Mometrix

Algebra 1

Student Edition Textbook

Written and edited by Matthew Bowling

Printed in the United States of America

This paper meets the requirements of ANSI/NISO Z39.48-1992 (Permanence of Paper).

Paperback
ISBN 13: 978-1-5167-2747-6
ISBN 10: 1-5167-2747-9

TABLE OF CONTENTS

INTRODUCTION (READ THIS FIRST!) 1
LESSONS 1
LESSON QUIZ AND CHAPTER QUIZ ANSWERS 1
OVERCOMING TEST ANXIETY 1

CHAPTER 1: SOLVING EQUATIONS 2
1.1 SOLVING ONE-STEP EQUATIONS 2
1.2 SOLVING TWO-STEP EQUATIONS 7
1.3 SOLVING EQUATIONS USING THE DISTRIBUTIVE PROPERTY 12
1.4 SOLVING EQUATIONS WITH VARIABLES ON BOTH SIDES 17
1.5 SIMPLIFYING ALGEBRAIC EXPRESSIONS 23
1.6 SOLVING EQUATIONS BY COMBINING LIKE TERMS 27
1.7 SOLVING MULTI-STEP EQUATIONS 32
1.8 SOLVING EQUATIONS INVOLVING ABSOLUTE VALUE 39
1.9 SOLVING FORMULAS FOR SPECIFIC VARIABLES 44
1.10 WRITING AND SOLVING EQUATIONS INVOLVING DIRECT VARIATION 50
CHAPTER 1 QUIZ 56

CHAPTER 2: FUNCTIONS AND SEQUENCES 60
2.1 EVALUATING FUNCTIONS 60
2.2 DOMAIN AND RANGE OF LINEAR FUNCTIONS 64
2.3 MANIPULATING LINEAR FUNCTIONS 75
2.4 WRITING FUNCTIONS FOR REAL-WORLD PROBLEMS 83
2.5 ARITHMETIC SEQUENCES 87
2.6 GEOMETRIC SEQUENCES 91
CHAPTER 2 QUIZ 95

CHAPTER 3: LINEAR EQUATIONS 101
3.1 SLOPE AND RATE OF CHANGE 101
3.2 GRAPHING LINEAR EQUATIONS 111
3.3 FINDING THE SLOPE-INTERCEPT FORM EQUATION OF A LINE 128
3.4 FINDING THE POINT-SLOPE FORM EQUATION OF A LINE 136
3.5 FINDING THE STANDARD FORM EQUATION OF A LINE 140
3.6 CONVERTING BETWEEN STANDARD FORM AND SLOPE-INTERCEPT FORM 146
3.7 WRITING LINEAR EQUATIONS 151
3.8 PARALLEL AND PERPENDICULAR LINES AND THE AXES 161
3.9 EQUATIONS OF PARALLEL LINES 168
3.10 EQUATIONS OF PERPENDICULAR LINES 174
3.11 CALCULATING CORRELATION COEFFICIENT 181
3.12 ASSOCIATION, CORRELATION, AND CAUSATION 192
CHAPTER 3 QUIZ 196

CHAPTER 4: INEQUALITIES 202
4.1 INEQUALITY BASICS 202
4.2 SOLVING ONE-STEP INEQUALITIES 207
4.3 SOLVING MULTI-STEP INEQUALITIES 215
4.4 SOLVING COMPOUND INEQUALITIES 223

4.5 Solving Inequalities Involving Absolute Values 230
4.6 Writing Linear Inequalities 237
4.7 Graphing Linear Inequalities 246
Chapter 4 Quiz 266

Chapter 5: Systems of Equations 274
5.1 The Graphing Method 274
5.2 The Substitution Method 294
5.3 The Elimination Method 312
5.4 Writing Systems of Linear Equations 319
5.5 Graphing Solutions to Systems of Linear Inequalities 337
Chapter 5 Quiz 362

Chapter 6: Polynomials 370
6.1 Adding and Subtracting Polynomials 370
6.2 Multiplying Monomials 376
6.3 Dividing Monomials 382
6.4 Multiplying Polynomials by Monomials 388
6.5 Multiplying Binomials 393
6.6 Dividing Polynomials by Monomials 399
6.7 Dividing Polynomials by Binomials 404
6.8 Factoring Using Common Monomial Factors 412
6.9 Factoring Trinomials of the Form $x2 + bx + c$ 417
6.10 Factoring the Difference of Two Squares 421
6.11 AC Method 426
6.12 Multistep Factoring 432
Chapter 6 Quiz 437

Chapter 7: Algebraic Fractions 441
7.1 Simplifying Algebraic Fractions 441
7.2 Solving Proportions Containing Algebraic Fractions 448
7.3 Multiplying Algebraic Fractions 455
7.4 Dividing Algebraic Fractions 464
7.5 Adding and Subtracting Algebraic Fractions 472
7.6 Adding and Subtracting Algebraic Fractions with Binomial Denominators 480
7.7 Solving Equations Involving Algebraic Fractions 487
Chapter 7 Quiz 494

Chapter 8: Quadratic Functions 499
8.1 Domain and Range of Quadratic Functions 499
8.2 Writing Equations Given Vertex and a Point 514
8.3 Factoring Quadratic Equations 520
8.4 Writing Quadratic Functions Given Solutions and Graphs 525
8.5 Solving Quadratic Equations Using Square Roots 546
8.6 Completing the Square 552
8.7 Converting Between Standard and Vertex Forms 560
8.8 The Quadratic Formula 566
8.9 The Discriminant 574
8.10 Graphing Quadratic Functions 581
8.11 Manipulating Quadratic Functions 604
Chapter 8 Quiz 618

CHAPTER 9: EXPONENTIAL FUNCTIONS 626
9.1 DOMAIN AND RANGE OF EXPONENTIAL FUNCTIONS 626
9.2 WRITING EXPONENTIAL FUNCTIONS 634
9.3 GRAPHING EXPONENTIAL FUNCTIONS AND DETERMINING KEY FEATURES 641
CHAPTER 9 QUIZ 659
CHAPTER 10: RADICAL EXPRESSIONS 670
10.1 SIMPLIFYING RADICAL EXPRESSIONS 670
10.2 SOLVING EQUATIONS INVOLVING RADICALS 674
10.3 ADDING AND SUBTRACTING RADICAL EXPRESSIONS 681
10.4 MULTIPLYING RADICAL EXPRESSIONS 686
10.5 DIVIDING RADICAL EXPRESSIONS 691
10.6 RATIONALIZING THE DENOMINATOR 696
10.7 SIMPLIFYING RADICAL EXPRESSIONS WITH BINOMIAL DENOMINATORS 702
CHAPTER 10 QUIZ 708
PRACTICE TEST #1 713
ANSWER KEY AND EXPLANATIONS 731
PRACTICE TESTS #2 AND #3 761
LESSON AND CHAPTER QUIZ ANSWER KEY 762
LESSON QUIZZES 762
CHAPTER QUIZZES 767
HOW TO OVERCOME TEST ANXIETY 768
ADDITIONAL BONUS MATERIAL 774

Introduction (Read This First!)

Lessons

This guide is broken down by chapter and by lesson. Each lesson includes a set of quiz questions to put into practice the concepts you just read about. There is also a chapter quiz at the end of each chapter. The lesson quizzes and chapter quizzes can be taken on paper or on almost any internet-connected device via our online **interactive testing interface**, which can be accessed by visiting the bonus page for this product at **mometrix.com/bonus948/algebra1tb** (or via the QR code) and clicking on **Lesson Quizzes** or **Chapter Quizzes**.

Lesson Quiz and Chapter Quiz Answers

An answer key for the quizzes is located at the end of this book, but in an effort to keep the book to a manageable length, we've opted to provide the complete worked out explanations for the quiz questions electronically only. So if you prefer to take the quizzes on paper, you'll want to visit the link below to see the **worked out answer explanations**. Visit the bonus page for this product at **mometrix.com/bonus948/algebra1tb** (or via the QR code) and click on either **Lesson Quiz Answers** or **Chapter Quiz Answers** to see how to work out the answer to each question. Both PDFs have a clickable table of contents for easy navigation to the right quiz.

Overcoming Test Anxiety

If you struggle with test anxiety, we strongly encourage you to check out our recommendations for how you can overcome it. Test anxiety is a formidable foe, but it can be beaten, and we want to make sure you have the tools you need to defeat it.

Chapter 1: Solving Equations

1.1 Solving One-Step Equations

Linear equations are a major topic in algebra. Mastering linear equations begins with a solid foundation in solving one-step equations. Once you have mastered one-step equations, you can tackle two-step equations and then multi-step equations. Solving linear equations is a skill you will want to become confident in as you navigate the world of algebra.

As you would expect, one-step equations can be solved in one step. This means that it only takes one operation (addition, subtraction, multiplication, or division) to solve for the variable.

Solving Equations Using Addition and Subtraction

Solving one-step equations is very useful in many real-world scenarios. For example, if a rectangle has a length of 10 feet, and the length of the rectangle is 7 feet longer than the width, a one-step equation can be set up in order to solve for the width of the rectangle. The equation $10 = w + 7$ states that the length of the rectangle is equal to the width plus 7. The only value for w that makes both sides of the equation the same (balanced) is $w = 3$. The width of the rectangle is 3 feet. "Guess and check" is a valid strategy for solving one-step equations; however, there is a more efficient method that will save you time. This strategy is known as **inverse operations**.

In the previous example, the equation $10 = w + 7$ was set up to represent the relationship between the length and width of a rectangle. The objective of inverse operations is to isolate the variable. Remember, a **variable** is simply a letter that represents an unknown value. In this case, the variable w represents the width of the rectangle. To isolate the w, we need to get the w on its own. Right now, the w is not isolated because there is a $+7$ next to it. To get rid of this $+7$, we need to perform its inverse. The inverse of adding 7 is subtracting 7. We need to subtract 7 from *both* sides of the equation in order to keep the equation **balanced** (same amount on both sides). When 7 is subtracted from both sides of the equation $10 = w + 7$ becomes $3 = w$.

$$10 - 7 = w + 7 - 7$$
$$3 = w$$

The $+7$ is now gone, because $+7$ and -7 combine to equal 0. The equation now states $3 = w$, which means that the width is 3 feet.

This process works for all one-step equations involving addition and subtraction. For example, the equation $9 = x - 4$ can be solved by isolating the variable x. Right now, the x is not isolated because there is a –4 next to it. This –4 is referred to as a **constant**, which is simply a number on its own in an equation (no variable attached to it). To get rid of the –4, we need to do the inverse of subtracting 4. The inverse of subtracting 4 is adding 4. When 4 is added to both sides of the equation $9 = x - 4$ becomes $13 = x$.

$$9 + 4 = x - 4 + 4$$
$$13 = x$$

The important thing to remember is that when solving one-step equations with addition or subtraction, the goal is to *isolate the variable*. To achieve this, we need to locate the constant that is next to the variable and either add or subtract the constant to eliminate it from that side of the equation. If the constant is positive, subtract the constant from both sides. If the constant is negative, then add the constant to both sides.

Solving Equations Using Multiplication and Division

Just like with solving one-step equations using addition and subtraction, to solve a one-step equation using multiplication or division, you perform the inverse operation on both sides. The opposite of multiplication is division, so if the variable is multiplied by a number (called a coefficient) divide by the coefficient on both sides of the equation. The opposite of division is multiplication, so if the variable is divided by a number, multiply both sides of the equation by that number.

For example, in this equation the variable is being multiplied by a constant. This means you will need to use the inverse of multiplication to solve it.

$$5x = 40$$

The inverse of multiplying by 5 is dividing by 5. Divide both sides by 5 to get $x = 8$.

In this equation, the variable is being divided by a constant. So, you will need to use the inverse of division to solve this equation.

$$\frac{x}{7} = -3$$

The inverse of dividing by 7 is multiplying by 7. Multiply both sides by 7 to get $x = -21$. Remember, the goal of solving any equation is to isolate the variable. In one-step equations this requires you to do the inverse operation to get the variable on its own.

Review Video: Solving One-Step Equations
Visit mometrix.com/academy and enter code: 777004

Examples

Example 1:

Solve for the variable x.

$$x - 9 = 40$$

Answer:

To isolate the variable, add 9 to both sides of the equation.

$$x - 9 + 9 = 40 + 9$$
$$x = 49$$

Example 2:

Solve for the variable y.

$$-\frac{1}{2} + y = 2$$

Answer:

To isolate the variable, add $\frac{1}{2}$ to both sides of the equation.

$$-\frac{1}{2}+\frac{1}{2}+y=2+\frac{1}{2}$$
$$y=2\frac{1}{2}$$

Example 3:

Solve for the variable y.

$$\frac{y}{17}=-2$$

Answer:

To isolate the variable, multiply both sides of the equation by 17.

$$\frac{y}{17}\times 17=-2\times 17$$

Remember, a negative number times a positive number equals a negative number, so $-2\times 17=-34$.

$$y=-34$$

Example 4:

Solve for the variable z.

$$2z=70$$

Answer:

To isolate the variable, divide both sides of the equation by 2.

$$\frac{2z}{2}=\frac{70}{2}$$
$$z=35$$

Practice Questions

1. The equation $x-5=12$ is given. What is the value of x?

a. $x=6$
b. $x=7$
c. $x=17$
d. $x=60$

2. The equation $x+7=25$ is given. Solve for x.

a. $x=18$
b. $x=30$
c. $x=32$
d. $x=175$

3. What is the value of x in the equation $\frac{x}{5} = -6$?

a. $x = -30$
b. $x = -11$
c. $x = 11$
d. $x = 30$

4. The equation $-3x = 45$ is given. What is the value of x?

a. $x = -135$
b. $x = -15$
c. $x = 15$
d. $x = 135$

5. The equation $y - 14 = 63$ is given. What is the value of y?

a. $y = 49$
b. $y = 57$
c. $y = 71$
d. $y = 77$

6. What is the value of y in the equation $y + 27 = -89$?

a. $y = -116$
b. $y = -62$
c. $y = 62$
d. $y = 116$

7. The equation $\frac{y}{17} = 22$ is given. What is the value of y?

a. $y = 5$
b. $y = 39$
c. $y = 110$
d. $y = 374$

8. The equation $-15y = 375$ is given. Solve for y.

a. $y = -5{,}625$
b. $y = -25$
c. $y = 25$
d. $y = 5{,}625$

9. The equation $w - 57 = -279$ is given. Solve for w.

a. $w = -336$
b. $w = -222$
c. $w = 222$
d. $w = 336$

10. The equation $w + 109 = -73$ is given. What is the value of w?

a. $w = -182$
b. $w = -36$
c. $w = 36$
d. $w = 182$

11. The equation $\frac{w}{92} = 5$ is given. What is the value of w?

a. $w = 18$
b. $w = 87$
c. $w = 97$
d. $w = 460$

12. What is the value of w in the equation $-30w = 240$?

a. $w = -7{,}200$
b. $w = -8$
c. $w = 210$
d. $w = 60$

13. Kusha and Zola pick 37 mangoes from a tree. If Kusha picked 22 mangoes, how many mangoes did Zola pick?

a. 15 mangoes
b. 19 mangoes
c. 55 mangoes
d. 59 mangoes

14. Alex is taking a math test that requires one hour to complete. There are two sections of the test. Alex completes the first section in 41 minutes. Write a one-step equation to represent the time Alex spends on the two tests, then solve for the amount of time that Alex has left to complete the second section of the test.

a. $60 = 14 + y$, 46 minutes for section two
b. $60 = 41 + y$, 19 minutes for section two
c. $40 = 61 + y$, 21 minutes for section two
d. $41 = 60 + y$, 19 minutes for section two

15. Joan is selling raffle tickets at a basketball game for $8 dollars each. How many tickets does she need to sell in order to make $440? Write a one-step equation in order to solve for the number of tickets and then solve for the number of tickets.

a. $440 = 55x$, 8 tickets need to be sold
b. $8x = 440$, 65 tickets need to be sold
c. $550 = 8x$, 50 tickets need to be sold
d. $440 = 8x$, 55 tickets need to be sold

1.2 Solving Two-Step Equations

A one-step equation requires one operation to solve for the variable. For example, the equation $4x = 20$ requires only one step (division by 4) to solve for x: $x = 5$. However, many equations will require more than one operation to solve.

Two-step equations are very common in algebra, and they are used to model many real-world scenarios. For example, the expression $20x + 15$ could represent the money you spent at a bookstore. $20x$ could represent spending \$20 on each book you purchased, and +15 could represent spending \$15 on a magazine. If you spend \$75 total, the equation would become $20x + 15 = 75$.

Solving two-step equations generally follows two steps:

Step 1: Move the variables to one side of the equation and the constants on the other side.

Step 2: Do the necessary operation to isolate the variable.

For example, consider the equation from the previous example:

$$20x + 15 = 75$$

Step 1: Move the variables to one side and the constants to the other side. For this problem, do that by subtracting 15 from both sides.

$$20x + 15 - 15 = 75 - 15$$
$$20x = 60$$

Step 2: Isolate the variable. For this problem, do that by dividing both sides by 20.

$$\frac{20x}{20} = \frac{60}{20}$$
$$x = 3$$

Since x is the number of books, you purchased 3 books.

Now consider an equation with negatives:

$$-3x - 8 = 40$$

Step 1: Move the variables to one side and the constants to the other side by adding 8 to both sides.

$$-3x - 8 + 8 = 40 + 8$$
$$-3x = 48$$

Step 2: Isolate the variable by dividing both sides by -3.

$$\frac{-3x}{-3} = \frac{48}{-3}$$
$$x = -16$$

Sometimes step 2 (isolating the variable) will require you to multiply instead of divide. For example, consider this equation:

$$5 + \frac{x}{6} = 13$$

Step 1: Move the variables to one side and the constants to the other side. In this example, this involves subtracting 5 from both sides.

$$5 + \frac{x}{6} - 5 = 13 - 5$$
$$\frac{x}{6} = 8$$

Step 2: Isolate the variable by multiplying both sides by 6. Remember, a fraction bar always represents division.

$$\frac{x}{6} \times 6 = 8 \times 6$$
$$x = 48$$

Review Video: Solving Two-Step Equations Visit mometrix.com/academy and enter code: 868774

EXAMPLES

EXAMPLE 1:

Solve for the variable y.

$$3y - 8 = 7$$

Answer:

Step 1: Add 8 to both sides of the equation.

$$3y - 8 + 8 = 7 + 8$$
$$3y = 15$$

Step 2: Divide both sides by 3.

$$\frac{3y}{3} = \frac{15}{3}$$
$$y = 5$$

EXAMPLE 2:

Solve for the variable x.

$$\frac{x + 5}{2} = 10$$

Answer:

Step 1: Since both the x and the $+5$ are divided by 2, we need to multiply both sides by 2 before we can subtract the 5.

$$\frac{x+5}{2} \times 2 = 10 \times 2$$
$$x + 5 = 20$$

Step 2: Subtract 5 from both sides.

$$x + 5 - 5 = 20 - 5$$
$$x = 15$$

Example 3:

Solve for the variable x.

$$\frac{2}{3}x + 10 = 4$$

Answer:

Step 1: Subtract 10 from both sides.

$$\frac{2}{3}x + 10 - 10 = 4 - 10$$
$$\frac{2}{3}x = -6$$

Step 2: Now we want to divide both sides by $\frac{2}{3}$, which is the same as multiplying both sides by its reciprocal, $\frac{3}{2}$.

$$\frac{3}{2} \times \frac{2}{3}x = -6 \times \frac{3}{2}$$
$$x = \frac{-18}{2}$$
$$x = -9$$

Practice Questions

1. Solve for the value of x that will make the following equation true: $2x + 4 = 8$.

a. $x = 4$
b. $x = 1$
c. $x = 2$
d. $x = 2.5$

2. Solve for the value of x that satisfies the equation $-4x - 3 = 13$.

a. $x = 4$
b. $x = 3$
c. $x = -5$
d. $x = -4$

3. Solve for the value of x that makes the following equation true: $10x + 40 = 100$.

a. $x = 6$
b. $x = 5$
c. $x = 4$
d. $x = 8$

4. Solve the following two-step equation for z: $6z - 2 = 10$.

a. $z = 2$
b. $z = 3$
c. $z = 4$
d. $z = 6$

5. Solve the following equation for x: $\frac{x}{5} - 3 = 2$.

a. $x = 1$
b. $x = 25$
c. $x = 15$
d. $x = 30$

6. Solve for n in the following equation: $-n + 5 = -17$.

a. $n = 22$
b. $n = -12$
c. $n = -22$
d. $n = 12$

7. Solve the following equation for the value of x that will make it true: $\frac{x}{3} - 7 = 11$.

a. $x = \frac{7}{3}$
b. $x = 77$
c. $x = \frac{4}{3}$
d. $x = 54$

8. Solve for x in the following two-step equation: $-3x - 2 = -1$.

a. $x = \frac{1}{3}$
b. $x = -1$
c. $x = -\frac{1}{3}$
d. $x = 3$

9. Solve for x in the equation $9x + 36 = 72$.

a. $x = \frac{4}{3}$
b. $x = 6$
c. $x = 8$
d. $x = 4$

10. Solve for x in the equation $-7x + 4 = -10$.

a. $x = \frac{6}{7}$
b. $x = -2$
c. $x = -\frac{6}{7}$
d. $x = 2$

11. Solve for x in the equation $8x + 1 = -21$.

a. $x = \frac{5}{2}$
b. $x = -\frac{11}{4}$
c. $x = -3$
d. $x = -4$

12. Solve for x given the equation $-\frac{3}{5}x + 8 = 23$.

a. $x = -25$
b. $x = \frac{14}{5}$
c. $x = -9$
d. $x = 9$

13. Danielle is interested in buying some blouses that are on sale for \$6.00 each. She has a budget of \$32, and she also has a \$10-off coupon. Write an equation from the information given and determine how many blouses Danielle can buy.

a. $6x = 22$, and Danielle can buy 3 blouses.
b. $6x - 10 = 32$, and Danielle can buy 7 blouses.
c. $10x - 6 = 32$, and Danielle can buy 3 blouses.
d. $6x - 10 = 32$, and Danielle can buy 8 blouses.

14. Billy rides his bicycle 5 miles each morning that he works his paper route. One morning this week, he rode an extra mile to visit with his grandparents. At the end of the week, he had ridden 21 miles. How many mornings did he deliver papers? Support your answer with an equation.

a. 3 mornings; $6x + 3 = 21$
b. 4 mornings; $21x - 5 = 22$
c. 4 mornings; $5x + 1 = 21$
d. 4 mornings; $6x + 3 = 21$

15. Nora earns \$4 per hour at her waitressing job and today received \$29 in tips. From her shift today, she earned a total of \$53 from both tips and hourly wages. Write an equation from the information given and determine how many hours Nora worked today.

a. $4x + 29 = 53$; Nora worked 6 hours.
b. $29x + 4 = 53$; Nora worked 2 hours.
c. $53x - 29 = 4$; Nora worked 2 hours.
d. $4x - 29 = 53$; Nora worked 9 hours.

1.3 Solving Equations Using the Distributive Property

The **distributive property** states that when you multiply a sum, like 4 multiplied by $(3 + 6)$, you can simply multiply each term in the sum, or *summand*, separately and then add the products. This means that $4(3 + 6)$ is the same thing as $4 \times 3 + 4 \times 6$.

The distributive property is often used when solving equations. For example, consider this equation:

$$14 + 2(4x - 3) = 40$$

The distributive property can be used to solve for the variable x. Notice the 2 that is multiplied by $(4x - 3)$. $2(4x - 3)$ can be simplified by multiplying (distributing) the 2 into both terms within the parentheses. Essentially, we need to multiply $2(4x)$ and $2(-3)$ and then add them together. Don't forget to include the sign ($+$ or $-$) that is in front of each term. In this case, the $-$ needs to stick with the 3. After we distribute, we have:

$$14 + 8x - 6 = 40$$

From here, we can arrange and combine all of the variables on one side of the equation and all of the constants on the other side. Start by adding 14 and -6.

$$8 + 8x = 40$$

Then, subtract 8 from both sides.

$$8 + 8x - 8 = 40 - 8$$
$$8x = 32$$

Now divide both sides by 8.

$$\frac{8x}{8} = \frac{32}{8}$$
$$x = 4$$

Let's look at an equation that requires the distributive property on *both* sides of the equation. Consider the equation:

$$3(x + 4) = 2(x - 1)$$

Notice that on the left side we need to distribute the 3, and on the right side we need to distribute the 2.

$3(x + 4)$ becomes $3(x) + 3(4) = 3x + 12$.

$2(x - 1)$ becomes $2(x) + 2(-1) = 2x - 2$.

$3(x + 4) = 2(x - 1)$ now looks like this:

$$3x + 12 = 2x - 2$$

From here, we can move all of the variables to one side and all of the constants to the other side. First, subtract $2x$ from both sides of the equation.

$$3x + 12 - 2x = 2x - 2 - 2x$$
$$x + 12 = -2$$

Then, subtract 12 from both sides.

$$x + 12 - 12 = -2 - 12$$
$$x = -14$$

Remember to pay close attention to the sign ($+$ or $-$) that is in front of each term when using the distributive property. For example, when distributing $-3(-4x + 5)$, the first step would be to multiply $-3(-4x)$ which equals $12x$. The second step would be to multiply $-3(5)$ which equals -15. It's important to remember the rules for multiplying negative numbers. When multiplying two negative numbers, the answer will be positive. When multiplying a positive number and a negative number, the answer will be negative.

$$-3(-4x + 5) = 12x - 15$$

> **Review Video: Solving Equations Using the Distributive Property**
> Visit mometrix.com/academy and enter code: 765499

EXAMPLES

EXAMPLE 1:

Simplify the following expression.

$$4(7x - 3) - 12$$

Answer:

Start by multiplying (distributing) the 4 by both terms in the parentheses.

$$4(7x) + 4(-3) - 12$$

$$28x - 12 - 12$$

Now, combine like terms to simplify.

$$28x - 24$$

EXAMPLE 2:

Solve for the variable x.

$$-2(9 + x) = 24$$

Answer:

Start by multiplying (distributing) the -2 by both terms in the parentheses.

$$(-2)(9) + (-2)(x) = 24$$
$$-18 - 2x = 24$$

Now add 18 to both sides so that all of the constant terms will be together on one side of the equation.

$$-18 - 2x + 18 = 24 + 18$$
$$-2x = 42$$

Now divide both sides by –2 to solve for x.

$$\frac{-2x}{-2} = \frac{42}{-2}$$
$$x = -21$$

EXAMPLE 3:

Solve for the variable y.

$$4(y - 2) - 2y = 5(y - 4)$$

Answer:

Distribute the 4:

$$4(y) + 4(-2) - 2y = 5(y - 4)$$
$$4y - 8 - 2y = 5(y - 4)$$

Distribute the 5:

$$4y - 8 - 2y = 5(y) + 5(-4)$$
$$4y - 8 - 2y = 5y - 20$$

Combine like terms on the left side of the equation.

$$(4y - 2y) - 8 = 5y - 20$$
$$2y - 8 = 5y - 20$$

Move all of the variables to one side and all of the constants to the other side. First, subtract $2y$ from both sides of the equation.

$$2y - 2y - 8 = 5y - 2y - 20$$
$$-8 = 3y - 20$$

Then, add 20 to both sides.

$$-8 + 20 = 3y - 20 + 20$$
$$12 = 3y$$

Divide both sides by 3.

$$\frac{12}{3} = \frac{3y}{3}$$
$$y = 4$$

Practice Questions

1. Simplify the expression $3(2x - 8)$.

a. $6x - 8$
b. $2x - 24$
c. $5x - 11$
d. $6x - 24$

2. Simplify the expression $4(x + 13)$.

a. $x + 52$
b. $4x + 52$
c. $5x + 52$
d. $5x + 17$

3. Simplify the expression $6(3x + 9)$.

a. $x + 54$
b. $9x + 15$
c. $18x + 54$
d. $18x + 15$

4. Simplify the expression $5(4x - 7)$.

a. $9x - 12$
b. $9x - 35$
c. $20x - 12$
d. $20x - 35$

5. The equation $2(3y + 6) = 60$ is given. What is the value of y?

a. $y = 8$
b. $y = 12$
c. $y = 16$
d. $y = 24$

6. What is the value of y in the equation $8(y + 8) = -56$?

a. $y = -15$
b. $y = -5$
c. $y = 5$
d. $y = 15$

7. The equation $-3(4y - 9) = 303$ is given. What is the value of y?

a. $y = -28$
b. $y = -23$
c. $y = 23$
d. $y = 28$

8. The equation $4(-4y + 9) = 180$ is given. Solve for y.

a. $y = -18$
b. $y = -9$
c. $y = 14$
d. $y = 27$

9. The equation $5(2w + 4) = 8(w + 9)$ is given. Solve for w.

a. $w = 5$
b. $w = 13$
c. $w = 26$
d. $w = 46$

10. The equation $6(3w - 5) = 4(6w + 3)$ is given. What is the value of w?

a. $w = -7$
b. $w = -3$
c. $w = 3$
d. $w = 7$

11. The equation $-7(3w + 6) = -2(3w - 84)$ is given. What is the value of w?

a. $w = -21$
b. $w = -14$
c. $w = 14$
d. $w = 21$

12. Solve for the variable n.

$$6(n + 3) = -2(n + 31)$$

a. $n = -10$
b. $n = 5.6$
c. $n = -4$
d. $n = -5$

13. The length of a rectangular garden is $x + 5$ feet, and the width of the garden is x feet. What is the perimeter of the rectangular garden?

a. $2x + 5$ feet
b. $2x + 10$ feet
c. $4x + 7$ feet
d. $4x + 10$ feet

14. Tima and 4 of her friends go to the movies. The ticket for the movie costs x dollars. They each buy a small popcorn for y dollars and a small drink for z dollars. Which expression shows the total cost of Tima and her friends going to the movies?

a. $4xyz$ dollars
b. $5xyz$ dollars
c. $4x + 4y + 4z$ dollars
d. $5x + 5y + 5z$ dollars

15. Mr. Jonas buys 15 pens, pencils, and erasers that cost \$1.15, \$0.79, and \$1.25 each, respectively. How much does Mr. Jonas spend in total on the supplies for his students?

a. \$11.85
b. \$17.25
c. \$18.75
d. \$47.85

1.4 Solving Equations with Variables on Both Sides

Some equations are very short and can be solved in one step. For example, $3x = 12$ can be solved simply by dividing both sides of the equation by 3 so that $x = 4$. However, many equations you encounter in algebra will require two or more steps to solve for the variable. Many equations will have variables and constants on both sides.

Equations with variables and constants on both sides of the equation can be solved by following these two steps:

Step 1: Move all of the variables to one side of the equation and all of the constants to the other side (using inverse operations) and combine like terms.

Step 2: Use an inverse operation to isolate the variable.

For example, the following equation has variables on both sides. Use the two-step process to solve for x.

$$-20x = 9 - 2x$$

Step 1: Move all of the variables to one side and the constants to the other side. In this example, we do that by adding $2x$ to both sides of the equation.

$$-20x + 2x = 9 - 2x + 2x$$
$$-18x = 9$$

Step 2: Isolate the variable using an inverse operation. The inverse of multiplying by –18 is dividing by –18. Divide both sides by –18.

$$\frac{-18x}{-18} = \frac{9}{-18}$$
$$x = -\frac{1}{2}$$

Some equations will have variables and constants on both sides. Use the same two-step process to solve for x.

$$-17x - 3 = -6x + 52$$

Step 1: Get all of the variables to one side and the constants to the other side. In this step, it does not matter which side you choose for the variables and constants, as long as they are on opposite sides. It also does not matter if the variables are moved first or the constants. For this example, start by adding $6x$ to both sides. Then, add 3 to both sides.

$$-17x - 3 + 6x = -6x + 52 + 6x$$
$$-11x - 3 = 52$$
$$-11x - 3 + 3 = 52 + 3$$
$$-11x = 55$$

Step 2: Isolate the variable using an inverse operation. The inverse of multiplying by –11 is dividing by –11. Divide both sides by –11.

$$\frac{-11x}{-11} = \frac{55}{-11}$$

$$x = -5$$

Review Video: Solving Equations with Variables on Both Sides
Visit mometrix.com/academy and enter code: 402497

EXAMPLES

EXAMPLE 1:

Solve for the variable x.

$$10x + 15 = 7x + 45$$

Answer:

Move the variables to one side and the constants to the other side. Subtract $7x$ and 15 from both sides of the equation.

$$\begin{gathered} 10x + 15 - 7x = 7x + 45 - 7x \\ 3x + 15 = 45 \\ 3x + 15 - 15 = 45 - 15 \\ 3x = 30 \end{gathered}$$

Note that 15 could have been subtracted first and then $7x$ and the result would have been the same.

Divide both sides by 3.

$$\begin{gathered} \frac{3x}{3} = \frac{30}{3} \\ x = 10 \end{gathered}$$

EXAMPLE 2:

Solve for the variable y.

$$5(y + 1) = 2(y + 10)$$

Answer:

Distribute the 5 on the left side and the 2 on the right side.

$$5y + 5 = 2y + 20$$

Move the variables to one side and the constants to the other side. Subtract $2y$ and 5 from both sides of the equation.

$$\begin{gathered} 5y + 5 - 2y = 2y + 20 - 2y \\ 3y + 5 = 20 \\ 3y + 5 - 5 = 20 - 5 \\ 3y = 15 \end{gathered}$$

Divide both sides by 3.

$$\begin{gathered} \frac{3y}{3} = \frac{15}{3} \\ y = 5 \end{gathered}$$

Example 3:

Solve for the variable n.

$$9n - 5 = \frac{1}{4}(16n + 60)$$

Answer:

Distribute the $\frac{1}{4}$ into the set of parentheses.

$$9n - 5 = 4n + 15$$

Move the variables to one side and the constants to the other side. Do this by subtracting $4n$ from both sides and adding 5 to both sides.

$$\begin{aligned} 9n - 5 - 4n &= 4n + 15 - 4n \\ 5n - 5 &= 15 \\ 5n - 5 + 5 &= 15 + 5 \\ 5n &= 20 \end{aligned}$$

Notice that when $4n$ was subtracted, it was placed at the ends of each side. Because of the commutative property of addition, it could also be written like this: $9n - 4n - 5 = 4n - 4n + 15$. Sometimes students prefer to write it this way to clearly see which terms need to be combined. Either way of writing it is correct.

Divide both sides by 5.

$$\begin{aligned} \frac{5n}{5} &= \frac{20}{5} \\ n &= 4 \end{aligned}$$

Example 4:

Solve for the variable x.

$$\frac{3x}{7} + 14 = 3 + 2x$$

Answer:

We want to move the variables to one side and the constants to the other side. We could start by subtracting 14 from both sides of the equation, then subtracting $2x$ from both sides. This will require subtracting $2x$ from $\frac{3x}{7}$, which means we would need to find a common denominator. We could avoid this by eliminating the fraction before we begin work. To do so we need to multiply both sides of the equation by the denominator, 7. We will refer to this process of eliminating a fraction as "clearing the fraction."

$$7\left(\frac{3x}{7} + 14\right) = 7(3 + 2x)$$

Distribute the 7 on both sides of the equation.

$$3x + 98 = 21 + 14x$$

Now we will move the variables to one side and the constants to the other side. Start by subtracting 98 from both sides of the equation, then subtract $14x$ from both sides.

$$3x + 98 - 98 = 21 + 14x - 98$$
$$3x = -77 + 14x$$
$$3x - 14x = -77 + 14x - 14x$$
$$-11x = -77$$

Divide both sides by −11.

$$\frac{-11x}{-11} = \frac{-77}{-11}$$

Practice Questions

1. Solve for the variable m.

$$9m + 4 = 7m + 12$$

a. $m = 4$
b. $m = 2$
c. $m = -4$
d. $m = 3.5$

2. Solve for the variable x.

$$2x - 7 = 5x + 8$$

a. $x = -5$
b. $x = 5.5$
c. $x = 2.5$
d. $x = 5$

3. The equation $2x + 3 = x + 16$ is given. What is the value of x?

a. $x = 8$
b. $x = 13$
c. $x = 19$
d. $x = 21$

4. The equation $4x - 8 = x + 19$ is given. Solve for x.

a. $x = 9$
b. $x = 10$
c. $x = 27$
d. $x = 30$

5. What is the value of x in the equation $\frac{5x}{7} = -18 + 2x$?

a. $x = -12$
b. $x = -7$
c. $x = 14$
d. $x = 36$

6. The equation $-6x + 7 = x + 56$ is given. What is the value of x?

a. $x = -9$
b. $x = -7$
c. $x = 7$
d. $x = 9$

7. The equation $-8y + 5 = -2y + 53$ is given. What is the value of y?

a. $y = -8$
b. $y = -6$
c. $y = 6$
d. $y = 8$

8. What is the value of y in the equation $3y - 12 = -5y + 28$?

a. $y = 2$
b. $y = 5$
c. $y = 8$
d. $y = 20$

9. The equation $-\frac{y}{2} - 26 = -3y + 49$ is given. What is the value of y?

a. $y = 25$
b. $y = 30$
c. $y = 46$
d. $y = 150$

10. The equation $-14y - 33 = -12y + 29$ is given. Solve for y.

a. $y = -37$
b. $y = -31$
c. $y = 31$
d. $y = 37$

11. The equation $6w + 16 = 4w - 30$ is given. Solve for w.

a. $w = -23$
b. $w = -7$
c. $w = 7$
d. $w = 23$

12. The equation $-7w - 52 = -3w + 20$ is given. What is the value of w?

a. $w = -18$
b. $w = -8$
c. $w = 8$
d. $w = 18$

13. Two pools at a swim club are being drained at the same time. Pool 1 has 2,100 gallons of water and is draining at 50 gallons per minute. Pool 2 has 3,600 gallons of water and is draining at 80 gallons per minute. After how many minutes, *t*, will the two pools have the same amount of water?

a. 43 minutes
b. 50 minutes
c. 54 minutes
d. 71 minutes

14. Car rental A charges a $15 daily fee and $0.20 per mile. Car rental B charges a $10 daily fee and $0.40 per mile. After how many miles, *m*, in one day will the cost of the two cars be the same?

a. 8.3 miles
b. 25 miles
c. 41.7 miles
d. 125 miles

15. A sign store charges $122 for a sign and $4 per letter to print on the sign. A second sign store charges $158 for a sign and $2.50 per letter to print on the sign. How many letters, *p*, would it take for the two signs to cost the same?

a. 6 letters
b. 14 letters
c. 24 letters
d. 43 letters

1.5 Simplifying Algebraic Expressions

Simplifying an expression means rewriting an expression in a more condensed form. For example, $5x + 4x + 3x + 2x + x$ is an expression that can be simplified to $15x$. The skill of simplifying expressions is a concept that is used very frequently in algebra.

When simplifying expressions, it is important to combine like terms. Combining **like terms** means combining terms that have the same variable raised to the same power. For example, $3x^2$ and $5x^2$ are considered like terms because they have the same variable raised to the same power. Constants are always like terms (the variable is raised to the power 0).

Like terms: $4xy$ and $55xy$, $3t$ and $-5t$, $4x^2$ and $4x^2$

Not like terms: $4x$ and $4y$, $3xz$ and $3xz^2$, $5t$ and $5y$

The process of combining like terms is used to simplify expressions. For example, consider the expression $(4x + 30y + 2xy) + (9x - 5y + 4xy)$. When the like terms are combined, the expression becomes a lot shorter (simplified). Make sure not to forget the symbol (+ or −) that is in front of each term. $(4x + 9x) + (30y - 5y) + (2xy + 4xy)$ simplifies to $13x + 25y + 6xy$.

Let's look at an example using subtraction. Consider the expression $(7x^2 + 2x - 1) - (3x^2 + 3x - 8)$. The subtraction symbol between the sets of parentheses needs to be distributed into each term in the second set of parentheses. This means that $-(3x^2 + 3x - 8)$ becomes $-3x^2 - 3x + 8$.

Now we can combine like terms, and the expression looks like this:

$$(7x^2 - 3x^2) + (2x - 3x) + (-1 + 8)$$

When all of the like terms are combined, the result is $4x^2 - x + 7$.

Review Video: Simplifying Algebraic Expressions with Parentheses
Visit mometrix.com/academy and enter code: 850843

Examples

Example 1:

Simplify the following expression:

$$(14x + 5xy - 7) + (9x + 2xy + 3)$$

Answer:

Combine like terms (same variable raised to the same power). Don't forget the sign (+ or −) that is in front of each term.

$$(14x + 5xy - 7) + (9x + 2xy + 3)$$

$$(14x + 9x) + (5xy + 2xy) + (-7 + 3)$$

$$23x + 7xy - 4$$

Example 2:

Simplify the following expression:

$$(25xz + 13z + 1) + (5xz - 3z - 1)$$

Answer:

Combine like terms (same variable raised to the same power). Don't forget the sign (+ or −) that is in front of each term.

$$(25xz + 13z + 1) + (5xz - 3z - 1)$$

$$(25xz + 5xz) + (13z - 3z) + (1 - 1)$$

$$30xz + 10z$$

EXAMPLE 3:

Simplify the following expression:

$$(45x^2 - 3y + 5) - (30x^2 - 3y + 4)$$

Answer:

Combine like terms (same variable raised to the same power). Don't forget the sign (+ or −) that is in front of each term. Notice that the second set of parentheses is being subtracted from the first, so make sure to distribute the minus sign before combining like terms.

$$(45x^2 - 3y + 5) - (30x^2 - 3y + 4)$$

$$(45x^2 - 3y + 5) + (-30x^2 - (-3y) - 4)$$

$$(45x^2 - 30x^2) + (-3y - (-3y)) + (5 - 4)$$

$$(45x^2 - 30x^2) + (-3y + 3y) + (5 - 4)$$

$$15x^2 + 1$$

PRACTICE QUESTIONS

1. What is $(3m + n + 6) + (8m + 3n - 8)$ in its most simplified form?

a. $5m + 3n - 2$
b. $5m + 4n - 14$
c. $11m + 4n - 2$
d. $11m + 3n - 14$

2. Simplify the expression $(6m - 4n - 9) + (-2m - 3n + 2)$.

a. $4m - n - 11$
b. $4m - 7n - 7$
c. $8m - n - 7$
d. $8m - 7n - 11$

3. Which expression shows the most simplified form of the expression $(7m + 3n - 5) - (3m + 5n + 8)$?

a. $4m - 2n - 13$
b. $4m + 8n - 3$
c. $11m - 2n - 3$
d. $11m + 8n - 13$

4. Which shows the expression $(-4xy + 2x + 3) - (xy + x - 9)$ in its most simplified form?

a. $-5xy + x + 12$
b. $-5xy + 3x - 6$
c. $-3xy + x + 12$
d. $-3xy + 3x - 6$

5. What is $(y + x + 10) - (5y - 6x + 3)$ in its most simplified form?

a. $-6y - 5x + 13$
b. $-4y + 7x + 7$
c. $4y - 5x + 7$
d. $6y + 7x + 13$

6. Which shows the expression $(5x - 3y - 5) + (-2x - 8y - 6)$ in its most simplified form?

a. $3x - 11y - 11$
b. $3x - 5y - 1$
c. $7x - 11y - 1$
d. $7x - 5y - 11$

7. Which expression shows the most simplified form of the expression $(9y + 3xy + 8) + (-3y - xy - 2)$?

a. $6y + 4xy + 10$
b. $6y + 2xy + 6$
c. $12y + 2xy + 10$
d. $12y + 4xy + 6$

8. Simplify the expression $(7y - 4xy - 5) - (-3y - 2xy - 2)$.

a. $4y - 6xy - 7$
b. $4y - 2xy - 3$
c. $10y - 6xy - 7$
d. $10y - 2xy - 3$

9. Which expression shows the most simplified form of the expression $(-3u - 4v + 1) + (-5u + v + 4)$?

a. $-8u - 5v + 3$
b. $-8u - 3v + 5$
c. $-2u - 5v + 5$
d. $-2u - 3v + 3$

10. Which shows the expression $(-6w + 3vw - 9) - (-2w - 4vw + 6)$ in its most simplified form?

a. $-8w - vw - 15$
b. $-8w + 7vw - 3$
c. $-4w - vw - 3$
d. $-4w + 7vw - 15$

11. Which expression shows the most simplified form of the expression $(8w - 5vw + 2) + (-3w - 6vw - 7)$?

a. $5w - 11vw - 5$
b. $5w - vw - 9$
c. $11w - 11vw - 5$
d. $11w - vw - 9$

12. Simplify the following expression: $(14xy + 3z - 8) - (2xy + 5z - 4)$.

a. $12xy + 2z + 4$
b. $12x - 2xz - 4$
c. $12xy - 2z + 1$
d. $12xy - 2z - 4$

13. What is the perimeter of a rectangle with length $2x + 3y - 5$ meters and width $x + y - 2$ meters?

a. $3x + 4y - 7$ meters
b. $3x + 4y + 7$ meters
c. $6x + 8y - 14$ meters
d. $6x + 8y + 14$ meters

14. Amazing Bakery buys $10y + 2x + 5$ pounds of flour and uses $3y - 5x + 3$ pounds of flour for cakes. How many pounds of flour does the bakery have left?

a. $7y - 3x + 8$ pounds
b. $7y + 7x + 2$ pounds
c. $13y - 3x + 2$ pounds
d. $13y + 7x + 8$ pounds

15. Mary travels $5y + 3x + 12$ miles from home to school and $4y - 7x + 9$ miles from school to work. How many total miles does Mary travel from home to work?

a. $y - 4x + 3$ miles
b. $y + 4x + 21$ miles
c. $9y - 4x + 21$ miles
d. $9y + 4x + 3$ miles

1.6 Solving Equations by Combining Like Terms

Combining like terms refers to adding or subtracting terms with the same variable raised to the same exponent. Constants are considered like terms (the power of the exponent is zero), constant multiples of x are like terms, constant multiples of x^2 are like terms and so on. Combining like terms therefore reduces sets of like terms to a single term. The main advantage of doing this is simplifying the equation. Often, combining like terms can be done as the first step in solving an equation, though it can also be done later, such as after distributing terms in a product.

For example, consider the equation $2(x+3)+3(2+x+3)=-4$. The 2 and the 3 in the second set of parentheses are like terms, and we can combine them, yielding $2(x+3)+3(x+5)=-4$. Now we can carry out the multiplication implied by the parentheses, distributing the outer 2 and 3 accordingly: $2x+6+3x+15=-4$. The $2x$ and the $3x$ are like terms, and we can add them together: $5x+6+15=-4$. Now, the constants 6, 15, and -4 are also like terms, and we can combine them as well: subtracting 6 and 15 from both sides of the equation, we get $5x=-4-6-15$, or $5x=-25$, which simplifies further to $x=-5$.

Review Video: Solving Equations by Combining Like Terms
Visit mometrix.com/academy and enter code: 668506

Examples

Example 1:

Solve for x.

$$4x+2x-x=20+5$$

Answer:

Combine the like terms.

$$4x+2x-x=20+5$$
$$5x=25$$

Divide both sides by 5.

$$\frac{5x}{5}=\frac{25}{5}$$
$$x=5$$

Example 2:

Solve for y.

$$15y-y=80+4$$

Answer:

Combine the like terms.

$$15y-y=80+4$$
$$14y=84$$

Divide both sides by 14.

$$\frac{14y}{14} = \frac{84}{14}$$
$$y = 6$$

EXAMPLE 3:
Solve for x.

$$2x = 14 + 3 + 1$$

Answer:

Combine like terms on the right side of the equation.

$$2x = 14 + 3 + 1$$
$$2x = 18$$

Divide both sides by 2.

$$\frac{2x}{2} = \frac{18}{2}$$
$$x = 9$$

EXAMPLE 4:
Solve for x.

$$10x + 10 - 2 + 5x = 7x - 2 + 7x$$

Answer:

Combine *like terms* on the left and right sides of the equation.

$$10x + 10 - 2 + 5x = 7x - 2 + 7x$$
$$15x + 8 = 14x - 2$$

Move the variables to one side and the constants to the other side by subtracting $14x$ from both sides and subtracting 8 from both sides.

$$15x - 14x + 8 = 14x - 14x - 2$$
$$x + 8 = -2$$
$$x + 8 - 8 = -2 - 8$$
$$x = -10$$

EXAMPLE 5:
Solve for x.

$$8\left(\frac{1}{2}x + 1\right) = 6\left(4 - \frac{2}{3}x\right)$$

Answer:

Distribute the 8 and the 6 into the sets of parentheses.

$$8\left(\frac{1}{2}x+1\right)=6\left(4-\frac{2}{3}x\right)$$
$$4x+8=24-4x$$

Move the variables to one side and the constants to the other side by adding $4x$ to both sides and subtracting 8 from both sides. Then, combine like terms.

$$4x+4x+8-8=24-8-4x+4x$$
$$8x=16$$

Divide both sides by 8.

$$\frac{8x}{8}=\frac{16}{8}$$
$$x=2$$

Practice Questions

1. Solve for the variable x.

$$14x-1=111$$

a. $x=9$
b. $x=8$
c. $x=7$
d. $x=6$

2. Solve for the variable x.

$$3x-2.5x=6.5-10.5$$

a. $x=-6$
b. $x=-8$
c. $x=8$
d. $x=4$

3. Solve for the variable t.

$$-5t+3t+1=-14$$

a. $t=6$
b. $t=6.5$
c. $t=7$
d. $t=7.5$

4. Solve for the variable m.

$$3m-6+9m=30m+9$$

a. $m=-\frac{5}{6}$
b. $m=\frac{5}{6}$
c. $m=-\frac{4}{9}$
d. $m=-\frac{3}{5}$

5. Solve for the variable x.

$$-7x + 10x - 12 = 3$$

a. $x = 5.5$
b. $x = 6$
c. $x = 5$
d. $x = 6.5$

6. Solve for the variable n.

$$-1 + 3n = -7 - 6n$$

a. $n = 2$
b. $n = 3$
c. $n = \frac{2}{3}$
d. $n = -\frac{2}{3}$

7. The terms in each list are all considered like terms except for which list?

a. $2x, -7x, 2x$
b. $5xy, 9xy, xy$
c. $8x^2, 14x^2, -2x^2$
d. $45x, 45y, 45z$

8. Solve for the variable y.

$$3y + 5 - 2y = 6y - 10$$

a. $y = 3$
b. $y = 5$
c. $y = -2$
d. $y = -7$

9. Solve for the variable k.

$$-2k - 3 = -2(2k + 1)$$

a. $k = 3.5$
b. $k = 2.5$
c. $k = \frac{1}{2}$
d. $k = \frac{4}{5}$

10. Solve for the variable x.

$$-2(-2x + 3) + 4x = -2(6 - 3x) + 3x$$

a. $x = 3$
b. $x = 5$
c. $x = 8$
d. $x = 6$

11. Solve for the variable y.

$$-\frac{1}{2}(10y-2)+3=14$$

a. $y=-2$
b. $y=2$
c. $y=\frac{1}{2}$
d. $y=-\frac{1}{4}$

12. Solve for the variable n.

$$5(n+3)-n=2(n-4)+3$$

a. $n=-10$
b. $n=8$
c. $n=-13$
d. $n=9$

13. Ruby is working on solving equations in math class. Ruby's teacher puts the following problem on the board:

$$6+4w=4w-3+30w-6$$

The teacher tells the class to simplify the equation by combining like terms before solving for w. What should Ruby's simplified equation be?

a. $30w=-15$
b. $38w=-15$
c. $15=30w$
d. $15=38w$

14. In 2010, the cost of a laptop was $(-4c^2+450)$ dollars. In 2020, the cost of a laptop was $(-8c^2+600)$ dollars. To represent the difference in laptop costs, Marcus subtracts the 2010 cost from the 2020 cost: $(-8c^2+600)-(-4c^2+450)$. Simplify Marcus's expression by applying the distributive property and combining like terms.

a. $-12c^2+150$
b. $-4c^2+150$
c. $-12c^2+1{,}050$
d. $-4c^2+1{,}050$

15. Eddie and his friends are going to a water park that charges \$7 to enter the park and then \$2 per ride. He purchases lunch at the park for \$12. At the end of the day Eddie has spent \$31. How many rides did he go on?

a. 6 rides
b. 4 rides
c. 11 rides
d. 8 rides

1.7 Solving Multi-Step Equations

Equations are generally categorized as one-step, two-step, or multi-step. As the name suggests, **multi-step equations** are solved using more than two steps, and they require an organized and consistent approach. The goal of solving an equation is to isolate the variable. In order to get a variable on its own, skills such as distributing, combining like terms, and inverse operations will be used.

Sometimes multi-step equations can appear complex, but there is a step-by-step process that can be used to stay on track. When presented with a multi-step equation, consider following these four steps:

1. Clear the fraction
2. Distribute
3. Combine like terms
4. Isolate the variable

Consider the equation $\frac{2\left(x+\frac{1}{2}\right)}{3} = 5$.

1. Start by clearing the fraction. The fraction shows division by 3 on the left side, and the inverse of this is multiplication by 3. Multiply both sides by 3, and the equation becomes $2(x + \frac{1}{2}) = 15$.
2. Use the distributive property. Multiply the 2 into the set of parentheses. $2(x + \frac{1}{2}) = 15$ becomes $2x + 1 = 15$.
3. Combine like terms by moving the variables to one side and the constants to the other side. $2x + 1 = 15$ becomes $2x = 14$.
4. The last step is to isolate the variable using inverse operations. The inverse of multiplying by 2 is dividing by 2. Divide both sides by 2 so that $2x = 14$ becomes $x = 7$.

Not all multi-step equations will require *all* of these steps. If that is the case, then just skip unnecessary step and move on to the next one.

For example, consider the equation $2(3x - 1) = 2x + 6$.

1. Skip step 1 because there is no fraction.
2. Distribute the 2 into the set of parentheses. $2(3x - 1) = 2x + 6$ becomes $6x - 2 = 2x + 6$.
3. Combine like terms so that all variables are on one side and all constants are on the other side. Either side will work because the outcome will be the same either way. $6x - 2 = 2x + 6$ becomes $4x = 8$.
4. Isolate the variable using inverse operations. Divide both sides by 4 so that $4x = 8$ becomes $x = 2$.

Review Video: Solving Multi-Step Equations
Visit mometrix.com/academy and enter code: 555445

EXAMPLES

EXAMPLE 1:

Solve for x: $\frac{-8+x}{3} = -6$.

Answer:

Multiply both sides of the equation by 3.

$$3 \times \frac{-8 + x}{3} = -6 \times 3$$
$$-8 + x = -18$$

Add 8 to both sides.

$$-8 + x + 8 = -18 + 8$$
$$x = -10$$

EXAMPLE 2:

Solve for x: $-4x - 5 + x = 7(7 - 7x) + 6(1 + 6x)$.

Answer:

Distribute the 7 and the 6 into the sets of parentheses.

$$-4x - 5 + x = 49 - 49x + 6 + 36x$$

Combine like terms on both sides of the equation.

$$-3x - 5 = -13x + 55$$

Move the variables to one side and the constants to the other side.

$$-3x - 5 + 5 = -13x + 55 + 5$$
$$-3x = -13x + 60$$
$$-3x + 13x = -13x + 60 + 13x$$
$$10x = 60$$

Divide both sides by 10.

$$\frac{10x}{10} = \frac{60}{10}$$
$$x = 6$$

EXAMPLE 3:

Solve for x: $7x - (3x + 5) - 8 = \frac{1}{2}(8x + 20) - 7x + 5$.

Answer:

Distribute on both sides.

$$7x - 3x - 5 - 8 = 4x + 10 - 7x + 5$$

Combine like terms on both sides.

$$4x - 13 = -3x + 15$$

Move the variables to one side and the constants to the other side.

$$\begin{aligned} 4x - 13 + 13 &= -3x + 15 + 13 \\ 4x &= -3x + 28 \\ 4x + 3x &= -3x + 3x + 28 \\ 7x &= 28 \end{aligned}$$

Divide both sides by 7.

$$\frac{7x}{7} = \frac{28}{7}$$
$$x = 4$$

EXAMPLE 4:

Solve for the variable t: $-3(2t - 3) = 25 - 8t$.

Answer:

Distribute the –3 into the set of parentheses.

$$-6t + 9 = 25 - 8t$$

Move the variables to one side and the constants to the other side.

$$\begin{aligned} -6t + 9 - 9 &= 25 - 8t - 9 \\ -6t &= -8t + 16 \\ -6t + 8t &= -8t + 8t + 16 \\ 2t &= 16 \end{aligned}$$

Divide both sides by 2.

$$\frac{2t}{2} = \frac{16}{2}$$
$$t = 8$$

EXAMPLE 5:

Solve for the variable x.

$$\frac{2}{3}(x + 57) = 178$$

Answer:

Clear the fraction by multiplying both sides of the equation by 3.

$$3 \times \frac{2}{3}(x + 57) = 178 \times 3$$
$$2(x + 57) = 534$$

Distribute the 2 into the set of parentheses.

$$2x + 114 = 534$$

Move the variables to one side and the constants to the other side.

$$2x + 114 - 114 = 534 - 114$$
$$2x = 420$$

Divide both sides by 2.

$$\frac{2x}{2} = \frac{420}{2}$$
$$x = 210$$

Practice Questions

1. Solve for the variable x.

$$-3(4x + 3) + 4(6x + 1) = 43$$

a. $x = 7$
b. $x = 6$
c. $x = 5$
d. $x = 4$

2. Solve for the variable x.

$$3(x + 12) - x = 8$$

a. $x = 17$
b. $x = -14$
c. $x = 14$
d. $x = -17$

3. Solve for the variable x.

$$\frac{3}{10}(x + 2) = 12$$

a. $x = \frac{3}{8}$
b. $x = 21$
c. $x = 38$
d. $x = 3\frac{1}{2}$

4. Solve for the variable x.

$$-\frac{1}{2}(x + 2) + 3x = -1$$

a. $x = -3$
b. $x = -1$
c. $x = \frac{1}{2}$
d. $x = 0$

5. Solve for the variable *y*.

$$3y - 5 = -8(6 + 5y)$$

a. $y = \frac{1}{4}$
b. $y = -6$
c. $y = -1$
d. $y = 48$

6. Solve for the variable *y*.

$$2(y + 2) - 5 = 3(y + 1)$$

a. $y = -4$
b. $y = -5$
c. $y = \frac{1}{2}$
d. $y = 14$

7. Solve for the variable *y*.

$$-6y + 3 - 7 = 20$$

a. $y = -4$
b. $y = 6$
c. $y = 4$
d. $y = 5$

8. Solve for the variable *y*.

$$5(y - 3) = 6y + 10$$

a. $y = -13$
b. $y = 18$
c. $y = -25$
d. $y = 25.5$

9. Solve for the variable *n*.

$$-3(n - 6) + 4(n + 1) = 7n - 10$$

a. $n = 7$
b. $n = 12$
c. $n = \frac{3}{16}$
d. $n = \frac{16}{3}$

10. Solve for the variable *n*.

$$\frac{n + 4}{3} = 2$$

a. $n = 2$
b. $n = 10$
c. $n = 1$
d. $n = 9$

11. Solve for the variable *n*.

$$5(1 + 4n) + 2n = 27$$

a. $n = 11$
b. $n = 1\frac{1}{2}$
c. $n = 1$
d. $n = 1.1$

12. Solve for the variable *n*.

$$2(5n - 8) = -6(n - 8)$$

a. $n = 3.5$
b. $n = 4$
c. $n = 2$
d. $n = -8$

13. If the triangle and the square have the same perimeter, then what is the value of *x*? Set up a multi-step equation to help you solve for *x*.

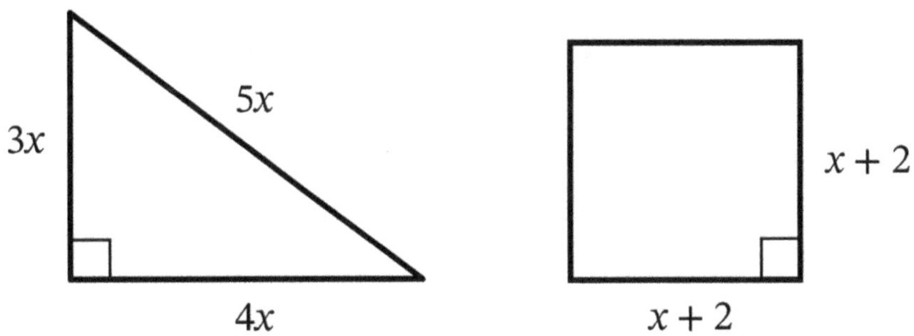

a. $x = 4$
b. $x = 1$
c. $x = 2$
d. $x = 1.5$

14. Mrs. Rodríguez has 8 busses available for a class field trip. She has 315 students attending the field trip, and 11 of these students say they don't need to ride the bus because they can get a ride from a family member. Set up an equation to solve for the number of students that will be on each bus. Each bus will carry the same number of students.

a. 38 students per bus
b. 26 students per bus
c. 21 students per bus
d. 33 students per bus

15. John wants to calculate the number of minutes he spent on the phone this month. He pays $30 per month plus $0.07 per minute, and his phone bill this month was $59.40. Set up a multi-step equation to solve for the number of minutes John spent on the phone this month.

a. 310 minutes
b. 200 minutes
c. 505 minutes
d. 420 minutes

1.8 Solving Equations Involving Absolute Value

When solving an equation with an absolute value, the first step is to isolate the absolute value term. We then consider two possibilities: when the expression inside the absolute value is positive or when it is negative. Absolute value is always a non-negative number because absolute value represents a distance from zero. You cannot have a negative distance, so the value will always be non-negative. In the case of the expression being positive, the expression in the absolute value equals the expression on the other side of the equation. In the case of the expression being negative, it equals the additive inverse of that expression—the expression times negative one. We consider each case separately and solve for the missing variable.

For instance, consider solving $|2x - 1| + x = 5$ for x. We can first isolate the absolute value by moving the x to the other side by subtracting x from both sides:

$$\begin{aligned} |2x - 1| + x - x &= 5 - x \\ |2x - 1| &= 5 - x \end{aligned}$$

Now, we have two possibilities. First, that $2x - 1$ is positive:

$$2x - 1 = 5 - x$$

Rearrange and combine like terms to solve for this value of x:

$$\begin{aligned} 2x - 1 + x &= 5 - x + x \\ 3x - 1 &= 5 \\ 3x - 1 + 1 &= 5 + 1 \\ 3x &= 6 \\ \frac{3x}{3} &= \frac{6}{3} \\ x &= 2 \end{aligned}$$

The other possibility is that $2x - 1$ is negative:

$$2x - 1 = -(5 - x)$$

Rearrange and combine like terms to solve for a second value of x:

$$\begin{aligned} 2x - 1 &= -5 + x \\ 2x - 1 - x &= -5 + x - x \\ x - 1 &= -5 \\ x - 1 + 1 &= -5 + 1 \\ x &= -4 \end{aligned}$$

Therefore, the two solutions for this absolute value equation are $x = 2$ or $x = -4$.

Note that the absolute value of a sum or difference applies to the sum or difference as a whole, not to the individual terms; in general, $|2x - 1|$ is not equal to $|2x + 1|$ or to $|2x| - 1$.

Review Video: Solving Absolute Value Equations
Visit mometrix.com/academy and enter code: 501208

Examples

Example 1:

Solve for x: $|x| = 9$.

Answer:

We have two possibilities. First, x may be positive:

$$x = 9$$

The other possibility is that x is negative:

$$x = -9$$

Therefore, the two solutions for this absolute value equation are $x = 9$ or $x = -9$.

Example 2:

Solve for x: $|x| + 2 = 3$.

Answer:

First, isolate the absolute value part of the equation by subtracting 2 from both sides.

$$|x| + 2 - 2 = 3 - 2$$
$$|x| = 1$$

We have two possibilities. First, that x may be positive:

$$x = 1$$

The other possibility is that x is negative:

$$x = -1$$

$$x = 1 \text{ or } x = -1$$

Example 3:

Solve for x: $|2x - 3| - 4 = 3$.

Answer:

Isolate the absolute value portion of the equation by adding 4 to both sides. Then set up a positive and negative outcome.

$$|2x - 3| - 4 + 4 = 3 + 4$$
$$|2x - 3| = 7$$

$$2x - 3 = 7$$
$$2x - 3 + 3 = 7 + 3$$
$$2x = 10$$
$$\frac{2x}{2} = \frac{10}{2}$$

$$2x - 3 = -7$$
$$2x - 3 + 3 = -7 + 3$$
$$2x = -4$$
$$\frac{2x}{2} = \frac{-4}{2}$$

$x = 5$ $x = -2$

EXAMPLE 4:

Solve for x.

$$|-2x + 7| = 25$$

Answer:

Isolate the absolute value portion of the equation, and then set up a positive and negative outcome.

$$-2x + 7 = 25$$
$$-2x + 7 - 7 = 25 - 7$$
$$-2x = 18$$
$$\frac{-2x}{-2} = \frac{18}{-2}$$
$$x = -9$$

$$-2x + 7 = -25$$
$$-2x + 7 - 7 = -25 - 7$$
$$-2x = -32$$
$$\frac{-2x}{-2} = \frac{-32}{-2}$$
$$x = 16$$

PRACTICE QUESTIONS

1. Solve for the variable x given the equation $|x| = 15$.

a. $x = 15$ or $x = -15$
b. $x = 15$
c. $x = 15$ or $x = 51$
d. $x = 0$ or $x = -15$

2. Solve for the variable x in the equation $|x| = 7$.

a. $x = 3$ or $x = -3$
b. $x = 7$ or $x = -7$
c. $x = 7$ or $x = -77$
d. $x = 7$ or $x = 0$

3. The equation $|x| - 3 = 7$ is given. What is the value of x?

a. $x = 3$ or $x = -3$
b. $x = 4$ or $x = -4$
c. $x = 10$ or $x = -10$
d. $x = 11$ or $x = -11$

4. The equation $|x| + 5 = 9$ is given. Solve for x.

a. $x = 4$ or $x = -4$
b. $x = 5$ or $x = -5$
c. $x = 13$ or $x = -13$
d. $x = 14$ or $x = -14$

5. Solve for the variable x given the equation $|2x - 4| = 4$.

a. $x = 0$
b. $x = 4$
c. $x = 4$ and $x = 0$
d. $x = 3$ and $x = 0$

6. Solve for the variable x given the equation $|2x + 6| = 10$.

a. $x = -2$ and $x = 8$
b. $x = -8$
c. $x = 2$
d. $x = 2$ and $x = -8$

7. What is the value of x in the equation $4|x + 2| - 6 = 18$?

a. $x = 1$ or $x = -5$
b. $x = 4$ or $x = -8$
c. $x = 14$ or $x = -18$
d. $x = 18$ or $x = -22$

8. The equation $|2x - 8| + 3 = 21$ is given. What is the value of x?

a. $x = 9$ or $x = -1$
b. $x = 12$ or $x = -4$
c. $x = 13$ or $x = -5$
d. $x = 16$ or $x = -8$

9. The equation $6|3y - 9| = 90$ is given. What is the value of y?

a. $y = 2$ or $y = -8$
b. $y = 3$ or $y = -3$
c. $y = 8$ or $y = -2$
d. $y = 35$ or $y = -29$

10. What is the value of y in the equation $2|2y - 4| - 3 = 17$?

a. $y = 3$ or $y = -3$
b. $y = 7$ or $y = -3$
c. $y = 9$ or $y = -5$
d. $y = 12$ or $y = -8$

11. The equation $|6y + 3| - 12 = 9$ is given. What is the value of y?

a. $y = -3$ or $y = -4$
b. $y = 1$ or $y = -1$
c. $y = 3$ or $y = -4$
d. $y = 4$ or $y = -4$

12. The equation $3|y - 5| - 16 = 11$ is given. Solve for y.

a. $y = -14$ or $y = 4$
b. $y = 14$ or $y = -4$
c. $y = 20$ or $y = -10$
d. $y = 35$ or $y = -25$

13. A cereal company has a product with 700 g of granola. The scale accepts any bag that is within 20 g of the required weight. What are the maximum and minimum weights that the scale will accept?

a. Max: 680 g, Min: 660 g
b. Max: 720 g, Min: 680 g
c. Max: 740 g, Min: 700 g
d. Max: 760 g, Min: 720 g

14. James scores an average of 38 points per game, with the game points being within 7 points of his game average. What is the lowest possible score for James during a game?

a. 19
b. 26
c. 31
d. 45

15. A real estate company is selling a house that is worth $280,000. They are willing to accept any offer that is within 3% of the value of the house. What is the least amount of money that the real estate company can accept for the house?

a. $196,400
b. $271,600
c. $288,400
d. $364,600

1.9 Solving Formulas for Specific Variables

Many formulas in math and science will have more than one variable. For example, the formula for the perimeter of a rectangle can be expressed as $P = 2(L + W)$, where P represents perimeter, L represents length, and W represents width. The formula is currently solved for P, but it can be manipulated to solve for any of the variables. The rules that apply to solving multi-step equations will also apply to manipulating a formula to isolate the variable of your choice. If we want to solve for the variable L in the perimeter formula, $P = 2(L + W)$, we can treat the formula as a multi-step equation. Start by dividing both sides by 2.

$$\frac{P}{2} = \frac{2(L + W)}{2}$$
$$\frac{P}{2} = L + W$$

Now, subtract W from both sides.

$$\frac{P}{2} - W = L + W - W$$
$$\frac{P}{2} - W = L$$

Consider the formula $V = \frac{D}{T}$, where V represents velocity, D represents distance, and T represents time. Currently, the formula is solved for V, but it can be manipulated to solve for D or T. For example, the variable T can be isolated by first multiplying both sides by T.

$$T \times V = \frac{D}{T} \times T$$
$$VT = D$$

Now divide both sides by V.

$$\frac{VT}{V} = \frac{D}{V}$$
$$T = \frac{D}{V}$$

D can be isolated by multiplying both sides by T.

$$T \times V = \frac{D}{T} \times T$$
$$VT = D$$

Another widely used formula is the simple interest formula. The formula states that $I = prt$, where I represents interest, p represents principal, r represents rate, and t represents time in years. For example, the interest you will earn over 5 years, on a \$10,000 principal, with an interest rate of 3.5% can be calculated using the formula. $I = prt$ would become $I = (10{,}000)(0.035)(5)$, which simplifies to 1,750. This means that after 5 years you will earn \$1,750. However, sometimes it is helpful to use the simple interest formula to solve for p, r, or t. In this case, the formula needs to be manipulated to isolate the desired variable. $I = prt$ can become $t = \frac{I}{pr}$, or $r = \frac{I}{pt}$, or $p = \frac{I}{rt}$ depending on the variable you are trying to isolate. For example, the time it will take to earn \$5,000 interest on a principal amount of \$20,000 with an interest rate of 2.5% can be calculated using the formula $t = \frac{I}{pr}$. This becomes $t = \frac{5,000}{(20,000)(0.025)}$, which simplifies to $t = 10$ years.

When manipulating a formula or equation to isolate a specific variable, make sure to utilize *inverse operations*, just as we did for equations with one variable. For example, when solving for the variable z in the equation $y = 2x + 3z$, start by subtracting $2x$ from both sides of the equation.

$$y - 2x = 2x + 3z - 2x$$
$$y - 2x = 3z$$

Now solve for z by dividing both sides by 3.

$$\frac{y - 2x}{3} = \frac{3z}{3}$$
$$\frac{y - 2x}{3} = z$$

Review Video: Solving Equations for Specific Variables
Visit mometrix.com/academy and enter code: 130695

EXAMPLES

EXAMPLE 1:

Solve for the variable x in the equation $y = 4x - 8z$.

Answer:

Use inverse operations to isolate the variable x. Start by adding $8z$ to both sides.

$$y + 8z = 4x - 8z + 8z$$
$$y + 8z = 4x$$

Now divide both sides by 4.

$$\frac{y + 8z}{4} = \frac{4x}{4}$$
$$\frac{y}{4} + \frac{8z}{4} = x$$
$$\frac{y}{4} + 2z = x$$

EXAMPLE 2:

Solve for the variable k in the equation $r = 5x + 3k$.

Answer:

Use inverse operations to isolate the variable k. Start by subtracting $5x$ from both sides.

$$r - 5x = 5x + 3k - 5x$$
$$r - 5x = 3k$$

Now divide both sides by 3.

$$\frac{r - 5x}{3} = \frac{3k}{3}$$

$$\frac{r}{3} - \frac{5x}{3} = k \text{ or } \frac{r}{3} - \frac{5}{3}x = k$$

PRACTICE QUESTIONS

1. Given the formula $e = mc^2$, solve for m.

a. $m = \frac{c^2}{e}$
b. $m = \frac{e}{c^2}$
c. $m = \frac{e}{c}$
d. $m = ec^2$

2. The area of a circle can be calculated using the formula $A = \pi r^2$. Given this formula, solve for the circle's radius, r.

a. $r = \frac{A}{\pi}$
b. $r = \frac{A}{\pi^2}$
c. $r = A\sqrt{\frac{1}{\pi}}$
d. $r = \sqrt{\frac{A}{\pi}}$

3. From the Pythagorean theorem, $a^2 + b^2 = c^2$, solve for c.

a. $c = a + b$
b. $c = a^2 + b^2$
c. $c = \sqrt{a^2} + \sqrt{b^2}$
d. $c = \sqrt{a^2 + b^2}$

4. From the Pythagorean theorem, $a^2 + b^2 = c^2$, solve for b.

a. $b = c - a$
b. $b = \sqrt{c^2 - a^2}$
c. $b = c^2 - a^2$
d. $b = \sqrt{c^2} - \sqrt{a^2}$

5. Given the equation $4x + 2y - 8 = 4z$, solve for the variable y.

a. $y = -4x + 4z + 4$
b. $y = -2x - 2z + 8$
c. $y = -2x + 2z + 4$
d. $y = -4x + 4z + 8$

6. In physics, the term momentum refers to how difficult it will be to stop a moving object. Momentum is calculated using the formula $p = mv$, where p is momentum, m is the object's mass, and v is the object's velocity or speed. Rearrange this formula to solve for v.

a. $v = mp$
b. $v = \frac{m}{p}$
c. $v = p - m$
d. $v = \frac{p}{m}$

7. For any triangle, the sum of its three interior angles must equal 180°. This fact can be written into a formula as $A + B + C = 180°$. Rearrange this formula to get B by itself.

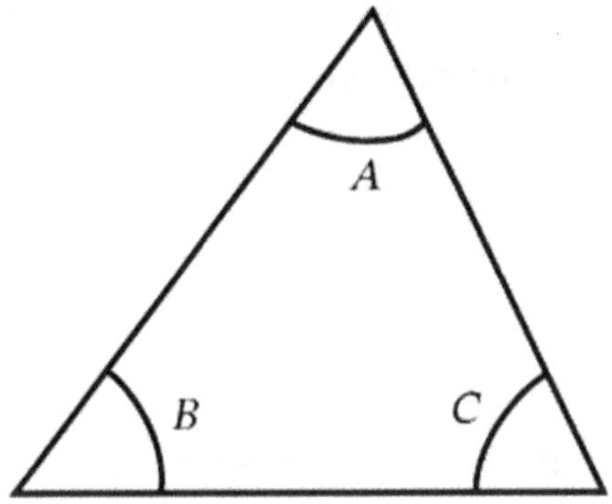

$$A + B + C = 180°$$

a. $B = 180° + A + C$
b. $B = 180° - A - C$
c. $B = 180° - AC$
d. $B = AC - 180$

8. In thermodynamics, scientists study heat and other forms of energy and how they move. One important formula used in thermodynamics is $Q = mc\Delta T$, where an object's mass m, specific heat c, and change in temperature ΔT are multiplied together to calculate heat transfer Q. Rearrange the formula to solve for ΔT.

a. $\Delta T = \frac{Q}{mc}$
b. $\Delta T = \frac{mc}{Q}$
c. $\Delta T = mcQ$
d. $\Delta T = Q - mc$

9. Use the formula $Q = mc\Delta T$ to solve for m.

a. $m = Qc\Delta T$
b. $m = Q - c - \Delta T$
c. $m = \frac{cQ}{\Delta T}$
d. $m = \frac{Q}{c\Delta T}$

10. The volume of a sphere can be found using the formula $V = \frac{4}{3}\pi r^3$. Solve for r.

a. $r = \sqrt[3]{\frac{3}{4\pi}V}$
b. $r = \sqrt[3]{\frac{4\pi}{3}V}$
c. $r = V\sqrt[3]{\frac{3}{4\pi}}$
d. $r = \frac{\sqrt[3]{V}}{\frac{4}{3}\pi}$

11. The formula for converting Fahrenheit to Celsius is $C = \frac{5}{9}(F - 32)$. Derive the formula for converting Celsius to Fahrenheit from this formula by solving for F.

a. $F = \frac{9}{5}(C + 32)$
b. $F = \frac{9}{5}C + 32$
c. $F = \frac{5}{9}C + 32$
d. $F = \frac{9}{5}C - 32$

12. One of the Pythagorean identities in trigonometry is $1 + tan^2\theta = sec^2\,\theta$. From this identity, solve for $\tan^2\theta$.

a. $tan^2\theta = sec\,\theta - 1$
b. $tan^2\theta = sec^2\theta + 1$
c. $tan^2\theta = \sqrt{sec\theta} - \sqrt{1}$
d. $tan^2\theta = sec^2\theta - 1$

13. Scientists are sometimes interested in the densities of varying objects, which is the ratio of an object's mass to its size. This relationship is given in the formula $\rho = \frac{m}{V}$, where the Greek letter ρ (pronounced "row") is equal to the object's mass divided by the object's volume. A university physics lab has a block of copper with density $\rho = 8.96\ g/cm^3$ and volume $V = 10\ cm^3$. Rearrange the density formula to solve for m, and then determine the copper block's mass.

a. $m = 0.896$ g
b. $m = 1.116$ g
c. $m = 89.6$ g
d. $m = 18.96$ g

14. Morgan recently opened a savings account and put \$200 (p) in it. Her bank pays interest at the rate of 2.5% (r) each year. If Morgan wants to earn \$10 (i) in interest at this bank without adding or removing money from the account, how long will she have to wait? (Hint: rearrange the formula $i = prt$ to solve for t, then substitute in the known values of i, p, and r.)

a. $t = 4$ years
b. $t = 50$ years
c. $t = \frac{1}{2}$ years
d. $t = 2$ years

15. Mr. Ledger is an accountant for the Make It Count accounting firm, and he has the following financial information (in thousands) for this fiscal year:

Total assets $(TA) = 850$
Total liabilities $(TL) = 425$
Common stock $(CS) = 175$
Retained Earnings $(RE) = ???$

Mr. Ledger is still waiting to receive final information about this year's retained earnings, but he can determine that value by using the accounting equation: $TA = TL + CS + RE$. Rearrange this formula and solve for the value of RE.

a. $RE = 250$
b. $RE = 600$
c. $RE = 1{,}100$
d. $RE = 1{,}450$

1.10 Writing and Solving Equations Involving Direct Variation

Direct variation describes a relationship between two variables, where the value of one depends directly on the value of the other. In equation form, direct variation is described as $y = kx$. The equation describes a relationship between x and y, where y varies *directly* with x. In other words, the value of y will change depending on what the value of x is. Since there is no constant added to kx, this means the graph of a direct function will always pass through the origin. The k in $y = kx$ is called the constant, or the **constant of proportionality**. The value k can be used to determine the rate of change.

$y = 7x$ is an example of a direct variation equation. The value of y will vary directly with the value of x. If x is 3, then y is 21. If x is 9, then y is 63. There is a direct relationship between the value of x and y. The constant of proportionality, or k, in this equation is 7.

Often when you are solving direct variation problems, you are looking for the value of k, or you are using the value of k to find x or y.

For example, find the constant of proportionality, k, in the following scenario: a car travels 320 miles in 4 hours.

The ratio between distance traveled and time should be constant. The equation states that $y = kx$, and in this scenario x (independent variable) is hours, and y (dependent variable) is miles. The equation $y = kx$ becomes $320 = k(4)$. Now solve for k by dividing both sides by 4.

$$\frac{320}{4} = \frac{k(4)}{4}$$
$$80 = k$$

Sometimes you will be writing direct variation equations based on data. For example, the data in the table below shows x- and y-values, where the value of y depends on the value of x. As x increases, y also increases. The data can be used to determine the value of k. The formula $y = kx$ becomes $12 = k(6)$ when an ordered pair from the table below is plugged into the formula. $12 = k(6)$ simplifies to $k = 2$.

x	y
6	12
7	14
8	16

Direct variation equations can include positive and negative values for k, as well as fractional and decimal values. More examples of direct variation equations include: $y = -2x$, $y = \frac{1}{4}x$, $y = (0.43)x$, and $y = -1.1x$.

Review Video: Direct Variation
Visit mometrix.com/academy and enter code: 547565

EXAMPLES

EXAMPLE 1:

The amount of money spent on a monthly gym membership (y) varies directly to the number of months of enrollment (x). If gym memberships are \$24 per month, what is the total cost for seven months?

Answer:

$y = kx$ becomes $y = 24(7)$, which simplifies to $y = 168$. The total cost for a seven-month gym membership is $168.

Example 2:

Write and solve a direct variation equation based on the following scenario.

When lightning strikes, you can estimate your distance from the strike based on the number of seconds it takes to hear the thunder after seeing the lighting. For every 1 mile you are away from the storm, it takes 5 seconds for you to hear thunder. In this experiment, the constant of proportionality is equal to 5 ($k = 5$).

If you see a lightning strike and then hear thunder 25 seconds later, approximately how many miles are you from the strike?

Answer:

$$y = kx$$

k is 5.

y is 25 seconds.

x is the unknown number of miles.

$$25 = 5x$$
$$\frac{25}{5} = \frac{5x}{5}$$
$$x = 5$$

The distance from the lightning strike is approximately 5 miles.

Example 3:

Write and solve a direct variation equation based on the following scenario.

A plant grows 4 inches each week, which represents the constant of proportionality ($k = 4$). If the plant is currently 6 weeks old, how tall is the plant?

Answer:

$$y = kx$$

k is 4.

y is the unknown height in inches.

x is 6 weeks.

$$y = 4(6)$$
$$y = 24$$

The height of the plant is 24 inches.

Example 4:

Determine whether the information given in the table below represents an example of direct variation. If it does, find the value of k.

x	y
2	8
4	16
6	24

Answer:

To see this, use the formula $k = \frac{y}{x}$, and check whether k is the same from data point to data point.

First, the table tells us that when $x = 2$, $y = 8$. The value of k for this data point is then $k = \frac{8}{2} = 4$.

Next, when $x = 4$, $y = 165$. The value of k for this data point is $k = \frac{16}{4} = 4$.

Finally, when $x = 6$, $y = 24$. The final value of k is $k = \frac{24}{6} = 4$.

Since k is the same at all data points, this is an example of direct variation. And of course, $k = 4$.

Practice Questions

1. Write a direct variation equation where $k = 5$.

a. $5 = yx$
b. $y = 5k$
c. $y = 5x$
d. $x = 5y$

2. Determine which of the following equations is an example of a direct variation equation.

a. $y = 2x + 1$
b. $y = 4x - 4$
c. $y = 12x$
d. $y = -3x + \frac{2}{3}$

3. Write a direct variation equation where $k = \frac{3}{5}$.

a. $y = \frac{3}{5} + x$
b. $x = \frac{3}{5}y$
c. $y = \frac{3}{5}x$
d. $y = \frac{3}{5x}$

4. Which of the following equations is not an example of direct variation?

a. $y = -10x$
b. $y = x$
c. $y = \frac{1}{2}x$
d. $y = 2x + 2$

5. Write a direct variation equation where $k = -3$, and evaluate the equation when $x = -1$.

a. $y = 3x, y = -3$
b. $y = -3x, y = 3$
c. $y = 3x, y = 3$
d. $y = -3x, y = -3$

6. Write a direct variation equation where $k = 12$, and evaluate the equation at $x = 5$.

a. $y = 12x, y = 60$
b. $y = 12 + x, y = 17$
c. $y = 12x + 12, y = 72$
d. $y = \frac{12}{x}, y = \frac{12}{5}$

7. For problems involving direct variation, we have the equation $y = kx$. What is the formula used for direct variation problems when you are given x and y and want to solve for k?

a. $x = yk$
b. $k = x + y$
c. $k = \frac{y}{x}$
d. $x = \frac{k}{y}$

8. Determine whether the information given in the table below represents an example of direct variation. If it does, find the value of k.

x	y
1	-5
3	-15
5	-25

a. This data does not represent an example of direct variation.
b. This data represents an example of direct variation and $k = 5$.
c. This data represents an example of direct variation and $k = -\frac{1}{5}$.
d. This data represents an example of direct variation and $k = -5$.

9. Determine whether the information given in the table below represents an example of direct variation. If it does, find the value of k.

x	y
10	20
12	36
14	42

a. This data does not represent an example of direct variation.
b. This data represents an example of direct variation and $k = 2.5$.
c. This data represents an example of direct variation and $k = 3$.
d. This data represents an example of direct variation and $k = 2$.

10. Determine whether the information given in the table below represents an example of direct variation. If it does, write the direct variation equation that corresponds to the data.

x	y
3	-2
6	-4

x	y
9	-6

a. This data does not represent an example of direct variation.
b. This data represents an example of direct variation and $y = -\frac{3}{2}x$.
c. This data represents an example of direct variation and $y = -\frac{2}{3}x$.
d. This data represents an example of direct variation $y = \frac{3}{2} + x$.

11. Determine whether the information given in the table below represents an example of direct variation. If it does, find the value of k.

x	y
-4	16
-2	8
0	0
2	-8

a. This data does not represent an example of direct variation.
b. This data represents an example of direct variation and $k = -\frac{1}{4}$.
c. This data represents an example of direct variation and $k = 4$.
d. This data represents an example of direct variation and $k = -4$.

12. Fill in the following table, given that the data are an example of direct variation and $k = 2$.

x	y
	-9
	-2
	5
	12

a. $x = -18, -4, 10, 24$
b. $x = -4.5, -1, 2.5, 6$
c. $x = -9, -2, 5, 12$
d. $x = 4.5, 1, -2.5, -6$

13. On average, adults typically consume about 2,000 calories per day. Let x represent "number of days" and write a direct variation equation describing how many calories an adult consumes on average given a certain number of days.

a. $y = 2{,}000x$
b. $y = 2{,}000 + x$
c. $y = 2x + 1{,}000$
d. $y = \frac{x}{2{,}000}$

14. Chase is driving 60 mph and still has 2.5 hours of driving on his trip. Using his speed, write a direct variation equation and solve to determine how many miles he has left to drive.

a. $y = 60x$; 125 miles
b. $y = 60x$; 150 miles
c. $y = 25x$; 180 miles
d. $y = 2.5x$; 125 miles

15. Casey is getting ready to bake cookies for a bake sale. If the recipe calls for half a cup of flour for a dozen cookies, which of the following is a direct variation equation that represents how many cups of flour she will need for each dozen cookies she bakes?

a. $y = \frac{1}{2}x$
b. $y = 2x$
c. $y = x$
d. $y = \frac{1}{3}x$

Chapter 1 Quiz

1. Find the value of x in the equation $29 + x = 44$.

a. $x = 15$
b. $x = 16$
c. $x = 25$
d. $x = 73$

2. The kindergarten section of Westchester School District has 12 classrooms. If each classroom has 16 students, how many kindergarteners are in the entire school district?

a. 28 students
b. 192 students
c. 200 students
d. 212 students

3. Find the value of n in the equation $\frac{n}{5} + 8 = 12$.

a. $n = 4$
b. $n = 20$
c. $n = 9$
d. $n = 25$

4. Emily is renting a paddle boat for the day. It costs 14 dollars to rent, plus 3 dollars for every hour she has the boat. If Emily paid a total of 32 dollars, how many hours did she rent the paddle boat for?

a. 2 hours
b. 10 hours
c. 14 hours
d. 6 hours

5. Find the value of q in the equation $4(q - 4) = 16$.

a. $q = 7$
b. $q = 8$
c. $q = 3$
d. $q = 12$

6. A rectangular planter has a width of 7 inches and a length of $(2x + 4)$ inches. If the area of the planter is 70 in^2, what is the value of x?

a. 1 inch
b. 2 inches
c. 3 inches
d. 4 inches

7. Find the value of p in the equation $5p = 3p + 16$.

a. $p = -2$
b. $p = -8$
c. $p = 8$
d. $p = 2$

8. 24 more than 5 times Jonathan's age is the same as 7 times his age. How old is Jonathan?

a. 10 years old
b. 2 years old
c. 24 years old
d. 12 years old

9. Simplify the equation by combining like terms:

$$8x + 4 + x + 2y - 1 = 23$$

a. $8x + 2y = 23$
b. $8x + x + 2y = 20$
c. $9x + 2y = 23$
d. $9x + 2y = 20$

10. Jack is finding the perimeter of a rectangle using the formula $2l + 2w = P$. The length of the rectangle is $(12b + 5)$ inches, and the width is 8 inches. The perimeter of the rectangle is $(4b + 126)$ inches. Jack uses these values to write the following equation for the perimeter of the rectangle:

$$2(12b + 5) + 2(8) = 4b + 126$$

What is the value of b?

a. $b = 4$
b. $b = 5$
c. $b = 6$
d. $b = 7$

11. Determine the value of x when $|x| + 12 = 15$.

a. $x = \pm 3$
b. $x = -3$
c. $x = \frac{1}{3}$
d. $x = 3$

12. On opening night of new galleries, Francine's Art Studio typically has far more guests than on normal business days. The number of guests on any given day is given by x in the equation $2|x - 95| + 10 = 150$. Solve for x and determine how many guests are present on the busiest opening night and how many are present on the slowest day.

a. 150 guests on the busiest opening night and 95 guests on the slowest day
b. 190 guests on the busiest opening night and 140 guests on the slowest day
c. 105 guests on the busiest opening night and 20 guests on the slowest day
d. 165 guests on the busiest opening night and 25 guests on the slowest day

13. Simplify the following expression:

$$(5x + 3xy - 2) + (7x + 2xy - 9)$$

a. $12x + 5xy - 11$
b. $-2x + xy + 7$
c. $12x + xy - 11$
d. $15xy - 4x - 9$

14. Andy opens a bank account and deposits $50 initially, and then $15 every week. How many weeks will it take Andy to have $500 in his savings account?

a. 27 weeks
b. 28 weeks
c. 29 weeks
d. 30 weeks

15. Find the value of x in the equation $3(2x - 5) - 4x = 49$.

a. $x = 32$
b. $x = 64$
c. $x = 17$
d. $x = 27$

16. A rectangular garden has a length of 5 feet and a width of $(2x - 4)$ feet. If the area of the garden is $(2x + 4)$ square feet, what is the value of x? Write an equation representing this scenario and solve to find the value of x.

a. $x = 6$
b. $x = 4$
c. $x = 8$
d. $x = 3$

17. Solve the formula $y = 3x + 2z$ for z.

a. $z = \frac{2}{3x-y}$
b. $z = \frac{2}{y-3x}$
c. $z = \frac{3x-y}{2}$
d. $z = \frac{y-3x}{2}$

18. The formula for the perimeter of a parallelogram is $P = 2l + 2w$, where P stands for perimeter, l stands for length, and w stands for width. Solve the formula for w.

A. $w = \frac{2}{2l-P}$
b. $w = \frac{2}{P-2l}$
c. $w = \frac{2l-P}{2}$
d. $w = \frac{P-2l}{2}$

19. If y varies directly with x at a constant variation of 3, find y when $x = 108$.

a. $y = 3{,}108$
b. $y = 324$
c. $y = 111$
d. $y = 36$

20. Roberto's weekly pay is directly related to the number of hours he works at the hardware store. His pay is \$360 for 24 hours of work. How much will Roberto's weekly pay be if he works 40 hours? Write and solve an equation to represent this scenario.

a. \$350
b. \$384
c. \$400
d. \$600

Chapter 2: Functions and Sequences

2.1 Evaluating Functions

Expressions, equations, and functions are similar terms, with slightly different meanings. Remember, an **expression** is a math phrase with numbers, variables, and operations. For example, $4x + 15y + 2z^2$ is an expression that can be evaluated if the value of each variable is known. If the values for x, y, and z are provided, they can be plugged into the expression to evaluate. Let's say $x = 7, y = 2$, and $z = 3$. To evaluate the expression simply plug in each value. $4x + 15y + 2z^2$ becomes $4(7) + 15(2) + 2(3)^2$, which simplifies to 76.

Functions are equations that have an input (x) and an output (y), where every input has exactly one output. For example, $y = 2x$ is a function. When a value is plugged into the input (x), the result will be exactly one output (y). Plug in 5 for x, and the output is 10. Plug in –3 for x, the output is –6. Functions can be written using function notation.

Function notation uses $f(x)$ instead of y to indicate the outputs. Essentially, $f(x)$ represents y. $f(x)$ means that y will depend on the value of x. The $f(x)$ represents the dependent variable (output), and x represents the independent variable (input).

Functions are evaluated by plugging in values for x (input) and determining what the result is for $f(x)$ (output). These input/output pairs form **ordered pairs**. When a large number of these (x, y) pairs are plotted on a coordinate plane, the resulting line or shape is a graphical representation of the function.

For example, consider the function $f(x) = x^2$. Plugging in values for x will result in output values for $f(x)$. Plugging in –2 for x results in $f(x) = (-2)^2$, which simplifies to $f(x) = 4$. This means that the point $(-2,4)$ will be located on the graph of the function $f(x) = x^2$.

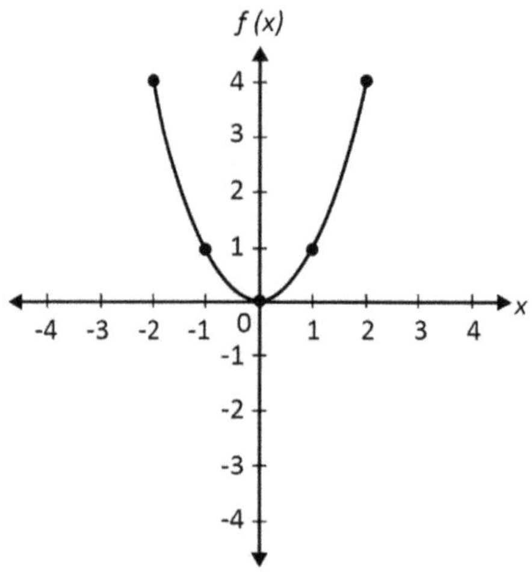

x	f(x)
-2	4
-1	1
0	0
1	1
2	4

Review Video: Evaluating Functions
Visit mometrix.com/academy and enter code: 588515

EXAMPLES

EXAMPLE 1

Evaluate the function $f(x) = 7x - 5$ when $x = 6$.

ANSWER

Plug in 6 for x.

$$f(6) = 7(6) - 5$$

Multiply, then subtract.

$$f(6) = 42 - 5 = 37$$

EXAMPLE 2

Simplify the expression $3x - 5y + \frac{1}{2}z$ when $x = 2$, $y = 4$, and $z = \frac{1}{2}$.

ANSWER

$$3(2) - 5(4) + \frac{1}{2}\left(\frac{1}{2}\right)$$

Start simplifying by multiplying.

$$6 - 20 + \frac{1}{4}$$

Add and subtract to get the final answer.

$$-13\frac{3}{4}$$

PRACTICE QUESTIONS

1. The function $f(x) = -3x - 8$ is given. What is $f(5)$?

a. $f(5) = -23$
b. $f(5) = -7$
c. $f(5) = 7$
d. $f(5) = 23$

2. The function $f(x) = -5x + 18$ is given. What is $f(-4)$?

a. $f(-4) = -38$
b. $f(-4) = -2$
c. $f(-4) = 2$
d. $f(-4) = 38$

3. The function $f(x) = 4x - 16$ is given. What is $f(9)$?

a. $f(9) = 3$
b. $f(9) = 20$
c. $f(9) = 29$
d. $f(9) = 52$

4. The function $g(x) = 7x + 25$ is given. What is $g(-10)$?

a. $g(-10) = -45$
b. $g(-10) = -42$
c. $g(-10) = 42$
d. $g(-10) = 45$

5. The function $g(x) = 5x - 33$ is given. What is $g(-6)$?

a. $g(-6) = -63$
b. $g(-6) = -44$
c. $g(-6) = 44$
d. $g(-6) = 63$

6. The function $g(x) = 9x + 26$ is given. What is $g(-7)$?

a. $g(-7) = -90$
b. $g(-7) = -37$
c. $g(-7) = 37$
d. $g(-7) = 90$

7. The function $g(x) = -10x + 31$ is given. What is $g(12)$?

a. $g(12) = -151$
b. $g(12) = -89$
c. $g(12) = 89$
d. $g(12) = 151$

8. Evaluate the function $f(x) = 3(x + 8) - 16$ when $x = 6$.

a. $f(6) = 1$
b. $f(6) = 298$
c. $f(6) = 26$
d. $f(6) = 34$

9. Evaluate the function $f(x) = x^2 + 2x + 4$ when $x = 10$.

a. $f(10) = 124$
b. $f(10) = 44$
c. $f(10) = 126$
d. $f(10) = 234$

10. If $f(x) = x^4 + 2x^3$, find the value of $f(3) + 2f(4)$.

a. 2,519
b. 521
c. 519
d. 903

11. Michelle works at an appliance shop. She makes \$15 per hour, plus an additional \$30 for each sale she makes. On Tuesday, Michelle made 5 sales, so she earned an extra \$150. Michelle's total pay for Tuesday can be represented by the function $f(x) = 15x + 150$, where x represents the number of hours worked. How much did Michelle earn altogether if she worked for 8 hours on Tuesday?

a. \$173
b. \$260
c. \$308
d. \$270

12. A gym membership costs \$15 per week. The total cost of the membership is represented by the function $f(x) = 15x$, with x representing the number of weeks in the membership. The gym also rents out lockers for an additional \$3 per week. The total cost of a membership with a locker rental is represented by the function $g(x) = 18x$. What is the difference between the cost of a 20-week membership with a locker rental and a 20-week membership without a locker rental?

a. \$40
b. \$60
c. \$30
d. \$20

13. When traveling, Maya gets paid a daily per diem of \$38 for food and \$0.58 per mile she travels using her car. If Maya travels to a city that is 75 miles away, what is her pay for traveling to and from the city in one day?

a. \$81.50
b. \$119.50
c. \$125.00
d. \$163.00

14. A kayak rental company charges \$9 per hour and a \$15 flat daily fee to rent a kayak. How much would it cost to rent a kayak for 5 hours?

a. \$24
b. \$60
c. \$84
d. \$136

15. A gym charges a one-time registration fee of \$50 and \$6 per week for membership. How much would someone pay for a 12-week gym membership fee?

a. \$56
b. \$68
c. \$122
d. \$606

2.2 Domain and Range of Linear Functions

A **linear function** has an independent variable and a dependent variable. The dependent variable y changes at a constant rate when the value of the independent variable x changes. This consistent relationship between independent and dependent variables is what creates a straight line when graphed. Linear functions can be represented using equations, tables, and graphs.

The **domain** of a function refers to all of the possible x-values, or input values, for which the function is defined. The **range** of a function refers to all of the possible y-values, or output values, produced by the function.

Linear functions such as $y = 2x - 3$, can be graphed using the slope and y-intercept. In this case, the line will extend in both directions indefinitely. The possible x- and y-values will both be infinite. This means that both the domain and range are the set of all real numbers, or D: $(-\infty, \infty)$, R: $(-\infty, \infty)$.

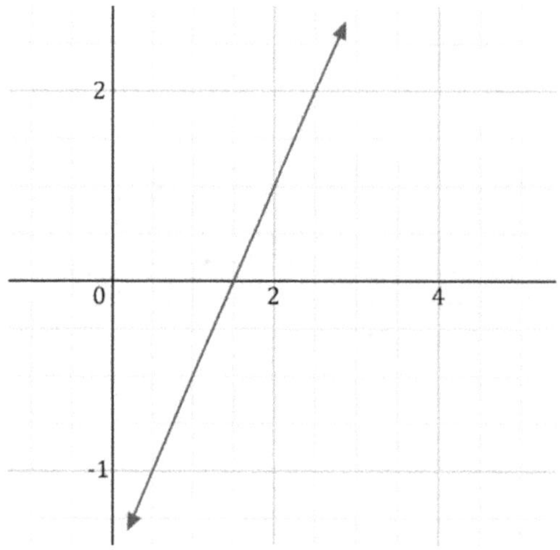

However, some functions will not have a domain and range that is the set of all real numbers. For example, linear inequalities will often have a domain and range that is more limited. **Linear inequalities** are simply linear functions that incorporate an inequality symbol such as $<$, $>$, $\leq$, or $\geq$.

For example, a linear inequality is graphed below. The domain can be determined by identifying all x-values that satisfy the graph. In this case, the domain is any value less than 1 ($x < 1$). The range can be determined by identifying all y-values that satisfy the graph. In this case, the only value of y that satisfies the graph is $y = -2$.

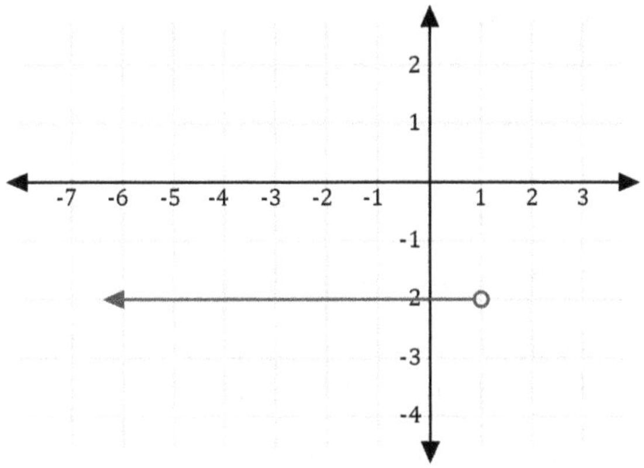

If the graph of a linear inequality shows a solid, or "closed," circle, it indicates the symbols $\leq$ or $\geq$. If the graph shows a hollow, or "open," circle, it indicates the symbols $<$ or $>$.

When identifying domain and range from a table, the x-values will represent the domain, and the y-values will represent the range. For example, in the table below, the domain is $\{-6, -1, 1, 5, 7\}$ and the range is $\{-4, -1, 2, 4\}$. Repeat values do not need to be listed more than once.

x	y
-6	2
−1	−1
1	4
5	4
7	−4

There will be times when the domain and range need to be identified, but an equation, graph, or table are not provided. When determining the domain and range from a written description or real-world situation, start by identifying the independent and dependent variables. For example, if your job pays $40 per hour, and you work 38 hours per week, the *independent variable* is the number of hours you work, and the *dependent variable* is the income per week. That is, your pay, y, depends on the number of hours you work, x.

Let's say you worked 38 hours in week 1, 40 hours in week 2, 35 hours in week 3, 40 hours in week 4, and 39 hours in week 5. The domain of hours worked over the 5 weeks would be D: $\{35, 38, 39, 40\}$. The range would be calculated by multiplying the number of hours by 40 (your pay per hour). The range would be R: $\{1{,}400, 1{,}520, 1{,}560, 1{,}600\}$.

The important thing to remember is that linear functions have independent and dependent variables. One variable *depends* on the other, and this consistent relationship of a constant increase in the dependent variable from one independent variable to the next is what creates a straight line when graphed. Sometimes the lines will extend indefinitely, and sometimes the lines will appear as segments. The independent variable, x, represents the domain, or input values. The dependent variable, y, represents the range, or output values.

Examples

Example 1:

Determine the domain and range of the linear function:

$$y = 3x - 1$$

Answer

The domain of the function is all of the possible x-values.

The range of the function is all of the possible y-values.

To find the domain and range, we need to determine the lowest and highest possible x- and y-values. This is easily identified if the function is graphed.

Graph the y-intercept, and then use the slope to build the line.

$$y = 3x - 1$$

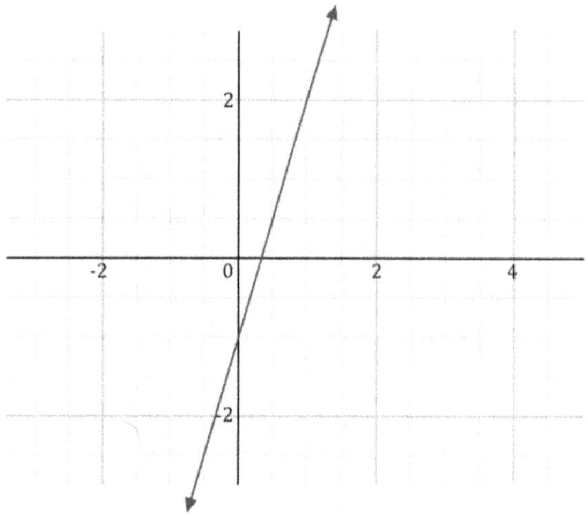

The x-values extend infinitely in the positive and negative directions. This means that the domain of the function is $(-\infty, \infty)$.

The y-values extend infinitely in the positive and negative directions as well. This means that the range of the function is also $(-\infty, \infty)$.

EXAMPLE 2:

Determine the domain and range of the graph below.

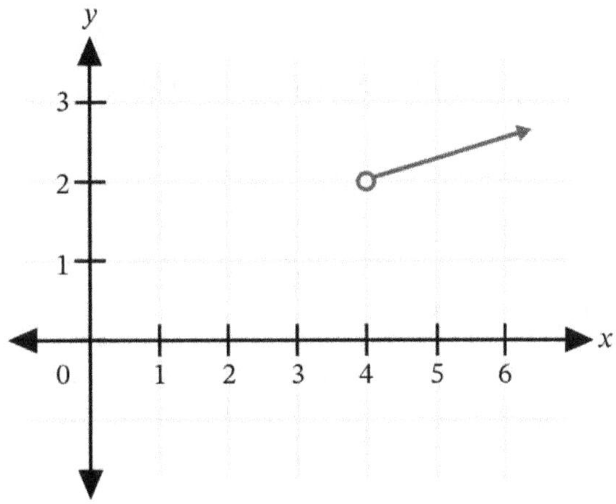

ANSWER

The domain of the function is all of the possible x-values.

The range of the function is all of the possible y-values.

To find the domain and range, we need to determine the lowest and highest possible x- and y-values. On this graph, there is an open circle at one end and an arrow on the other end. This implies that the x-values can be any value greater than 4, and the y-values can be any value greater than 2.

Domain (x-values): $x > 4$

Range (y-values): $y > 2$

EXAMPLE 3:

Determine the domain and range of the table below.

x	y
5	3
2	5
4	7
3	9
1	12

ANSWER

The domain is all of the possible x-values, and the range is all of the possible y-values.

Domain: $\{1,2,3,4,5\}$

Range: $\{3,5,7,9,12\}$

EXAMPLE 4:

A diver is currently 33 feet deep on a dive adventure. The graph below illustrates how pressure increases the deeper you dive. Express the domain and range of the scenario that is graphed.

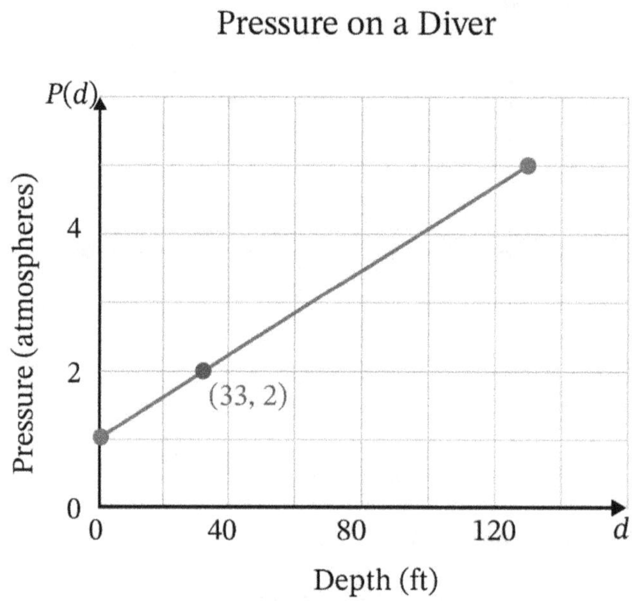

ANSWER

The domain (all possible x-values) is any depth from 0 to 130. The range (all possible y-values) is any pressure greater than or equal to 1 and less than 5.

Domain: $0 \leq d \leq 130$

Range: $1 \leq P \leq 5$

EXAMPLE 5:

Determine the range values for the function $f(x) = 2x + 1$ for the domain values $\{1,2,3\}$.

ANSWER

Since domain represents input and range represents output, to find the range values when given domain values, simply substitute each domain value into the function for x and solve for $f(x)$.

$$f(1) = 2(1) + 1 = 2 + 1 = 3$$

$$f(2) = 2(2) + 1 = 4 + 1 = 5$$

$$f(3) = 2(3) + 1 = 6 + 1 = 7$$

Therefore, the range values for the function $f(x) = 2x + 1$ for the domain values $\{1,2,3\}$ are $\{3,5,7\}$.

PRACTICE QUESTIONS

1. Determine the domain and range of the linear function graphed below.

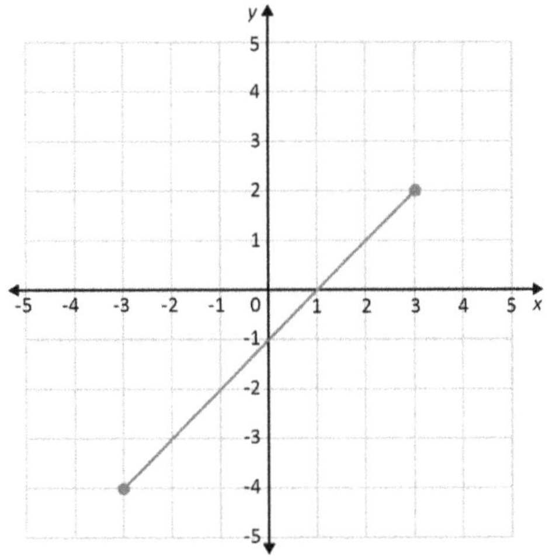

a. Domain: $-3 < y < 3$
Range: $-4 < x < 2$
b. Domain: $-3 \leq x \leq 3$
Range: $-4 \leq y \leq 2$
c. Domain: $-3 \leq x \leq 3$
Range: $-4 \leq y \leq 2$
d. Domain: $-4 \leq y \leq 2$
Range: $-3 \leq x \leq 3$

2. Determine the domain and range of the linear function graphed below.

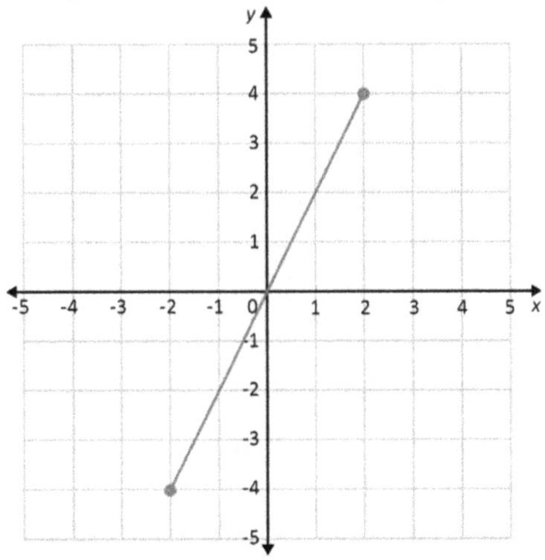

a. Domain: $-2 \leq y < 2$
Range: $-4 < x \leq 4$
b. Domain: $-2 \leq x \leq 2$
Range: $-4 \leq y \leq 4$
c. Domain: $-4 < y < 4$
Range: $-2 < x < 2$
d. Domain: $-2 < x < 2$
Range: $-4 < y < 4$

3. Determine the domain and range of the linear function graphed below.

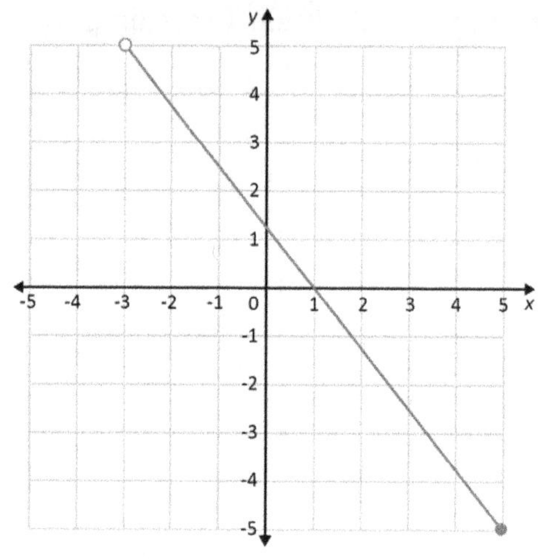

a. Domain: $-3 < x \leq 5$
   Range: $-5 \leq y < 5$
b. Domain: $-5 \leq y < 5$
   Range: $-3 < x \leq 5$
c. Domain: $-3 \leq x < 5$
   Range: $-5 < y \leq 5$
d. Domain: $-3 \leq y \leq 5$
   Range: $-5 \leq x \leq 5$

4. Determine the domain and range of the linear function graphed below.

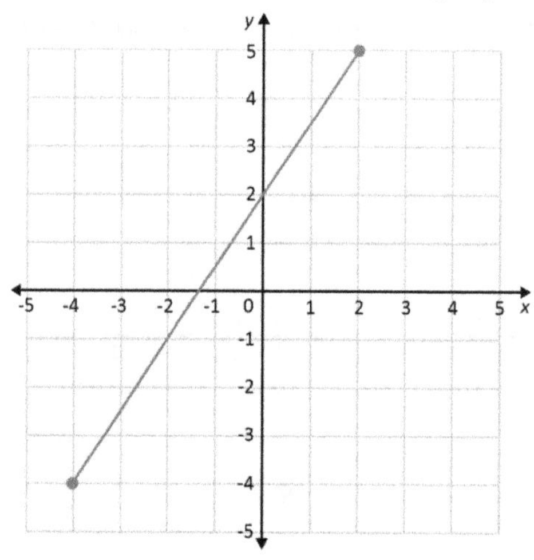

a. Domain: $-4 \leq y < 2$
   Range: $-4 < x \leq 5$
b. Domain: $-4 < x < 2$
   Range: $-4 < y < 5$
c. Domain: $-4 \leq y \leq 5$
   Range: $-4 \leq x \leq 2$
d. Domain: $-4 \leq x \leq 2$
   Range: $-4 \leq y \leq 5$

5. Here is a table of values that represents a linear function. What is the domain of the linear function?

x	y
-1	-8
2	-2
5	4
7	8

a. $\{-1,2,5,7\}$
b. $\{-8,-2,4,8\}$
c. $\{-8,-1,-2\}$
d. $\{2,4,5,7,8\}$

6. Here is a table of values that represents a linear function. What is the range of the linear function?

x	y
-3	0
-1	2
2	5
4	7

a. $\{-3,-1,2,4\}$
b. $\{0,2,5,7\}$
c. $\{-1,-3\}$
d. $\{0,2,2,4,5,7\}$

7. The graph of the linear function shows the cost of grapes as a function of the amount of grapes purchased in a grocery store. What is the domain of the linear function?

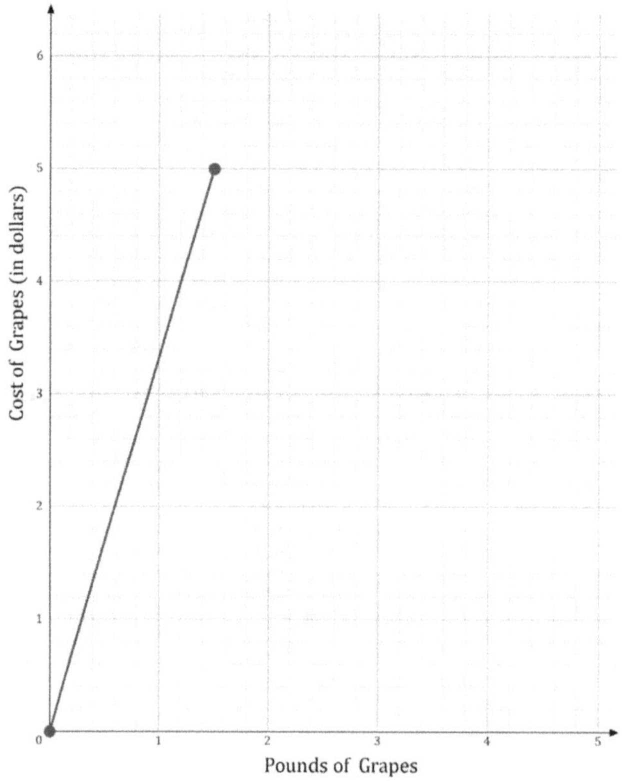

a. $0 \leq x \leq 1.5$
b. $0 \leq x \leq 5$
c. $0 \leq x \leq 6$
d. $1.5 \leq x \leq 5$

8. Use the linear function $f(x) = 4x + 1$ to determine a correct set of possible domain and range values.

a. D: $\{2, 4, 6, 8\}$, R: $\{-3, 5, 7, 8\}$
b. D: $\{1, 2, 3, 4\}$, R: $\{-5, -9, -13, -17\}$
c. D: $\{1, 2, 3, 4\}$, R: $\{5, 9, 13, 17\}$
d. D: $\{1, 3, 5, 7\}$, R: $\{5, 9, 11, 17\}$

9. Which domain and range values would satisfy the linear function $f(x) = -3$?

a. (1,3), (0,0), (1,4)
b. (−1,3), (0,0), (1, −3)
c. (−1,2), (0,5), (1, −3)
d. (1,3), (0,0), (1,9)

10. If the domain of the linear function $f(x) = 0.5x + 2$ is D: $\{-1, 0, 1, 2\}$, what is the range of the function?

a. R: $\{1.5, 2.5, 3.5, 4.5\}$
b. R: $\{-1.5, 2, -2.5, 3\}$
c. R: $\{1, 2, 3, 4\}$
d. R: $\{1.5, 2, 2.5, 3\}$

11. A bowling alley charges 8 dollars for use of the lane, plus an additional 3 dollars for each shoe rental. This relationship can be represented with the function $y = 3x + 8$. In this equation, x stands for the pairs of shoes rented, and y stands for the total cost to bowl. Determine reasonable domain and range values for this scenario.

a. Domain: $\{0, 1, 2, 3, 4\}$
   Range: $\{0, 10, 12, 14, 16\}$
b. Domain: $\{0, 1, 2, 3, 4\}$
   Range: $\{8, 11, 14, 17, 20\}$
c. Domain: $\{-8, -11, -14, -17, -20\}$
   Range: $\{0, 1, 2, 3, 4\}$
d. Domain: $\{0, -1, -2, -3, -4\}$
   Range: $\{6, 11, 13, 15, 17\}$

12. Hannah gets paid \$12 an hour to babysit. She can use the equation $y = 12x$ to determine her pay for each babysitting job, where x is the number of hours she babysits and y is her total pay for that job. If Hannah can only work 6 hours a day, what is the range of the function $y = 12x$?

a. $0 \leq y \leq 6$
b. $0 \leq y \leq 72$
c. $6 \leq y \leq 12$
d. $6 \leq y \leq 72$

13. Sam gets paid \$9 an hour working at Soccer Stop. He uses the equation $y = 9x$ to calculate his weekly pay, where x is the number of hours he worked and y is his total pay. If Sam's paycheck is always between \$135 and \$198, what is the domain of the function $y = 9x$?

a. $15 \leq x \leq 22$
b. $15 \leq x \leq 135$
c. $22 \leq x \leq 198$
d. $135 \leq x \leq 198$

14. An online video game rental company charges a monthly membership fee of \$15 and \$2 for each game that is rented. Dexter uses the linear function $y = 2x + 15$ to calculate his monthly bill, where x is the number of games he rents in one month and y is his total monthly bill. What is the range of the linear function if Dexter can only afford to rent a maximum of 8 games per month?

a. $0 \leq y \leq 15$
b. $0 \leq y \leq 31$
c. $15 \leq y \leq 31$
d. $17 \leq y \leq 31$

15. An online clothing rental company charges a monthly membership fee of $20 and $5 for each piece of clothing that is rented. Mike uses the linear function $y = 5x + 20$ to calculate his monthly bill, where x is the total pieces of clothing that he rented for the month and y is his monthly bill. What is the domain of the linear function if Mike's monthly bill is between $80 and $95, inclusive?

a. $\{12\}$
b. $\{15\}$
c. $\{12,13,14,15\}$
d. $\{13,14,15,16\}$

2.3 Manipulating Linear Functions

A **function manipulation**, also called a **transformation**, is when the graph of the original function, $f(x)$, gets moved on the coordinate plane or is changed in size. Two types of linear function manipulations are translations and dilations.

Translations of Functions

A **translation** is when the graph of a function slides in a specific direction on the coordinate plane. A translation can occur either vertically, where the graph of the function is shifted up or down along the y-axis, or horizontally, where the graph of the function is shifted left or right along the x-axis. When translations occur, the graph of the function does not change in shape or size, and it is not rotated or reflected.

Vertical Translations

Vertical translations happen when the graph of a function is shifted either up or down by k units. To shift the graph of the parent function $f(x)$ up, add k units to the equation for $f(x)$ to get $g(x) = f(x) + k$. This results in k getting added to all y-coordinates on the graph of the function. For example, $g(x) = f(x) + 2$ moves the graph $f(x)$ up 2 units on the coordinate plane.

To shift the graph of the parent function $f(x)$ down, subtract k units from $f(x)$ to get $g(x) = f(x) - k$. Doing so results in k getting subtracted from all y-coordinates on the graph of the function. For instance, $g(x) = f(x) - 4$ moves the graph $f(x)$ down 4 units on the coordinate plane.

Horizontal Translations

Horizontal translations occur when the graph of a function is shifted either left or right by h units. To shift the graph of the parent function $f(x)$ to the left, add h units to x inside of parentheses to get $g(x) = f(x + h)$. Adding h inside of parentheses shifts the graph in a negative direction, to the left. When a function manipulation occurs inside the parentheses, the graph moves in the opposite direction of the sign given in the function. For example, $g(x) = f(x + 3)$ shifts the graph $f(x)$ left 3 units on the coordinate plane.

To shift the graph of the parent function $f(x)$ to the right, subtract h units from x inside the parentheses to get $g(x) = f(x - h)$. Subtracting h inside of parentheses shifts the graph in a positive direction, to the right. This is because the function manipulation occurs before the input is evaluated, causing the graph to move in the opposite direction of the sign given. For example, $g(x) = f(x - 5)$ shifts the graph of $f(x)$ right 5 units on the coordinate plane.

Dilations of Functions

Another way in which linear functions are manipulated is with dilations. A **dilation** is when the size of a function's graph is changed. When a dilation results in a larger image, it is called an enlargement, and the graph of the original function is stretched. When a dilation results in a smaller image, it is called a reduction, and the graph of the original function is compressed. The **scale factor** of a dilation determines how much larger or smaller the new graph will be in comparison to the parent function.

Vertical Dilations

A **vertical dilation** is when the graph of a function is either stretched or compressed in a vertical direction. To stretch or compress the graph of the parent function $f(x)$ vertically, multiply $f(x)$ by a to get $g(x) = af(x)$. Doing so results in all y-coordinates on the graph of the function getting multiplied by a. If the scale factor, a, is greater than 1, then the graph of the parent function will stretch vertically, moving away from the x-axis. If a is a number between 0 and 1, then the graph of the parent function will compress vertically, moving in toward the x-axis. For example, $g(x) = 6f(x)$ stretches the graph of the parent function vertically by a factor of 6 because each y-coordinate gets multiplied by 6. On the other hand, $g(x) = \frac{1}{6}f(x)$ compresses the graph of the parent function vertically by a factor of $\frac{1}{6}$ because each y-coordinate gets multiplied by $\frac{1}{6}$.

Horizontal Dilations

A **horizontal dilation** is when the graph of a function is either stretched or compressed in a horizontal direction. To stretch or compress the graph of the parent function $f(x)$ horizontally, multiply x by b inside the parentheses to get $g(x) = f(bx)$. Doing so results in all x-coordinates on the graph of the function getting multiplied by b. A horizontal dilation occurs inside the parentheses, so the x-values will change before they are evaluated by the function. If b is a number between 0 and 1, the graph of the parent function will stretch horizontally, moving away from the y-axis. If b is greater than 1, the graph of the parent function will compress horizontally, moving in toward the y-axis. For example, $g(x) = f(7x)$ compresses the graph of the parent function horizontally by a factor of $\frac{1}{7}$. On the other hand, $g(x) = f\left(\frac{1}{7}x\right)$ stretches the graph of the parent function horizontally by a factor of 7.

Rules for Function Manipulations

Function	Type of Manipulation	Result
$f(x) + k$	Vertical translation	k is added to all y-coordinates to shift $f(x)$ up k units
$f(x) - k$	Vertical translation	k is subtracted from all y-coordinates to shift $f(x)$ down k units
$f(x + h)$	Horizontal translation	h is subtracted from all x-coordinates to shift $f(x)$ left h units
$f(x - h)$	Horizontal translation	h is added to all x-coordinates to shift $f(x)$ right k units
$af(x)$	Vertical dilation	y-coordinates are multiplied by a to compress vertically when $0 < a < 1$, or stretch vertically when $a > 1$
$f(bx)$	Horizontal dilation	x-coordinates are multiplied by b to compress horizontally when $k > 1$, or stretch horizontally when $0 < k < 1$

Examples

Example 1:

Write a function, g, that represents the graph of $f(x) = x$ after it is stretched vertically by a factor of 2.

Answer

$$g(x) = 2f(x)$$

This function manipulation is a vertical dilation. To represent a vertical dilation by a factor of 2, multiply $f(x)$ by 2. Doing so results in all y-coordinates being multiplied by 2. In the graph shown, the point $(1,1)$ on the parent function becomes the point $(1,2)$ on the graph of $g(x) = 2f(x)$.

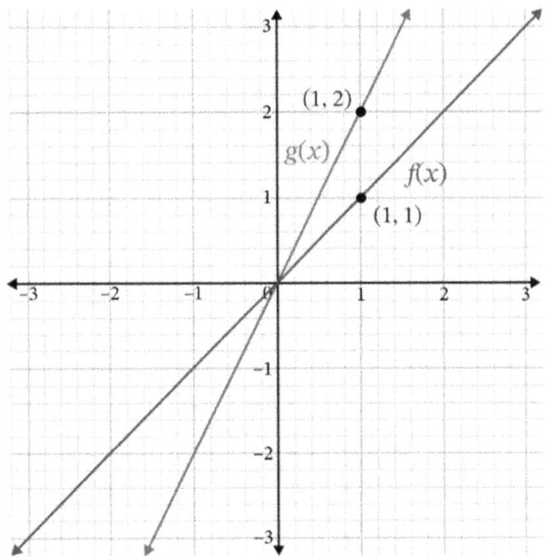

Example 2:

Write a function, g, that represents the graph of $f(x) = x$ after it is shifted 4 units up.

Answer

$$g(x) = f(x) + 4$$

This function manipulation is a vertical translation. To represent a vertical translation that shifts 4 units up, add 4 to $f(x)$. Doing so results in 4 being added to all y-coordinates. In the graph shown, the point $(0,0)$ on the parent function becomes the point $(0,4)$ on the graph of $g(x) = f(x) + 4$.

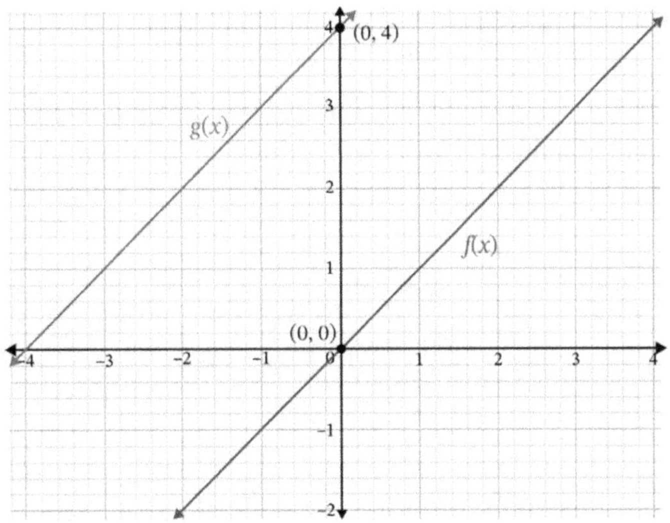

Example 3:

Write a function, g, that represents the graph of $f(x) = x$ after it is shifted 3 units to the right.

Answer

$$g(x) = f(x - 3)$$

This function manipulation is a horizontal translation. To represent a horizontal translation that shifts 3 units to the right, subtract 3 from x within parentheses. In the graph shown, the point (0,0) on the parent function becomes the point (3,0) on the graph of $g(x) = f(x - 3)$.

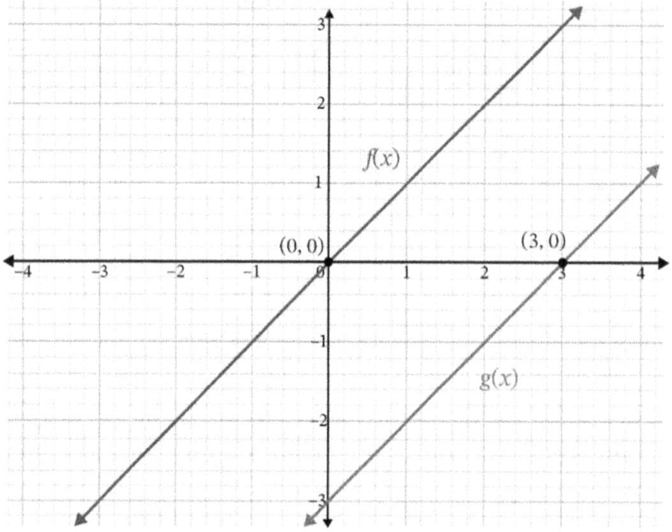

Example 4:

Write a function, g, that represents the graph of $f(x) = x$ after it is compressed horizontally by a factor of $\frac{1}{5}$.

Answer

$$g(x) = f(5x)$$

This function manipulation is a horizontal dilation. To represent a dilation that is compressed horizontally by a factor of $\frac{1}{5}$, the reciprocal of the fraction, $\frac{5}{1} = 5$, becomes the value of b in the equation $f(bx)$. In the graph shown, the point (5,5) on the parent function becomes the point (1,5) on the graph of $g(x) = f(5x)$.

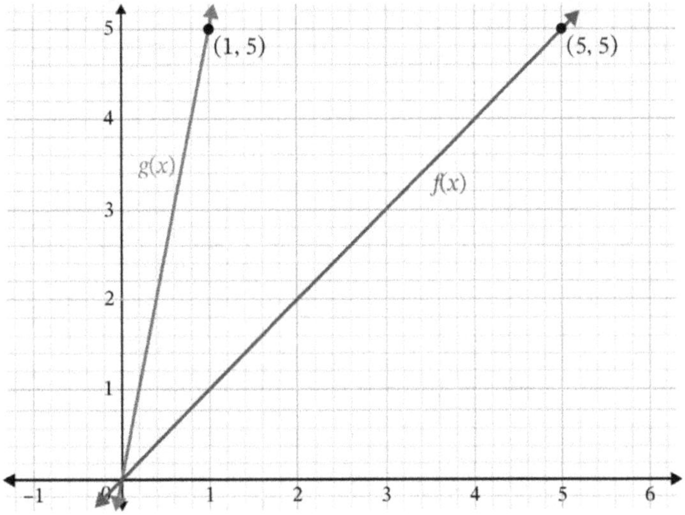

PRACTICE QUESTIONS

1. Which function represents the graph of $f(x) = x$ after it is stretched vertically by a factor of 3?

a. $g(x) = 3f(x)$
b. $g(x) = \frac{1}{3}f(x)$
c. $g(x) = f(x) + 3$
d. $g(x) = f(x + 3)$

2. Which function represents the graph of $f(x) = x$ after it is shifted 2 units up?

a. $g(x) = f(x + 2)$
b. $g(x) = f(x) + 2$
c. $g(x) = f(x) - 2$
d. $g(x) = f(x - 2)$

3. Which function represents the graph of $f(x) = x$ after it is shifted four units to the right?

a. $g(x) = f(x + 4)$
b. $g(x) = f(x - 4)$
c. $g(x) = f(x) + 4$
d. $g(x) = f(x) - 4$

4. Which function represents the graph of $f(x) = x$ after it is compressed horizontally by a factor of $\frac{1}{3}$?

a. $g(x) = f(3x)$
b. $g(x) = 3f(x)$
c. $g(x) = f\left(\frac{1}{3}x\right)$
d. $g(x) = \frac{1}{3}f(x)$

5. The coordinate plane below shows the graph of the parent function $f(x) = x$ (shown as a solid line), and a manipulation in which the parent function is compressed vertically by a factor of $\frac{1}{2}$ (shown as a dashed line). Which function represents this dilation?

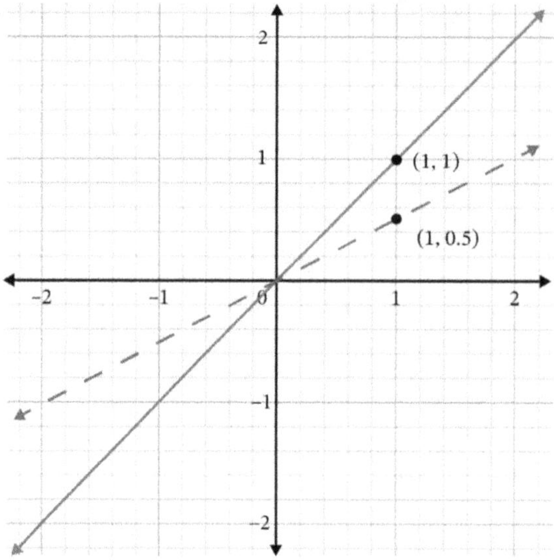

a. $g(x) = f(x) + \frac{1}{2}$
b. $g(x) = f\left(\frac{1}{2}x\right)$
c. $g(x) = \frac{1}{2}f(x)$
d. $g(x) = 2f(x)$

6. Which function represents the graph of $f(x) = x$ after it is shifted 5 units down?

a. $g(x) = f(x + 5)$
b. $g(x) = f(x) + 5$
c. $g(x) = f(x) - 5$
d. $g(x) = f(x - 5)$

7. Which function represents the graph of $f(x) = x$ after it is shifted five units to the left?

a. $g(x) = f(x) + 5$
b. $g(x) = f(x + 5)$
c. $g(x) = f(x) - 5$
d. $g(x) = f(x - 5)$

8. The graph of $f(x) = x$ is translated vertically 3 units up to create the graph of $g(x)$. What is the function of the translated graph?

a. $g(x) = f(x) + 3$
b. $g(x) = f(x) - 3$
c. $g(x) = f(x + 3)$
d. $g(x) = f(x - 3)$

9. The graph of $f(x) = x$ is translated vertically 4 units down to create the graph of $g(x)$. What is the function of the translated graph?

a. $g(x) = f(x) + 4$
b. $g(x) = f(x) - 4$
c. $g(x) = f(x + 4)$
d. $g(x) = f(x - 4)$

10. The graph of $f(x) = x$ is translated horizontally to the right by 2 units to create the graph of $g(x)$. What is the function of the translated graph?

a. $g(x) = f(x) + 2$
b. $g(x) = f(x) - 2$
c. $g(x) = f(x + 2)$
d. $g(x) = f(x - 2)$

11. The graph of $f(x) = x$ is translated horizontally to the left 5 units to create the graph of $g(x)$. What is the function of the translated graph?

a. $g(x) = f(x) + 5$
b. $g(x) = f(x) - 5$
c. $g(x) = f(x + 5)$
d. $g(x) = f(x - 5)$

12. The graph of $f(x) = x$ is stretched vertically by a factor of p to create the graph of $g(x)$. What is the function of the stretched graph?

a. $g(x) = f(x) + p$
b. $g(x) = f(x) - p$
c. $g(x) = pf(x)$
d. $g(x) = \frac{f(x)}{p}$

13. A skating rink rents ice skates for $1 per hour. The function $f(x) = x$ represents the hourly rental fee for x hours. The skating rink decides to add a $6 fee per rental to cover cleaning and maintenance expenses. The function that represents the new cost for skate rental is $g(x) = f(x) + 6$. How does the graph of this transformation compare to the graph of the parent function, $f(x)$?

a. The graph of $g(x)$ is a translation of $f(x)$, with a vertical shift up 6 units.
b. The graph of $g(x)$ is a translation of $f(x)$, with a vertical shift down 6 units.
c. The graph of $g(x)$ is a translation of $f(x)$, with a horizontal shift right 6 units.
d. The graph of $g(x)$ is a dilation of $f(x)$, vertically stretched by a factor of 6.

14. A law firm wants to create a smaller version of their logo to use on their letterhead. The edge of the original logo, represented by the function $f(x) = x$, needs to be compressed horizontally by a factor of $\frac{1}{4}$. Which function represents this dilation?

a. $g(x) = \frac{1}{4}f(x)$
b. $g(x) = 4f(x)$
c. $g(x) = f\left(\frac{1}{4}x\right)$
d. $g(x) = f(4x)$

15. Alex is a graphic designer and is working on creating a website. He needs to manipulate a linear image so it fits on the webpage. The original image, represented by the function $f(x) = x$, needs to be compressed vertically by a factor of $\frac{1}{4}$. Which function represents this dilation?

a. $g(x) = f(4x)$
b. $g(x) = 4f(x)$
c. $g(x) = f\left(\frac{1}{4}x\right)$
d. $g(x) = \frac{1}{4}f(x)$

2.4 Writing Functions for Real-World Problems

Many real-world scenarios can be represented using algebra. Writing functions to represent real-world situations is what makes algebra relevant and useful. As you learn about topics such as equation writing, the distributive property, the order of operations, simplifying expressions, and other skills, remember that these topics will eventually come together in a way that is relevant and useful in many real-world situations.

For example, if a taxi driver charges a flat rate of \$3 and then \$1.50 per mile, we can use algebra to calculate how much the driver earns on a 46-mile trip. Let x represent the *independent variable* and let y represent the *dependent variable*. In this scenario, the x-variable is the number of miles traveled, and the y-variable is the total amount of money earned because it *depends* on how many miles are traveled. We can use slope-intercept form, $y = mx + b$, to set up this equation, where m represents the slope, or "rate of change", and b represents the y-intercept, or the starting point. In this case, the rate of change is \$1.50 and the starting point is at \$3. Plug in $m = 1.5$ and $b = 3$ into the slope-intercept form, $y = mx + b$, to get $y = 1.5x + 3$. We can now plug in any value for x (the number of miles traveled) and calculate y (the total cost of the trip). For a 46-mile trip, plug in 46 for x. The equation $y = 1.5x + 3$ becomes $y = 1.5(46) + 3$, which simplifies to $y = 72$. So, a 46-mile trip will cost \$72.

Some functions will have a b-value of 0. For example, Frances wants to rent ice skates, and she has \$40 to spend. The skating rink charges \$8 per hour to rent the skates. How many hours can Frances afford to skate? The x-value in this scenario is the number of hours Frances rents the skates, and the y-value is the total cost (in this case \$40). In this scenario, we are solving for x (number of hours). The formula $y = mx + b$ becomes $40 = 8x + 0$. In this case, b is 0 because there is no additional flat rate or fee associated with the skate rental. Solve the equation $40 = 8x$ to find $x = 5$. This means that Frances can rent ice skates for 5 hours if she has \$40.

EXAMPLES

EXAMPLE 1:

Danny owns a tree trimming company. He charges each client a \$50 one-time consultation fee and then \$25 per hour for his services. How many hours will Danny need to work for one client in order to make \$500?

ANSWER

The slope is the rate of change, so the slope (m) will be how much Danny charges per hour, which is \$25. The y-intercept will be the one-time consultation fee, because that fee only happens once, not each hour he works.

$$y = 25x + 50$$

This equation represents \$25 per hour plus a \$50 one-time fee, where x represents the number of hours worked, and y represents the total amount of money earned. Substitute 500 for y in order to determine the total number of hours needed to earn \$500.

$$500 = 25x + 50$$

$$450 = 25x$$

$$18 = x$$

Danny will earn \$500 after working 18 hours for a client.

Example 2:

A diver jumps off a boat 10 feet above sea level and descends 5 feet every second. After 7 seconds, how far below sea level is the diver?

Answer

To create a function for this scenario, start by determining the rate of change. Since the diver is descending 5 feet every second, the rate of change (m) is –5. Then, determine the y-intercept for the function. Since the diver starts 10 feet above sea level, the y-intercept (b) is 10. Substitute these values into the slope-intercept equation ($y = mx + b$).

$$y = -5x + 10$$

Since we are asked how far below sea level the diver is after 7 seconds, substitute 7 for x and solve for y.

$$y = -5(7) + 10 = -35 + 10 = -25$$

Therefore, the diver is 25 feet below sea level after 7 seconds. Notice that a negative value of y gives a positive distance below sea level. A positive value of y would have given a positive distance above sea level.

Practice Questions

1. A car rental company charges a daily fee of $25 and $0.45 per mile driven. What is the total rental charge for one day if 150 miles was driven by the car?

a. $67.50
b. $92.50
c. $161.25
d. $175.45

2. A boat rental company charges a $15 daily fee and $4 per hour for renting a boat. Which equation can be used to calculate the total one-day rental fee, y, for renting the boat for x hours?

a. $y = 4x + 15$
b. $y = 15x + 4$
c. $x = 4y + 15$
d. $x = 15y + 4$

3. If purchasing a pound of apples costs $1.65, how much would 7 pounds of apples cost?

a. $8.65
b. $10.30
c. $11.55
d. $11.95

4. If a gallon of gas costs $2.85, how many gallons of gas can Ian buy if he only has $30? (Round your answer to the nearest hundredth.)

a. 5.26 pounds
b. 7.38 pounds
c. 10.53 gallons
d. 21.45 gallons

5. Alliya pays a monthly membership fee of $8 for an online audio book rental app that charges $3 per audio book. How many books did Alliya rent in one month if her total bill is $20?

a. 3 books
b. 4 books
c. 5 books
d. 6 books

6. Josh works at a coffee shop downtown and is paid $14 per hour. The coffee shop pays the city $35 a month for each employee to park downtown and then deducts the amount from the employee's paycheck at the end of the month. What is Josh's monthly pay if he works 164 hours in one month?

a. $2,085
b. $2,261
c. $2,296
d. $2,331

7. Grey works at a boutique where she is paid $17 an hour. The boutique provides parking and lunch each day Grey works, then at the end of the month the boutique deducts $100 from Grey's paycheck. How many hours did Grey work if her monthly pay is $2,960?

a. 168 hours
b. 174 hours
c. 180 hours
d. 185 hours

8. Zola has a savings account balance of $500. Since she has a new job, she has decided to save $150 a month from her paycheck and deposit it into her savings account. How much money will Zola have in her savings account after 8 months?

a. $650
b. $1,700
c. $4,150
d. $5,200

9. Nadya is saving up for a trip to Europe. She has $1,200 in her savings account. She saves $200 per month from her paycheck. How many months will it take Nadya to reach $4,000?

a. 3 months
b. 4 months
c. 10 months
d. 14 months

10. Sarah's phone company charges her a monthly flat fee of $30 and $0.03 per sent text message per month. How much is Sarah's phone bill if she sends 400 text messages in one month?

a. $12
b. $18
c. $42
d. $48

11. A certain phone company charges a monthly flat fee of $15 plus $0.05 per sent text message per month. How many text messages would someone have to send for their monthly bill to be $55?

a. 800 text messages
b. 900 text messages
c. 1,100 text messages
d. 1,400 text messages

12. Lora has a savings account balance of $2,400. She wants to save money to buy a new car. She decides to put $300 from each monthly paycheck into the savings account. Which equation can be used to model this situation where x is the number of months that Lora has been saving money and y is the total amount of money in the savings account?

a. $y = 300x + 2{,}400$
b. $y = 2{,}400x + 300$
c. $x = 300y + 2{,}400$
d. $x = 2{,}400y + 300$

13. An online music store charges a monthly membership fee of $10 and $0.20 per song that is downloaded. Which equation can be used to model this situation where x is the number of songs downloaded and y is the total monthly bill?

a. $x = 0.20y + 10$
b. $y = 0.20x + 10$
c. $y = 10x + 0.20$
d. $x = 10y + 0.20$

14. Lex charges $25 to mow a lawn. He spends $80 a week on fuel for his lawn mower. Which equation can be used to model this situation where x is the number of lawns Lex mows in one week and y is the total amount of money that he makes in one week?

a. $x = 25y - 80$
b. $x = 80y - 25$
c. $y = 25x - 80$
d. $y = 80x - 25$

15. A musician pays $15 a month to sell their music on an app that pays them $0.89 each time one of their songs is downloaded. Which equation can be used to model this situation where x is the number of songs downloaded in one month and y is the total amount of money paid to the musician by the app?

a. $y = 0.89x - 15$
b. $y = 15x - 0.89$
c. $y = 0.89x + 15$
d. $y = 15x + 0.89$

2.5 Arithmetic Sequences

An **arithmetic sequence** is a sequence where each term increases or decreases by a constant amount. This constant amount is called the **common difference**. For increasing arithmetic sequences, the common difference will always be positive, and for decreasing arithmetic sequences, the common difference will always be negative. For example, 10, 20, 30, 40 ... is an arithmetic sequence because each term increases by 10, and it is considered an increasing arithmetic sequence because its common difference is positive. Sequences such as 3, 6, 12, 24... are not arithmetic because each term does not increase or decrease by the same amount.

When working with sequences, you will often be asked to determine the nth term. The nth term can be any term in the sequence. For example, if you are trying to find the 50th term in the sequence 2, 4, 6, 8..., the 50th term is considered the nth term when $n = 50$. Finding the 50th term in a sequence seems like an enormous task. However, there is a formula for finding the nth term of an arithmetic sequence that makes the task much more manageable. The explicit formula is: $a_n = a_1 + d(n - 1)$, where n represents the place of the term you are looking for, a_1 represents the first term in the sequence, and d represents the common difference (the amount of increase or decrease).

Consider the sequence 3, 9, 15, 21, ...

We know this is an arithmetic sequence because there is a common difference of 6. In other words, you get the next number by adding 6 to the previous number. If we want to know the 35th term in the sequence, n is 35. At this point we have enough information to find the 35th term using the explicit formula for an arithmetic sequence.

In this sequence, $a_1 = 3$ and $d = 6$. So, $a_n = a_1 + d(n - 1)$ becomes $a_{35} = 3 + (6)(35 - 1)$ when the values are plugged in. $a_{35} = 3 + (6)(35 - 1)$ simplifies to 207, so the 35th term in the sequence is 207.

The formula also works for negative terms. For example, the formula can be used to determine the 99th term in the following sequence.

$$-4, -9, -14, -19, \ldots$$

In this sequence, $a_1 = -4$ and $d = -5$.

The formula $a_n = a_1 + d(n - 1)$ becomes $a_{99} = -4 + (-5)(99 - 1)$. This simplifies to - 494. The 99th term in the sequence is - 494.

Review Video: Arithmetic Sequences
Visit mometrix.com/academy and enter code: 676885

Examples

Example 1:

Find the 10th term in the arithmetic sequence.

$$2, 4, 6, 8, \ldots$$

Answer

For shorter smaller term numbers, the sequence can be written out to find the specific term. The common difference between each term is 2.

2, 4, 6, 8, 10, 12, 14, 16, 18, 20

The 10th term in the sequence is 20.

Example 2:

Use the formula for the nth term of an arithmetic sequence to find the 16th term in the following arithmetic sequence.

$$2, 10, 18, 26, \ldots$$

Answer

The explicit formula for finding the nth term in an arithmetic sequence is $a_n = a_1 + d(n-1)$, where a_1 represents the first term in the sequence, n represents the number of the term you are looking for, and d represents the common difference. In this sequence, $a_1 = 2$ and $d = 8$.

The formula becomes $a_{16} = 2 + (8)(16-1)$, which simplifies to $a_{16} = 122$.

Example 3:

Use the explicit formula for the nth term of an arithmetic sequence to determine the value of the 50th term in the following sequence.

$$6, 12, 18, 24, \ldots$$

Answer

The explicit formula for finding the nth term in an arithmetic sequence is $a_n = a_1 + d(n-1)$, where a_1 represents the first term in the sequence, n represents the number of the term you are looking for, and d represents the common difference. In this sequence, $a_1 = 6$ and $d = 6$.

To find when $n = 50$ the formula $a_n = a_1 + \mathrm{d}(n-1)$ becomes $a_{50} = 6 + (6)(50-1)$, which simplifies to $a_{50} = 300$.

Practice Questions

1. Which statement is true about the common difference for an increasing arithmetic sequence?

a. The common difference is always zero.
b. The common difference is always negative.
c. The common difference is always positive.
d. The common difference can be positive or negative.

2. Which statement is true about the common difference for a decreasing arithmetic sequence?

a. The common difference is always zero.
b. The common difference is always negative.
c. The common difference is always positive.
d. The common difference can be positive or negative.

3. Here is an arithmetic sequence: 5, 8, 11, 14, ... What is the next term in the sequence?

a. 16
b. 17
c. 18
d. 19

4. Here is an arithmetic sequence: 43, 37, 31, 25, ... What is the next term in the sequence?

a. 17
b. 18
c. 19
d. 20

5. Here is an arithmetic sequence: 3, 12, 21, 30, ... What is the 7th term in the sequence?

a. 57
b. 58
c. 59
d. 60

6. Here is an arithmetic sequence: 62, 55, 48, 41, ... What is the 6th term in the sequence?

a. 25
b. 26
c. 27
d. 28

7. Here is an arithmetic sequence: 2, 17, 32, ... What are the next three terms in the sequence?

a. 47, 62, 77
b. 50, 65, 70
c. 47, 52, 57
d. 50, 65, 80

8. Here is an arithmetic sequence: 88, 83, 78, ... What are the next three terms in the sequence?

a. 73, 78, 83
b. 72, 67, 62
c. 73, 68, 63
d. 72, 77, 82

9. Here is an arithmetic sequence: 7, 12, 17, 22, ... What is the formula for the sequence?

a. $a_n = 5 + (n - 1)$
b. $a_n = 7 + (n - 1)$
c. $a_n = 5 + 7(n - 1)$
d. $a_n = 7 + 5(n - 1)$

10. Here is an arithmetic sequence: 54, 50, 46, 42, ... What is the formula for the sequence?

a. $a_n = 4 - 54(n - 1)$
b. $a_n = 54 - 4(n - 1)$
c. $a_n = 4 + 54(n - 1)$
d. $a_n = 54 + 4(n - 1)$

11. The formula for an arithmetic sequence is $a_n = 107 - 13(n - 1)$. What is the 5th term of the sequence?

a. 13
b. 55
c. 94
d. 107

12. The formula for an arithmetic sequence is $a_n = 38 + 9(n - 1)$. What is the 3rd term of the sequence?

a. 9
b. 38
c. 56
d. 65

13. The table below shows the relationship between the number of sides of a polygon and the sum of the interior angles of that polygon. What is the formula for the arithmetic sequence that represents this relationship where $n = 1$ when the polygon has 3 sides?

Number of sides	3	4	5
Sum of the interior angles	180°	360°	540°

a. $a_n = 180(n - 1)$
b. $a_n = 360(n - 1)$
c. $a_n = 180 + 180(n - 1)$
d. $a_n = 180 + 360(n - 1)$

14. Michaela takes a new job with a starting salary of $50,000 per year. She receives an annual raise of $2,500. Assuming Michaela receives the same raise each year, what will her annual salary be in six years?

a. $52,500
b. $57,500
c. $60,000
d. $62,500

15. Sabrina is training for a marathon. She runs 20 minutes on day one, 35 minutes on day two, and 50 minutes on day three. If she continues to increase her minutes at the same rate each day, how many minutes will she run on the 8th day of her training?

a. 75 minutes
b. 90 minutes
c. 105 minutes
d. 125 minutes

2.6 Geometric Sequences

A **geometric sequence** is formed by multiplying or dividing each previous term by a **common ratio**. For example, 2, 4, 8, 16, ... is a geometric sequence with a common ratio of 2. Each term is multiplied by 2 to find the next term.

Real-world examples of geometric sequences can be observed all around us. For example, if you toss a basketball and let it bounce away from you, it will gradually bounce lower and lower until it stops bouncing altogether. You will notice that each bounce height is a fraction of the previous bounce's height. If you measure each bounce height, each measurement is a constant fraction of its remaining energy, so the sequence of heights is geometric.

To find the nth term of a geometric sequence we can use the explicit formula: $a_n = a_1 \cdot r^{n-1}$.

a_n represents the nth term in the sequence.

a_1 represents the first term in the sequence.

r represents the common ratio.

n represents the term number.

For example, the nth term formula can be used to find the 7th term in the following geometric sequence: 4, 12, 36, 108, ...

In the sequence, $a_1 = 4$ and $r = 3$. So, $a_n = a_1 \cdot r^{n-1}$ becomes $a_n = (4)(3)^{7-1}$, which simplifies to $a_n = (4)(3)^6$, which is equivalent to $a_7 = (4)(729) = 2{,}916$. The 7th term in the sequence is 2,916.

The same formula works in a geometric sequence where terms are decreasing. For example, in the following sequence, the common ratio is a fraction. $-3, 1, -\frac{1}{3}, \frac{1}{9}, \ldots$

Each new term in the sequence is calculated by dividing by -3, which is the same thing as multiplying by $-\frac{1}{3}$. The common ratio, r, is $-\frac{1}{3}$. If we are looking for the 5th term, the nth term formula $a_n = a_1 \times r^{n-1}$ becomes $a_5 = (-3)\left(-\frac{1}{3}\right)^{5-1}$. This simplifies to $a_5 = (-3)\left(-\frac{1}{3}\right)^4$ and then $a_5 = -\frac{1}{27}$.

Examples

Example 1:

Find the 8th term in the following geometric sequence.

2, 4, 8, 16, ...

Answer

When asked for a term close to what is in the original sequence, it is sometimes easy enough to list out the terms using the pattern. Multiply each previous term by 2 to find the next term.

2, 4, 8, 16, 32, 64, 128, <u>256</u>

The 8th term in the sequence is 256.

EXAMPLE 2:

Write a formula that can be used to find the 9th term in the following geometric sequence.

$$2, 8, 32, \ldots$$

ANSWER

The explicit formula for finding the nth term in a geometric sequence is $a_n = a_1 \cdot r^{n-1}$, where n represents the place of the term you are looking for, a_1 represents the first term in the sequence, and r represents the common ratio. In the given sequence $a_1 = 2$ and $r = 4$. So, $a_n = a_1 \cdot r^{n-1}$ becomes $a_9 = 2 \cdot (4)^{9-1}$, which simplifies to 131,072.

EXAMPLE 3:

Use the explicit formula for the nth term of a geometric sequence to determine the value of the 11th term in the following sequence.

$$3, 12, 48, \ldots$$

ANSWER

The explicit formula for finding the nth term in a geometric sequence is $a_n = a_1 \cdot r^{n-1}$, where n represents the place of the term you are looking for, a_1 represents the first term in the sequence, and r represents the common ratio. In the given sequence $a_1 = 3$ and $r = 4$. So, $a_n = a_1 \cdot r^{n-1}$ becomes $a_{11} = 3 \cdot (4)^{11-1}$, which simplifies to 3,145,728.

PRACTICE QUESTIONS

1. Find the sixth term in the following geometric sequence:

$$4, 8, 16, 32, \ldots$$

a. 64
b. 256
c. 164
d. 128

2. Find the fifth term in the following geometric sequence:

$$2, 6, 18, \ldots$$

a. 162
b. 218
c. 256
d. 180

3. Find the seventh term in the following geometric sequence:

$$4{,}000, 2{,}000, 1{,}000, 500, \ldots$$

a. 62.5
b. 8,000
c. 40
d. 125

4. Find the seventh term in the following geometric sequence:

$$2, 10, 50, 250, \ldots$$

a. 25,625
b. 16,650
c. 22,250
d. 31,250

5. Use the geometric sequence formula to write an equation describing the following sequence:

$$8, 32, 128, \ldots$$

a. $a_n = 8 \cdot 4^{n-1}$
b. $a_n = 8 \cdot 8^{n-1}$
c. $a_n = 4 \cdot 8^{n-1}$
d. $a_n = 8 \cdot 16^{n-1}$

6. Write an equation to find the twelfth term of the following geometric sequence:

$$9, 36, 144, 576, \ldots$$

a. $a_{12} = 9 \cdot 4^{12-1}$
b. $a_{12} = 4 \cdot 12^{9}$
c. $a_{12} = 9 \cdot 6^{12-1}$
d. $a_{12} = 9 \cdot 36^{12-1}$

7. Write an equation to find the 20th term of the following geometric sequence:

$$100, 50, 25, \ldots$$

a. $a_{20} = 100 \cdot (2)^{20-1}$
b. $a_{20} = 100 \cdot \left(\frac{1}{2}\right)^{20-1}$
c. $a_{20} = 100 \cdot (50)^{20-1}$
d. $a_{20} = 50 \cdot (2)^{20-1}$

8. Write an equation to find the 100th term of the following geometric sequence:

$$\frac{1}{4}, \frac{1}{2}, 1, \ldots$$

a. $a_{100} = \frac{1}{4} \cdot \left(\frac{1}{2}\right)^{100-1}$
b. $a_{100} = \frac{1}{4} \cdot 2^{100-1}$
c. $a_{100} = \frac{1}{2} \cdot 2^{100-1}$
d. $a_{100} = \frac{1}{4} \cdot \left(\frac{1}{4}\right)^{100-1}$

9. Find the fifth term in the following geometric sequence:

$$3, 6, 12, \ldots$$

a. $a_5 = 48$
b. $a_5 = 36$
c. $a_5 = 16$
d. $a_5 = 54$

10. Find the fourth term in the following geometric sequence:

$$1, 7, \ldots$$

a. $a_4 = 350$
b. $a_4 = 149$
c. $a_4 = 288$
d. $a_4 = 343$

11. Find the sixth term in the following geometric sequence:

$$10, 20, 40, \ldots$$

a. $a_6 = 800$
b. $a_6 = 3{,}200$
c. $a_6 = 320$
d. $a_6 = 160$

12. Find the eighth term in the following geometric sequence:

$$5, 15, \ldots$$

a. $a_8 = 10{,}935$
b. $a_8 = 2{,}560$
c. $a_8 = 12{,}550$
d. $a_8 = 8{,}515$

13. Kevin is working on a school fundraiser and invites 10 people to attend. Each of those ten people invite five more people to attend. Then, each of *those* people invite five additional people to attend. If this pattern continues, how many people will be invited in the fourth "wave" of invitations? (Kevin's original ten invites is the first "wave".)

a. $a_4 = 5{,}000$
b. $a_4 = 3{,}500$
c. $a_4 = 1{,}250$
d. $a_4 = 2{,}250$

14. While eating an apple, Quinn notices that there are five seeds in it. From these seeds, she considers planting five apple trees, which could produce a total of 50 apples with five seeds each. If she planted all of *those* seeds, how many seeds would she have from that generation of trees?

a. 12,500 seeds
b. 2,500 seeds
c. 10,000 seeds
d. 850 seeds

15. In the medical science laboratory, Alan is studying a type of bacteria which can double in number each day. On Monday, there were 150 bacteria cells. How many cells should there be on Friday?

a. 15,500 cells
b. 2,400 cells
c. 3,000 cells
d. 2,250 cells

Chapter 2 Quiz

1. What is an arithmetic sequence?

a. A sequence of numbers where each term is found by multiplication.
b. A sequence of numbers with a common difference between consecutive terms.
c. A sequence of numbers with no common difference between consecutive terms.
d. A sequence of numbers where each term is found by division.

2. The graph of the linear function $y = 2x + 1$ is shown below. Determine the domain and range of this function.

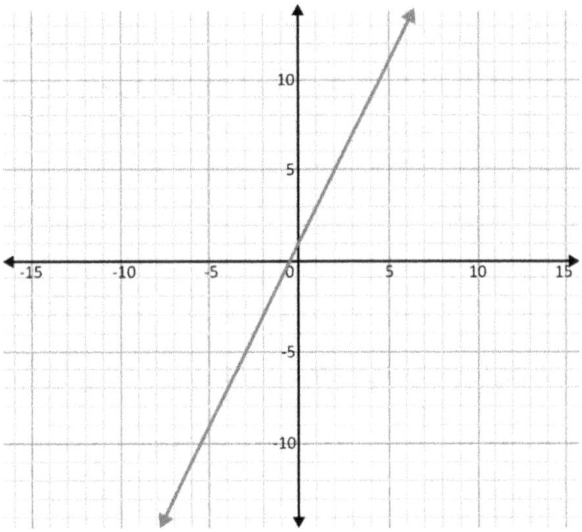

a. Domain: $(-\infty, \infty)$; Range: $(-\infty, \infty)$
b. Domain: $(2, \infty)$; Range: $(2, \infty)$
c. Domain: $(-\infty, 2)$; Range: $(-\infty, 2)$
d. Domain: $(-2,2)$; Range: $(-2,2)$

3. If $f(x) = x^3 + 3x^2$, then what is the value of $f(5) + 3f(2)$?

a. 260
b. 520
c. 770
d. 206

4. What is a geometric sequence?

a. An ordered set of numbers in which each subsequent term is found by addition.
b. An ordered set of numbers in which each subsequent term is found by subtraction.
c. An ordered set of numbers in which each term is a fixed multiple of the number that comes before it.
d. An ordered set of numbers in which each term is not a fixed multiple of the number that comes before it.

5. Which function represents the graph of $f(x) = x$ after it is shifted two units to the right?

a. $g(x) = f(2x)$
b. $g(x) = 2f(x)$
c. $g(x) = f(x + 2)$
d. $g(x) = f(x - 2)$

6. Which function represents the graph of $f(x) = x$ after it is shifted three units up?

a. $g(x) = f(x + 3)$
b. $g(x) = f(x) + 3$
c. $g(x) = f(x) - 3$
d. $g(x) = f(x - 3)$

7. Use the formula $a_n = a_1 \cdot r^{(n-1)}$ to find the 8th term in the geometric sequence shown below.

$$3, 9, 27, 81, \ldots$$

a. 6,561
b. 6,165
c. 2,190
d. 2,187

8. If $f(x) = 3x + 2$, then find the value of $f(20) - f(10)$.

a. 634
b. 94
c. 32
d. 30

9. Determine the domain and range of the linear function graphed below.

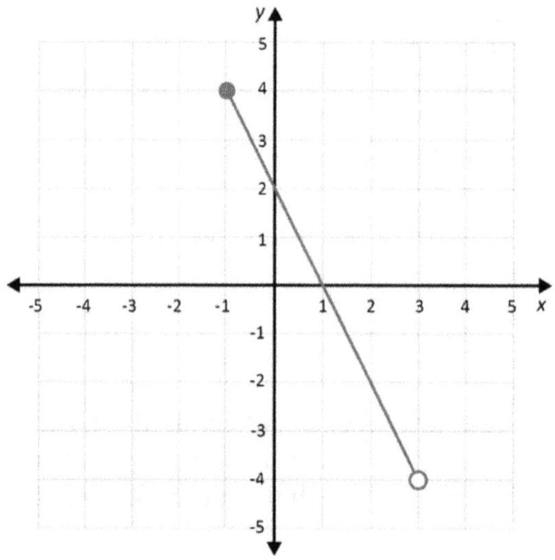

a. Domain: $\{-4 \leq y < 4\}$; Range: $\{-1 < x \leq 3\}$
b. Domain: $\{-1 < x \leq 3\}$; Range: $\{-4 \leq y < 4\}$
c. Domain: $\{-1 \leq x < 3\}$; Range: $\{-4 < y \leq 4\}$
d. Domain: $\{-4 < y \leq 4\}$; Range: $\{-1 \leq x < 3\}$

10. Use the formula $a_n = a_1 + d(n - 1)$ to find the 12th term in the arithmetic sequence below.

$$90, 86, 82, 78, 74, \ldots$$

a. 54
b. 50
c. 46
d. 42

11. Sarah is training for a bike race. According to her training schedule, she will increase her overall distance by four miles each week. She rides her bike for 7 miles in the first week, 11 miles in the second week, and 15 miles during the third week.

$$7, 11, 15, \ldots$$

Assuming Sarah continues increasing her overall distance by four miles each week, which formula best represents the number of miles she rides on her n^{th} week of training?

a. $a_n = 3n - 4$
b. $a_n = 3n + 4$
c. $a_n = 4n - 3$
d. $a_n = 4n + 3$

12. An online video subscription service charges \$12 per month, plus an additional \$3 per movie that gets rented. The relationship between the number of movies rented and the total charge per month can be expressed with the function $y = 3x + 12$. In this equation, x stands for the number of movies rented, and y stands for the total charge per month. Determine a reasonable domain and range for this scenario.

a. Domain: $\{-1, -2, -3, -4, -5\}$; Range: $\{0, 12, 15, 18, 21\}$
b. Domain: $\{0, 1, 2, 3, 4\}$; Range: $\{0, 12, 15, 18, 21\}$
c. Domain: $\{0, 1, 2, 3, 4\}$; Range: $\{12, 15, 18, 21, 24\}$
d. Domain: $\{-1, -2, -3, -4, -5\}$; Range: $\{12, 15, 18, 21, 24\}$

13. If $f(x) = x^2 - 1$, then solve $f(4) \cdot f(3)$.

a. 12
b. 158
c. 120
d. 143

14. Which explicit formula matches the geometric sequence given below?

$$2, 8, 32, 128, \ldots$$

a. $a_n = 4 \cdot 2^{n-1}$
b. $a_n = 2 \cdot 4^{n-1}$
c. $a_n = 4 \cdot 2^{1-n}$
d. $a_n = 2 \cdot 4^{1-n}$

15. A bowling alley rents shoes for $1 an hour. The function $f(x) = x$ represents the hourly rental fee for x hours. The bowling alley decides to add an additional $5 fee for cleaning. The function that represents the new cost for shoe rental at the bowling alley is $g(x) = f(x) + 5$. How does the graph of the transformation $g(x)$ compare to the graph of the parent function $f(x)$?

a. The graph of $g(x)$ is a translation of $f(x)$, with a horizontal shift of 5 units to the left.
b. The graph of $g(x)$ is a translation of $f(x)$, with a vertical shift of 5 units up.
c. The graph of $g(x)$ is a dilation of $f(x)$, vertically stretched by a factor of 5.
d. The graph of $g(x)$ is a dilation of $f(x)$, horizontally compressed by a factor of $\frac{1}{5}$.

16. If $f(x) = 2x^2 + 1$ and $g(x) = 2x - 7$, then what is the value of $f(-3) - g(10)$?

a. 32
b. -24
c. -8
d. 6

17. Ruby grows strawberries on her produce farm. During her first year, she planted 5 seeds. The next year, she planted 20 seeds. The year after that, she planted 80 seeds. Ruby plans to continue expanding her strawberry farm at the same rate each year.

$$5, 20, 80, \ldots$$

Which explicit formula best represents Ruby's growing strawberry farm?

a. $a_n = 4 \cdot 5^{(1-n)}$
b. $a_n = 5 \cdot 4^{(1-n)}$
c. $a_n = 4 \cdot 5^{(n-1)}$
d. $a_n = 5 \cdot 4^{(n-1)}$

18. In some locations the current temperature can be approximated by listening and counting cricket chirps. Write a linear function that provides a reasonable fit to the data displayed in the graph below. Approximate the temperature if 60 chirps are heard.

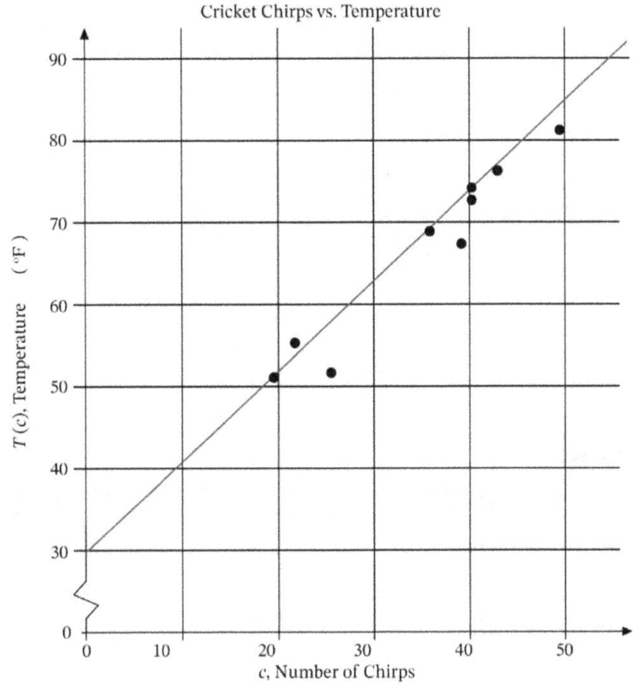

a. $f(x) = \frac{5}{4}x$, 60 chirps indicate a temperature of approximately 100 °F.
b. $f(x) = \frac{5}{4}x + 30$, 60 chirps indicate a temperature of approximately 105 °F.
c. $f(x) = \frac{5}{4}x + 30$, 60 chirps indicate a temperature of approximately 90 °F.
d. $f(x) = 8x + 30$, 60 chirps indicate a temperature of approximately 100 °F.

19. Marc is setting up a spaghetti bridge experiment. He places long pieces of dried spaghetti across the gap between two tables. Marc is interested in seeing the correlation between the number of pieces of spaghetti, and the number of marbles they can hold in a cup.

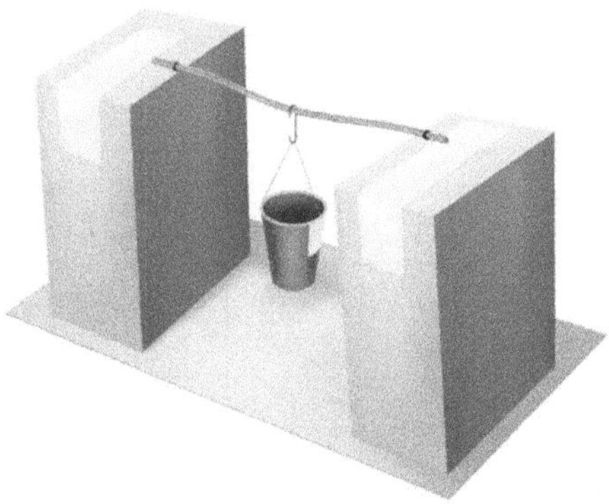

Marc's data is shown below.

Number of Pieces of Spaghetti	Number of Marbles to Break Spaghetti
1	18
2	40
3	56
4	70

Write a linear function that provides a reasonable fit for the data and use the function to estimate the number of marbles that 20 pieces of spaghetti can hold.

a. $f(x) = 18x$, 20 pieces of spaghetti can hold approximately 360 marbles.
b. $f(x) = 9x$, 20 pieces of spaghetti can hold approximately 180 marbles.
c. $f(x) = 20x$, 20 pieces of spaghetti can hold approximately 200 marbles.
d. $f(x) = 5x$, 20 pieces of spaghetti can hold approximately 3,600 marbles.

20. At the Village Inn Pizza Parlor, you can get a large pepperoni pizza for $20 and then every additional topping costs $0.75. Write a linear function to model the situation, and then determine the price of a pizza with 4 extra toppings.

a. $f(x) = 4x + 20$, a pizza with 4 additional toppings will cost $26.
b. $f(x) = 20x + 0.75$, a pizza with 4 additional toppings will cost $28.
c. $f(x) = 0.75x + 20$, a pizza with 4 additional toppings will cost $23.
d. $f(x) = 0.75 + 4$, a pizza with 4 additional toppings will cost $27.

Chapter 3: Linear Equations

3.1 Slope and Rate of Change

Rate of change, or slope, can be represented graphically, algebraically, or tabularly. The **slope** represents the vertical change divided by the horizontal change "rise over run." The slope is sometimes referred to as the "gradient" of a line. Essentially, the slope represents the steepness of a line. In slope-intercept form, $y = mx + b$, m represents the slope of the line.

Calculating the Slope from a Graph

When a graph is provided, the slope can be determined by locating two points on the line and counting the "rise over run" from one point to the other. For example, the graph on the left shows a rise of 4 and a run of 3, making the slope $m = \frac{4}{3}$. The graph on the right shows a rise of −3 and a run of 3, making the slope $m = \frac{-3}{3}$, or −1.

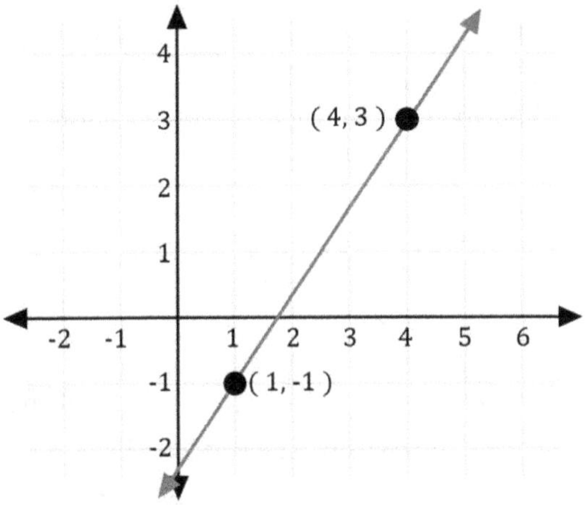

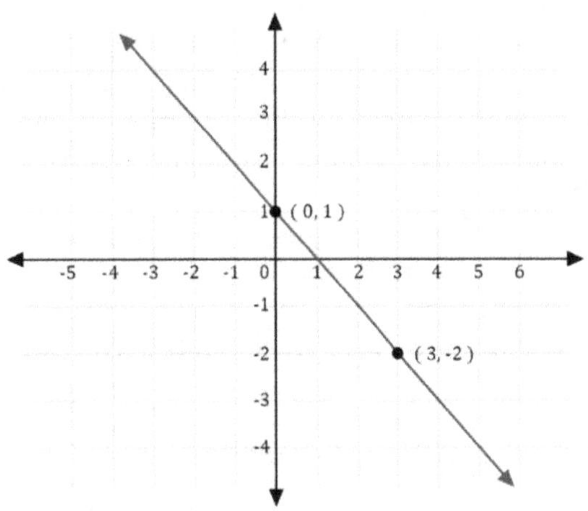

A line with a positive slope will extend up and to the right. A line with a negative slope will extend down and to the right.

Calculating the Slope Algebraically

Slope can be determined algebraically. If you are provided with a linear equation, you can manipulate the equation into slope-intercept form to identify the slope, m. For example, the equation $2y - 3 = 8x + 7$ can be manipulated algebraically to reveal the rate of change, or slope. The equation $2y - 3 = 8x + 7$ needs to be in the form $y = mx + b$ to determine the value of m. Use inverse operations to isolate the variable y. Start by adding 3 to both sides.

$$2y - 3 = 8x + 7$$
$$2y - 3 + 3 = 8x + 7 + 3$$
$$2y = 8x + 10$$

Now divide both sides by 2.

$$\frac{2y}{2} = \frac{8x + 10}{2}$$
$$y = 4x + 5$$

The slope, or rate of change, is 4.

Calculating the Slope Tabularly

Calculating slope using a table requires at least two ordered pairs (x, y). To calculate the rate of change from a table, use the **slope formula**: $m = \frac{y_2 - y_1}{x_2 - x_1}$. This formula divides the vertical change by the horizontal change, which produces the slope. For example, in the table below we see four ordered pairs: (0,0), (2, −2), (4, −4), and (6, −6). Select any two ordered pairs from the table to plug into the formula. If the first two ordered pairs are used, (0,0) and (2, −2), then the formula becomes $m = \frac{-2-0}{2-0} = \frac{-2}{2} = -1$. The slope of the line that this table produces is $m = -1$.

x	y
0	0
2	-2
4	-4
6	-6

The **rate of change** of a line can be used to represent real-world scenarios. We can use linear equations, graphs, and tables to determine the rate of change to make predictions about the future. For example, the table below shows the relationship between the number of inches of rainfall and the duration of a storm. Finding the rate of change will allow us to make predictions about the number of inches of rainfall for longer storms.

Duration of the storm (hrs), *x*	0	1	2	3
Amount of rain (in), *y*	0	0.5	1	1.5

The slope in this scenario can be found using the formula $m = \frac{y_2 - y_1}{x_2 - x_1}$. When the x- and y-values are plugged in, the formula becomes $m = \frac{1-0.5}{2-1}$, which simplifies to $m = \frac{1}{2}$. Now that we know the slope, we can make predictions about the rainfall for other storms. If there is a storm that lasts 5 hours, there will be $5 \times \frac{1}{2} = \frac{5}{2}$ inches of rain. This means that a storm that lasts 5 hours will produce approximately $2\frac{1}{2}$ inches of rain.

Understanding rate of change, equations, tables, and graphs is important because these concepts can be applied to many real-world situations, to make accurate predictions.

Review Video: Finding the Slope of a Line
Visit mometrix.com/academy and enter code: 766664

EXAMPLES

EXAMPLE 1:

Find the slope using the slope formula.

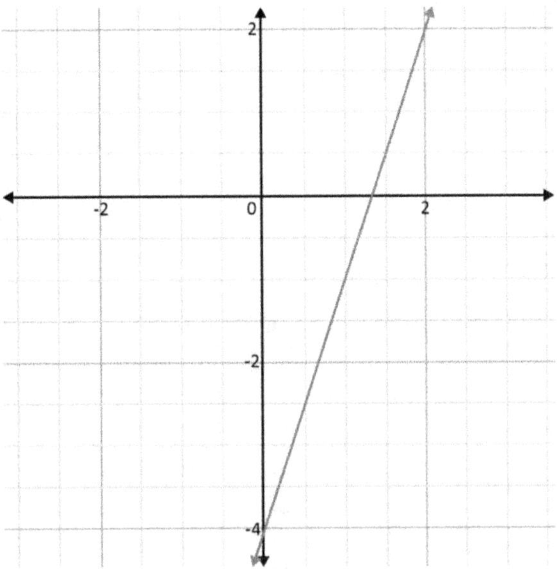

Answer:

Slope formula: $m = \frac{y_2 - y_1}{x_2 - x_1}$

Select two points on the line. Plug their x- and y-values into the formula. For example, the points $(0, -4)$ and $(2, 2)$ are on the line, so plug these values into the formula to find the slope.

$$m = \frac{y_2 - y_1}{x_2 - x_1} = \frac{2 - (-4)}{2 - 0} = \frac{6}{2} = 3$$

The slope of the line is 3.

EXAMPLE 2:

Determine the slope of the graphed line $y = 5$.

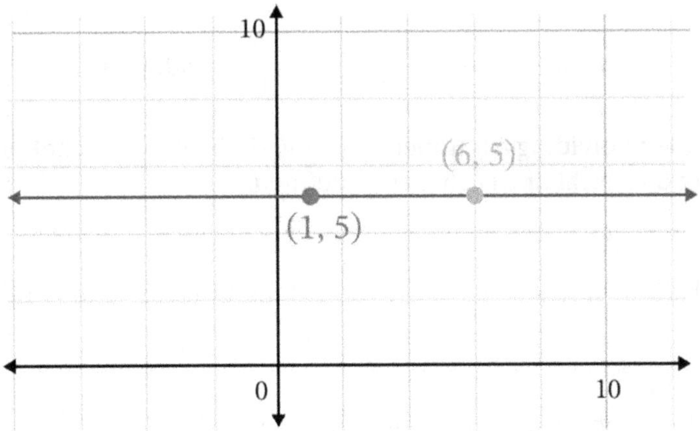

Answer:

The slope is the "steepness" of a line. Slope is calculated as vertical change over horizontal change, or $m = \frac{\text{rise}}{\text{run}}$. In a horizontal line, the slope is 0 because the line does not "rise" at all. This can be confirmed using the slope formula.

$$m = \frac{y_2 - y_1}{x_2 - x_1} = \frac{5 - 5}{1 - 6} = \frac{0}{-5} = 0$$

The slope of the line is zero.

Example 3:

Determine the slope of the graphed line $x = 2$.

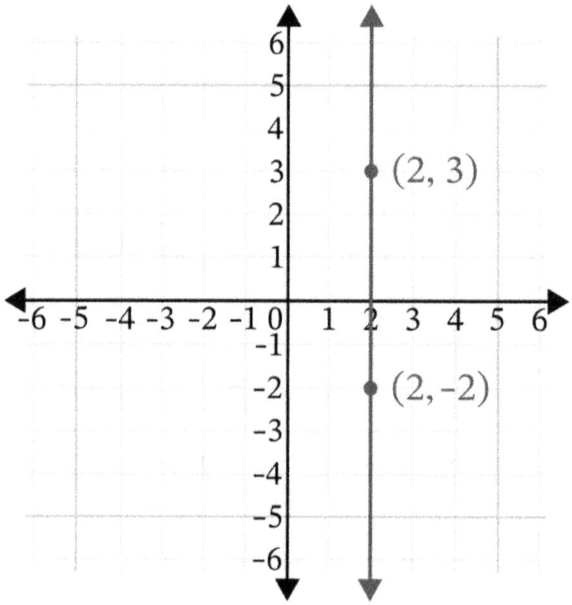

Answer:

The slope is the "steepness" of a line. Slope is calculated as vertical change over horizontal change, or $m = \frac{\text{rise}}{\text{run}}$. In a vertical line, the slope is undefined because the line does not "run" at all. This can be confirmed using the slope formula.

$$m = \frac{y_2 - y_1}{x_2 - x_1} = \frac{3 - (-2)}{2 - 2} = \frac{5}{0} = \text{undefined}$$

Because the slope results in dividing a number by 0, and division by 0 is not possible, the slope is undefined. The slope of a vertical line is always undefined.

Example 4:

Determine if the graph of the line $y = -\frac{1}{4}x + 4$ will extend upward or downward.

Answer:

Positive slopes will extend "upward" (bottom left to upper right).

Negative slopes will extend "downward" (upper left to bottom right).

The equation $y = -\frac{1}{4}x + 4$ has a slope of $-\frac{1}{4}$, so the graphed line will slope downward.

EXAMPLE 5:

Determine the slope of the equation of the line $y = \frac{1}{2}x + 1$.

Answer:

In slope-intercept form, $y = mx + b$, the variable m represents the slope. In the equation $y = \frac{1}{2}x + 1$, the slope is $\frac{1}{2}$.

PRACTICE QUESTIONS

1. The graph of a linear function passes through the points $(7, 1)$ and $(-2, 3)$. What is the slope of this line?

a. $\frac{9}{2}$
b. $-\frac{9}{2}$
c. $\frac{2}{9}$
d. $-\frac{2}{9}$

2. The table below represents some points on the graph of a linear function.

x	y
0	4
2	12
4	20
6	28
8	36

What is the slope of this linear function?

a. 4
b. 8
c. $\frac{1}{4}$
d. $\frac{1}{2}$

3. What is the slope of a line with the equation $3x + 2y = 5$?

a. $\frac{2}{3}$
b. -3
c. 3
d. $-\frac{3}{2}$

4. The graph of $y = -1.5$ is shown on the coordinate plane below. What is the slope of this line?

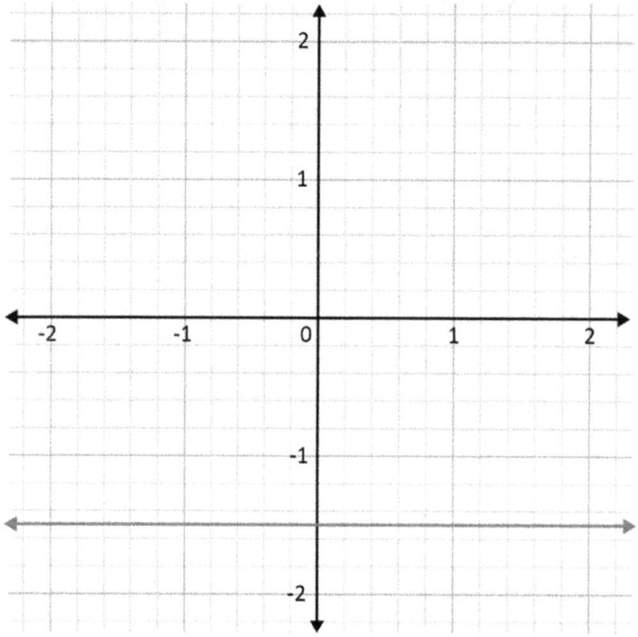

a. 0
b. -1.5
c. 1.5
d. Undefined

5. What is the slope of a line with the equation $8x + 3y = -9$?

a. $\frac{8}{3}$
b. -8
c. $-\frac{8}{3}$
d. 8

6. The graph of a linear equation is shown on the coordinate plane below. What is the slope of this line?

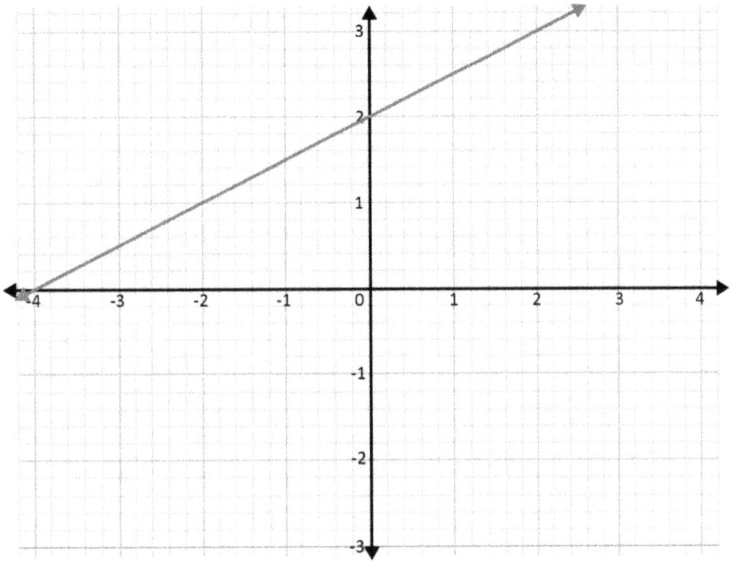

a. 2
b. -2
c. $\frac{1}{2}$
d. $-\frac{1}{2}$

7. The graph of $x = 4$ is shown on the coordinate plane below. What is the slope of this line?

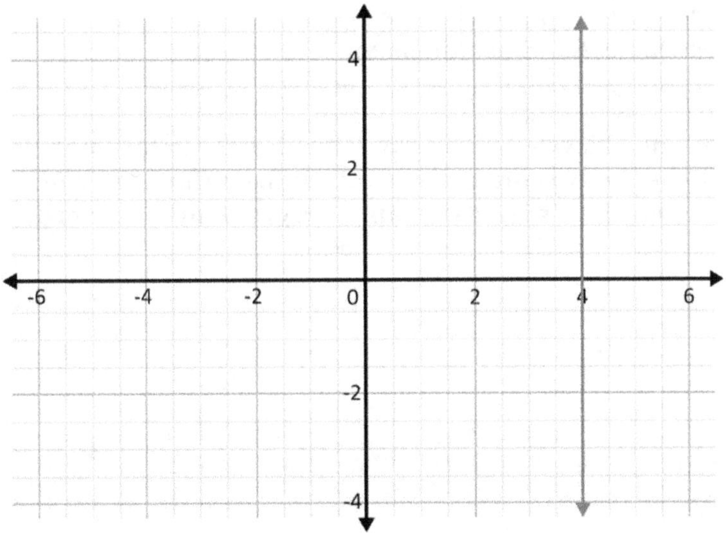

a. 0
b. Undefined
c. 4
d. -4

8. The graph of $y = 2$ is shown on the coordinate plane below. What is the slope of this line?

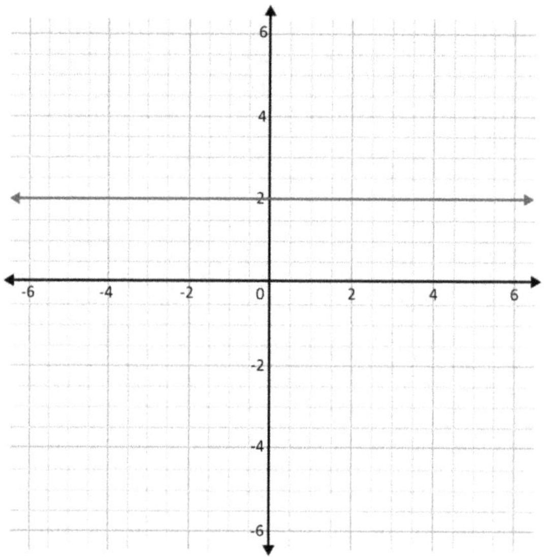

a. 0
b. 2
c. −2
d. Undefined

9. A taxi driver charges a base fee of \$2.70, plus \$1.44 for each kilometer traveled. This relationship is represented by the equation $y = 1.44x + 2.70$. Which of the following best describes the rate of change?

a. The taxi driver charges \$4.14 per kilometer.
b. The taxi driver charges \$2.70 per kilometer.
c. The taxi driver charges \$1.44 per kilometer.
d. The taxi driver charges \$1.44 for every 2 kilometers traveled.

10. Maria decides to rent a bicycle for the day. The bicycle rental company charges a flat fee of \$8, as well as \$1.50 per hour rented. Given this information, determine the direction in which the line would slope if this relationship were graphed on a coordinate plane.

a. Downward
b. Upward
c. Zero
d. Undefined

11. The Chatterley family went on a road trip. The table below shows the relationship between the amount of time they spent driving in hours (x) and the distance traveled in miles (y).

Time Driving in Hours (x)	Distance Traveled in Miles (y)
0	0
2	40
4	80
6	120

What is the rate of change of the distance traveled per hour?

a. 120 miles per hour
b. 80 miles per hour
c. 40 miles per hour
d. 20 miles per hour

12. A savings account balance is modeled by the linear function shown on the graph below.

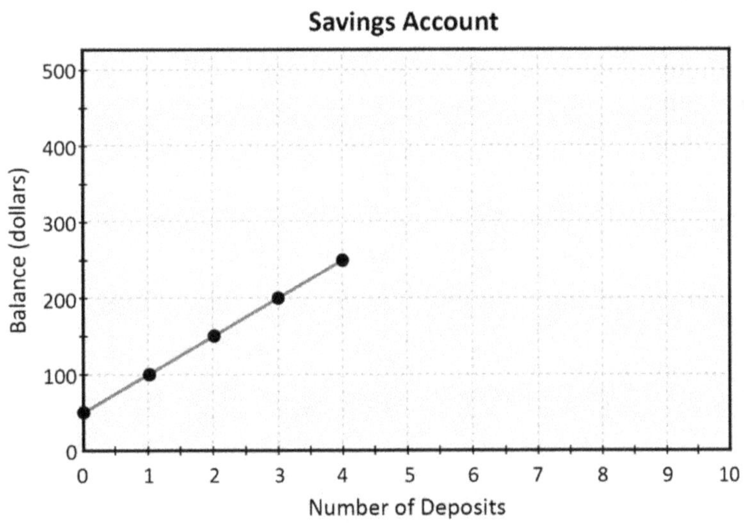

What is the rate of change of the balance per deposit?

a. \$5 per deposit
b. \$0.50 per deposit
c. \$100 per deposit
d. \$50 per deposit

13. Grace runs around a 100-meter track. She runs one warm-up lap, then proceeds with her workout. She uses the linear equation $y = 100 + 4.5x$ to describe the distance, y, she runs x seconds after her warm-up lap. Which of the following best describes the rate of change of the linear equation?

a. Grace runs 4.5 meters in 2 seconds.
b. Grace runs 4.5 meters in 1 second.
c. Grace runs 9 meters in 1 second.
d. Grace runs 100 meters in 1 second.

14. The table below represents some points on the graph of a linear function.

x	y
1	-14
3	-6
5	2
7	10

What is the rate of change of this function?

a. -4
b. 2
c. 4
d. 8

15. Shannon's employer pays an hourly rate plus a monthly stipend of 100 dollars. The equation $y = 100 + 30x$ can be used to determine the total pay, y, that Shannon would be paid for x hours of work in a month. What is the rate of change of Shannon's total pay with respect to the number of hours worked?

a. 30 dollars per hour
b. 60 dollars per hour
c. 70 dollars per hour
d. 100 dollars per hour

3.2 Graphing Linear Equations

One algebraic skill that is necessary to master is the ability to make graphs from equations. Many of the equations you encounter will be in slope-intercept form or standard form.

Slope-intercept form: $y = mx + b$, where m is the slope and b is the y-intercept.

Standard form: $Ax + By = C$, where A, B, and C are real numbers and $A > 0$.

If you are given an equation in slope-intercept form, and you need to graph the equation, start by identifying the y-intercept, and then use the slope to locate additional points. For example, consider the equation $y = 3x - 5$. The slope is 3 or $\frac{3}{1}$ and the y-intercept is –5. A y-intercept of –5 means the graphed line will intersect the y-axis at $(0, -5)$. Plot this point on the coordinate plane, and then use the slope to locate additional points. The slope is $\frac{3}{1}$, which means the vertical change is 3 and the horizontal change is 1 ("rise" of 3 and "run" of 1). From the point $(0, -5)$ "rise" 3 units and "run" 1 unit to the right. This new point $(1, -2)$ will be on the graphed line. The process of "rise" 3 and "run" 1 can be repeated to plot multiple points on the line; however, only 2 points are required to form the line.

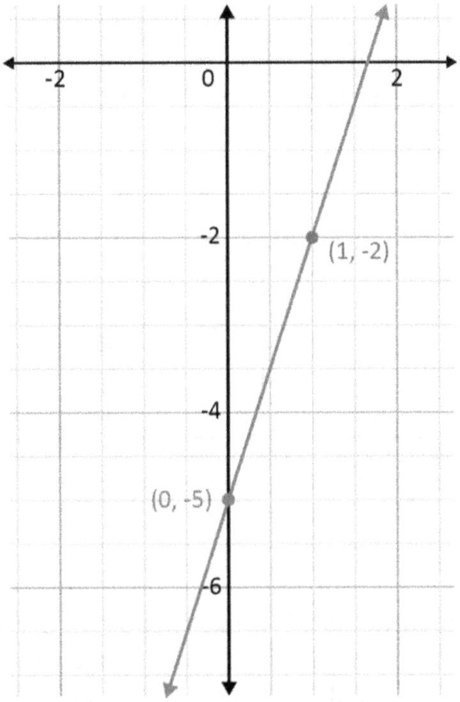

The equations for vertical and horizontal lines will look a little different. A vertical line will have the same x-value, regardless of the y-value. The equation for a vertical line will simply be the location of the x-intercept. The x-intercept is also sometimes referred to as the zero of a function. For example, $x = 3$ is a vertical line passing through the x-axis at 3. Regardless of the y-value, the x-value will always be 3.

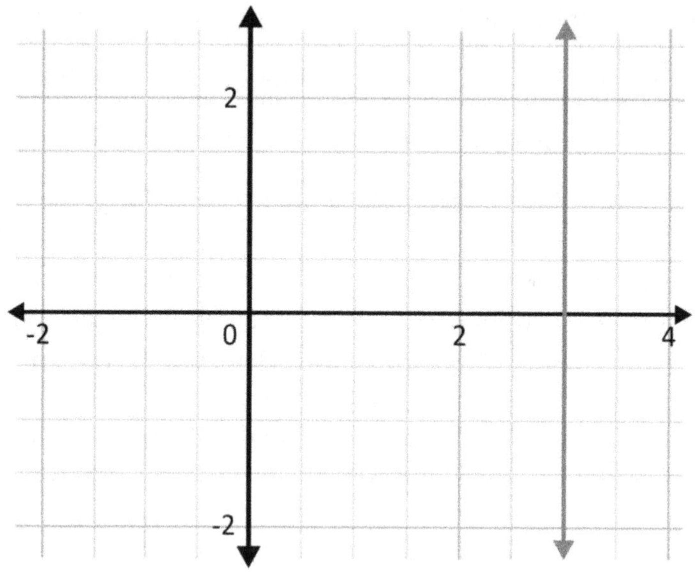

Similarly, a horizontal line will have only one y-value, regardless of what the x-value is. For example, the equation $y = 4$ describes a horizontal line that passes through the y-axis at 4.

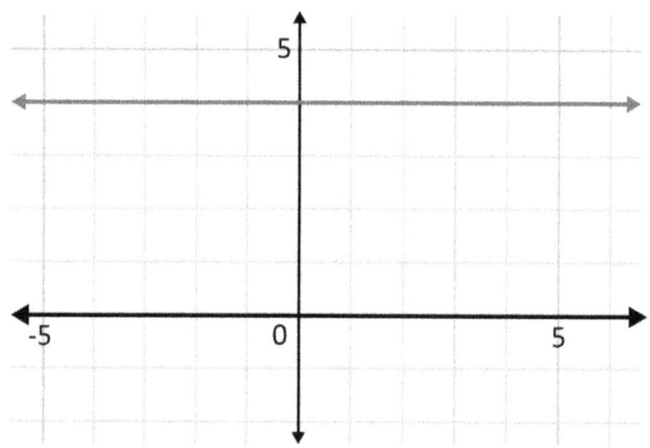

Some equations will be presented in standard form. When graphing an equation in standard form, $Ax + By = C$, you can locate the x- and y-intercepts, and then connect the two points. For example, consider the equation $2x + 3y = 6$. Find the x-intercept by plugging in 0 for x.

$$\begin{aligned} 2(0) + 3y &= 6 \\ 3y &= 6 \\ y &= 2 \end{aligned}$$

This means that the y-intercept is $(0, 2)$. Now plug in 0 for y.

$$\begin{aligned} 2x + 3(0) &= 6 \\ 2x &= 6 \\ x &= 3 \end{aligned}$$

This means that the x-intercept is $(3, 0)$. Knowing the x- and y-intercepts is enough information to create a line. Plot the two intercepts and connect the points.

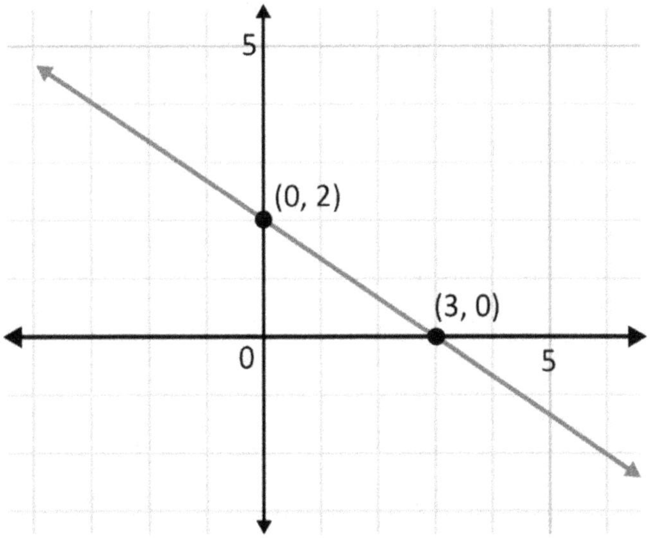

Review Video: Graphing Linear Equations
Visit mometrix.com/academy and enter code: 479576

EXAMPLES

EXAMPLE 1:

Graph the following equation by using the slope and y-intercept: $y = \frac{1}{2}x - 3$.

Answer:

Start by identifying the y-intercept of the graphed line. The equation $y = \frac{1}{2}x - 3$ is in slope-intercept form, $y = mx + b$, where b represents the y-intercept. This means that -3 is the y-intercept. Plot a point at the y-intercept $(0, -3)$ and then use the slope to find the next point. The variable m identifies the slope, so in the equation $y = \frac{1}{2}x - 3$, the slope is $\frac{1}{2}$. A slope of $\frac{1}{2}$ represents a "rise" of 1 and a "run" of 2. From the y-intercept, move "up" 1 and "over" 2, and plot this point $(2, -2)$.

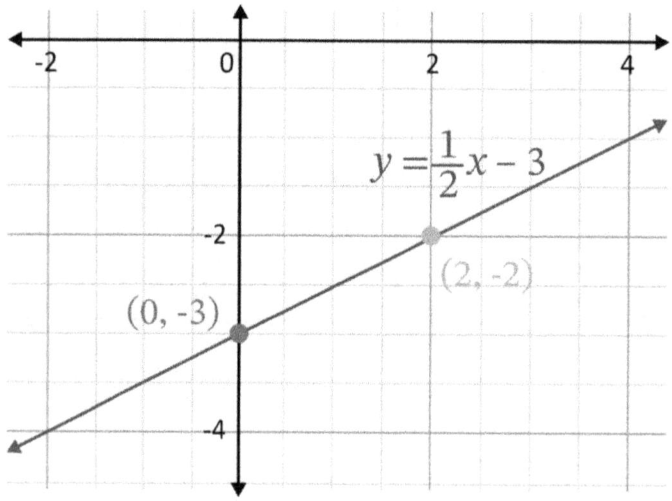

EXAMPLE 2:

Graph the following equation: $y = 5$.

Answer:

The graph of a horizontal line will have only one y-value. For example, $y = \frac{1}{3}$, $y = -2$, and $y = 7$ are all horizontal lines.

The equation $y = 5$ forms a horizontal line where the value of y for every value of x is 5.

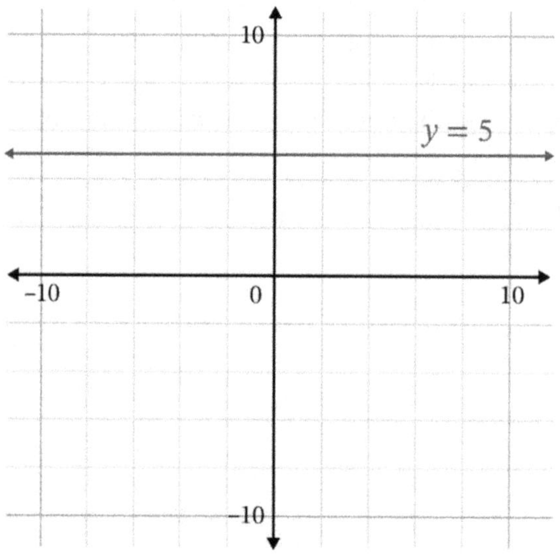

EXAMPLE 3:

Find the x- and y-intercepts to graph the following equation in standard form: $4x + 2y = 57$.

Answer:

To find the x-intercept, plug in 0 for y.

$$\begin{aligned} 4x + 2(0) &= 57 \\ 4x &= 57 \\ x &= 14.25 \end{aligned}$$

The x-intercept is $(14.25,0)$.

To find the y-intercept in standard form, plug in 0 for x.

$$\begin{aligned} 4(0) + 2y &= 57 \\ 2y &= 57 \\ y &= 28.5 \end{aligned}$$

The y-intercept is $(0,28.5)$.

To graph the line, plot the x- and y-intercepts and connect the points.

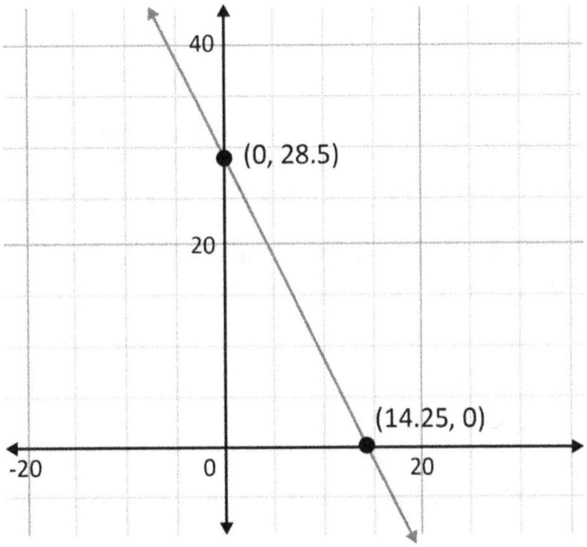

EXAMPLE 4:

Locate the x-intercept, y-intercept, and zero of the line below.

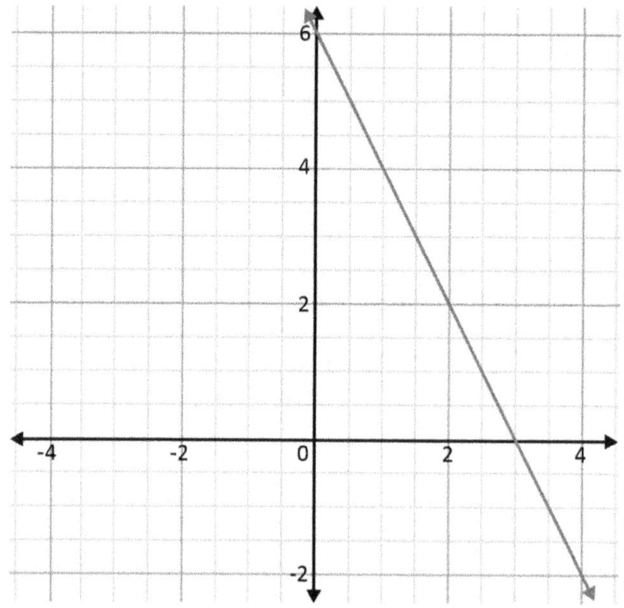

Answer:

The x-intercept is the point where the graphed line intersects the x-axis: (3,0).

The y-intercept is the point where the graphed line intersects the y-axis: (0,6).

The "zero" of a line is the value of x that makes the equation equal to zero. This is the x-intercept, because at the x-intercept the value of y is zero. The x-intercept of this graphed line is $(3, 0)$. So, the zero of the graph is $x = 3$.

PRACTICE QUESTIONS

1. Which graph represents the linear equation $y = 2x - 3$?

a. 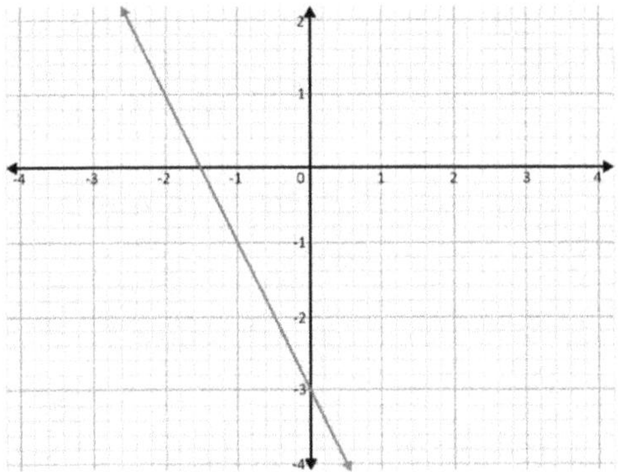

c. 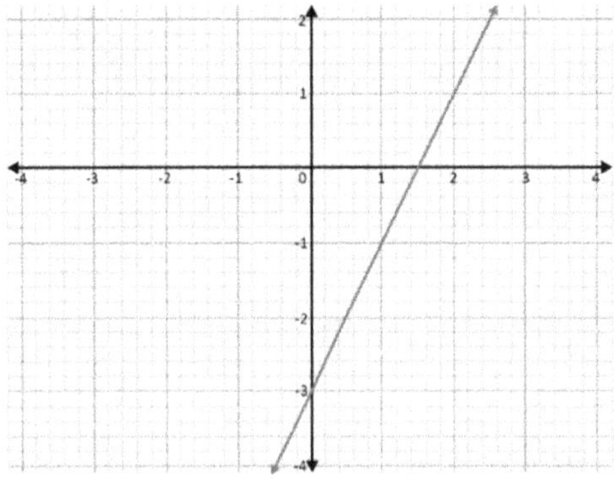

b. 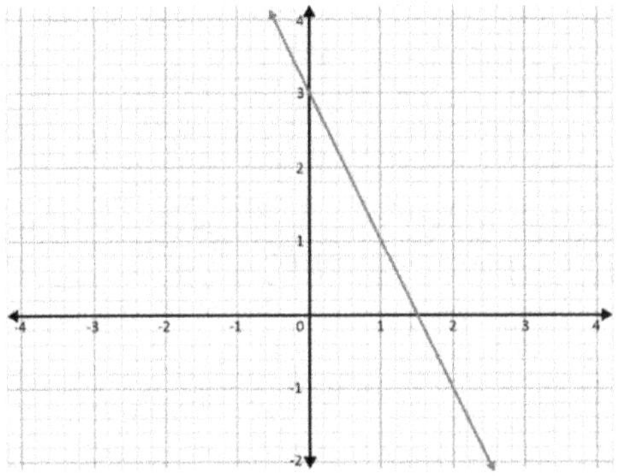

d. 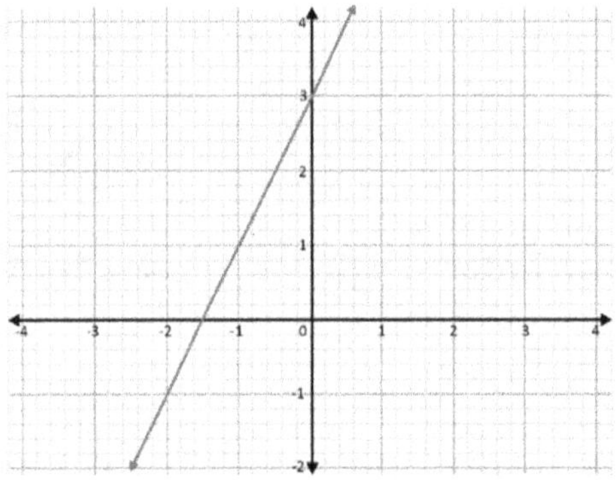

2. The graph of a linear equation is shown on the coordinate plane below.

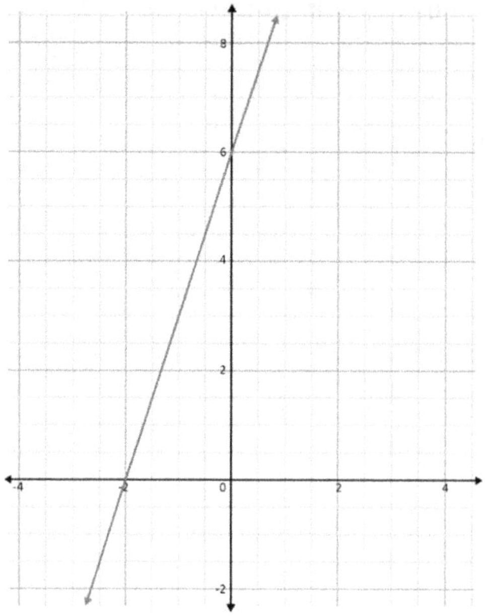

Which ordered pair best represents the location of the x-intercept?

a. $(6, 0)$
b. $(-2, 0)$
c. $(0, -2)$
d. $(0, 6)$

3. The graph of a linear equation is shown on the coordinate plane below.

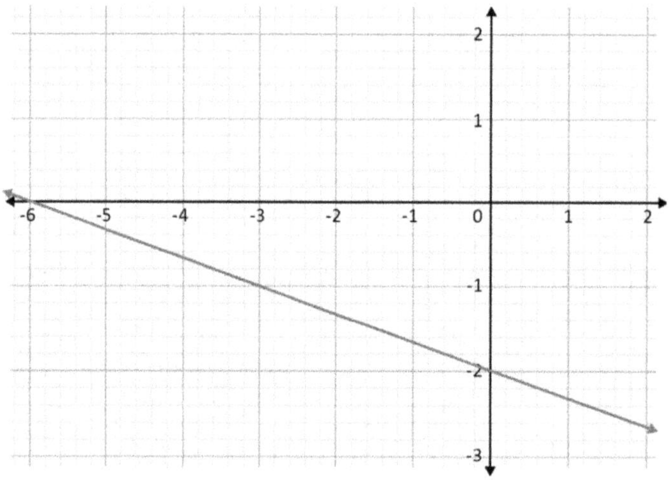

Which ordered pair best represents the location of the y-intercept?

a. $(-2, 0)$
b. $(0, -2)$
c. $(-6, 0)$
d. $(0, -6)$

4. The graph of a linear equation is shown on the coordinate plane below.

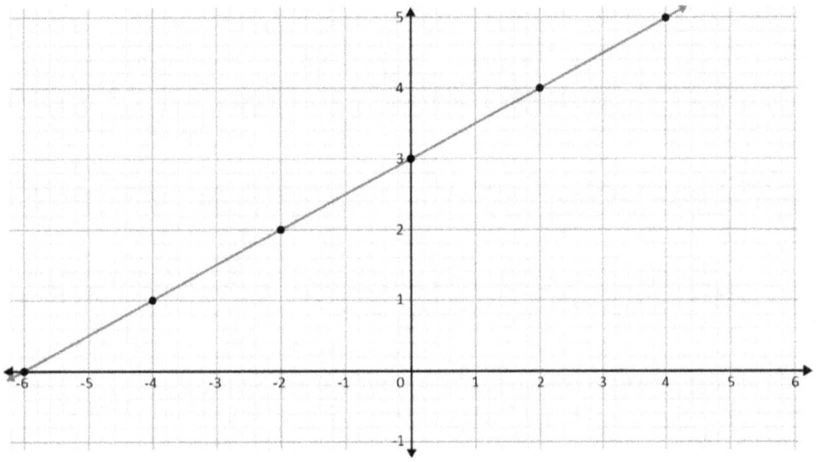

What is the slope of this line?

a. $m = \frac{1}{2}$
b. $m = 2$
c. $m = -\frac{1}{2}$
d. $m = -2$

5. Which graph represents the linear equation $x = 4$?

a. 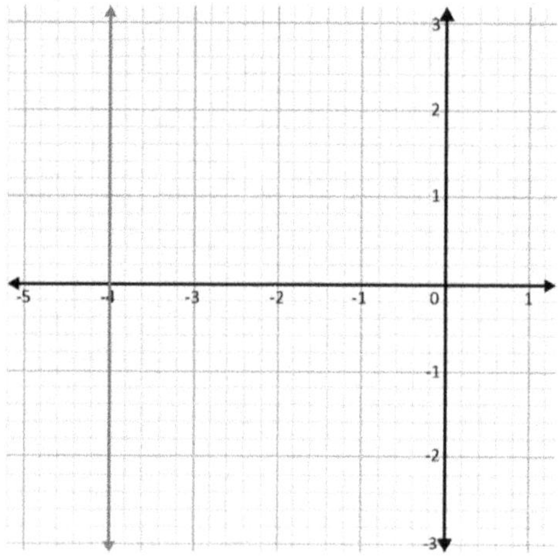

c. 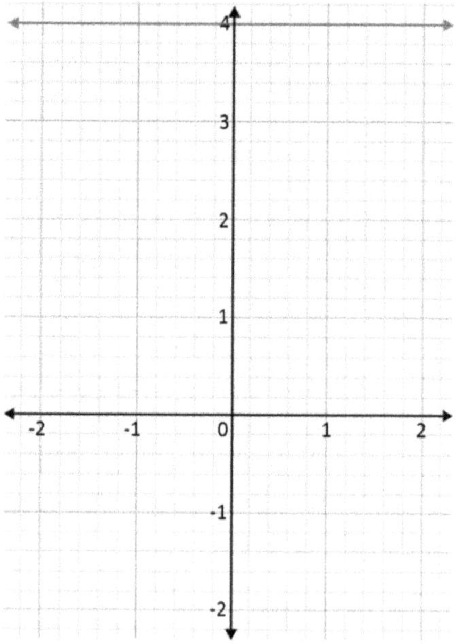

b. 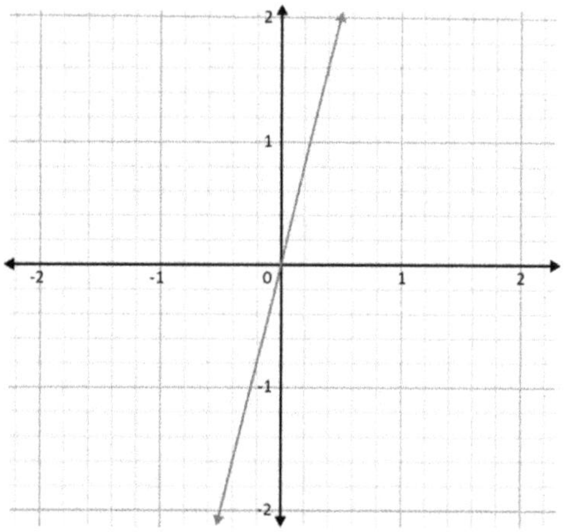

d. 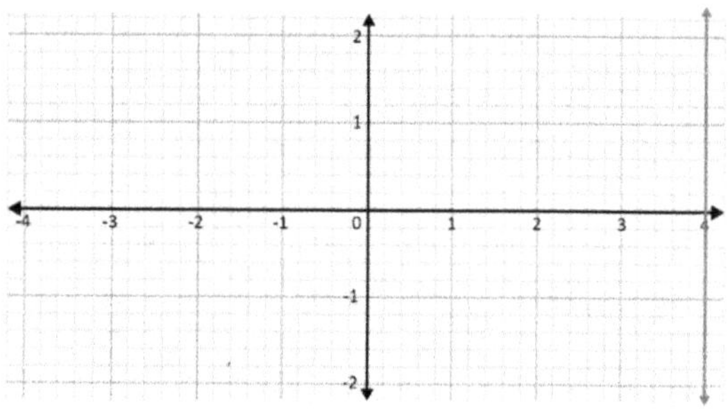

6. Which graph represents the linear equation $2x + 4y = 8$?

a.

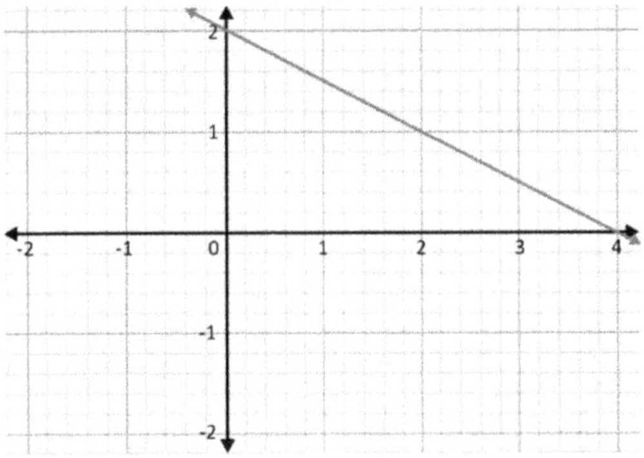

c.

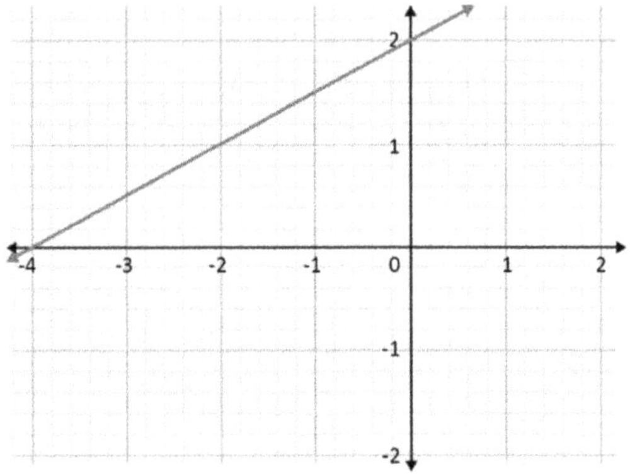

b.

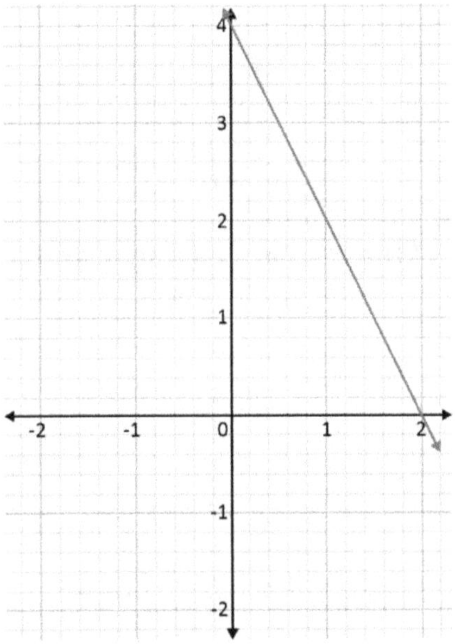

d.

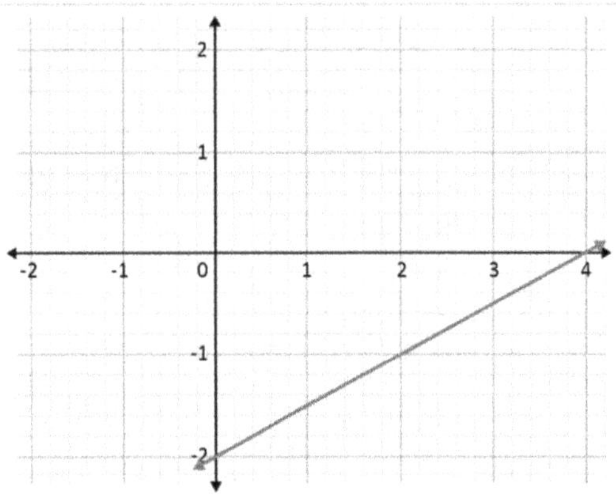

7. The graph of a linear function is shown below.

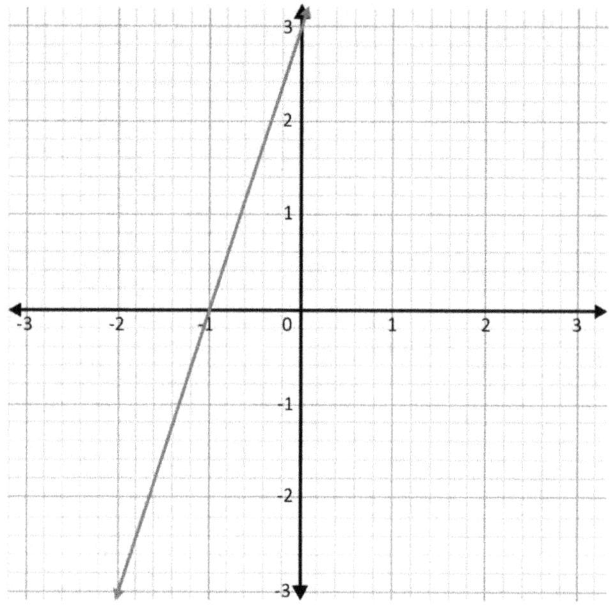

Which coordinate pair best represents the location of the zero of this linear function?

a. $(0, -1)$
b. $(3, 0)$
c. $(-1, 0)$
d. $(0, 3)$

8. Which graph represents the linear equation $y = -6$?

a. 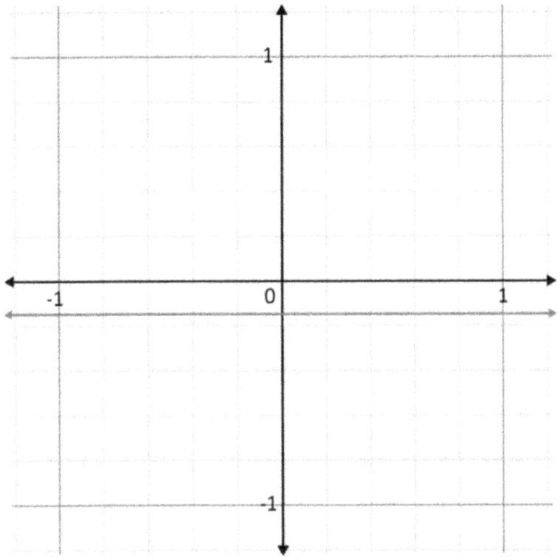

c. 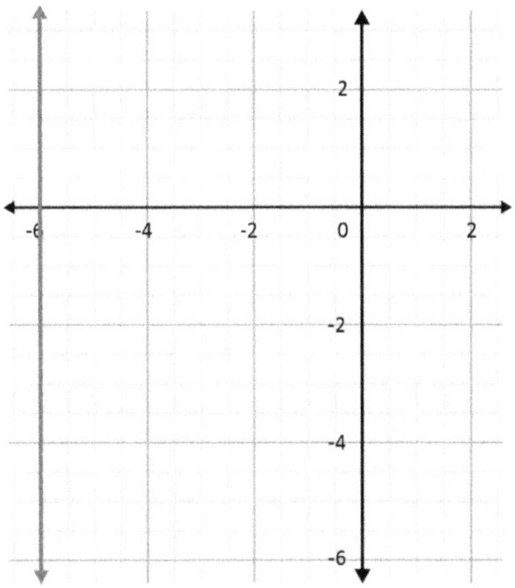

b. 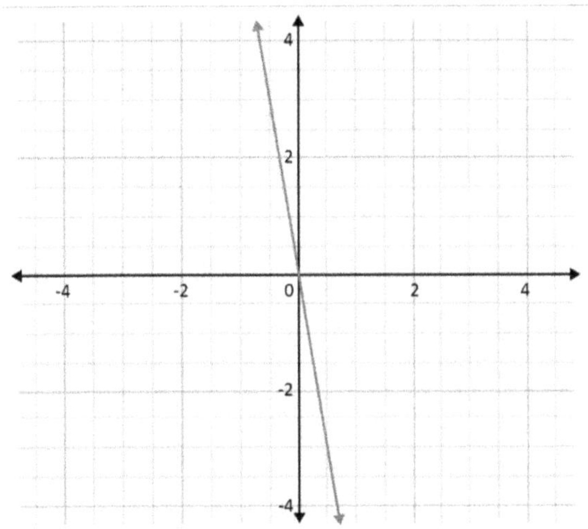

d. 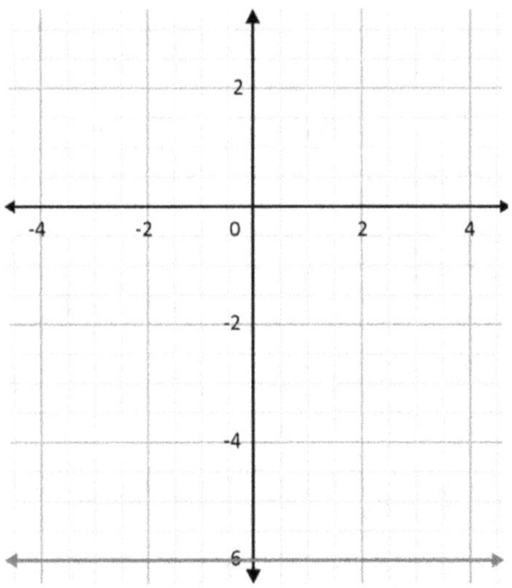

9. What is the y-intercept of the linear equation $4x + 2y = 6$?

a. $(0, 4)$
b. $(0, 2)$
c. $(0, 3)$
d. $\left(0, 1\frac{1}{2}\right)$

10. The graph of the linear equation $3x + 4y = 1$ is shown on the coordinate plane below.

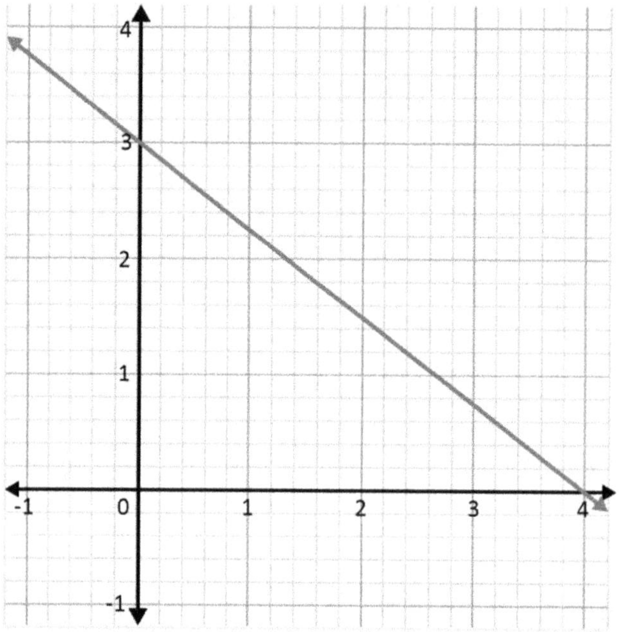

Which ordered pair best represents the location of the y-intercept?

a. $(3, 0)$
b. $(0, 3)$
c. $(4, 0)$
d. $(0, 4)$

11. Which graph represents the linear equation $-6x + 3y = 15$?

a.

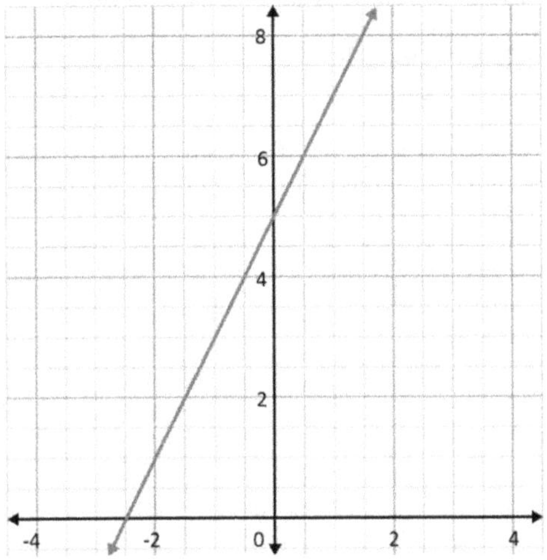

c.

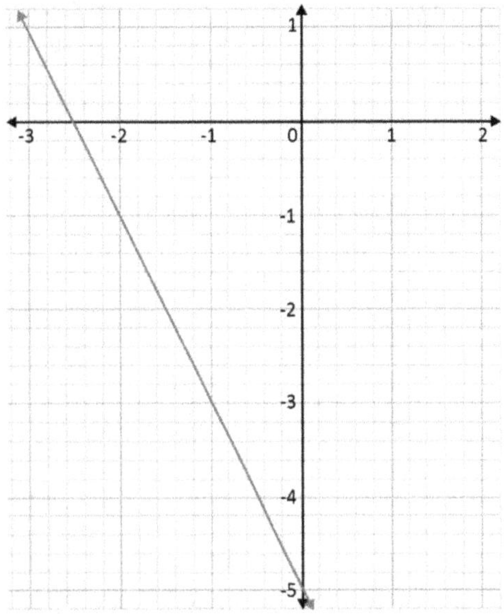

b.

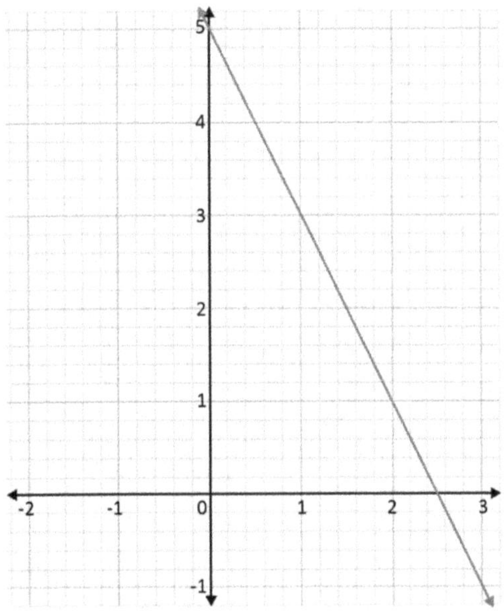

d.

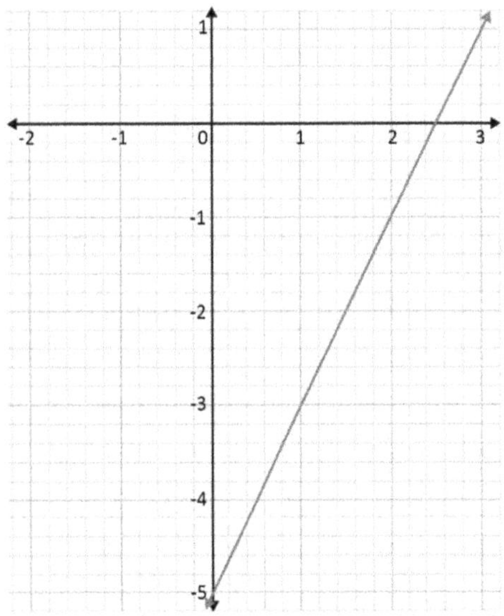

12. What is the x-intercept of the linear equation $2x + 4y = 12$?

a. $(6, 0)$
b. $(4, 0)$
c. $(3, 0)$
d. $(2, 0)$

13. Mike and Christina are graphing linear equations and identifying their slopes. After graphing the equation $y = 3.5$, they disagree about the slope of the line. Mike thinks the slope is 0, and Christina thinks the slope is undefined. Who is correct?

a. Christina is correct because a vertical line always has an undefined slope.
b. Christina is correct because a slope with a decimal is always undefined.
c. Mike is correct because a vertical line always has a slope of 0.
d. Mike is correct because a horizontal line always has a slope of 0.

14. The graph below shows how Matthew's savings account balance has changed over the course of a year. He opened his savings account with $250, and after 12 months, he had $2,650 in his savings account. How much money does Matthew save each month?

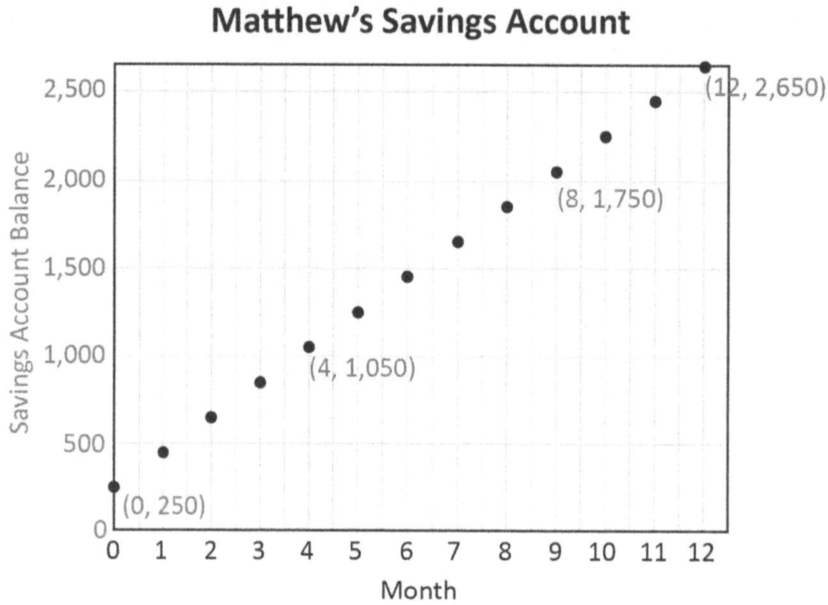

a. Matthew spent $200 each month.
b. Matthew saved $1,200 each month.
c. Matthew saved $200 each month.
d. Matthew saved $250 each month.

15. Marissa goes for a jog with her dog every morning. The distance jogged can be modeled by the equation $d = \frac{1}{10}t$, where d is the distance jogged in miles, and t is the time jogged in minutes. Which graph best represents the relationship between minutes and distance?

a. 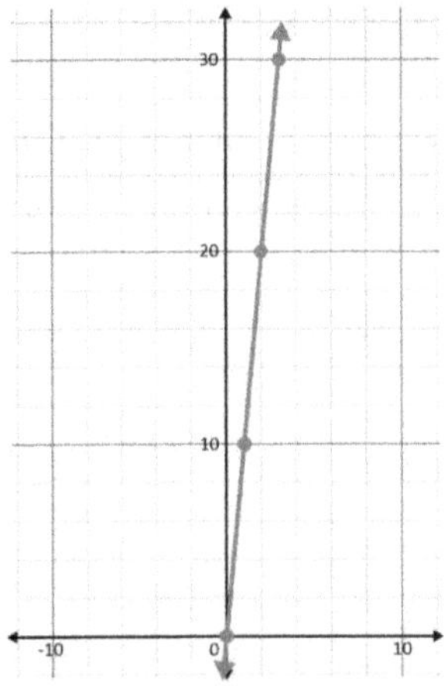

b. 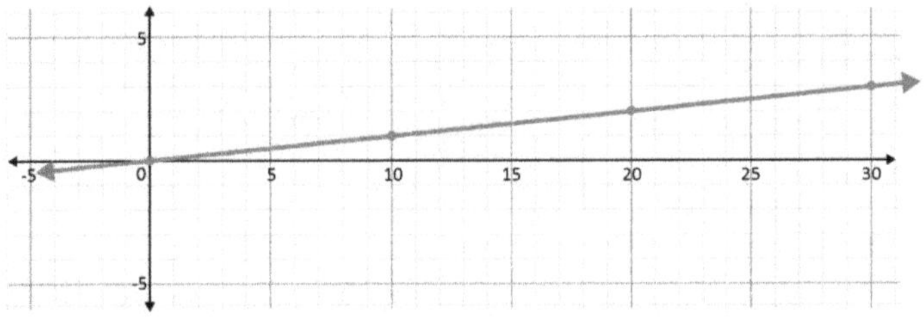

c. 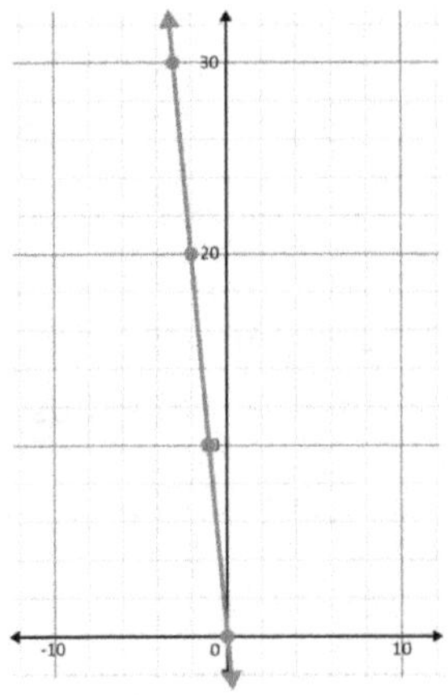

d. 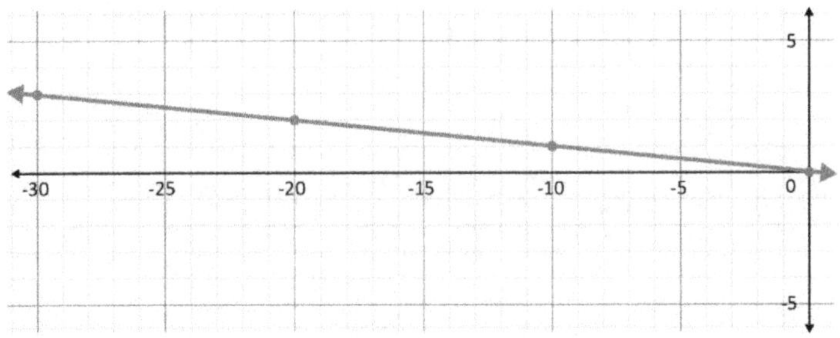

3.3 Finding the Slope-Intercept Form Equation of a Line

Slope-intercept form provides two very useful pieces of information when graphing linear equations. As the name suggests, the slope, m, and y-intercept, b, are provided. The y-intercept is the location that the line crosses the y-axis. The slope is "rise over run" or $\frac{\text{vertical change}}{\text{horizontal change}}$.

Slope-intercept form is $y = mx + b$, where m is the slope, and b is the y-intercept. For example, $y = 5x - 2$ has a slope of 5, or $\frac{5}{1}$, and a y-intercept of –2. The graphed line will cross the y-axis at –2, and from this point the line will rise up 5 units and run right 1 unit.

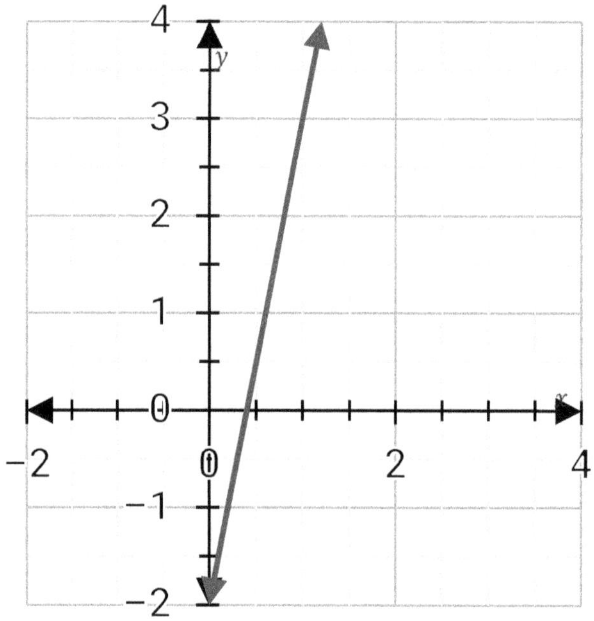

When writing an equation from a graph, identify the y-intercept and the slope. For example, in the graph below, the y-intercept is 3, and the slope is $\frac{1}{4}$, a rise of 1 and a run of 4. Take these values and place them in the slope-intercept formula for m and b. The equation of the line $y = mx + b$ becomes $y = \frac{1}{4}x + 3$.

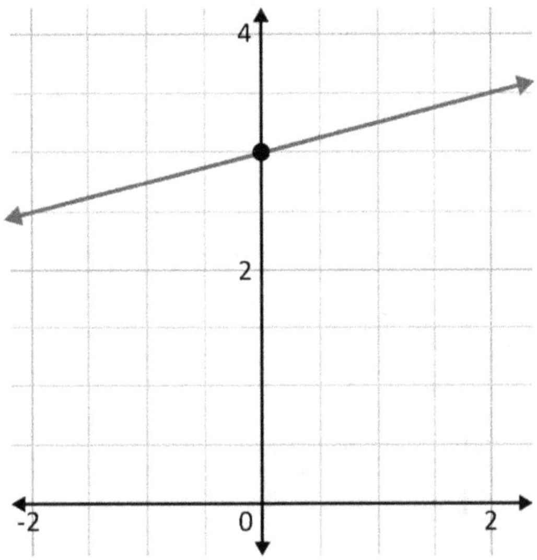

You can also write the equation of a line in slope-intercept form if you are provided two points on a line. For example, if you know the points $(5, -6)$ and $(10, -9)$, you can write the equation of the line.

m is found using the slope formula: $m = \frac{y_2 - y_1}{x_2 - x_1}$.

b is found by plugging in values for x, y, and m, and solving for b.

Start by finding m.

$$m = \frac{y_2 - y_1}{x_2 - x_1} = \frac{-6 - (-9)}{5 - 10} = -\frac{3}{5}$$

Then, find b by plugging in the values from one of the provided ordered pairs, like $(10, -9)$.

$$\begin{aligned} -9 &= -\frac{3}{5}(10) + b \\ -9 &= -6 + b \\ -3 &= b \end{aligned}$$

Now you have the slope, m, and y-intercept, b. Plug m and b into the equation.

$$y = mx + b$$
$$y = -\frac{3}{5}x - 3$$

The points $(0, -3)$ and $(10, -9)$ are on the line that is graphed from the equation $y = -\frac{3}{5}x - 3$.

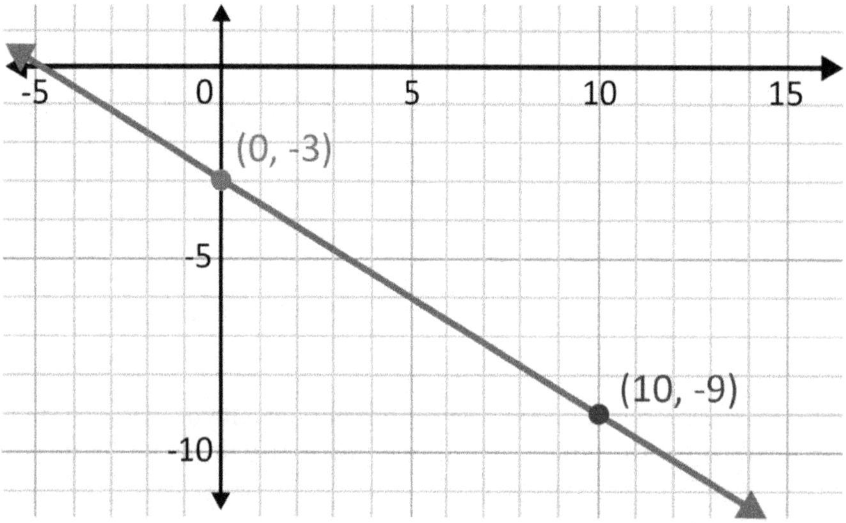

Review Video: Finding the Slope-Intercept Equation
Visit mometrix.com/academy and enter code: 429758

EXAMPLES

EXAMPLE 1:

A line has a slope of 3 and passes through the point (3,4). Write the equation in slope-intercept form and then graph the equation.

Answer:

When writing an equation in slope-intercept form, m and b need to be identified. Remember, m represents the slope, and b is the y-intercept. We know that the slope is 3, and we know the point $(3, 4)$ is on the graphed line. Plug these three values into the slope-intercept equation to solve for b.

$$y = mx + b$$
$$4 = 3(3) + b$$

Now solve for b.

$$4 = 9 + b$$
$$4 - 9 = 9 + b - 9$$
$$-5 = b$$

Now we have values for m and b, so plug these into the formula.

$$y = mx + b$$

$$y = 3x - 5$$

To graph an equation that is in slope-intercept form, start by making a point at the y-intercept, and then use the slope to find more points on the line. In this case, make a point at the y-intercept $(0, -5)$, and then use a "rise over run" of $\frac{3}{1}$ to find another point on the line. From $(0, -5)$ "rise" up 3 and "run" right 1. The result is the point $(1, -2)$. Draw a line extending through these two points.

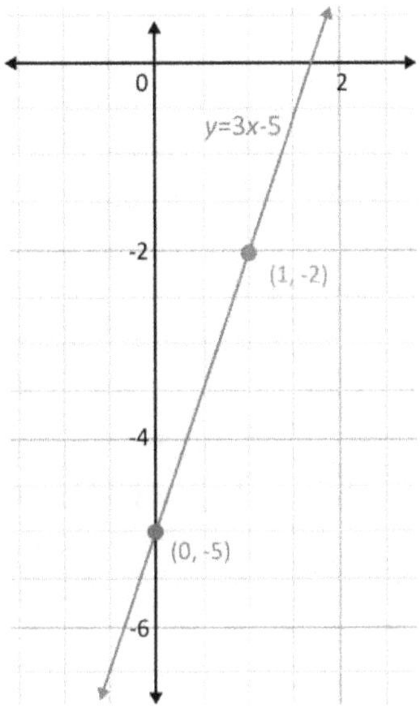

Example 2:

A line has a slope of $\frac{3}{4}$, and passes through the point $(3, 2)$. Write the equation for the line in slope-intercept form and then graph the line.

Answer:

Slope-intercept form, $y = mx + b$, requires a value for m and a value for b. In this case, we know that the value of m is $\frac{3}{4}$. We also know that the point (3,2) falls on the line of the graph. Plug in the slope and the point (3,2) into the equation to solve for b.

$$y = mx + b$$
$$2 = \frac{3}{4}(3) + b$$
$$2 = \frac{9}{4} + b$$
$$2 = 2\frac{1}{4} + b$$
$$-\frac{1}{4} = b$$

Now that the values for m and b are identified, plug them into the formula.

$$y = mx + b$$
$$y = \frac{3}{4}x - \frac{1}{4}$$

To graph the equation $y = \frac{3}{4}x - \frac{1}{4}$, make a point at the y-intercept. In this case, the y-intercept is $-\frac{1}{4}$. Plot the point $\left(0, -\frac{1}{4}\right)$ and then from this point "rise" up 3 and "run" right 4 because the slope is $\frac{3}{4}$. The new point is $\left(4, 2\frac{3}{4}\right)$. Draw a line extending through these two points.

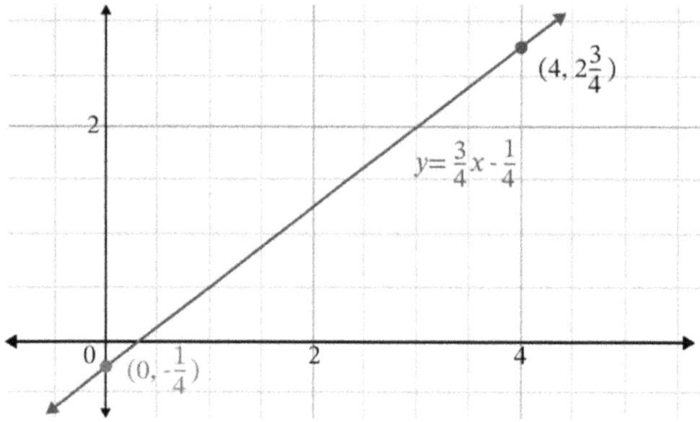

Practice Questions

1. Write the equation for the line that has a slope of -2 and passes through the point $(6, -3)$.

a. $y = -2x + 9$
b. $y = 2x$
c. $y = -x + 2$
d. $y = 2x + 9$

2. Write the equation for the line that has a slope of 2 and passes through the point $(-1, -6)$.

a. $y = 2x - 4$
b. $y = \frac{1}{2}x - 4$
c. $y = 2x + 4$
d. $y = 2x - \frac{1}{4}$

3. Write the equation for the line that has a slope of 2 and passes through the point $(-3, 5)$.

a. $y = 2x - 1$
b. $y = 2x + 11$
c. $y = -2x + 1$
d. $y = 2x + \frac{1}{11}$

4. Write the equation for the line that has a slope of -1 and passes through the point $(-2, 5)$.

a. $y = -x + 3$
b. $y = -\frac{1}{2}x + 3$
c. $y = -x - 3$
d. $y = -x + 2$

5. Write the equation for the line that has a slope of -3 and passes through the point $(1, 1)$.

a. $y = -3x$
b. $y = -3x - 4$
c. $y = 3x + 4$
d. $y = -3x + 4$

6. **Write the equation for the line that has a slope of $-\frac{5}{3}$ and passes through the point $(4, -1)$.**

a. $y = 3x + 17$
b. $y = \frac{5}{3}x - \frac{17}{3}$
c. $y = -\frac{5}{3}x + \frac{17}{3}$
d. $y = -5x - \frac{17}{3}$

7. **Write an equation for the line that passes through the two points $(-4, -2)$ and $(1, 3)$.**

a. $y = 2x + 2$
b. $y = x - 2$
c. $y = x + 2$
d. $x = y + 2$

8. **Write an equation for the line that passes through the two points $(-3, 3)$ and $(3, -1)$.**

a. $y = -\frac{2}{3}x + \frac{1}{3}$
b. $y = -\frac{2}{3}x - 1$
c. $y = \frac{2}{3}x + 1$
d. $y = -\frac{2}{3}x + 1$

9. **Write an equation for the line that passes through the two points $(1, 6)$ and $(3, 2)$.**

a. $y = \frac{1}{2}x + 8$
b. $y = -2x + 8$
c. $y = -2x + 0.5$
d. $y = 0.2x + 8$

10. **Write an equation for the line that passes through the two points $(-5, 2)$ and $(-1, 0)$.**

a. $y = -\frac{1}{2}x - \frac{1}{2}$
b. $y = -x - \frac{1}{2}$
c. $y = -\frac{1}{2}x - 1$
d. $x = -\frac{1}{2}y - \frac{1}{2}$

11. **Write an equation for the line that passes through the two points $(3, 2)$ and $(9, 7)$.**

a. $y = \frac{1}{2}x - \frac{1}{2}$
b. $y = \frac{5}{6}x + \frac{1}{2}$
c. $y = \frac{5}{6}x - \frac{1}{2}$
d. $y = -\frac{5}{6}x - 2$

12. **Write an equation for a line that passes through the two points $(1, -7)$ and $(3, -15)$.**

a. $y = 0.4x - 3$
b. $y = -4x + 1$
c. $y = -\frac{4x}{3}$
d. $y = -4x - 3$

13. Evan is emptying out a fish tank. The tank starts out with 175 gallons of water and gradually empties at a steady rate. The graph below shows gallons of water on the y-axis and time (in hours) on the x-axis. The point $(3, 100)$ represents 100 gallons of water after 3 hours. The point $(7, 0)$ represents 0 gallons of water after 7 hours. Which equation below represents the rate of water emptying?

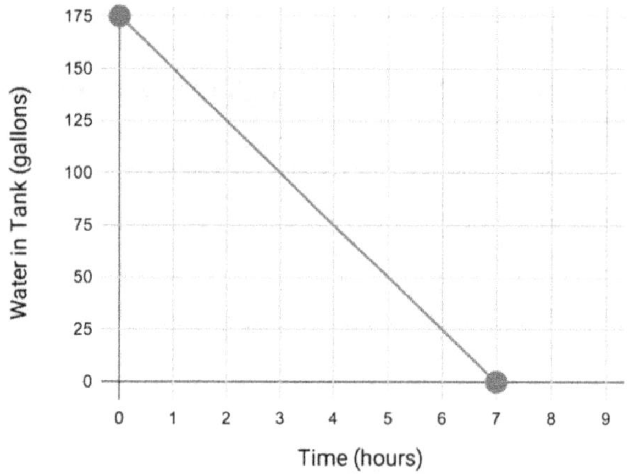

a. $y = 25x - 175$
b. $y = -25x + 175$
c. $y = 25x + 175$
d. $y = -25x + 7$

14. Gloria is raising a flag to the top of the flagpole in front of her school. The flag starts at a height of 3 feet and gradually raises as she pulls the ropes attached to the pole. The point $(2, 6)$ represents the flag height of 6 feet after 2 seconds. Write an equation in slope-intercept form based on the scenario.

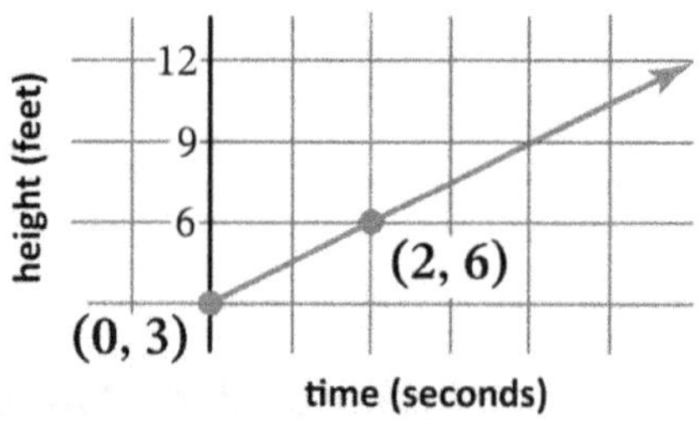

a. $y = \frac{2}{3}x + 3$
b. $y = \frac{3}{2}x + 3$
c. $y = 2x + 3$
d. $y = \frac{3}{2}x + 2$

15. Carl charges $45 per hour as a plumber. On a graph, the point $(0, 100)$ represents the one-time fee of $100 that Carl charges before any hours of work have been completed. Write an equation in slope-intercept form where x is the hours Carl works on a job and y is the total amount Carl charges for the job.

a. $y = 45x + 100$
b. $y = 45x$
c. $y = 4.5x + 100$
d. $y = 45x - 100$

3.4 Finding the Point-Slope Form Equation of a Line

Point–slope form, $y - y_1 = m(x - x_1)$, is one way to write a linear equation. Point-slope form is very convenient when you are locating points on a line if the slope and one point on the line are provided, or if two points on the line are provided.

If you are provided the slope and one point on the line, you can write an equation in point-slope form. Simply plug in the slope, m, and the ordered pair (x, y) into the formula. For example, if the slope is 2 and the point is $(1,3)$, the formula $y - y_1 = m(x - x_1)$ becomes $y - 3 = 2(x - 1)$. The equation can then be simplified to be expressed in standard form or slope–intercept form. If the equation $y - 3 = 2(x - 1)$ is simplified, it becomes $y - 3 = 2x - 2$, which is equivalent to $2x - y = -1$ in standard form, or $y = 2x + 1$ in slope-intercept form.

If you need to write an equation in point-slope form and two points are provided, you will follow a similar process. Consider the two points $(0, -2)$ and $(3,4)$.

Start by using the slope formula to identify the slope. The slope formula is $m = \frac{y_2 - y_1}{x_2 - x_1}$. Plug in the two points, $(0, -2)$ and $(3,4)$, and solve for m.

$$m = \frac{y_2 - y_1}{x_2 - x_1} = \frac{4 - (-2)}{3 - 0} = \frac{6}{3} = 2$$

The slope of the line is 2.

To write the equation in point-slope form, plug in the values for x_1, y_1, and m. The formula $y - y_1 = m(x - x_1)$ becomes $y - 4 = 2(x - 3)$ or $y - (-2) = 2(x - 0)$ in point-slope form.

> **Review Video: Finding the Point-Slope Form Equation**
> Visit mometrix.com/academy and enter code: 537903

EXAMPLES

EXAMPLE 1:

Write an equation of a line in point-slope form that has a slope of $\frac{1}{2}$ and passes through the point $(3,7)$.

Answer:

Point-slope form: $y - y_1 = m(x - x_1)$

x_1 and y_1 represent a point on the line. m represents the slope.

If the slope is $\frac{1}{2}$ and a point on the line is $(3,7)$, substitute $\frac{1}{2}$ for m and $(3,7)$ for (x_1, y_1).

$$y - y_1 = m(x - x_1)$$
$$y - 7 = \frac{1}{2}(x - 3)$$

Therefore, the equation in point-slope form is $y - 7 = \frac{1}{2}(x - 3)$.

EXAMPLE 2:

Write an equation of a line in point-slope form where $(3, -2)$ and $(1,0)$ are points on the line.

Answer:

Point-slope form requires the slope and at least one point to be provided. In this case, the slope is not provided, but two points on the line are known: $(3,-2)$ and $(1,0)$. Use the slope formula and these two points to identify the slope.

$$m = \frac{y_2 - y_1}{x_2 - x_1} = \frac{0-(-2)}{1-3} = \frac{2}{-2} = -1$$

Now the slope and at least one point on the line are known. Plug in one of the points and the slope into the point-slope formula. Using the point $(3,-2)$, the formula $y - y_1 = m(x - x_1)$ becomes $y + 2 = -1(x-3)$. Using the point $(1,0)$, the formula $y - y_1 = m(x - x_1)$ becomes $y = -1(x-1)$.

Practice Questions

1. Write an equation in point-slope form for a line that has a slope of $\frac{1}{2}$ and passes through the point $(-3,6)$.

a. $y = \frac{1}{2}(x+3)$
b. $y + 6 = \frac{1}{2}(x-3)$
c. $y - 3 = \frac{1}{2}(x-3)$
d. $y - 6 = \frac{1}{2}(x+3)$

2. Write an equation in point-slope form for a line that has a slope of 2 and passes through the point $(-3,1)$.

a. $y - 1 = 2(x-3)$
b. $y + 1 = 2(x+3)$
c. $x - 1 = 2(y+3)$
d. $y - 1 = 2(x+3)$

3. Write an equation in point-slope form for a line that has a slope of $\frac{3}{2}$ and passes through the point $(0,2)$.

a. $y - 2 = \frac{3}{2}(x-0)$
b. $y + 2 = \frac{3}{2}(x-1)$
c. $x - 2 = \frac{3}{2}(y-0)$
d. $y - \frac{3}{2} = 2(x-0)$

4. When writing an equation in point-slope form, $y - y_1 = m(x - x_1)$, the variable m represents the ______, and the variables x_1 and x_2 represent ________________.

a. slope/y-intercept
b. y-intercept/a point on the line
c. slope/a point on the line
d. an ordered pair/another point on the line

5. A line goes through the point $(1, 4)$. If the slope of the same line is 5, what is the equation in point-slope form?

a. $y - 4 = 5(x - 1)$
b. $y = 5x + 4$
c. $y + 4 = 5(x + 1)$
d. $-4x + 5x = -1$

6. Write an equation in point-slope form for a line that has a slope of 2 and passes through the point $\left(\frac{1}{2}, 1\right)$.

a. $y - 1 = 2(x - 1)$
b. $y - 1 = 2\left(x - \frac{1}{2}\right)$
c. $y + 1 = \frac{1}{2}(x - 2)$
d. $y = 2\left(x - \frac{1}{2}\right)$

7. Which of the following is an equation of a line in point-slope form that passes through the points $(0, -2)$ and $(3, 4)$?

a. $y = 2(x - 3)$
b. $y - 4 = 4(x + 3)$
c. $y + 4 = (x - 3)$
d. $y - 4 = 2(x - 3)$

8. Which of the following is an equation in point-slope form that passes through the points $(-3, 5)$ and $(2, 8)$?

a. $y + 8 = \frac{3}{5}(x + 2)$
b. $y = \frac{3}{5}x - 2$
c. $y - 2 = \frac{3}{5}(x - \frac{3}{5})$
d. $y - 8 = \frac{3}{5}(x - 2)$

9. Which of the following is an equation of a line in point-slope form that passes through the points $(-2, -2)$ and $(4, 1)$?

a. $y - 1 = x - 4$
b. $y - 1 = \frac{1}{2}(x + 4)$
c. $y = \frac{1}{2}(x - 4)$
d. $y - 1 = \frac{1}{2}(x - 4)$

10. What is the correct way to write an equation in point-slope form for the two points $(-2, -3)$ and $(4, -2)$?

a. $y + 2 = \frac{1}{6}(x - 2)$
b. $y + 3 = \frac{1}{6}(x + 2)$ or $y + 2 = 2(x - 2)$
c. $y + 2 = \frac{1}{6}(x - 4)$ or $y + 3 = \frac{1}{6}(x + 2)$
d. $y = \frac{1}{6}x + 2$

11. What is an equation of a line in point-slope form that passes through the points $(3, 3)$ and $(-3, -3)$.

a. $y - 3 = (x - 3)$
b. $y = x - 3$
c. $3y - 3x = 1$
d. $y - 3 = (x + 3)$

12. What is an equation of a line in point-slope form that passes through the points $(4, -7)$ and $(1, 2)$.

a. $y = -3(x - 1)$
b. $y - 2 = 3(x - 1)$
c. $y - 2 = -3\left(x - \frac{1}{3}\right)$
d. $y - 2 = -3(x - 1)$

13. Eliza is making necklaces to sell at her community market. Eliza wants to sell larger orders of necklaces, so her new sales strategy is to sell two necklaces for \$10 and then \$3 for every necklace after that. She hopes this will encourage customers to buy multiple necklaces. Express the scenario in point-slope form where x is the number of necklaces in each order and y is the cost of the order.

a. $y + 10 = 3(x + 2)$
b. $y - 10 = 3(x - 2)$
c. $y - 2 = 3(x - 10)$
d. $y - 1 = 10(x - 2)$

14. Gregory is on a work trip, and the company that he works for allows him to use a pre-loaded company debit card to spend \$25 per day for his lunch expenses. On the 4th day of his trip, he has \$175 left on the debit card. Express the scenario in point-slope form where x is the day of the trip and y is the amount of money on the debit card.

a. $y = -25x + 175$
b. $y = -25x - 4$
c. $y - 4 = -25(x - 4)$
d. $y - 175 = -25(x - 4)$

15. Jamie and Julia disagree on the correct way to write an equation in point-slope form if the slope is 3 and a point on the line is $(-2, 5)$. Jamie is confident that the equation would be $y - 5 = 3(x - 2)$. However, Julia states that the equation is $y - 5 = 3(x + 2)$. Who is correct?

a. Jamie and Julia are both correct.
b. Jamie and Julia are both incorrect.
c. Jamie is correct: $y - 5 = 3(x - 2)$.
d. Julia is correct: $y - 5 = 3(x + 2)$.

3.5 Finding the Standard Form Equation of a Line

Standard form, $Ax + By = C$, is one way to express a linear equation, where A, B, and C are integers, and $A > 0$. The benefit of writing an equation in standard form is that it makes it much easier to identify the x- and y-intercepts. In standard form, simply plug in 0 for x to find the y-intercept, and plug in 0 for y to find the x-intercept. For example, we can easily identify the x- and y-intercept for the following equation in standard form: $10x + 5y = 10$.

Plug in 0 for x to find the y-intercept.

$$\begin{aligned} 10(0) + 5y &= 10 \\ 5y &= 10 \\ y &= 2 \end{aligned}$$

Plug in 0 for y to find the x-intercept.

$$\begin{aligned} 10x + 5(0) &= 10 \\ 10x &= 10 \\ x &= 1 \end{aligned}$$

A graph can be created if the x- and y-intercepts are known. Plot the x- and y-intercepts, and then connect the two points.

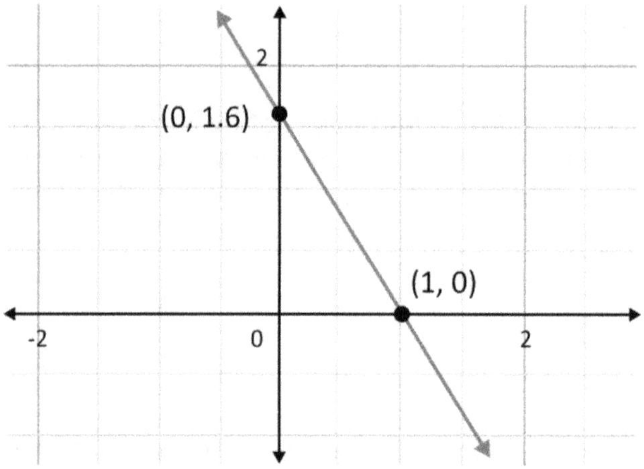

Writing an equation in standard form when two points on the line are provided is an important skill in algebra. This can be accomplished in three steps.

Consider the two points $(1,6)$ and $(3,-4)$. The three-step process can be used to write an equation in standard form.

Step 1: Use the slope formula to identify the slope (m).

$$m = \frac{y_2 - y_1}{x_2 - x_1}$$
$$m = \frac{-4 - 6}{3 - 1}$$
$$m = -\frac{10}{2} = -5$$

The slope of the line is –5.

Step 2: Find the y-intercept.

Now use the slope and one point to find the y-intercept. Plug in x, y, and m into the slope-intercept equation $y = mx + b$. If you use the point (1,6), the equation becomes $6 = -5(1) + b$. Solve for b.

$$\begin{aligned} 6 &= -5(1) + b \\ 6 &= -5 + b \\ 11 &= b \end{aligned}$$

The equation in slope-intercept form is $y = -5x + 11$.

Step 3: Rewrite the equation in standard form.

Now rewrite the equation $y = -5x + 11$ in standard form. The x- and y-terms need to be on one side and the constant on the other side. Add $5x$ to both sides.

$$5x + y = 11$$

If you are provided the slope and one point, you can use the point-slope formula to write the equation of the line. The point-slope formula is $y - y_1 = m(x - x_1)$. For example, consider a slope of –2 and the point $(4, -3)$. We can use the point–slope formula to write the equation of the line.

Plug in –2 for the slope, m, and plug in $(4, -3)$ into x_1 and y_1.

$$y - (-3) = -2(x - 4)$$

Now, simplify the equation.

$$y + 3 = -2x + 8$$

Now move the x- and y-terms to the left side and the constant to the right side.

$$2x + y = 5$$

The equation is now in standard form.

Review Video: Finding the Standard Form of a Linear Equation
Visit mometrix.com/academy and enter code: 416857

Review Video: Slope-Intercept and Point-Slope Forms
Visit mometrix.com/academy and enter code: 113216

Examples

Example 1:

Write the equation of a line in standard form with a slope of $\frac{2}{3}$ that passes through the point (3,4).

Answer:

Start by writing the equation in point-slope form since the slope and one point are provided.

$$y - y_1 = m(x - x_1)$$

$$y - 4 = \frac{2}{3}(x - 3)$$

Now rewrite the equation $y - 4 = \frac{2}{3}(x - 3)$ in standard form. We need the x- and y-terms to be on one side and the constant to be on the other side. Start by distributing the 2.

$$y - 4 = \frac{2}{3}x - 6$$

Now add 4 to both sides, and then subtract $2x$ from both sides.

$$y = \frac{2}{3}x - 2$$
$$-\frac{2}{3}x + y = -2$$

Remember, A cannot be negative or a fraction in standard form.

Get rid of the negative by dividing both sides of the equation by -1.

$$\frac{2}{3}x - y = 2$$

Get rid of the fraction by multiplying both sides of the equation by the denominator of the fraction, 3.

$$3\left(\frac{2}{3}x - y\right) = (2)(3)$$
$$2x - 3y = 6$$

EXAMPLE 2:

Write an equation in slope-intercept form that runs through the points $(3,7)$ and $(0,-2)$.

Answer:

Start by finding the slope using the slope formula.

$$m = \frac{y_2 - y_1}{x_2 - x_1} = \frac{-2 - 7}{0 - 3} = \frac{-9}{-3} = 3$$

The slope is 3. Now, determine the y-intercept. The y-intercept is $(0,-2)$, so the value for b is -2.

Now, use the slope and y-intercept to write the equation in slope-intercept form.

$$y = 3x - 2$$

PRACTICE QUESTIONS

1. Write an equation in standard form that has a slope of -3 and passes through the point $(1,5)$.

a. $9x + y = 8$
b. $3x + y = 9$
c. $x + 3y = 8$
d. $3x + y = 8$

2. Write an equation in standard form that has a slope of $-\frac{1}{3}$ and passes through the point $(4, -1)$.

a. $3x - y = \frac{1}{3}$
b. $x + 3y = 1$
c. $x + y = 1$
d. $\frac{1}{3}x - y = \frac{1}{3}$

3. Write an equation in standard form that has a slope of 2 and passes through the point $(3, 5)$.

a. $\frac{1}{2}x - y = 2$
b. $2x + y = -1$
c. $-3x - y = 2$
d. $2x - y = 1$

4. Write an equation in standard form that has a slope of 5 and passes through the point $(6, -9)$.

a. $5x - y = 39$
b. $-5x + y = 39$
c. $5x - y = 24$
d. $-5x + y = 24$

5. Write an equation in standard form that has a slope of $-\frac{3}{7}$ and passes through the point $(1, -2)$.

a. $\frac{1}{3}x + \frac{1}{7}y = -11$
b. $3x - 7y = 11$
c. $3x + 7y = -11$
d. $3x + 7y = -\frac{1}{3}$

6. The example below shows a conversion from point-slope form to standard form. Identify the error.

Point: (2,3) and Slope $= 4$
Step 1: $y - 3 = 4(x - 2)$
Step 2: $y - 3 = 4x - 8$
Step 3: $-4x + y = -5$

a. The constant, -5, needs to be on the left side of the equation in step 3.
b. The constant cannot be negative in step 3.
c. A cannot be negative in step 3.
d. There is no error.

7. Write an equation in standard form that passes through the two points $(2, 1)$ and $(-1, 3)$.

a. $2x - 3y = 9$
b. $2x + 3y = 7$
c. $3x + 2y = -7$
d. $y = 3x + 2y$

8. Write an equation in standard form that passes through the two points $(-1,3)$ and $(4,-2)$.

a. $2x + 4y = -2$
b. $x + 3y = -3$
c. $x + y = 2$
d. $2x + y = 2$

9. Write an equation in standard form that passes through the two points $(-1,-5)$ and $(-4,1)$.

a. $2x + 3y = \frac{1}{7}$
b. $-2x + y = 7$
c. $-7x + y = 3$
d. $2x + y = -7$

10. Write an equation in standard form that passes through the two points $(-1,5)$ and $(0,8)$.

a. $3x - y = -8$
b. $\frac{1}{3}x - y = -8$
c. $3x + y = 8$
d. $x - y = 8$

11. Write an equation in standard form that passes through the two points $(1,3)$ and $(3,7)$.

a. $5x - y = -1$
b. $4x - y = -1$
c. $3x - y = -1$
d. $2x - y = -1$

12. Write an equation in standard form that passes through the two points $(-2,3)$ and $(-4,-5)$.

a. $3x - y = 11$
b. $4x + 3y = -11$
c. $4x - y = -11$
d. $2x - y = 11$

13. Your school needs to raise \$5,000 for a new soccer stadium. A fundraiser is selling \$200 student tickets and \$300 adult tickets for a concert to raise money for the new stadium. Write an equation in standard form that could be used to determine the number of student and adult tickets that need to be sold to reach the \$5,000 goal exactly.

a. $20.0x + 30.0y = 5{,}000$
b. $200x - 300y = 5{,}000$
c. $200x + 300y = 5{,}000$
d. $250x + 350y = 5{,}000$

14. Jamie sells snacks during half-time at a basketball game. He sells pizza slices for \$1 and bags of popcorn for 50¢. After the game, he has earned a total of \$200. Write an equation in standard form that can be used to determine the number of pizza slices, x, and popcorn bags, y, he sold.

a. $2x + 0.5y = 200$
b. $2x + y = 400$
c. $y + 0.5x = 2.00$
d. $x + 5y = 400$

15. The following equation is in standard form: $2x + 5y = 50$. Match the equation with one of the written descriptions below.

a. Robin spent \$2 on ice cream and \$5 on fries. Robin has \$50 left over.
b. April earned \$5 per hour pulling weeds out of the garden. She worked for 50 hours and took a 2-hour break.
c. Susan spent \$50 on flowers at the nursery. Each outdoor plant cost \$2 and each indoor plant cost \$6.
d. Dave has \$50 dollars to spend on lunch. Drinks cost \$2 each and sandwiches cost \$5 each.

3.6 Converting Between Standard Form and Slope-Intercept Form

Equations can be presented in many forms. Standard form and slope-intercept form are two commonly used forms of linear equations. Writing an equation in a different form does not change the outcome of the equation, it simply rearranges the parts and pieces of the equation, often telling you different things about the function.

Standard Form: $Ax + By = C$
Slope-Intercept Form: $y = mx + b$

Graphing an equation that is presented in standard form involves more steps than graphing an equation in slope-intercept form. Converting equations from standard form to slope-intercept form is a valuable skill to master because it can make graphing much simpler.

The benefit of converting an equation into slope-intercept form is that the slope and y-intercept are very easy to identify. For example, $y = 2x + 5$ is in slope-intercept form where x and y represent variables, the coefficient in front of the x represents the slope, m, and the constant represents the y-intercept, b. Graphing becomes quick and straightforward when the slope and y-intercept are easily identified.

When converting an equation from standard form to slope-intercept form, solve for y. For example, this equation is in standard form.

$$6x + 2y = 24$$

To convert this into slope-intercept form, start by subtracting $6x$ from both sides.

$$2y = -6x + 24$$

Now divide both sides by 2.

$$y = -3x + 12$$

The equation is now in slope-intercept form. The slope is –3, and the y-intercept is 12. In this form, the equation is ready to be graphed.

Consider the equation $2x + 3y = 9$. Converting the equation into slope-intercept form will make the graphing process much quicker. Solve for y using inverse operations. Start by subtracting $2x$ from both sides.

$$3y = -2x + 9$$

Now, divide both sides by 3.

$$y = -\frac{2}{3}x + 3$$

At this point, the slope and y-intercept are clearly identifiable. The slope is $-\frac{2}{3}$ and the y-intercept is 3.

We can also convert an equation from slope-intercept form to standard form. To convert to standard form, rearrange the slope-intercept equation, and keep in mind the special rules about standard form equations. Remember, A, B, and C must be integers, and A must be positive.

Convert this equation to standard form.

$$y = \frac{1}{2}x - 7$$

Start by subtracting $\frac{1}{2}x$ from both sides of the equation.

$$-\frac{1}{2}x + y = -7$$

Then, check to make sure the rules are followed. $-\frac{1}{2}$ is not an integer, and it is not positive. Eliminate this by multiplying the entire equation by –2.

$$x - 2y = 14$$

The equation is now in standard form.

> **Review Video: How to Convert Between Standard and Slope-Intercept Forms**
> Visit mometrix.com/academy and enter code: 916601

EXAMPLES

EXAMPLE 1:

Convert the equation $3x + 2y = 8$ from standard form to slope-intercept form.

Answer:

When converting from standard form to slope-intercept form, solve for the variable y. When y is isolated, the equation is in slope-intercept form: $y = mx + b$.

Start by subtracting $3x$ from both sides.

$$2y = -3x + 8$$

Now, divide both sides by 2.

$$y = -\frac{3}{2}x + 4$$

EXAMPLE 2:

Convert the equation $y = 3x + 4$ from slope-intercept form to standard form.

Answer:

In standard form, the x- and y-terms are on the left side of the equation and the constant is on the right side. When converting from slope-intercept form to standard form simply move the x- and y-terms to one side and the constant to the other side using inverse operations.

Subtract $3x$ from both sides.

$$-3x + y = 4$$

Remember, equations in standard form should not have a negative value for A, so multiply both sides of the equation by –1.

$$3x - y = -4$$

Practice Questions

1. Convert the following equation from standard form to slope-intercept form: $6x + 2y = 4$.

a. $y = 3x - 2$
b. $y = -3x + 2$
c. $y = -\frac{1}{3}x + 2$
d. $y = -2x + \frac{1}{2}$

2. Convert the following equation from standard form to slope-intercept form: $3x - 2y = 21$.

a. $y = 4x - \frac{21}{2}$
b. $y = \frac{3}{2}x - \frac{21}{2}$
c. $y = \frac{3}{2}x - 21y$
d. $y = \frac{21}{2}x - 4$

3. Convert the following equation from standard form to slope-intercept form: $8x + 4y = 16$.

a. $y = 2x + 4$
b. $y = -2x + 4$
c. $y = -2x - 4$
d. $y = -4x + 2$

4. Convert the following equation from standard form to slope-intercept form: $x + y = 7$.

a. $y = x + 7$
b. $y = x$
c. $y = -x + 7$
d. $y = x - 7$

5. Convert the following equation from standard form to slope-intercept form: $10x - 7y = -35$.

a. $y = \frac{10}{7}x + 5$
b. $y = x + 5$
c. $y = \frac{10}{7}x$
d. $y = 1.7x + 5$

6. Convert the following equation from standard form to slope-intercept form: $2x - 3y = -6$.

a. $x = \frac{2}{3}y + 2$
b. $y = \frac{2}{3}x + 2$
c. $y = \frac{2}{3}x$
d. $y = \frac{2}{3}x + \frac{1}{2}$

7. When converting from standard form to slope-intercept form, which variable should you isolate using inverse operations?

a. x
b. y
c. A
d. B

8. Convert the equation $y = 5x - 9$ from slope-intercept form to standard form.

a. $-9x + y = -5$
b. $9x - y = 5$
c. $5x - y = 9$
d. $-5x + y = -9$

9. Below is a step-by-step process for converting from standard form to slope-intercept form. Identify the step that has an error.

Step 1: $6x + 3y = 1$
Step 2: $3y = -6x + 1$
Step 3: $y = -2x + 1$

a. Error in Step 1
b. Error in Step 2
c. Error in Step 3
d. There is no error.

10. Convert the following equation from slope-intercept form to standard form: $y = \frac{2}{3}x + 3$.

a. $3x - 2y = -9$
b. $-\frac{2}{3}x + y = 3$
c. $-3y + 2x = 9$
d. $2x - 3y = -9$

11. Convert the following equation from slope-intercept form to standard form: $y = -\frac{3}{8}x + 6$.

a. $3x + 8y = 48$
b. $6x + 3y = -8$
c. $3x - 8y = -48$
d. $\frac{3}{8}x + y = 6$

12. Convert the following equation from slope-intercept form to standard form: $y = \frac{12}{5}x + \frac{44}{5}$.

a. $12x - 5y = -44$
b. $-y + 12x = 44$
c. $-\frac{12}{5}x + y = \frac{44}{5}$
d. $-12x + y = 44$

13. Ryan is converting from slope-intercept form to standard form. His original equation is $y = 6x - 3$. Ryan starts by subtracting $6x$ from both sides of the equation, which results in $-6x + y = -3$. What is Ryan's next step?

a. Add 3 to both sides.
b. Divide both sides by –6.
c. Subtract y from both sides.
d. Multiply both sides by –1.

14. Lauren and Jose disagree on whether standard form or slope-intercept form is more convenient for graphing. Lauren says that it is best to convert equations from slope-intercept to standard form and then graph. Jose says that it is best to convert equations from standard form to slope-intercept form and then graph. Who is correct?

a. Lauren is correct.
b. Jose is correct.
c. Neither Lauren nor Jose is correct.
d. Lauren and Jose are both correct.

15. Marci is at a bookstore buying books and bookmarks for her family members. The books she purchases cost \$20 each and the bookmarks cost \$1 each. Marci spends \$80 at the bookstore. The equation that represents this scenario is $20x + y = 80$. Convert the equation to slope-intercept form.

a. $y = \frac{1}{80}x + \frac{1}{4}$
b. $x = -\frac{1}{20}y + 4$
c. $y = -20x + 80$
d. $y = -80x + 20$

3.7 Writing Linear Equations

A general rule when writing equations from tables, graphs, or written descriptions, is that two things are required: slope and y-intercept. With these two pieces of information, an equation can be written in slope-intercept form: $y = mx + b$. The process looks a little different for tables, graphs, and written descriptions, but overall, the goal is to find the slope, m, and y-intercept, b, so that the equation can be written.

Writing Equations from a Table

x	y
3	-42
4	-56
5	-70
6	-84

Step 1: Identify the slope

Slope can be found using the slope formula: $m = \frac{y_2-y_1}{x_2-x_1}$. Select any two pairs of (x, y) values and plug them into the formula. Using the first two ordered pairs, $(3, -42)$ and $(4, -56)$, the formula becomes:

$$m = \frac{y_2 - y_1}{x_2 - x_1} = \frac{-56 - (-42)}{4 - 3} = \frac{-14}{1} = -14$$

The slope is $m = -14$.

Step 2: Identify the y-intercept

The y-intercept is the location on a graph where the line intersects the y-axis. In a table, this is located at the value $x = 0$. When x is zero, y represents the y-intercept. The value of $x = 0$ is not provided, but we can use the slope, $m = -14$, to work our way backwards to $x = 0$. As the value of x decreases by 1, the value of y increases by 14. When $x = 2, y = -28$. When $x = 1, y = -14$. When $x = 0, y = 0$. In this case, the y-intercept is at the origin $(0,0)$. This means that $b = 0$.

Another way to find the y-intercept is by substituting a point into the partial equation and solving for b. Since the slope is known, the equation is:

$$y = -14x + b$$

Substitute any point on the line for (x, y) in the equation and solve for b.

$$\begin{aligned} -42 &= -14(3) + b \\ -42 &= -42 + b \\ b &= 0 \end{aligned}$$

Notice that this method gives the same value for b as working backwards.

Now that the slope and y-intercept are identified, the equation can be written.

$$\begin{aligned} y &= -14x + 0 \\ y &= -14x \end{aligned}$$

WRITING EQUATIONS FROM A GRAPH

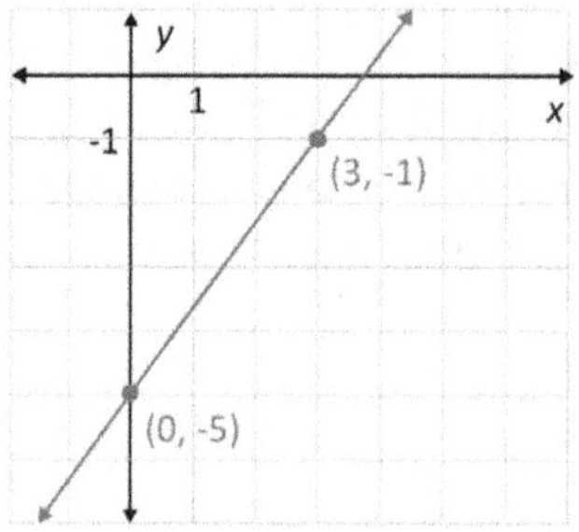

The process of writing an equation from a graphed line is very similar to the process for a table. Again, the objective is to identify the slope and y-intercept.

Step 1: Identify the slope

The slope can be found a couple different ways. The first way to find the slope is by counting the rise and the run. Remember, slope is the ratio $\frac{\text{rise}}{\text{run}}$. In this example, the graph rises 4 from one point to the next and runs to the right 3. This gives a slope of $\frac{4}{3}$.

Another way to determine the slope is by using the slope formula: $m = \frac{y_2 - y_1}{x_2 - x_1}$. Select any two ordered pairs on the graphed line, and plug them into the formula. For example, if the points $(0, -5)$ and $(3, -1)$ are selected, the formula becomes:

$$m = \frac{y_2 - y_1}{x_2 - x_1} = \frac{-1 - (-5)}{3 - 0} = \frac{4}{3}$$

The slope of the line is $m = \frac{4}{3}$.

Step 2: Identify the y-intercept

To determine the y-intercept, locate where the line intersects the y-axis. In this case, the line crosses the y-axis at $(0, -5)$, which means the y-intercept is $b = -5$.

Now the slope and y-intercept are identified, so the equation can be written.

$$y = \frac{4}{3}x - 5$$

WRITING EQUATIONS FROM WRITTEN DESCRIPTIONS

When writing linear equations that represent written descriptions, identify the slope and y-intercept within the scenario. For example, consider the following scenario. You earn \$20 per hour working a job, and you also earn a \$40 bonus when the job is complete. In this case, the slope is 20, because for every hour, you earn \$20. The y-intercept will be $b = 40$, because this amount is not based on the number of hours worked rather it is a separate flat payment. Now that the slope, $m = 20$, and the y-intercept, $b = 40$, are identified, the equation for how much you make can be written. The equation will be $y = 20x + 40$, where x represents the number of hours you worked, and y represents the total amount of money made.

EXAMPLES

EXAMPLE 1:

Use the table of values to write a linear equation.

x	y
0	3
2	11
4	19
6	27
8	35

Answer:

The two pieces of information required to write an equation in slope-intercept form, $y = mx + b$, from a table are the slope, m, and the y-intercept, b. The y-intercept occurs when $x = 0$, which is the location where the line crosses the y-axis. So, locate the ordered pair with $x = 0$ in the table. In this table, the line crosses the y-axis at (0,3). This means that the value of b is 3.

Use the slope formula to calculate the slope, m. Choose two ordered pairs in the table and plug these into the slope formula.

$$m = \frac{y_2 - y_1}{x_2 - x_1} = \frac{11 - 3}{2 - 0} = \frac{8}{2} = 4$$

The slope of the line is $m = 4$.

Now, plug in 4 for m and 3 for b.

$$y = mx + b$$
$$y = 4x + 3$$

EXAMPLE 2:

Write an equation in slope-intercept form to match the graph below.

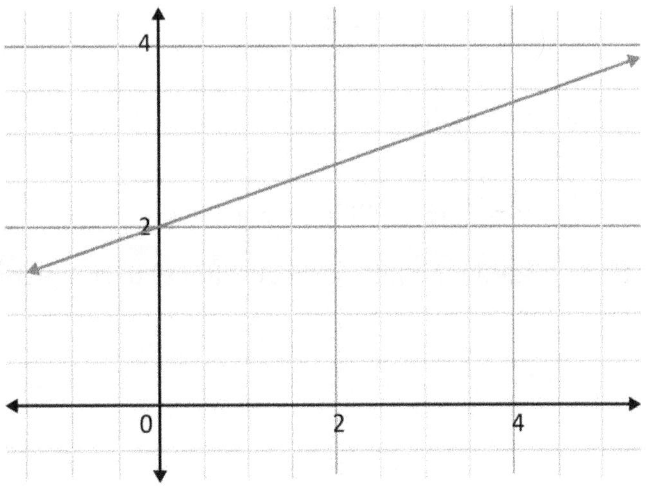

Answer:

Two pieces of information are needed to write an equation in slope-intercept form, $y = mx + b$ from a graph, the slope, m, and the y-intercept, b.

The y-intercept of this graph is 2. This is the location where the line intersects the y-axis.

Now we need to identify the value of the slope, m. The slope is the vertical change divided by the horizontal change $\left(\frac{\text{rise}}{\text{run}}\right)$. The slope can be calculated using the slope formula. Find two points on the graph such as (0,2) and (3,3).

$$m = \frac{y_2 - y_1}{x_2 - x_1} = \frac{3 - 2}{3 - 0} = \frac{1}{3}$$

The slope is $\frac{1}{3}$.

Now that the y-intercept, $b = 2$, and the slope, $m = \frac{1}{3}$, are known, the equation for the graph can be written.

$$y = \frac{1}{3}x + 2$$

EXAMPLE 3:

James charges $55 per hour as an electrician. He also charges a one-time $80 pre-inspection fee before beginning work on a project. How much would he be paid for a job that took 3 hours to complete?

Answer:

Start by expressing this scenario as a linear equation in slope-intercept form. The slope, or rate of increase, in this scenario is $55 dollars per hour, where x represents the number of hours, and y represents the total amount charged to the client. The $80 fee occurs before the hourly rate starts. This represents a one-time fee, so $b = 80$.

$$m = 55$$
$$b = 80$$
$$y = 55x + 80$$

To find out how much he would be paid for a job that took 3 hours to complete, substitute 3 for the value of x and solve for y.

$$y = 55(3) + 80 = 165 + 80 = 245$$

Therefore, James would be paid $245 for a job that took 3 hours to complete.

Practice Questions

1. Write a linear equation based on the table of values.

x	y
-2	-4
-1	-1
0	2
1	5
2	8

a. $y = 3x + 2$
b. $y = 2x + 3$
c. $y = \frac{1}{3}x - 2$
d. $y = -3x - 2$

2. Write a linear equation based on the table of values.

x	y
1	6
2	12
3	18

a. $y = 6x$
b. $y = 0.6x$
c. $y = \frac{1}{6}x$
d. $x = 6y$

3. Use the table below to write a linear equation.

x	y
0	-7
1	-5
2	-3
3	-1
4	1

a. $y = -2x - 7$
b. $y = \frac{1}{2}x - 7$
c. $y = 7x - 2$
d. $y = 2x - 7$

4. Write a linear equation that represents the data table below.

x	y
1	11
2	15
3	19
4	23
5	27

a. $y = 4x$
b. $y = 2x + 7$
c. $y = 4x + 7$
d. $y = \frac{1}{4}x + 7$

5. Write a linear equation that describes the graph.

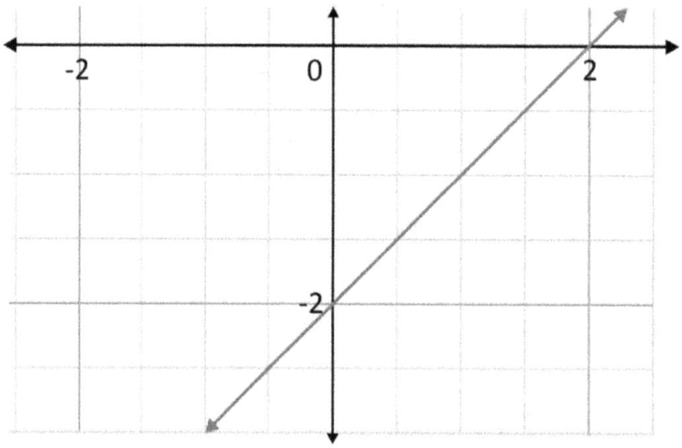

a. $y = x + 2$
b. $y = x - 2$
c. $x = y - 2$
d. $y = 2x - 2$

6. Write a linear equation based on the graph.

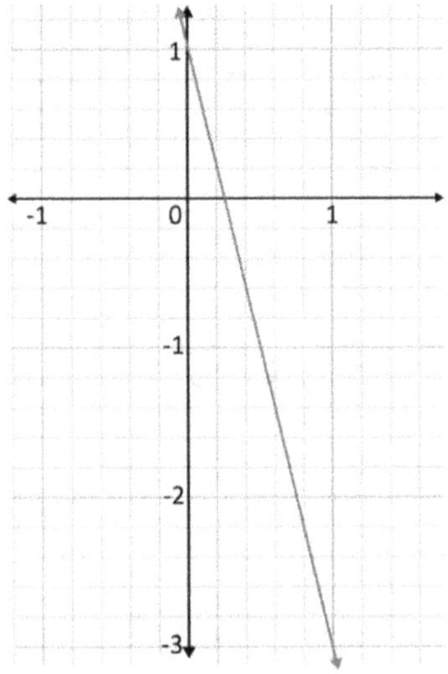

a. $y = 5x + \frac{1}{4}$
b. $y = 4x - 1$
c. $y = -4x + 1$
d. $y = \frac{1}{4}x + 3$

7. Write a linear equation that describes the graph.

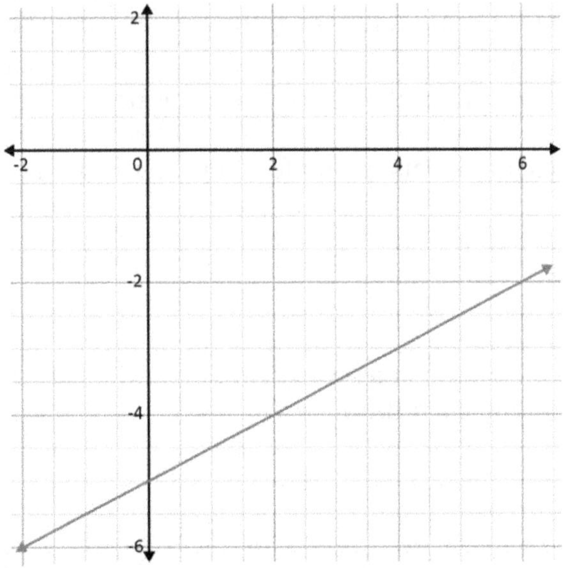

a. $y = 2x - 5$
b. $y = \frac{1}{2}x + 5$
c. $y = \frac{1}{2}x - 5$
d. $y = -\frac{1}{2}x + 5$

8. Write a linear equation that describes the graph.

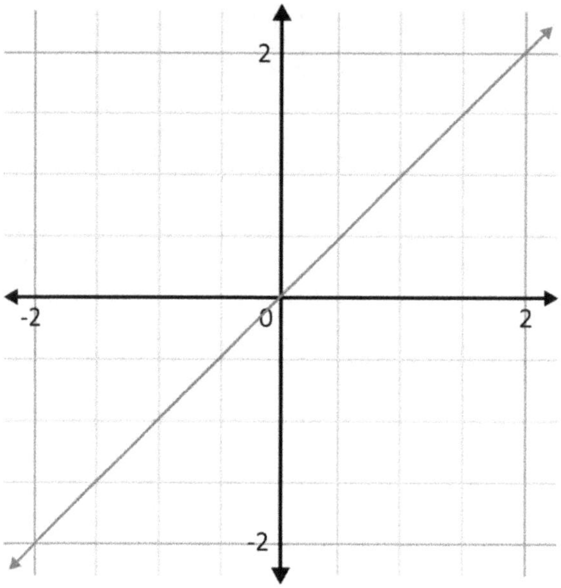

a. $y = x$
b. $y = \frac{1}{2}x$
c. $y = 1x + 1$
d. $y = x + 1$

9. Write a linear equation that represents the following description:

Damion is arranging a sailboat tour for his vacation. The sailboat company charges a one-time fee of $150 to reserve the boat and $25 per person for the tour. Damion has 6 family members joining him on the vacation. Write a linear equation to calculate the total cost of the sailing tour for Damion and his family.

a. $x = 25y + 150$
b. $y = 25x - 150$
c. $y = 5x + 150$
d. $y = 25x + 150$

10. Michael is hiking up a mountain at a steady rate of 3 miles per hour. How many miles has Michael traveled if he hikes for 2.5 hours?

a. 6.25 miles
b. 7.5 miles
c. 8.5 miles
d. 0.75 miles

11. Margaret can bake 42 cookies in 2 hours. If she bakes cookies for x hours and then eats 3 of the cookies, how many cookies remain (y)?

a. $y = 21x + 3$
b. $y = 21x - 3$
c. $y = 42x - 3$
d. $y = 42x + 3$

12. Edgar owns a car wash that charges $24 per wash. Today, one of Edgar's most valued customers was so pleased with the quality of service that he left a $50 tip! How much has Edgar's car washing business earned after 20 washes and the tip?

a. $150
b. $265
c. $530
d. $1,024

13. A parking garage charges a flat rate of $2 and then $0.50 for every hour a car is parked there. If Andrea parks her car in the garage at 2:00 p.m. and then leaves the garage at 7:00 p.m., how much is her bill?

a. $4.50
b. $5.00
c. $4.50
d. $6.50

14. Susan has been recording the number of t-shirts she sells each week. She currently sells 8 shirts per week. As she continues to grow her small business, her goal is to sell at least 50 t-shirts after 7 weeks. Write a linear equation based on the graph and determine if Susan will meet her goal.

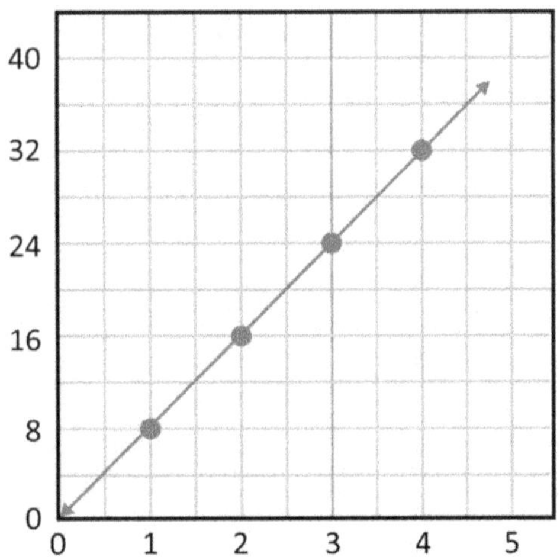

a. $y = 8x + 7$; no, Susan will not meet her goal.
b. $y = 8x + 1$; yes, Susan will meet her goal.
c. $y = x + 8$; no, Susan will not meet her goal.
d. $y = 8x$; yes, Susan will meet her goal.

15. Erika owns a yoga studio. The table below shows the number of enrolled members at her studio each year. Erika wants to write a linear equation from the table so that she can predict the number of members she will have after 7 years of business. Using an equation that describes the table, how many members will Erika have after 7 years?

Time (years)	0	1	2	3	4
Number Enrolled	8	11	14	17	20

a. 23
b. 26
c. 29
d. 31

3.8 Parallel and Perpendicular Lines and the Axes

Parallel lines are lines that extend indefinitely and never intersect. Parallel lines look similar to railroad tracks and always remain the same distance apart.

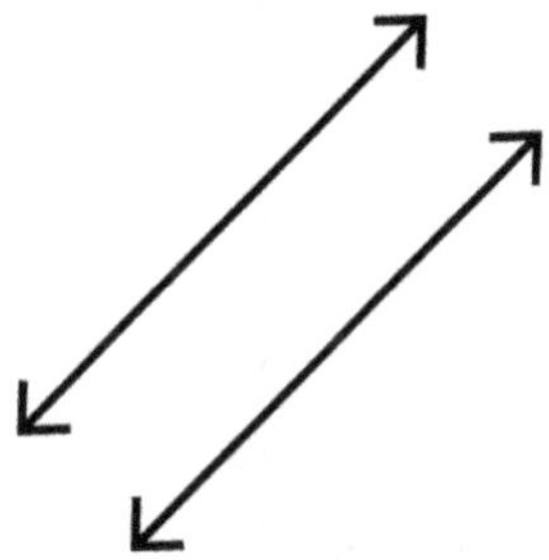

In contrast, **perpendicular lines** intersect at a 90-degree angle. It is important to know how to graph parallel and perpendicular lines in algebra. For example, graphing a line that is parallel or perpendicular to the x- or y-axis.

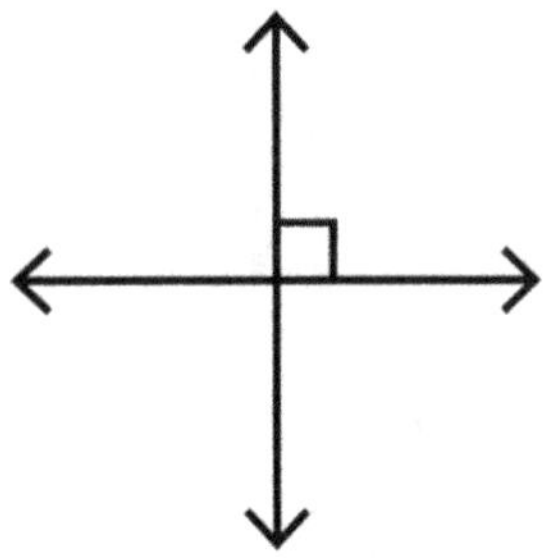

Horizontal lines will be parallel to the x-axis.

If an equation has a fixed value for y, such as $y = 3$, then the line will be parallel to the x-axis. This is because the y-value will always be 3 for any value of x. If x is 9, y is 3, represented by the point (9,3). If x is –100, y is still 3, represented by the point $(-100,3)$.

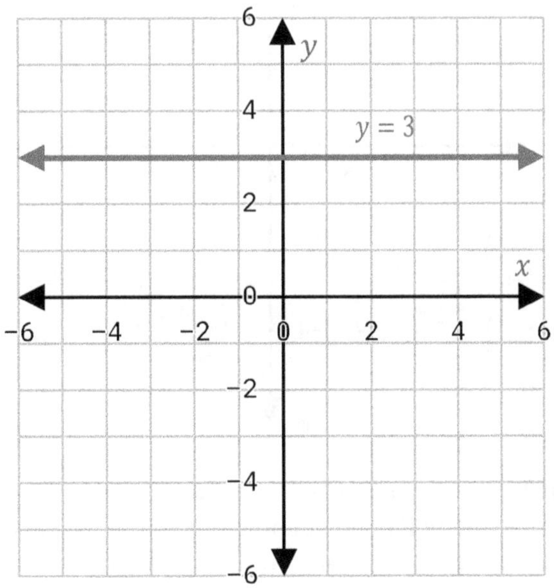

Any equation that shows y equal to one value will be parallel to the x-axis. For example, $y = 6$, $y = -2.5$, and $y = 1$ are all parallel to the x-axis.

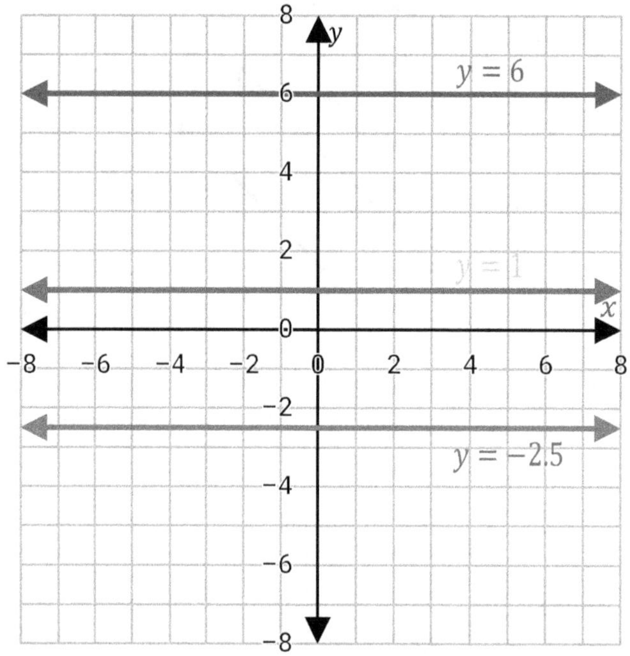

Vertical lines will be perpendicular to the x-axis.

If an equation has a fixed value for x, such as $x = -5$, then the line will be perpendicular to the x-axis.

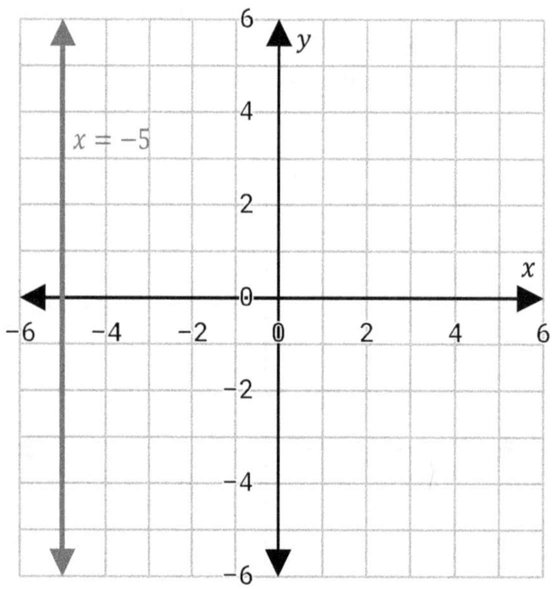

Since there is only one value for x, regardless of the value of y, plotted points will form a line that extends vertically. This vertical line will be perpendicular to the x-axis because the x-axis is a horizontal line. The lines will intersect to form a 90-degree angle. For example, $x = 2$, $x = -2.4$, and $x = 100$ will all form lines perpendicular to the x-axis.

Vertical lines will be parallel to the y-axis.

The y-axis is a vertical line. This means that there is only one value for x and an infinite number of values for y. When writing an equation that is parallel to the y-axis, simply choose a value for x. For example, $x = 2$ is a vertical line parallel to the y-axis.

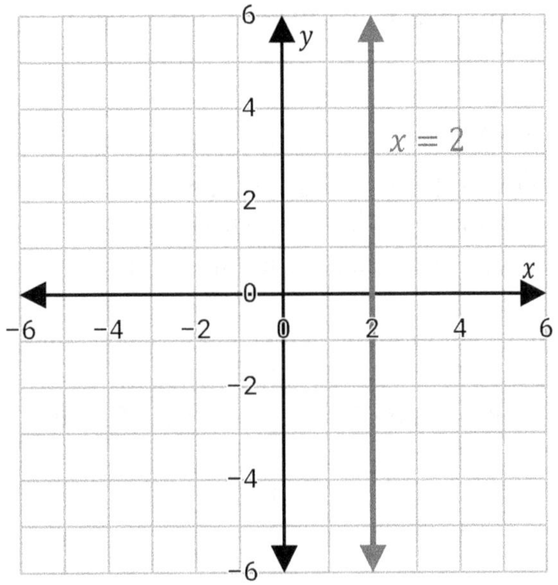

The lines $x = 8, x = -50, x = 1$ are all examples of lines that are parallel to the y-axis.

Horizontal lines will be perpendicular to the y-axis.

Lines will be perpendicular to the y-axis if they are horizontal lines. Horizontal lines have one value for y and an infinite number of values for x. For example, $y = 10$ will form a horizontal line extending left and right through the y-value of 10. Regardless of the x-value, the y-value will always remain 10.

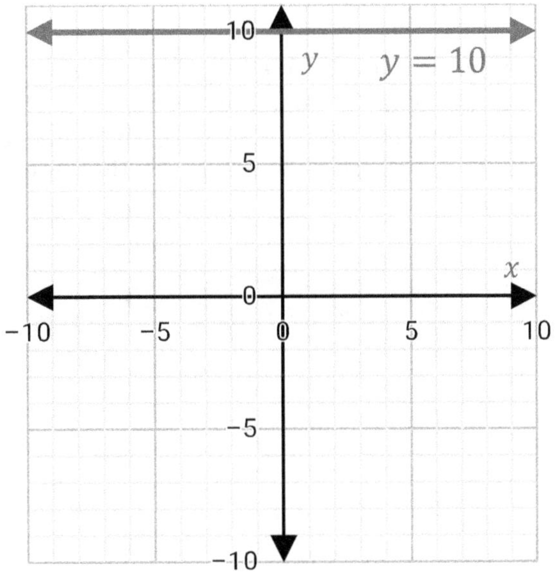

Horizontal lines will be perpendicular to the y-axis. This means that any equation such as $y = 1$, $y = 30$, and $y = -2$ will be perpendicular to the y-axis because they are horizontal.

The slope of a horizontal line is zero.

Remember, slope is calculated as the "rise over run", or the vertical change divided by the horizontal change. A horizontal line has no vertical change (0), so all slope values will be zero. For example, a slope of $\frac{0}{4}$ is 0.

The slope of a vertical line is "undefined."

Since slope is calculated as vertical change divided by horizontal change, and a vertical line has no horizontal change, the denominator of the fraction will always be zero. We can't divide by zero, so the slope is considered "undefined". For example, a slope of $\frac{4}{0}$ is "undefined."

EXAMPLES

EXAMPLE 1:

Write an equation that represents the line. What is the slope of the line?

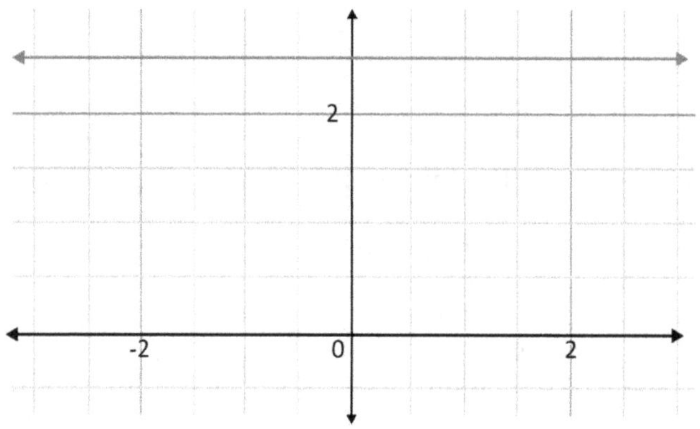

Answer:

The slope of the x-axis is 0, and the slope of the line is 0. The x-axis and the line have the same slope, which means they are parallel. Slope is calculated as the vertical change divided by the horizontal change, or "rise" over "run." There is no "rise" for any points on the line, so the slope is 0. The equation $y = 2.5$ tells us that regardless of the value of x, the value of y will always be 2.5.

EXAMPLE 2:

Write an equation that represents the line. What is the slope of the line?

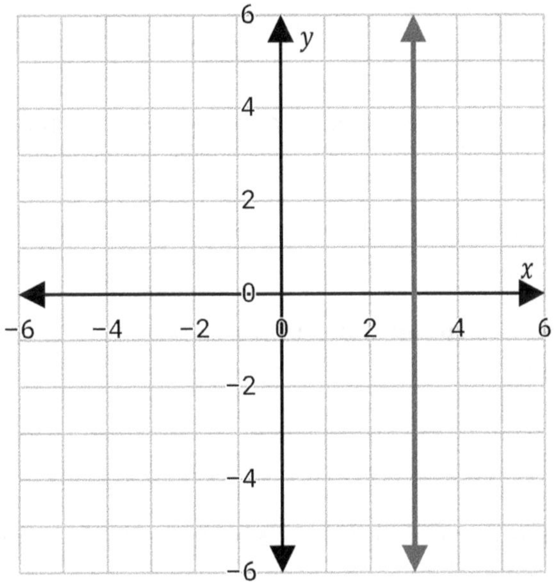

Answer:

The slope is calculated by dividing the vertical change by the horizontal change. The line has no "run," and it "rises" forever. This means that the slope is undefined. The line intersects the x-axis at a 90-degree angle, forming perpendicular lines. The slopes of perpendicular lines will be negative inverses of each other. The equation $x = 3$ tells us that regardless of the value of y, the value of x will always be 3.

PRACTICE QUESTIONS

1. Which equation of a line is parallel to the x-axis, and what is its slope?

a. $y = 8; m = 0$
b. $y = 8x; m = 0$
c. $x = 8; m = \text{undefined}$
d. $x = 8y; m = \text{undefined}$

2. Which equation of a line is parallel to the y-axis, and what is its slope?

a. $y = 4; m = 0$
b. $y = 4x; m = 0$
c. $x = 4; m = \text{undefined}$
d. $x = 4y; m = \text{undefined}$

3. Which equation of a line is perpendicular to the x-axis, and what is its slope?

a. $y = 2x; m = 0$
b. $x = 2y; m = 0$
c. $y = 2; m = \text{undefined}$
d. $x = 2; m = \text{undefined}$

4. Which equation of a line is perpendicular to the y-axis, and what is its slope?

a. $x = 1; m = 0$
b. $x = 1; m =$ undefined
c. $y = 1; m = 0$
d. $y = 1; m =$ undefined

5. Which equation of a line is parallel to the y-axis, and what is its slope?

a. $y = -3x; m = 0$
b. $y = -3; m = 0$
c. $x = -3y; m =$ undefined
d. $x = -3; m =$ undefined

6. Which equation of a line is perpendicular to the x-axis, and what is its slope?

a. $x = 12; m =$ undefined
b. $y = 12; m =$ undefined
c. $x = 12; m = 0$
d. $y = 12; m = 0$

7. Which equation of a line is parallel to the x-axis, and what is its slope?

a. $x = 7.5y; m =$ undefined
b. $x = 7.5; m =$ undefined
c. $y = 7.5x; m = 0$
d. $y = 7.5; m = 0$

8. Which equation of a line is perpendicular to the y-axis, and what is its slope?

a. $y = 6; m = 0$
b. $y = 6x; m = 0$
c. $x = 6; m =$ undefined
d. $x = 6y; m =$ undefined

9. Which equation of a line is perpendicular to the x-axis, and what is its slope?

a. $y = -1; m =$ undefined
b. $x = -1; m =$ undefined
c. $y = -1; m = 0$
d. $x = -1; m = 0$

10. Which equation of a line is perpendicular to the y-axis, and what is its slope?

a. $y = -10; m =$ undefined
b. $y = -10; m = 0$
c. $x = -10; m =$ undefined
d. $x = -10; m = 0$

11. Which equation of a line is parallel to the y-axis, and what is its slope?

a. $y = -5.5; m =$ undefined
b. $y = -5.5x; m = 0$
c. $x = -5.5; m =$ undefined
d. $x = -5.5; m = 0$

12. Which equation of a line is parallel to the x-axis, and what is its slope?

a. $y = -11x; m =$ undefined
b. $y = -11; m = 0$
c. $x = -11y; m =$ undefined
d. $x = -11; m = 0$

13. A hiker is using a coordinate plane to analyze a map of a state park. The map shows several different hiking trails within the park. The green hiking trail runs along the x-axis, and the red hiking trail runs perpendicular to the green hiking trail. The green and red hiking trails intersect at point $(11.4, 0)$ on the coordinate plane. What is the linear equation of the red hiking trail, and what is its slope?

a. $x = 11.4; m =$ undefined
b. $y = 11.4; m =$ undefined
c. $x = 11.4; m = 0$
d. $y = 11.4; m = 0$

14. A truck driver is using a coordinate plane to analyze a map of his driving route. The map shows freeways in San Francisco. Nimitz Freeway runs along the y-axis, and MacArthur Freeway runs parallel to Nimitz Freeway. What could be the linear equation for MacArthur Freeway, and what is its slope?

a. $y = 2.8x; m = 0$
b. $y = 2.8; m = 0$
c. $x = 2.8y; m =$ undefined
d. $x = 2.8; m =$ undefined

15. A Pennsylvania Railroad line runs perpendicular to Main Street. When examined on a coordinate plane, the railroad line runs along the x-axis, and Main Street intersects the railroad at point $(-9, 0)$. What is the linear equation for Main Street, and what is its slope?

a. $x = -9; m = 0$
b. $y = -9; m = 0$
c. $x = -9; m =$ undefined
d. $y = -9; m =$ undefined

3.9 Equations of Parallel Lines

Writing the equation for a line that is parallel to another line is an important skill in algebra. Many times, you will be expected to write the equation for a line that is parallel to another line that passes through a provided point. For example, let's write the equation of a line that is parallel to the graphed line below and passes through the point (0,1).

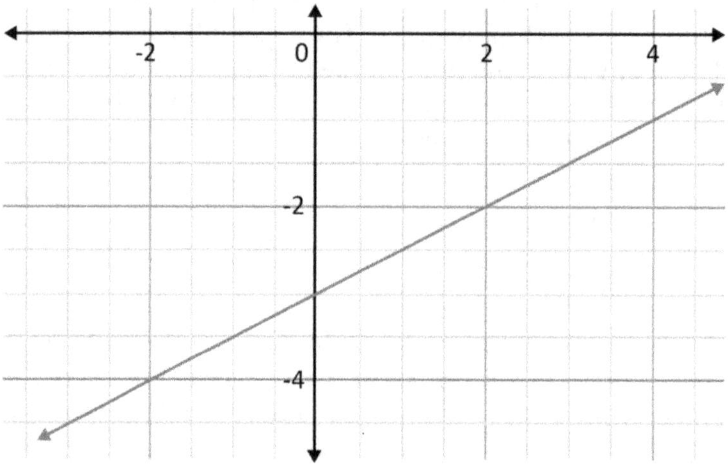

Start by identifying the relevant information:

The point that the new line needs to pass through: (0,1).

The slope of the provided line: $\frac{1}{2}$.

It is important to remember that parallel lines have the same slope.

The equation for the new parallel line can be found using a 2-step process:

Step 1: Recognize that the slope is $\frac{1}{2}$ for the original line as well as the new parallel line: $m = \frac{1}{2}$.

Step 2: Determine the new y-intercept, b, using the provided point (0,1). Plug in x, y, and m into the slope-intercept formula.

$$y = mx + b$$
$$1 = \frac{1}{2}(0) + b$$
$$1 = b$$

The slope and y-intercept of the new line are now known. The equation of the line that passes through the point (0,1) and is parallel to the provided line, is $y = \frac{1}{2}x + 1$.

The quick check to determine if two lines are parallel is to see if their slopes are the same. If the slopes are the same, the lines are either parallel or the same line.

Review Video: Equations of Parallel Lines
Visit mometrix.com/academy and enter code: 573410

EXAMPLES

EXAMPLE 1:

Determine if the two lines are parallel.

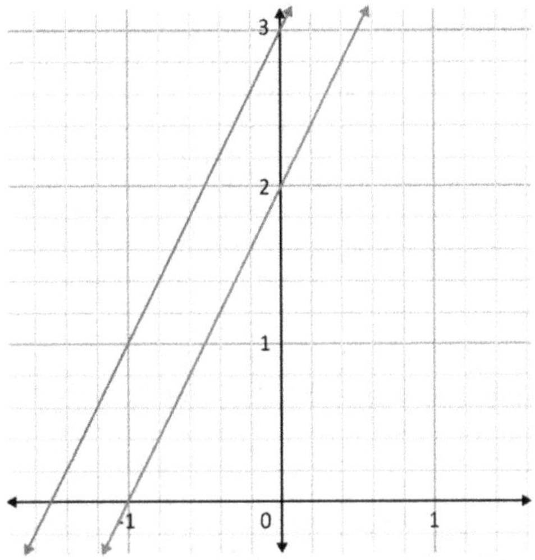

Answer:

Parallel lines will have the same slope. The slope of both lines is 2 ("rise" of 2, and "run" of 1). Therefore, the two lines are parallel.

EXAMPLE 2:

Determine if the two equations are parallel lines: $y = -3x - 1$ and $y = -3x + 1$.

Answer:

Equations that have the same slope, m, will be parallel. Both equations are in slope-intercept form, $y = mx + b$. The slope, m, in both equations is –3. Therefore, the equations form parallel lines.

EXAMPLE 3:

Write the equation of a line that is parallel to the line $y = \frac{1}{2}x + 1$, shown below, and runs through the point (2,3).

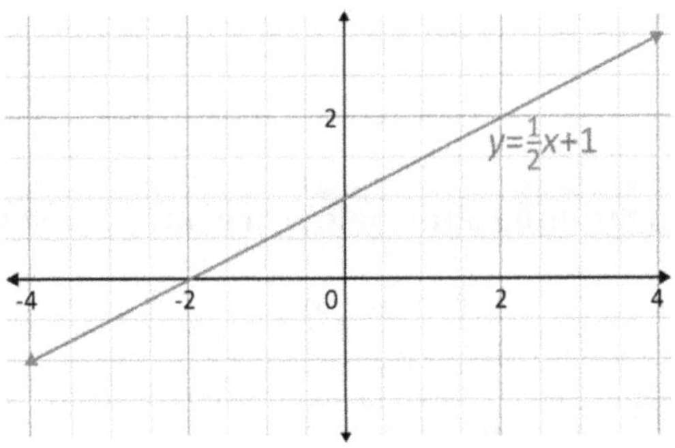

Answer:

Lines that are parallel will have the same slope, m. The slope of the given line is $\frac{1}{2}$. This means the slope of a line that is parallel to this line also needs to have a slope of $\frac{1}{2}$. Use the point (2,3) to determine the new y-intercept. Plug in (2,3) and m into the equation $y = mx + b$.

$$y = mx + b$$
$$3 = \frac{1}{2}(2) + b$$
$$3 = 1 + b$$
$$2 = b$$

The y-intercept of the new line is 2.

The equation of the line that is parallel to $y = \frac{1}{2}x + 1$ and runs through the point (2,3) is $y = \frac{1}{2}x + 2$.

Practice Questions

1. Which equation of a line passes through the point $(3, 2)$ and is parallel to the graph of $y = 6x + 4$?

a. $y = 6x + 20$
b. $y = 6x - 16$
c. $y = -6x - 16$
d. $y = \frac{1}{6}x - 16$

2. Determine whether the linear equations below are parallel. If so, what is their slope?

$$4x + y = 5$$
$$4x + y = -2$$

a. The lines are parallel. The slope of both lines is 4.
b. The lines are parallel. The slope of both lines is –4.
c. The lines are not parallel because their slopes are different.
d. The lines are not parallel because they are the same line.

3. Which equation of a line passes through the point $(7, 3)$ and is parallel to the graph of $4x + 2y = 8$?

a. $y = -4x + 4$
b. $y = -2x + 11$
c. $y = -2x + 17$
d. $y = -4x + 8$

4. Determine whether the linear equations below are parallel. If so, what is their slope?

$$y = \frac{1}{3}x + 4$$
$$3x + y = -4$$

a. The lines are parallel. The slope of both lines is –3.
b. The lines are parallel. The slope of both lines is $\frac{1}{3}$.
c. The lines are not parallel because they are the same line.
d. The lines are not parallel because their slopes are different.

5. Which equation of a line passes through the point $(1, 1)$ and is parallel to the graph of $3x + 5y = 10$?

a. $y = 3x + 1\frac{3}{5}$
b. $y = \frac{5}{3}x + \frac{2}{5}$
c. $y = -\frac{3}{5}x + 1\frac{3}{5}$
d. $y = -\frac{3}{5}x + \frac{2}{5}$

6. Which equation of a line passes through point $(4, 5)$ and is parallel to the graph of $8x + 2y = 12$?

a. $y = -4x + 21$
b. $y = -4x - 11$
c. $y = 8x + 21$
d. $y = -\frac{1}{4}x - 11$

7. Determine whether the linear equations below are parallel. If so, what is their slope?

$$2x + y = 4$$
$$2x + y = \frac{1}{4}$$

a. These lines are parallel. The slope of both lines is –2.
b. These lines are parallel. The slope of both lines is 2.
c. These lines are not parallel because their slopes are different.
d. The lines are not parallel because they are the same line.

8. Determine whether the linear equations below are parallel. If so, what is their slope?

$$-\frac{1}{4}x - y = -3$$
$$x + 4y = -32$$

a. The lines are not parallel because they are the same line.
b. The lines are not parallel because the slopes are different.
c. The lines are parallel. The slope of both lines is 4.
d. The lines are parallel. The slope of both lines is $-\frac{1}{4}$.

9. Which equation of a line passes through the point $(-6, 1)$ and is parallel to the graph of $9x + 3y = 1$?

a. $y = -3x + 19$
b. $y = -3x - 19$
c. $y = -3x - 17$
d. $y = -9x - 53$

10. Determine whether the linear equations below are parallel. If so, what is their slope?

$$\frac{2}{3}x - y = -4$$
$$\frac{3}{2}x + y = 4$$

a. The lines are parallel. The slope of both lines is $\frac{2}{3}$.
b. The lines are parallel. The slope of both lines is $-\frac{3}{2}$.
c. The lines are not parallel because they are the same line.
d. The lines are not parallel because their slopes are different.

11. Determine whether the linear equations below are parallel. If so, what is their slope?

$$\frac{3}{5}x - y = 2$$
$$3x - 5y = -5$$

a. The lines are parallel. The slope of both lines is 3.
b. The lines are parallel. The slope of both lines is $\frac{3}{5}$.
c. The lines are not parallel because their slopes are different.
d. The lines are not parallel because they are the same line.

12. Which equation of a line passes through the point $(8, 2)$ and is parallel to the graph of $x + 4y = 12$?

a. $y = -\frac{1}{4}x$
b. $y = -\frac{1}{4}x + 4$
c. $y = 4x + 4$
d. $y = 4x - 4$

13. When graphed on the coordinate plane, one bike path has the equation $4x - y = 1$, and another bike path has the equation $\frac{1}{4}x - y = 1$. Determine whether these two bike paths are parallel to one another. If so, what is their slope?

a. The lines are not parallel because their slopes are different.
b. The lines are not parallel because they are the same line.
c. The lines are parallel. The slope of both lines is 4.
d. The lines are parallel. The slope of both lines is $\frac{1}{4}$.

14. When graphed on a coordinate plane, the New Jersey Turnpike has the equation $\frac{1}{2}x + y = -3$. Interstate 295 runs parallel to the New Jersey Turnpike and passes through the point $(12, 3)$. What is the equation of the line for Interstate 295 when graphed on the same coordinate plane?

a. $y = \frac{1}{2}x + 3$
b. $y = \frac{1}{2}x - 3$
c. $y = -\frac{1}{2}x + 9$
d. $y = -\frac{1}{2}x - 3$

15. When graphed on a coordinate plane, a linear portion of the Schuylkill River has the equation $2x - 6y = -30$. A running trail runs parallel to this portion of the Schuylkill River and passes through the point $(15, 13)$. What is the equation of the line for the running trail?

a. $y = -3x + 8$
b. $y = -3x + 18$
c. $y = \frac{1}{3}x + 8$
d. $y = \frac{1}{3}x + 18$

3.10 Equations of Perpendicular Lines

Lines that are perpendicular, meaning they intersect at a 90° angle, will have slopes that are reciprocals with opposite signs. Numbers that are reciprocals of one another will multiply to 1. For example, $\frac{1}{3}$ and 3, or $\frac{2}{5}$ and $\frac{5}{2}$. A simple way to remember how to take the reciprocal of a number is to flip the fraction. Since perpendicular slopes also need opposite signs, some examples are 2 and $-\frac{1}{2}$, -4 and $\frac{1}{4}$.

When graphing a line that is perpendicular to another line, the slope and y-intercept need to be identified. In many algebra courses, you will be expected to write the equation of a line that is perpendicular to another line that passes through a provided point. For example, writing an equation for a line that is perpendicular to the graphed line below and passes through the point (3,1).

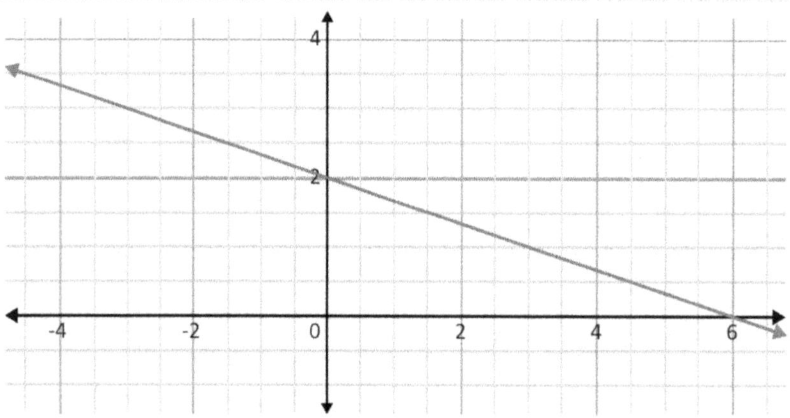

The new perpendicular line can be found using a 3-step process:

Step 1: Use the graph to determine the slope of the line. In this case, the slope is $-\frac{1}{3}$.

Step 2: Recognize that the slope of the new perpendicular line will be the reciprocal of $-\frac{1}{3}$ with the opposite sign. This means that the new slope will be 3.

Step 3: Determine the y-intercept, b, by using the provided point (3,1). Plug in x, y, and m into the slope-intercept formula.

$$\begin{aligned} y &= mx + b \\ 1 &= 3(3) + b \\ 1 &= 9 + b \\ -8 &= b \end{aligned}$$

Now that the slope and the y-intercept are known, the equation of the new perpendicular line can be written. The equation $y = 3x - 8$ will pass through the point (3,1), and form a line that is perpendicular to $y = -\frac{1}{3}x + 2$.

Remember, if two lines have slopes that are negative-reciprocals of each other, then the lines are perpendicular.

Review Video: Equations of Perpendicular Lines
Visit mometrix.com/academy and enter code: 837833

EXAMPLES

EXAMPLE 1:

Determine if the two lines are perpendicular.

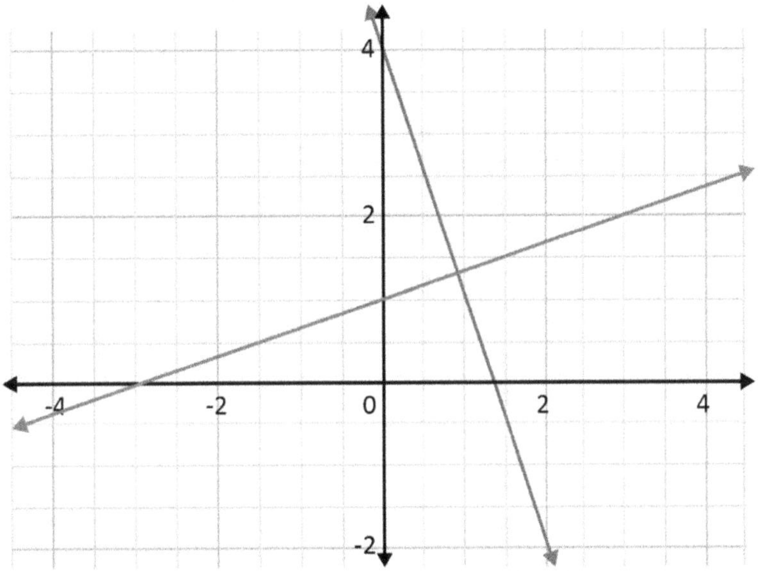

Answer:

Lines are perpendicular if one slope is the negative reciprocal of the other. The slope of the increasing line is $\frac{1}{3}$, and the slope of the decreasing line is –3. Since –3 is the negative reciprocal of $\frac{1}{3}$, the lines are perpendicular.

EXAMPLE 2:

Determine if the two equations form perpendicular lines: $y = 2x + 1$ and $y = -\frac{1}{2}x + 4$.

Answer:

If one slope is the negative reciprocal of the other slope, then the lines are perpendicular. The equation $y = 2x + 1$ has a slope of 2. The equation $y = -\frac{1}{2}x + 4$ has a slope of $-\frac{1}{2}$. Since $-\frac{1}{2}$ is the negative reciprocal of 2, the lines are perpendicular.

EXAMPLE 3:

Write the equation of a line that runs through the point (1,4) and is perpendicular to the graphed line.

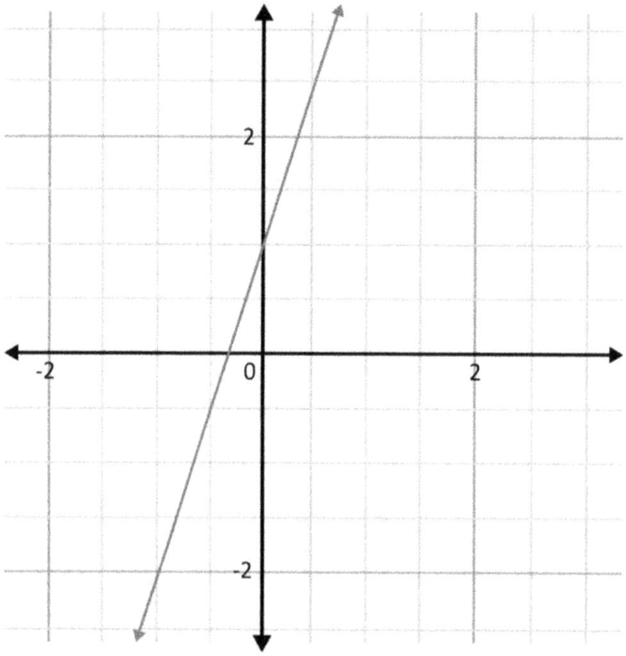

Answer:

The equation for the graphed line is $y = 3x + 1$. The slope is 3 and the y-intercept is 1. To create a line that is perpendicular to this line, the new slope needs to be the reciprocal of 3 with the opposite sign. In other words, the new slope will be the negative reciprocal of 3, which is $-\frac{1}{3}$.

At this point, we know that the slope of the new line is $-\frac{1}{3}$ and that the line will pass through the point (1,4). Now we just need to identify the y-intercept, b, of this new line, to write the equation of the line.

To find b, plug in (1,4) and the slope of $-\frac{1}{3}$ into the equation $y = mx + b$, and then solve for b.

$$
\begin{aligned}
y &= mx + b \\
4 &= -\frac{1}{3}(1) + b \\
4 &= -\frac{1}{3} + b \\
4\frac{1}{3} &= b
\end{aligned}
$$

This means that the y-intercept of the new perpendicular line is $4\frac{1}{3}$.

Now the slope, $m = -\frac{1}{3}$ and the y-intercept, $b = 4\frac{1}{3}$, are known, so the equation can be written: $y = -\frac{1}{3}x + 4\frac{1}{3}$.

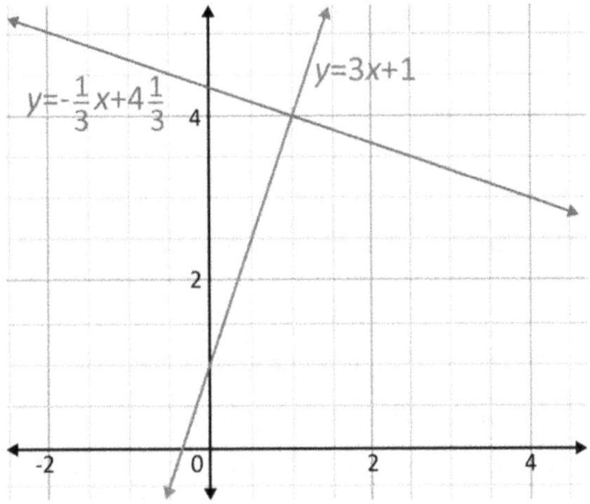

EXAMPLE 4:

Find the equation of a line that contains the point $(-1,3)$ and is perpendicular to the graph of $2x + 3y = 8$.

Answer:

Convert the equation to slope-intercept form to determine the slope of the line.

$$2x + 3y = 8$$
$$3y = -2x + 8$$
$$y = -\frac{2}{3}x + \frac{8}{3}$$

The slope of this line is $-\frac{2}{3}$, so the slope of a perpendicular line is $\frac{3}{2}$.

Next, substitute the coordinate point and slope into the slope-intercept equation and solve for b.

$$3 = \frac{3}{2}(-1) + b$$
$$3 = -\frac{3}{2} + b$$
$$\frac{9}{2} = b$$

Now that we know the slope and y-intercept, we can write the equation for the perpendicular line.

$$y = \frac{3}{2}x + \frac{9}{2}$$

PRACTICE QUESTIONS

1. Which equation of a line contains the point $(2, 5)$ and is perpendicular to the graph of $y = 4x + 3$?

a. $y = -4x + 5\frac{1}{2}$
b. $y = -4x + 4\frac{1}{2}$
c. $y = -\frac{1}{4}x + 5\frac{1}{2}$
d. $y = -\frac{1}{4}x + 4\frac{1}{2}$

2. Which statement best describes the relationship between the linear equations below?

$$\frac{1}{3}x + y = 2$$
$$y = 3x - 8$$

a. The linear equations are parallel. The slope of both lines is $\frac{1}{3}$.
b. The linear equations are perpendicular. The slope of both lines is 3.
c. The linear equations are perpendicular. Their slopes are opposite reciprocals of one another.
d. The linear equations are neither parallel nor perpendicular.

3. Which equation of a line contains the point $(2, -1)$ and is perpendicular to the graph of $5x + y = 1$?

a. $y = -\frac{1}{5}x + 1\frac{2}{5}$
b. $y = \frac{1}{5}x - 1\frac{2}{5}$
c. $y = -\frac{1}{5}x - \frac{3}{5}$
d. $y = \frac{1}{5}x - \frac{3}{5}$

4. Which statement best describes the relationship between the linear equations below?

$$7x - y = -2$$
$$x + 7y = 21$$

a. The linear equations are parallel. The slope of both lines is 7.
b. The linear equations are neither parallel nor perpendicular.
c. The linear equations are perpendicular. Their slopes are opposite reciprocals of one another.
d. The linear equations are perpendicular. The slope of both lines is $-\frac{1}{7}$.

5. Which equation of a line contains the point $(-4, -7)$ and is perpendicular to the graph of $4x + 7y = 3$?

a. $y = -\frac{1}{4}x - 7$
b. $y = -\frac{1}{4}x + 14$
c. $y = \frac{7}{4}x - 14$
d. $y = \frac{7}{4}x$

6. Which equation of a line contains the point $(-5, 3)$ and is perpendicular to the graph of $2x - y = 3$?

a. $y = -\frac{1}{2}x + \frac{1}{2}$
b. $y = -2x + 5\frac{1}{2}$
c. $y = -\frac{1}{2}x - \frac{1}{2}$
d. $y = -2x - 5\frac{1}{2}$

7. Which statement best describes the relationship between the linear equations below?

$$2x - y = 8$$
$$4x - y = 8$$

a. The linear equations are neither parallel nor perpendicular because they are the same line.
b. The linear equations are parallel because their slopes are the same.
c. The linear equations are perpendicular. Their slopes are opposite reciprocals of one another.
d. The linear equations are not perpendicular. Their slopes are not opposite reciprocals of one another.

8. Which statement best describes the relationship between the linear equations below?

$$\frac{1}{2}x - y = 9$$
$$4x - 2y = 22$$

a. The linear equations are perpendicular. Their slopes are opposite reciprocals of one another.
b. The linear equations are not perpendicular. Their slopes are not opposite reciprocals of one another.
c. The linear equations are neither parallel nor perpendicular because they are the same line.
d. The linear equations are parallel because their slopes are equal to one another.

9. Which equation of a line contains the point $(4, -4)$ and is perpendicular to the graph of $4x + y = -2$?

a. $y = \frac{1}{4}x - 3$
b. $y = \frac{1}{4}x - 5$
c. $y = 4x + 5$
d. $y = 4x + 3$

10. Which statement best describes the relationship between the linear equations below?

$$\frac{3}{2}x + y = 4$$
$$-\frac{2}{3}x + y = 1$$

a. The linear equations are parallel. The slope of both lines is $-\frac{2}{3}$.
b. The linear equations are neither parallel nor perpendicular.
c. The linear equations are perpendicular. The slope of both lines is $\frac{3}{2}$.
d. The linear equations are perpendicular. Their slopes are opposite reciprocals of one another.

11. Which statement best describes the relationship between the linear equations below?

$$4x + y = 7$$
$$8x + 2y = 3$$

a. The linear equations are parallel to one another. Their slopes are the same.
b. The linear equations are neither parallel nor perpendicular to one another.
c. The linear equations are not perpendicular. Their slopes are not opposite reciprocals of one another.
d. The linear equations are perpendicular. Their slopes are opposite reciprocals of one another.

12. Which equation of a line contains the point $(0, -4)$ and is perpendicular to the graph of $3x + y = -12$?

a. $y = -3x - 4$
b. $y = 3x + 4$
c. $y = \frac{1}{3}x - 4$
d. $y = \frac{1}{3}x + 4$

13. When graphed on a coordinate plane, the red subway line has the equation $2x - 5y = 10$, and the blue subway line has the equation $5x + 2y = 12$. Determine whether these two subway lines are perpendicular to one another. Choose the statement that best describes their relationship.

a. The subway lines are perpendicular. Their slopes are opposite reciprocals of one another.
b. The subway lines are perpendicular. The slope of both lines is $\frac{2}{5}$.
c. The subway lines are parallel. The slope of both lines is $-\frac{5}{2}$.
d. The subway lines are neither parallel nor perpendicular.

14. When graphed on a coordinate plane, Broad Street has the equation $x + y = 3$. Market Street runs perpendicular to Broad Street and passes through the point $(-3, 2)$. What is the linear equation for Market Street?

a. $y = x - 1$
b. $y = x + 5$
c. $y = -x + 1$
d. $y = -x - 5$

15. When graphed on a coordinate plane, one side of a rectangle has the equation $2x - y = 2$. An adjacent side of the rectangle runs perpendicular to the first side and passes through the point $(-4, 0)$. What is the equation of the line for the adjacent side of the rectangle?

a. $y = -\frac{1}{2}x + 2$
b. $y = -\frac{1}{2}x - 2$
c. $y = -2x + 2$
d. $y = 2x - 2$

3.11 Calculating Correlation Coefficient

What Is a Correlation Coefficient?

A **correlation coefficient** is a statistical measurement of the strength of a linear relationship between two quantitative variables or sets of data. It is a singular number that shows the strength of a statistical relationship. There are several different types of correlation coefficients, and the most common of these is Pearson product-moment correlation. This is often simply referred to as Pearson's, or Pearson's r.

How Are Correlation Coefficients Interpreted?

The correlation coefficient is represented by the variable r, where r is a number ranging from –1.0 to 1.0.

Positive Correlation

A correlation coefficient of 1.0 indicates a perfect positive linear association between two variables. In other words, for every positive increase in one variable, there is a proportionally positive increase in the other. For instance, there is an almost perfect positive correlation between the number of hours spent driving and the distance traveled at a constant speed. As the amount of time driving increases, the distance traveled also increases at a fixed proportion. When this data is plotted on the coordinate plane, it shows a nearly perfect straight line that points in a positive direction.

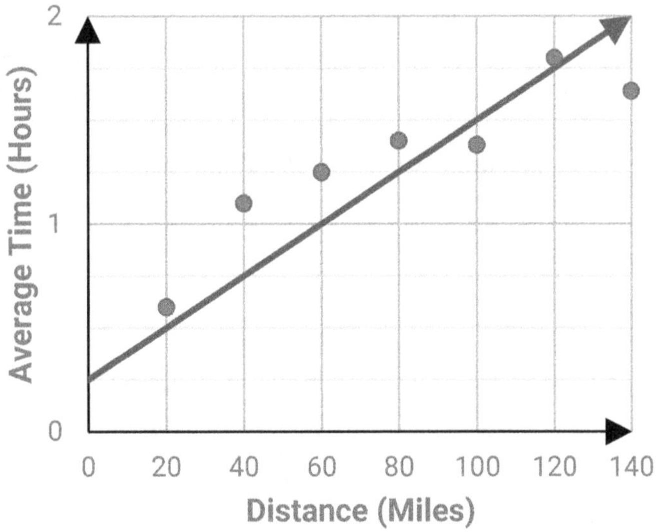

Negative Correlation

A correlation coefficient of –1.0 indicates a perfect negative linear association between two variables. In other words, for every positive increase in one variable, there is a proportionally negative decrease in the other. For instance, there is an almost perfect negative correlation between the number of hours driving at a constant speed and the distance to the destination. As the time spent driving increases, the distance to the destination decreases at a fixed proportion. When this data is plotted on the coordinate plane, it shows a nearly perfect straight line that points in a negative direction.

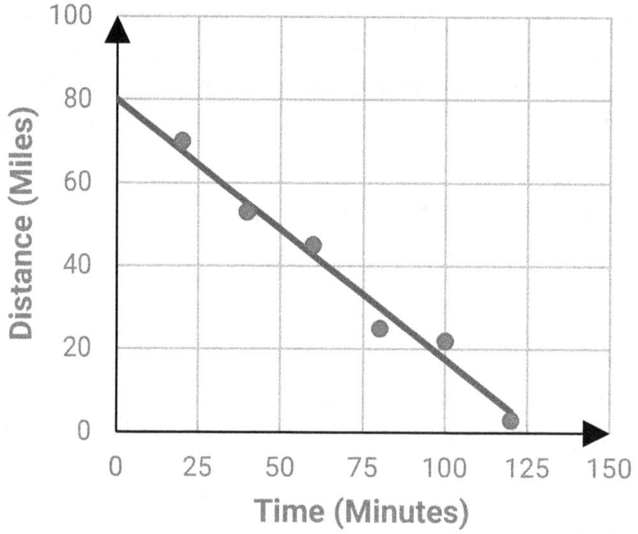

No Correlation

A correlation coefficient of 0 indicates no association or relationship between the two variables in question. In other words, for every positive increase in one variable, there is no pattern of increases or decreases in the other variable. There is no connection between the two sets of data. For instance, there is no correlation between a student's shoe size and their grade point average (GPA). As shoe size increases, there is no pattern of increase or decrease in GPA. When this data is plotted on the coordinate plane, it is scattered, with no linear pattern present.

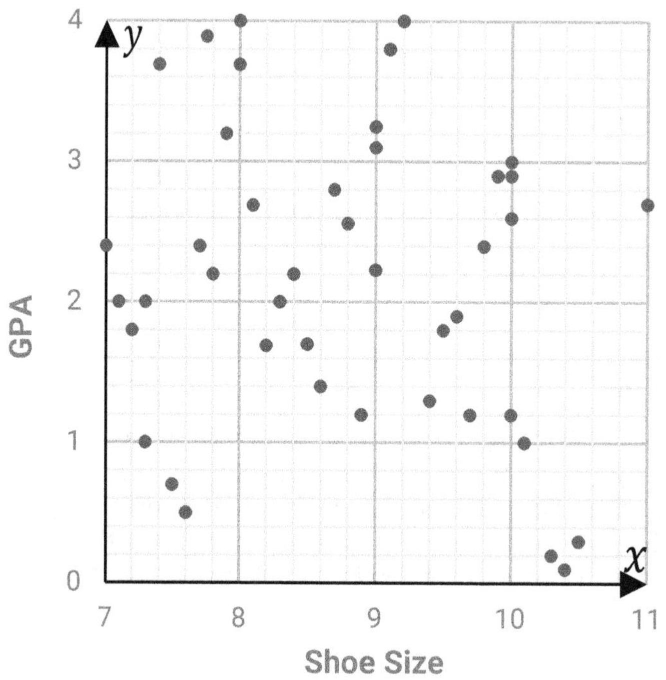

Absolute Value

The absolute value of a correlation coefficient shows the strength of the relationship between the two variables. The larger the absolute value, the stronger the relationship. For example, a correlation coefficient of –0.85 indicates a stronger negative correlation than a correlation coefficient of –0.39 because its absolute value is higher. Similarly, a correlation coefficient of 0.78 indicates a stronger positive correlation than a correlation coefficient of 0.25 because its absolute value is higher.

The chart below can help interpret correlation coefficients:

Correlation coefficient	Interpretation
-1.0	Perfect negative linear correlation
-0.75	Strong negative linear correlation
-0.50	Moderate negative linear correlation
-0.25	Weak negative linear correlation
0	No correlation
0.25	Weak positive linear correlation
0.50	Moderate positive linear correlation
0.75	Strong positive linear correlation
1.0	Perfect positive linear correlation

How Is the Correlation Coefficient Calculated?

The correlation coefficient (r) is calculated by dividing the covariance by the product of the standard deviations of each variable. **Covariance** is a measure of how much two variables change in the context of one another. **Standard deviation** is a measure of how much a data point deviates from the average of the data set.

The formula for finding the correlation coefficient is shown below:

$$r = \frac{\sum(x_i - \bar{x})(y_i - \bar{y})}{\sqrt{\sum(x_i - \bar{x})^2 \sum(y_i - \bar{y})^2}}$$

Calculating the correlation coefficient by hand can be done, but it is a tedious and time-consuming process. Instead, many mathematicians use technology, such as graphing calculators. By simply inputting each variable's data set into a graphing calculator, the correlation coefficient for two quantitative variables can be identified and interpreted.

Steps for finding the correlation coefficient using a TI-84 graphing calculator.

1. Turn on your calculator's diagnostics
2. Enter your data: STAT and then EDIT
3. Fill out List 1 and List 2 with your x- and y-values
4. STAT and then CALCULATE
5. Select #4.LinReg and then ENTER
6. You will see the correlation coefficient as $r =$ _____

For example, the data set below shows height in cm, x, and arm span in cm, y. The correlation coefficient will describe the strength of the relationship between these two variables. Use the 6 steps to find the correlation coefficient on a calculator.

Height (cm)	160	167	170	161	175	168	173	181	176	184
Arm Span (cm)	162	167	168	162	178	165	170	179	176	183

When the x-values (height) and the y-values (arm span) are entered into List 1 and List 2, r can be calculated. In this case, $r = 0.965$. This is a very strong positive relationship.

Examples

Example 1:

Calculate the correlation coefficient for the following scenario comparing the relationship between the number of days a student misses school to the score on their final exam.

Number of days absent: 10, 12, 2, 0, 8, 5

Score on the final exam: 70, 65, 96, 94, 75, 82

Answer:

The correlation coefficient can be found using a graphing calculator.

Finding the correlation coefficient on a graphing calculator:

Be sure you turn your diagnostic on.

Step 1: Stat

Step 2: Edit

Step 3: Enter

Step 4: L1 and L2 (enter the data for both variables)

Step 5: Stat

Step 6: LinReg - Enter

From here it tells you the value of $r = -0.980$.

r represents the correlation coefficient, and the value of r is always between -1 and 1. 1 represents a strong positive correlation, -1 represents a strong negative correlation, and 0 represents no correlation. As the number of absences increase, the score decreases. $r = -0.980$ is a strong negative correlation.

EXAMPLE 2:

Calculate the correlation coefficient for the following relationship between x and y.

x: 1, 2, 3, 4, 5, 6

y: 2, 4, 7, 9, 12, 14

Answer:

The correlation coefficient can be found using a graphing calculator.

Finding the correlation coefficient on a graphing calculator:

Be sure you turn your diagnostic on.

Step 1: Stat

Step 2: Edit

Step 3: Enter

Step 4: L1 and L2 (enter the data for both variables)

Step 5: Stat

Step 6: LinReg - Enter

From here it tells you the value of $r = 0.998$.

r represents the correlation coefficient, and the value of r is always between –1 and 1. 1 represents a strong positive correlation, –1 represents a strong negative correlation, and 0 represents no correlation. As x increases, y increases. $r = 0.998$ is a strong positive correlation.

EXAMPLE 3:

Calculate and interpret the correlation coefficient, rounded to the nearest hundredth, of the two sets of data shown.

x	1	2	3	4	5
y	65	74	71	82	81

Answer:

Use a graphing calculator to determine the correlation coefficient of this set of data.

Be sure you turn your diagnostic on.

Step 1: Stat

Step 2: Edit

Step 3: Enter

Step 4: L1 and L2 (enter the data for both variables)

Step 5: Stat

Step 6: LinReg - Enter

The correlation coefficient, rounded to the nearest hundredth, is 0.89. Since 0.89 is a positive number, it indicates a positive association. Since 0.89 is very close to 1.0, a correlation coefficient of 0.89 indicates a strong positive linear association. When this data is plotted on the coordinate plane, it shows a nearly perfect straight line that points in a positive direction.

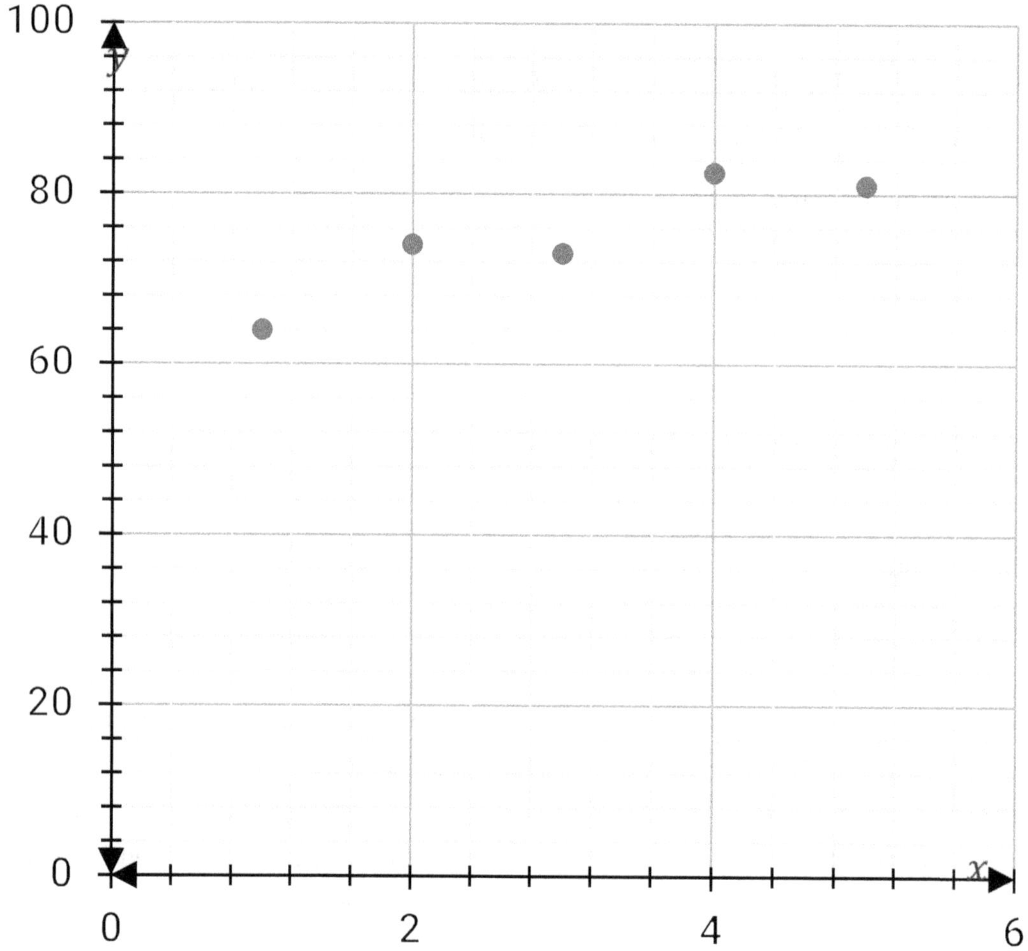

Example 4:

Calculate and interpret the correlation coefficient, rounded to the nearest hundredth, of the two sets of data shown.

x	64	85	91	42	18
y	72	61	71	54	80

Answer:

Use a graphing calculator to determine the correlation coefficient of this set of data.

Be sure you turn your diagnostic on.

Step 1: Stat

Step 2: Edit

Step 3: Enter

Step 4: L1 and L2 (enter the data for both variables)

Step 5: Stat

Step 6: LinReg - Enter

The correlation coefficient of this set of data, rounded to the nearest hundredth, is –0.26. Since –0.26 is negative and very close to 0, a correlation coefficient of –0.26 indicates a weak negative linear association. When this data is plotted on the coordinate plane, its line is mostly scattered but it points in a negative direction.

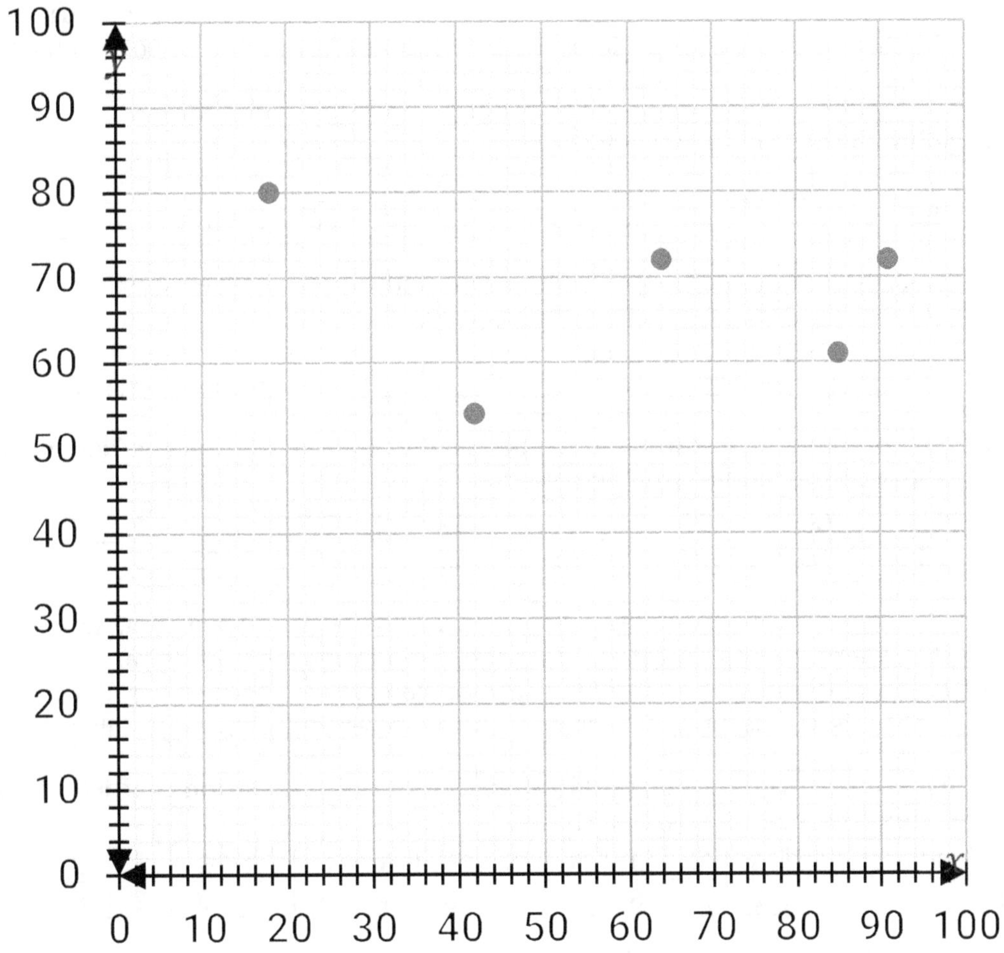

PRACTICE QUESTIONS

1. Given the data below, which statement best describes the correlation coefficient, rounded to the nearest hundredth, of the two variables?

x	64	73	70	75	63	66	69	70	75
y	140	190	185	190	165	160	160	170	190

a. $r \approx -0.85$, there is a strong negative linear association between the two variables.
b. $r \approx 0.85$, there is a strong positive linear association between the two variables.
c. $r \approx 0.15$, there is a weak positive linear association between the two variables.
d. $r = 0$, there is no association between the two variables.

2. Given the data below, which statement best describes the correlation coefficient, rounded to the nearest hundredth, of the two variables?

x	1	3	0.5	0.75	1.5
y	4	1	4	1.5	5

a. $r = 0$, there is no correlation between the two variables.
b. $r = -1$, there is a perfect negative linear association between the two variables.
c. $r \approx -0.48$, there is a moderate negative linear correlation between the two variables.
d. $r \approx 0.48$, there is a moderate positive linear correlation between the two variables.

3. Given the data below, which statement best describes the correlation coefficient, rounded to the nearest hundredth, of the two variables?

x	1	2	3	4	5	6	7	8	9
y	40	36	32	28	24	20	16	12	8

a. $r \approx -0.92$, there is a strong negative linear association between the two variables.
b. $r = 0$, there is no association between the two variables.
c. $r = 1$, there is a perfect positive linear association between the two variables.
d. $r = -1$, there is a perfect negative linear association between the two variables.

4. Given the data below, which statement best describes the correlation coefficient, rounded to the nearest hundredth, of the two variables?

x	25.5	22.5	30	26.5	34	29.5	25	26	26
y	25	30	36	20	23	28	25	16	14

a. $r \approx 0.15$, there is a weak positive linear association between the two variables.
b. $r \approx -0.15$, there is a weak negative linear association between the two variables.
c. $r \approx 0.85$, there is a strong positive linear association between the two variables.
d. $r = -1$, there is a perfect negative linear association between the two variables.

5. Given the data below, which statement best describes the correlation coefficient, rounded to the nearest hundredth, of the two variables?

x	1	1	2	2	3	3	4	4	5	5	6	6	7	7	8	8	9	9
y	3	7	2	6	2	3	8	4	7	3	4	8	2	3	2	6	3	7

a. $r = 1$, there is a perfect positive linear association between the two variables.
b. $r = 0$, there is no association between the two variables.
c. $r = -1$, there is a perfect negative linear association between the two variables.
d. $r \approx -0.09$, there is a weak negative linear association between the two variables.

6. Given the data below, which statement best describes the correlation coefficient, rounded to the nearest hundredth, of the two variables?

x	8	12	6	4	10	7	2	4	9
y	1	1	2	3	1	2	4	2	1

a. $r \approx -0.50$, there is a moderate negative linear association between the two variables.
b. $r \approx -0.89$, there is a strong negative linear association between the two variables.
c. $r \approx 0.89$, there is a strong positive linear association between the two variables.
d. $r = 0$, there is no association between the two variables.

7. Given the data below, which statement best describes the correlation coefficient, rounded to the nearest hundredth, of the two variables?

x	81	85	89	93	97	101	105	109	113
y	38.5	41	43.5	46	48.5	51	53.5	56	58.5

a. $r \approx -0.46$, there is a moderate negative linear association between the two variables.
b. $r = 0$, there is no linear association between the two variables.
c. $r = 1$, there is a perfect positive linear association between the two variables.
d. $r = -1$, there is a perfect negative linear association between the two variables.

8. Given the data below, which statement best describes the correlation coefficient, rounded to the nearest hundredth, of the two variables?

x	6	7	11	2	4	3	9	10	2
y	74	72	92	97	80	73	65	76	76

a. $r \approx -0.14$, there is a weak negative linear association between the two variables.
b. $r \approx 0.14$, there is a weak positive linear association between the two variables.
c. $r = 0$, there is no linear association between the two variables.
d. $r = 1$, there is a perfect positive linear association between the two variables.

9. Given the data below, which statement best describes the correlation coefficient, rounded to the nearest hundredth, of the two variables?

x	4	7	6	4	7	20	16	12
y	1	9	-1	6	10	19	20	-5

a. $r = 1$, there is a perfect positive linear association between the two variables.
b. $r = -1$, there is a perfect negative linear association between the two variables.
c. $r \approx -0.63$, there is a moderate negative linear association between the two variables.
d. $r \approx 0.63$, there is a moderate positive linear association between the two variables.

10. Given the data below, which statement best describes the correlation coefficient, rounded to the nearest hundredth, of the two variables?

x	5	5	7	6	5	7	4	3	8
y	8	4	13	11	5	13	3	2	12

a. $r \approx 0.92$, there is a strong positive linear association between the two variables.
b. $r \approx -0.81$, there is a strong negative linear association between the two variables.
c. $r \approx 0.18$, there is a weak positive linear association between the two variables.
d. $r = 0$, there is no association between the two variables.

11. Given the data below, which statement best describes the correlation coefficient, rounded to the nearest hundredth, of the two variables?

x	17	0	5	10	18	5	0	2	3
y	78	82	90	92	77	89	94	86	100

a. $r \approx -0.61$, there is a moderate negative linear correlation between the two variables.
b. $r \approx 0.61$, there is a moderate positive linear correlation between the two variables.
c. $r = 1$, there is a perfect positive linear association between the two variables.
d. $r = 0$, there is no correlation between the two variables.

12. Given the data below, which statement best describes the correlation coefficient, rounded to the nearest hundredth, of the two variables?

x	25	35	45	55	65	75	85	95	105
y	140	132	124	116	108	100	92	84	76

a. $r \approx -0.11$, there is a weak negative linear association between the two variables.
b. $r = 1$, there is a perfect positive linear association between the two variables.
c. $r = -1$, there is a perfect negative linear association between the two variables.
d. $r = 0$, there is no association between the two variables.

13. The table below shows how many hours each student spent studying for their final exam, as well as their final score on the exam out of 100%. Given this data set, which statement best describes the correlation coefficient, rounded to the nearest hundredth, of the two variables?

Name	Hours spent studying	Final exam score (out of 100%)
Abigail	1	60
Brian	1.5	80
Christina	2	75
David	3	80
Eloise	3.5	83
Flynn	4	90
Gianna	4.5	90
Hugo	5	85
Ira	5.5	90

a. $r \approx -0.83$, there is a strong negative linear association between hours spent studying and the final exam score.
b. $r \approx 0.84$, there is a strong positive linear association between hours spent studying and the final exam score.
c. $r \approx 0.16$, there is a weak positive linear association between hours spent studying and the final exam score.
d. $r = 0$, there is no association between hours spent studying and the final exam score.

14. The table below shows each student's height in inches, as well as the total number of books they read last year. Given this data set, which statement best describes the correlation coefficient, rounded to the nearest hundredth, of the two variables?

Name	Height (in inches)	Number of Books Read Last Year
Amma	68	20
Bella	66	22
Cathy	64	24
Dan	66	26
Eric	68	28

a. $r \approx 0.54$, there is a moderate positive linear correlation between height and the number of books read last year.
b. $r = 1$, there is a perfect positive linear association between height and the number of books read last year.
c. $r = -1$, there is a perfect negative linear association between height and the number of books read last year.
d. $r = 0$, there is no association between height and the number of books read last year.

15. The table below shows a paper company's total sales over the course of six years. Given this data set, which statement best describes the correlation coefficient, rounded to the nearest hundredth, of the two variables?

Year	Sales
2011	$80,000
2012	$76,000
2013	$77,000
2014	$74,000
2015	$70,000
2016	$72,000

a. $r = 1$, there is a perfect positive linear association between the year and total sales.
b. $r = -1$, there is a perfect negative linear association between the year and total sales.
c. $r \approx 0.91$, there is a strong positive linear association between the year and total sales.
d. $r \approx -0.91$, there is a strong negative linear association between the year and total sales.

3.12 Association, Correlation, and Causation

Three important relationships between variables are association, correlation, and causation. **Association** is a relationship between two variables. **Correlation** is the strength of the relationship between two variables. In other words, it is the measure of association. **Causation** is a relationship between two variables where one variable has a direct impact on the other variable. For example, a change in one variable causes a change in the other variable.

In everyday language, association, causation, and correlation are often mistakenly interchanged. It is important to be aware of the subtle differences between association and causation.

In statistics, an observed association between two variables does not necessarily imply causation. For example, consider the graph below.

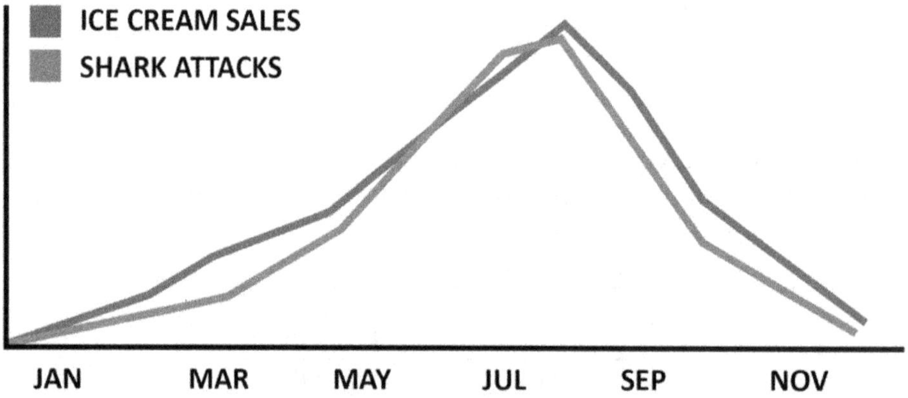

The graph clearly shows how ice cream sales and shark attacks are strongly correlated. The association is undeniable. As ice cream sales increase, shark attacks also increase! Does this imply causation? Not necessarily. An increase in ice cream sales does not *cause* an increase in shark attacks.

However, associated variables such as the amount of time weight training and gains in muscle mass, or the number of rainstorms and the number of flooding events, are examples of causation. One variable causes the other. The variables are *associated,* with a strong *correlation*, and it is very reasonable that one causes the other.

EXAMPLES

EXAMPLE 1:

For an hourly worker, does the relationship "hours worked and income earned" describe association, causation, or both?

Answer:

In this case, hours worked and income earned represent an association and causation relationship. Income and hours worked have a relationship where one variable directly impacts the other. As hours worked increases, income increases. As hours worked decreases, income decreases.

EXAMPLE 2:

Does the relationship "the more firemen sent to a fire, the more damage is done" describe association, causation, or both?

Answer:

In this case, more firemen will be sent to a fire if it is a large fire. So, on a graph it will appear that as the number of firemen increases, the amount of damage from the fire also increases. However, this association is not necessarily causation. The increase in damage is not due to the fact that more firemen are on the scene.

Practice Questions

1. Which of the following statements best describes a variable relationship that shows association but not causation?

a. The amount of flour needed for a cookie recipe and the total number of cookies the recipe yields
b. The amount of smoke and the size of a fire
c. The number of A's earned on a report card and overall grade point average
d. The number of cups of hot chocolate sold and pairs of mittens sold

2. Which of the following statements describes a variable relationship that best shows causation?

a. A secretary's typing speed in words per minute and the total amount of time it takes them to type 500 words
b. A toddler's age and the size of their vocabulary
c. The time it takes for a person to wash his or her hair and the total amount of shampoo needed
d. A middle school student's height and their score on a proficiency test

3. Which of the following statements best describes a variable relationship that shows association but not causation?

a. The amount of time spent exercising on a rowing machine and the number of calories burned while working out
b. Ounces of soup and the total number of servings
c. A sprinter's running speed in minutes per mile and the total amount of time it takes for them to run 1 mile
d. The day of the week and the number of airline tickets sold

4. Which of the following statements describes a variable relationship that best shows causation?

a. The unit rate of 1 pencil and the total cost of a box of 100 pencils
b. A person's age and the number of countries they've traveled to
c. A person's shoe size and their shirt size
d. The number of yards a football team gains on offense and the number of touchdowns scored

5. Which of the following statements describes a variable relationship that shows association but not causation?

a. The number of hours spent babysitting and the total amount of money paid for the babysitting job
b. A student's high school attendance record and their SAT score
c. The number of books purchased and the total amount of money spent at the bookstore
d. The number of steps taken and the total number of miles walked in a day

6. Which of the following statements describes a variable relationship that shows association but not causation?

a. A person's speed on a stationary bike and the total amount of time it takes for them to bike 10 miles
b. The number of hours of television a student watches per week and the number of homework assignments completed
c. The height of a child and their clothing size
d. The cost of bulk candy by the pound and the total amount spent on 5 pounds of candy

7. Which of the following statements describes a variable relationship that best shows causation?

a. The number of years of education a person completes and their earning potential
b. The number of sodas a person consumes each day and their body mass index
c. The number of pages in a photo album and the total number of pictures it can hold
d. The number of hours a student spends studying for a test and their test score

8. Which of the following statements describes a variable relationship that best shows causation?

a. The dimensions of a restaurant's outdoor patio and the total amount of square footage available on the patio for dining
b. The number of days a heating system runs throughout the year and the outdoor temperature
c. A person's age and their total amount of retirement savings
d. The total number of books a person has read and their score on an IQ test

9. Which of the following statements describes a variable relationship that shows association but not causation?

a. The percentage of battery power on a laptop and the total amount of time before it shuts down
b. The square footage of a person's home and their annual income
c. A person's walking speed and the total amount of time it takes for them to walk 3 miles
d. The cost of gas per gallon and the total amount spent to fill up a 12-gallon gas tank

10. Which of the following statements describes a variable relationship that best shows causation?

a. A student's SAT score and their college grade point average
b. The amount of time a person spends exercising and their total weight loss
c. The amount of gasoline put into a car and the distance it can travel
d. A company's annual sales and its total revenue

11. Which of the following statements describes a variable relationship that best shows causation?

a. An athlete's weight and their running speed
b. The amount of force exerted on a ball and the distance it travels over a flat surface
c. The number of cups of lemonade sold and the number of popsicles sold
d. A child's age and the number of states they have traveled to

12. Which of the following statements describes a variable relationship that shows association but not causation?

a. The number of extra treats a dog consumes and its weight gain
b. A student's number of correctly spelled words on a spelling test and their overall score
c. The number of cups of coffee consumed and the amount of time spent awake
d. The amount of wind present and the size of the waves in a body of water

13. Dave notices that the older he gets, the more time he has available to read. He wants to know if there is a relationship between a person's age and the total number of books they've read. Which of the following statements most likely best explains the relationship between these two variables?

a. There is a positive correlation and causation in the relationship between a person's age and the total number of books they've read.
b. There is no causation or association between a person's age and the total number of books they've read.
c. There is a positive correlation between a person's age and the total number of books they've read but this is not a causal relationship.
d. There is causation but no association between a person's age and the total number of books they've read.

14. Brooke is buying a fish tank for her aquarium. She wants to know if there is a relationship between the size of a fish tank and the number of gallons of water it can hold. Which of the following statements best explains the relationship between these two variables?

a. There is a positive correlation and causation in the relationship between the size of a fish tank and the number of gallons of water it can hold.
b. There is a negative correlation between the size of a fish tank and the number of gallons of water it can hold but this is not a causal relationship.
c. There is a positive correlation between the size of a fish tank and the number of gallons of water it can hold but this is not a causal relationship.
d. There is no causation or association between the size of a fish tank and the number of gallons of water it can hold.

15. Ben is working on his free throw percentage. He wants to know if there is a relationship between the number of free throw shots taken and the total number of points scored. Which of the following statements best explains the relationship between these two variables?

a. There is no causation or association between the number of free throw shots taken and the total number of points scored.
b. There is a positive correlation and causation in the relationship between the number of free throw shots taken and the total number of points scored.
c. There is causation but no association between the number of free throw shots taken and the total number of points scored.
d. There is a positive correlation between the number of free throw shots taken and the total number of points scored but this is not a causal relationship.

Chapter 3 Quiz

1. The graph of a linear equation is shown on the coordinate plane below.

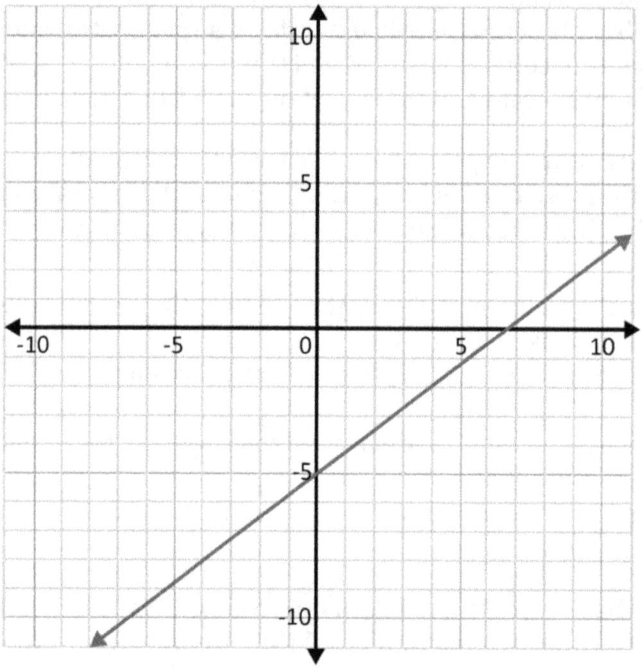

Which ordered pair represents the location of the *y*-intercept?

a. $\left(\frac{1}{5}, 0\right)$
b. $(0,5)$
c. $(0,-5)$
d. $(-5,0)$

2. Jack and Isla are graphing linear equations and identifying their slopes. After graphing the equation $x = -2$, they disagree on the slope of the line. Jack thinks the slope is 0, and Isla thinks the slope is undefined. Who is correct?

a. Jack is correct because a vertical line always has a slope of 0.
b. Jack is correct because a horizontal line always has an undefined slope.
c. Isla is correct because the slope is –2.
d. Isla is correct because a vertical line always has an undefined slope.

3. The graph of a linear function passes through points $(3, -20)$ and $(5, 8)$. What is the slope of this line?

a. 6
b. -6
c. 14
d. $\frac{1}{14}$

4. Jake is visiting Washington, D.C. and needs to rent a taxi to get from the airport to his hotel. The taxi company charges a flat fee of $3.50 for using the taxi, as well as $2.16 per mile traveled. Given this information, determine the direction in which the line would slope if this relationship were graphed on a coordinate plane with miles traveled on the x-axis and total charge on the y-axis.

a. Downward
b. Upward
c. Zero
d. Undefined

5. What is the slope-intercept equation of a line that passes through the point $(2, 1)$ and has a slope of $m = 3$?

a. $y = 2x + 3$
b. $y = 3x + 2$
c. $-3x + y = -5$
d. $y = 3x - 5$

6. Monica's toy store is advertising for an upcoming sale. It costs \$0.50 for each flyer printed and \$30 for ad space in a local newspaper. Which graph shows the total advertising cost (y) in relation to the number of flyers printed (x)?

a.

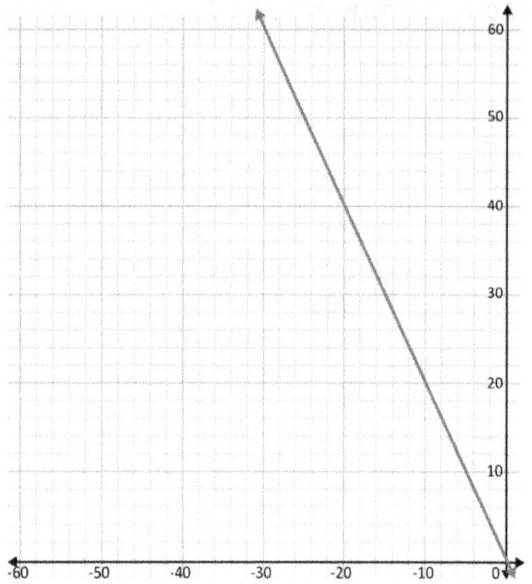

c.

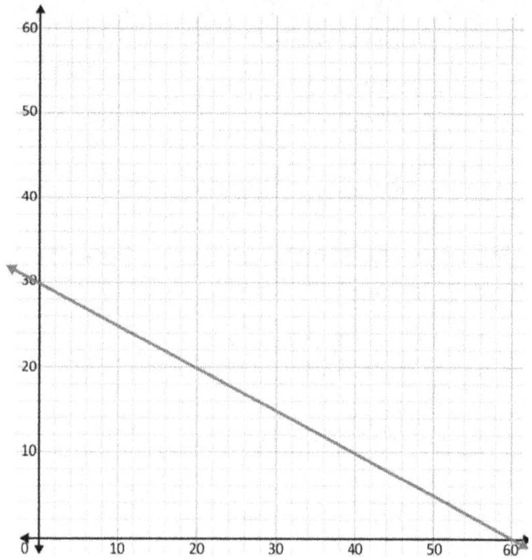

b.

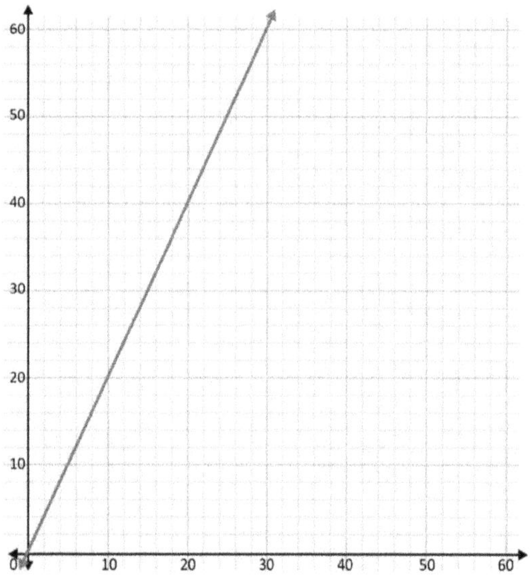

d.

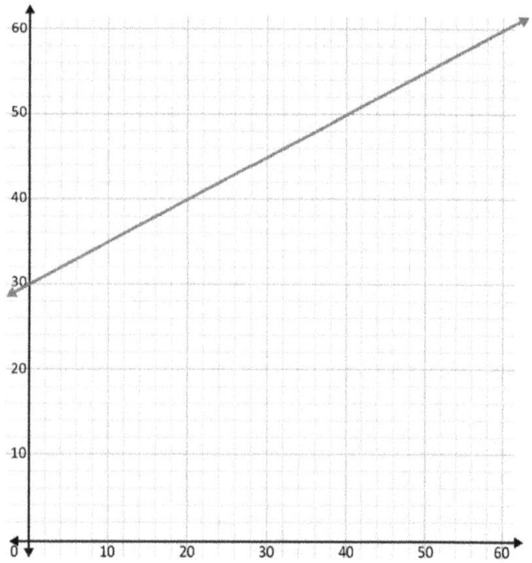

7. Write the equation of a linear function in point-slope form that has a slope of $m = 4$ and contains the point $(3, 17)$.

a. $y - 17 = 4(x - 3)$
b. $y - 3 = 4(x - 17)$
c. $y = 4x + 5$
d. $-4x + y = 5$

8. Justin spends an average of \$3 each day on snacks. After 8 days he has \$86 left. There is a linear relationship between the amount of money Justin has left and the number of days that have passed. Write an equation for this linear function in point-slope form.

a. $y - 8 = 3(x - 86)$
b. $y - 86 = 3(x - 8)$
c. $y - 86 = -3(x - 8)$
d. $y - 8 = -3(x - 86)$

9. What is the equation of a linear function in standard form that has a slope of $m = 5$ and contains the point $(2, 4)$?

a. $5x - y = 6$
b. $5x - y = -6$
c. $-5x + y = 6$
d. $5x - y = 14$

10. Glenville Middle School hosts an annual math-a-thon. During the first year, 35 students participated. During the third year, 57 students participated. There is a linear relationship between the years and the number of students who participated in the math-a-thon. The graph of this linear function passes through points $(1, 35)$ and $(3, 57)$. Write an equation in standard form that represents this linear function.

a. $11x - y = 24$
b. $24x - 11y = 35$
c. $11x - y = -24$
d. $11x + 24 = y$

11. Write a linear equation in standard form that represents the table of values shown below.

x	1	2	3	4
y	5	7	11	14

a. $3x + y = -2$
b. $2x + y = 3$
c. $2x - y = -3$
d. $y = 2x - 3$

12. Montgomery Township High School's spirit club is selling t-shirts and hoodies to raise money for a senior class trip. T-shirts, x, cost \$10 each, and hoodies, y, cost \$25 each. The spirit club's sponsor asked the club to raise \$1,000. The treasurer of the spirit club wants to know how many t-shirts and how many hoodies the club needs to sell to meet its goal of \$1,000. Write an equation in two variables that represents this scenario.

a. $10x + 25y = 1{,}000$
b. $1{,}000x + 25y = 10$
c. $25x - 10y = 1{,}000$
d. $1{,}000x - 25y = 10$

13. Which equation of a line contains the point $(2, 7)$ and is parallel to the graph of $y = 3x - 4$?

a. $y = 3x - 7$
b. $y = 3x - 1$
c. $y = 3x + 1$
d. $y = \frac{1}{3}x + 1$

14. When graphed on the coordinate plane, a hiking trail has the equation $2x - y = -1$, and a set of train tracks has the equation $2x + y = 4$. Determine whether the hiking trail and the train tracks are parallel to one another. If so, what is their slope?

a. These lines are parallel. The slope of both lines is 2.
b. These lines are not parallel because their slopes are different.
c. These lines are parallel. The slope of both lines is –2.
d. These lines are not parallel because the slope of the line $2x + y = 4$ is undefined.

15. Which equation of a line contains the point $(7, 10)$ and is perpendicular to the graph of $y = \frac{1}{2}x - 9$?

a. $y = 2x - 4$
b. $y = 2x + 24$
c. $y = -2x - 4$
d. $y = -2x + 24$

16. When graphed on a coordinate plane, Point Breeze Avenue has the equation $x + 3y = 3$. Wharton Street runs perpendicular to Point Breeze Avenue and passes through the point $(4, 1)$. What is the equation of the line for Wharton Street?

a. $y = \frac{1}{3}x - 11$
b. $y = -\frac{1}{3}x + 13$
c. $y = 3x - 11$
d. $y = 3x + 13$

17. Which equation of a line is parallel to the x-axis, and what is its slope?

a. $y = 2x; m = 0$
b. $y = 2; m = 0$
c. $x = 2y; m = \text{undefined}$
d. $x = 2; m = \text{undefined}$

18. Given the data below, which statement best describes the correlation coefficient, rounded to the nearest hundredth, of the daily high temperatures and the number of cups of lemonade sold?

High Temperature (Fahrenheit)	Cups of Lemonade Sold
54	338
82	784
57	330
80	740
65	411
75	615
65	450
72	453

a. $r \approx -0.83$, there is a strong negative linear association between the two variables.
b. $r \approx 0.94$, there is a strong positive linear association between the two variables.
c. $r \approx 0.36$, there is a weak positive linear association between the two variables.
d. $r = 0$, there is no association between the two variables.

19. Change the linear equation $6x - y = 8$ from standard form to slope-intercept form.

a. $y = 6x + 8$
b. $y = 6x - 8$
c. $y = -6x - 8$
d. $y = -6x + 8$

20. Which of the following statements describes a relationship that shows association but not causation?

a. The number of hours worked and the amount of money earned in a paycheck.
b. The number of rain clouds in the sky and the amount of total rainfall in a region.
c. The number of bottles of sunscreen sold and the number of popsicles sold.
d. The number of additional treats a cat consumes and its weight gain.

Chapter 4: Inequalities

4.1 Inequality Basics

In algebra and other upper-level fields of math, you will commonly find yourself working with mathematical expressions that do not equal each other. The statement comparing such expressions with symbols such as $<$ (less than) or $>$ (greater than) is called an **inequality**. An example of an inequality is $7x > 5$. To solve for x, divide both sides by 7 and the solution is shown to be $x > \frac{5}{7}$. Graphs of the solution set of inequalities are represented on a number line. Open circles are used to show that an expression approaches a number but is never quite equal to that number.

Conditional inequalities are those with certain values for the variable that will make the condition true and other values for the variable where the condition will be false. **Absolute inequalities** can have any real number as the value for the variable to make the condition true, while there is no real number value for the variable that will make the condition false. Solving inequalities is done by following the same rules for solving equations with the exception that when multiplying or dividing by a negative number the direction of the inequality sign must be flipped or reversed. **Compound inequalities** are situations where two inequality statements apply to the same variable expression. Example: $-c < ax + b < c$.

Graphing Simple Inequalities

To graph a simple inequality, we first mark on the number line the value that signifies the end point of the inequality. If the inequality is strict (involves a less than or greater than symbol), we use an open circle; if it is not strict (less than or equal to or greater than or equal to), we use a solid circle. We then fill in the part of the number line that satisfies the inequality: to the left of the marked point for less than (or less than or equal to), to the right for greater than (or greater than or equal to).

For example, we would graph the inequality $x < 5$ by putting an open circle at 5 and filling in the part of the line to the left.

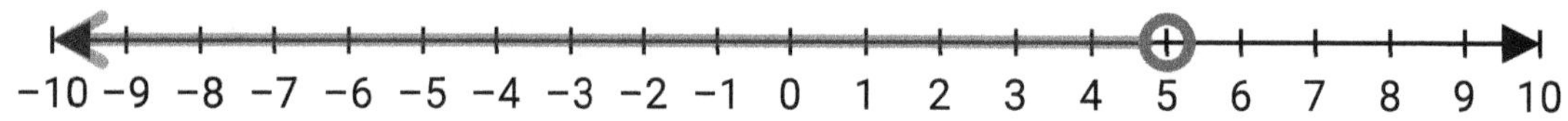

Review Video: Inequalities
Visit mometrix.com/academy and enter code: 347842

Examples

Example 1:

Convert the inequality to word form: $x \leq 7$.

Answer:

$x \leq 7$ is "x is less than or equal to seven"

EXAMPLE 2:

Graph $x < 10$ on a number line.

Answer:

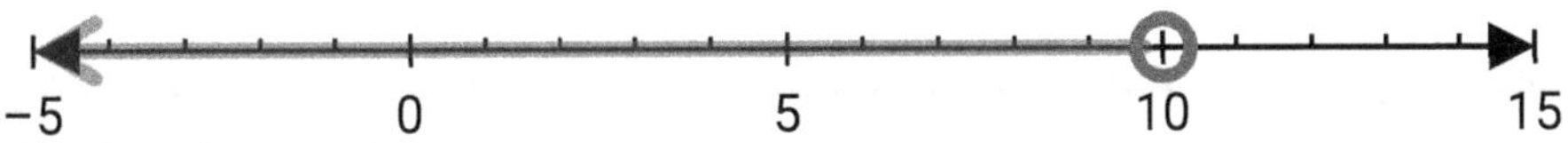

- < and > have an "open circle."
- ≤ and ≥ have a "closed circle/dot."
- The arrow extends to the left when less than or less than or equal to.
- The arrow extends to the right when greater than or greater than or equal to.

PRACTICE QUESTIONS

1. Which inequality symbol means less than?

a. $>$
b. $\leq$
c. $<$
d. $\geq$

2. Translate the phrase below into an algebraic inequality:

x is less than or equal to 4

a. $x \geq 4$
b. $x < 4$
c. $x > 4$
d. $x \leq 4$

3. Translate the algebraic inequality below into a phrase:

$x \geq 10$

a. x is greater than 10
b. x is less than 10
c. x is greater than or equal to 10
d. x is less than or equal to 10

4. What is the meaning of the inequality symbol $>$?

a. Less than
b. Greater than
c. Less than or equal to
d. Greater than or equal to

5. Which number line shows the solution set for $x > 8$?

a. 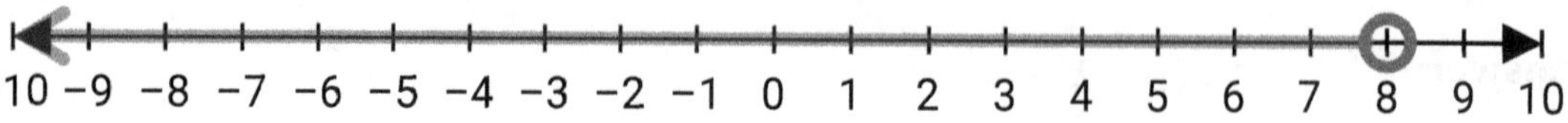

b. 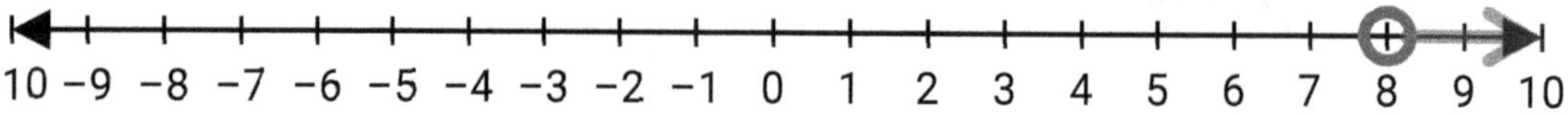

c. 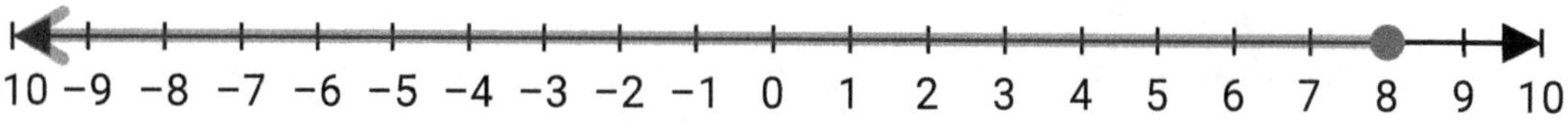

d. 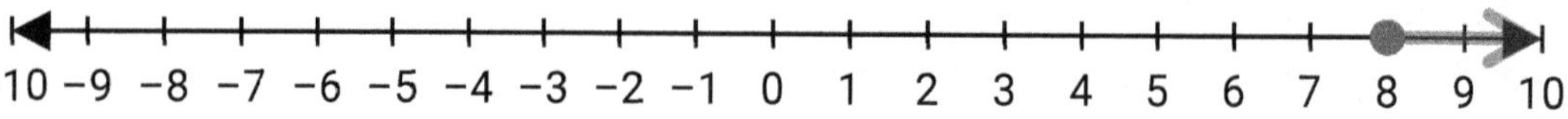

6. Which inequality statement matches the solution set graphed on the number line below?

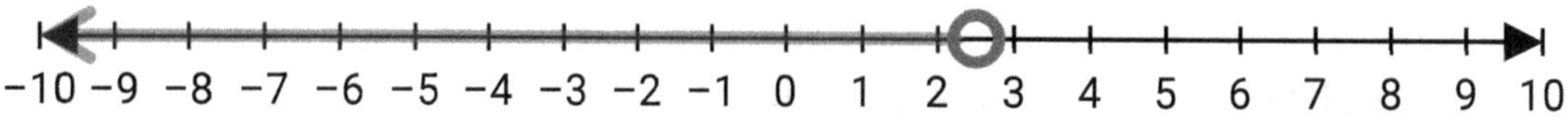

a. $x < 2.5$
b. $x > 2.5$
c. $x \leq 2.5$
d. $x \geq 2.5$

7. Which inequality symbol means greater than or equal to?

a. $>$
b. $<$
c. $\geq$
d. $\leq$

8. Translate the phrase below into an algebraic inequality:

x is less than 3

a. $x \geq 3$
b. $x \leq 3$
c. $x > 3$
d. $x < 3$

9. Which number line shows the solution set for $x \leq$?

a. 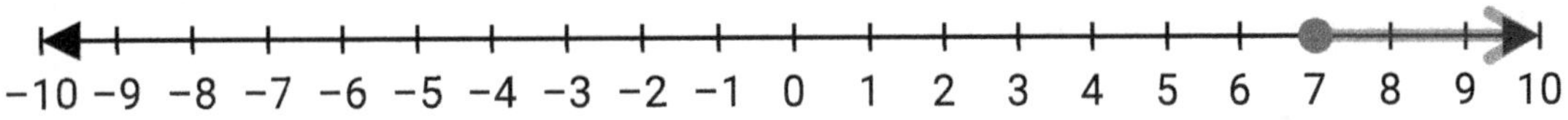

b. 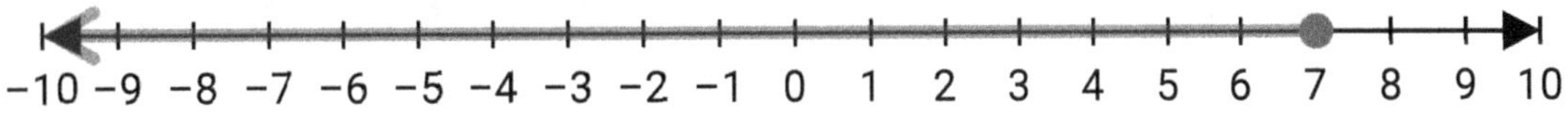

c. 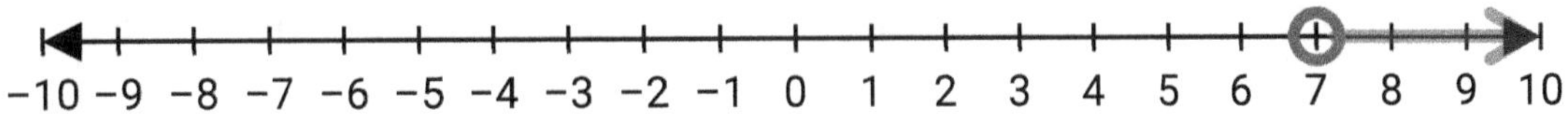

d. 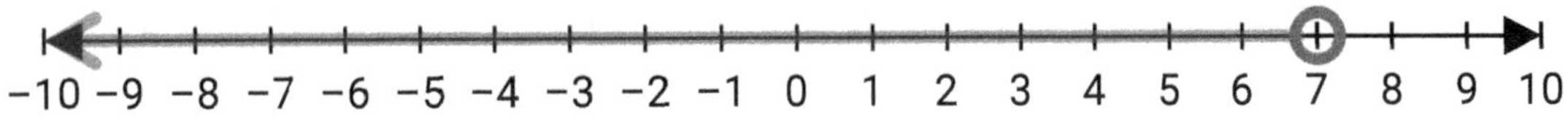

10. What is the meaning of the inequality symbol $\leq$?

a. Greater than
b. Less than
c. Greater than or equal to
d. Less than or equal to

11. Translate the algebraic inequality below into a phrase:

$$x \geq -2$$

a. x is greater than or equal to –2
b. x is greater than –2
c. x is less than or equal to –2
d. x is less than –2

12. Which inequality statement matches the solution set graphed on the number line below?

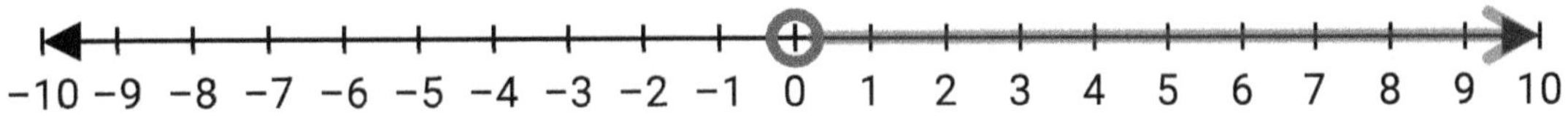

a. $x < 0$
b. $x > 0$
c. $x \leq 0$
d. $x \geq 0$

13. Jane is analyzing her finances. She wants to have at least \$200 in her savings account, and she makes a note of this by writing an inequality where x represents the amount of money in her savings account. Which inequality best represents this statement?

a. $x \leq 200$
b. $x \geq 200$
c. $x < 200$
d. $x > 200$

14. Breanna wants to go to an arcade with her friends. Breanna's mom tells her that she can go if she agrees to spend no more than 20 dollars. Which number line best represents the solution set for this scenario?

a. 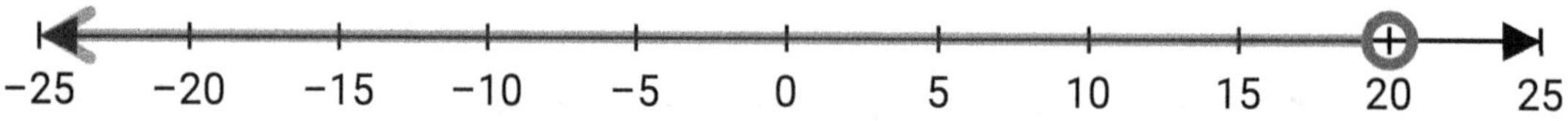

b. 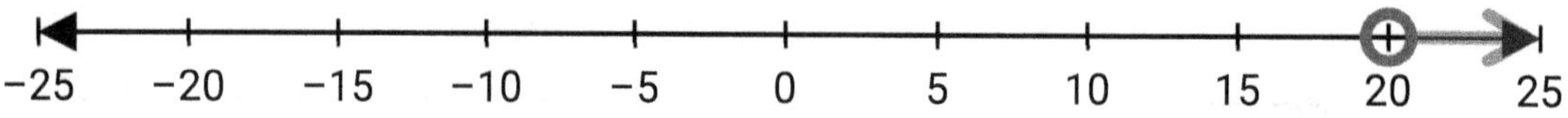

c. 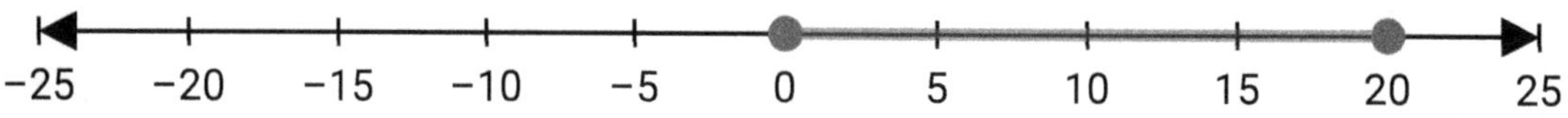

d. 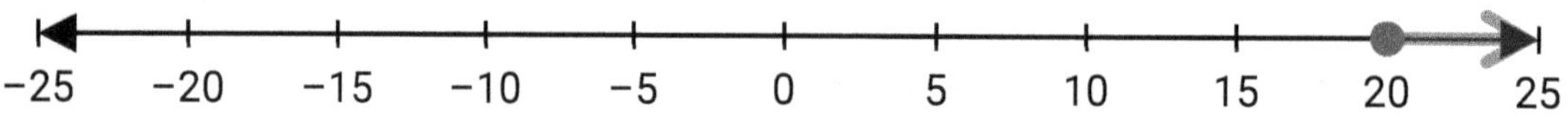

15. Cathy is looking at the weather forecast for the upcoming week. Saturday's average temperature will be colder than 65 °F. Which inequality best represents Saturday's temperature, *x*?

a. $x \leq 65$
b. $x \geq 65$
c. $x < 65$
d. $x > 65$

4.2 Solving One-Step Inequalities

Inequalities are used in many real–world situations. For example, if you have \$20, you can purchase a pair of shoes that are *less than or equal to* \$20. This can be expressed mathematically as $x \leq 20$. Or perhaps you are wondering if you are tall enough to ride the new rollercoaster at your local amusement park. The sign posted on the ride states that you must be at least 48 inches tall. This means that you can only enter the ride if your height is *greater than or equal to* 48 inches. This can be expressed as $h \geq 48$.

Solving inequalities is very similar to solving equations. Just like equations, the goal of solving inequalities is to isolate the variable. This is achieved by using inverse operations.

For example, consider the four inequalities below.

Example 1:

$$x - 3 \leq 9$$

The inverse of subtraction is addition, so add 3 to both sides.

$$x - 3 + 3 \leq 9 + 3$$
$$x \leq 12$$

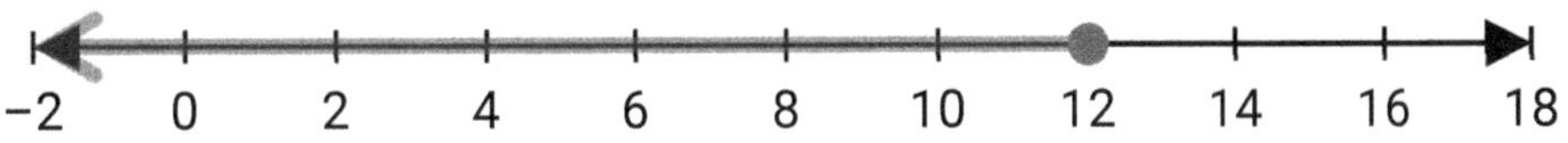

Example 2:

$$4 \geq x + 1$$

The inverse of addition is subtraction, so subtract 1 from both sides.

$$4 - 1 \geq x + 1 - 1$$
$$3 \geq x$$

The inequality sign can be flipped if the positions of the values on either side of the inequality sign are flipped as well. This is the proper way to write inequalities because it is often easier to interpret "x is less than or equal to 3" than "3 is greater than or equal to x."

$$x \leq 3$$

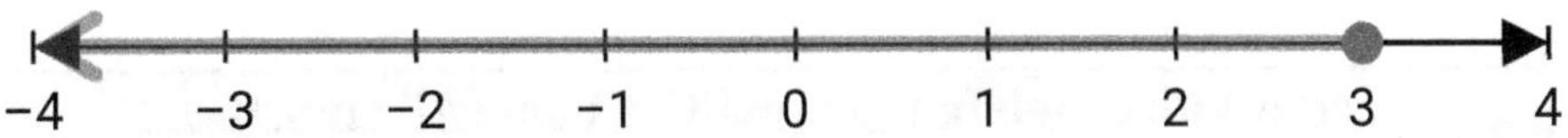

Example 3:

$$2x < 10$$

The inverse of multiplication is division, so divide both sides by 2.

$$\frac{2x}{2} < \frac{10}{2}$$
$$x < 5$$

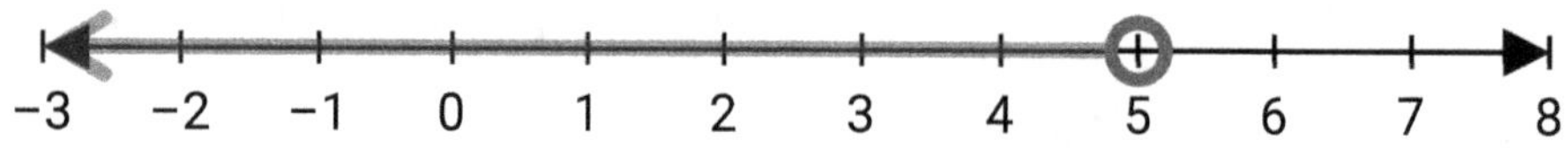

Example 4:

$$5 > \frac{x}{4}$$

The inverse of division is multiplication, so multiply both sides by 4.

$$5 \times 4 > \frac{x}{4} \times 4$$
$$20 > x$$

Reverse the inequality to write it in a more proper form.

$$x < 20$$

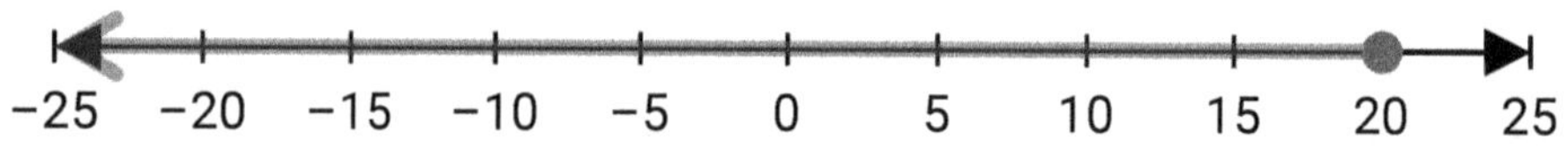

An important rule to remember:

When you multiply or divide by a negative, you need to *reverse* the inequality sign.

For example, to solve the inequality $-6x < 30$, you need to divide both sides by -6. Dividing by a negative requires the sign to be reversed. $\frac{-6x}{-6} < \frac{30}{-6}$ becomes $x > -5$.

Similarly, the inequality $\frac{x}{-5} > -4$ is solved by multiplying both sides by -5. Multiplying by a negative requires the sign to be reversed. $\frac{x}{-5} \cdot (-5) > -4 \cdot (-5)$ becomes $x < 20$.

Review Video: Solving Inequalities Using All Four Basic Operations
Visit mometrix.com/academy and enter code: 401111

EXAMPLES

EXAMPLE 1:

Solve the inequality for m, and graph the solution on a number line.

$$m - 3 < 5$$

Answer:

Solve the inequality as though it is a one-step equation. Isolate m by adding 3 to both sides.

$$m - 3 + 3 < 5 + 3$$
$$m < 8$$

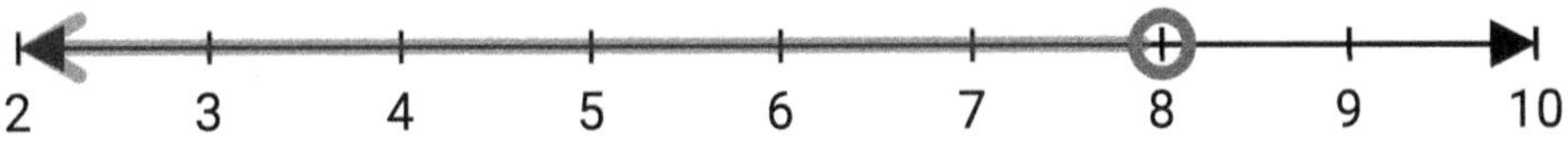

EXAMPLE 2:

Solve the inequality for y, and graph the solution on a number line.

$$9 + y \geq -2$$

Answer:

Solve the inequality as though it is a one-step equation. Isolate y by subtracting 9 from both sides.

$$9 + y - 9 \geq -2 - 9$$
$$y \geq -11$$

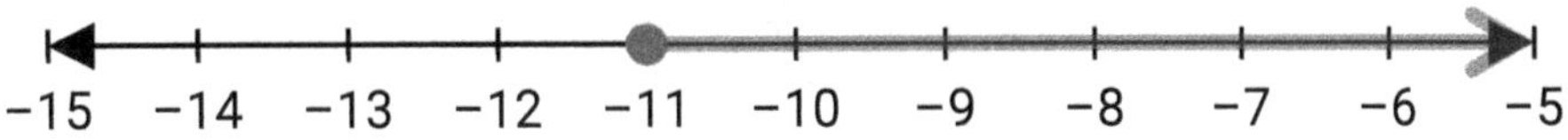

EXAMPLE 3:

Solve the inequality for z, and graph the solution on a number line.

$$\frac{z}{9} \leq 19$$

Answer:

Solve the inequality as though it is a one-step equation. Isolate z by multiplying both sides by 9.

$$\frac{z}{9} \cdot 9 \leq 19 \cdot 9$$
$$x \leq 171$$

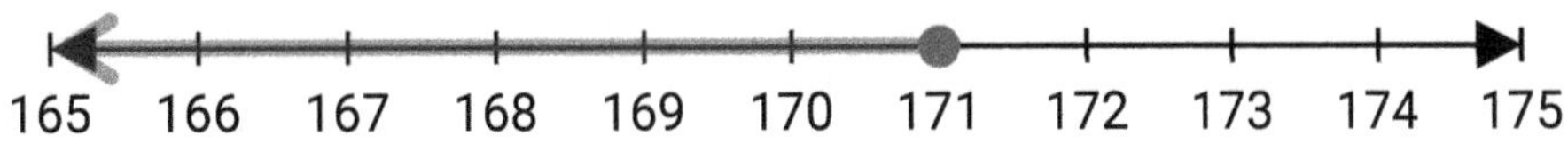

EXAMPLE 4:

Solve the inequality for n, and graph the solution on a number line.

$$-7n > 42$$

Answer:

Solve the inequality as though it is a one-step equation. Isolate n by dividing both sides by –7.

$$\frac{-7n}{-7} > \frac{42}{-7}$$

Remember, when you multiply or divide by a negative, you need to *reverse* the inequality sign.

$$n < -6$$

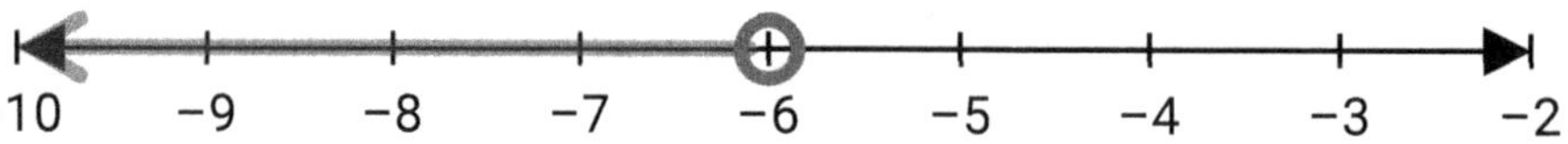

PRACTICE QUESTIONS

1. Solve the inequality:

$$x + 8 < 12$$

a. $x < 4$
b. $x < 20$
c. $x > 4$
d. $x > -4$

2. Solve the inequality:

$$x - 15 > 3$$

a. $x > 18$
b. $x > 12$
c. $x = 18$
d. $x > -18$

3. Solve the inequality:

$$5x \leq 70$$

a. $x \leq 65$
b. $x \leq 350$
c. $x \geq -14$
d. $x \leq 14$

4. Solve the inequality:

$$\frac{x}{3} \geq 6$$

a. $x \leq -18$
b. $x \geq 9$
c. $x \geq 2$
d. $x \geq 18$

5. Solve the inequality:

$$x - 2 < 12.5$$

a. $x < 10.5$
b. $x < 14.5$
c. $x \leq 14.5$
d. $x < 25$

6. Solve and graph the inequality:

$$7x > 133$$

a. $x < -19$

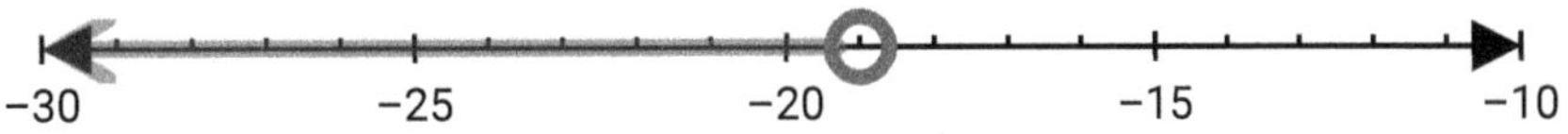

b. $x > 140$

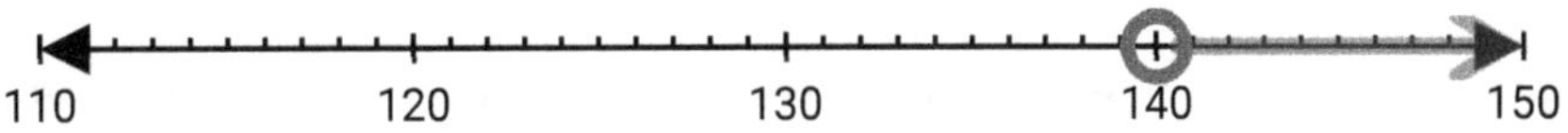

c. $x > 19$

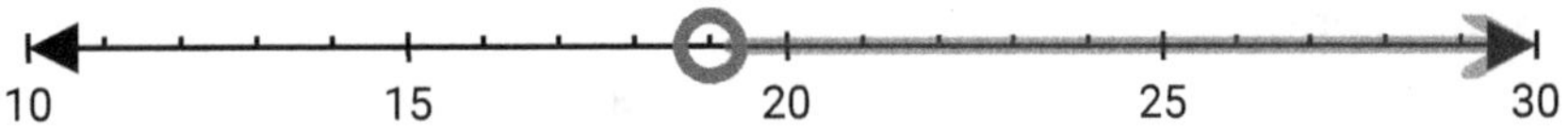

d. $x > 125$

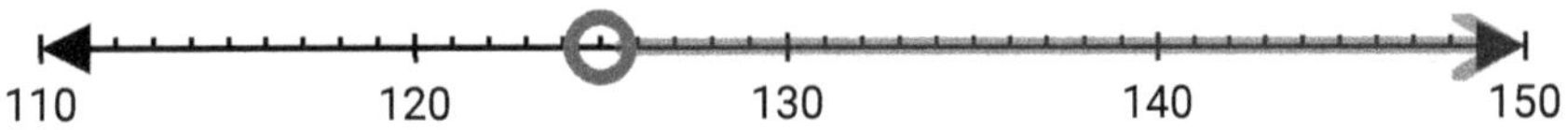

7. Solve the inequality:

$$\frac{x}{12} \leq 2$$

a. $x \leq 6$
b. $x \leq 24$
c. $x \leq 14$
d. $x \leq 10$

8. Solve and graph the inequality:

$$x + 9.25 \geq 21.25$$

a. $x \leq 30.5$

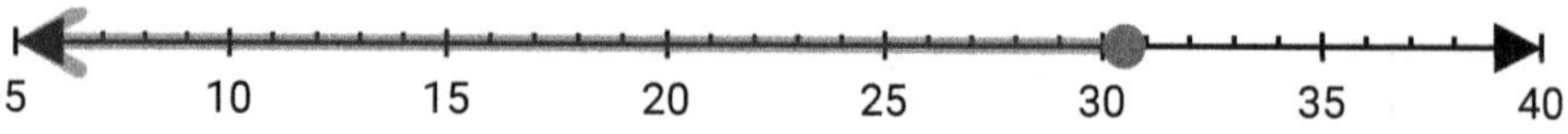

b. $x \leq 12$

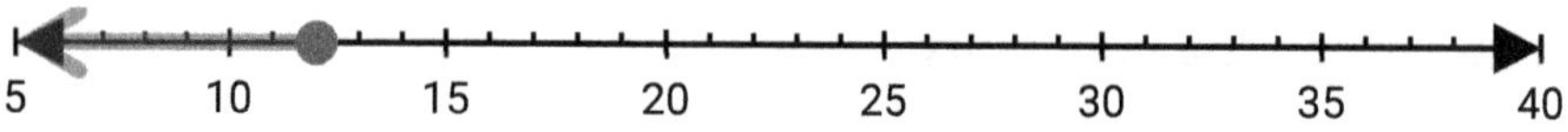

c. $x \geq 30.5$

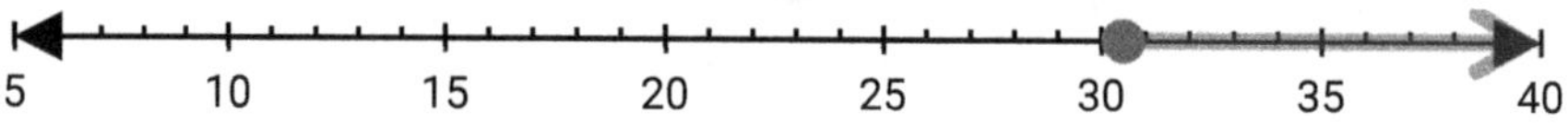

d. $x \geq 12$

9. Solve the inequality:

$$-4x < 52$$

a. $x \geq -208$
b. $x < -208$
c. $x > -13$
d. $x < -13$

10. Solve the inequality:

$$\frac{x}{18} > 11$$

a. $x > 29$
b. $x > -7$
c. $x > 198$
d. $x > \frac{11}{18}$

11. Solve and graph the inequality:

$$x + 54 \leq -3$$

a. $x \geq 57$

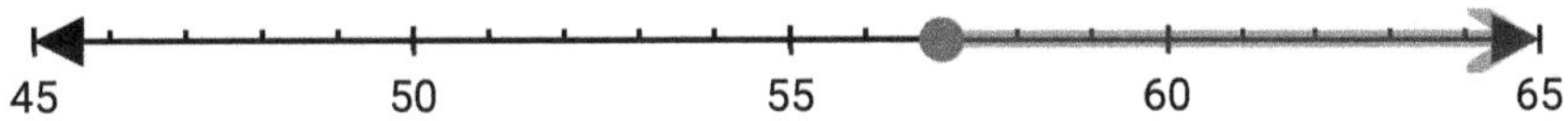

b. $x \leq -57$

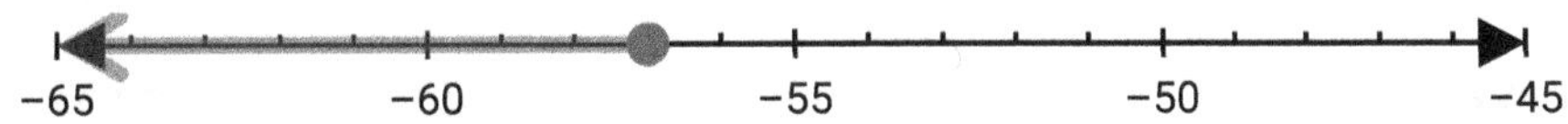

c. $x \leq 51$

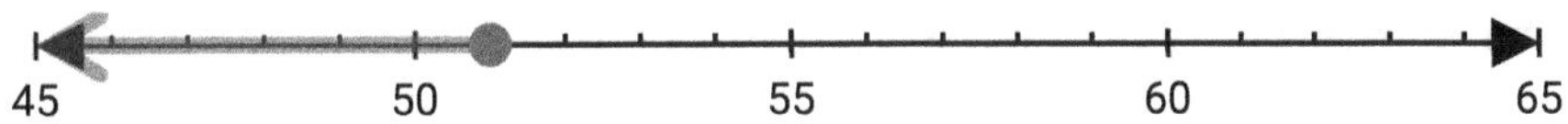

d. $x \geq -51$

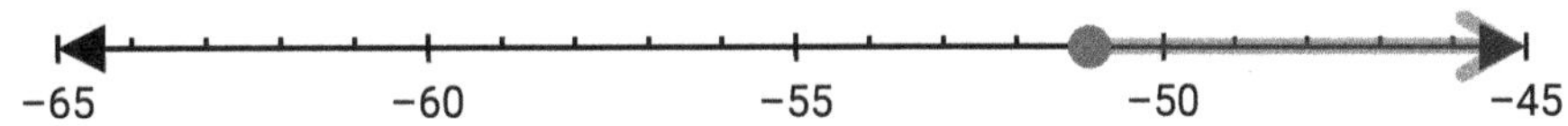

12. Solve the inequality:

$$x - 15 \geq 22$$

a. $x < 7$
b. $x < 37$
c. $x \geq 7$
d. $x \geq 37$

13. Stephanie baked a batch of cookies for her friends. She split the cookies into three equal groups and noted that each friend will receive more than a dozen cookies. Based on this information, choose the inequality statement and graph that represent the total number of cookies Stephanie could have baked.

a. $x \geq 36$

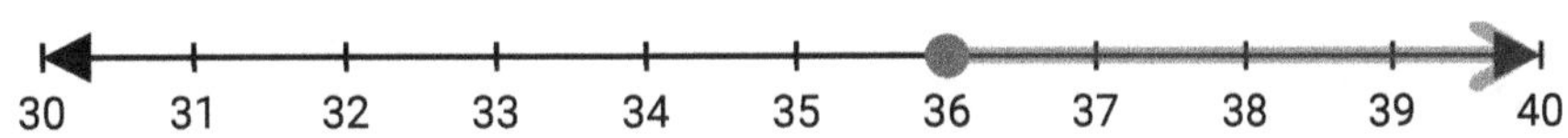

b. $x > 36$

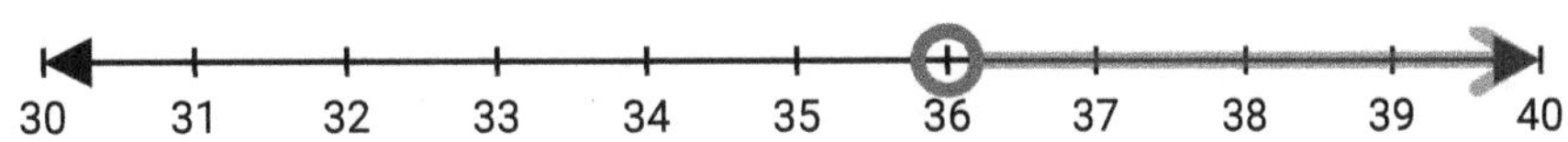

c. $x < 36$

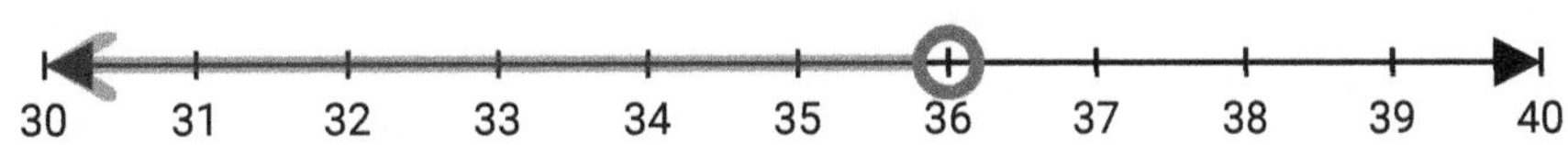

d. $x \leq 36$

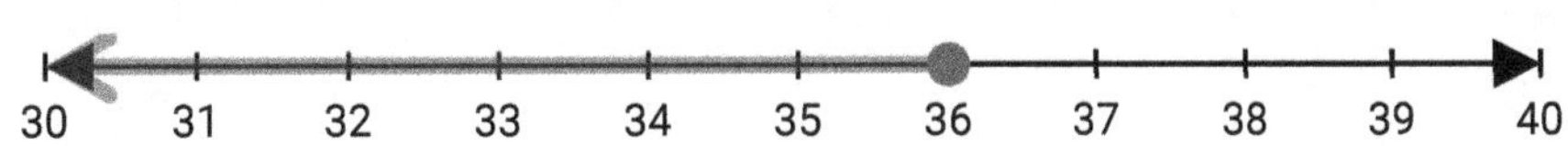

14. Alex's lemonade stand must sell more than 45 cups of lemonade to make a profit. He has already sold 30 cups. Based on this information, which inequality best represents the number of cups of lemonade, x, Alex still needs to sell?

a. $x > 15$
b. $x < 15$
c. $x \geq 15$
d. $x \leq 15$

15. Amaya went to the gas station and filled up her car's 12-gallon gas tank. Her gas tank uses 4 gallons of gas per hour when driving on the interstate. Based on this information, which inequality statement and graph best represents the number of hours, x, Amaya can drive on the interstate before she runs out of gas.

a. $x \geq 3$ 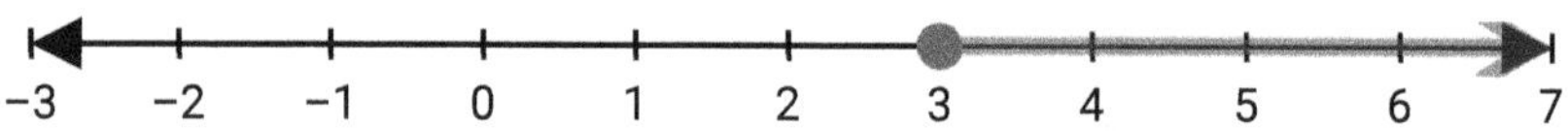

b. $0 \leq x \leq 3$ 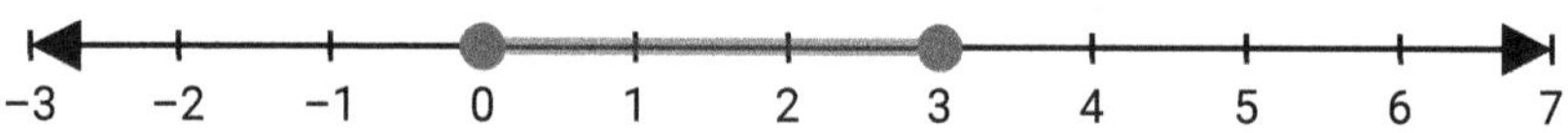

c. $x > 3$ 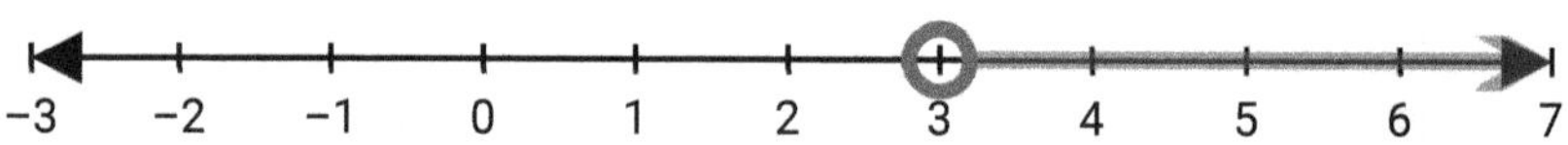

d. $x < 3$ 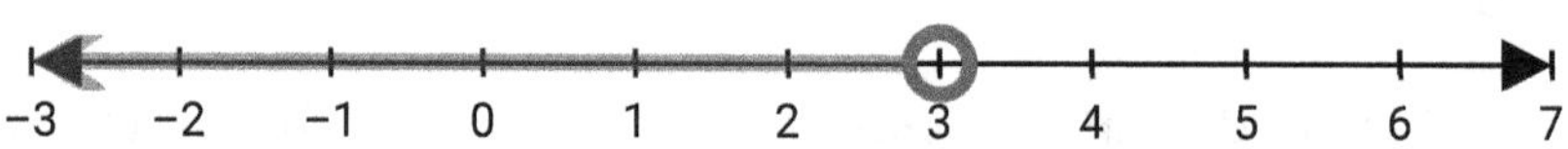

4.3 Solving Multi-Step Inequalities

Solving multi-step inequalities is very similar to solving multi-step equations. The objective is to isolate the variable, and this is achieved by using inverse operations.

Multi-step inequalities can be used to describe many real-world situations. For example, suppose you go to a restaurant and want to buy a few pizzas and one salad. Pizzas cost \$20 and a salad costs \$15. If you have \$150 to spend, the inequality $20p + 15 \leq 150$ states that the amount of money you spend on pizzas and salad needs to be less than or equal to \$150. If you solve for p, you will be able to see the maximum number of pizzas you can afford to buy.

$$20p + 15 - 15 \leq 150 - 15$$
$$20p \leq 135$$
$$\frac{20p}{20} \leq \frac{135}{20}$$
$$p \leq 6.75$$

In this case, $p \leq 6.75$, which means the number of pizzas you order needs to be less than or equal to 6.75 pizzas, so realistically you can afford to buy 6 pizzas and one salad.

Some multi-step inequalities will require you to use the distributive property. Again, treat the inequalities as though they are simply equations. Use inverse operations to isolate the variable. Consider the inequalities below.

$$5x + 2 > 4(x + 1)$$

Start by distributing the 4 on the right side.

$$5x + 2 > 4x + 4$$

Collect variables on one side and constants on the other side. Start by subtracting 2 from both sides.

$$5x + 2 - 2 > 4x + 4 - 2$$
$$5x > 4x + 2$$

Then, subtract $4x$ from both sides.

$$5x - 4x > 4x + 2 - 4x$$
$$x > 2$$

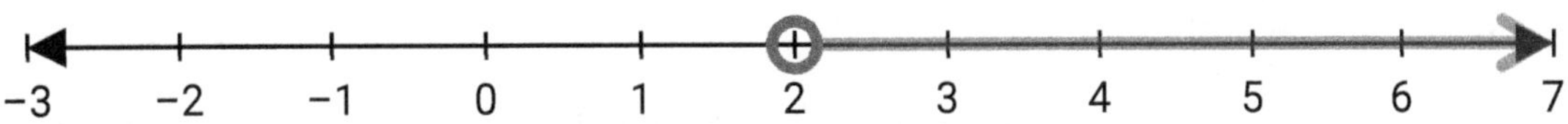

Here's another example.

$$4(x - 6) + 2 \geq x - 4$$

Start by distributing the 4.

$$4x - 24 + 2 \geq x - 4$$

Collect variables on one side and constants on the other side. First, combine like terms on the left side.

$$4x - 22 \geq x - 4$$

Then, add 22 to both sides.

$$4x - 22 + 22 \geq x - 4 + 22$$
$$4x \geq x + 18$$

Next, subtract x from both sides.

$$4x - x \geq x + 18 - x$$
$$3x \geq 18$$

Finally, divide both sides by 3.

$$\frac{3x}{3} \geq \frac{18}{3}$$
$$x \geq 6$$

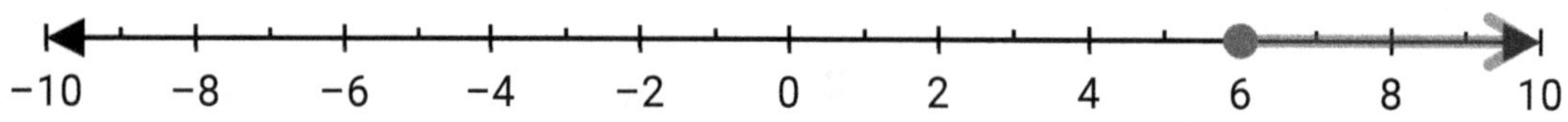

Remember, if you multiply or divide by a negative number, you need to reverse the inequality symbol. Solve the inequality below.

$$-4z - 12 < 2(z + 5)$$

Start by distributing the 2 on the right side.

$$-4z - 12 < 2z + 10$$

Collect variables on one side and constants on the other side. Start by adding 12 to both sides.

$$-4z - 12 + 12 < 2z + 10 + 12$$
$$-4z < 2z + 22$$

Then, subtract $2z$ from both sides.

$$-4z - 2z < 2z + 22 - 2z$$
$$-6z < 22$$

Finally, divide both sides by −6 and flip the sign.

$$\frac{-6z}{-6} < \frac{22}{-6}$$
$$z > -\frac{11}{3}$$

Review Video: Solving Multi-Step Inequalities
Visit mometrix.com/academy and enter code: 347842

EXAMPLES

EXAMPLE 1:

Solve for x.

$$11 + 2x \leq 3(2 - x)$$

Answer:

Treat the inequality as though it is an equation, and work to isolate the variable x. Start by distributing the 3.

$$11 + 2x \leq 6 - 3x$$

Now, collect the x's on one side of the inequality and the constants on the other side. Add $3x$ to both sides.

$$11 + 5x \leq 6$$

Then, subtract 11 from both sides.

$$5x \leq -5$$

Now divide both sides by 5.

$$x \leq -1$$

EXAMPLE 2:

Solve for y.

$$3y - 8 < 5y + 2$$

Answer:

Treat the inequality as though it is an equation, and work to isolate the variable y. Start by collecting the variables on one side and all of the constants to the other side. Subtract $3y$ from both sides.

$$-8 < 2y + 2$$

Then, subtract 2 from both sides.

$$-10 < 2y$$

Now divide both sides by 2.

$$-5 < y$$

If the entire inequality is flipped, the inequality sign can be flipped to put it in a form that is easier to read.

$$y > -5$$

Example 3:

Three more than four times a real number, x, is at least as large as six times the number minus nine. What is are the solutions for x?

Answer:

First, come up with an inequality for the question. Three more than four times a real number, x, can be represented by the expression $4x + 3$, and six times the number minus nine can be represented by $6x - 9$. Since $4x + 3$ is at least as large as $6x - 9$, that means it is greater than or equal to $6x - 9$.

$$4x + 3 \geq 6x - 9$$

From here, solve for x. Start by subtracting $6x$ from both sides.

$$-2x + 3 \geq -9$$

Then, subtract 3 from both sides.

$$-2x \geq -12$$

Finally, divide both sides by –2. Remember to flip the inequality symbol when dividing by a negative number.

$$x \leq 6$$

The real number x must be less than or equal to 6.

Practice Questions

1. Solve the inequality: $9x + 15 \leq -12$.

a. $x \leq -3$
b. $x \geq 3$
c. $x \leq \frac{1}{3}$
d. $x \leq -\frac{1}{3}$

2. Solve the inequality: $5x + 6 > 14 + 3x$.

a. $x < -4$
b. $x > 4$
c. $x < -8$
d. $x > 3$

3. Solve the inequality: $3 - 2x < 13 - 7x$.

a. $x < 2$
b. $x > -2$
c. $x > -10$
d. $x < 1$

4. Solve the inequality: $2(x - 5) \geq 6x + 18$.

a. $x \geq -\frac{28}{5}$
b. $x \leq -\frac{13}{4}$
c. $x \geq 7$
d. $x \leq -7$

5. Solve the inequality: $x - 18 < -4(x - 3)$.

a. $x > 15$
b. $x < 3$
c. $x < 6$
d. $x > -6$

6. Solve the inequality: $15 < -6(x + 1) - x$.

a. $x > \frac{21}{2}$
b. $x < -\frac{7}{3}$
c. $x < -3$
d. $x > 3$

7. Solve the inequality: $6x + 4(2x - 1) > 2 + 3(x + 9)$.

a. $x < -3$
b. $x > 1$
c. $x < -4$
d. $x > 3$

8. Graph the solution to the inequality: $7x - 8 > 2x + 17$.

a.
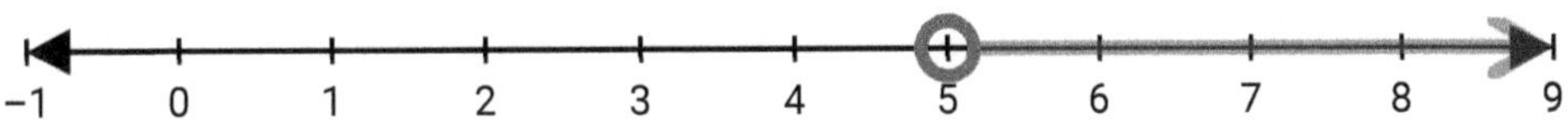

b.
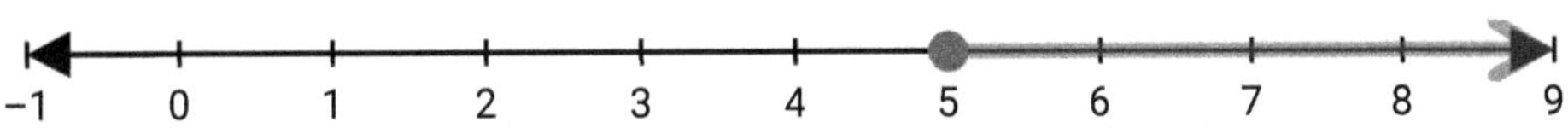

c.
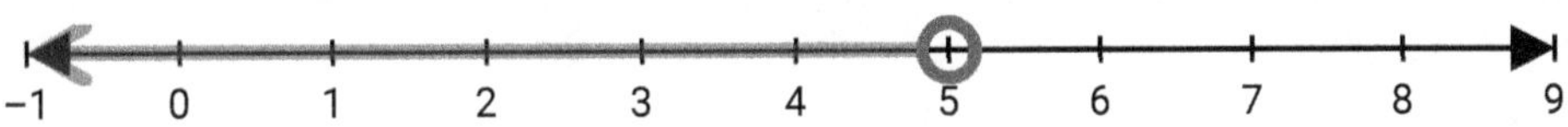

d.
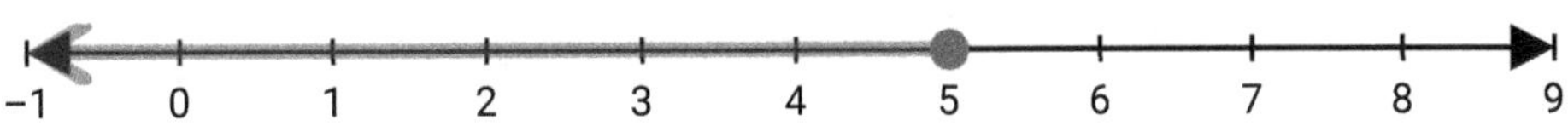

9. Graph the solution to the inequality: $4 - 3(x + 5) \leq 4x + 3$.

a. 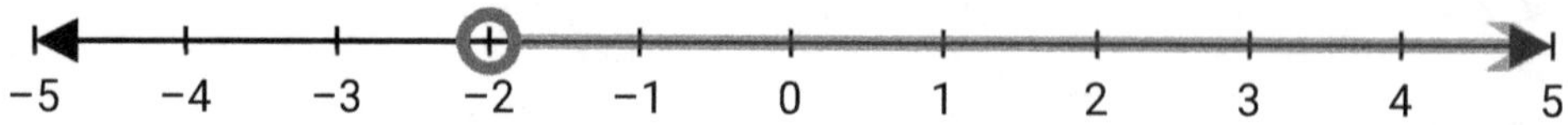

b. 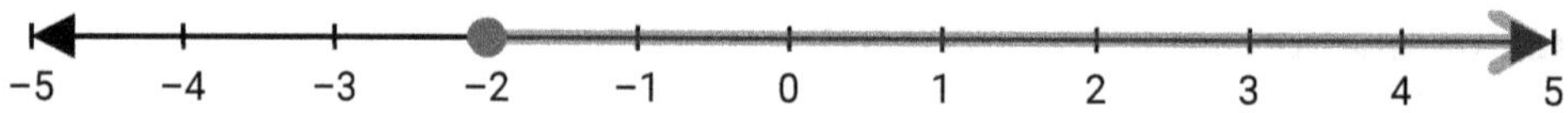

c. 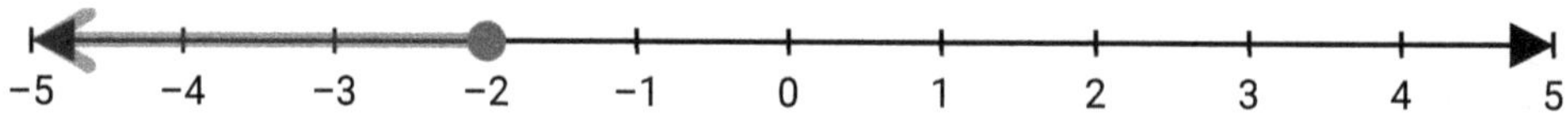

d. 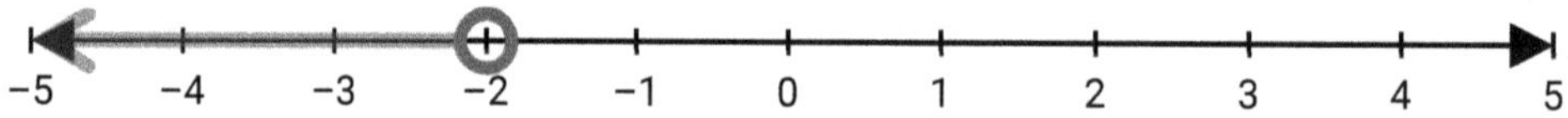

10. Graph the solution to the inequality: $14 - 2x < -2(5x + 9)$.

a. 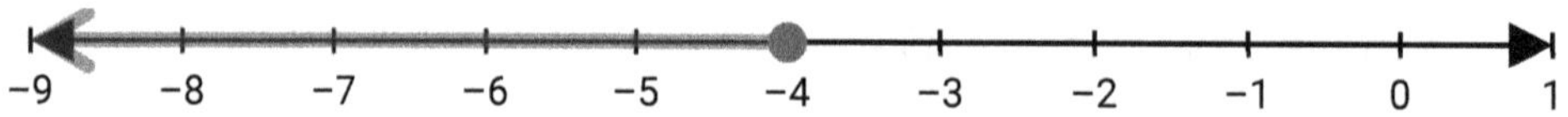

b. 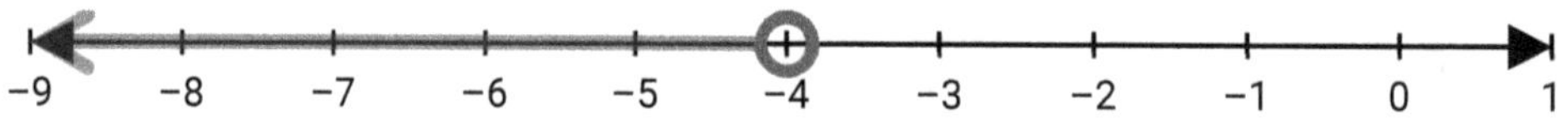

c. 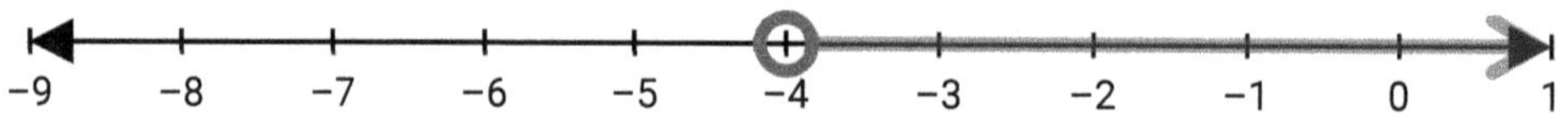

d. 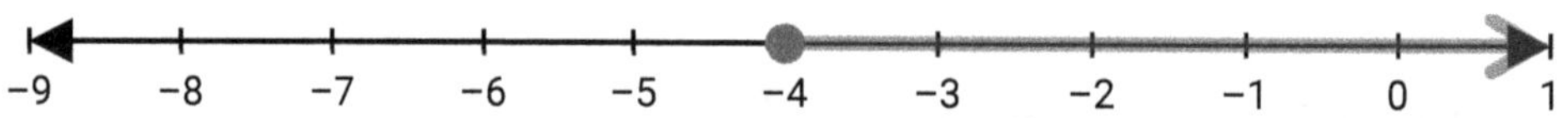

11. Graph the solution to the inequality: $2(x-3) \geq 4x+8$.

a. 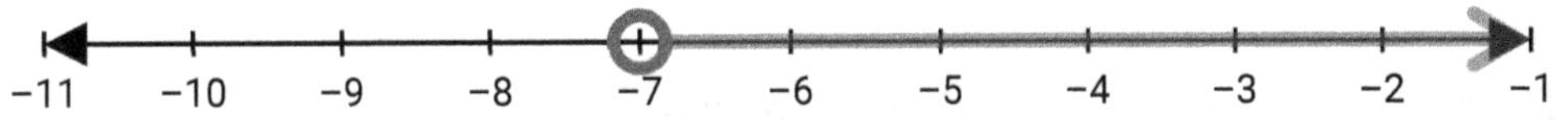

b. 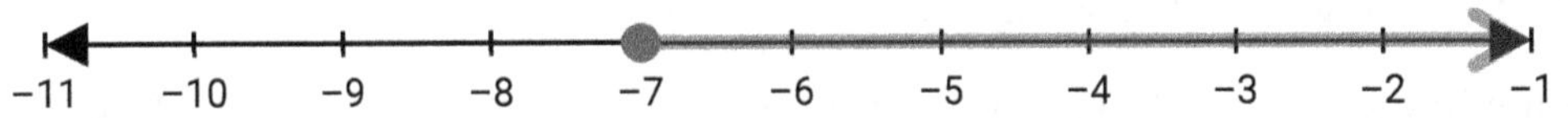

c. 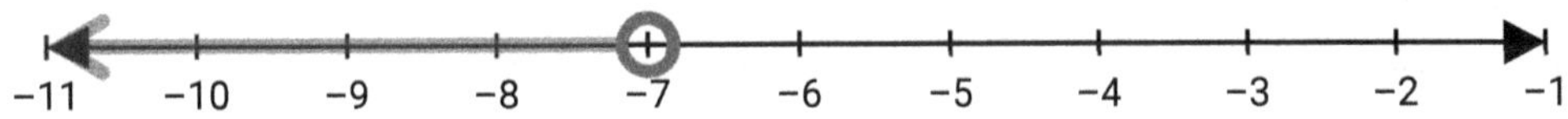

d. 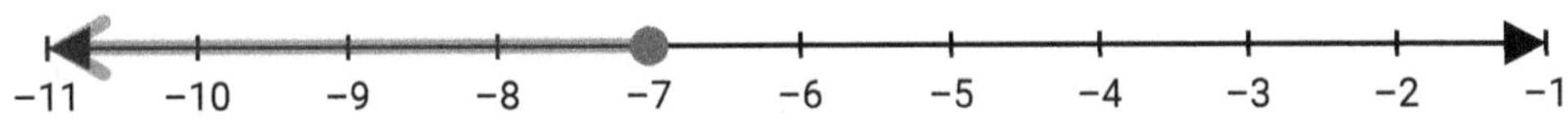

12. Solve and graph the following inequality: $2x-1<7$.

a. 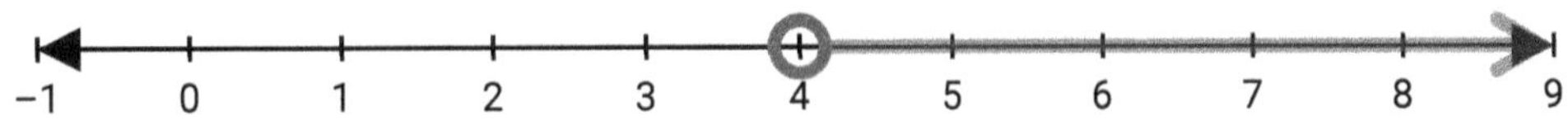

b. 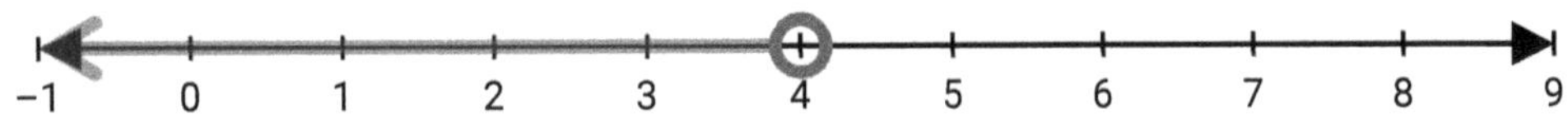

c. 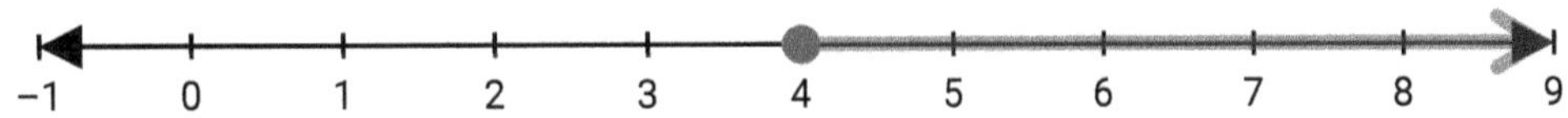

d. 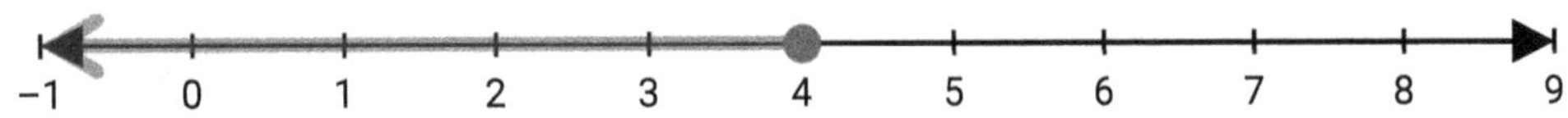

13. The cost, in dollars, for a company to make x number of calculators can be represented by the expression $25x+1,000$. If the cost is not to exceed \$2,500, what is the maximum number of calculators that can be made?

a. 90
b. 100
c. 140
d. 60

14. Twelve more than three times a real number, x, is at least as large as the sum of four times the number and ten. Which of the following statements must be true about the real number?

a. The real number must be less than or equal to 2.
b. The real number must be greater than or equal to 2.
c. The real number must be strictly less than 2.
d. The real number must be strictly greater than 2.

15. The perimeter of a rectangle is less than or equal to 80 meters. If the length of the rectangle is four more than twice its width, what is the maximum the width can be?

a. 10 meters
b. 6 meters
c. 12 meters
d. 14 meters

4.4 Solving Compound Inequalities

A **compound inequality** is an inequality that consists of two inequalities combined with "and" or "or." For instance, "$x + 1 < 2$ or $x + 1 > 3$" is a compound inequality, as is "$2x \geq 4$ and $2x \leq 6$." An *and* inequality is called a **conjunction** and can be written more compactly by having one inequality on each side of the common part: "$2x \geq 1$ and $2x \leq 6$," can also be written as $1 \leq 2x \leq 6$. An *or* inequality is called a **disjunction**.

If the two parts of an *and* inequality overlap, then the inequality has a solution; otherwise, no numbers satisfy the inequality. On the other hand, if the two parts of an *or* inequality overlap, then *all* numbers satisfy the inequality.

Solving a compound inequality requires solving each part separately. For example, given the compound inequality "$x + 1 < 2$ or $x + 1 > 3$," the first inequality, $x + 1 < 2$, is equivalent to $x < 1$, and the second part, $x + 1 > 3$, is equivalent to $x > 2$, so the whole compound inequality can be written as "$x < 1$ or $x > 2$." Similarly, $1 \leq 2x \leq 6$ can be solved by dividing each term by 2, yielding $\frac{1}{2} \leq x \leq 3$.

Graphing Compound Inequalities

To graph a compound inequality, start by plotting the endpoints of each inequality on the number line. For an *or* inequality, we then fill in the appropriate side of the line for each inequality. Typically, the two component inequalities do not overlap, which means the shaded part is *outside* the two points. However, sometimes they do overlap. In these instances, choose the option that is more restrictive. For an *and* inequality, we instead fill in the part of the line that makes both inequalities true.

For the inequality "$x \leq -3$ or $x > 4$," we first put a closed circle at -3 and an open circle at 4. We then fill the parts of the line outside these circles since the solution set doesn't overlap.

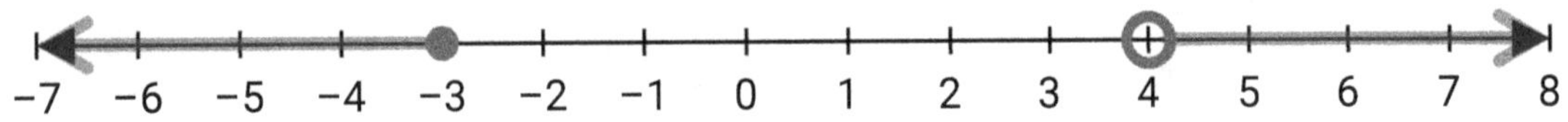

Review Video: Compound Inequalities
Visit mometrix.com/academy and enter code: 786318

Examples

Example 1:

Solve the compound inequality for x.

$$-4 < 2x - 8 \leq 10$$

Answer:

$$-4 < 2x - 8 \leq 10$$

Add 8 to all three parts of the inequality.

$$-4 + 8 < 2x - 8 + 8 \leq 10 + 8$$
$$4 < 2x \leq 18$$

Divide each of the three parts by 2.

$$\frac{4}{2} < \frac{2x}{2} \leq \frac{18}{2}$$
$$2 < x \leq 9$$

Compound inequalities are sometimes called *and* inequalities. This *and* inequality encompasses solutions for x that land between 2 and 9 (including 9). On a graph, this looks like a line connecting 2 and 9, with an open circle on 2 and a closed circle on 9.

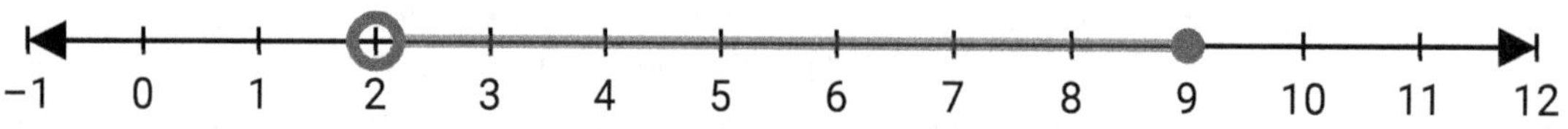

EXAMPLE 2:

Solve the compound inequality for x.

$$3x - 4 < -13 \text{ or } -2x - 6 \leq -4$$

Answer:

Solve each inequality separately:

$$3x - 4 < -13$$
$$3x - 4 + 4 < -13 + 4$$
$$3x < -9$$
$$\frac{3x}{3} < -\frac{9}{3}$$
$$x < -3$$

$$-2x - 6 \leq -4$$
$$-2x - 6 + 6 \leq -4 + 6$$
$$-2x \leq 2$$
$$\frac{-2x}{-2} \leq \frac{2}{-2}$$
$$x \geq -1$$

$$x < -3 \text{ or } x \geq -1$$

An *or* inequality has two groups of solutions. In this case, solutions for x are less than -3, *or* greater than or equal to -1.

Example 3:

Solve the compound inequality: $3x - 4 \leq 14$ and $6x - 9 > -15$

Answer:

With an *and* compound inequality, the solution must be true for both statements. Solve both inequalities.

$3x - 4 \leq 14$	and	$6x - 9 > -15$	Write down both inequalities side by side.
$3x - 4 + 4 \leq 14 + 4$	and	$6x - 9 + 9 > -15 + 9$	Add the constants to both sides of each inequality.
$3x \leq 18$	and	$6x > -6$	Simplify each inequality.
$\frac{3x}{3} \leq \frac{18}{3}$	and	$\frac{6x}{6} > \frac{-6}{6}$	Divide both sides of the inequality by the coefficients.
$x \leq 6$	and	$x > -1$	Simplify each inequality.

For the compound inequality to have a solution, x must be both greater than -1 and less than or equal to 6. The solution can be written either of the two following ways.

$$x \leq 6 \text{ and } x > -1$$

$$-1 < x \leq 6$$

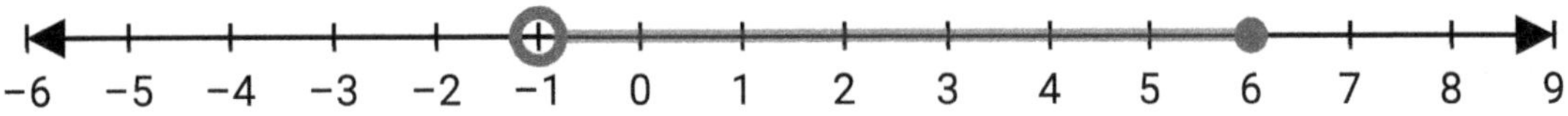

Example 4:

The difference of 2 times a real number, x, and 9 is either less than or equal to -7 or greater than 17. What is the range of values the number can be?

Answer:

The expression, the difference of 2 times a real number, x, and 9, can be written as $2x - 9$. Since the expression is less than or equal to -7 or greater than 17, we have the following compound inequality.

$$2x - 9 \leq 7 \text{ or } 2x - 9 > 17$$

Solve each of the inequalities separately.

$2x - 9 \leq 7$	$2x - 9 > 17$	Write down both inequalities side by side.
$2x - 9 + 9 \leq 7 + 9$	$2x - 9 + 9 > 17 + 9$	Add the constant to both sides of each inequality.
$2x \leq 16$	$2x > 26$	Simplify each inequality.
$\frac{2x}{2} \leq \frac{16}{2}$	$\frac{2x}{2} > \frac{26}{2}$	Divide both sides of the inequality by the coefficient.
$x \leq 8$	$x > 13$	Simplify each inequality.

Therefore, the range of values is $x \leq 8$ or $x > 13$.

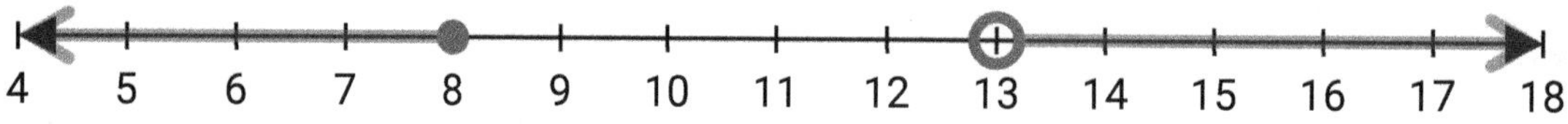

Practice Questions

1. Solve the compound inequality: $3x + 5 > 11$ or $2x + 3 \leq -1$.

a. $x \leq -2$ or $x \leq 2$
b. $x > -2$ or $x \leq 2$
c. $x \leq -2$ or $x > 2$
d. $x \geq -2$ or $x < 2$

2. Solve the compound inequality: $4x - 6 > 18$ and $3x - 10 \leq 20$.

a. $6 < x \leq 10$
b. $6 \leq x < 10$
c. $10 \leq x < 6$
d. $6 > x \geq 10$

3. Solve the combined inequality: $5x - 8 \leq 17$ or $2x + 4 \geq -2$.

a. $x \leq -3$ or $x \leq 5$
b. $x \geq 5$ or $x \leq -3$
c. $x \leq -3$ or $x \geq 5$
d. $x \leq 5$ or $x \geq -3$

4. Solve the compound inequality: $21 < 6x - 9$ and $-4x + 1 > -35$.

a. $5 < x < 9$
b. $9 < x < 5$
c. $5 > x > 9$
d. $9 > x < 5$

5. Solve the compound inequality: $-3x + 5 > 12$ or $5 \leq 2x + 6$.

a. $x < -\frac{7}{3}$ or $x \leq -\frac{1}{2}$
b. $x < -\frac{7}{3}$ or $x \geq -\frac{1}{2}$
c. $x > -\frac{7}{3}$ or $x \leq -\frac{1}{2}$
d. $x > -\frac{7}{3}$ or $x \geq -\frac{1}{2}$

6. Solve the compound inequality: $6x + 9 \geq 20$ and $2x - 3 \leq -14$.

a. $\frac{11}{6} \leq x \leq -\frac{11}{2}$
b. $-\frac{11}{2} \leq x \leq \frac{11}{6}$
c. All real numbers
d. No solution

7. Solve the compound inequality: $-4x + 11 \geq 27$ or $10 > -x + 17$.

a. $x \leq -4$ or $x < 7$
b. $x \geq 4$ or $x < -7$
c. $x \leq -4$ or $x > 7$
d. $x \geq -4$ or $x < 7$

8. Solve the compound inequality: $13 < -3x + 19$ and $4 \geq 1 - 3x$.

a. $-1 < x \leq 2$
b. $-1 \leq x < 2$
c. $-1 \geq x > 2$
d. $-1 > x < 2$

9. Graph the solution to the compound inequality: $2x + 1 < 5$ or $x + 2 > 7$.

a. 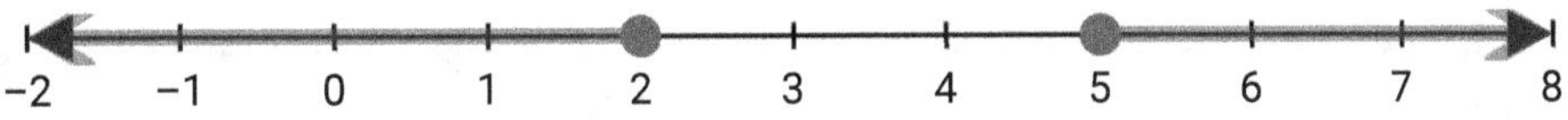

b. 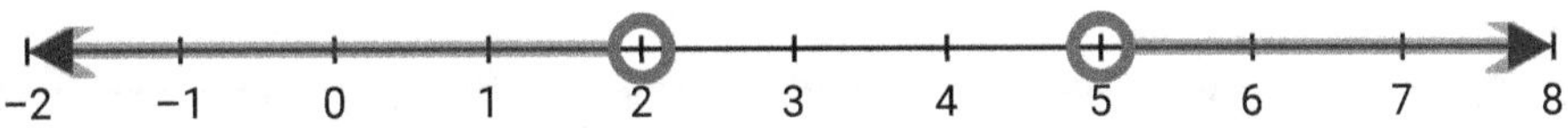

c. 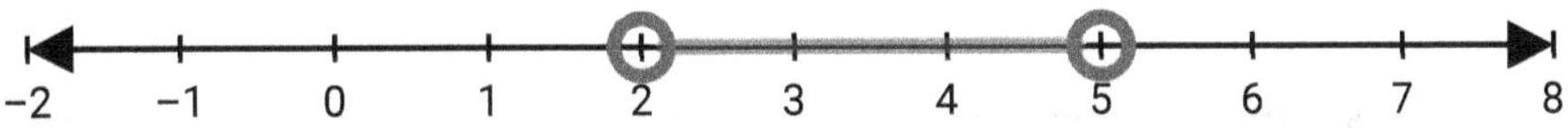

d. 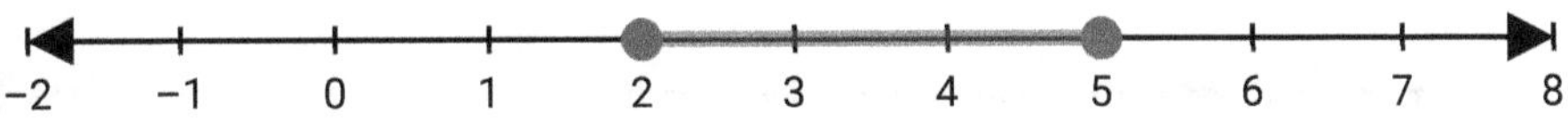

10. Graph the solution to the compound inequality: $4x + 3 \geq -9$ and $2x - 9 < 3$.

a. 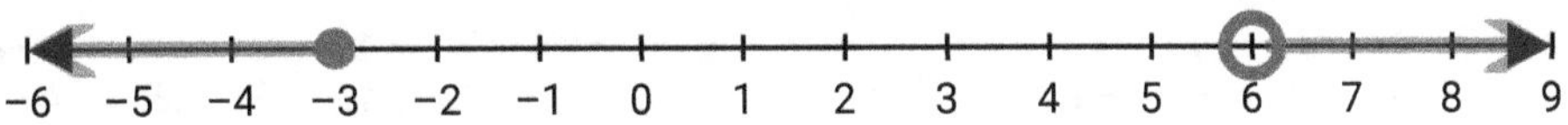

b. 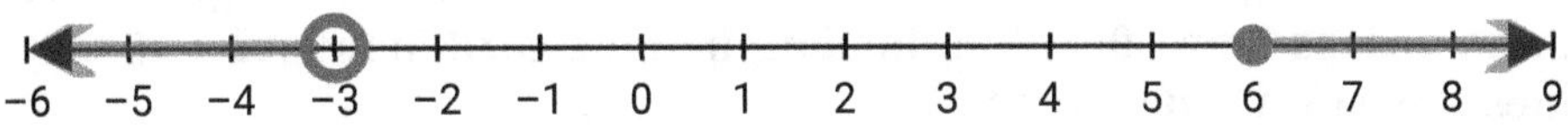

c. 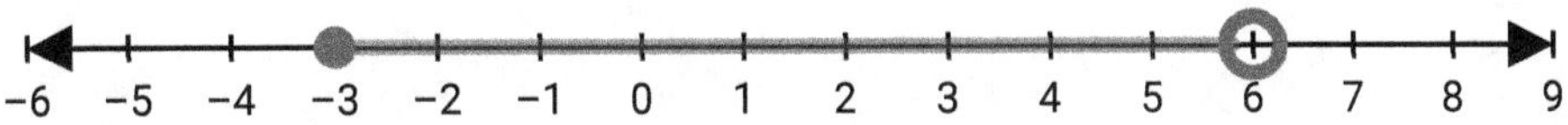

d. 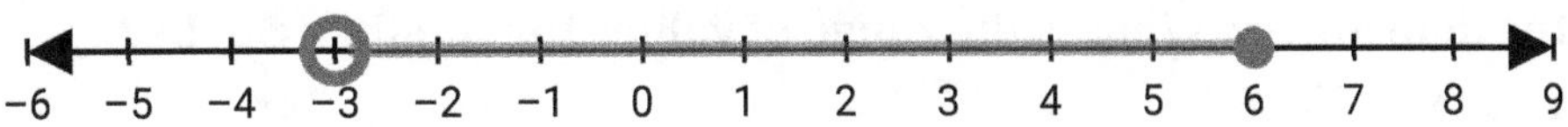

11. Graph the solution to the compound inequality: $-2x - 5 > 13$ or $9 - 6x \leq -3$.

a. 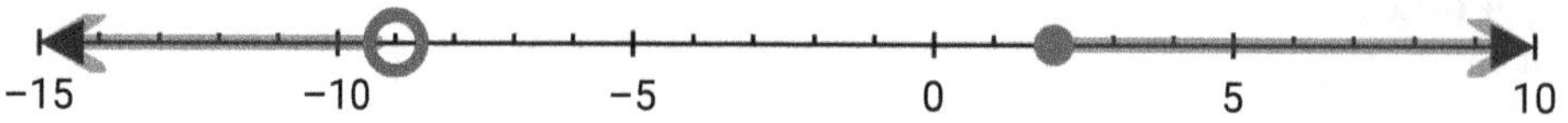

b. 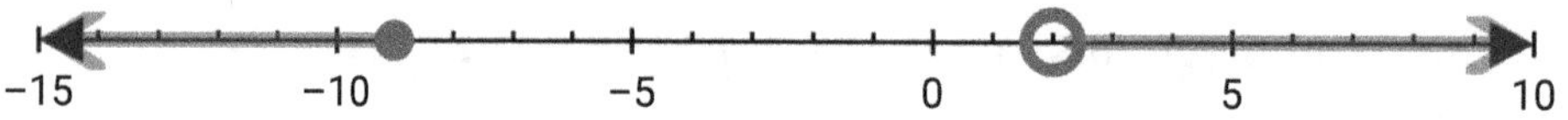

c. 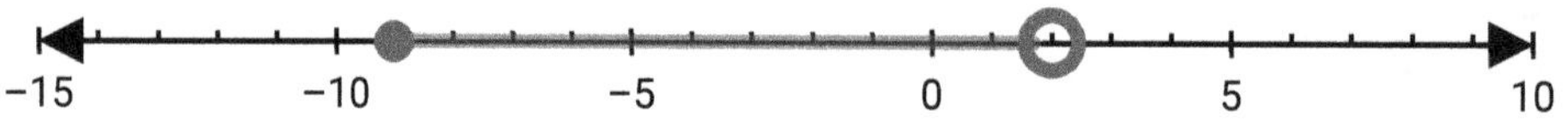

d. 

12. Graph the solution to the compound inequality: $3x - 5 \geq 7$ and $-9 < 2x + 1$.

a. 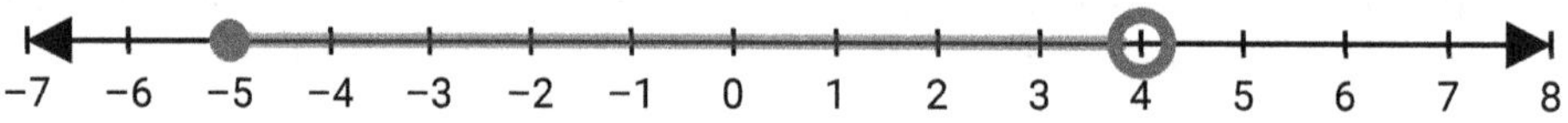

b. 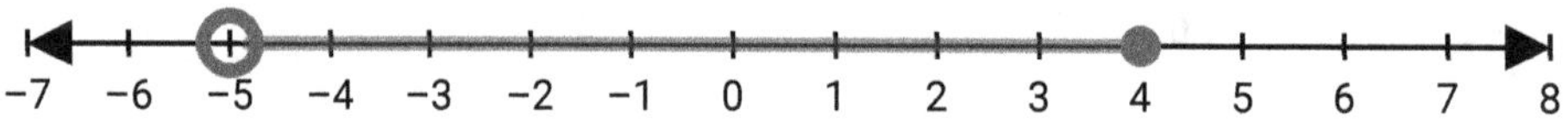

c. 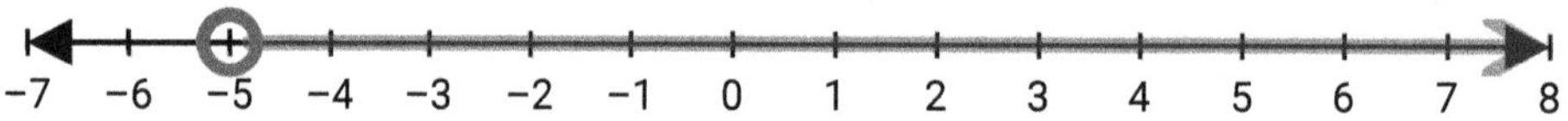

d. 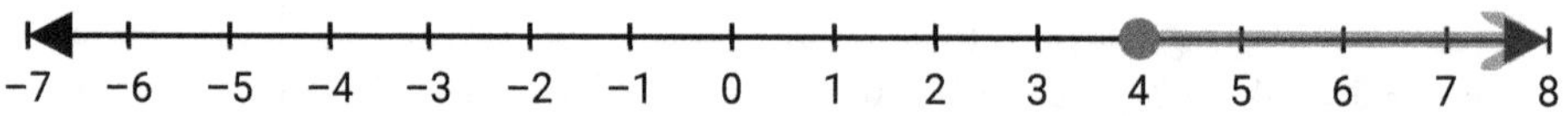

13. A gardener is building a rectangular garden. The perimeter of the garden can be no more than 70 feet and no less than 60 feet. If the length of the garden must be 25 feet, what is the range of values the width can be in feet?

a. $5 \geq x \geq 10$
b. $5 \leq x \leq 10$
c. $10 > x < 5$
d. $5 \geq x \leq 10$

14. The sum of three times a real number, x, and fifteen is either less than or equal to forty-five or greater than ninety. What is the range of values the number can be?

a. $x \leq 10$ or $x < 25$
b. $x \geq 25$ or $x < 10$
c. $x \leq 10$ or $x > 25$
d. $x \geq 10$ or $x < 25$

15. The total cost, C, that a company incurs for producing x number of ear buds in dollars is:

$$C(x) = 10x + 50$$

If the company can afford a cost that is no less than \$1,000 and no more than \$7,000, what is the range of the number of ear buds it can afford to produce?

a. $95 \leq x \leq 695$
b. $95 < x < 695$
c. $95 > x \leq 695$
d. $95 \geq x \geq 695$

4.5 Solving Inequalities Involving Absolute Values

To solve an inequality involving an absolute value, first isolate the absolute value bars. Then proceed to treat the two cases separately, as with an absolute value equation, but flip the inequality symbol in the case where the expression in the absolute value is negative (since that involves multiplying both sides by -1.) The two cases are then combined into a compound inequality. If the absolute value is on the greater side of the inequality, then it is an *or* compound inequality, if on the lesser side, then it's an *and* inequality.

Consider the inequality $2 + |x - 1| \geq 3$. We can isolate the absolute value bars by subtracting 2 from both sides.

$$|x - 1| \geq 1$$

Since we are working with an absolute value inequality, we're left with two cases. We have the case where we remove the absolute value bars, $x - 1 \geq 1$, and we have the case where the value in the absolute value bars is negative, $x - 1 \leq -1$. Note that in the latter, negative case, the inequality is flipped. $x - 1 \geq 1$ is equivalent to $x \geq 2$, and $x - 1 \leq -1$ is equivalent to $x \leq 0$. Since in the inequality $|x - 1| \geq 1$ the absolute value is on the greater side, the two cases combine into an *or* compound inequality, so the final, solved inequality is "$x \leq 0$ or $x \geq 2$."

GRAPHING INEQUALITIES INCLUDING ABSOLUTE VALUES

An inequality with an absolute value can be converted to a compound inequality. To graph the inequality, first convert it to a compound inequality, and then graph the compound inequality normally. If the absolute value is on the greater side of the inequality, we end up with an *or* inequality. We plot the endpoints of the inequality on the number line and fill in the part of the line outside those points. If the absolute value is on the lesser side of the inequality, we end up with an *and* inequality. We plot the endpoints of the inequality on the number line and fill in the part of the line between those points.

For example, solve the inequality $|x + 1| \geq 4$. Start by breaking it apart into two separate inequalities and solve each one separately.

$$x + 1 \geq 4 \qquad x + 1 \leq -4$$

$$x \geq 3 \qquad x \leq -5$$

Therefore, the inequality becomes $x \geq 3$ or $x \leq -5$. We place closed circles at the points 3 and -5 and fill in where each inequality is true.

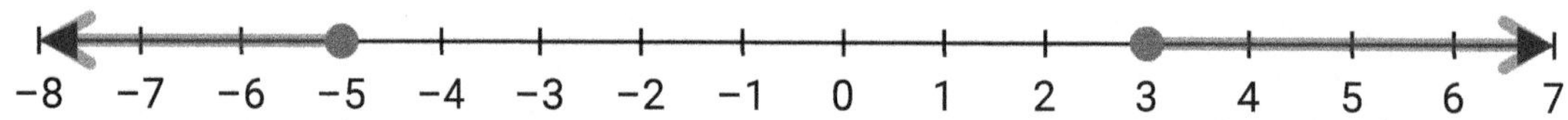

Review Video: Solving Absolute Value Inequalities
Visit mometrix.com/academy and enter code: 997008

EXAMPLES

EXAMPLE 1:

Solve the inequality.

$$|x| > 4$$

Answer:

Absolute value represents how far you are from 0. Distance cannot be negative, which is why absolute value is always positive. $|x| > 4$ is asking for the values of x that are farther than 4 units away from 0. Visualize this on a number line. Numbers greater than 4 or less than –4 will be farther than 4 units from 0.

$$x < -4 \text{ or } x > 4$$

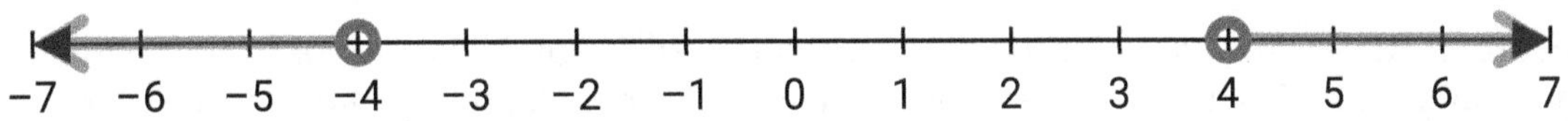

EXAMPLE 2:

Solve the inequality.

$$2|x| + 1 > 11$$

Answer:

Start by isolating the absolute value. Subtract 1 from both sides and then divide by 2.

$$\begin{aligned} 2|x| + 1 &> 11 \\ 2|x| &> 10 \\ |x| &> 5 \end{aligned}$$

This means that x is any value that has a greater distance from 0 than 5. Visualize a number line. Values greater than 5, and less than –5 are farther than 5 units from 0. This means that the absolute value of x is any value less than –5 and any value greater than 5: $x < -5$ or $x > 5$.

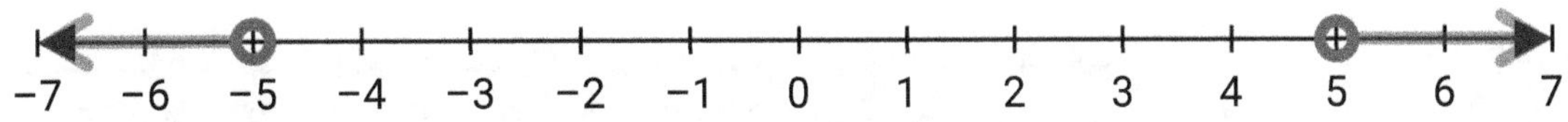

Example 3:

Solve the inequality.

$$|x + 5| > 8$$

Answer:

This expression says: Whatever is inside of the absolute value sign is a greater distance than 8 from 0. This means that $x + 5$ can be greater than 8 and $x + 5$ can be less than −8 (because both are farther than 8 units from 0). Solve both inequalities for x.

$$x + 5 > 8 \qquad x + 5 < -8$$

$$x > 3 \qquad x < -13$$

Therefore, $x > 3$ or $x < -13$.

Example 4:

A carpenter is measuring wood for a table. The table needs to be 78 inches long and can differ from this by no more than 0.5 inches. Using x as the length of the table, set up and solve an inequality with an absolute value to determine what lengths are allowed.

Answer:

Let x be the length of the table. The difference between this length and 78 inches, with the direction of the difference being irrelevant, is $|x - 78|$. The length of the table can only differ from 78 inches by 0.5 inches, so we can express this as the inequality $|x - 78| \le 0.5$. To solve an inequality of the form $|\text{expression}| \le a$ (where $a > 0$), we remove the absolute value by translating the inequality into the form $-a \le \text{expression} \le a$.

$$-0.5 \le x - 78 \le 0.5$$

Add 78 to all three parts of the inequality.

$$77.5 \le x \le 78.5$$

Therefore, the length of the table can be between 77.5 inches and 78.5 inches.

Practice Questions

1. Solve the inequality $|x| < 4$ and graph the solution on a number line, if possible.

a. $-4 < x < 4$

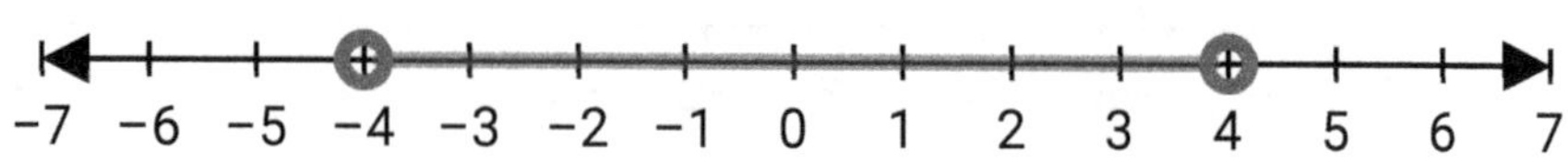

b. $x < -4$ or $x > 4$

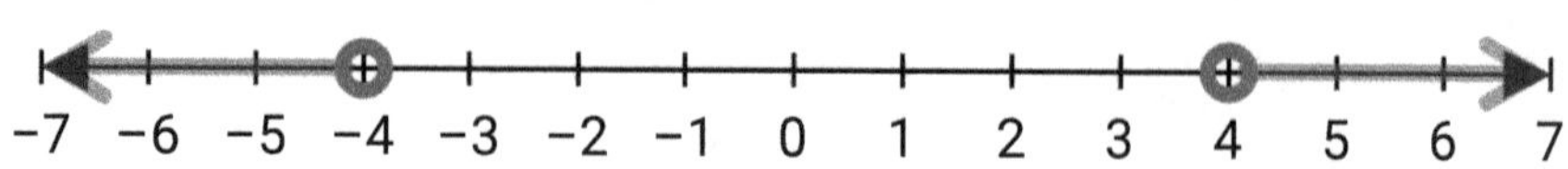

c. All real numbers

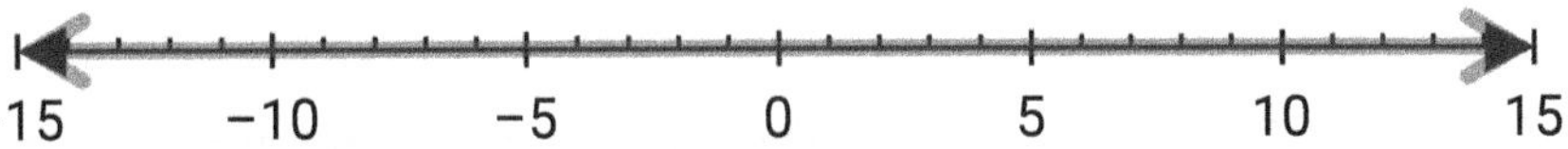

d. This inequality has no solution.

2. Solve the inequality $|x| \geq 3$ and graph the solution on a number line, if possible.

a. $-3 \leq x \leq 3$

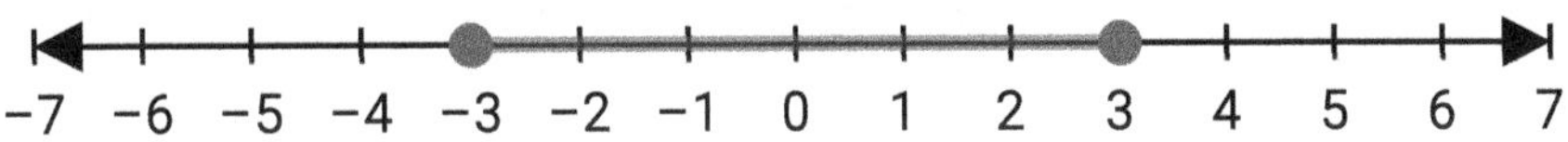

b. $x \leq -3$ or $x \geq 3$

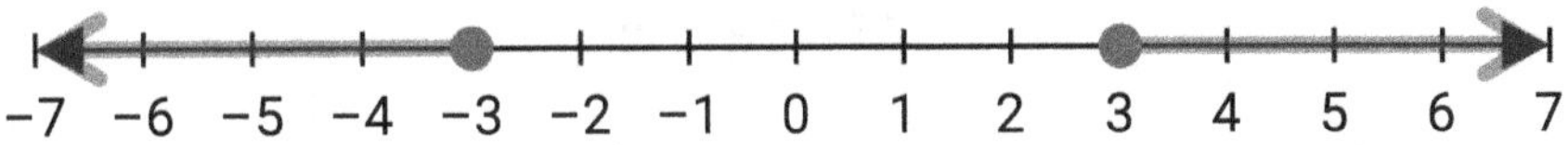

c. All real numbers

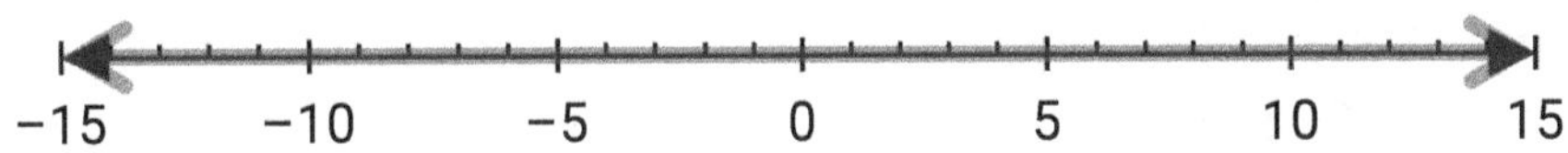

d. This inequality has no solution.

3. Solve the inequality $|x| < -2$ and graph the solution on a number line, if possible.

a. $-2 < x < 2$

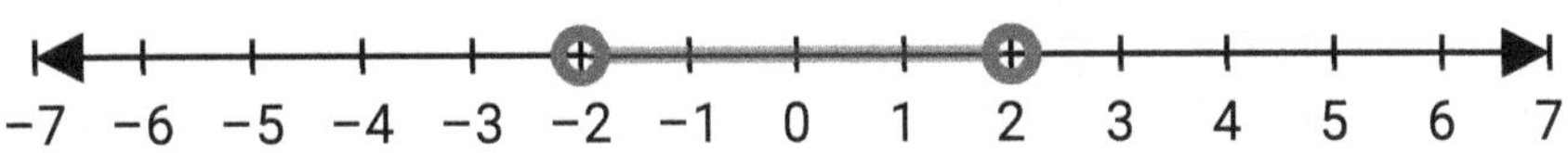

b. $x < -2$ or $x > 2$

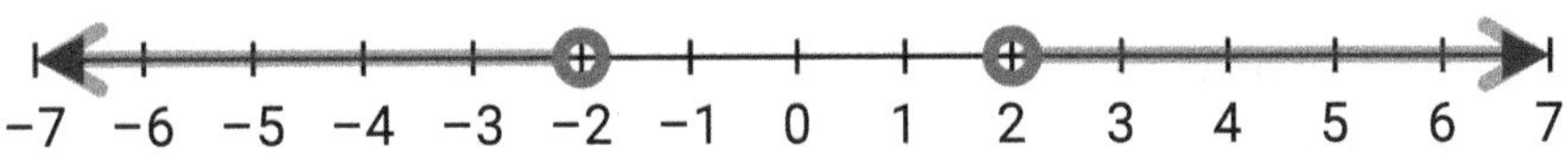

c. All real numbers

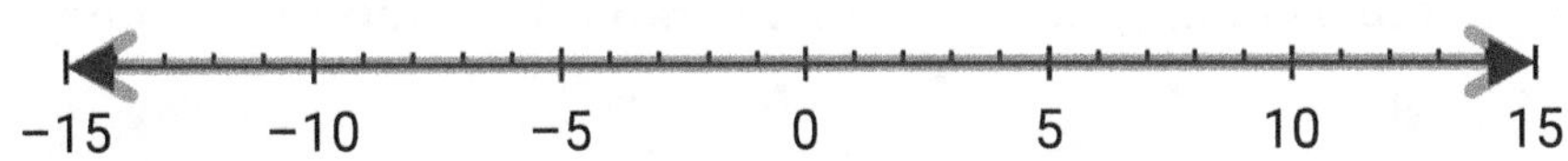

d. This inequality has no solution.

4. Solve the inequality $|x| + 5 > 6$.

a. $x < -11$ or $x > 11$
b. $x < -1$ or $x > 1$
c. $-11 < x < 11$
d. $-1 < x < 1$

5. Solve the inequality $|x| - 3 \leq -1$ and graph the solution on a number line, if possible.

a. $-2 \leq x \leq 2$

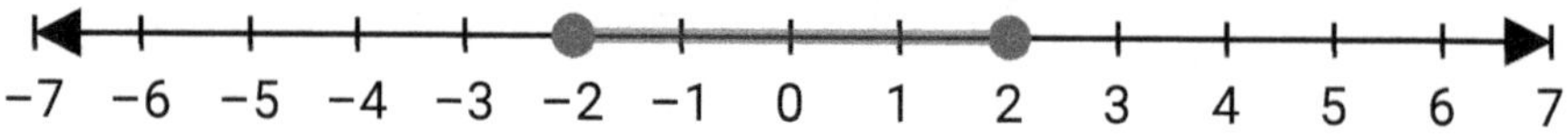

b. $x \leq -2$ or $x \geq 2$

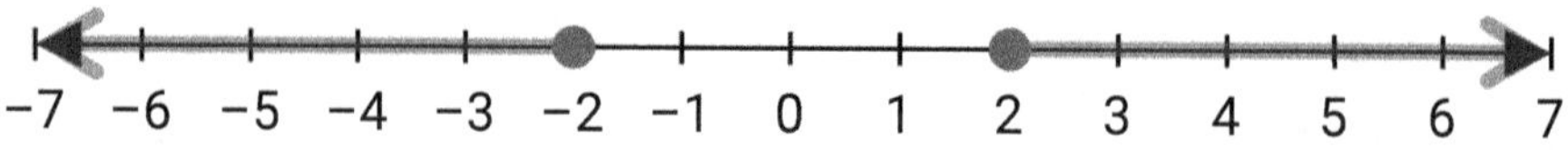

c. $-4 \leq x \leq 4$

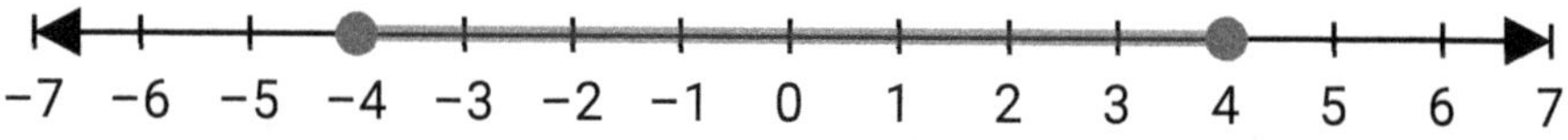

d. This inequality has no solution.

6. Solve the inequality $-4|x| \leq -6$.

a. $-2 \leq x \leq 2$
b. $-\frac{3}{2} \leq x \leq \frac{3}{2}$
c. $x \leq -\frac{3}{2}$ or $x \geq \frac{3}{2}$
d. This inequality has no solution.

7. Solve the inequality $2|x| - 7 \leq -1$.

a. $-\frac{13}{2} \leq x \leq \frac{13}{2}$
b. $x \leq -\frac{13}{2}$ or $x \geq \frac{13}{2}$
c. $-3 \leq x \leq 3$
d. $x \leq -3$ or $x \geq 3$

8. Solve the inequality $-3|x| + 1 \leq 13$.

a. $-4 \leq x \leq 4$
b. $x \leq -4$ or $x \geq 4$
c. $x \leq -\frac{15}{3}$ or $x \geq \frac{15}{3}$
d. All real numbers

9. Solve the inequality $|x - 3| > 1$ and graph the solution on a number line if possible.

a. $2 < x < 4$

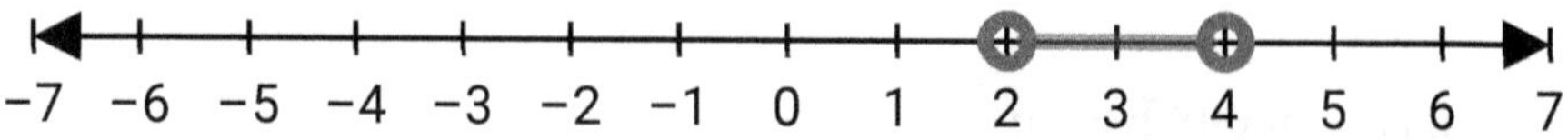

b. $x < 2$ or $x > 4$

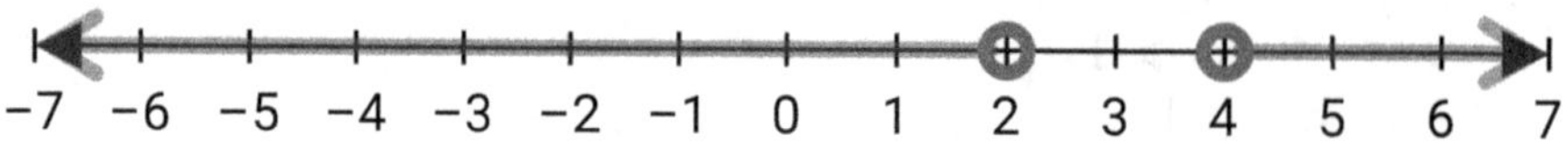

c. $x > 4$

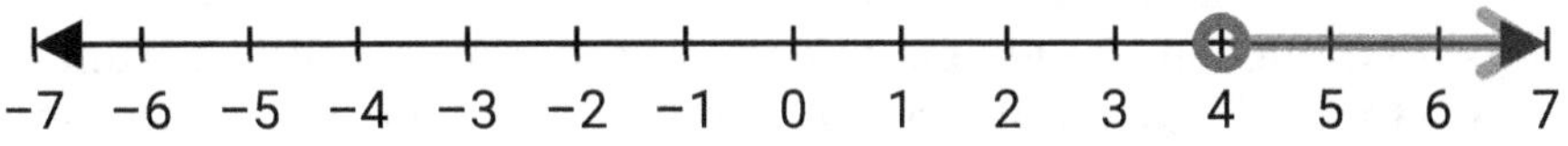

d. All real numbers

10. Solve the inequality $|4x + 12| < 8$.

a. $x < -14$ or $x > -10$
b. $-14 < x < -10$
c. $x < -5$ or $x > -1$
d. $-5 < x < -1$

11. Solve the inequality $|x - 1| \leq 3$.

a. $x \leq -2$ or $x \geq 4$
b. $-2 \leq x \leq 4$
c. $x \leq 2$
d. $-4 \leq x \leq 2$

12. Solve the inequality $|2x + 1| > 6$.

a. $-\frac{7}{2} < x < \frac{5}{2}$
b. $x < -\frac{7}{2}$ or $x > \frac{5}{2}$
c. $-4 < x < 2$
d. $x < -4$ or $x > 2$

13. The food safety guidelines at a restaurant specify that if food is being held for serving, the temperature of the food must differ from 90 °F by more than 50 °F, whether higher or lower. Using x as the holding temperature, set up and solve an inequality with an absolute value to determine what holding temperatures are allowed.

a. $x < -40$ °F or $x > 140$ °F
b. -40 °F $< x < 140$ °F
c. $x < 40$ °F or $x > 140$ °F
d. 40 °F $< x < 140$ °F

14. A statistician says he has collected data showing that the mean height of men (adult males) in the United States in the late twentieth century was 69 inches and that ninety-five percent of men had heights within 5.8 inches of this mean. Using x as a man's height, set up and solve an inequality with an absolute value to determine the heights of ninety-five percent of men in the United States at this time.

a. $x \leq 65.55$ inches or $x \geq 72.45$ inches
b. $x \leq 63.2$ inches or $x \geq 74.8$ inches
c. 65.55 inches $\leq x \leq 72.45$ inches
d. 63.2 inches $\leq x \leq 74.8$ inches

15. A theme park has a water slide in which groups of riders ride small boats from the top of the slide to the bottom. The slide operator has a large scale at the top of the slide to weigh each group of riders before they enter the boat. If a group is too heavy or too light, then the boat is unstable. So, the weight of each group must not differ from the ideal weight of 650 lb. by more than 120 lb. Using x as the weight of a group, set up and solve an inequality with an absolute value to determine the weights that are safe.

a. $-120 \text{ lb} \leq x \leq 120 \text{ lb}$
b. $0 \text{ lb} \leq x \leq 120 \text{ lb}$
c. $x \leq 530 \text{ lb}$ or $x \geq 770 \text{ lb}$
d. $530 \text{ lb} \leq x \leq 770 \text{ lb}$

4.6 Writing Linear Inequalities

Writing Inequalities from Graphs

When you are writing an inequality from a graph, it is important to determine if the line is dashed or solid and if the shaded region is above or below the line. These two pieces of information will allow you to select the correct symbol.

Graphs with dashed lines use the symbols $<$ and $>$. A dashed line means that the points on the line are not part of the solution.

Graphs with solid lines use the symbols $\leq$ and $\geq$. A solid line means that the points on the line are part of the solution.

When the graph is shaded above the line, the "greater than" or "greater than or equal to" symbol will be used.

When the graph is shaded below the line, the "less than" or "less than or equal to" symbol will be used.

Once the correct symbol has been determined, write the inequality as though you are writing an equation. Find the slope and y-intercept.

For example, the graph below shows a slope of $\frac{3}{2}$, and a y-intercept of $\frac{3}{2}$. The line is dashed, and the shaded region is below the line. With these four pieces of information, we can write the inequality to describe the graph: $y < \frac{3}{2}x + \frac{3}{2}$.

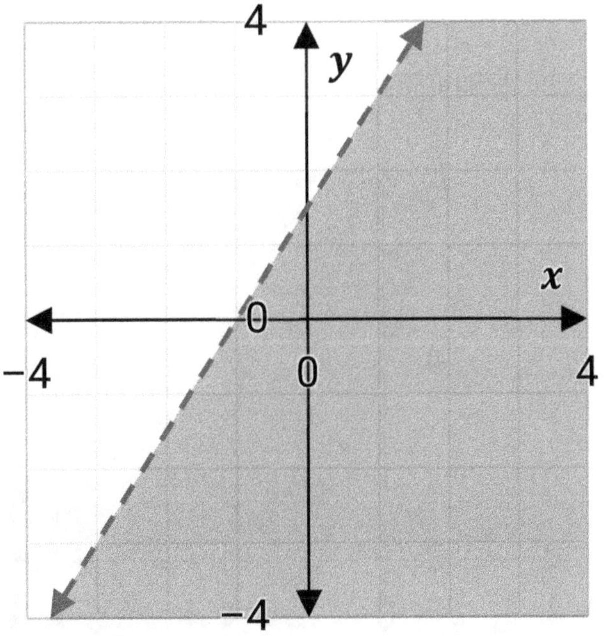

Writing Inequalities from Written Descriptions

There are four descriptions to look for when you are provided with a written description of an inequality.

"x is greater than…": $x >$

"x is less than…": $x <$

"x is greater than or equal to…": $x \geq$

"x is less than or equal to…": $x \leq$

It is unlikely that you will encounter these descriptions word for word. Likely, they will be embedded into real-world scenarios. For example, *Debby found more than 30 seashells on the beach*. This statement can be expressed as $s > 30$ because the number of seashells, s, she found is greater than and not equal to 30.

Examples

Example 1:

A local bakery sells muffins for $7 each and macarons for $3 each. They must sell more than $280 worth of muffins and macarons to meet their goal this week. Write an inequality that represents the amount of muffins they need to sell, x, and the amount of macarons they need to sell, y, in order to meet their goal.

Answer:

In this scenario, x stands for the number of muffins the bakery needs to sell, and y stands for the number of macarons they need to sell. Since each muffin costs $7, use the term $7x$ to represent the amount of money made from selling muffins. Since each macaron costs $3, use the term $3y$ to represent the amount of money made from selling macarons. The sum of these two amounts needs to be greater than $280, so the inequality that represents this scenario is $7x + 3y > 280$.

Example 2:

Write an inequality to describe the graph.

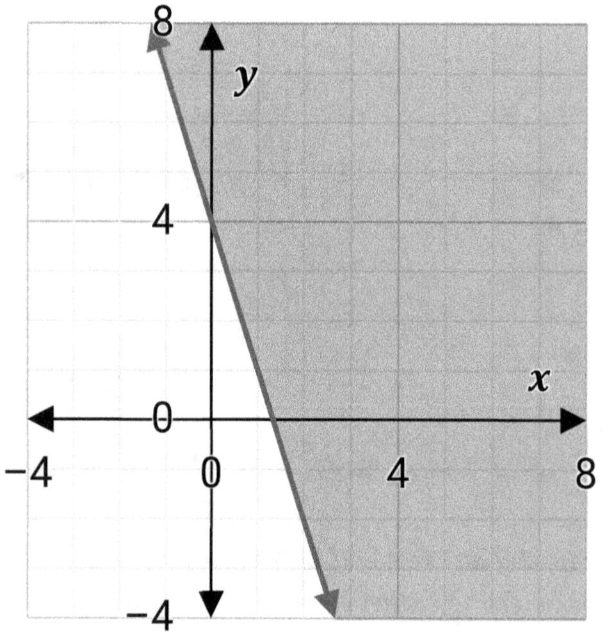

Answer:

To find the equation of the line, find the slope and the y-intercept.

$$\text{Slope} = \frac{1-4}{1-0} = \frac{-3}{1} = -3$$
$$y - \text{intercept} = 4$$

Equation of the line: $y = -3x + 4$

Determine the sign: $>, <, \geq,$ or $\leq$

The shaded region is above the line, and the line is solid. This means that the sign is $\geq$. The inequality is $y \geq -3x + 4$.

EXAMPLE 3:

Use the written description "y is less than or equal to the difference of twice a number, x, and 3" to write a linear inequality.

Answer:

The description "y is less than or equal to" is expressed as $y \leq$. The description "the difference of twice a number, x, and 3" is expressed as $2x - 3$. The inequality $y \leq 2x - 3$ means that any value less than or equal to $2x - 3$ is a solution.

PRACTICE QUESTIONS

1. Which inequality statement can be represented by the graph shown below?

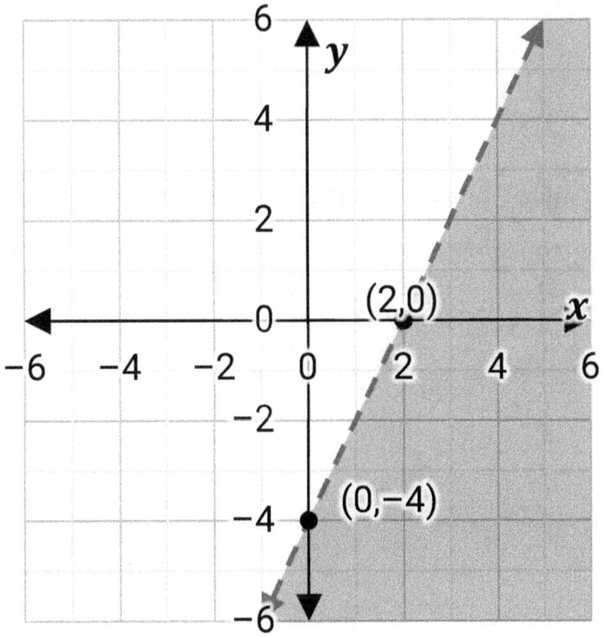

a. $y < 2x - 4$
b. $y < 2x + 4$
c. $y < -2x - 4$
d. $y \leq 2x - 4$

2. The drama club at a school is putting on a play to raise money. They hope to raise \$520 to cover the cost of the play and the end of school trip. An adult ticket costs \$6, and a child ticket costs \$2. Which inequality statement represents all the possible combinations of adult tickets, x, and child tickets, y, that will help them raise enough money?

a. $6x + 2y \geq 520$
b. $6x + 2y \leq 250$
c. $2x + 6y \geq 520$
d. $2x + 6y \leq 520$

3. A company creates paint by number kits. The company starts with a \$400 investment and earns \$25 from each sale they make. If they need to operate on less money than they have, which of the following inequalities represents the amount of money they have to operate from, y, based on the number of kits they sell, x?

a. $25x + y > 400$
b. $y > -25x + 400$
c. $25x - y < 400$
d. $y < 25x + 400$

4. Which inequality statement can be represented by the graph shown below?

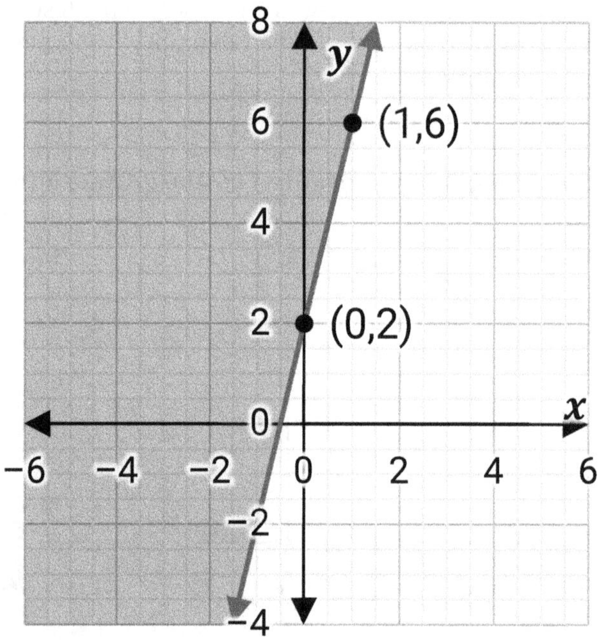

a. $y \geq 4x + 2$
b. $y > 4x + 2$
c. $y \geq -4x + 2$
d. $y \geq 4x - 2$

5. Sarah has a food delivery service. She charges $10 for delivery plus 50 cents per mile and allows for customers to give her tips. Which of the following inequalities represents the amount of money she earns, y, for a delivery that is x miles away?

a. $y < 0.5x + 10$
b. $y > 10x - 0.5$
c. $y \geq 0.5x + 10$
d. $y \leq 10x + 0.5$

6. Which inequality statement can be represented by the graph shown below?

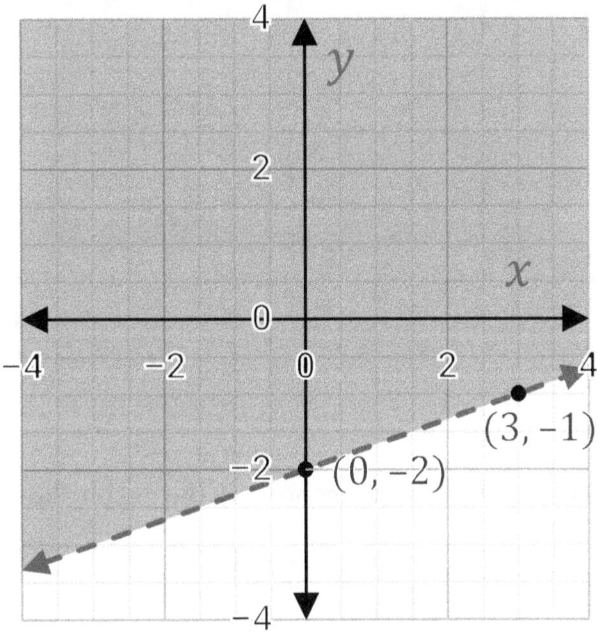

a. $y \leq \frac{1}{3}x - 2$
b. $y > \frac{1}{3}x - 2$
c. $y > 3x - 2$
d. $y > \frac{1}{3}x + 2$

7. Jackson runs a lawn mowing business where he earns $19 per hour he spends mowing lawns. Some weeks there are too many lawns for him to mow on his own, so he pays his brother $14 per hour to help him. If Jackson has never earned more than $600 in a week, which of the following inequalities represents the combination of hours that he has mowed, x, and the hours his brother has mowed, y?

a. $19x + 14y \geq 600$
b. $19x - 14y \leq 600$
c. $-19x + 14y < 600$
d. $19x - 14y > 600$

8. Which inequality statement can be represented by the graph shown below?

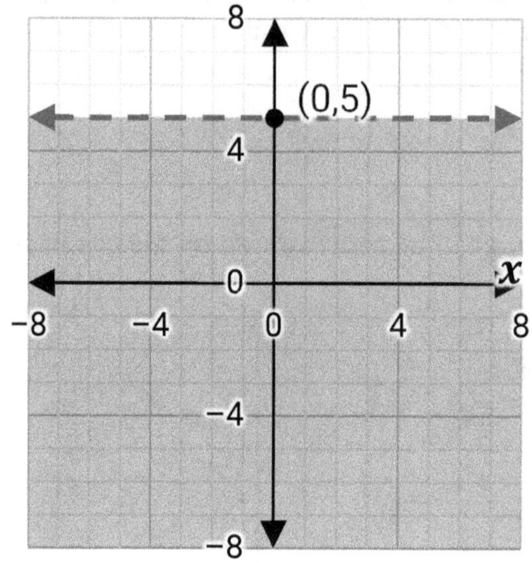

a. $y < 5$
b. $y > 5$
c. $y < -5$
d. $x < 5$

9. Which inequality statement can be represented by the graph shown below?

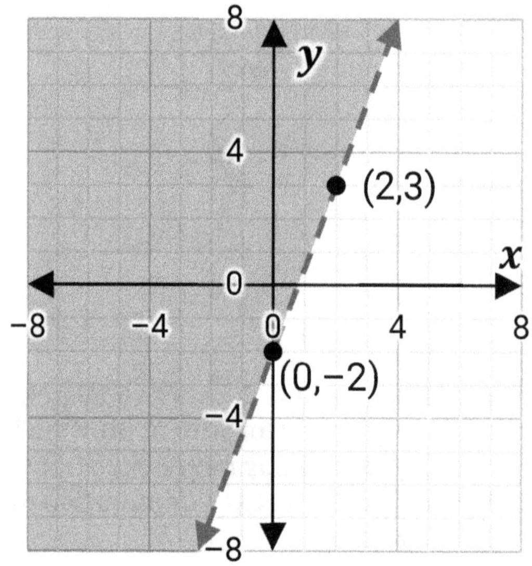

a. $y \leq \frac{5}{2}x + 2$
b. $y > \frac{5}{2}x + 2$
c. $y \geq \frac{5}{2}x - 2$
d. $y > \frac{5}{2}x - 2$

10. Which inequality statement can be represented by the graph shown below?

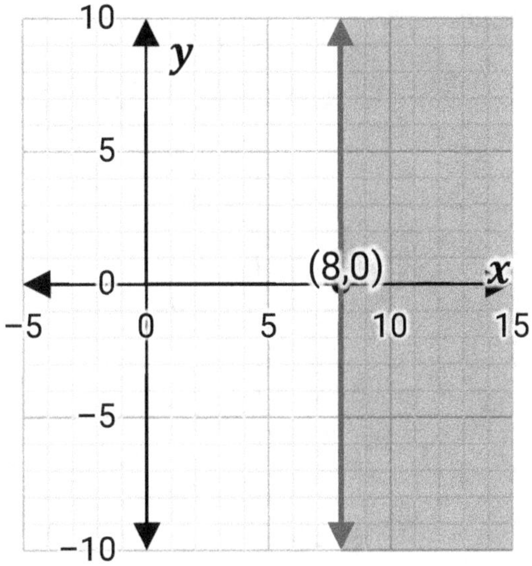

a. $x < 8$
b. $x > 8$
c. $x \geq 8$
d. $x \leq 8$

11. Which inequality statement can be represented by the graph shown below?

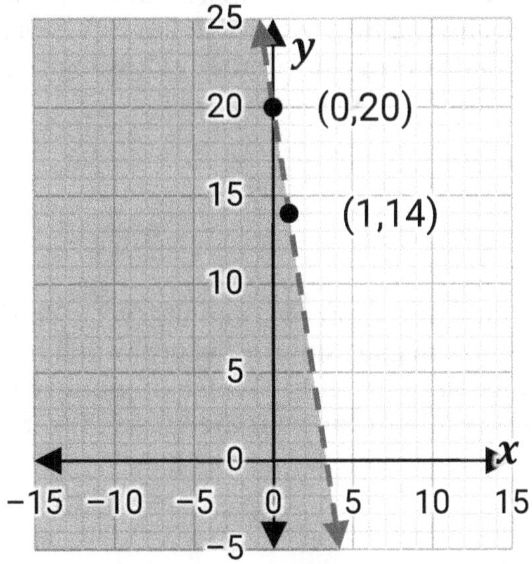

a. $y < 6x - 20$
b. $y < -6x + 20$
c. $y \leq -6x - 20$
d. $y \leq 6x + 20$

12. Which inequality statement can be represented by the graph shown below?

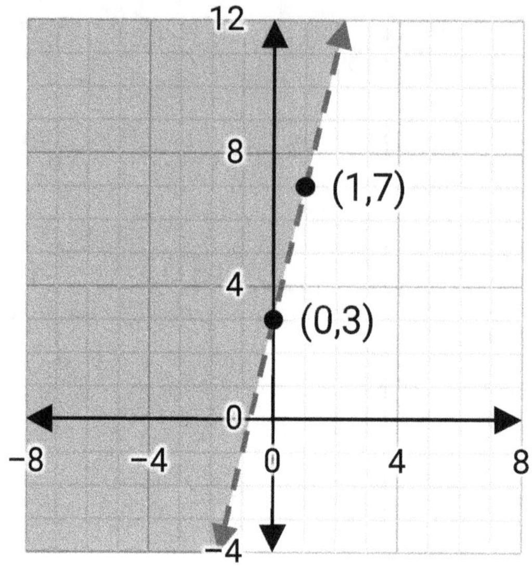

a. $y < 3x - 4$
b. $y > 3x + 4$
c. $y \geq 4x + 3$
d. $y > 4x + 3$

13. Kelly is at a basketball game with her friends. She plans to buy some hotdogs and sodas for the group. Each hotdog costs 2 dollars, and each soda costs 4 dollars. Kelly wants to know how many hotdogs and sodas she can buy with 40 dollars. Which inequality statement represents all possible combinations of the number of hotdogs, *x*, and the number of sodas, *y*, that Kelly can buy for no more than 40 dollars?

a. $4x + 40y \geq 2$
b. $40x - 2y > 4$
c. $2x + 4y \leq 40$
d. $2x - 4y < 40$

14. Glenn is visiting Philadelphia and wants to rent a bicycle. The bike costs $12 for a daily rental plus $0.15 for each minute of use. Glenn wants to rent a bicycle for as long as he can afford, but he can't spend more than *y* dollars. Which inequality statement represents the maximum number of minutes, *x*, Glenn can rent the bicycle for in terms of *y* dollars?

a. $y \leq 0.15x + 12$
b. $y > 0.15x + 12$
c. $y < 12x - 0.15$
d. $y \geq 12x - 0.15$

15. A coffee shop is getting new tables. Each small table seats three people, and each large table seats six people. The owner of the coffee shop wants to know how many small and large tables she should purchase. The maximum capacity of the coffee shop is 30 people. Which inequality statement represents all possible combinations of the number of large tables, x, and the number of small tables, y, that the coffee shop owner can seat for a maximum of 30 people?

a. $3x - 30y \geq 6$
b. $30x + 6y < 3$
c. $6x + 3y \leq 30$
d. $6x + 3y > 30$

4.7 Graphing Linear Inequalities

To graph an inequality in two variables, we first graph the border of the inequality. This means graphing the equation that we get if we replace the inequality sign with an equal sign. If the inequality is strict ($>$ or $<$), we graph the border with a dashed or dotted line; if it is not strict ($\geq$ or $\leq$), we use a solid line. We can then test any point not on the border to see if it satisfies the inequality. If it does, we shade in that side of the border; if not, we shade in the other side. As an example, consider $y > 2x + 2$. To graph this inequality, we first graph the border, $y = 2x + 2$. Since it is a strict inequality, we use a dashed line.

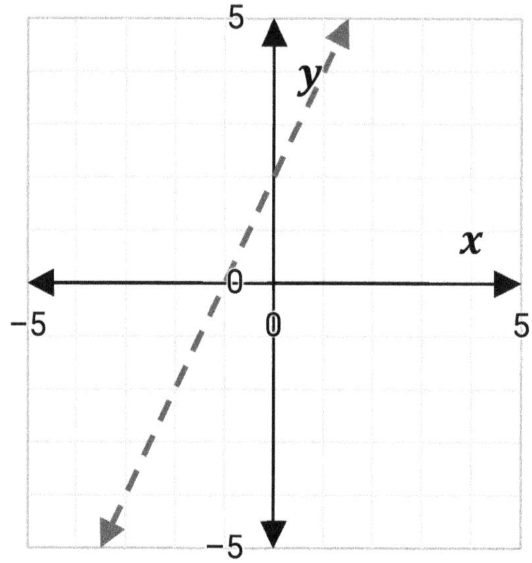

Then, we choose a test point. This can be any point not on the border; in this case, we will choose the origin, (0,0). This makes the calculation easy and is generally a good choice unless the border passes through the origin. Putting this into the original inequality, we get $0 > 2(0) + 2$, i.e., $0 > 2$. This is *not* true, so we shade in the side of the border that does *not* include the point (0,0).

Another way to determine which direction to shade is to look at the inequality in slope intercept form. If the inequality includes $y >$ or $y \geq$, then shade above the line, the area where all values of y are greater than the line $2x + 2$. If the inequality includes $y <$ or $y \leq$, then shade below the line, the area where all values of y are less than $2x + 2$. For the inequality $y > 2x + 2$, we see that we shade above the line.

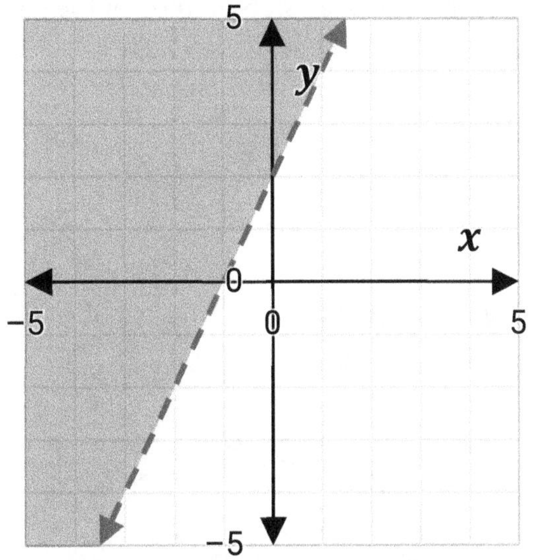

Review Video: Graphing Solutions to Linear Inequalities
Visit mometrix.com/academy and enter code: 391281

EXAMPLES

EXAMPLE 1:

Graph the linear inequality.

$$3x - 4y > 12$$

Answer:

Start by finding the x- and y-intercepts by first turning the inequality into an equation. Then, plug in 0 for x to find the y-intercept, and plug in 0 for y to find the x-intercept.

$$\begin{aligned} 3x - 4(0) &= 12 \\ 3x &= 12 \\ x &= 4 \end{aligned}$$

x-intercept: $(4, 0)$

$$\begin{aligned} 3(0) - 4y &= 12 \\ -4y &= 12 \\ y &= -3 \end{aligned}$$

y-intercept: $(0, -3)$

Now graph the x- and y-intercepts using a dashed line. Remember, $<$ and $>$ are graphed with dashed lines. Convert the inequality to slope-intercept form.

$$\begin{aligned} 3x - 4y &> 12 \\ -4y &> -3x + 12 \end{aligned}$$

$$y < \frac{3}{4}x - 3$$

Shade the lower region of the graph because in slope-intercept the equation states $y < \frac{3}{4}x - 3$ which says "y is *less than* $\frac{3}{4}x - 3$."

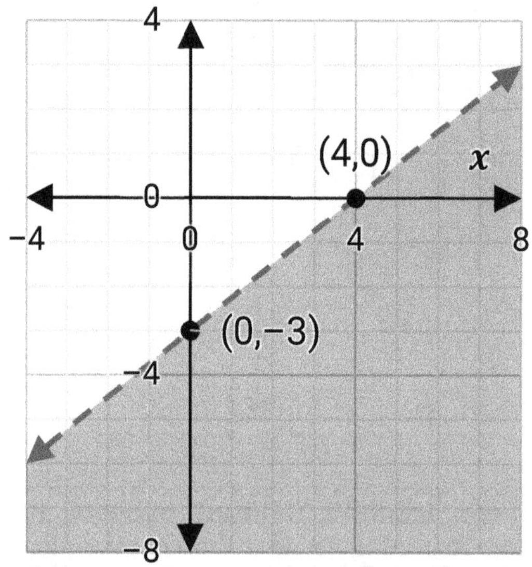

EXAMPLE 2:

Graph the linear inequality.

$$y < \frac{3}{2}x + 2$$

Answer:

The slope is $\frac{3}{2}$ and the y-intercept is 2. The line will be "dashed" not solid because the symbol is $<$. The equation of the line in slope-intercept form is $y = \frac{3}{2}x + 2$. Graph this as a dashed line. Shade in the below the line because the inequality states that y is less than the expression.

Remember, symbols < and > have a dashed line, which means that the line is not a part of the solution. The symbols ≥ and ≤ are graphed with solid lines, which means that the line is a part of the solution.

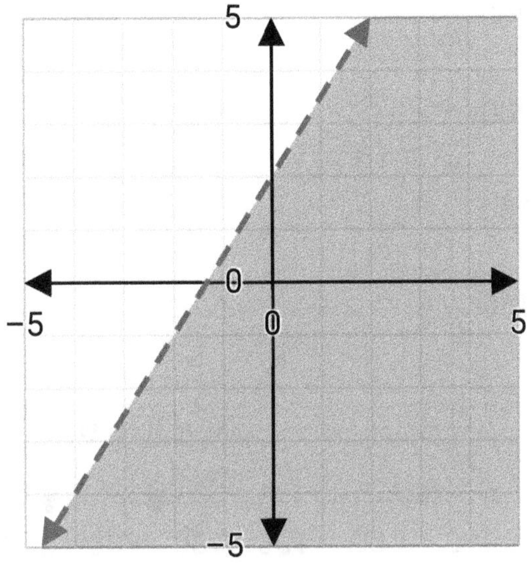

EXAMPLE 3:

There are x adults and y children in attendance at a play. If there are never more than 175 people in attendance, what graph represents the combination of the number of adults and children in attendance?

Answer:

The sum of the number of adults, x, and the number of children, y, in attendance at the play is never more than 175, so we have the following inequality.

$$x + y \leq 175$$

To graph the inequality, first graph the line $x + y = 175$ by finding the x- and y-intercepts and drawing a line through them.

x-intercept	y-intercept
$x + 0 = 175$	$0 + y = 175$
$x = 175$	$y = 175$

To sketch the graph of the line, plot the intercepts on a coordinate plane and draw a line through them. Since the inequality includes the equal sign, the line graphed is a solid line to indicate the points on it are a part of the solution set.

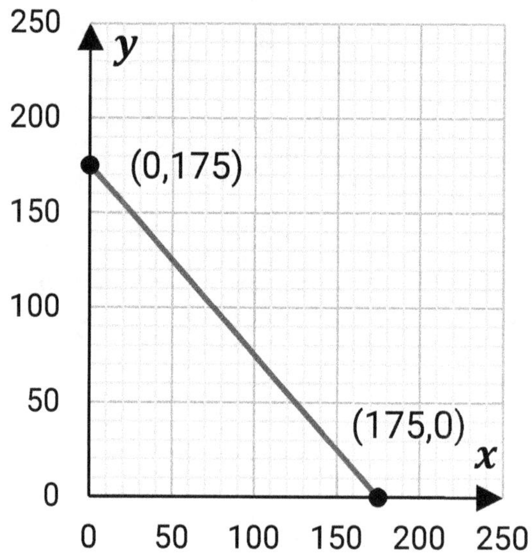

Shade the portion of the graph that satisfies the inequality. Convert the inequality to slope-intercept form.

$$x + y \leq 175$$

$$y \leq -x + 175$$

In slope intercept form, the direction the graph is shaded can be determined by the inequality symbol. Since the inequality contains the less than or equal to symbol ($\leq$), the graph of the inequality is shaded below the line.

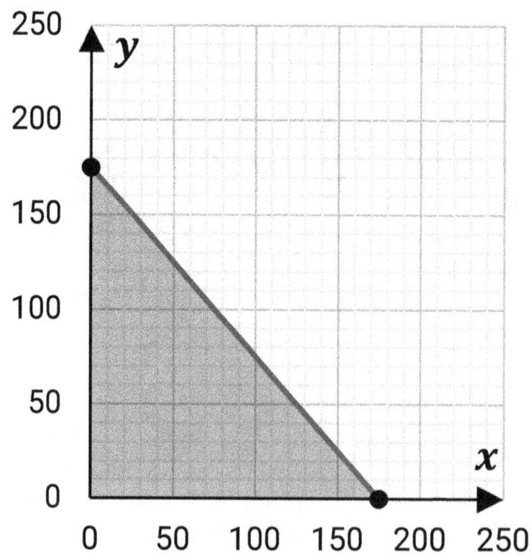

Thus, the combination of the number of adults, x, and the number of children, y, in attendance is any point on the line or shaded region of the graph.

PRACTICE QUESTIONS

1. Graph the inequality: $3x + 2y \geq 12$.

a. 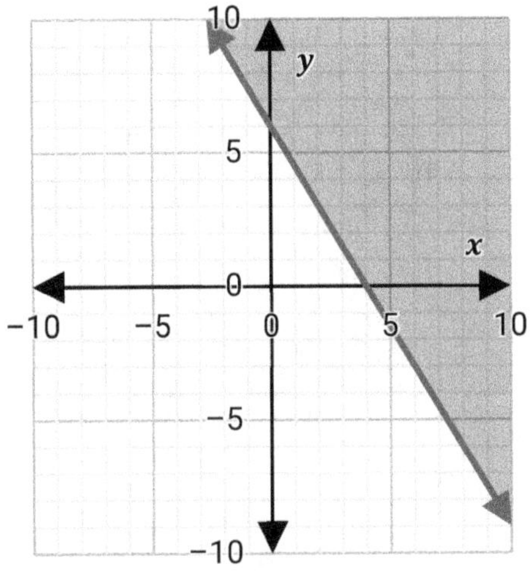

b. 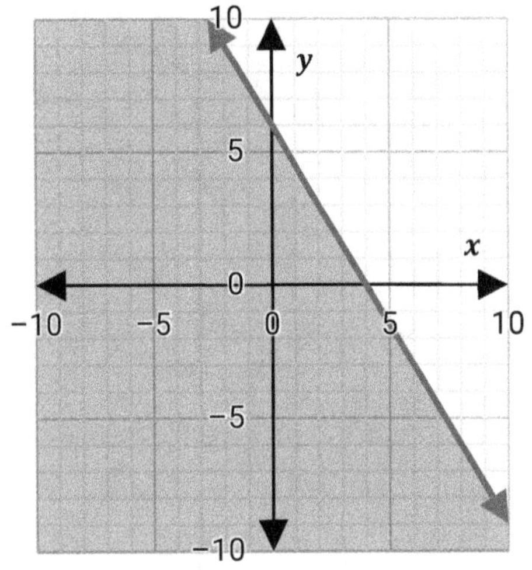

c. 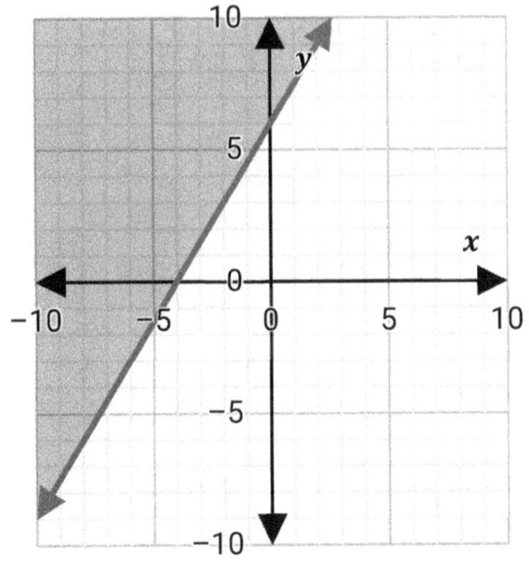

d. 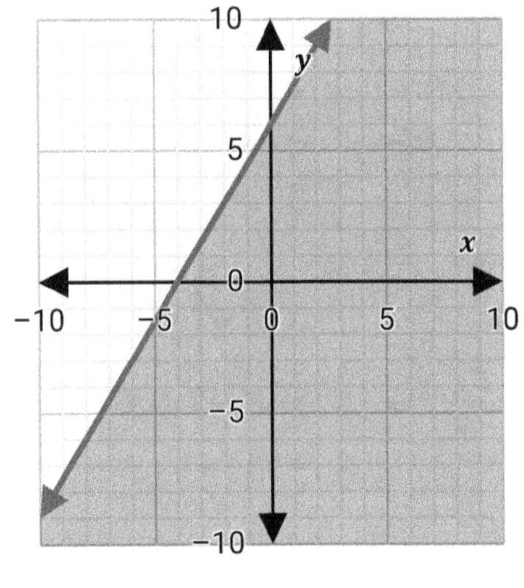

2. Graph the inequality: $x + 3y \leq 6$.

a. 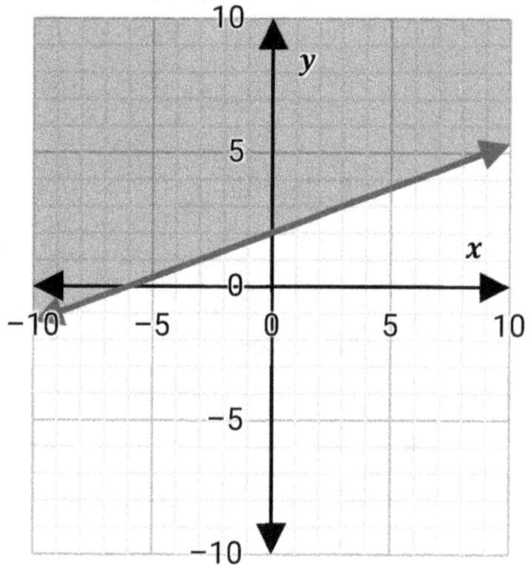

b. 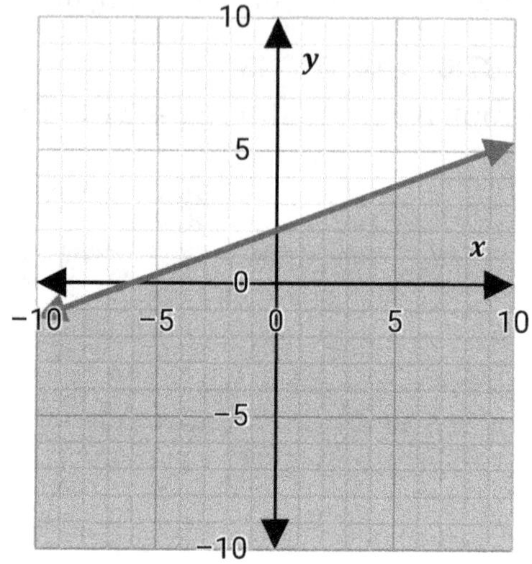

c. 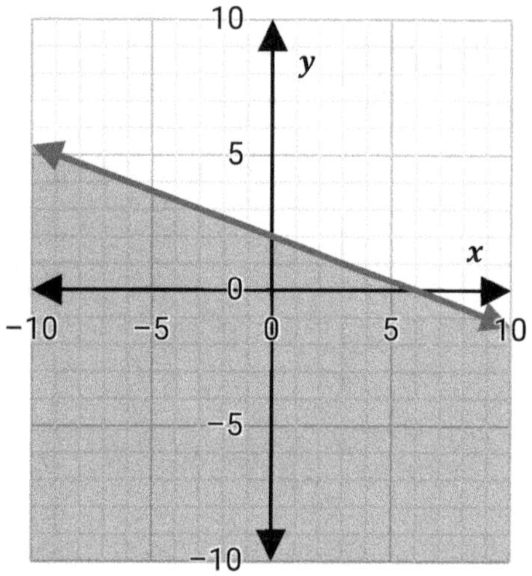

d. 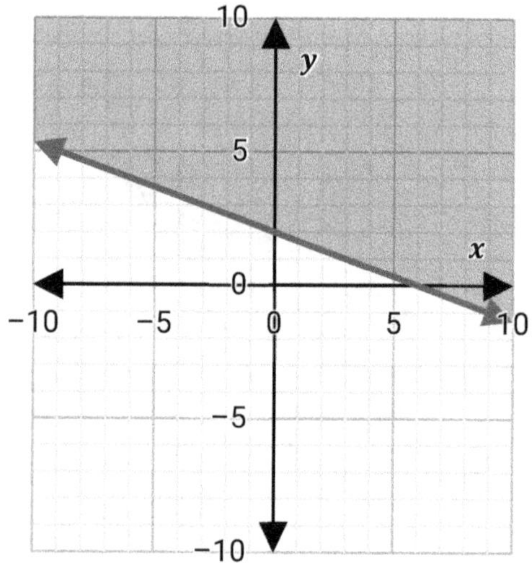

3. Solve the inequality: $5x - 10y < 4$.

a. 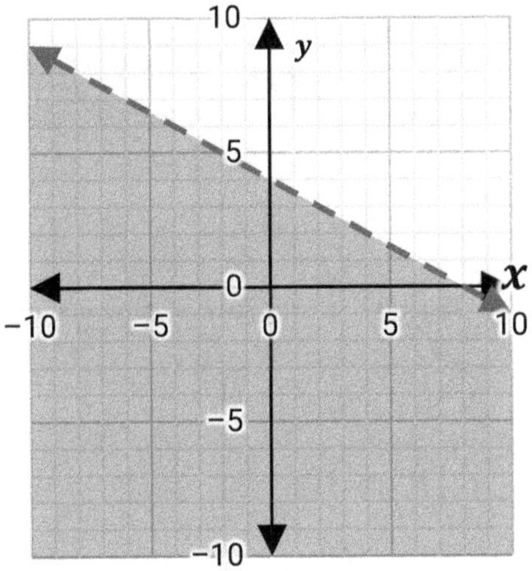

b. 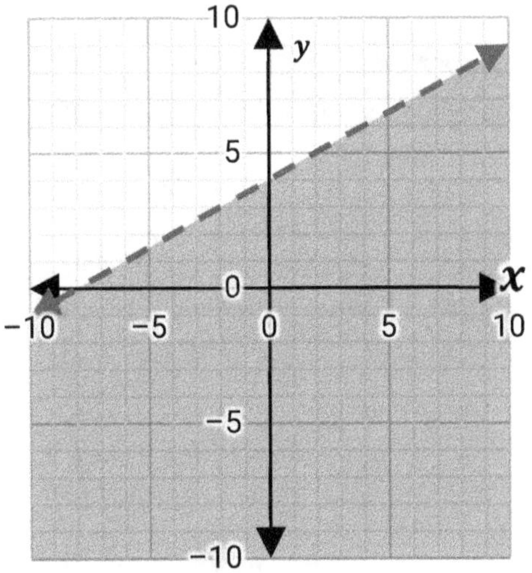

c. 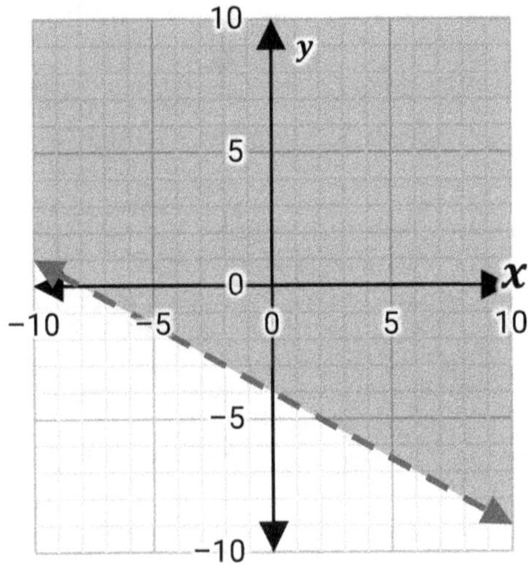

d. 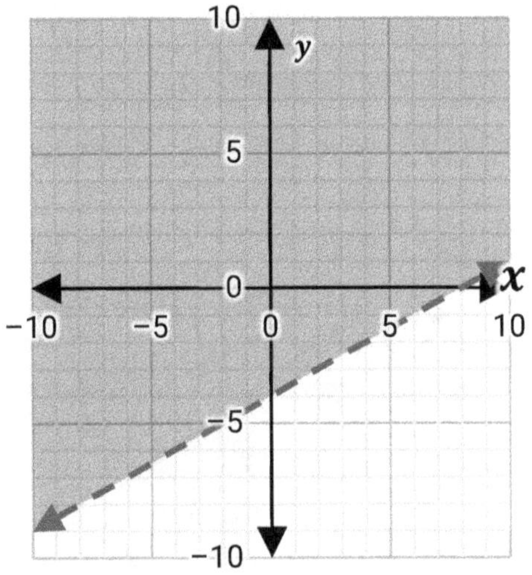

4. Graph the inequality: $\frac{1}{3}x + \frac{1}{2}y >$.

a.

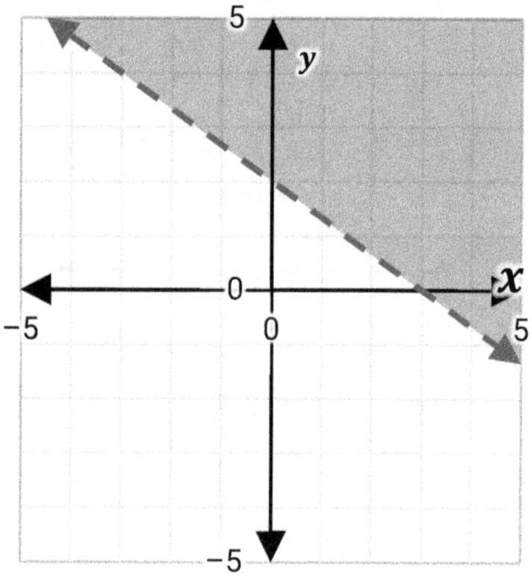

b.

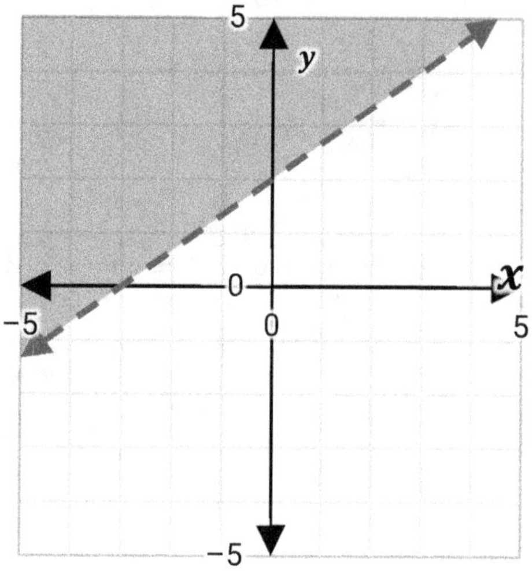

c.

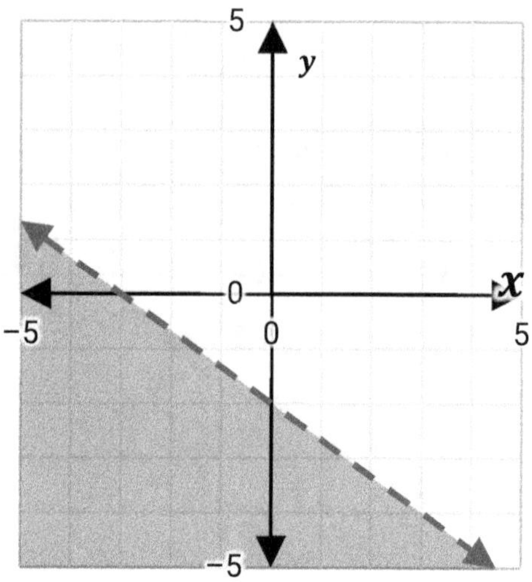

d.

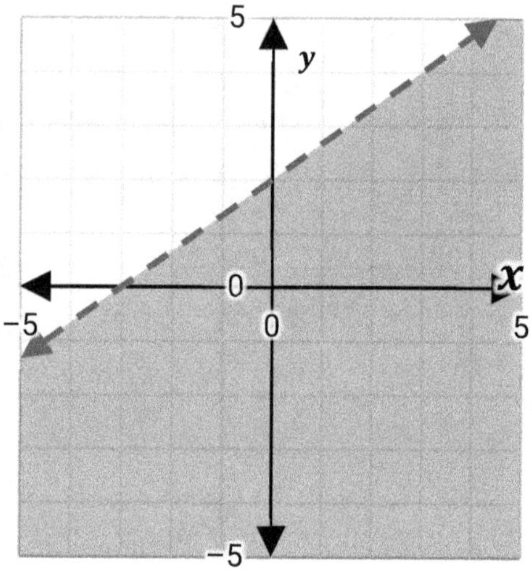

5. Graph the inequality: $2x - \frac{2}{3}y \leq 1$.

a.
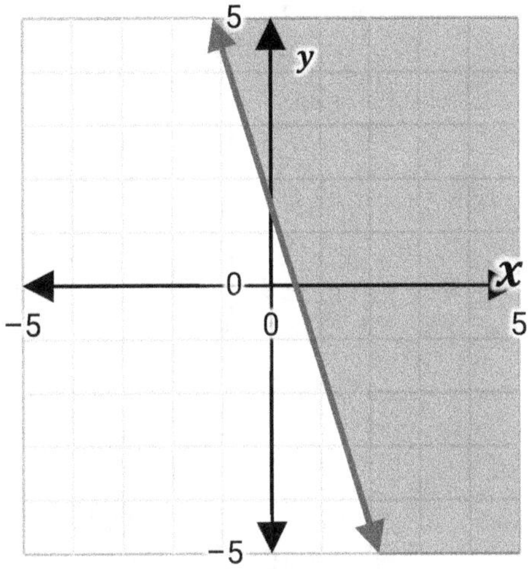

b.
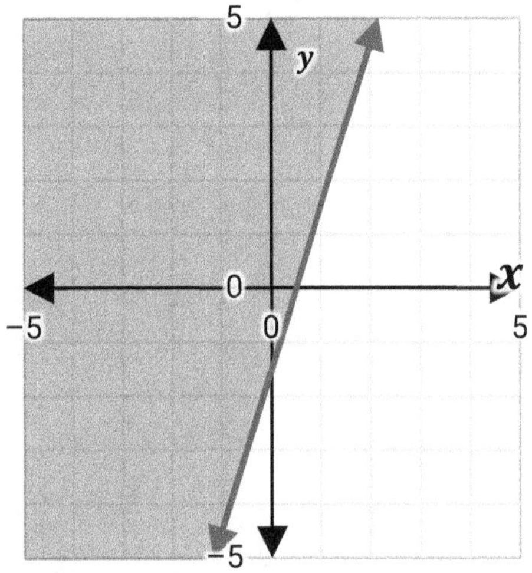

c.
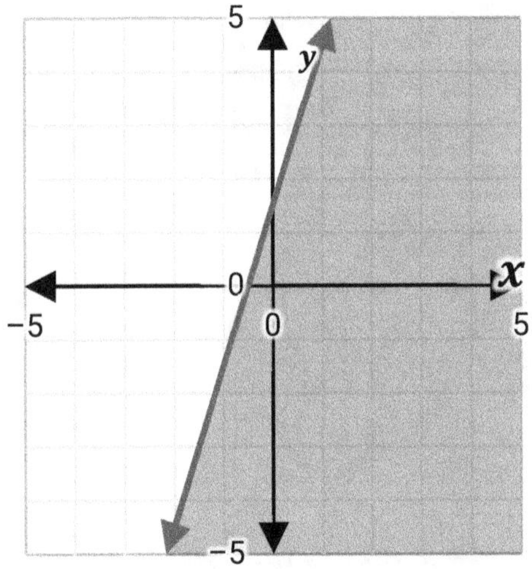

d.
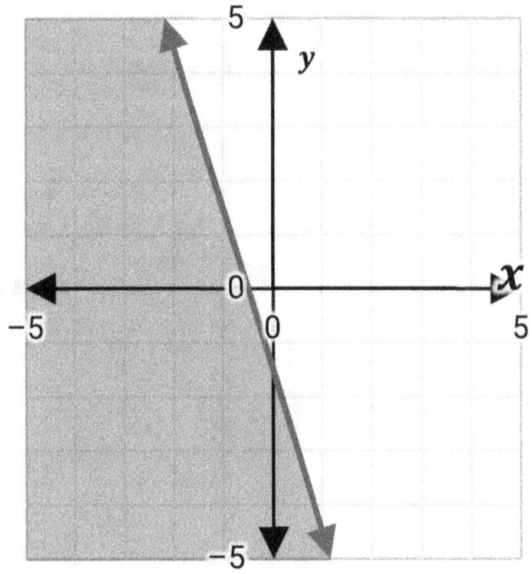

6. Graph the inequality: $-3x + \frac{3}{4}y < 6$.

a. 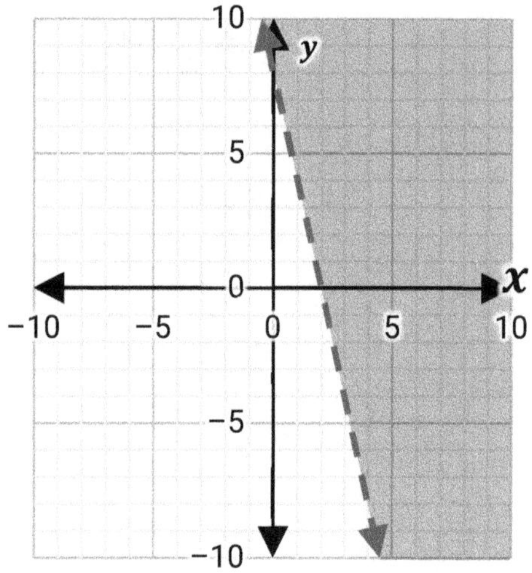

b. 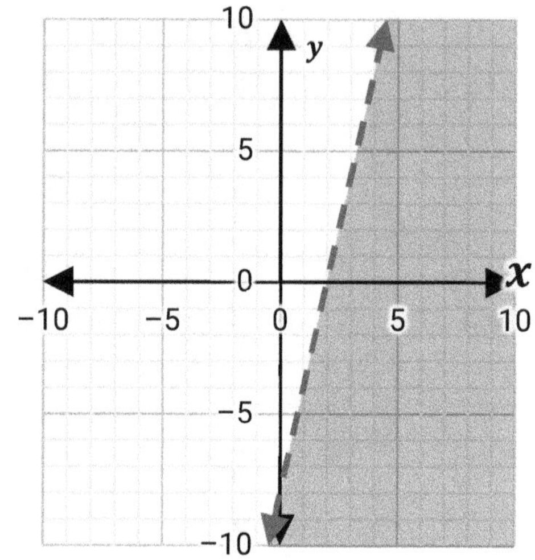

c. 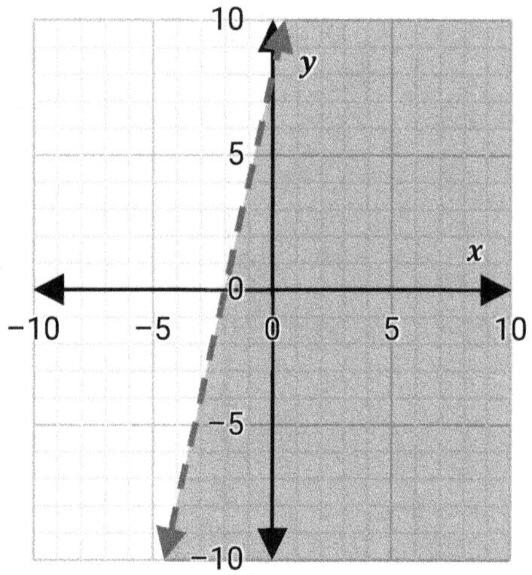

d. 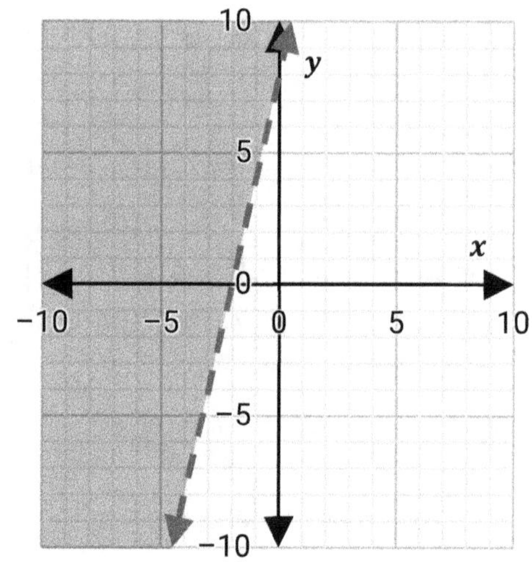

7. Graph the inequality: $y > \frac{2}{3}x + 1$.

a.
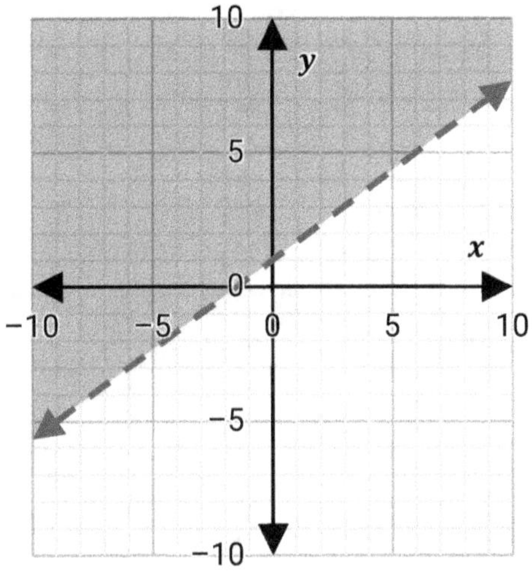

b.
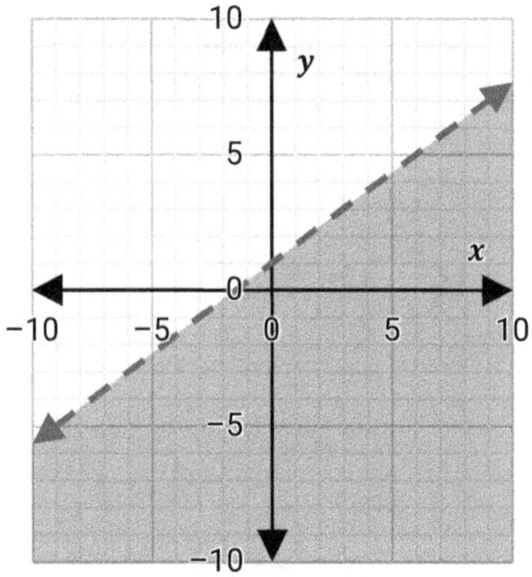

c.
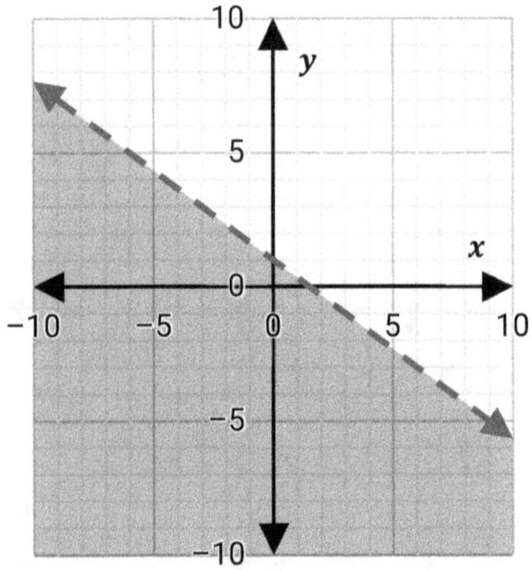

d.
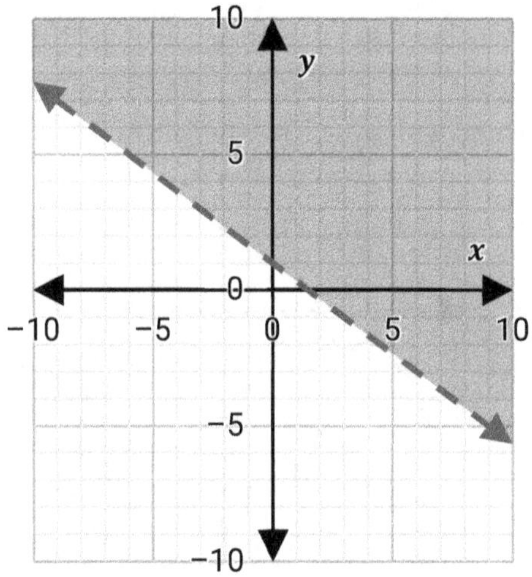

8. Graph the inequality: $y \leq \frac{3}{5}x - 2$.

a.

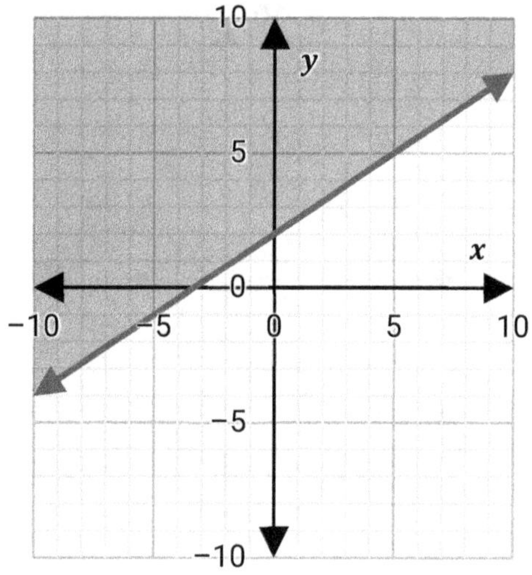

b.

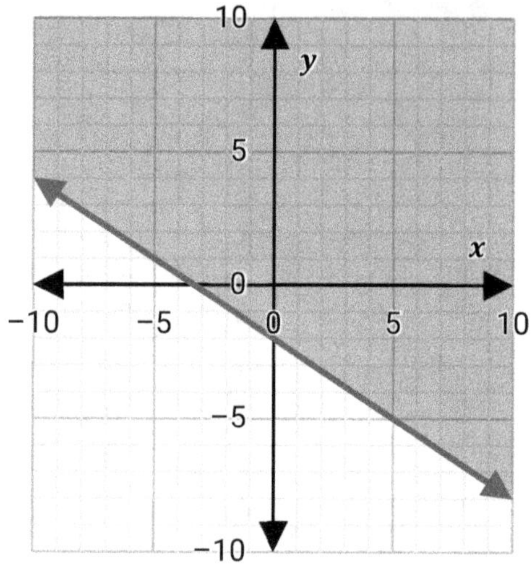

c.

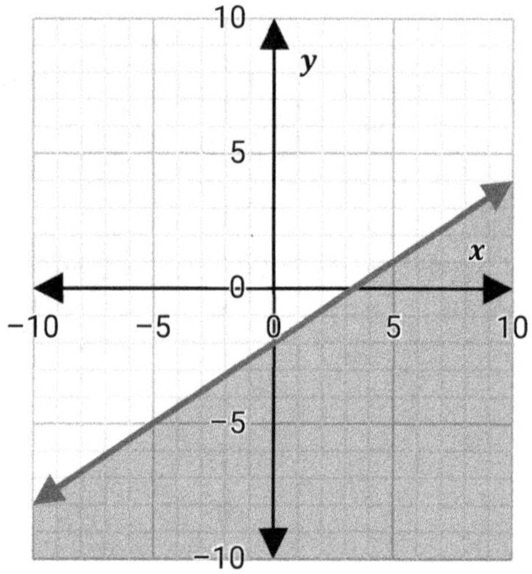

d.

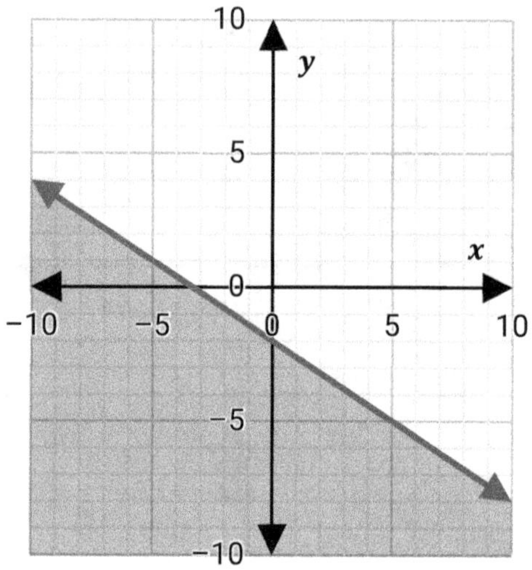

9. Graph the inequality: $y \geq -\frac{4}{7}x + 2$.

a.

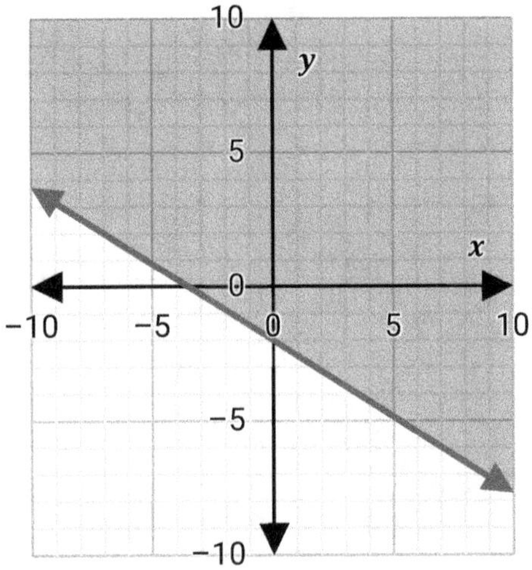

b.

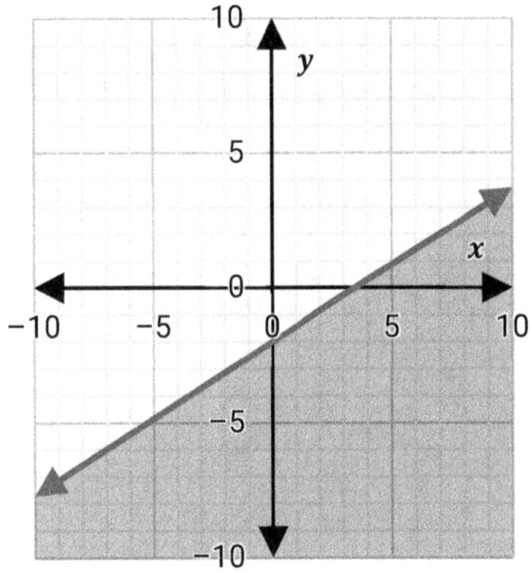

c.

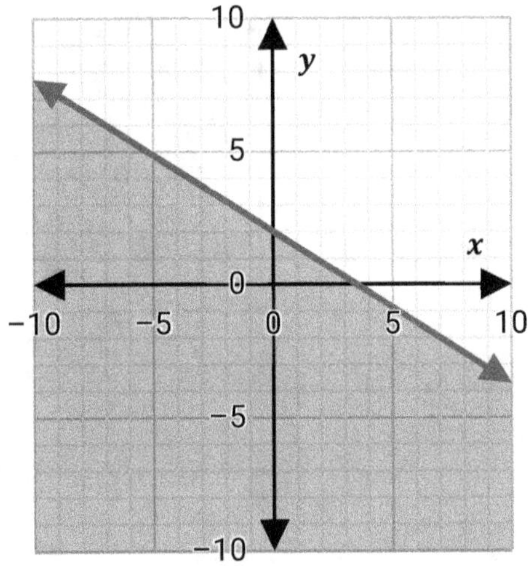

d.

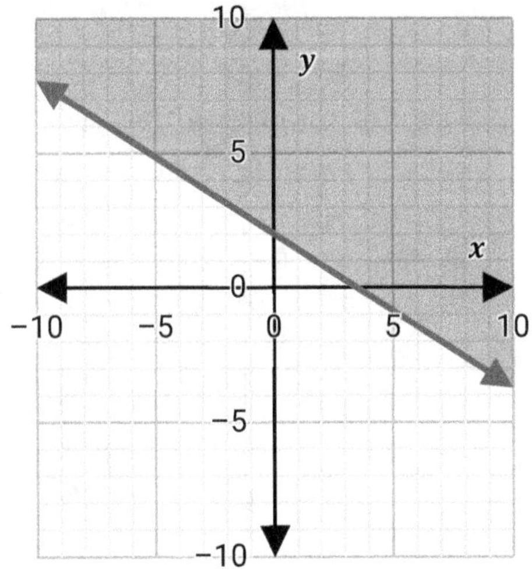

10. Graph the inequality: $y \leq -\frac{5}{2}x - 3$.

a.

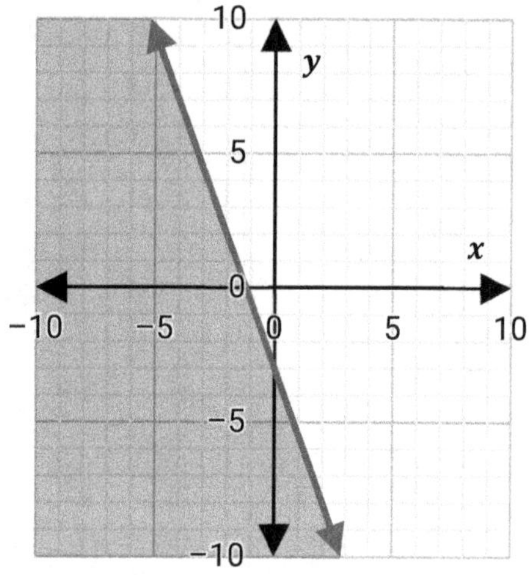

b.

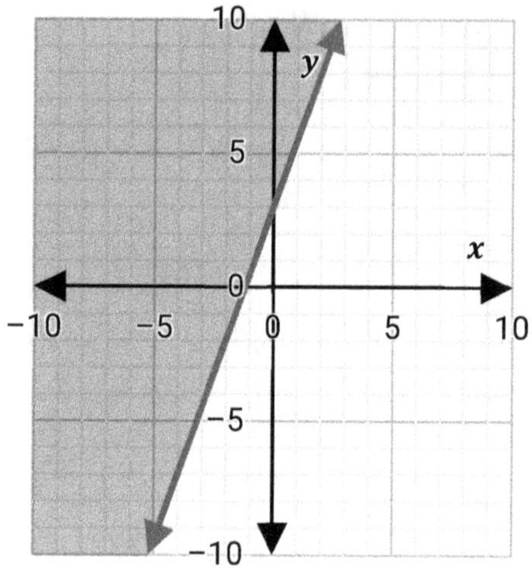

c.

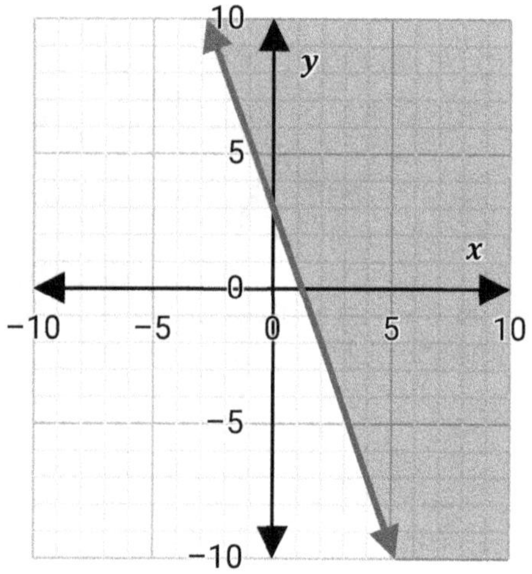

d.

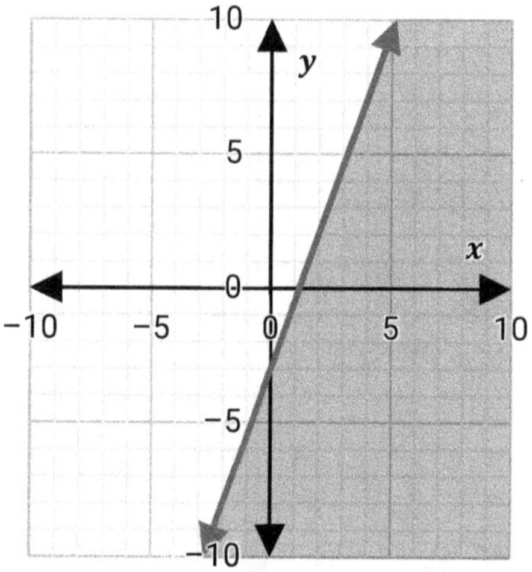

11. Graph the inequality: $y \leq 4x$.

a.

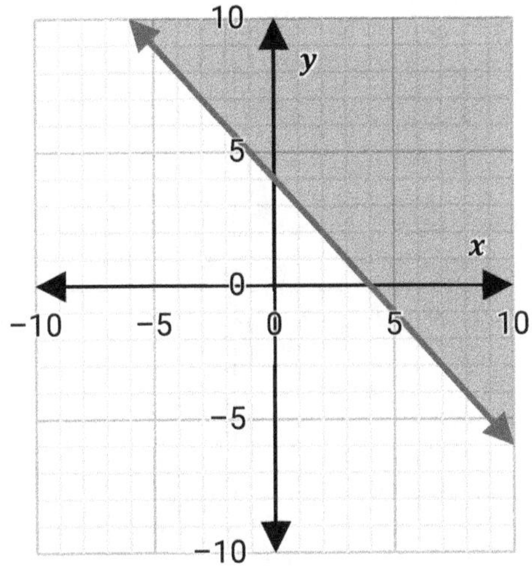

b.

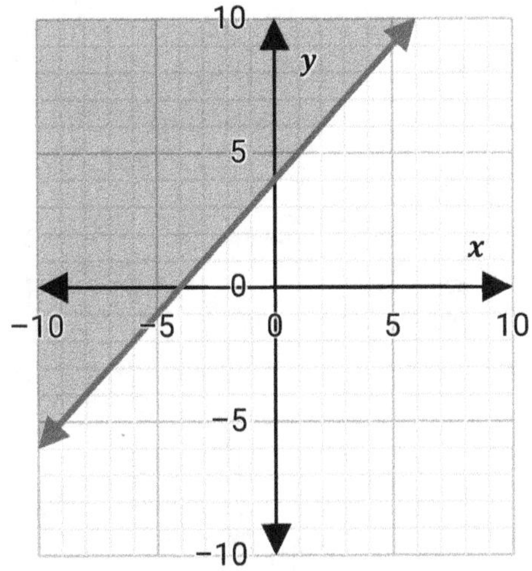

c.

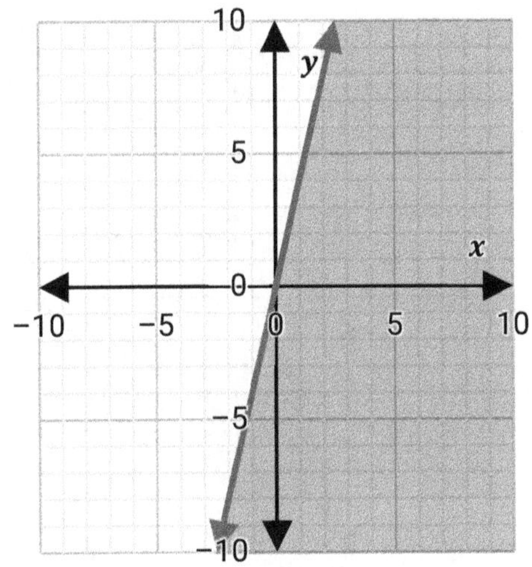

d.

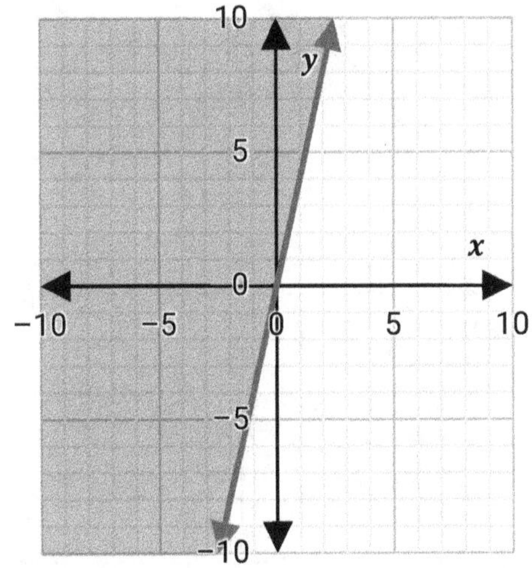

12. Graph the inequality: $y < -3x + 4$.

a.

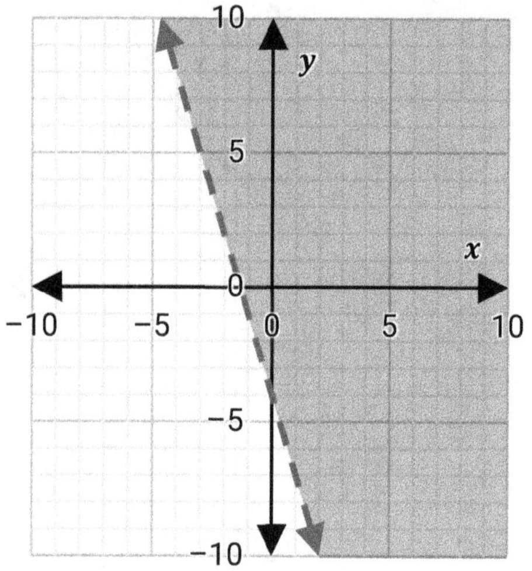

b.

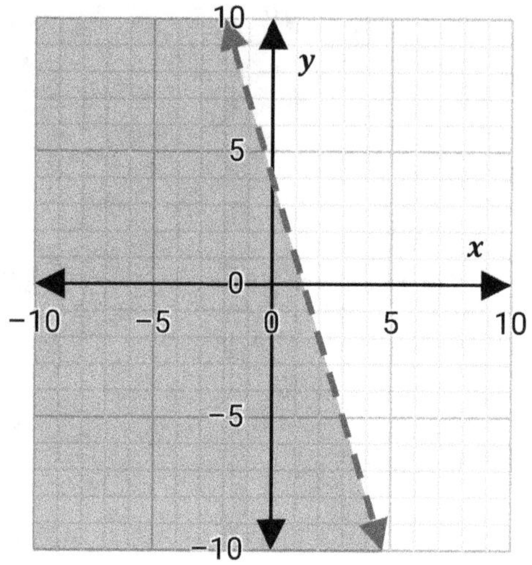

c.

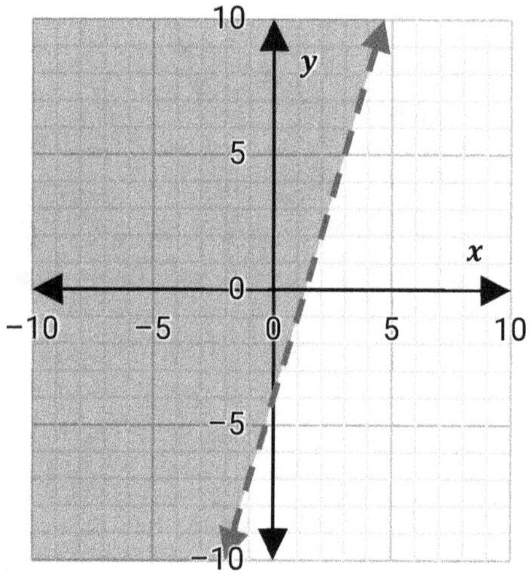

d.

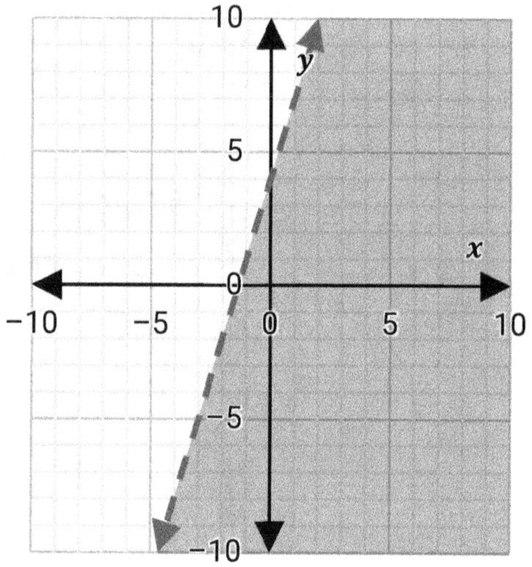

13. There are x number of male employees and y number of female employees that work for a company. If the company can have no more than 65 employees working for it at any given time, which graph represents the combination for the number male and female employees that work for the company?

a.

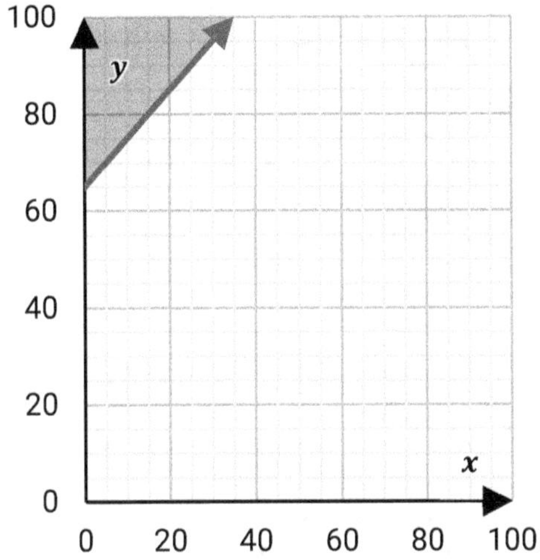

b.

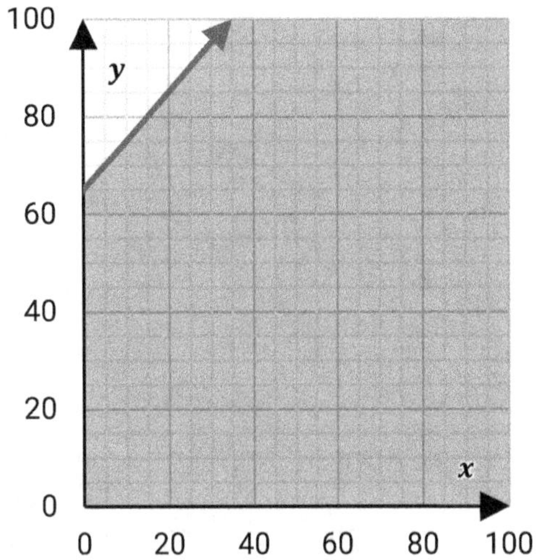

c.

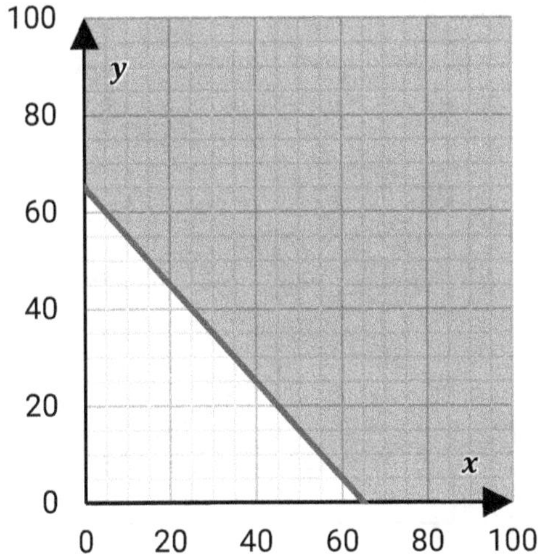

d.

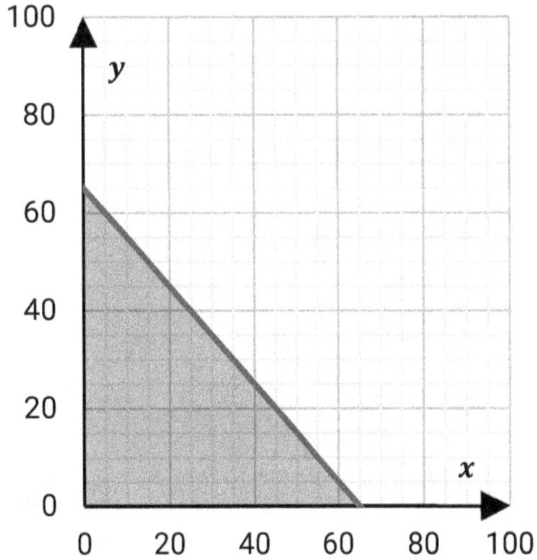

14. A department store purchases shirts for $5.00 each and purchases a pair of pants for $10.00 each. If x represents the number of shirts the store purchases and y represents the number of pants the store purchases, which graph represents the combination of shirts and pants the store can purchase so the total cost does not exceed $500.00?

a.

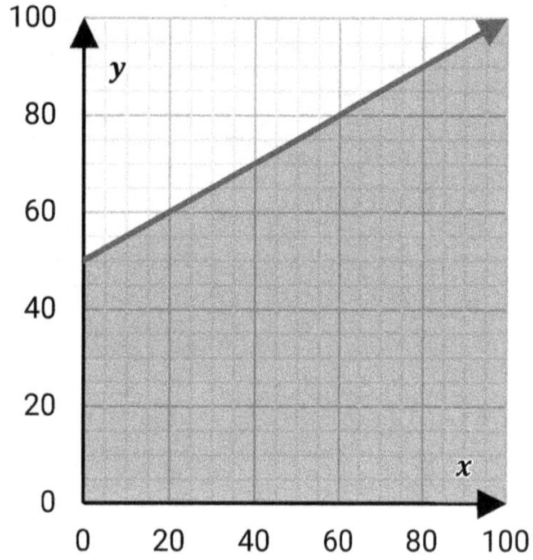

b.

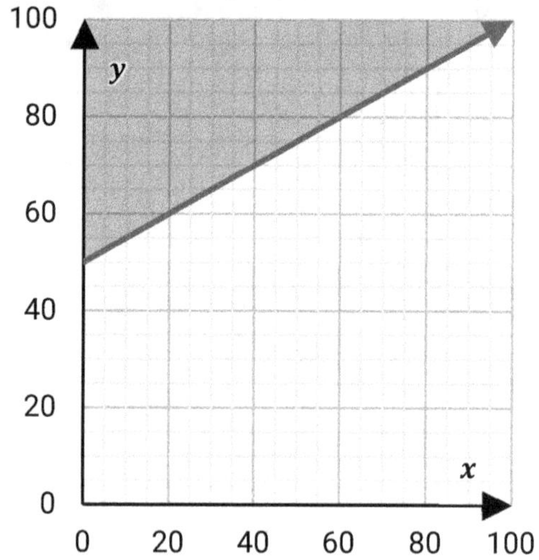

c.

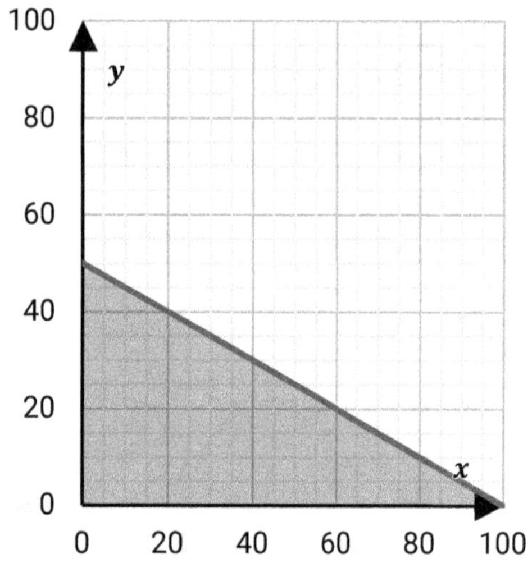

d.

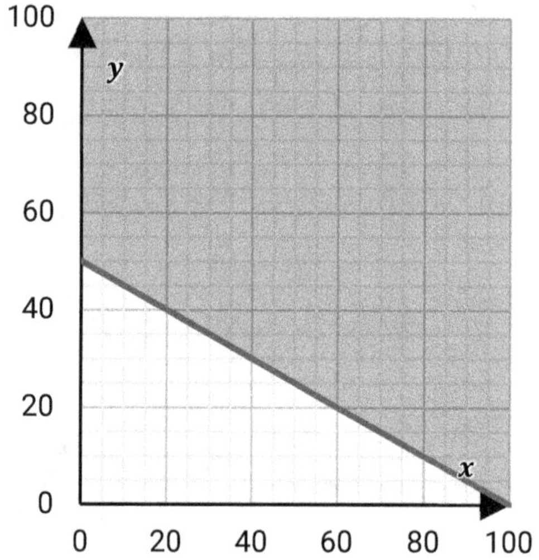

15. Let x and y be two numbers. If y is greater than five more than three times x, which graph represents the different combinations for x and y?

a. 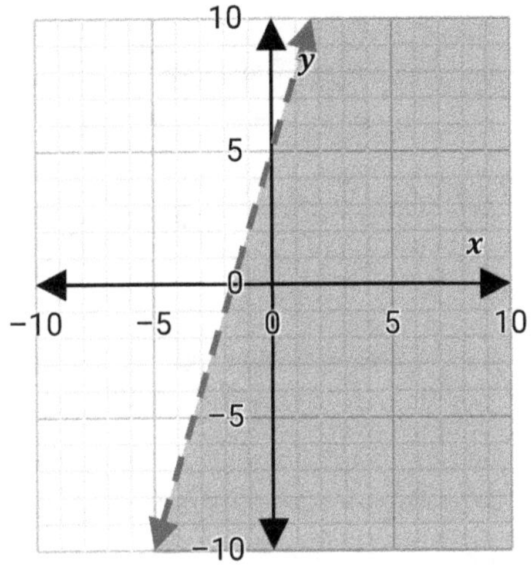

b. 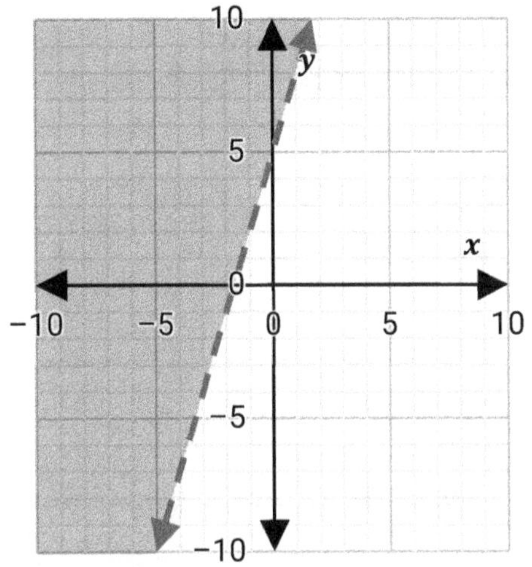

c. 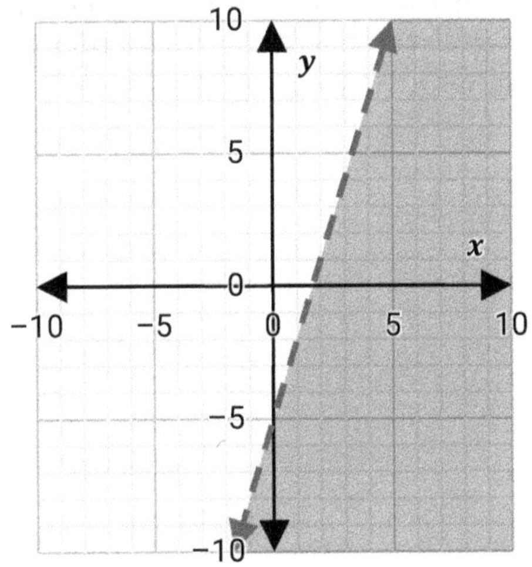

d. 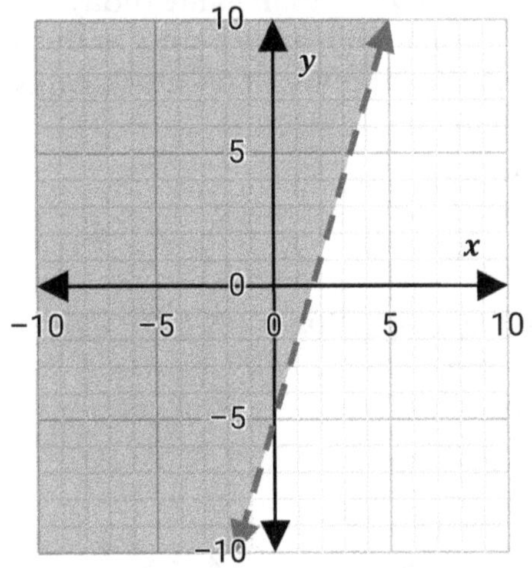

Chapter 4 Quiz

1. Which choice shows the correct way to graph the inequality $x < -2$?

a. 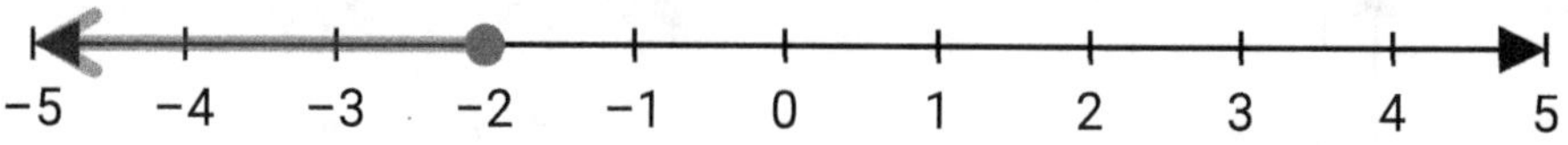

b. 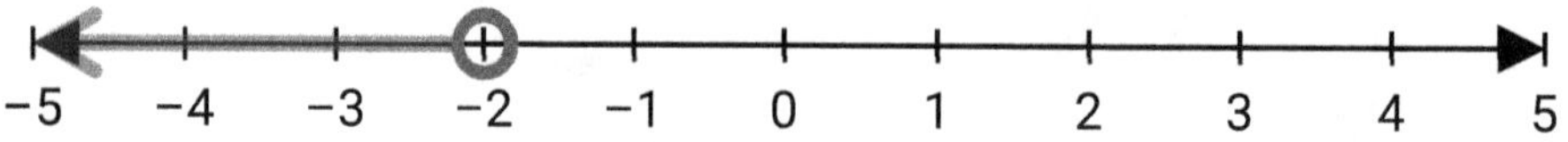

c. 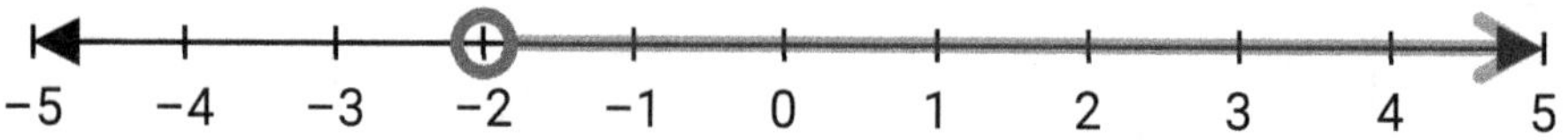

d. 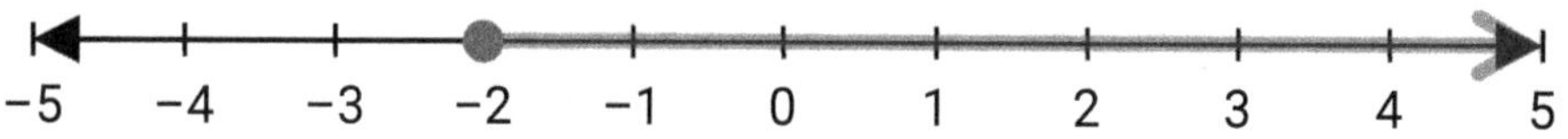

2. Aaron has a basketball game today. Aaron's father says that he will buy him an ice cream cone if he scores at least 6 points in the game. Express the scenario as an inequality where x represents the number of points Aaron scores in his game.

a. $x \le 6$
b. $x \ge 6$
c. $x < 6$
d. $x > 6$

3. Solve the inequality: $x + 5 > 3$.

a. $x > 2$
b. $x < 2$
c. $x > 8$
d. $x > -2$

4. Solve the inequality: $x - 4 < 1$.

a. $x < 3$
b. $x < 5$
c. $x < -3$
d. $x > 5$

5. Jesse is trying to save at least \$450 to buy a new bicycle. She's mowing one lawn every day and makes \$30 per lawn. Write an inequality that represents her savings goal, with x representing the number of lawns she mows.

a. $450x \ge 30$
b. $450x \le 30$
c. $30x \ge 450$
d. $30x \le 450$

6. Which inequality is represented by the graph below?

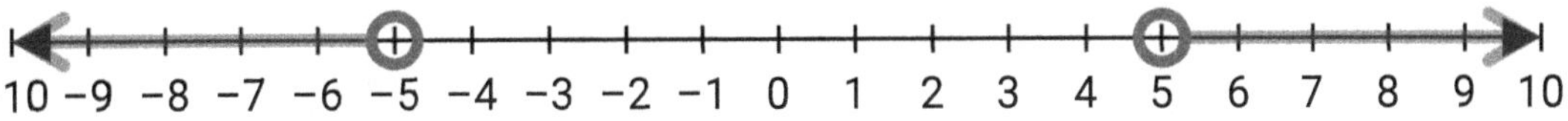

a. $x \leq 5$
b. $x < 5$
c. $|x| \geq 5$
d. $|x| > 5$

7. Solve the inequality:

$$3|x| + 7 > 9$$

a. $x < -\frac{2}{3}$ or $x > \frac{2}{3}$
b. $x > -\frac{2}{3}$ or $x < \frac{2}{3}$
c. $x < -\frac{16}{3}$ or $x > \frac{2}{3}$
d. $x > -\frac{16}{3}$ or $x < \frac{2}{3}$

8. A company that manufactures ping pong balls needs to ship boxes that contain 100 ping pong balls, plus or minus 5 balls. The absolute value inequality that represents this scenario is $|x - 100| \leq 5$, where x represents the acceptable number of ping pong balls in each box. Based on the allowable error of 5 balls, which compound inequality statement represents the number of ping pong balls that could be in each box?

a. $95 \leq x \leq 105$
b. $95 \geq x$ or $x \leq 105$
c. $95 < x < 105$
d. $95 > x$ or $x < 105$

9. A local amusement park charges $10 per adult ticket and $2 per child ticket. If they made at least $670 one day, which inequality represents the combination of adult tickets sold, x, and child tickets sold, y?

a. $10x + 2y \leq 670$
b. $10x + 2y \geq 670$
c. $2x + 10y \geq 670$
d. $2x + 10y \leq 670$

10. The shaded half-plane represents the solution set of a linear inequality. Find the inequality.

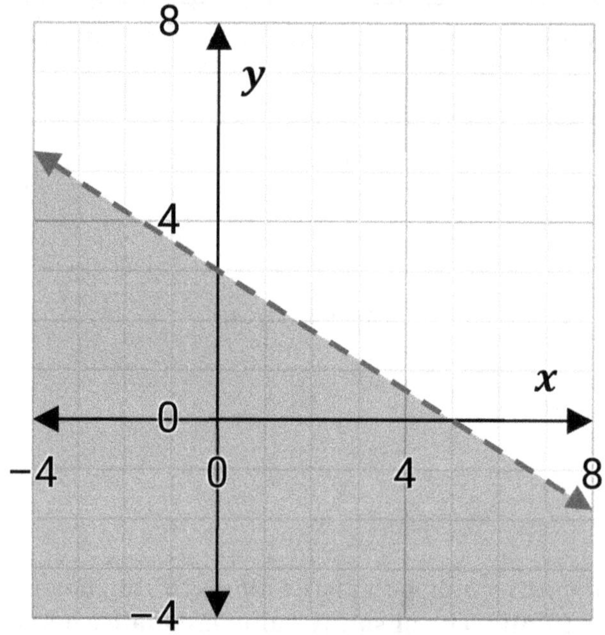

a. $3x + 5y < 15$
b. $3x - 5y < -15$
c. $3x - 5y > -15$
d. $3x + 5y > 15$

11. You manufacture candles by hand in a rented crafting space that costs you \$100 per day. The materials to produce a single candle cost you \$5, and you can sell the candle for \$8. If x is the number of candles you make in a day and y is the number you sell, write an inequality expressing the values of x and y that allow you to make a profit on a particular day. Note that it is possible to sell more candles than you make in a day since you can sell candles left over from previous days.

a. $8y + 5x \geq 100$
b. $8y + 5x > 100$
c. $8y - 5x \geq 100$
d. $8y - 5x > 100$

12. Graph the following inequality: $3x - 4y > 12$.

a.

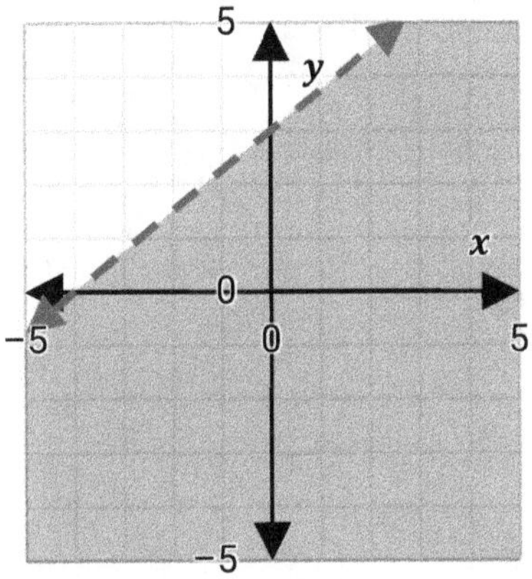

c.

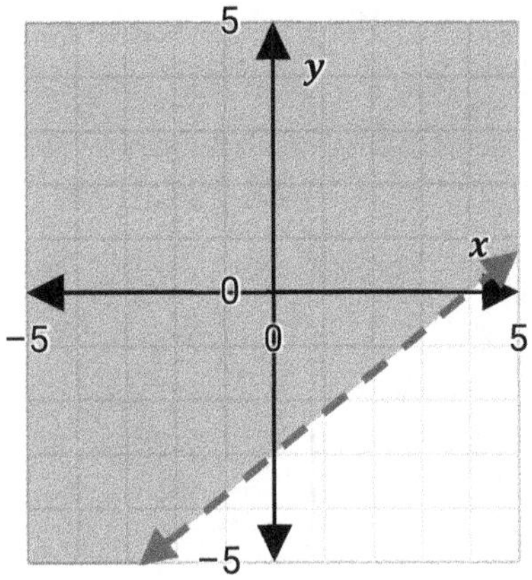

b.

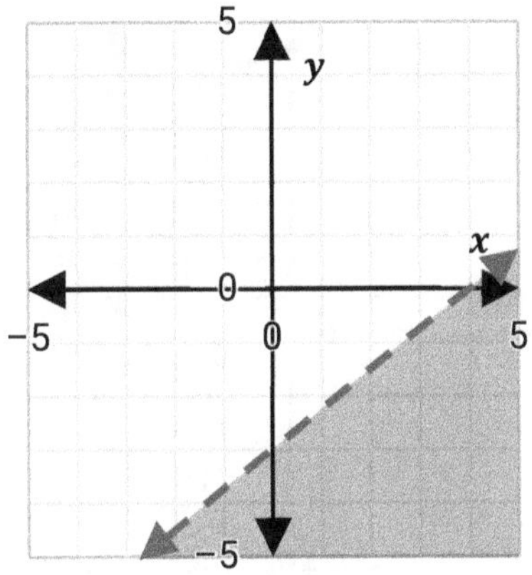

d.

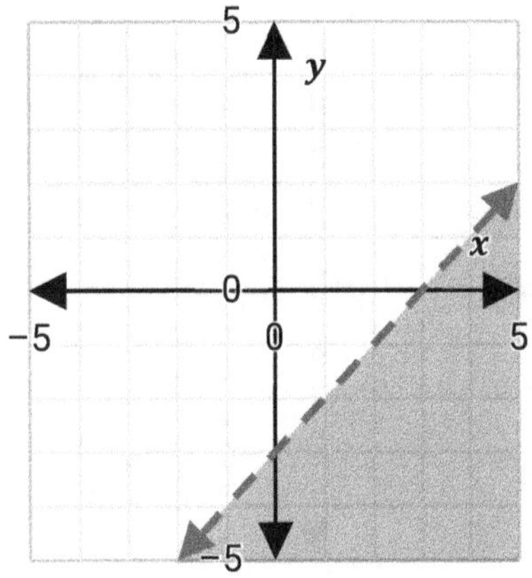

13. Graph the following inequality: $-7x + 6y < 6$.

a. 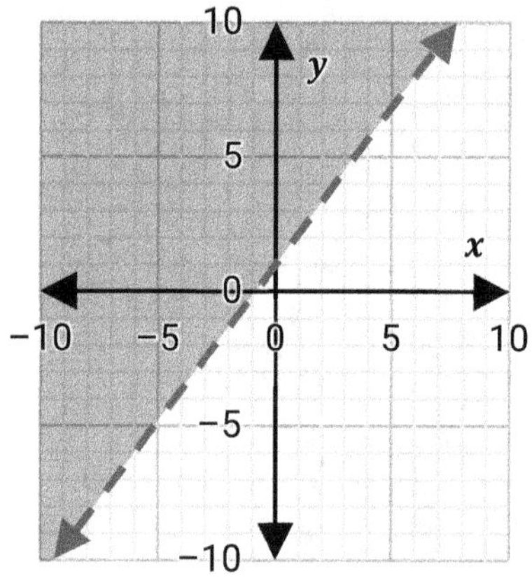

c. 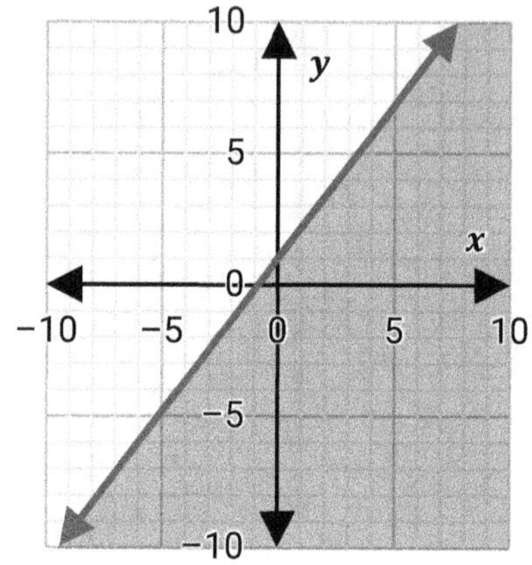

b. 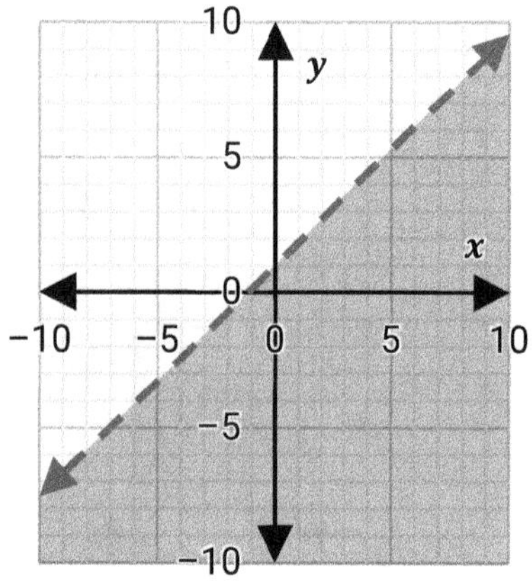

d. 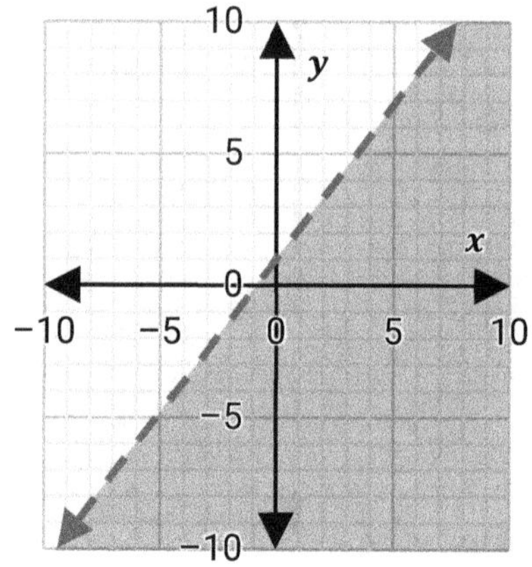

14. Danny has $1,000 in his savings account. He wants to have at least $300 left in this account after the end of the summer. He plans on only using the money in this account for groceries. If Danny withdraws $100 every week for groceries, how many weeks can he withdraw money for groceries? Express the scenario as an inequality, and then graph the solution.

a. $w \geq -7$

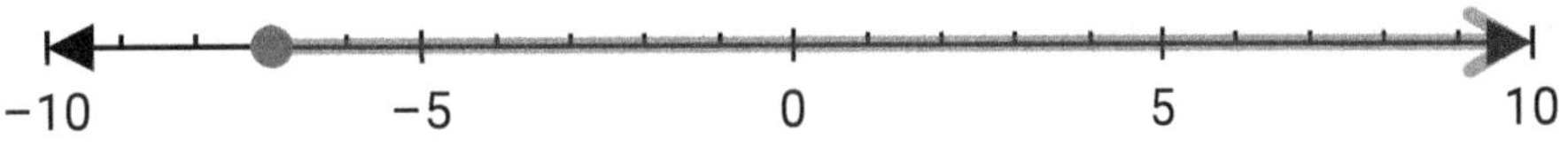

b. $w \geq 7$

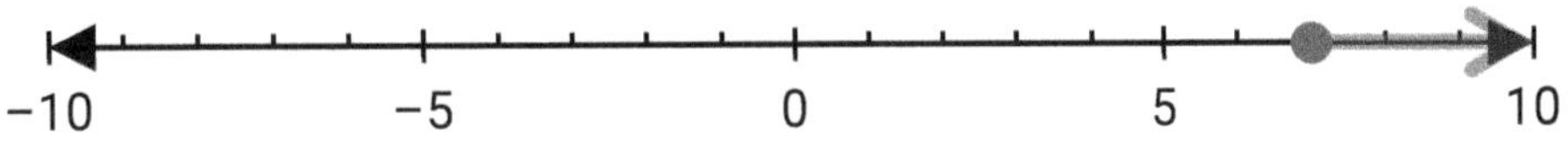

c. $w \leq -7$

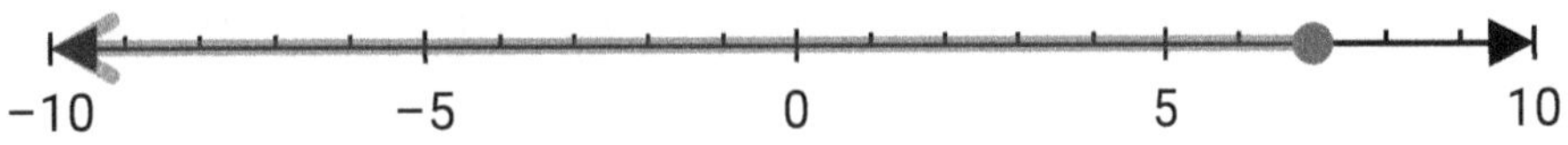

d. $0 \leq w \leq 7$

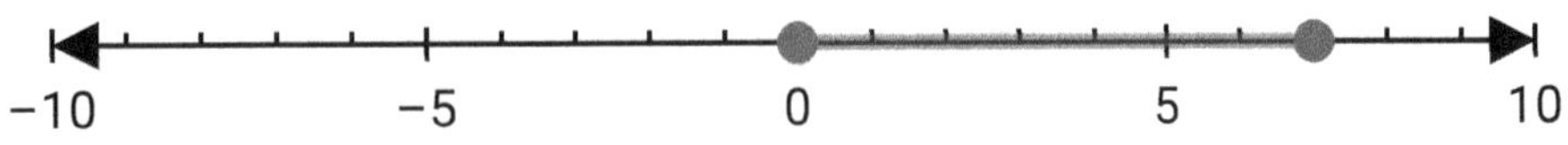

15. Solve the inequality and then graph your answer: $2x + 7 > 1$.

a. $x > -3$

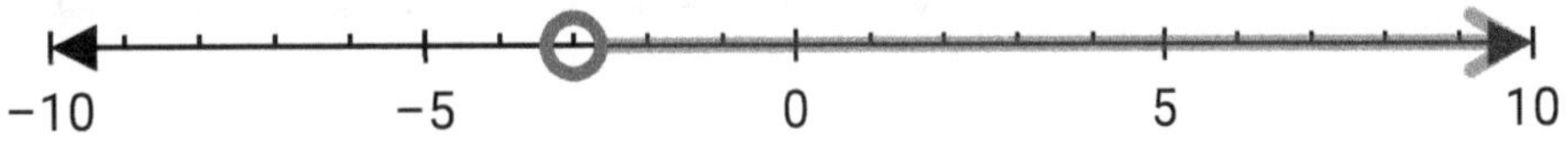

b. $x > -2$

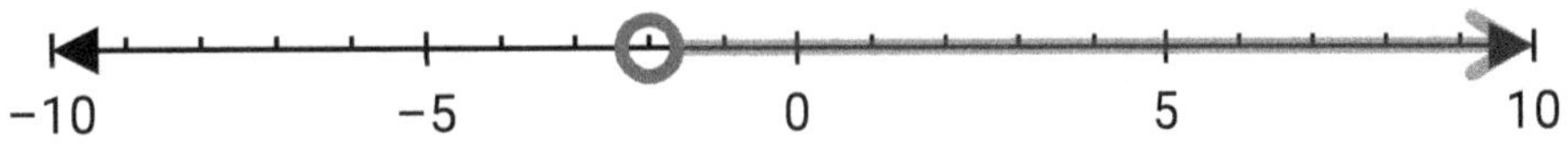

c. $x > -6$

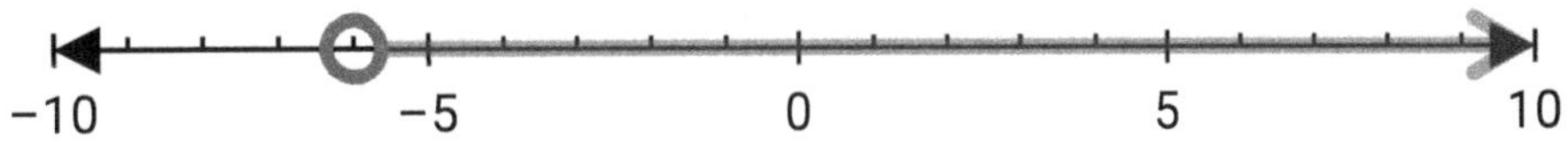

d. $x < -3$

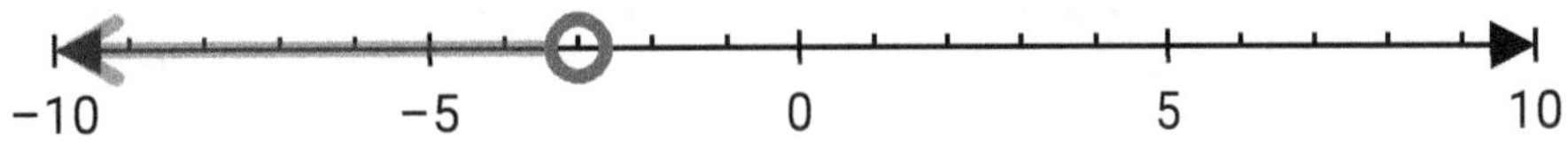

16. Solve the inequality: $5x + 2 \leq 2x + 17$.

a. $x \geq 5$
b. $x \leq 15$
c. $x \geq 15$
d. $x \leq 5$

17. Seven times a number decreased by 5 is less than two times the same number increased by 20. What values for the number make this true? Let x represent the number.

a. $x < 20$
b. $x < 25$
c. $x < 5$
d. $x < -5$

18. Solve the inequality and then graph your answer: $2x + 9 > 1$ and $3x - 13 < 2$.

a. $-4 < x < 5$

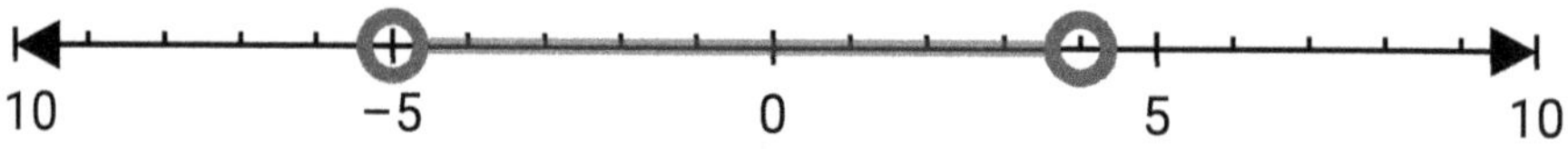

b. $-5 < x < 4$

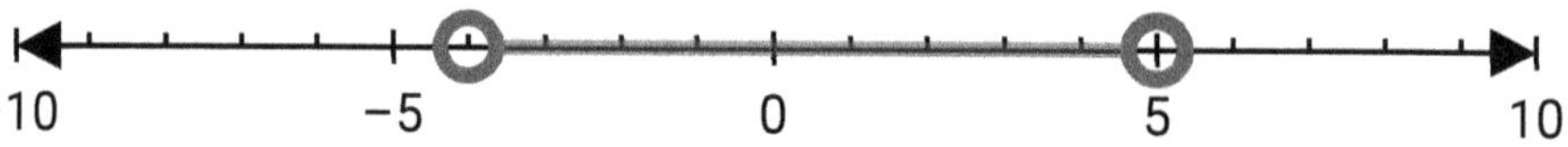

c. $x < 5$

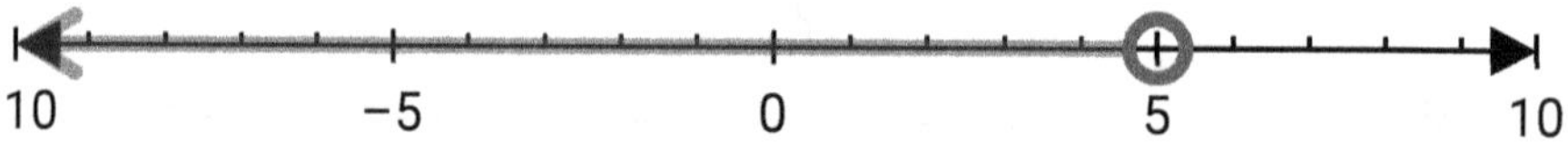

d. $-\infty < x < \infty$

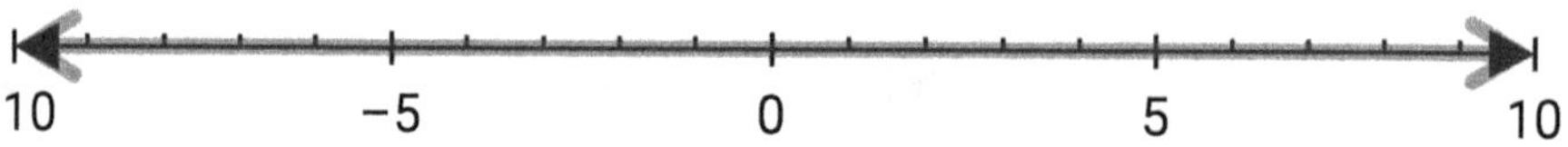

19. Solve the inequality and then graph your answer: $x + 5 \leq 1$ or $2x - 4 \geq 4$.

a. $x \leq -4 \text{ or } x \geq 4$

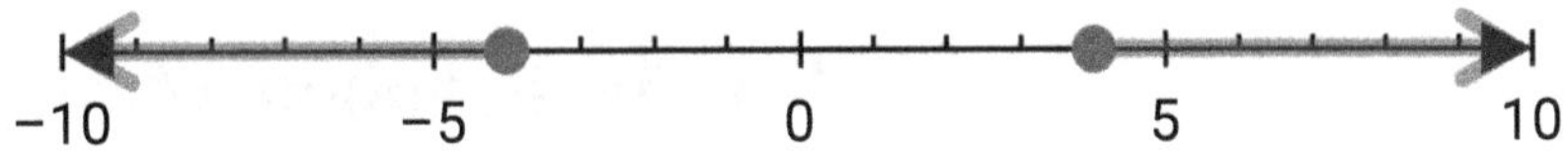

b. $x \leq -4$

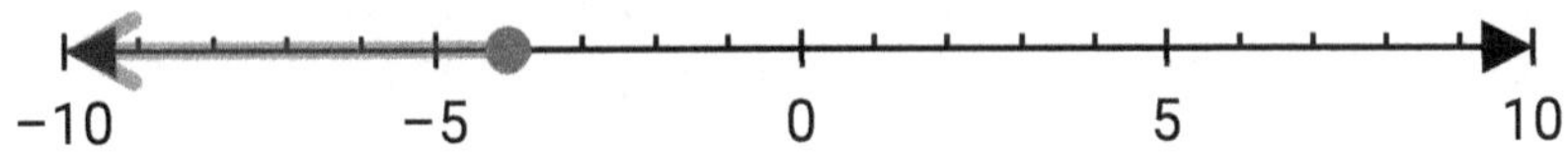

c. $x \geq 4$

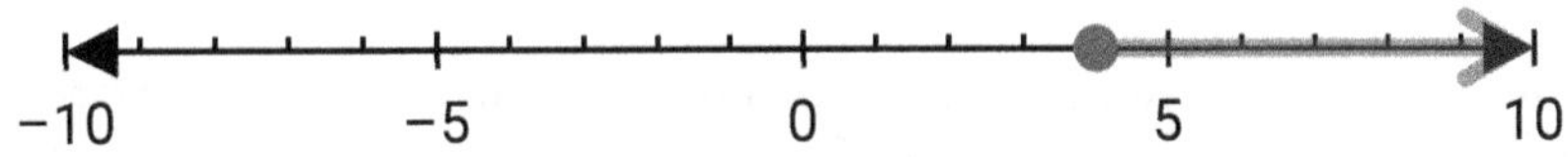

d. $-4 \leq x \leq 4$

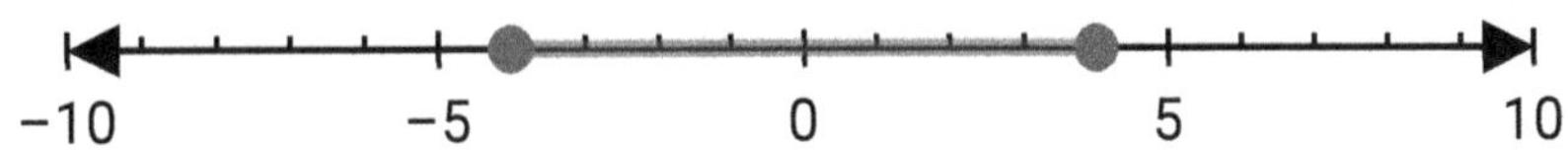

20. Solve the inequality: $3x + 7 < 19$ or $x + 7 > 5$.

a. $x < 4$ and $x > -2$
b. $x > 4$
c. $x > -2$
d. $-\infty < x < \infty$

Chapter 5: Systems of Equations

5.1 The Graphing Method

A **system of equations** is a set of two or more equations that can be solved to find their point of intersection or lack of a point of intersection. To solve a system of linear equations **graphically**, plot both equations on the same graph. The solution of the equations is the point where both lines cross. If the lines do not cross, they are parallel lines and the system has **no solution**. If the lines overlap, they are the same line and the system has **infinite solutions**.

For example, consider the following system of equations:

$$\begin{cases} y = 2x + 7 \\ y = -x + 1 \end{cases}$$

Since these equations are given in slope-intercept form, they can be graphed quickly. The y-intercepts of the lines are (0,7) and (0,1). The respective slopes are 2 and –1. The lines are graphed below.

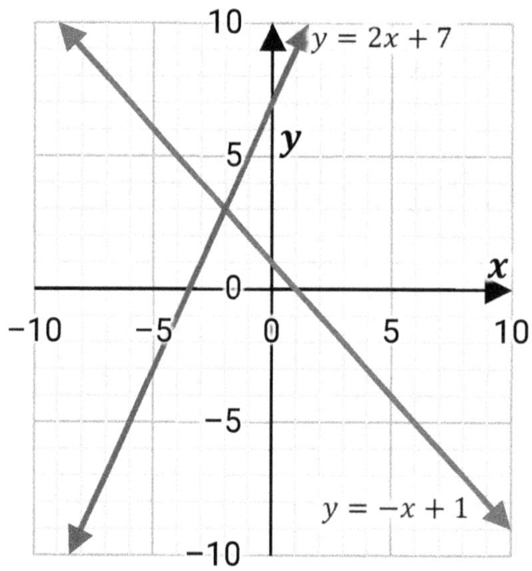

The two lines intersect at the point $(-2,3)$, so this is the solution to the system of equations.

Solving a system graphically is generally only practical if both coordinates of the solution are integers; otherwise, the intersection will lie between gridlines on the graph and the coordinates will be difficult or impossible to determine exactly. It also helps if, as in this example, the equations are in slope-intercept form so they can be graphed quickly. Otherwise, another method of solution is likely to be more useful.

Review Video: Solving Systems by Graphing
Visit mometrix.com/academy and enter code: 634812

Examples

Example 1:

Solve the system of linear equations by graphing.

$$\begin{cases} y = 2x - 4 \\ y = -2x + 8 \end{cases}$$

Answer

Graph both lines and identify where they intersect.

$y = 2x - 4$ has a y-intercept of –4 and a slope of 2.

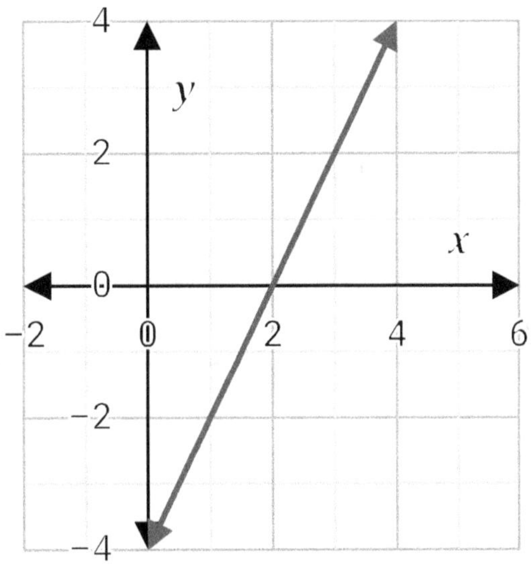

$y = -2x + 8$ has a y-intercept of 8 and a slope of –2.

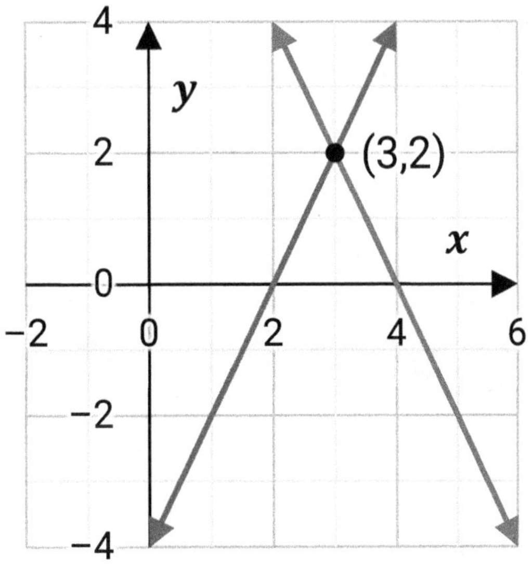

The lines intersect at (3,2), so this is the solution to the system of equations.

EXAMPLE 2:

Solve the system of linear equations by graphing.

$$\begin{cases} 4x - 3y = 4 \\ 2x + 3y = 11 \end{cases}$$

ANSWER

The equations are in standard form. Convert both equations to slope-intercept form so that they are easier to graph.

$$\begin{aligned} 4x - 3y &= 4 \\ -3y &= -4x + 4 \\ y &= \frac{4}{3}x - \frac{4}{3} \end{aligned}$$

$$\begin{aligned} 2x + 3y &= 11 \\ 3y &= -2x + 11 \\ y &= -\frac{2}{3}x + \frac{11}{3} \end{aligned}$$

Graph both equations.

$y = \frac{4}{3}x - \frac{4}{3}$ has a y-intercept of $-\frac{4}{3}$ and a slope of $\frac{4}{3}$.

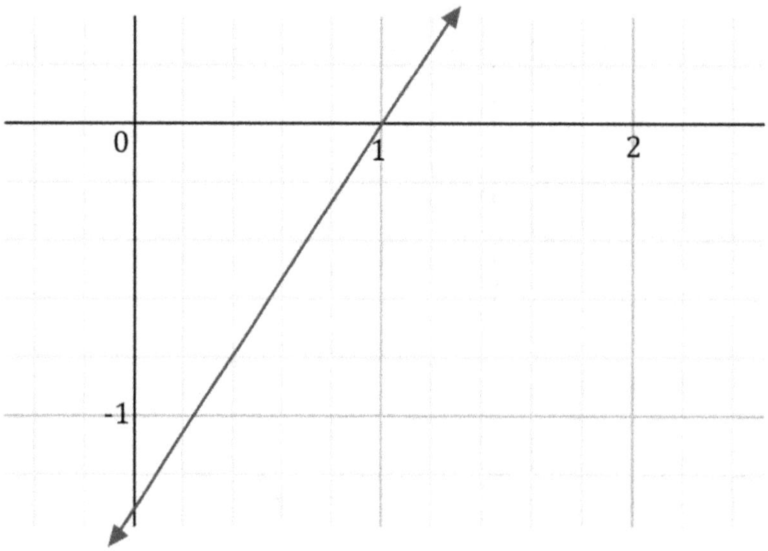

$y = -\frac{2}{3}x + \frac{11}{3}$ has a y-intercept of $\frac{11}{3}$ and a slope of $-\frac{2}{3}$.

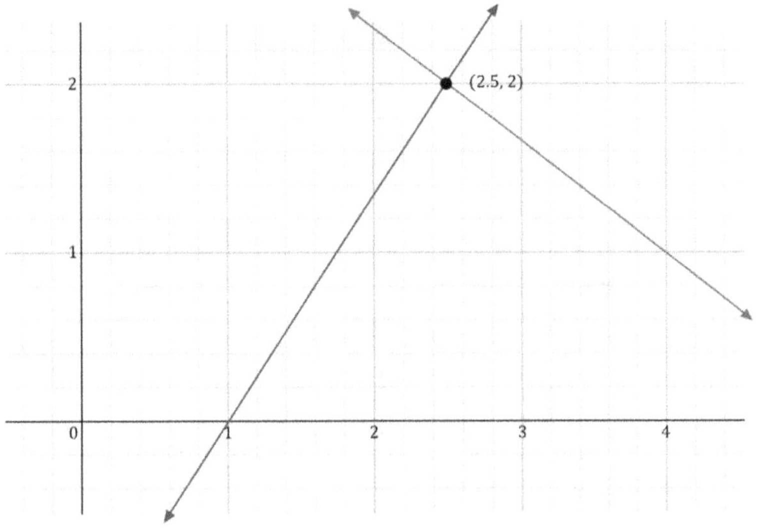

The lines intersect at (2.5,2), so this is the solution to the system of equations.

EXAMPLE 3:

Shane and Max both started new jobs working at different restaurants. Shane's income is \$10 per hour. Shane also received a one-time \$100 bonus as an incentive for signing his contract to work for the restaurant. Max's income is also \$10 per hour. Unfortunately, Max had to pay a one-time fee of \$50 to cover the mandatory background check and paperwork fee. How many hours will it take for Shane and Max to earn the exact same amount of money? Solve the system of equations using the graphing method.

ANSWER

Let x represent the number of hours worked and y represent the total amount of money earned.

$$\begin{cases} y = 10x + 100 \\ y = 10x - 50 \end{cases}$$

When the equations are graphed, the lines never intersect, meaning they are parallel. When two lines are parallel, this means their slopes are the same, and this indicates that there is no solution. In this case, "no solution" means that there is no specific number of hours of work that will produce the exact same amount of money for Shane and Max. In other words, there is no x and y pair that will match the constraints provided in the problem. Regardless of how many hours of work occur, the amount of money earned will never be exactly the same.

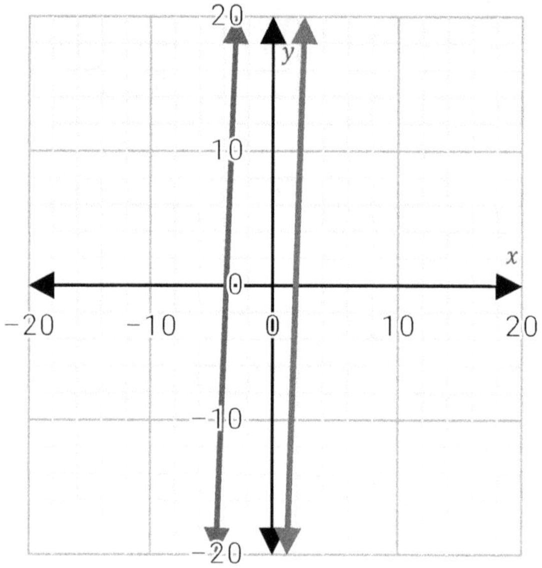

PRACTICE QUESTIONS

1. Solve the following system of equations using the graphing method: $\begin{cases} y = -\frac{1}{2}x + 4 \\ y = x + 1 \end{cases}$.

a. (2,3)

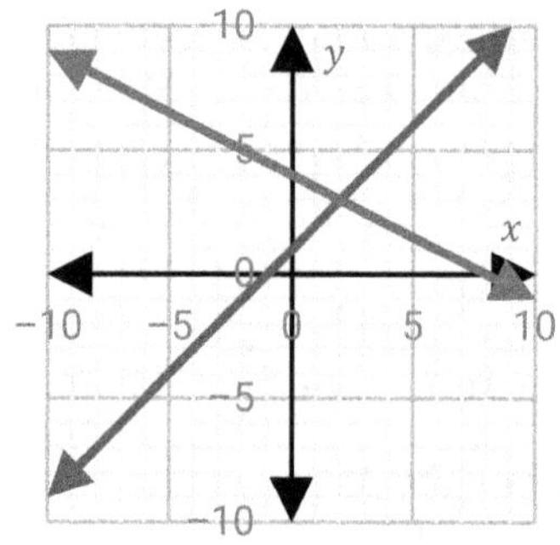

b. (6,7)

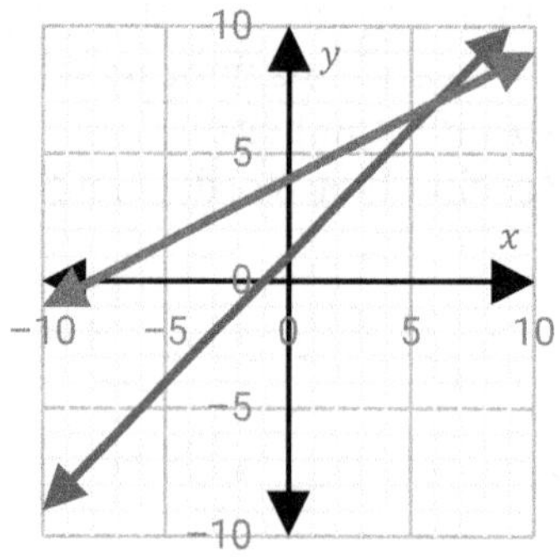

c. (3,2)

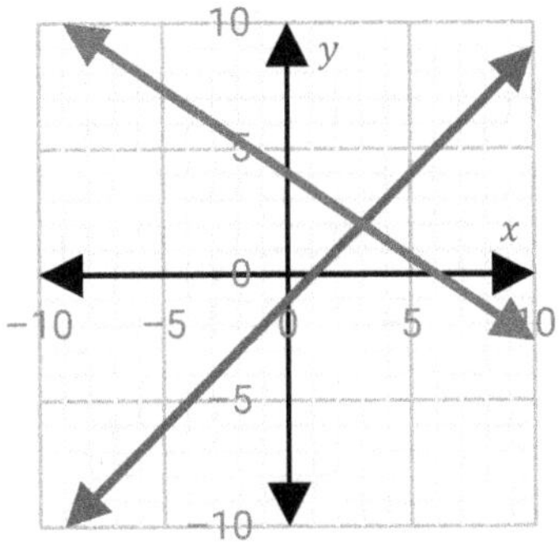

d. No solution

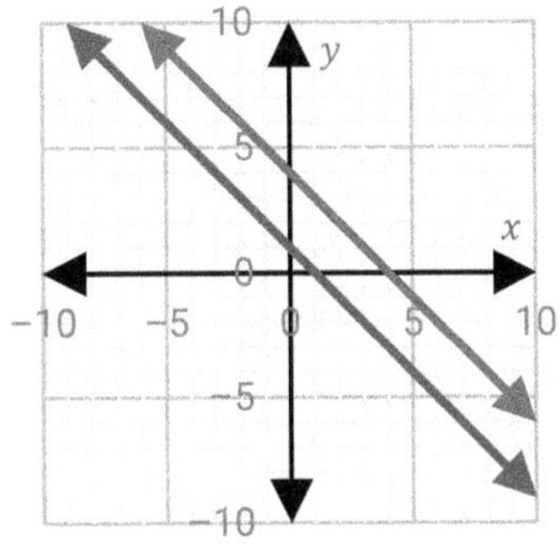

2. Solve the following system of equations using the graphing method: $\begin{cases} y = 2x + 3 \\ y = x + 5 \end{cases}$.

a. (−2,7)

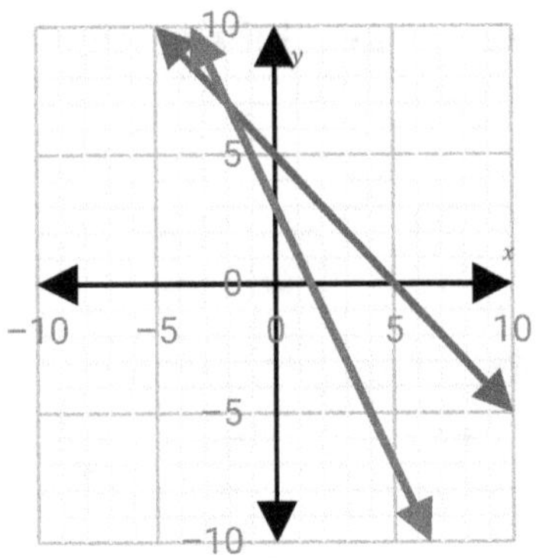

b. (1,5)

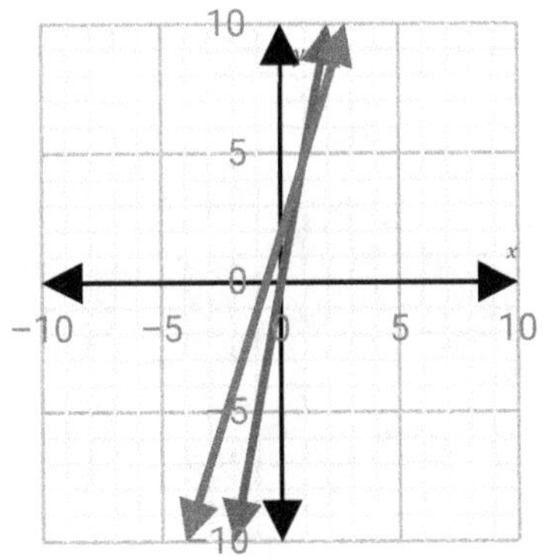

c. (2,7)

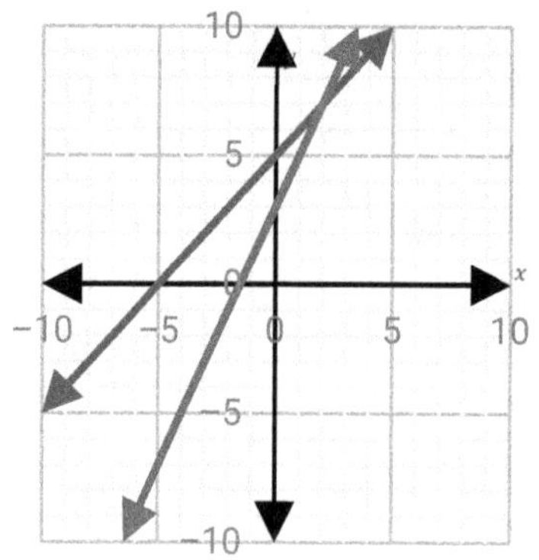

d. No solution

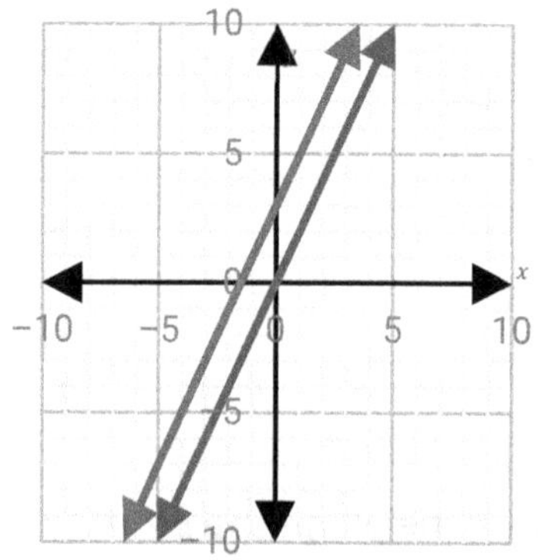

3. Solve the following system of equations using the graphing method: $\begin{cases} y = -2x + 4 \\ y = x - 2 \end{cases}$.

a. (−2,0)

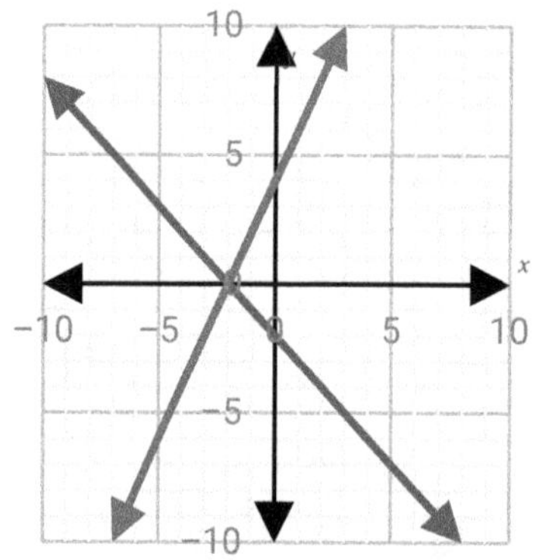

b. (1,2)

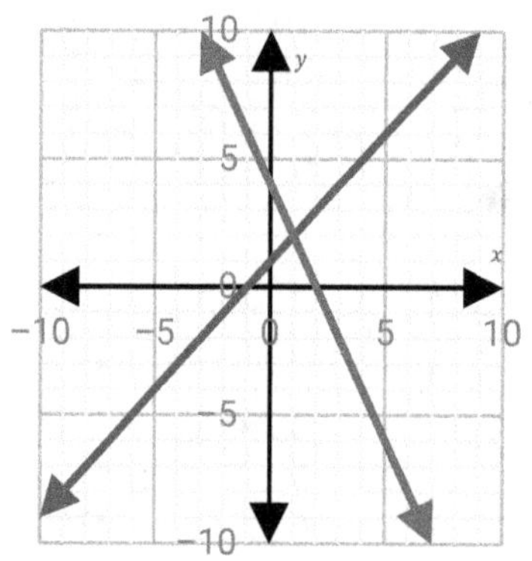

c. (2,0)

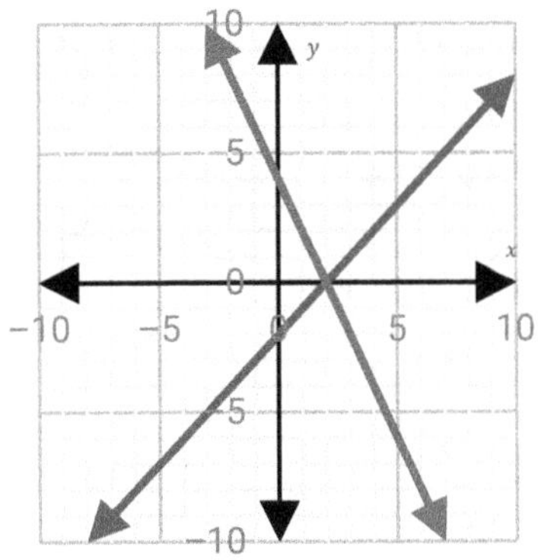

d. No solution

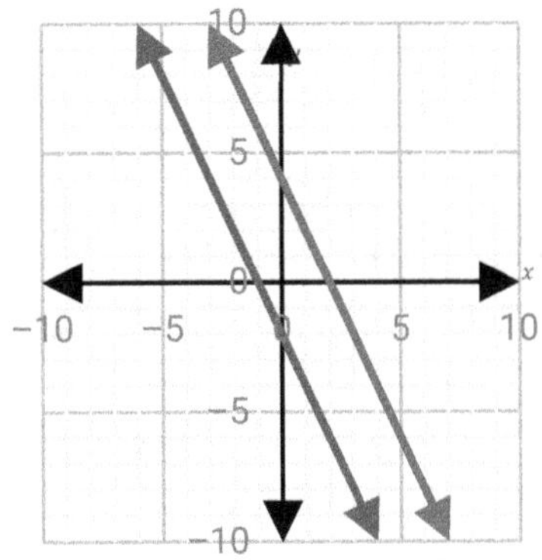

4. Solve the following system of equations using the graphing method: $\begin{cases} y = 2x - 1 \\ y = 3x \end{cases}$.

a. (1, −3)

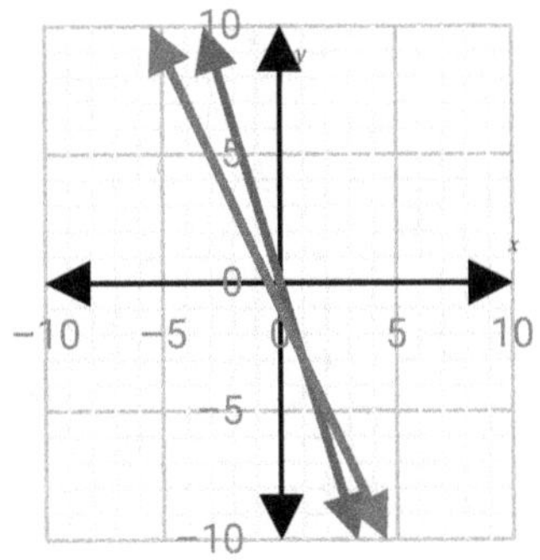

b. (−1, −3)

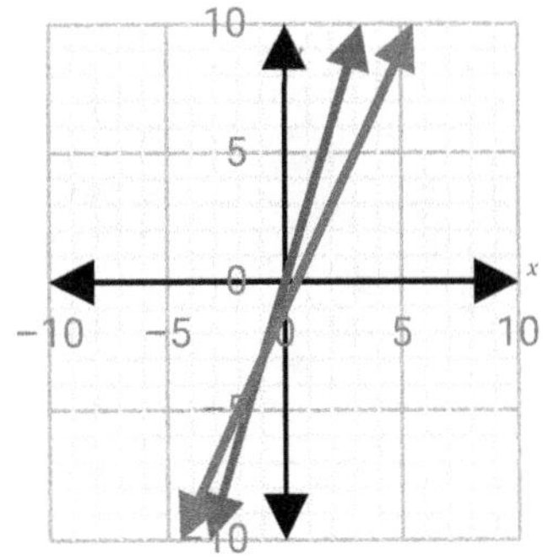

c. $(1, -3)$

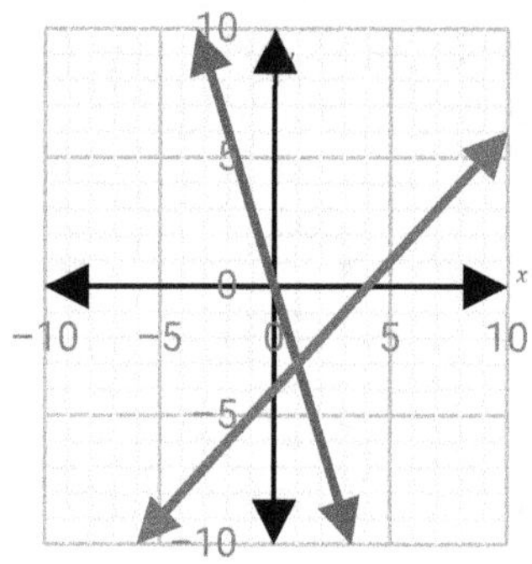

d. No solution

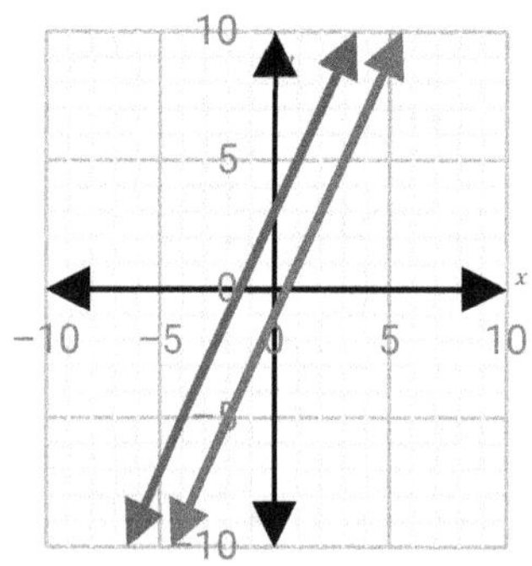

5. Solve the following system of equations using the graphing method: $\begin{cases} y = -3x + 5 \\ y = -4 \end{cases}$.

a. $(-4, -8)$

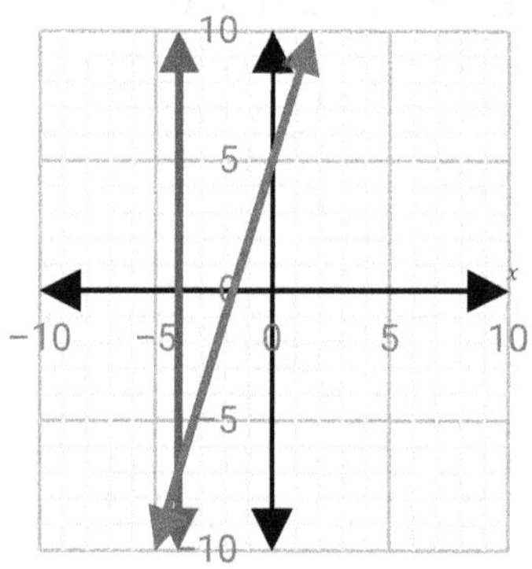

b. $(2, -2)$

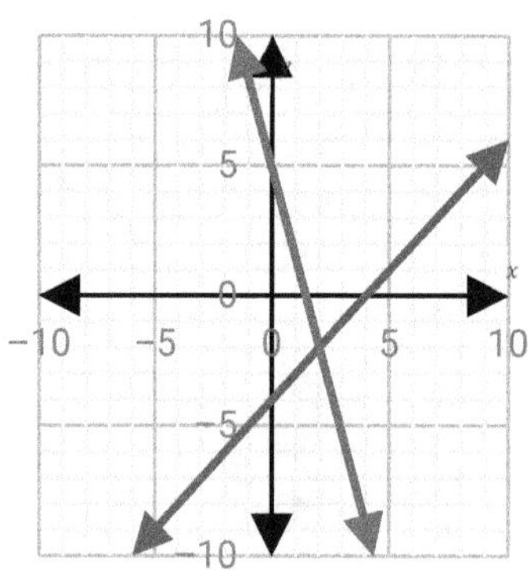

c. (3, −4)

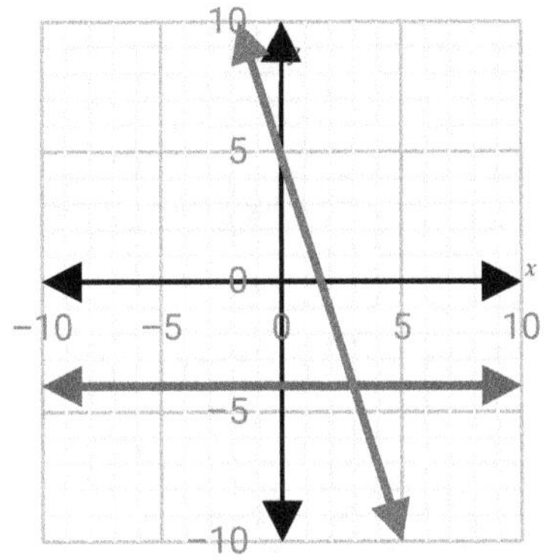

d. No solution

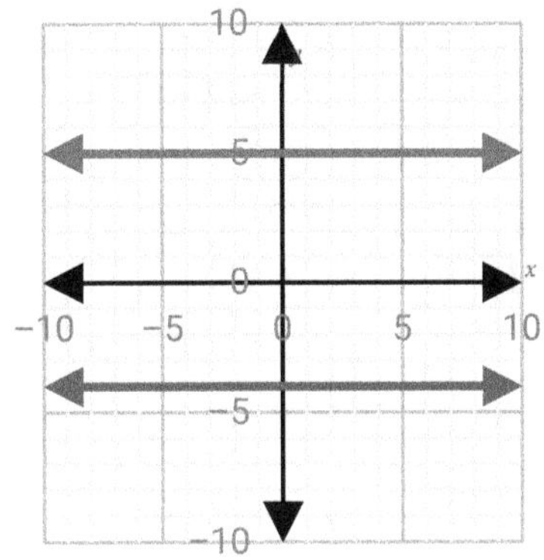

6. Solve the following system of equations using the graphing method: $\begin{cases} y = -\frac{2}{3}x - 2 \\ x = 3 \end{cases}$.

a. (−4, −8)

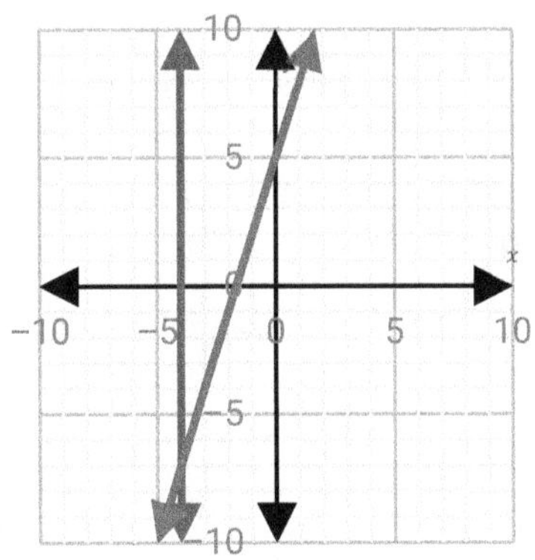

b. (3, −4)

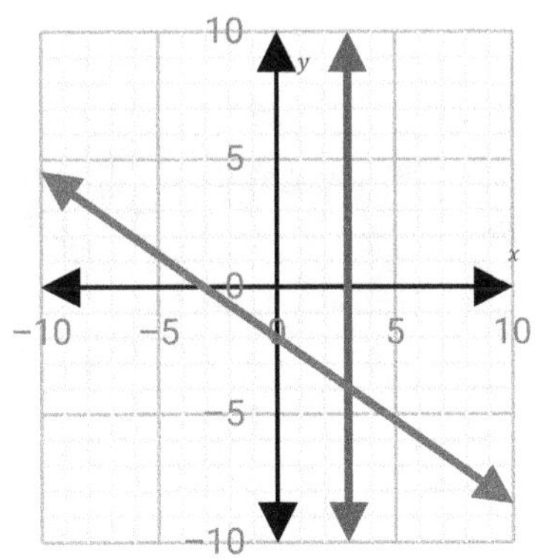

c. (3, −4)

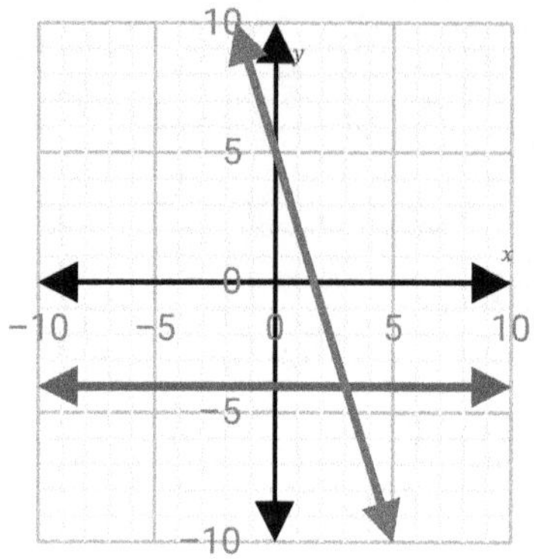

d. No solution

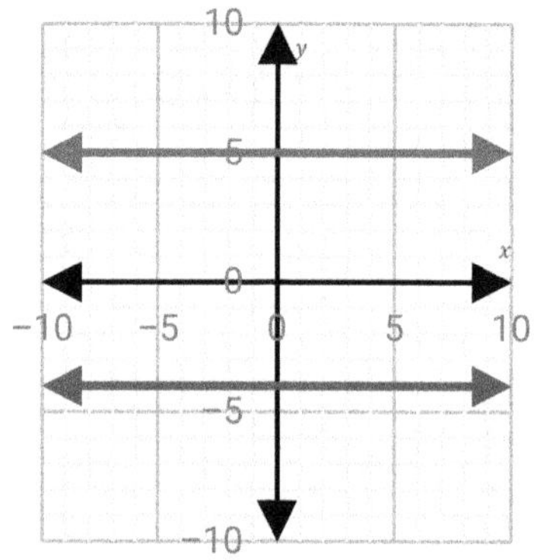

7. Solve the following system of equations using the graphing method: $\begin{cases} 3x + y = 5 \\ y = -3x - 2 \end{cases}$.

a. (1,1)

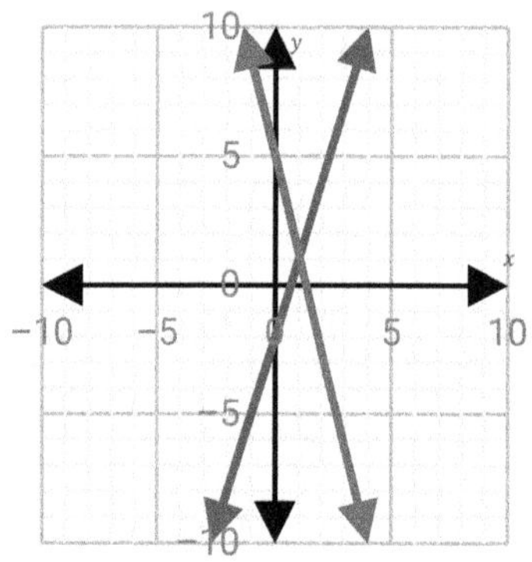

b. (0.5, −4)

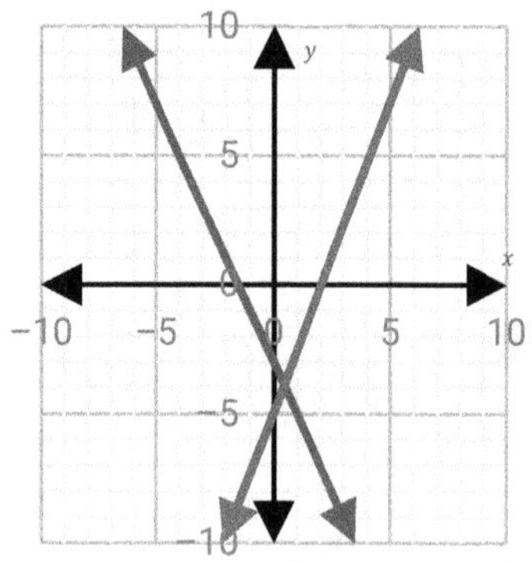

c. (1,2)

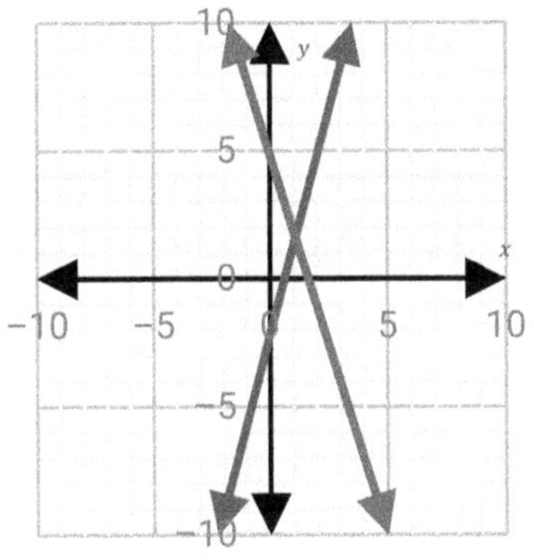

d. No solution

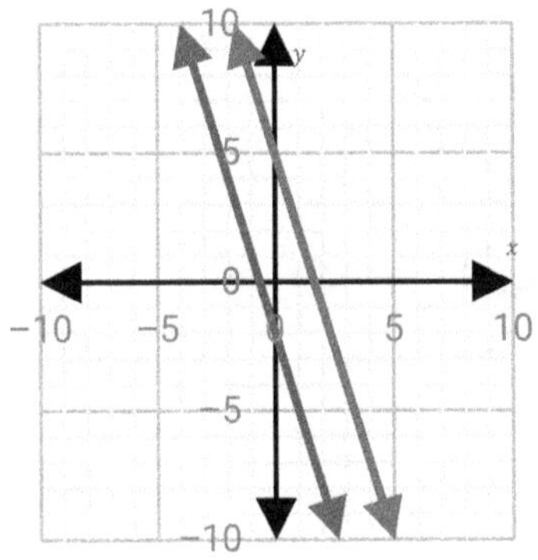

8. Solve the following system of equations using the graphing method: $\begin{cases} 2x - y = -1 \\ y = x - 1 \end{cases}$.

a. $(-2,-3)$

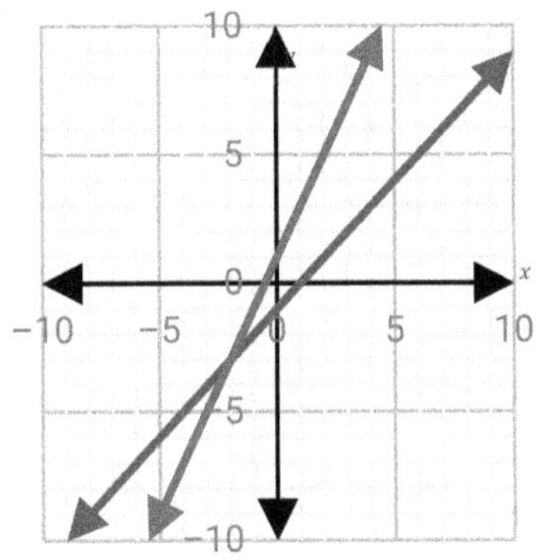

b. $(0,-1)$

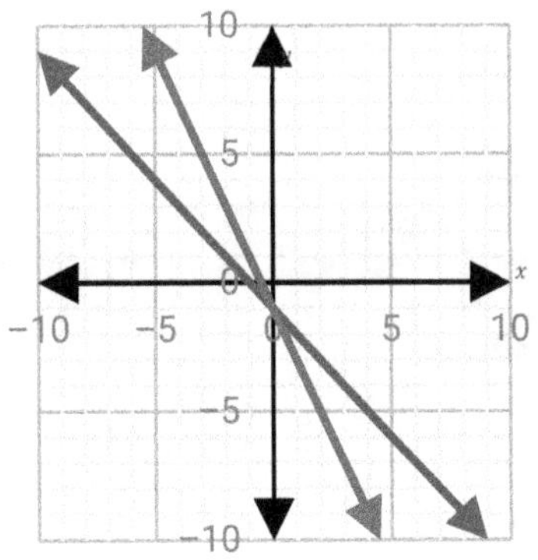

c. (2,3)

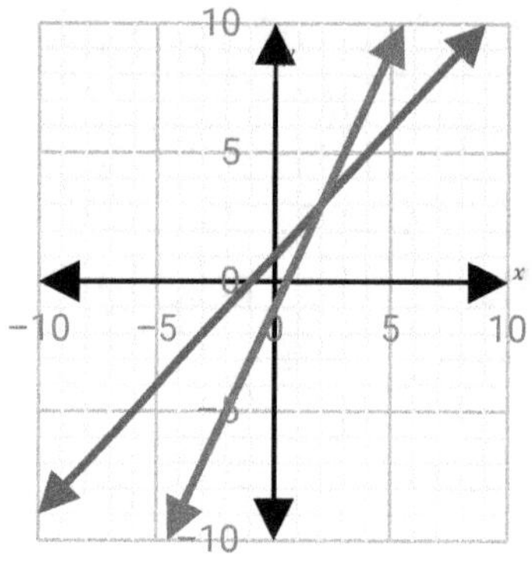

d. No solution

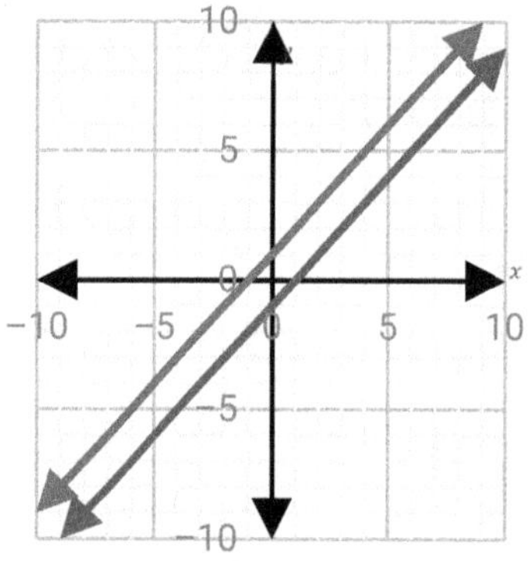

9. Solve the following system of equations using the graphing method: $\begin{cases} 2x - 3y = -3 \\ x = 3y - 9 \end{cases}$.

a. (−6,5)

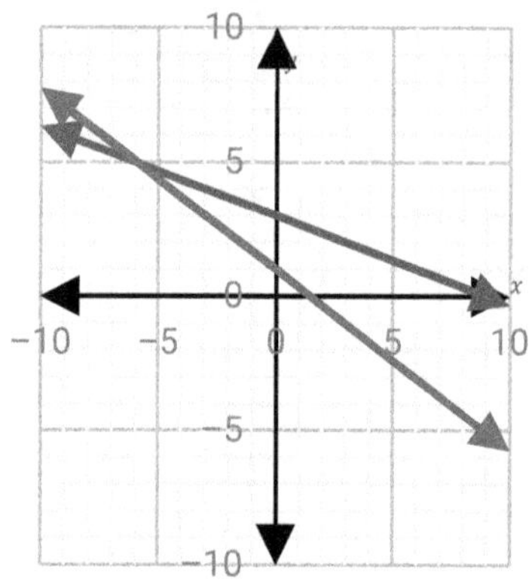

b. (6,5)

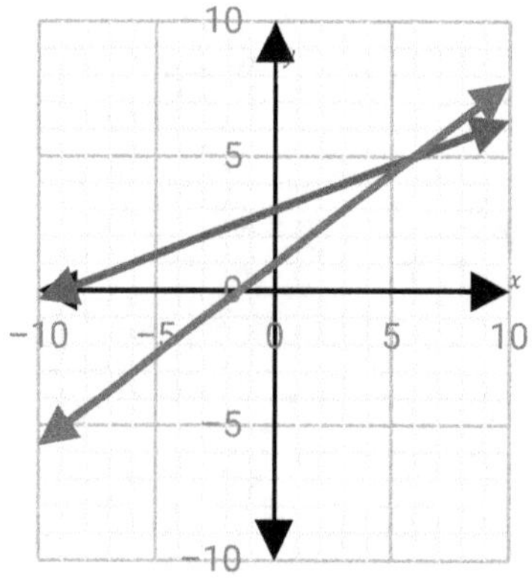

c. (−4,2)

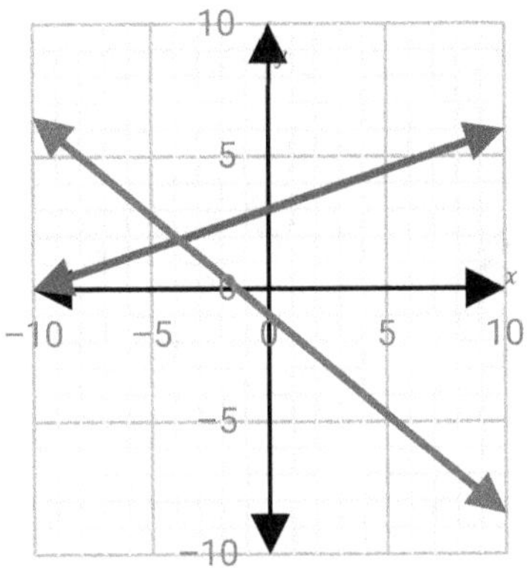

d. No solution

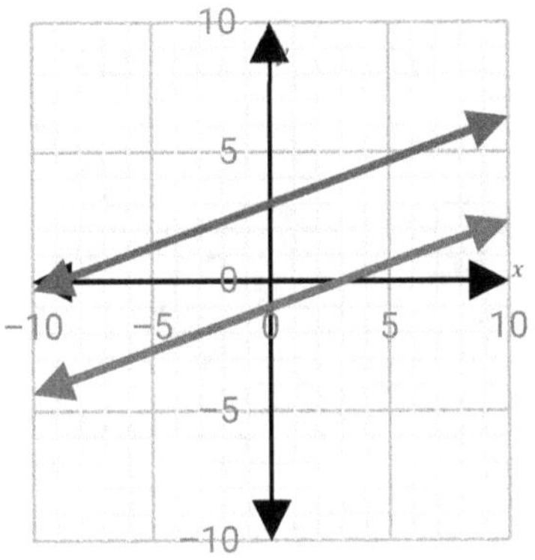

10. Solve the following system of equations using the graphing method: $\begin{cases} x + 4y = 12 \\ y = -\frac{1}{2}x + 1 \end{cases}$.

a. (−2,0)

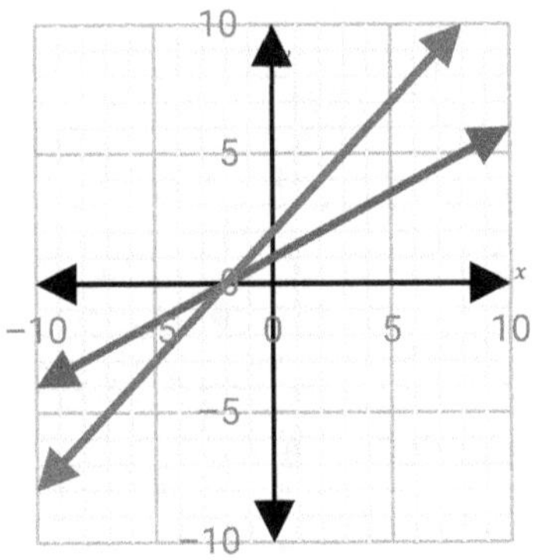

b. (8,5)

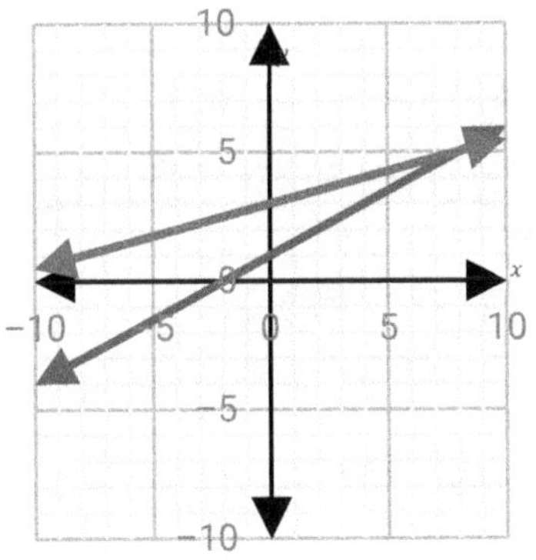

c. (−8,5)

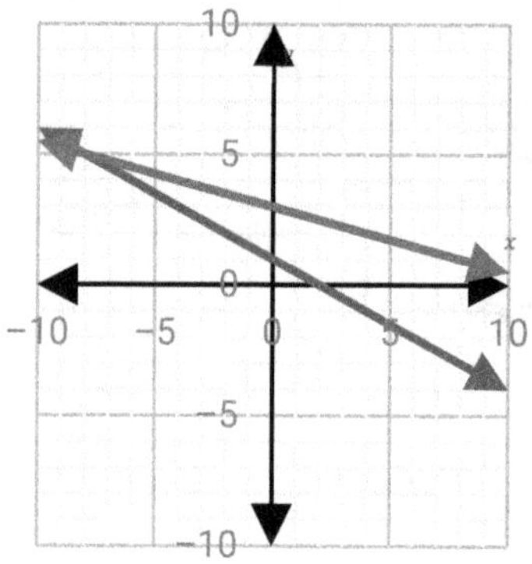

d. No solution

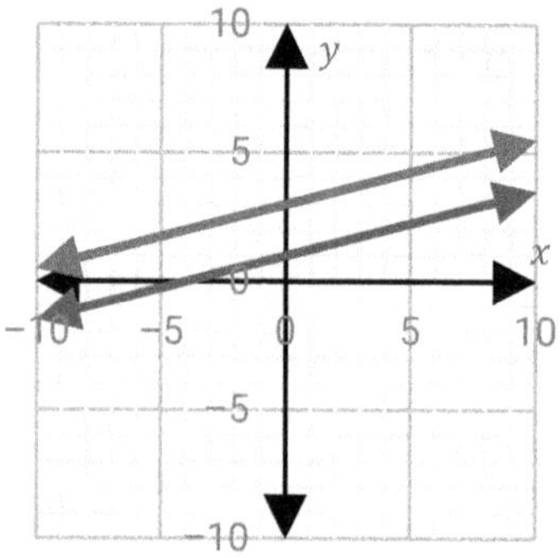

11. Solve the following system of equations using the graphing method: $\begin{cases} -4x - 2y = 6 \\ y = -\frac{4}{3}x - 5 \end{cases}$.

a. (3, −9)

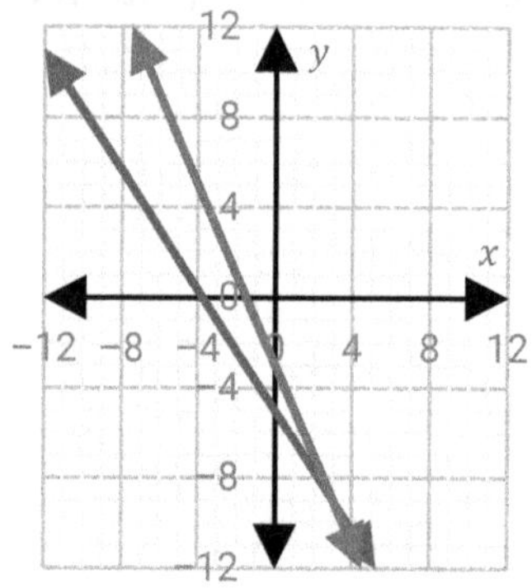

b. (−3, −9)

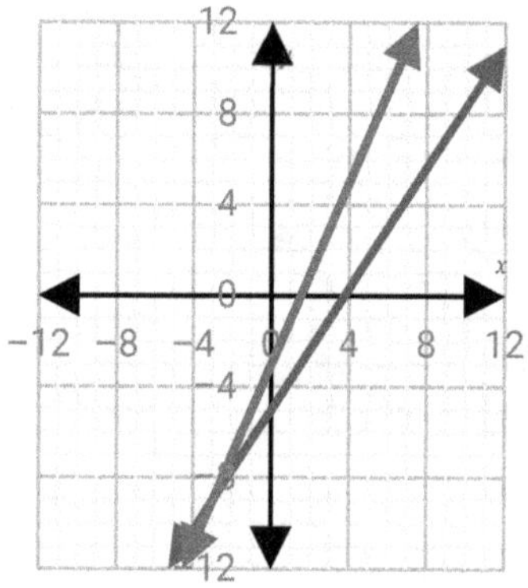

c. $(-0.5, -4)$

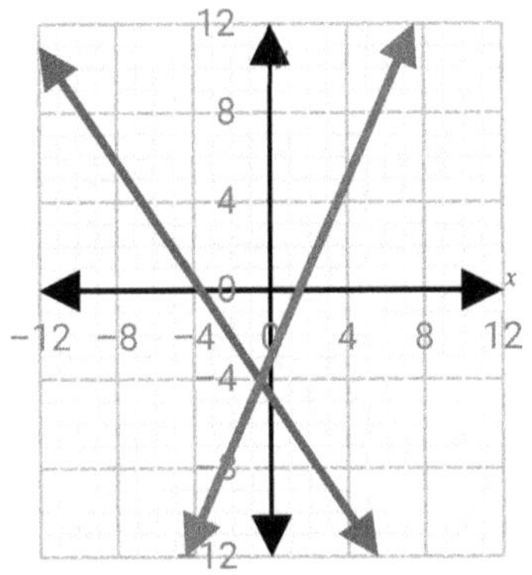

d. No solution

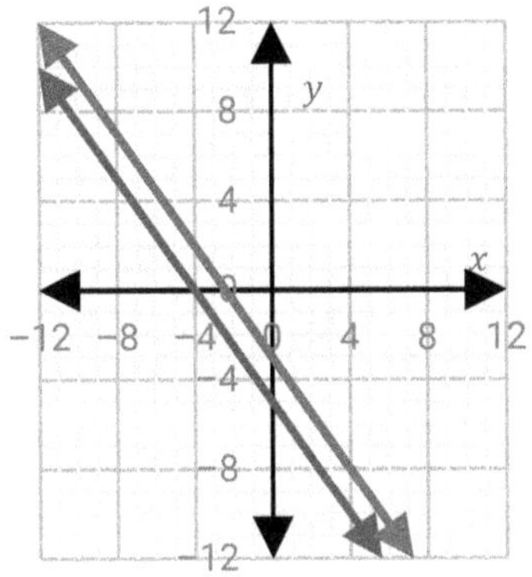

12. Solve the following system of equations using the graphing method: $\begin{cases} 6x - 3y = 9 \\ -4x + 2y = 14 \end{cases}$.

a. $(2.5, 2)$

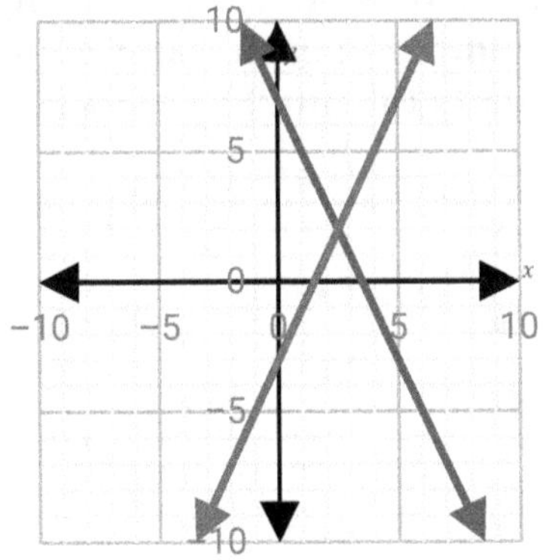

b. $(-2.5, 2)$

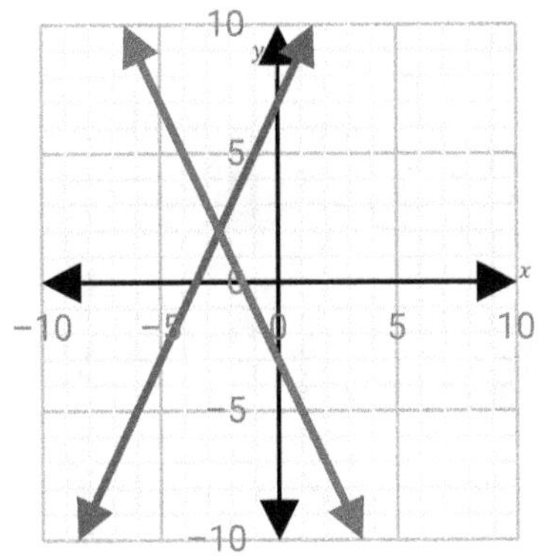

c. $(-1,-5)$

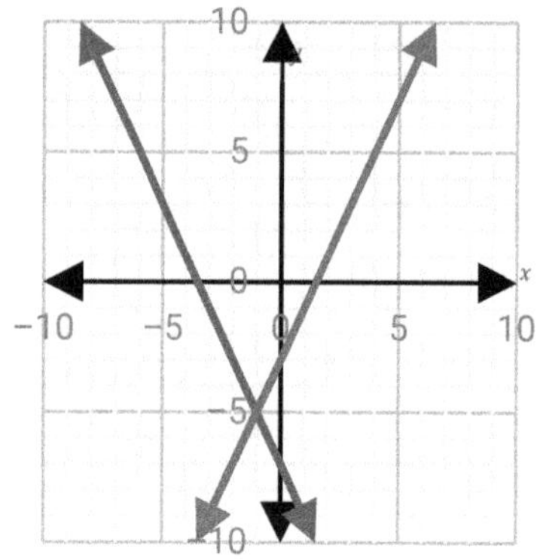

d. No solution

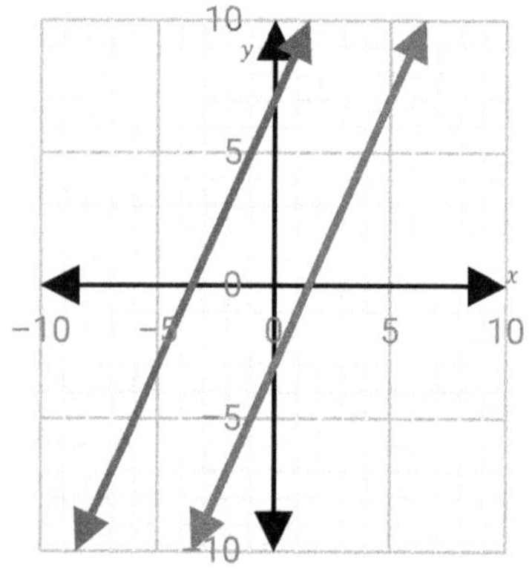

13. A store sells two types of picture frames. One frame sells for \$3.00 and the other one sells for \$4.00. The store sells 50 of the picture frames totaling \$180 on a single day. Which of the following is the graph of the solution for the number of each frame sold where x represents the number of \$3.00 frames sold and y represents the number of \$4.00 frames sold?

a. $(-20,30)$

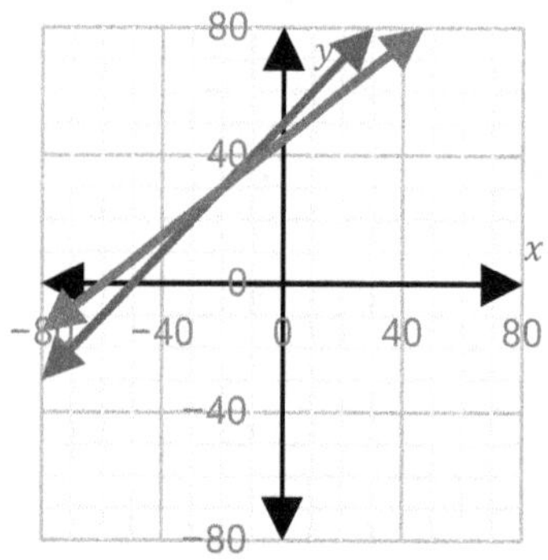

b. $(20,30)$

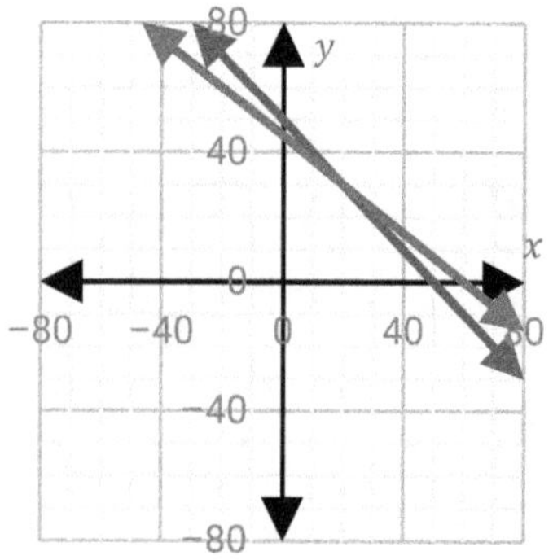

c. (10,40)

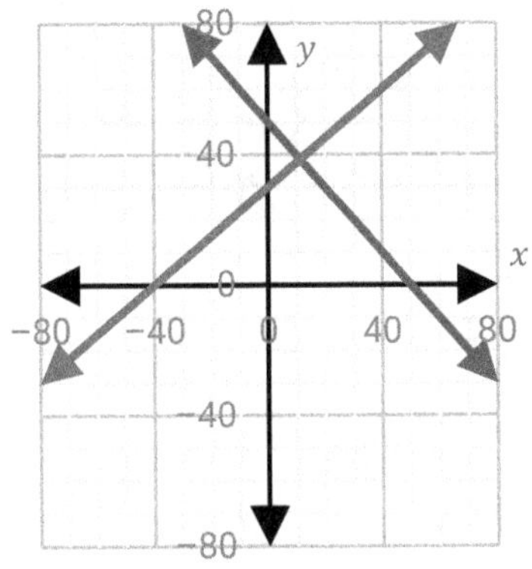

d. No solution

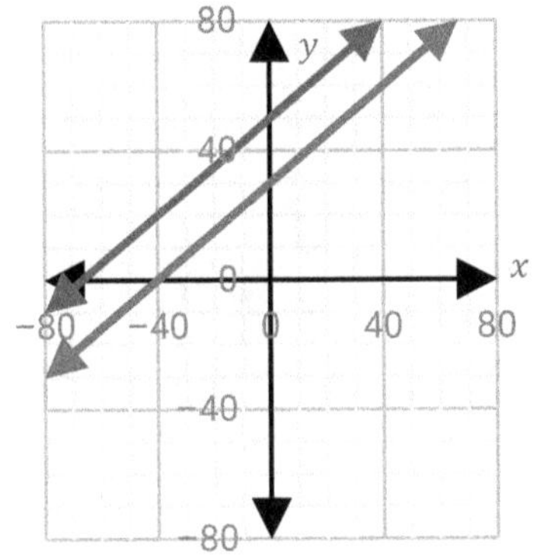

14. The sum of two numbers is −15. The one number is twice the other number. Which of the following is the graph for the solution of the two numbers?

a. (5,10)

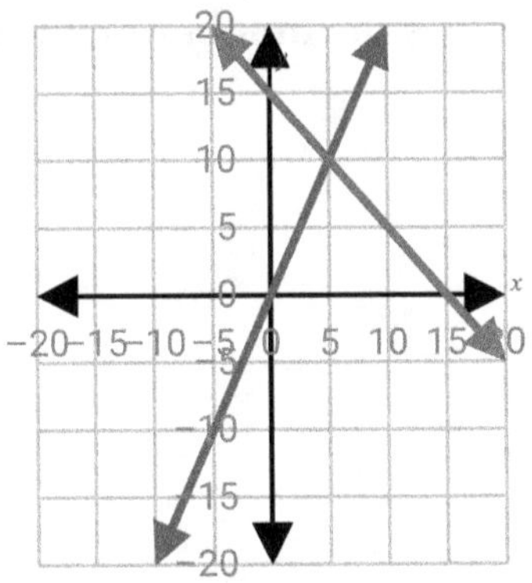

b. (5, −10)

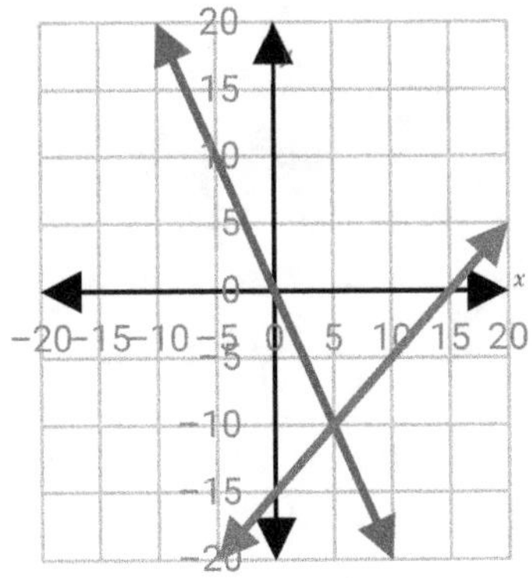

c. (−5, −10)

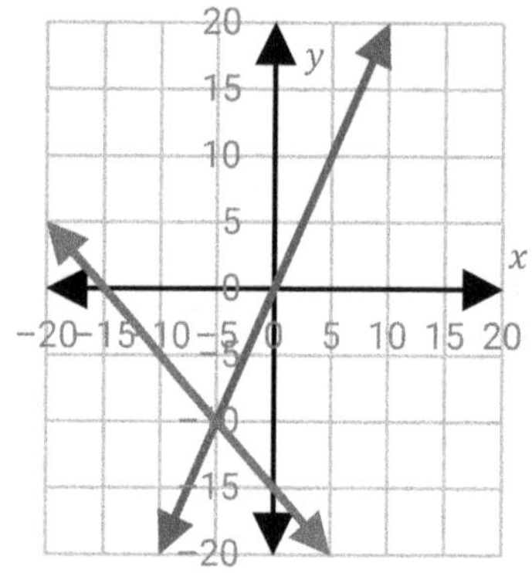

d. No solution

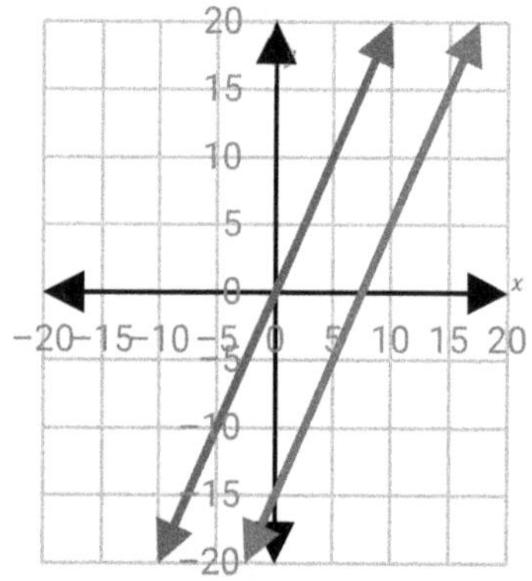

15. A paper bag contains a total of 20 nickels and dimes. The total amount of money in the bag is \$1.40. Which of the following is the graph of the solution for the number of each coin in the bag where x is the number of nickels and y is the number of dimes?

a. (−12,8)

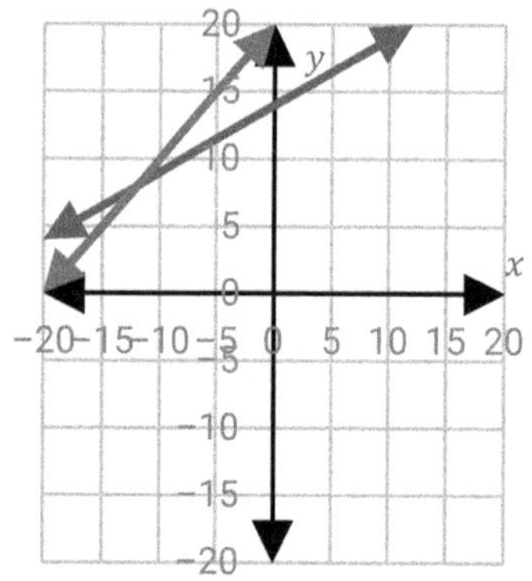

b. (12,8)

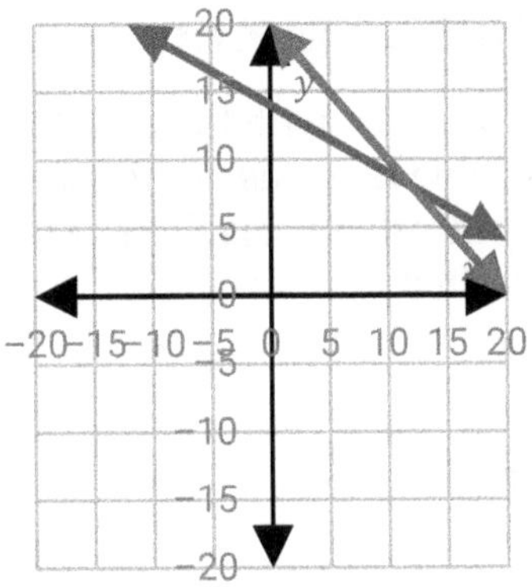

c. (8,12)

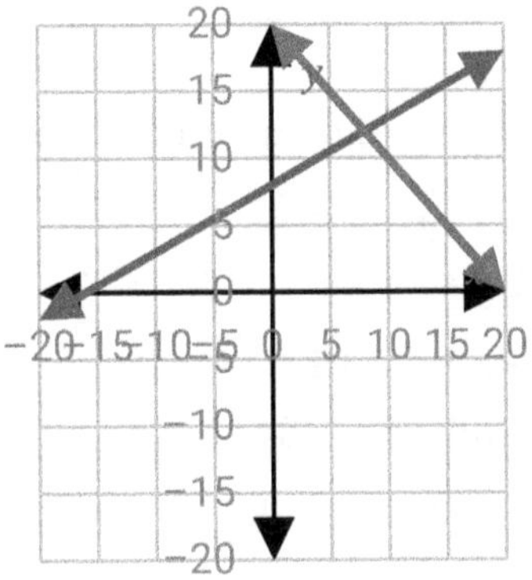

d. No solution

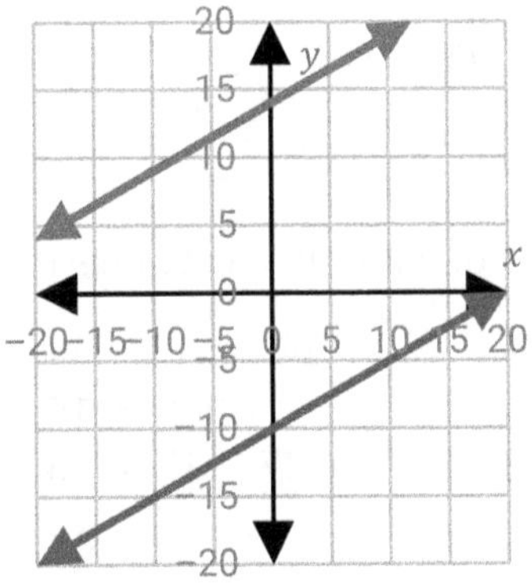

5.2 The Substitution Method

To solve a system of linear equations by **substitution**, start with the equation that requires fewer steps to solve for a single variable. Express this variable in terms of the other variable. Substitute this expression in the other equation and solve for the other variable. The solution should be expressed in the form (x, y). The answer can be checked by substituting the values into either of the original equations. Solve the following system of equations using substitution.

$$\begin{cases} x + 6y = 15 \\ 3x - 12y = 15 \end{cases}$$

Solve the first equation for x by subtracting $6y$ from both sides.

$$x = 15 - 6y$$

Substitute the expression $15 - 6y$ in place of x in the second equation and solve for y.

$$\begin{aligned} 3(15 - 6y) - 12y &= 15 \\ 45 - 18y - 12y &= 15 \\ 45 - 30y &= 15 \\ -30y &= -30 \\ y &= 1 \end{aligned}$$

Plug this value for y into the first equation to solve for x.

$$x = 15 - 6(1) = 15 - 6 = 9$$

Therefore, the solution is (9,1).

Check the solution by substituting the found values of x and y. The values must make both equations true.

$$\begin{aligned} (9) + 6(1) &= 15 \\ 9 + 6 &= 15 \\ 15 &= 15 \end{aligned} \qquad \begin{aligned} 3(9) - 12(1) &= 15 \\ 27 - 12 &= 15 \\ 15 &= 15 \end{aligned}$$

There are two other solution types you may come across when solving systems using the substitution method. A system of equations has no solution if the two lines are parallel. These lines will never cross. For example, $\begin{cases} 2x - 3y = 27 \\ -2x + 3y = 14 \end{cases}$ is an example of a system of equations that has no solution. The other solution type is a system of equations that has infinite solutions. A system of equations only has infinite solutions if the lines are exactly the same. For example, $\begin{cases} x - 2y = 11 \\ 2x - 4y = 22 \end{cases}$ is an example of a system of equations that has no solution. At first glance, the equations do not look identical; however, if you divide the second equation by 2, both equations become $x - 2y = 11$.

Review Video: The Substitution Method
Visit mometrix.com/academy and enter code: 565151

EXAMPLES

EXAMPLE 1:

Solve the system of linear equations using the substitution method.

$$\begin{cases} 7x + 6y = -9 \\ y = -2x + 1 \end{cases}$$

ANSWER

The second equation is already solved for y. Substitute the value of y, which is $-2x + 1$, for y in the first equation.

$$7x + 6y = -9$$
$$7x + 6(-2x + 1) = -9$$

Now, solve for x using the distributive property and inverse operations.

$$7x - 12x + 6 = -9$$
$$-5x + 6 = -9$$
$$-5x + 6 - 6 = -9 - 6$$
$$-5x = -15$$
$$\frac{-5x}{-5} = \frac{-15}{-5}$$
$$x = 3$$

Now, substitute $x = 3$ in either equation. Here we will substitute 3 for x in the second equation.

$$y = -2x + 1$$
$$y = -2(3) + 1$$
$$y = -6 + 1$$
$$y = -5$$

The solution of the system of equations is $x = 3$ and $y = -5$. This represents the ordered pair, $(3, -5)$, where the graphed lines intersect.

EXAMPLE 2:

Solve the system of linear equations using the substitution method.

$$\begin{cases} 2x - y = 1 \\ 4x + y = 8 \end{cases}$$

ANSWER

Solve one of the equations for one of the variables. In this case, solve the second equation for y.

$$4x + y = 8$$
$$y = -4x + 8$$

Now, substitute the value of y, which is $-4x + 8$, for y in the first equation.

$$2x - y = 1$$
$$2x - (-4x + 8) = 1$$

Now, solve for x.

$$\begin{aligned} 2x + 4x - 8 &= 1 \\ 6x - 8 &= 1 \\ 6x - 8 + 8 &= 1 + 8 \\ 6x &= 9 \\ \frac{6x}{6} &= \frac{9}{6} \\ x &= \frac{9}{6} \\ x &= 1.5 \end{aligned}$$

Now, substitute $x = 1.5$ into either equation. Here we will substitute $x = 1.5$ in the first equation.

$$\begin{aligned} 2x - y &= 1 \\ 2(1.5) - y &= 1 \\ 3 - y &= 1 \\ 3 - 3 - y &= 1 - 3 \\ -y &= -2 \\ \frac{-y}{-1} &= \frac{-2}{-1} \\ y &= 2 \end{aligned}$$

The solution of the system of equations is $x = 1.5$ and $y = 2$. The ordered pair $(1.5, 2)$ represents where the graphed lines intersect.

Example 3:

Ellen needs to mix part of a 10% acid solution with part of a 35% acid solution to create 2 liters of a 15% acid solution. How many liters of each solution should she mix? Use the substitution method to solve.

Answer

Assign a variable for each solution. Let x represent the amount of 10% acid solution and y represent the amount of 35% acid solution. The mixture requires some of solution x and some of solution y, for a total of 2 liters. Set up the first equation.

$$x + y = 2$$

The second equation will be related to the concentration of each acid. Acid x is 10% and acid y is 35%. Multiply the concentration by the number of liters. The equation needs to show how the concentration of one container plus the concentration of the other container produces a total concentration of 15%. Write the percentages in decimal form. The second equation becomes:

$$0.1x + 0.35y = 0.15$$

The number of liters of solution needs to be multiplied by the concentration of the solution. This means that the total concentration (0.15) needs to be multiplied by the number of liters. The equation will be set equal to 0.15(2), or 0.3.

$$0.1x + 0.35y = 0.3$$

Make the equation simpler by converting each decimal value to a whole number. Multiply each term by 100.

$$10x + 35y = 30$$

Now that a system of equations is established, the values of x and y can be calculated using the substitution method.

$$\begin{cases} x + y = 2 \\ 10x + 35y = 30 \end{cases}$$

Solve the simpler equation for one of the variables. For example, solve $x + y = 2$ for y.

$$y = 2 - x$$

Substitute this value of y into the other equation for y.

$$10x + 35(2 - x) = 30$$

Solve for the variable x using inverse operations. Start by distributing the 35 into the parentheses.

$$10x + 70 - 35x = 30$$

Now subtract 70 from both sides.

$$10x - 35x = -40$$

Combine the x terms.

$$-25x = -40$$

Divide both sides by –25.

$$x = 1.6$$

1.6 liters of the 10% solution is required. Use the value of x to solve for the value of y. Use the original equation $x + y = 2$.

$$1.6 + y = 2$$

Subtract 1.6 from both sides.

$$y = 0.4$$

This means that 0.4 liters of the 35% solution is required.

Ellen needs to use 1.6 liters of the 10% acid solution and 0.4 liters of the 35% acid solution.

PRACTICE QUESTIONS

1. Solve the system of equations using the substitution method: $\begin{cases} x = -5 \\ y = 2x + 8 \end{cases}$

a. $(-1,6)$
b. $(1,10)$
c. $(-2,-5)$
d. $(-5,-2)$

2. Solve the system of equations using the substitution method: $\begin{cases} y = 2x + 1 \\ y = x + 2 \end{cases}$

a. $(3,1)$
b. $(1,3)$
c. $(2,2)$
d. $(-1,1)$

3. Solve the system of equations using the substitution method: $\begin{cases} 3x + y = 10 \\ y = 2x \end{cases}$

a. $(3,4)$
b. $(-1,-2)$
c. $(2,4)$
d. $(4,2)$

4. Use the substitution method to simplify the system of equations: $\begin{cases} y = -3x \\ 2x + 3y = -7 \end{cases}$

a. $(2,-6)$
b. $(7,-7)$
c. $(3,-1)$
d. $(1,-3)$

5. Use the substitution method to solve the system of equations: $\begin{cases} y = \frac{1}{2}x - 1 \\ y = x - 4 \end{cases}$

a. $(-1,-5)$
b. $(2,0)$
c. $(6,2)$
d. $(2,6)$

6. Use the substitution method to solve the system of equations: $\begin{cases} y = x + 2 \\ x = \frac{1}{3}y + 4 \end{cases}$

a. $(7,9)$
b. $(9,7)$
c. $(6,6)$
d. $(1,3)$

7. Use the substitution method to solve the system of equations: $\begin{cases} x + 3y = 9 \\ 2x + y = -2 \end{cases}$

a. $(4, -3)$
b. $(-3,4)$
c. $(0,3)$
d. $(-1,0)$

8. Use the substitution method to solve the system of equations: $\begin{cases} 2x - y = 9 \\ x + 3y = 8 \end{cases}$

a. $(4, -1)$
b. $(-1,3)$
c. $(5,1)$
d. $(1,5)$

9. Use the substitution method to solve the system of equations: $\begin{cases} x + 4y = -6 \\ -2x + y = 3 \end{cases}$

a. $(2,1)$
b. $(2,7)$
c. $(-2, -1)$
d. $(-1, -2)$

10. Use the substitution method to solve the system of equations: $\begin{cases} -x - y = -12 \\ 2x + y = 16 \end{cases}$

a. $(8,4)$
b. $(4,8)$
c. $(-1,13)$
d. $(-1,18)$

11. Use the substitution method to solve the system of equations: $\begin{cases} x - 6y = -4 \\ 3x + 5y = 11 \end{cases}$

a. $(-3,4)$
b. $(-10,1)$
c. $(1,2)$
d. $(2,1)$

12. Use the substitution method to solve the system of equations: $\begin{cases} 2x + y = 10 \\ 4x + 2y = 6 \end{cases}$

a. $(2,6)$
b. $(1,1)$
c. Infinitely many solutions
d. No solution

13. For a school fundraiser, a science class sells chips and candy bars at a school's football game. The chips are sold for \$1.50 each and the candy bars are sold for \$1.25 each. The class raised a total of \$127.50 from selling a total of 90 bags of chips and candy bars. How many of each item were sold?

a. 60 bags of chips and 30 candy bars
b. 30 bags of chips and 60 candy bars
c. 45 bags of chips and 45 candy bars
d. 70 bags of chips and 20 candy bars

14. The sum of two numbers is 20. Three times the smaller number equals the larger number. What are the two numbers?

a. 6 and 14
b. 5 and 15
c. 4 and 12
d. No solution

15. An acid solution is 10% acid. A second acid solution containing 50% acid is mixed with the first solution. How many liters of each solution should be mixed to have 60 liters of a 40% acid solution?

a. 40 liters of the first solution and 20 liters of the second solution
b. 30 liters of the first solution and 3 liters of the second solution
c. 15 liters of the first solution and 45 liters of the second solution
d. 45 liters of the first solution and 15 liters of the second solution

5.2 The Substitution Method

1. D: To use the substitution method, first substitute the value for one variable in one equation into the other equation. Since we are given $x = -5$, substitute -5 into the equation for y.

$$y = 2(-5) + 8$$

Now, solve the second equation for y.

$$y = -10 + 8$$
$$y = -2$$

Write the answer as the ordered pair $(-5, -2)$.

2. B: To use the substitution method, substitute the value for one variable in one equation into the other equation. Substituting $2x + 1$ into the second equation for y, we get the following equation.

$$2x + 1 = x + 2$$

Now, solve the second equation for x.

$2x + 1 = x + 2$	Write the second equation with the substituted value.
$x + 1 = 2$	Subtract x from both sides of the equation.
$x = 1$	Subtract 1 from both sides of the equation.

Next, substitute the value for x into either of the two original equations to find the y-value.

$y = (1) + 2$	Substitute $x = 1$ into the second equation.
$y = 3$	Simplify the right-hand side of the equation.

Write the answer as the ordered pair (1,3).

3. C: To use the substitution method, first substitute the value for one variable in one equation into the other equation. Substituting $2x$ into the first equation for y, we get the following equation.

$$3x + 2x = 10$$

Now, solve the first equation for x.

$3x + 2x = 10$	Write the first equation with the substituted value.
$5x = 10$	Combine like terms on the left-hand side of the equation.
$x = 2$	Divide both sides of the equation by 5.

Next, substitute the value for x into either of the two original equations to find the y-value.

$y = 2(2)$	Substitute $x = 2$ into the second equation.
$y = 4$	Simplify the right-hand side of the equation.

Write the answer as the ordered pair (2,4).

4. D: To use the substitution method, first substitute the value for one variable in one equation into the other equation. Substituting $y = -3x$ into the second equation, we get the following equation.

$$2x + 3(-3x) = -7$$

Now, solve the second equation for x.

$2x + 3(-3x) = -7$	Write the second equation with the substituted value.
$2x - 9x = -7$	Multiply $-3x$ by 3.
$-7x = -7$	Combine like terms on the left-hand side of the equation.
$x = 1$	Divide both sides of the equation by -7.

Next, substitute the value for x into either of the two original equations to find the y-value.

$y = -3(1)$	Substitute $x = 1$ into the first equation.
$y = -3$	Simplify the right-hand side of the equation.

Write the answer as the ordered pair $(1, -3)$.

5. C: To use the substitution method, first substitute the value for one variable in one equation into the other equation. Substituting $y = \frac{1}{2}x - 1$ into the second equation, we get the following equation.

$$\frac{1}{2}x - 1 = x - 4$$

Now, solve the second equation for x.

$\frac{1}{2}x - 1 = x - 4$	Write the second equation with the substituted value.
$-\frac{1}{2}x - 1 = -4$	Subtract x from both sides of the equation.
$-\frac{1}{2}x = -3$	Add 1 to both sides of the equation.
$x = 6$	Multiply both sides of the equation by –2.

Next, substitute the value for x into either of the two original equations to find the y-value.

$y = (6) - 4$	Substitute $x = 6$ into the second equation.
$y = 2$	Simplify the right-hand side of the equation.

Write the answer as the ordered pair (6,2).

6. A: To use the substitution method, first substitute the value for one variable in one equation into the other equation. Substituting $x = \frac{1}{3}y + 4$ into the first equation, we get the following equation.

$$y = \left(\frac{1}{3}y + 4\right) + 2$$

Now, solve the first equation for y.

$y = \left(\frac{1}{3}y + 4\right) + 2$	Write the first equation with the substituted value.
$y = \frac{1}{3}y + 6$	Simplify the right-hand side of the equation.
$\frac{2}{3}y = 6$	Subtract $\frac{1}{3}y$ from both sides of the equation.
$y = 9$	Multiply both sides of the equation by $\frac{3}{2}$.

Next, substitute the value for y into either of the two original equations to find the x-value.

$(9) = x + 2$	Substitute $y = 9$ into the first equation.
$7 = x$	Subtract 2 from both sides of the equation.

Write the answer as the ordered pair (7,9).

7. B: To use the substitution method, first substitute the value for one variable in one equation into the other equation. Solving the first equation for x, we have the following equation.

$x + 3y = 9$	Write the first equation.

$x = 9 - 3y$	Subtract $3y$ from both sides of the equation to isolate x.

Substitute $x = 9 - 3y$ into the second equation.

$$2(9 - 3y) + y = -2$$

Now, solve the second equation for y.

$2(9 - 3y) + y = -2$	Write the second equation with the substituted value.
$2(9) + 2(-3y) + y = -2$	Use the distributive property for the left-hand side of the equation.
$18 - 5y = -2$	Combine like terms on the left-hand side of the equation.
$-5y = -20$	Subtract 18 from both sides of the equation.
$y = 4$	Divide both sides of the equation by –5.

Next, substitute the value for y into either of the two original equations to find the x-value.

$2x + (4) = -2$	Substitute $y = 4$ into the second equation.
$2x = -6$	Subtract 4 from both sides of the equation.
$x = -3$	Divide both sides of the equation by 2.

Write the answer as the ordered pair $(-3,4)$.

8. C: To use the substitution method, first substitute the value for one variable in one equation into the other equation. Solving the second equation for x, we have the following equation.

$x + 3y = 8$	Write the second equation.
$x = 8 - 3y$	Subtract $3y$ from both sides of the equation to isolate x.

Substitute $x = 8 - 3y$ into the first equation.

$$2(8 - 3y) - y = 9$$

Now, solve the first equation for y.

$2(8 - 3y) - y = 9$	Write the first equation with the substituted value.
$2(8) + 2(-3y) - y = 9$	Use the distributive property for the left-hand side of the equation.
$16 - 7y = 9$	Combine like terms on the left-hand side of the equation.
$-7y = -7$	Subtract 16 from both sides of the equation.
$y = 1$	Divide both sides of the equation by –7.

Next, substitute the value for y into either of the two original equations to find the x-value.

$x + 3(1) = 8$	Substitute $y = 1$ into the second equation.
$x + 3 = 8$	Simplify the left-hand side of the equation.
$x = 5$	Subtract 3 from both sides of the equation.

Write the answer as the ordered pair (5,1).

9. C: To use the substitution method, first substitute the value for one variable in one equation into the other equation. Solving the first equation for x, we have the following equation.

$x + 4y = -6$	Write the first equation.
$x = -6 - 4y$	Subtract $4y$ from both sides of the equation to isolate x.

Substitute $x = -6 - 4y$ into the second equation.

$$-2(-6 - 4y) + y = 3$$

Now, solve the second equation for y.

$-2(-6 - 4y) + y = 3$	Write the second equation with the substituted value.
$(-2)(-6) + (-2)(-4y) + y = 3$	Use the distributive property for the left-hand side of the equation.
$12 + 9y = 3$	Combine like terms on the left-hand side of the equation.
$9y = -9$	Subtract 12 from both sides of the equation.
$y = -1$	Divide both sides of the equation by 9.

Next, substitute the value for y into either of the two original equations to find the x-value.

$x + 4(-1) = -6$	Substitute $y = -1$ into the first equation.
$x = -2$	Add 4 to both sides of the equation.

Write the answer as the ordered pair $(-2, -1)$.

10. B: To use the substitution method, first substitute the value for one variable in one equation into the other equation. Solving the second equation for y, we have the following equation.

$2x + y = 16$	Write the second equation.
$y = 16 - 2x$	Subtract $2x$ from both sides of the equation.

Substitute $y = 16 - 2x$ into the first equation.

$$-x - (16 - 2x) = -12$$

Now, solve the first equation for x.

$-x - (16 - 2x) = -12$	Write the first equation with the substituted value.
$-x + (-1)(16) + (-1)(-2x) = -12$	Use the distributive property for the left-hand side of the equation.
$x - 16 = -12$	Combine like terms on the left-hand side of the equation.
$x = 4$	Add 16 to both sides of the equation.

Next, substitute the value for x into either of the two original equations to find the y-value.

$2(4) + y = 16$	Substitute $x = 4$ into the second equation.

$8 + y = 16$	Simplify the left-hand side of the equation.
$y = 8$	Subtract 8 from both sides of the equation.

Write the answer as the ordered pair (4,8).

11. D: To use the substitution method, first substitute the value for one variable in one equation into the other equation. Solving the first equation for x, we have the following equation.

$x - 6y = -4$	Write the first equation.
$x = -4 + 6y$	Add $6y$ to both sides of the equation.

Substitute $x = -4 + 6y$ into the second equation.

$$3(-4 + 6y) + 5y = 11$$

Now, solve the second equation for y.

$3(-4 + 6y) + 5y = 11$	Write the second equation with the substituted value.
$3(-4) + 3(6y) + 5y = 11$	Use the distributive property for the left-hand side of the equation.
$-12 + 23y = 11$	Combine like terms on the left-hand side of the equation.
$23y = 23$	Add 12 to both sides of the equation.
$y = 1$	Divide both sides of the equation by 23.

Next, substitute the value for y into either of the two original equations to find the x-value.

$x - 6(1) = -4$	Substitute $y = 1$ into the first equation.
$x - 6 = -4$	Simplify the left-hand side of the equation.
$x = 2$	Add 6 to both sides of the equation.

Write the answer as the ordered pair (2,1).

12. D: To use the substitution method, first substitute the value for one variable in one equation into the other equation. Solving the first equation for y, we have the following equation.

$2x + y = 10$	Write the first equation.
$y = 10 - 2x$	Subtract $2x$ from both sides of the equation.

Substitute $y = 10 - 2x$ into the second equation.

$$4x + 2(10 - 2x) = 6$$

Now, solve the second equation for y.

$4x + 2(10 - 2x) = 6$	Write the second equation with the substituted value.
$4x + 2(10) + 2(-2x) = 6$	Use the distributive property for the left-hand side of the equation.
$20 = 6$	Combine like terms on the left-hand side of the equation.

Since $20 = 6$ is a false statement, there is no solution to the system of equations.

13. A: Let x be the number of bags of chips sold and y be the number of candy bars sold at the fundraiser. The amount raised, in dollars, for selling the chips is $1.50x$ and the amount raised by selling the candy bars is $1.25y$. Since the total amount raised is $127.50, we have the following equation.

$$1.5x + 1.25y = 127.5$$

Since there was a total of 90 items sold, we also have the following equation.

$$x + y = 90$$

We can solve the following system of linear equations to find the number of each item sold.

$$\begin{cases} 1.5x + 1.25y = 127.5 \\ x + y = 90 \end{cases}$$

To use the substitution method, substitute the value for one variable in one equation into the other equation. Solve the second equation for y.

$x + y = 90$	Write the second equation.
$y = 90 - x$	Subtract x from both sides of the equation.

Substitute $y = 90 - x$ into the first equation.

$$1.5x + 1.25(90 - x) = 127.5$$

Now, solve the first equation for x.

$1.5x + 1.25(90 - x) = 127.5$	Write the first equation with the substituted value.
$1.5x + 1.25(90) + 1.25(-x) = 127.5$	Use the distributive property for the left-hand side of the equation.
$1.5x + 135 - 1.25x = 127.5$	Simplify the distribution on the left-hand side of the equation.
$0.25x + 112.5 = 127.5$	Combine like terms on the left-hand side of the equation.
$0.25x = 15$	Subtract 112.5 from both sides of the equation.
$x = 60$	Divide both sides of the equation by 0.25.

Next, substitute the value for x into either of the two original equations to find the y-value.

$(60) + y = 90$	Substitute $x = 60$ into the second equation.
$y = 30$	Subtract 60 from both sides of the equation.

Since $x = 60$ and $y = 30$, there were 60 bags of chips and 30 candy bars sold.

14. B: Let x be the smaller number and y be the larger number. The sum of the numbers is 20.

$$x + y = 20$$

Since the larger number is three times the smaller number, we also have the following equation.

$$y = 3x$$

We can solve the following system of linear equations to find each number.

$$\begin{cases} x + y = 20 \\ y = 3x \end{cases}$$

To use the substitution method, first substitute the value for one variable in one equation into the other equation.

$x + (3x) = 20$	Substitute $3x$ into the first equation.

Now, solve the first equation for x.

$x + (3x) = 20$	Write the first equation with the substituted value.
$4x = 20$	Combine like terms on the left-hand side of the equation.
$x = 5$	Divide both sides of the equation by 4.

Next, substitute the value for x into either of the two original equations to find the y-value.

$(5) + y = 20$	Substitute $x = 5$ into the first equation.
$y = 15$	Subtract 5 from both sides of the equation.

Since $x = 5$ and $y = 15$, the two numbers are 5 and 15.

15. C: Let x be the first acid solution and y be the second acid solution. Since the first acid solution is 10% acid, it has $0.10x$ liters of acid in it. The second solution is 50% acid, so it has $0.50y$ liters of acid in it. If the 60-liter solution is to contain 40% acid, it has 60 liters $\times$ 0.40 $=$ 24 liters of acid in it. Combining the two acid solutions we have the following equation.

$$0.1x + 0.5y = 24$$

Since we have a total of 60 liters by combining the two solutions, we also have the following equation.

$$x + y = 60$$

We can solve the following system of linear equations to find how much of each solution we need.

$$\begin{cases} 0.1x + 0.5y = 24 \\ x + y = 60 \end{cases}$$

To use the substitution method, first substitute the value for one variable in one equation into the other equation. Solving the second equation for y, we have:

$x + y = 60$	Write the second equation.
$y = 60 - x$	Subtract x from both sides of the equation.

Substitute $y = 60 - x$ into the first equation.

$$0.1x + 0.5(60 - x) = 24$$

Now, solve the first equation for x.

$0.1x + 0.5(60 - x) = 24$	Write the first equation with the substituted value.
$x + 5(60 - x) = 240$	Multiply both sides of the equation by 10 to clear out the decimals.
$x + 5(60) + 5(-x) = 240$	Use the distributive property for the left-hand side of the equation.
$x + 300 - 5x = 240$	Simplify the distribution on the left-hand side.
$-4x + 300 = 240$	Combine like terms on the left-hand side of the equation.
$-4x = -60$	Subtract 300 from both sides of the equation.
$x = 15$	Divide both sides of the equation by –4.

Next, substitute the value for x into either of the two original equations to find the y-value.

$15 + y = 60$	Substitute $x = 60$ into the second equation.
$15 + y - 15 = 60 - 15$	Subtract 15 from both sides of the equation.
$y = 45$	Simplify both sides of the equation.

Since $x = 15$ and $y = 45$, we need 15 liters of the first solution and 45 liters of the second solution.

Practice Questions

1. Solve the system of equations using the substitution method: $\begin{cases} x = -5 \\ y = 2x + 8 \end{cases}$

a. $(-1,6)$
b. $(1,10)$
c. $(-2,-5)$
d. $(-5,-2)$

2. Solve the system of equations using the substitution method: $\begin{cases} y = 2x + 1 \\ y = x + 2 \end{cases}$

a. $(3,1)$
b. $(1,3)$
c. $(2,2)$
d. $(-1,1)$

3. Solve the system of equations using the substitution method: $\begin{cases} 3x + y = 10 \\ y = 2x \end{cases}$

a. $(3,4)$
b. $(-1,-2)$
c. $(2,4)$
d. $(4,2)$

4. Use the substitution method to simplify the system of equations: $\begin{cases} y = -3x \\ 2x + 3y = -7 \end{cases}$

a. $(2, -6)$
b. $(7, -7)$
c. $(3, -1)$
d. $(1, -3)$

5. Use the substitution method to solve the system of equations: $\begin{cases} y = \frac{1}{2}x - 1 \\ y = x - 4 \end{cases}$

a. $(-1, -5)$
b. $(2,0)$
c. $(6,2)$
d. $(2,6)$

6. Use the substitution method to solve the system of equations: $\begin{cases} y = x + 2 \\ x = \frac{1}{3}y + 4 \end{cases}$

a. $(7,9)$
b. $(9,7)$
c. $(6,6)$
d. $(1,3)$

7. Use the substitution method to solve the system of equations: $\begin{cases} x + 3y = 9 \\ 2x + y = -2 \end{cases}$

a. $(4, -3)$
b. $(-3,4)$
c. $(0,3)$
d. $(-1,0)$

8. Use the substitution method to solve the system of equations: $\begin{cases} 2x - y = 9 \\ x + 3y = 8 \end{cases}$

a. $(4, -1)$
b. $(-1,3)$
c. $(5,1)$
d. $(1,5)$

9. Use the substitution method to solve the system of equations: $\begin{cases} x + 4y = -6 \\ -2x + y = 3 \end{cases}$

a. $(2,1)$
b. $(2,7)$
c. $(-2, -1)$
d. $(-1, -2)$

10. Use the substitution method to solve the system of equations: $\begin{cases} -x - y = -12 \\ 2x + y = 16 \end{cases}$

a. (8,4)
b. (4,8)
c. (−1,13)
d. (−1,18)

11. Use the substitution method to solve the system of equations: $\begin{cases} x - 6y = -4 \\ 3x + 5y = 11 \end{cases}$

a. (−3,4)
b. (−10,1)
c. (1,2)
d. (2,1)

12. Use the substitution method to solve the system of equations: $\begin{cases} 2x + y = 10 \\ 4x + 2y = 6 \end{cases}$

a. (2,6)
b. (1,1)
c. Infinitely many solutions
d. No solution

13. For a school fundraiser, a science class sells chips and candy bars at a school's football game. The chips are sold for $1.50 each and the candy bars are sold for $1.25 each. The class raised a total of $127.50 from selling a total of 90 bags of chips and candy bars. How many of each item were sold?

a. 60 bags of chips and 30 candy bars
b. 30 bags of chips and 60 candy bars
c. 45 bags of chips and 45 candy bars
d. 70 bags of chips and 20 candy bars

14. The sum of two numbers is 20. Three times the smaller number equals the larger number. What are the two numbers?

a. 6 and 14
b. 5 and 15
c. 4 and 12
d. No solution

15. An acid solution is 10% acid. A second acid solution containing 50% acid is mixed with the first solution. How many liters of each solution should be mixed to have 60 liters of a 40% acid solution?

a. 40 liters of the first solution and 20 liters of the second solution
b. 30 liters of the first solution and 3 liters of the second solution
c. 15 liters of the first solution and 45 liters of the second solution
d. 45 liters of the first solution and 15 liters of the second solution

5.3 The Elimination Method

To solve a system of equations using **elimination**, begin by rewriting both equations in standard form $Ax + By = C$. Check to see if the coefficients of one pair of like variables add to zero. If not, multiply one or both of the equations by a non-zero number to make one set of like variables add to zero. Add the two equations to solve for one of the variables. Substitute this value into one of the original equations to solve for the other variable. Check your work by substituting into the other equation. Now, let's look at solving the following system using the elimination method.

$$\begin{cases} 5x + 6y = 4 \\ x + 2y = 4 \end{cases}$$

If we multiply the second equation by –3, we can eliminate the y-terms.

$$\begin{aligned} 5x + 6y &= 4 \\ -3x - 6y &= -12 \end{aligned}$$

Add the equations together and solve for x.

$$\begin{aligned} 2x &= -8 \\ x &= \frac{-8}{2} = -4 \end{aligned}$$

Plug the value for x back in to either of the original equations and solve for y:

$$\begin{aligned} (-4) + 2y &= 4 \\ 2y &= 8 \\ y &= \frac{8}{2} = 4 \end{aligned}$$

Therefore, the solution is $(-4, 4)$. You can check your work by substituting the found values for x and y into both equations.

$$\begin{aligned} 5(-4) + 6(4) &= 4 \\ -20 + 24 &= 4 \\ 4 &= 4 \end{aligned} \qquad \begin{aligned} (-4) + 2(4) &= 4 \\ -4 + 8 &= 4 \\ 4 &= 4 \end{aligned}$$

Similar to solving systems using the graphing or substitution methods, solving systems using the elimination method can also result in no solution if the lines are parallel and infinite solutions if the lines are the same.

Review Video: The Elimination Method
Visit mometrix.com/academy and enter code: 449121

EXAMPLES

EXAMPLE 1:

Solve the system of equations using the elimination method.

$$\begin{cases} y + 8x = 21 \\ 11x - y = 17 \end{cases}$$

Answer

Align the equations so that one variable is "eliminated" when the two equations are added or subtracted. In this case, line up the x- and y-variables. Adding the equations will eliminate the y-terms.

$$\begin{array}{r} 8x + y = 21 \\ \underline{11x - y = 17} \end{array}$$

The y-variables will cancel out when the two equations are added.

$$\begin{array}{r} 8x + y = 21 \\ \underline{+\,11x - y = 17} \\ 19x = 38 \\ \frac{19x}{19} = \frac{38}{19} \\ x = 2 \end{array}$$

Now, plug in 2 for x in either equation in order to solve for y. Here we use the first equation.

$$\begin{array}{c} y + 8x = 21 \\ y + 8(2) = 21 \\ y + 16 = 21 \\ y + 16 - 16 = 21 - 16 \\ y = 5 \end{array}$$

The solution to the system is (2,5).

Example 2:

Solve the system of equations using the elimination method.

$$\begin{cases} 2x + y = 7 \\ 3x - 2y = -7 \end{cases}$$

Answer

As written, none of the variables will cancel out, or eliminate, if the two equations are added. However, if the first equation is multiplied by 2, $2y$ and $-2y$ will cancel out.

Multiply the first equation by 2.

$$\begin{array}{c} 2(2x + y = 7) \\ 4x + 2y = 14 \end{array}$$

Now, line up the equations and add them together.

$$\begin{array}{r} 4x + 2y = 14 \\ \underline{+3x - 2y = -7} \\ 7x = 7 \\ \frac{7x}{7} = \frac{7}{7} \\ x = 1 \end{array}$$

Now, plug in 1 for x in either equation to solve for y. Here we plug 1 in for x in the first equation.

$$\begin{aligned} 2x + y &= 7 \\ 2(1) + y &= 7 \\ 2 + y &= 7 \\ 2 + y - 2 &= 7 - 2 \\ y &= 5 \end{aligned}$$

The solution to the system is (1,5).

EXAMPLE 3:

Gloria has multiple hummingbird feeders on her ranch. She wants to make a large batch of sugar water so she can fill up all the feeders. She needs 30 liters of the sugar water mixture, and she needs the solution to have a sugar concentration of 30%. Gloria currently has two containers of sugar water. One container has a sugar concentration of 25% and the other container has a concentration of 40%. How many liters of each solution must Gloria mix to get the desired solution? Use the elimination method to solve.

ANSWER

Gloria needs to mix 20 liters of the 25% concentration solution with 10 liters of the 40% concentration solution.

Start by setting up a model to represent the scenario.

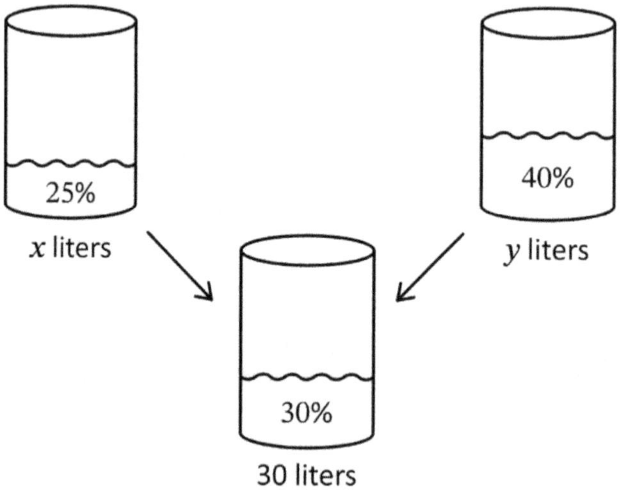

Use the model to write a system of equations. One equation will be related to concentration and one equation will be related to volume in liters.

Volume equation: The total volume of the solution will be the result of some of container x being combined with some of container y.

$$x + y = 30$$

Concentration equation: The equation needs to show how the concentration of one container plus the concentration of the other container produces a total concentration of 30%. Write the percentages in decimal form. The equation equals 9 liters because the concentration needs to be 30% of the total amount of liquid, and the number of liquid needs to be 30 liters: $0.3 \times 30 = 9$.

$$0.25x + 0.4y = 9$$

The system of equations:

$$\begin{cases} x + y = 30 \\ 0.25x + 0.4y = 9 \end{cases}$$

To solve the system using elimination, one of the variables needs to cancel out. This can be achieved by multiplying the top equation by –25, and the bottom equation by 100.

$$\begin{cases} -25x - 25y = -750 \\ 25x + 40y = 900 \end{cases}$$

Now, add the equations so the x-terms cancel out.

$$15y = 150$$

Solve for y by dividing both sides by 15.

$$y = 10$$

This means that 10 liters of the 40% concentration solution is required. Now, solve for x using either of the original equations.

$$\begin{aligned} x + y &= 30 \\ x + 10 &= 30 \\ x &= 20 \end{aligned}$$

Therefore, 20 liters will come from the 25% concentration solution. Gloria needs to mix 10 liters of the 40% solution with 20 liters of the 25% solution.

Practice Questions

1. Use the elimination method to solve the system of equations: $\begin{cases} 2x + y = 7 \\ -2x + 3y = 5 \end{cases}$

a. (2,3)
b. (3,2)
c. (1,5)
d. (−1,1)

2. Use the elimination method to solve the system of equations: $\begin{cases} 2x + y = 13 \\ x + y = 8 \end{cases}$

a. (4,4)
b. (4,5)
c. (5,3)
d. (3,5)

3. Use the elimination method to solve the system of equations: $\begin{cases} x + 3y = 7 \\ x - y = -1 \end{cases}$

a. (2,1)
b. (1,2)
c. (4,1)
d. (5,6)

4. Use the elimination method to solve the system of equations: $\begin{cases} 2x - 4y = -4 \\ x + 3y = 13 \end{cases}$

a. (6,12)
b. (1,4)
c. (3,4)
d. (4,3)

5. Use the elimination method to solve the system of equations: $\begin{cases} 3x - y = -8 \\ 2x + 2y = 8 \end{cases}$

a. $(-2,2)$
b. $(2,2)$
c. $(-1,5)$
d. $(5,-1)$

6. Use the elimination method to solve the system of equations: $\begin{cases} x + 5y = -4 \\ -3x - 4y = 1 \end{cases}$

a. $(1,-1)$
b. $(-1,1)$
c. $(6,-2)$
d. $(-3,2)$

7. Use the elimination method to solve the system of equations: $\begin{cases} -5x - y = 7 \\ 2x + 3y = 5 \end{cases}$

a. $(-1,-2)$
b. $(1,1)$
c. $(3,-2)$
d. $(-2,3)$

8. Use the elimination method to solve the system of equations: $\begin{cases} 4x + 2y = 6 \\ -8x - 4y = -12 \end{cases}$

a. (1,2)
b. (2,1)
c. No solution
d. Infinite solutions

9. Use the elimination method to solve the system of equations: $\begin{cases} 2x - 3y = -6 \\ 3x - 2y = 1 \end{cases}$

a. (4,3)
b. (3,4)
c. (0,2)
d. (1,2)

10. Use the elimination method to solve the system of equations: $\begin{cases} x + 4y = 5 \\ 3x - 6y = 6 \end{cases}$

a. $\left(3, \frac{1}{2}\right)$
b. $\left(\frac{1}{2}, 3\right)$
c. (1,1)
d. (4,1)

11. Use the elimination method to solve the system of equations: $\begin{cases} y = -\frac{1}{5}x - \frac{3}{5} \\ -2x + 5y = 9 \end{cases}$

a. $\left(5, \frac{2}{5}\right)$
b. $\left(-5, -\frac{1}{5}\right)$
c. $\left(-4, \frac{1}{5}\right)$
d. $\left(\frac{1}{5}, -4\right)$

12. Use the elimination method to solve the system of equations: $\begin{cases} 6x - 3y = 12 \\ y = 2x + 5 \end{cases}$

a. (3,2)
b. (1,7)
c. No solution
d. Infinite solutions

13. The sum of two numbers, *x* and *y* is -60, and two times *x* minus *y* is -15. What are *x* and *y*?

a. $x = -35$ and $y = -25$
b. $x = -25$ and $y = -35$
c. $x = -15$ and $y = -45$
d. $x = -9$ and $y = -3$

14. For a fundraiser, your school sells bags of cotton candy and cans of soda at a school play. Cotton candy is sold for $2.25 a bag and sodas are sold for $2.00 each. Your school raised a total of $325 and sold a total of 150 bags of cotton candy and cans of soda. How many of each item were sold?

a. 100 bags of cotton candy and 50 cans of soda
b. 50 bags of cotton candy and 100 cans of soda
c. 75 bags of cotton candy and 75 cans of soda
d. 80 bags of cotton candy and 70 cans of soda

15. An acid solution is 20% acid. A second acid solution containing 60% acid is mixed with the first solution. How many liters of each solution should be mixed to have 90 liters of a 30% acid solution?

a. 22.5 liters of the first solution and 67.5.5 liters of the second solution
b. 67.5 liters of the first solution and 22.5 liters of the second solution
c. 40 liters of the first solution and 50 liters of the second solution
d. 75 liters of the first solution and 15 liters of the second solution

5.4 Writing Systems of Linear Equations

A **system of linear equations** refers to two or more linear equations that are "working together." Graphically, this looks like two or more lines graphed on the same coordinate plane. When you solve a system of linear equations, you are identifying the point where the lines intersect, or if they intersect at all. There are three possible solutions when solving systems of linear equations: one solution (one point of intersection), no solution (parallel lines), or infinite solutions (lines that completely overlap).

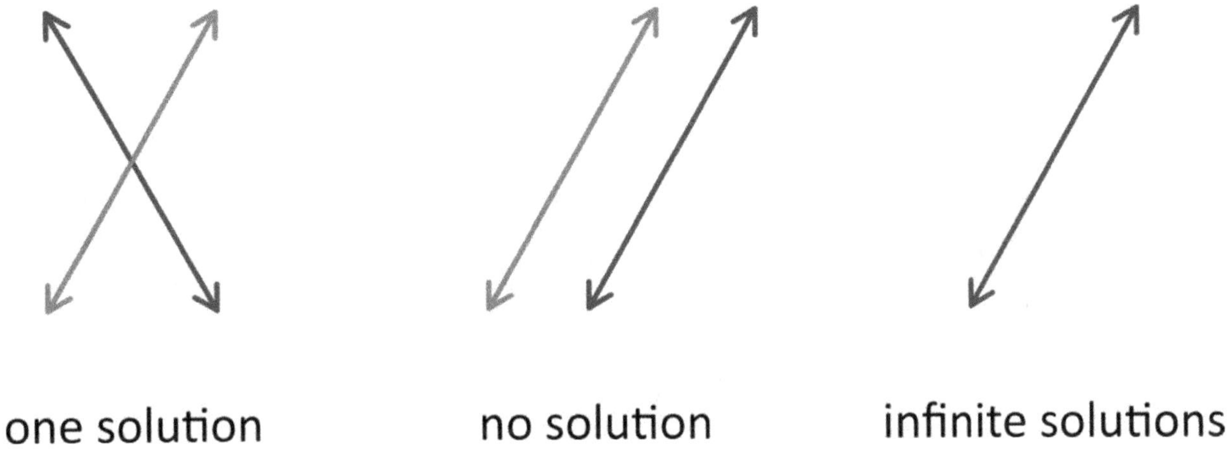

WRITING SYSTEMS FROM A GRAPH

When you write a system of equations, you are writing the two equations that are represented in the scenario. When a system of equations is presented graphically, start by determining the equation of each line. Remember, the y-intercept is the location where the line passes through the y-axis. The slope is the vertical change divided by the horizontal change, also known as "rise over run." For example, in the graph below, the thick line has a y-intercept of 1 and a slope of 3. The equation in slope-intercept form is $y = 3x + 1$. The thin line has a y-intercept of 2 and a slope of 2. The equation in slope-intercept form is $y = 2x + 2$. Now that the equations for both lines are identified, the system can be expressed as the following.

$$\begin{cases} y = 3x + 1 \\ y = 2x + 2 \end{cases}$$

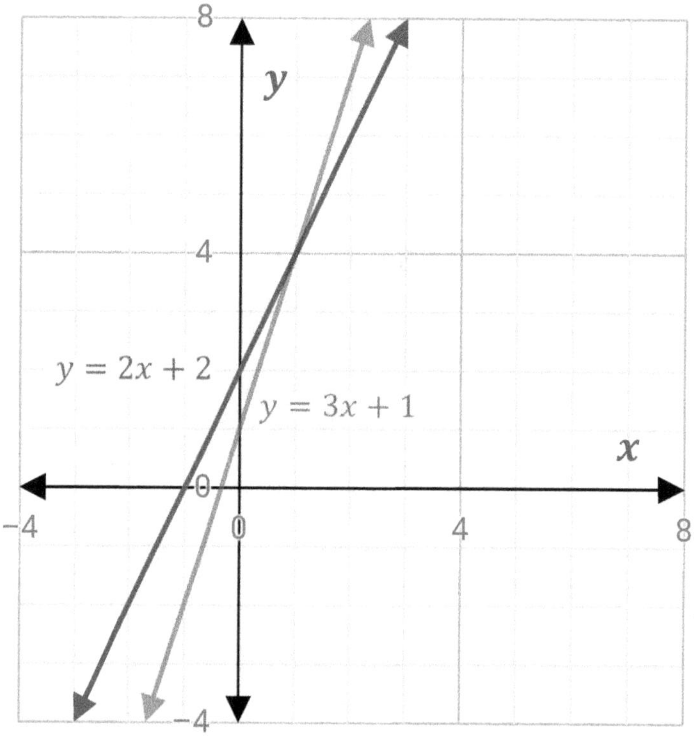

WRITING SYSTEMS FROM A TABLE

When writing a system of equations from a table, start by writing an equation to match the data in each table. Remember, the y-intercept can be easily identified in a table by locating $x = 0$ and then finding the corresponding y-value. The slope can be calculated using the slope-formula: $m = \frac{y_2 - y_1}{x_2 - x_1}$. If $x = 0$ is not given in the table, the equation can be written using the point-slope formula and then converted to slope-intercept form.

In the table below on the left, the y-intercept is –2 because $y = -2$ when $x = 0$, and the slope is $m = \frac{-2-(-1)}{0-(-2)} = \frac{-1}{2} = -\frac{1}{2}$. This means that the equation in slope-intercept form is $y = -\frac{1}{2}x - 2$. The table on the right has a y-intercept of 1, because $y = 1$ when $x = 0$, and a slope of $m = \frac{-3-9}{1-(-2)} =$

$\frac{-12}{3} = -4$. This means that the equation in slope-intercept form is $y = -4x + 1$. The two equations can be written together as a system of linear equations to describe the tables:

$$\begin{cases} y = -\frac{1}{2}x - 2 \\ y = -4x + 1 \end{cases}$$

x	y
-2	-1
-1	-1.5
0	-2
1	-2.5

x	y
-2	9
0	1
1	-3
4	-15

Writing Systems from a Written Description

Word problems that have two different variables with different quantities attached to each variable will often be expressed as a system of equations.

For example, let's say a group of friends goes out for ice cream on Friday. They buy 3 chocolate cones and 2 vanilla cones, and the total cost is $11. The ice cream was so good that the same group of friends went back the next day. This time the group orders 4 chocolate cones and 2 vanilla cones, and the total cost is $12.

When writing a system of equations for a written description, start by organizing the important information and then put the information in equation form.

3 chocolate cones and 2 vanilla cones for a total cost of $11 can be expressed as the equation $3c + 2v = 11$.

4 chocolate cones and 2 vanilla cones for a total cost of $12 can be expressed as the equation $4c + 2v = 12$.

The following system of equations represents the written description.

$$\begin{cases} 3c + 2v = 11 \\ 4c + 2v = 12 \end{cases}$$

Once you have created a system of equations, use any of the 3 solving methods to solve for your variables. This system is set up well for using the elimination method. Start by multiplying the second equation by −1.

$$\begin{cases} 3c + 2v = 11 \\ -4c - 2v = -12 \end{cases}$$

Now, add the equations together and solve for c.

$$-c = -1$$
$$c = 1$$

From here, substitute the value for c into either original equation and solve for v.

$$3(1) + 2v = 11$$
$$3 + 2v = 11$$

$$2v = 8$$
$$v = 4$$

Therefore, the solution to this system of equations is $c = 1$ and $v = 4$, meaning that each chocolate cone costs \$1 and each vanilla cone costs \$4.

EXAMPLES

EXAMPLE 1:

Write a system of equations that describes the tables below, and then solve the system.

x	y
0	-3
1	$-2\frac{1}{2}$
2	-2
3	$-1\frac{1}{2}$

x	y
0	-3
1	1
2	5
3	9

ANSWER

Write an equation for each table. The y-intercept is determined by locating $x = 0$ and then identifying the corresponding y-value.

The slope is determined by using the slope formula: $m = \frac{y_2 - y_1}{x_2 - x_1}$.

The table on the left shows a y-intercept of –3, because $y = -3$ when $x = 0$, and a slope of $m = \frac{-2-(-3)}{2-0} = \frac{1}{2}$. This results in the equation $y = \frac{1}{2}x = 3$.

The table on the right shows a y-intercept of –3, because $y = -3$ when $x = 0$, and a slope of $m = \frac{5-(-3)}{2-0} = \frac{8}{2} = 4$. This results in the equation $y = 4x - 3$.

The following system of equations matches the tables.

$$\begin{cases} y = \frac{1}{2}x - 3 \\ y = 4x - 3 \end{cases}$$

This system is set up well for using the elimination method. Start by multiplying the second equation by –1.

$$\begin{cases} y = \frac{1}{2}x - 3 \\ -y = -4x + 3 \end{cases}$$

Then, add the two equations and solve for x.

$$0 = -\frac{7}{2}x$$
$$0 = x$$

Next, substitute the value for x into either original equation and solve for y.

$$y = 4(0) - 3 = 0 - 3 = -3$$

Therefore, the solution to the system of equations is $(0, -3)$.

Example 2:

Write a system of equations that describes the graph below, and then solve the system of equations.

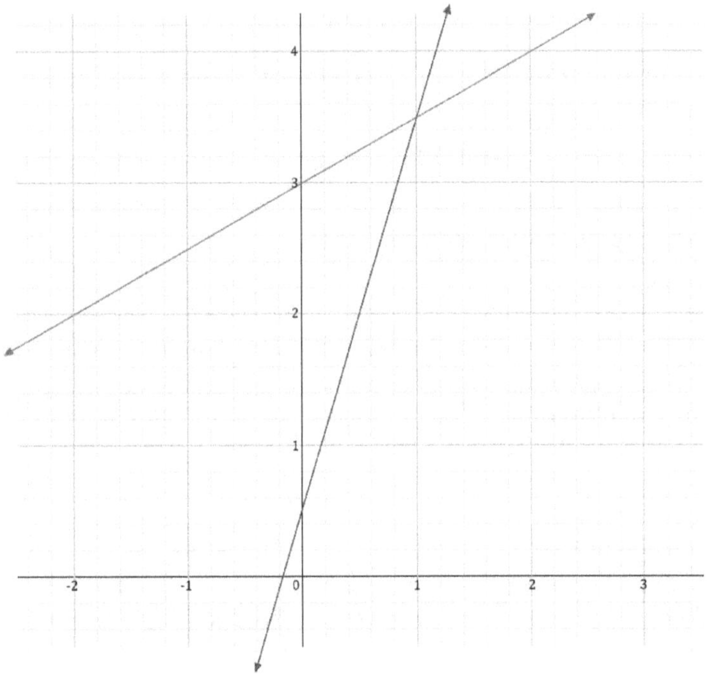

Answer

First Line: The y-intercept is 3 because the line passes through the point $(0,3)$, and the slope is $m = \frac{3-2}{0-(-2)} = \frac{1}{2}$: $y = \frac{1}{2}x + 3$.

Second Line: The y-intercept is $\frac{1}{2}$ because the line passes through the point $\left(0, \frac{1}{2}\right)$, and the slope is $m = \frac{2-\frac{1}{2}}{\frac{1}{2}-0} = \frac{\frac{3}{2}}{\frac{1}{2}} = \frac{3}{2} \times \frac{2}{1} = 3$: $y = 3x + \frac{1}{2}$.

The system of equations that matches the graph:

$$\begin{cases} y = \frac{1}{2}x + 3 \\ y = 3x + \frac{1}{2} \end{cases}$$

One way to solve this system of equations is by using the substitution method. Substitute $3x + \frac{1}{2}$ for y in the first equation and then solve for x.

$$\begin{aligned} 3x + \frac{1}{2} &= \frac{1}{2}x + 3 \\ \frac{5}{2}x + \frac{1}{2} &= 3 \\ \frac{5}{2}x &= \frac{5}{2} \\ x &= 1 \end{aligned}$$

Then, substitute this value for x into either original equation and solve for y.

$$y = \frac{1}{2}(1) + 3 = \frac{1}{2} + 3 = 3\frac{1}{2}$$

Therefore, the solution to the system of equations is $\left(1{,}3\frac{1}{2}\right)$.

EXAMPLE 3:

Write a system of equations that describes the scenario, and then solve the system of equations.

A rectangular garden has a length that is 7 feet longer than its width. The total perimeter of the garden is 72 feet. How long is the garden?

ANSWER

Write two equations that are both true for the scenario. The length of the garden is 7 feet longer than the width, which can be expressed by the equation $l = w + 7$. Since the perimeter of a rectangle can be found using the equation $2l + 2w = P$, the second equation for the system is $2l + 2w = 72$.

$$\begin{cases} l = w + 7 \\ 2l + 2w = 72 \end{cases}$$

This system of equations can be used to solve for the length and the width of the garden. Graphing, substitution, or elimination can be used to calculate the length and width.

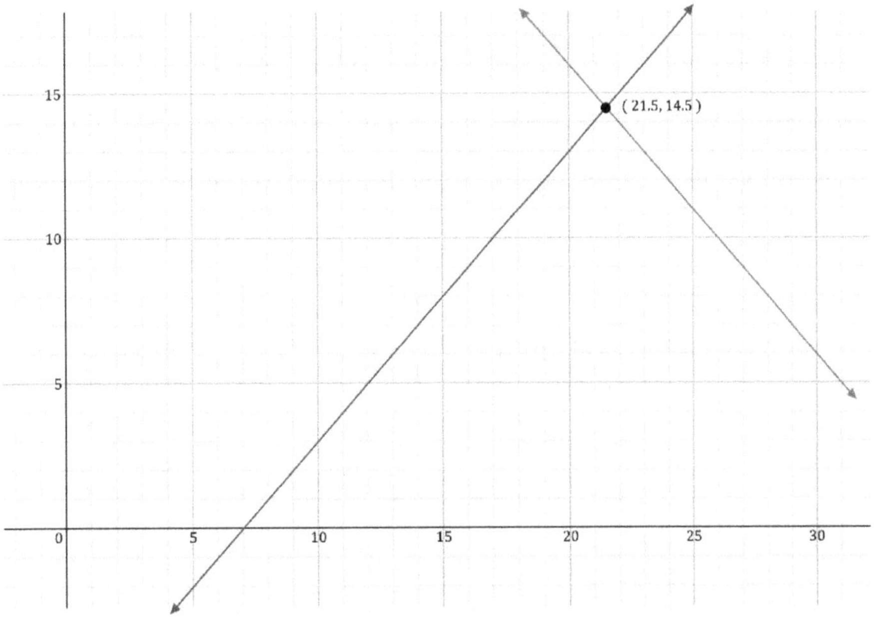

If the graphing method is used, the point of intersection is identified as (21.5,14.5). This means that the length of the garden is 21.5 feet and the width of the garden is 14.5 feet.

Example 4:

Write a system of linear equations in slope-intercept form based on the two tables below.

Table A

x	y
2	−5
8	−9
14	−13

Table B

x	y
2	4
3	5
4	6

Answer

Each table will produce a line when graphed. An equation can be used to describe this graph, and this equation can be found using point-slope form. As the name implies, point-slope form requires a point and a slope.

Each table will have its own equation.

Table A:

Start by finding the slope of the line. To determine the slope, use the slope formula.

$$m = \frac{y_2 - y_1}{x_2 - x_1}$$

Use the values in the table to substitute in values for x_1, x_2, y_1, and y_2. Substitute $(2, -5)$ and $(8, -9)$ into the slope formula.

$$m = \frac{-9 - (-5)}{8 - 2} = \frac{-4}{6} = -\frac{2}{3}$$

A point is required. Any point from the table can be selected. Here the point $(2, -5)$ is used.

Now that the slope and a point are determined, substitute both pieces of information into the point-slope formula.

$$y - y_1 = m(x - x_1)$$

The point $(2, -5)$ will be substituted in for x_1 and y_1, and the slope $-\frac{2}{3}$ will be substituted in for m.

$$y - (-5) = -\frac{2}{3}(x - 2)$$
$$y + 5 = -\frac{2}{3}(x - 2)$$

This equation can be converted to slope-intercept form by solving for y.

$$y + 5 = -\frac{2}{3}x + \frac{4}{3}$$
$$y = -\frac{2}{3}x - \frac{11}{3}$$

Table B:

Start by finding the slope of the line. To determine the slope, use the slope formula.

$$m = \frac{y_2 - y_1}{x_2 - x_1}$$

Use the values in the table to substitute in values for x_1, x_2, y_1, and y_2. Substitute (2,4) and (3,5) into the slope formula.

$$m = \frac{5 - 4}{3 - 2} = \frac{1}{1} = 1$$

A point is required. Any point from the table can be selected. Here the point (2,4) is used.

Now that the slope and a point are determined, substitute both pieces of information into the point-slope formula.

$$y - y_1 = m(x - x_1)$$

The point (2,4) will be substituted in for x_1 and y_1, and the slope of 1 will be substituted in for m.

$$y - 4 = 1(x - 2)$$
$$y - 4 = (x - 2)$$

This equation can be converted to slope-intercept form by solving for y.

$$y = x + 2$$

The system of equations built from the tables:

$$\begin{cases} y = -\frac{2}{3}x - \frac{11}{3} \\ y = x + 4 \end{cases}$$

PRACTICE QUESTIONS

1. Write a system of two linear equations given a table of values for each equation. Then, state the solution to the system

Table 1

x	y
-2	-4
0	2
2	8
4	14

Table 2

x	y
-2	-6
0	-4
2	-2
4	0

a. $\begin{cases} y = 3x + 2 \\ y = x - 4 \end{cases}$ and the solution is $(-3, -7)$.

b. $\begin{cases} y = 3x - 2 \\ y = x + 4 \end{cases}$ and the solution is $(2,4)$.

c. $\begin{cases} y = 3x \\ y = x + 2 \end{cases}$ and the solution is $(0,0)$.

d. $\begin{cases} y = x - 4 \\ y = 3x \end{cases}$ and the solution is $(-3, -9)$.

2. Write a system of two linear equations given a table of values for each equation. Then, state the solution to the system.

Table 1

x	y
-3	-3
-1	1
1	5
5	13

Table 2

x	y
-3	-2
-1	0
1	2
5	6

a. $\begin{cases} y = 3x \\ y = x \end{cases}$ and the solution is $(0,0)$.

b. $\begin{cases} y = 2x \\ y = 3x \end{cases}$ and the solution is $(3,6)$.

c. $\begin{cases} y = 2x + 3 \\ y = x + 1 \end{cases}$ and the solution is $(-2, -1)$.

d. $\begin{cases} y = 3x - 2 \\ y = x - 1 \end{cases}$ and the solution is $(-1, -5)$.

3. Write a system of two linear equations given a table of values for each equation. Then, state the solution to the system.

Table 1

x	y
−3	−5
−1	−1
2	5
5	11

Table 2

x	y
−3	15
−1	9
2	0
5	−9

a. $\begin{cases} x - y = 4 \\ 3x + y = 6 \end{cases}$ and the solution is (6,4).

b. $\begin{cases} 2x - y = -2 \\ x + y = 5 \end{cases}$ and the solution is (2,3).

c. $\begin{cases} 2x + y = -1 \\ 3x - y = 6 \end{cases}$ and the solution is (1, −3).

d. $\begin{cases} 2x - y = -1 \\ 3x + y = 6 \end{cases}$ and the solution is (1,3).

4. Write a system of two linear equations given a table of values for each equation. Then, state the solution to the system.

Table 1

x	y
−2	10
1	7
2	6
4	4

Table 2

x	y
−2	−14
1	−8
2	−6
4	−2

a. $\begin{cases} x - y = -8 \\ 2x - y = 10 \end{cases}$ and the solution is (−6, −2).

b. $\begin{cases} x + y = 8 \\ 2x - y = 10 \end{cases}$ and the solution is (6,2).

c. $\begin{cases} x + y = 8 \\ x + 2y = -10 \end{cases}$ and the solution is (26, −18).

d. $\begin{cases} 3x + 4y = 17 \\ 2x - 4y = -10 \end{cases}$ and the solution is (15,10).

5. Write a system of two linear equations given a table of values for each equation. Then, state the solution to the system.

Table 1

x	y
-5	1
1	9
4	13
7	17

Table 2

x	y
-5	-7
1	17
4	29
7	41

a. $\begin{cases} 2x + y = 14 \\ y = 5x - 8 \end{cases}$ and the solution is $(4,6)$.

b. $\begin{cases} 2x + y = 18 \\ y = 3x - 10 \end{cases}$ and the solution is $(6,3)$.

c. $\begin{cases} 4x + 3y = 23 \\ y = 4x - 13 \end{cases}$ and the solution is $(2,5)$.

d. $\begin{cases} 4x - 3y = -23 \\ y = 4x + 13 \end{cases}$ and the solution is $(-2,5)$.

6. Write a system of two linear equations given a table of values for each equation. Then, state the solution to the system.

Table 1

x	y
-2	$-\frac{7}{5}$
-1	-1
1	$-\frac{1}{5}$
3	$\frac{3}{5}$

Table 2

x	y
-2	-3
-1	$-\frac{7}{3}$
1	-1
3	$\frac{1}{3}$

a. $\begin{cases} 2x - 5y = 3 \\ 2x - 3y = 5 \end{cases}$ and the solution is $(4,1)$.

b. $\begin{cases} 2x + 5y = 3 \\ 2x - 3y = 5 \end{cases}$ and the solution is $(1,-1)$.

c. $\begin{cases} x + y = -2 \\ 2x + 3y = -5 \end{cases}$ and the solution is $(-1,-1)$.

d. $\begin{cases} 2x + 5y = 3 \\ x - y = -2 \end{cases}$ and the solution is $(1,3)$.

7. Write a system of two linear equations given the graphs for each equation. Then state the solution to the system.

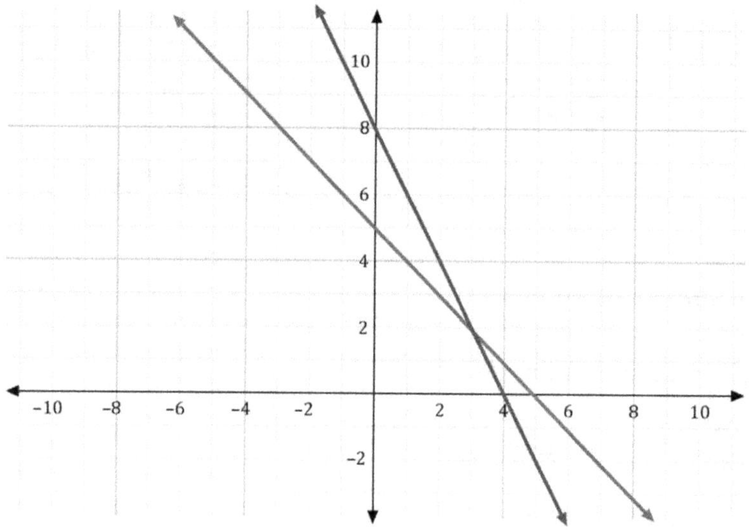

a. $\begin{cases} y = -x + 5 \\ y = -2x + 8 \end{cases}$ and the solution is (3,2).

b. $\begin{cases} y = x + 5 \\ y = 2x + 8 \end{cases}$ and the solution is (−3,2).

c. $\begin{cases} y = x - 5 \\ y = 2x - 8 \end{cases}$ and the solution is (3,2).

d. $\begin{cases} y = -x - 5 \\ y = -2x + 4 \end{cases}$ and the solution is (3, −2).

8. Write a system of two linear equations given the graphs for each equation. Then state the solution to the system.

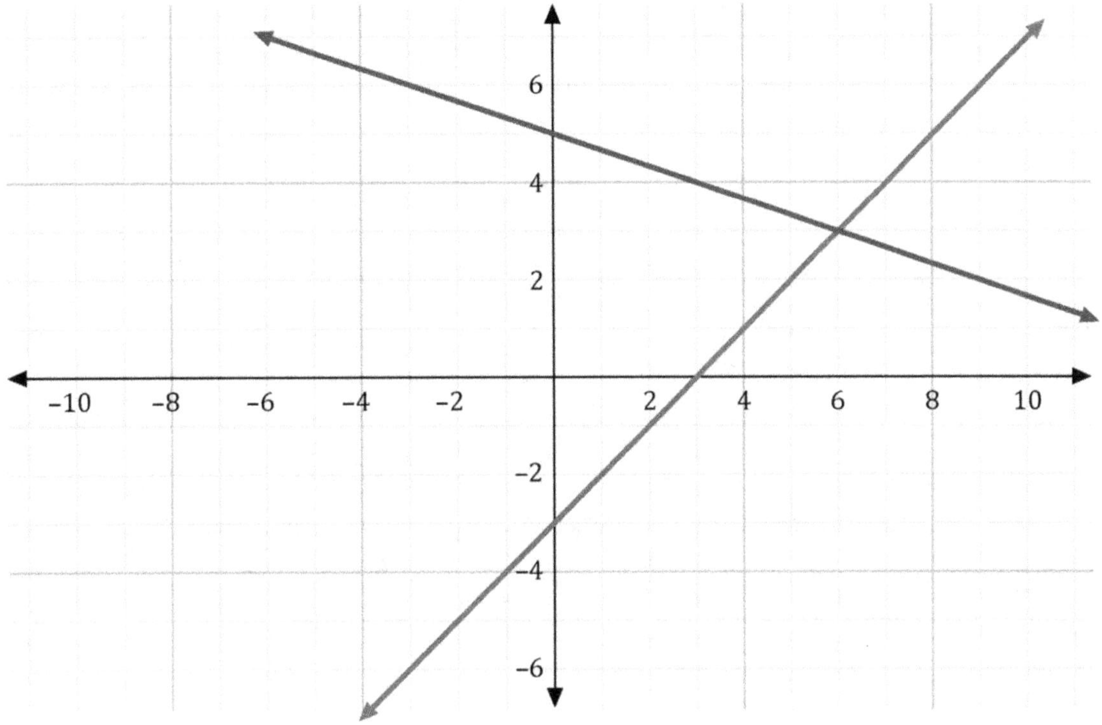

a. $\begin{cases} y = -x - 5 \\ y = \frac{1}{3}x - 5 \end{cases}$ and the solution is (6,3).

b. $\begin{cases} y = x - 3 \\ y = 3x + 5 \end{cases}$ and the solution is (3,6).

c. $\begin{cases} y = x - 3 \\ y = -\frac{1}{3}x + 5 \end{cases}$ and the solution is (6,3).

d. $\begin{cases} y = -x - 3 \\ y = \frac{1}{3}x + 5 \end{cases}$ and the solution is (−6,3).

9. Write a system of two linear equations given the graphs for each equation. Then state the solution to the system.

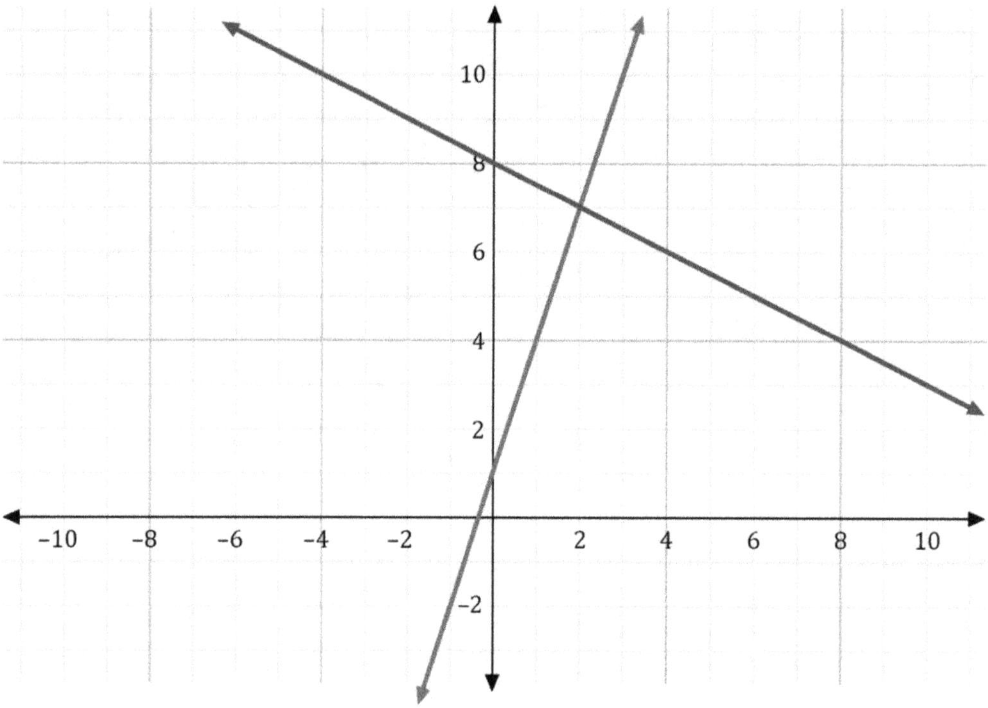

a. $\begin{cases} y = -3x + 1 \\ y = \frac{1}{2}x + 8 \end{cases}$ and the solution is $(-2,7)$.

b. $\begin{cases} y = 3x + 1 \\ y = -\frac{1}{2}x + 8 \end{cases}$ and the solution is $(2,7)$.

c. $\begin{cases} y = 3x + 1 \\ y = -2x - 8 \end{cases}$ and the solution is $(2,7)$.

d. $\begin{cases} y = \frac{1}{3}x - 1 \\ y = \frac{1}{2}x + 8 \end{cases}$ and the solution is $(-2,7)$.

10. Write a system of two linear equations given the graphs for each equation. Then state the solution to the system.

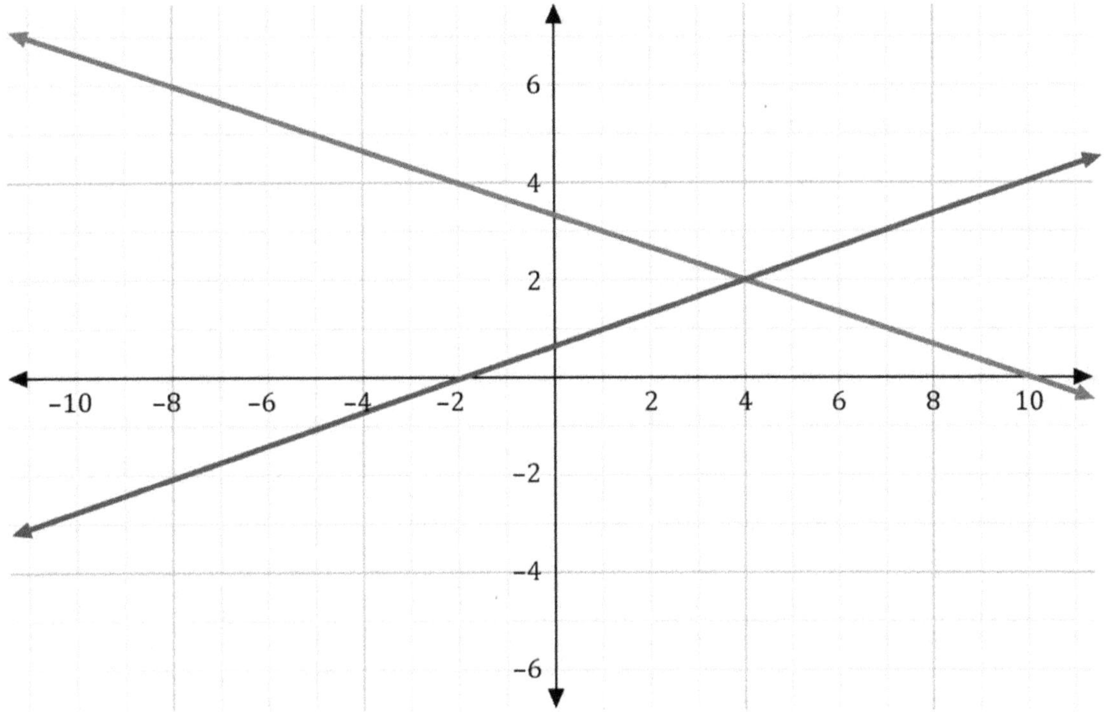

a. $\begin{cases} -x + 3y = 10 \\ x + 3y = 2 \end{cases}$ and the solution is $(-4,2)$.

b. $\begin{cases} 3x + y = 30 \\ 3x - y = -6 \end{cases}$ and the solution is $(4,2)$.

c. $\begin{cases} x - 3y = 10 \\ x + 3y = -2 \end{cases}$ and the solution is $(4,-2)$.

d. $\begin{cases} x + 3y = 10 \\ x - 3y = -2 \end{cases}$ and the solution is $(4,2)$.

11. Write a system of two linear equations given the graphs for each equation. Then state the solution to the system.

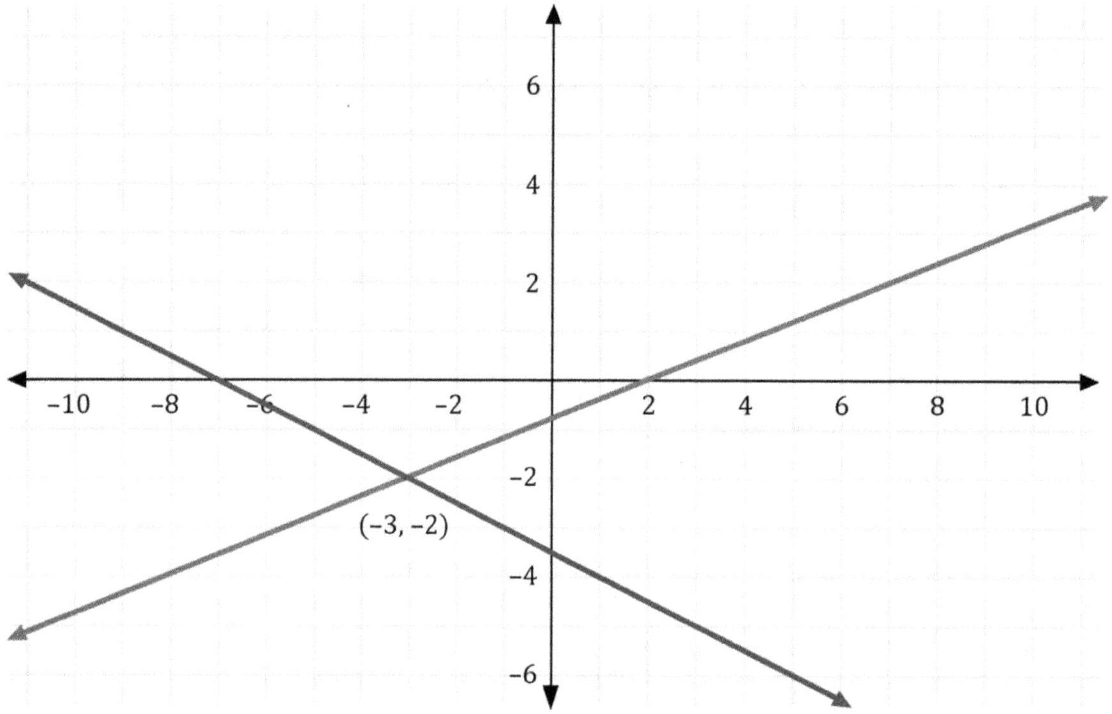

a. $\begin{cases} -2x + 5y = 4 \\ x + 3y = 6 \end{cases}$ and the solution is $(-3, -2)$.

b. $\begin{cases} x - y = 5 \\ x + 2y = -7 \end{cases}$ and the solution is $(3, -2)$.

c. $\begin{cases} 2x - 5y = 4 \\ x + 2y = -7 \end{cases}$ and the solution is $(-3, -2)$.

d. $\begin{cases} 2x + 5y = 4 \\ x - 2y = -7 \end{cases}$ and the solution is $(-3, 2)$.

12. Write a system of two linear equations given the graphs for each equation. Then state the solution to the system.

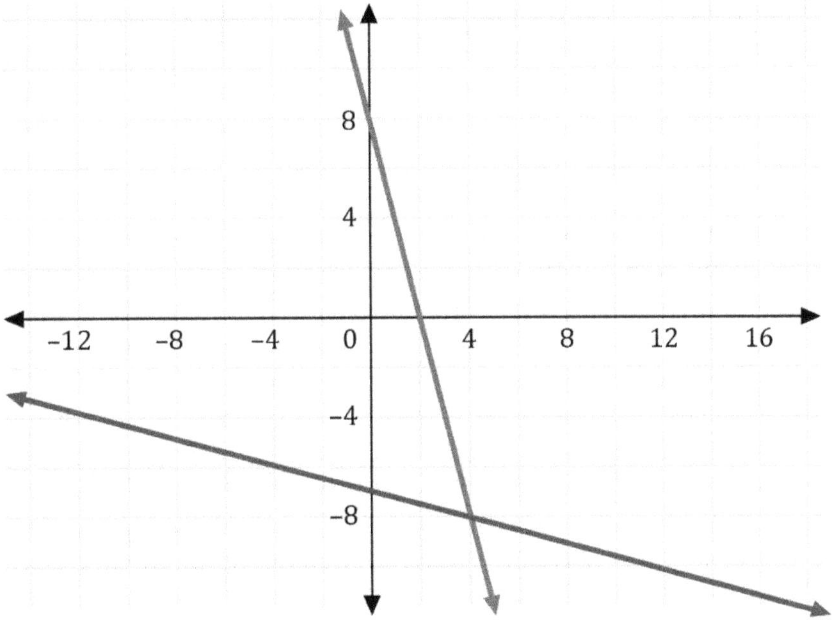

a. $\begin{cases} -4x + y = 8 \\ x - 4y = 28 \end{cases}$ and the solution is $(4, -8)$.

b. $\begin{cases} x + 4y = 8 \\ 4x + y = -28 \end{cases}$ and the solution is $(-8,4)$.

c. $\begin{cases} 4x - y = 8 \\ x - 4y = -28 \end{cases}$ and the solution is $(-4,8)$.

d. $\begin{cases} 4x + y = 8 \\ x + 4y = -28 \end{cases}$ and the solution is $(4, -8)$.

13. A small grocery store sold a total of 11 hygiene items one day—specifically packages of soap and bottles of shampoo. A package of soap sells for $3.00, and a bottle of shampoo sells for $8.00. The total sales for the soap and shampoo on that same day is $48.00. How many packages of soap and bottles of shampoo are sold that day?

a. 18 packages of soap and 7 bottles of shampoo
b. 8 packages of soap and 3 bottles of shampoo
c. 3 packages of soap and 8 bottles of shampoo
d. 24 packages of soap and 24 bottles of shampoo

14. A jar is filled with a total of 25 coins—specifically quarters and fifty-cent coins. The total value of the coins in the jar is $10.00. How many quarters and fifty-cent coins are in the jar?

a. 12 quarters and 13 fifty-cent coins
b. 15 quarters and 5 fifty-cent coins
c. 10 quarters and 15 fifty-cent coins
d. 10 quarters and 25 fifty-cent coins

15. The sum of two integers, x and y is –15, and x is 35 more than y. What are the values of x and y?

a. $x = 10$ and $y = -25$

b. $x = -10$ and $y = 25$

c. $x = 10$ and $y = 25$

d. $x = -10$ and $y = -25$

5.5 Graphing Solutions to Systems of Linear Inequalities

One way to solve a system of inequalities in two variables is by graphing. When looking at the graph of a system of inequalities, the solution to the system is where the two inequalities overlap.

For example, to solve the system $\begin{cases} y \geq x - 1 \\ y \leq -x \end{cases}$, start by graphing each inequality. Graph the inequality $y \geq x - 1$ by graphing the line $y = x - 1$ with a solid line and shade above the line because of the inequality symbol $\geq$.

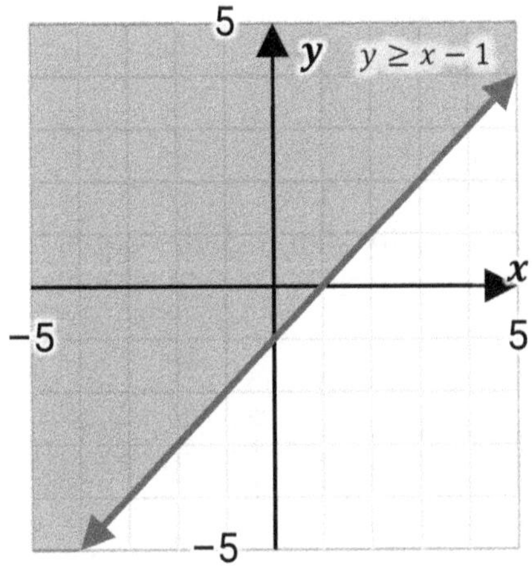

Next, graph the inequality $y \leq -x$. Start by graphing the equation $y = -x$ with a solid line and then shade below the line because of the inequality symbol $\leq$.

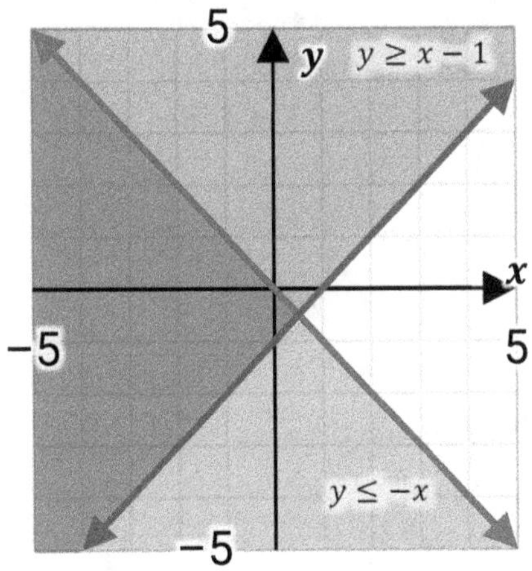

The solution to the system is the overlapping portion of the graph. Any point in this section is a solution to the system of inequalities. For example, $(-5,0)$ is a solution to the system of inequalities because it falls in the overlapping region of the graph. The point $(1,2)$ is not a solution to the system of inequalities because it only falls in one of the graph's shaded regions.

Review Video: Graphing Solutions to Linear Inequalities
Visit mometrix.com/academy and enter code: 391281

Examples

Example 1:

Graph the solution set for the system of inequalities:

$$\begin{cases} y \geq 2x - 1 \\ y \leq x + 1 \end{cases}$$

Answer

Graph each inequality as though it is an equation.

$$y \geq 2x - 1$$
$$y = 2x - 1$$

The y-intercept is –1, and the slope is 2.

The line will be solid because the inequality sign includes equal to. The shaded region will be above the line because y is isolated on one side of the inequality, and it is greater than or equal to the expression on the other side.

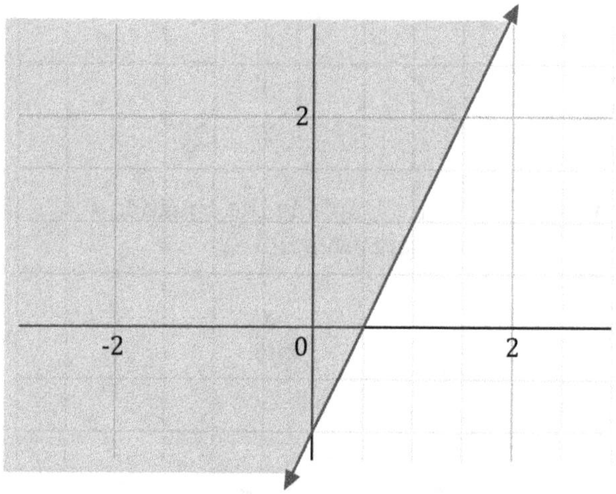

$$y \leq x + 1$$
$$y = x + 1$$

The y-intercept is 1, and the slope is 1.

The line will be solid because the inequality sign includes equal to, and the shaded region will be below the line because y is isolated on one side of the inequality, and it is less than or equal to the expression on the other side.

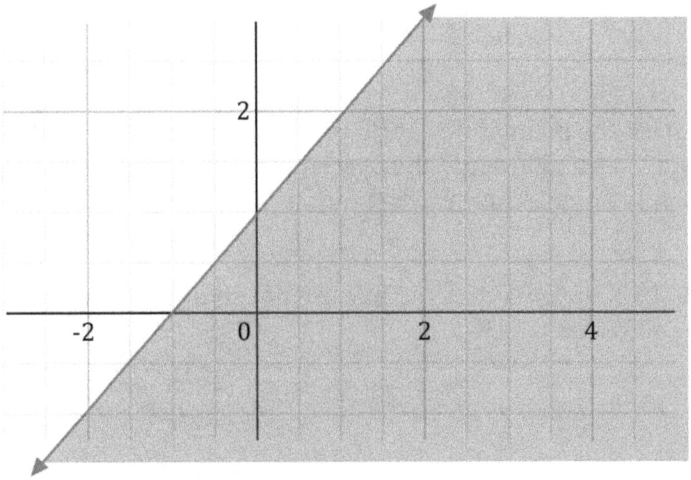

The overlapping shaded area represents the solution set. Any point that is located in this region is a solution.

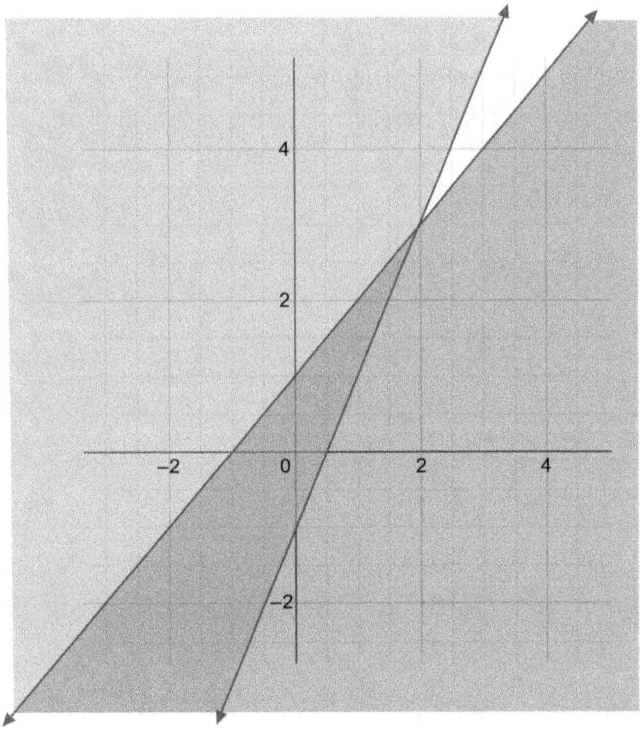

EXAMPLE 2:

Is the point (2,6) in the solution set for the system of inequalities below? Prove your answer by graphing.

$$\begin{cases} y > 2x + 1 \\ y < 6x + 5 \end{cases}$$

ANSWER

To graph the system, treat the inequalities as though they are equations. Identify the slope and y-intercept. Graph both lines and identify the overlapping area that is shared by both inequalities. This overlapping region is the solution set. If a point is in this region, then it is a solution to the system.

On a dashed line (< and >), the points on the line are not solutions. On a solid line (≤ and ≥), the points on the line are solutions.

$$y > 2x + 1$$

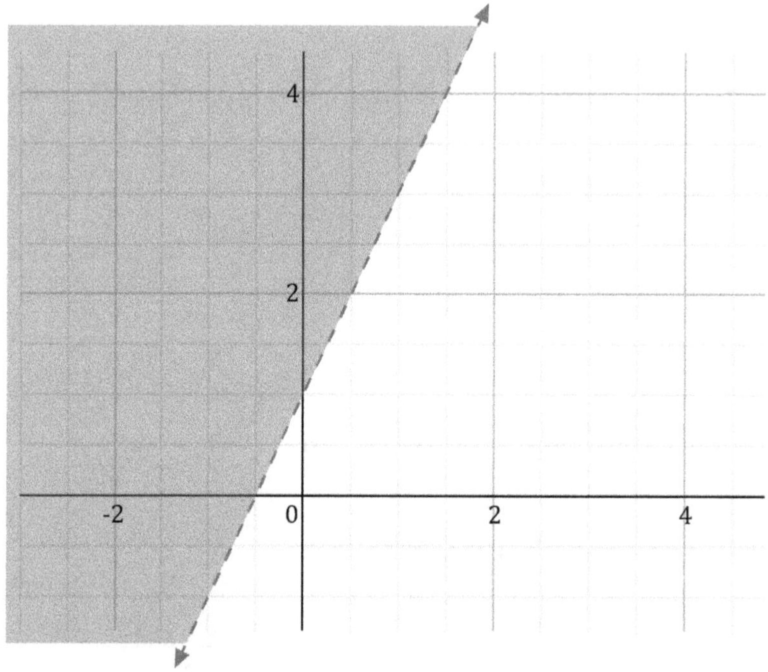

$y < 6x + 5$

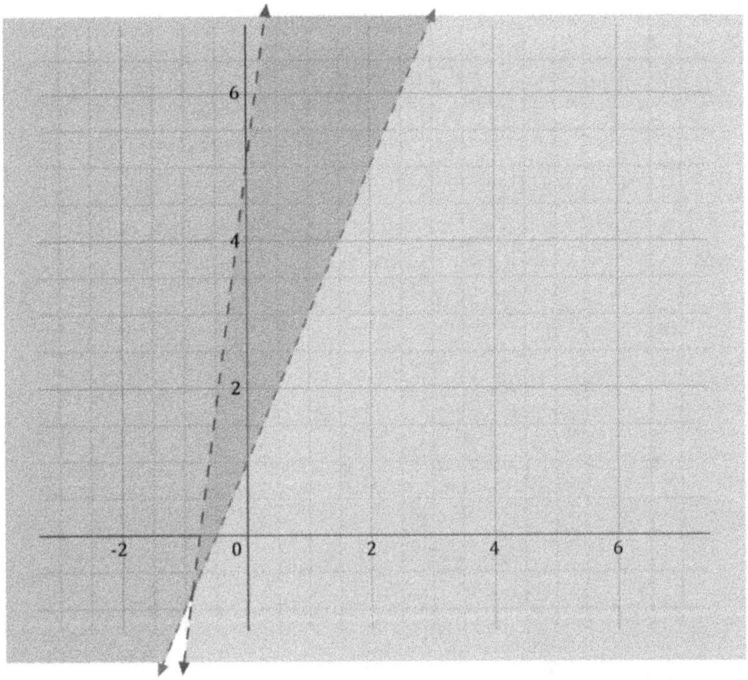

The solution set is the overlapping section, and the point (2,6) is in the solution set.

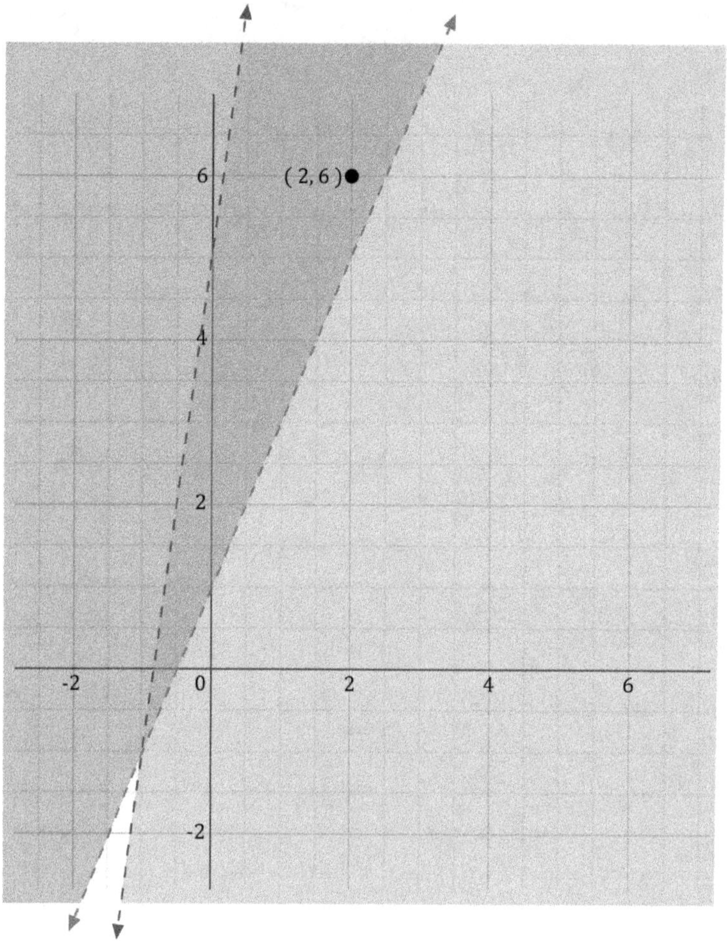

Example 3:

Is the point (2,3) a solution for the system of inequalities?

$$\begin{cases} y \geq 2x + 1 \\ y \geq x - 3 \end{cases}$$

Answer

For the point (2,3) to be a solution to the system, the x- and y-values need to satisfy both inequalities. Plug in 2 for x and 3 for y in both inequalities.

$$y \geq 2x + 1$$
$$(3) \geq 2(2) + 1$$
$$3 \geq 4 + 1$$
$$3 \geq 5$$

This is a false statement.

$$y \geq x - 3$$
$$(3) \geq (2) - 3$$
$$3 \geq -1$$

This is a true statement.

Only one statement is satisfied, which means the point (2,3) is not in the solution set for the system of inequalities.

EXAMPLE 4:

Solve the system of linear inequalities using a graph.

$$\begin{cases} x > -4 \\ y < 5 \end{cases}$$

ANSWER

There are four steps involved in solving systems of linear inequalities by graphing.

1. Graph both of the linear inequalities.
2. Decide if the lines will be dotted or solid.
3. Shade the appropriate section.
4. Note the area of intersection (the solution).

$$\begin{cases} x > -4 \\ y < 5 \end{cases}$$

Start by graphing $x > -4$. This will be a vertical line at $x = -4$. The symbols $<$ and $>$ require a dotted line. The symbols $\leq$ and $\geq$ require a solid line. $x > -4$ states that "x is greater than 4", which means that the possible x-values in the shaded region will be "greater" or "to the right" of the dotted line.

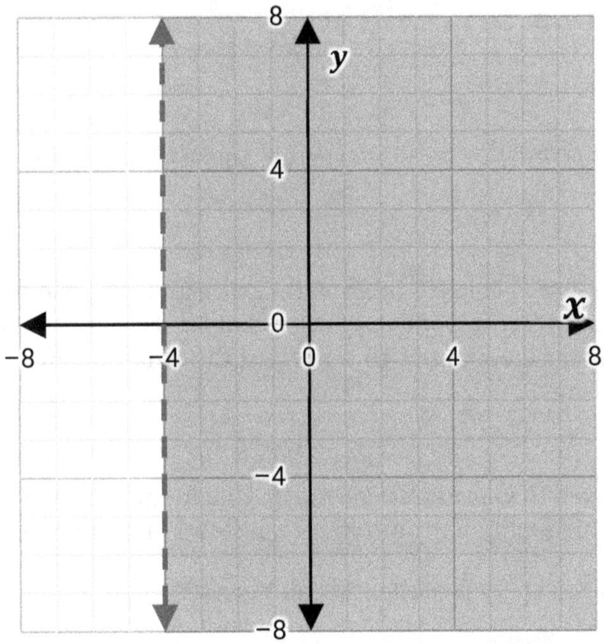

Now graph $y < 5$ on the same coordinate plane. This will be a horizontal line at $y = 5$. The symbols $<$ and $>$ require a dotted line. The symbols $\leq$ and $\geq$ require a solid line. $y < 5$ states that "y is less than 5", which means that the possible y-values in the shaded region will be "less than" or "below" of the dotted line. The shaded area of overlap represents all the possible solutions.

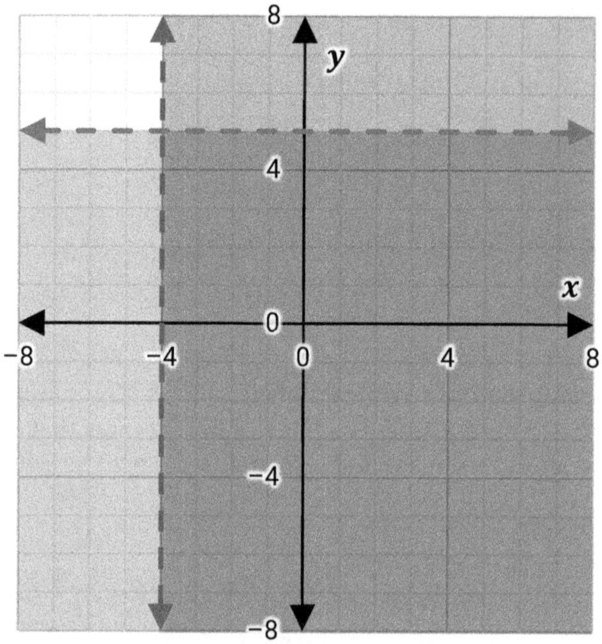

Remember, dotted lines indicate that solutions *do not* lie on the line. Solid lines indicate that solutions *do* lie on the line.

EXAMPLE 5:

Josie owns a food truck that sells chicken tacos and steak burritos. Josie charges \$12 for chicken tacos and \$30 for steak burritos. On Monday Josie sold at least 5 chicken tacos and earned at most \$240. Write a system of linear inequalities to represent the situation and solve the system by graphing.

ANSWER

Two inequalities will be used to express the scenario. One inequality will address the amount of food that is sold, and one inequality will address the amount of money earned.

Let x represent chicken tacos and y represent steak burritos.

Josie sold "at least 5 chicken tacos". This means that she sold exactly 5 tacos or more than 5 tacos. x represents tacos, so $x \geq 5$ states that the number of tacos sold is greater than or equal to 5.

Josie charges \$12 for chicken tacos and \$30 for steak burritos. Multiply the number of tacos by 12 and the number of burritos by 30, and then find the sum of these two values in order to determine the amount of money earned. The amount of money Josie earned is "at most" \$240. This means that the amount of money earned is exactly \$240 or less than \$240. $12x + 30y \leq 240$ represents the amount of money Josie earned.

The following system of inequalities represents the situation.

$$\begin{cases} x \geq 5 \\ 12x + 30y \leq 240 \end{cases}$$

In order to solve the system, graph both inequalities on the same coordinate plane. The solution will be the overlapping area of the graph.

Graph $x \geq 5$ with a solid line, and shade to the right of the line. The line will be vertical, through the x-axis located at 5.

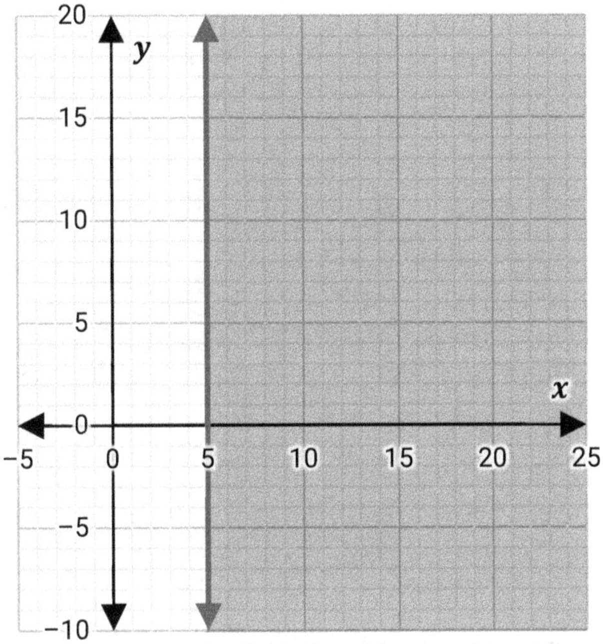

Graph $12x + 30y \leq 240$ by first determining the x- and y-intercepts.

To find the x-intercept, plug in 0 for y. Solve for x as though the inequality is an equation.

$$12x + 30(0) = 240$$
$$12x = 240$$
$$x = 20$$

The x-intercept is (20,0).

To find the y-intercept, plug in 0 for x. Solve for y as though the inequality is an equation.

$$12(0) + 30y = 240$$
$$30y = 240$$
$$y = 8$$

The y-intercept is (0,8).

Plot the points (20,0) and (0,8) on the same coordinate plane as $x \geq 5$. Connect the points to form a line. The line will be solid because the original inequality had a ≤ instead of a <, and the shading will occur below the line. This can be determined by choosing a test point, either above or below the line, and checking it in the original inequality. If the statement is true, then that side of the line is shaded; otherwise, shade the opposite side of the line. For example, use the point (0,0).

$$\begin{aligned} 12(0) + 30(0) &\leq 240 \\ 0 + 0 &\leq 240 \\ 0 &\leq 240 \end{aligned}$$

Since this statement is true, and the point is below the line, the shading will occur below the line.

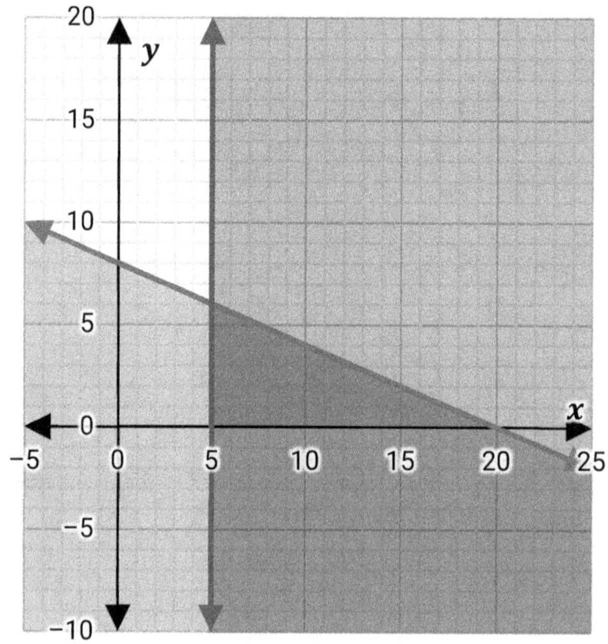

All possible (x, y) solutions are in the shaded overlapping area. For this specific problem, solutions will only be in the first quadrant. This is because customers cannot order negative amounts of food, and customers cannot be charged negative prices for food.

Any (x, y) pair located in the overlapping region is a solution. For example, the point (10,1) is in the "solution" region. This means that one possible solution to the problem is 10 chicken tacos and 1 steak burrito.

PRACTICE QUESTIONS

1. Graph the system $\begin{cases} x + y \geq 5 \\ 2x + y \leq 8 \end{cases}$, and determine if the point $(-2, 9)$ is in the solution set.

a. The given point is in the solution.

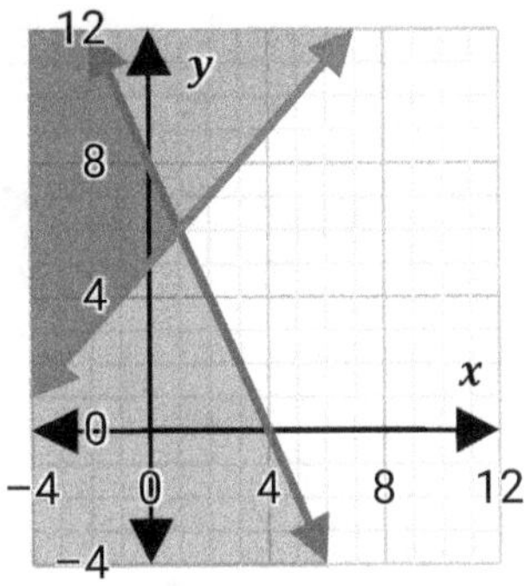

b. The given point is in the solution.

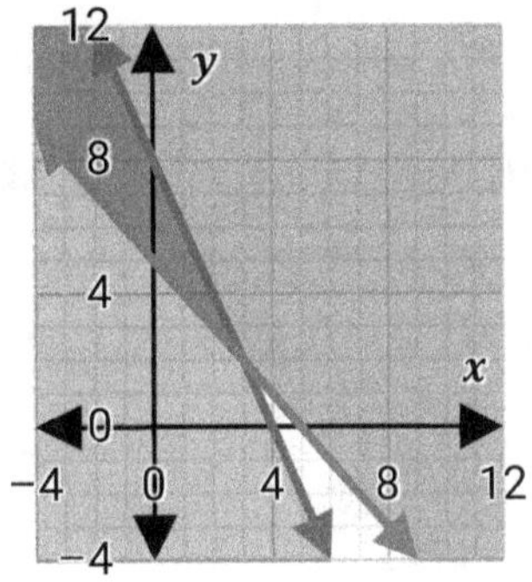

c. The given point is not in the solution.

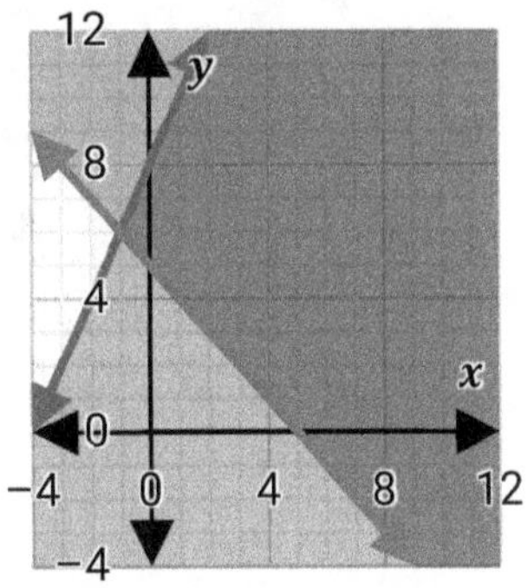

d. The given point is not in the solution.

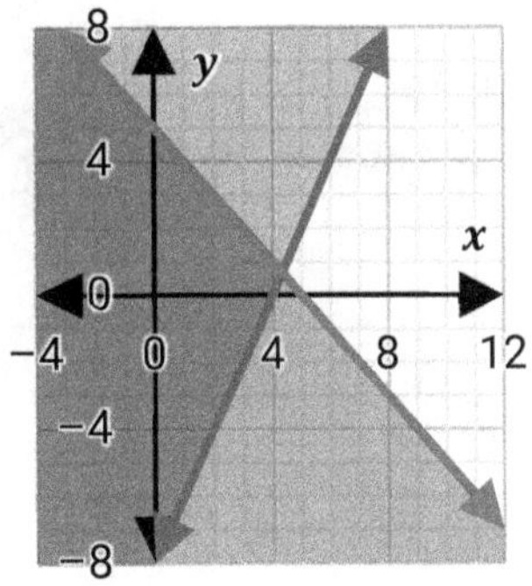

2. Graph the system $\begin{cases} 2x - y \le 4 \\ x + 3y \ge 6 \end{cases}$, and determine if the point $(3, 5)$ is in the solution set.

a. The given point is in the solution.

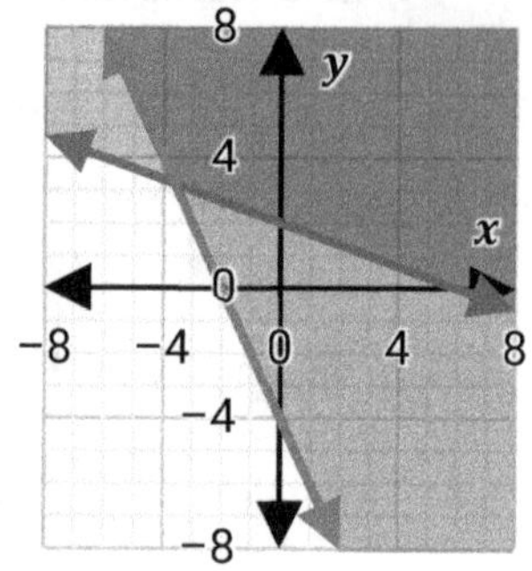

b. The given point is in the solution.

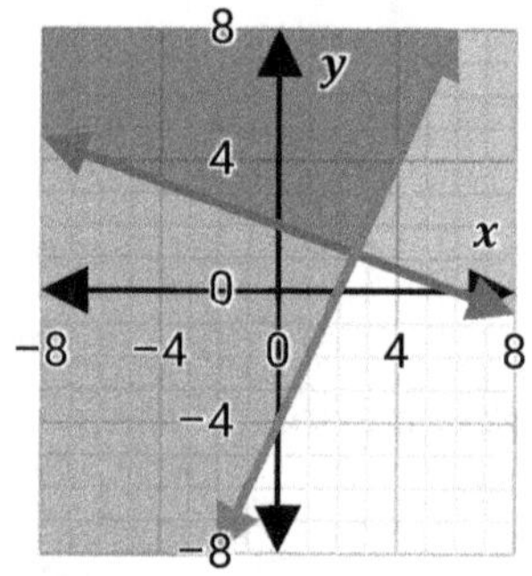

c. The given point is not in the solution.

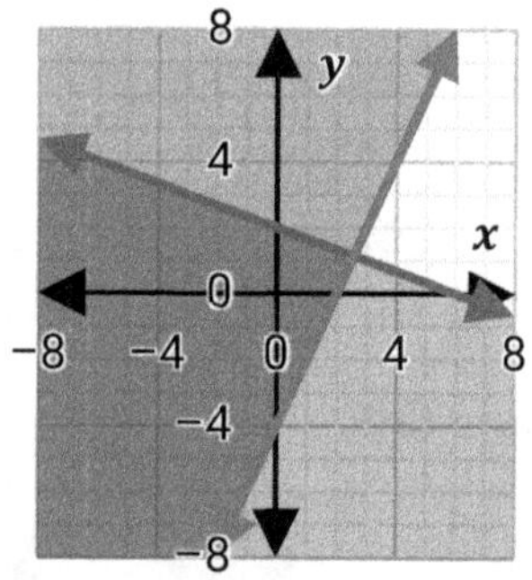

d. The given point is not in the solution.

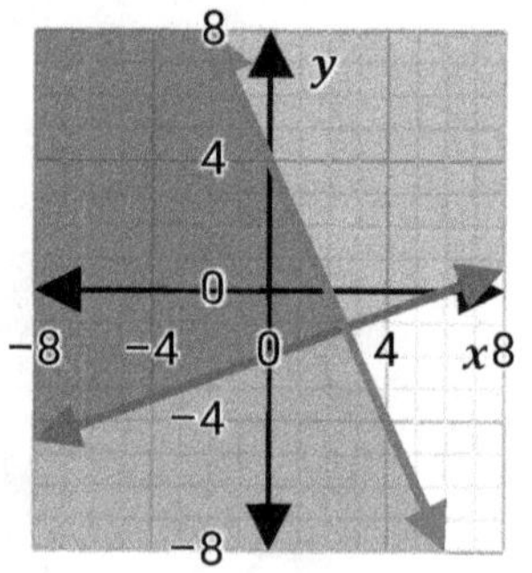

3. **Graph the system** $\begin{cases} 3x + y < 9 \\ x + 2y \geq 12 \end{cases}$**, and determine if the point** $(-3, 8)$ **is in the solution set.**

a. The given point is not in the solution.

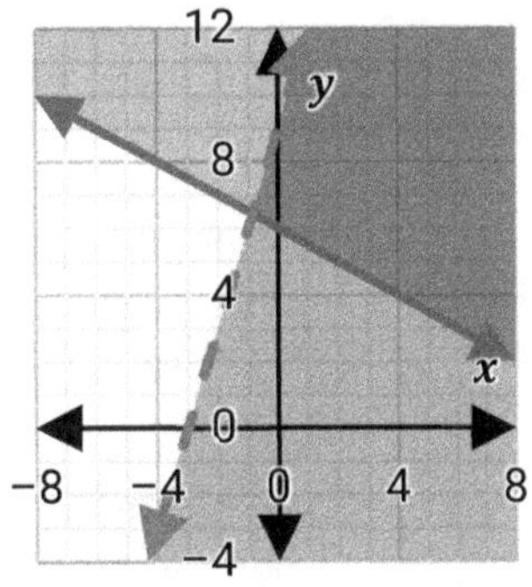

b. The given point is not in the solution.

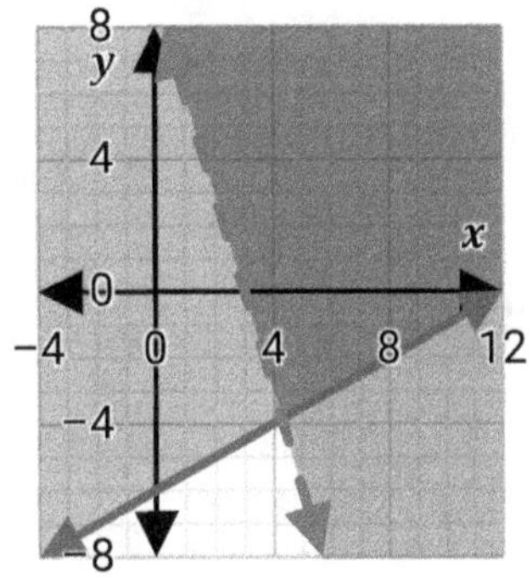

c. The given point is not in the solution.

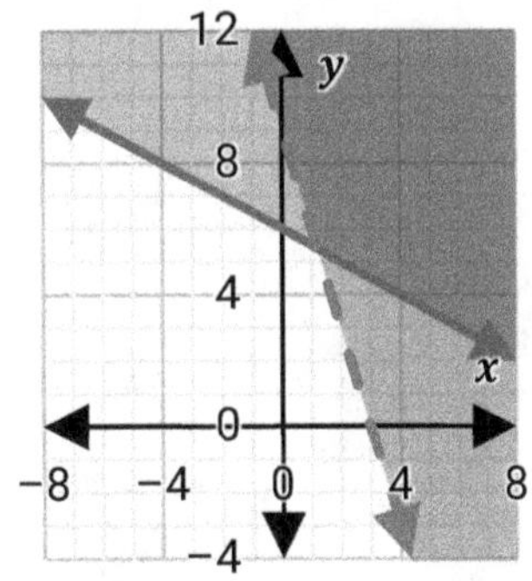

d. The given point is in the solution.

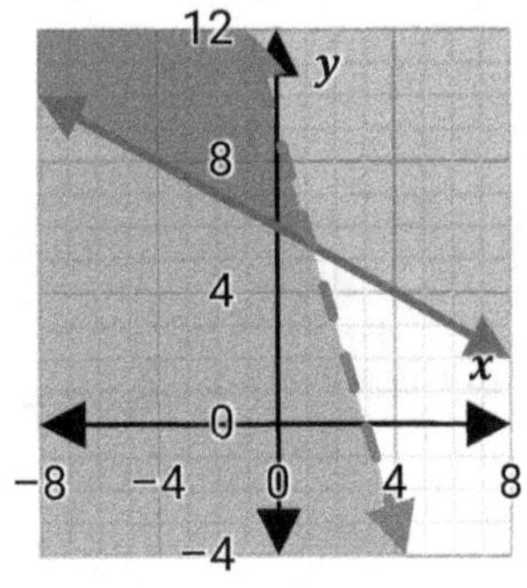

4. Graph the system $\begin{cases} x - 5y > 5 \\ 2x + 3y < 6 \end{cases}$**, and determine if the point** $(0, -3)$ **is in the solution set.**

a. The given point is in the solution.

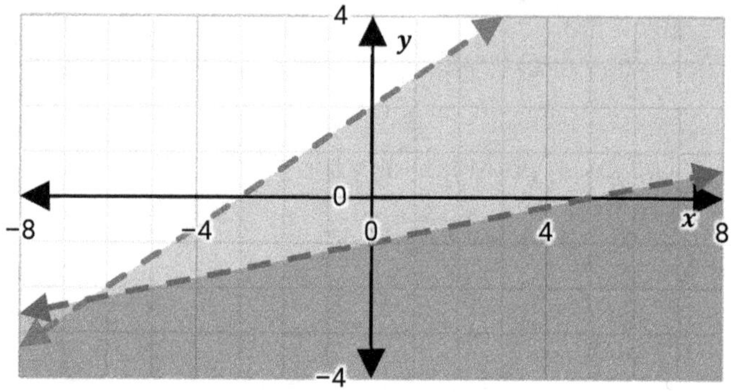

b. The given point is not in the solution.

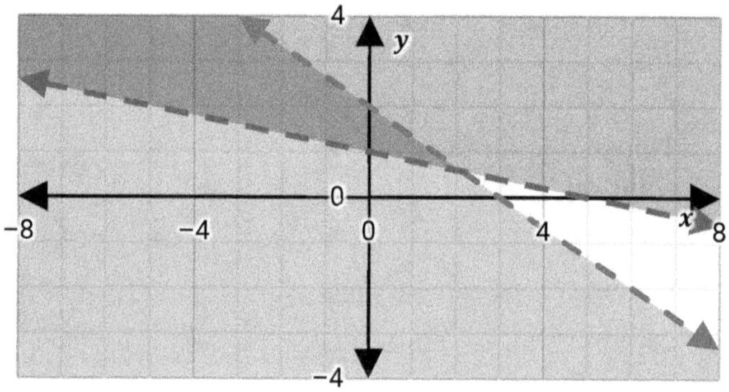

c. The given point is in the solution.

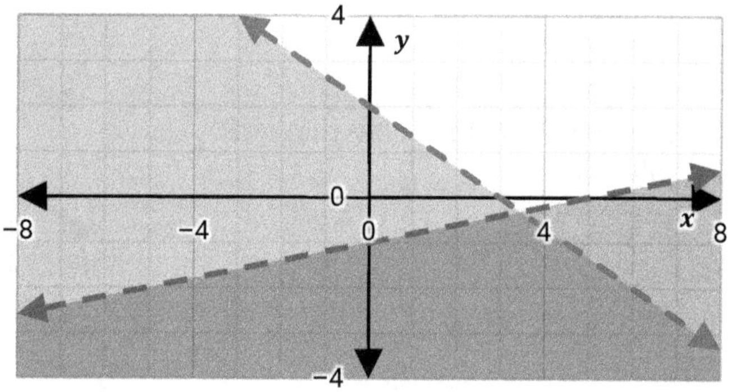

d. The given point is not in the solution.

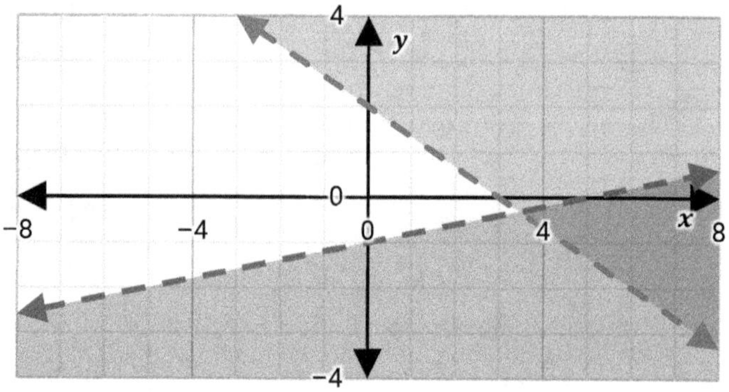

5. Graph the system $\begin{cases} 4x - y \leq 8 \\ x - y > 0 \end{cases}$, and determine if the point $(-2, 3)$ is included in the solution set.

a. The given point is in the solution.

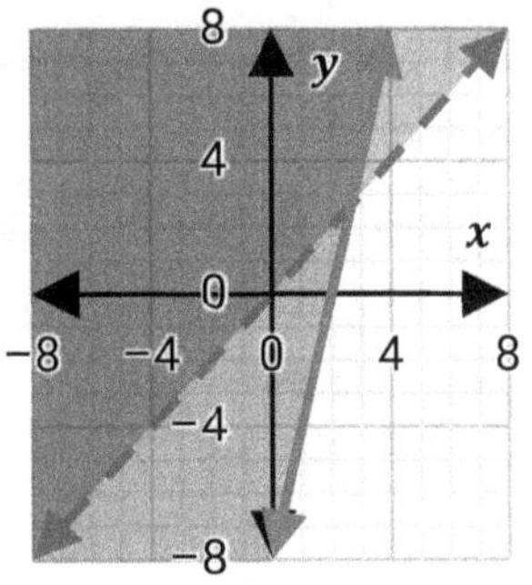

b. The given point is not in the solution.

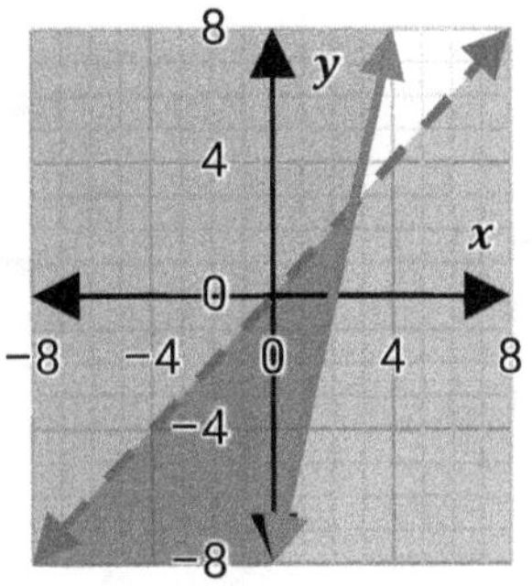

c. The given point is not in the solution.

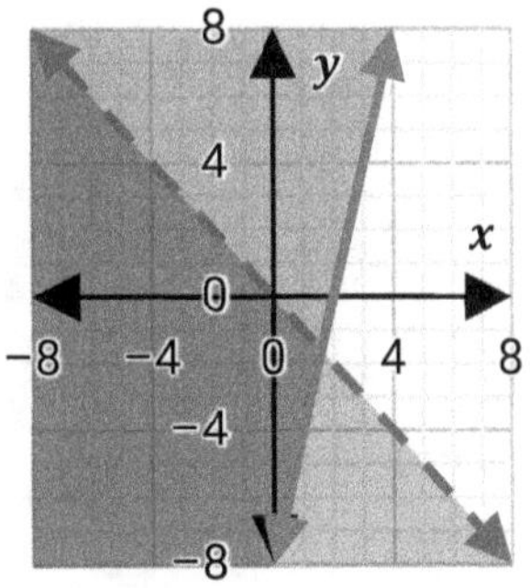

d. The given point is in the solution.

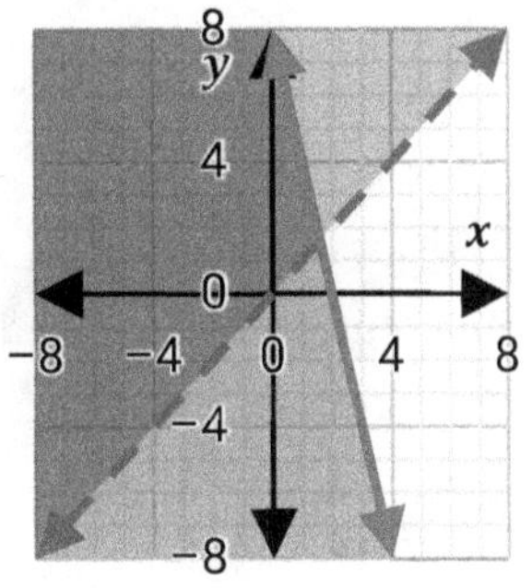

6. Graph the system $\begin{cases} 3x + 5y \geq 15 \\ 2x - 6y \leq 12 \end{cases}$, and determine if the point $(6, 2)$ is in the solution set.

a. The given point is in the solution.

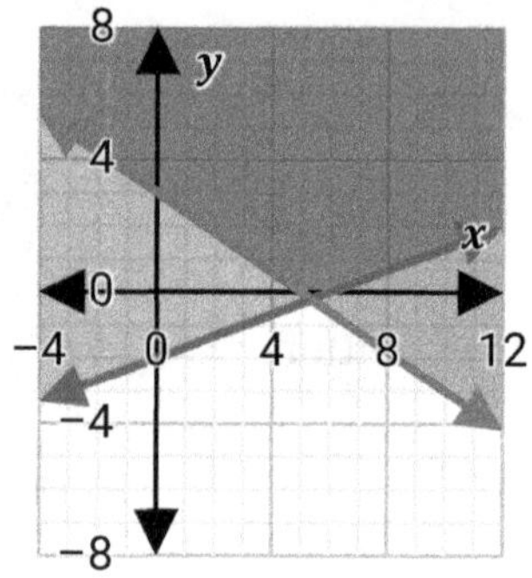

b. The given point is not in the solution.

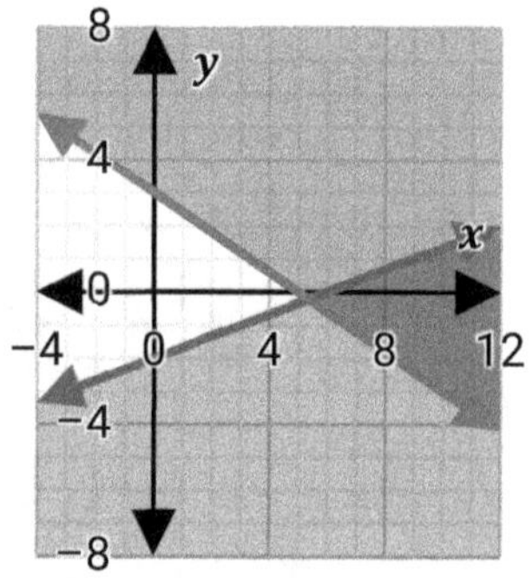

c. The given point is not in the solution.

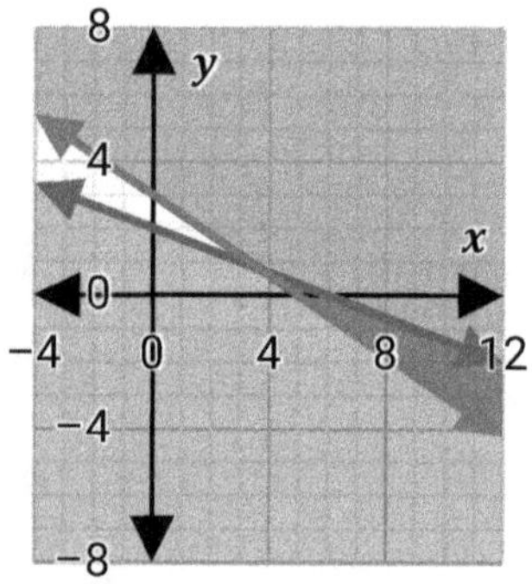

d. The given point is in the solution.

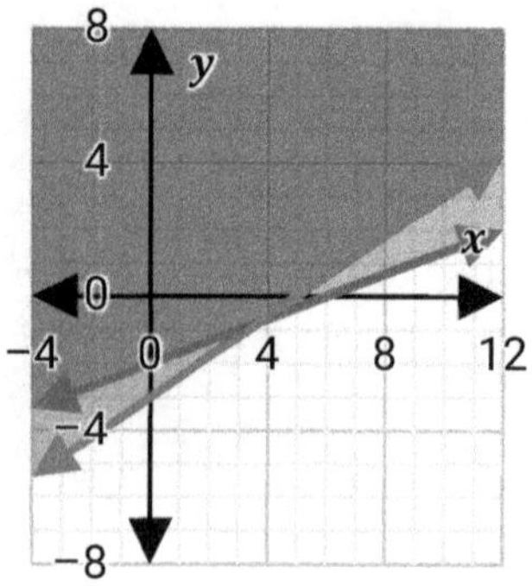

7. Graph the system $\begin{cases} y > 2x \\ x + y \geq -3 \end{cases}$**, and determine if the point** $(2, 4)$ **is in the solution set.**

a. The given point is not in the solution.

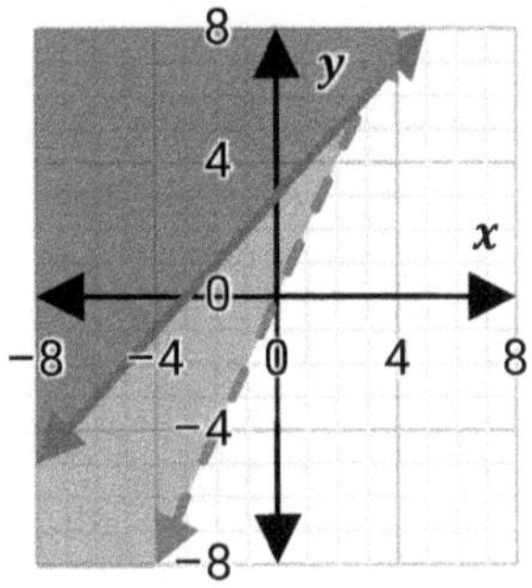

b. The given point is not in the solution.

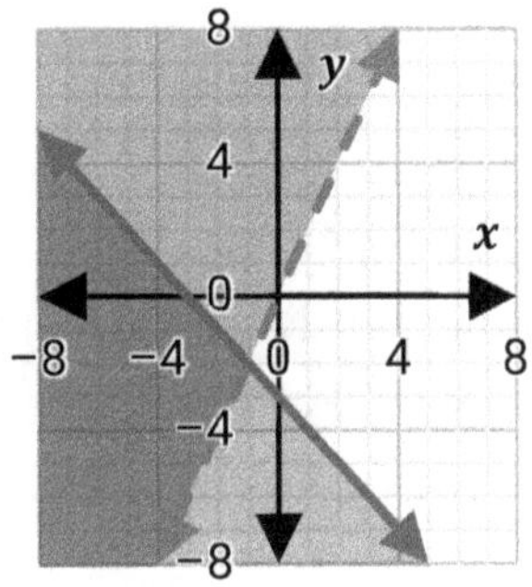

c. The given point is in the solution.

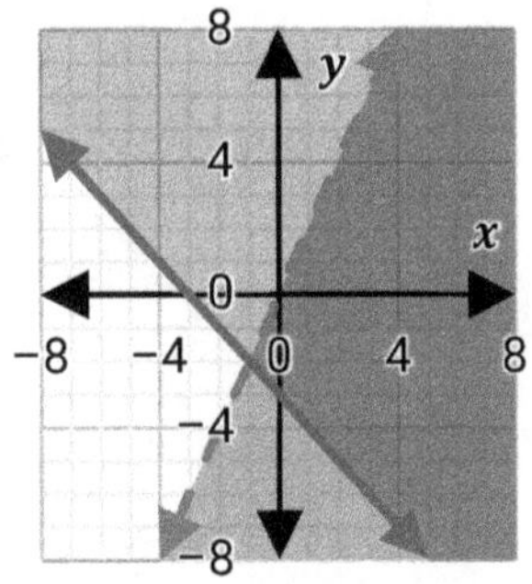

d. The given point is not in the solution.

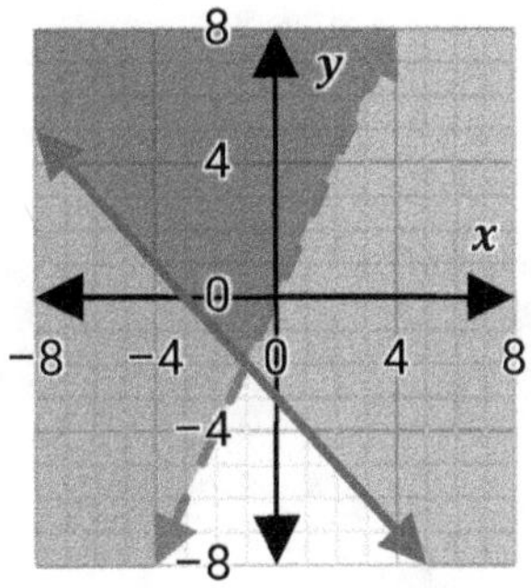

8. Graph the system $\begin{cases} 2x - y \geq 4 \\ y < \frac{2}{5}x \end{cases}$, and determine if the point $(2, 0)$ is in the solution set.

a. The given point is not in the solution.

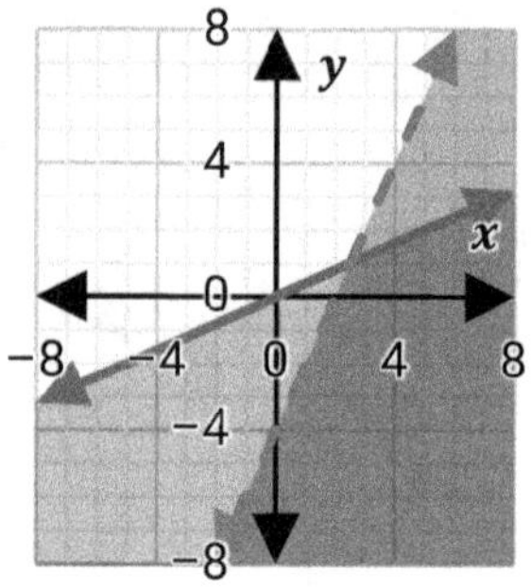

b. The given point is in the solution.

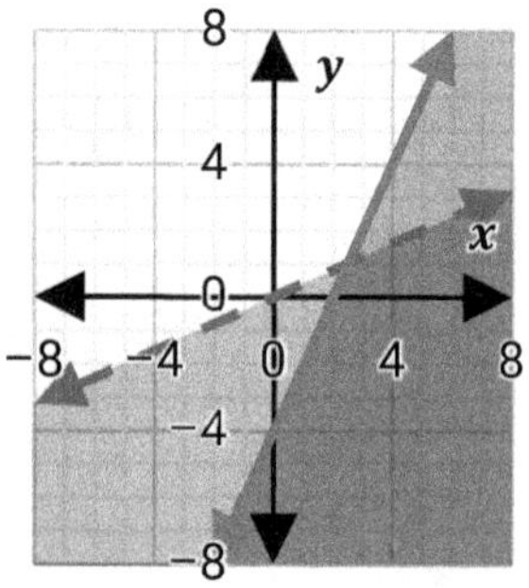

c. The given point is in the solution.

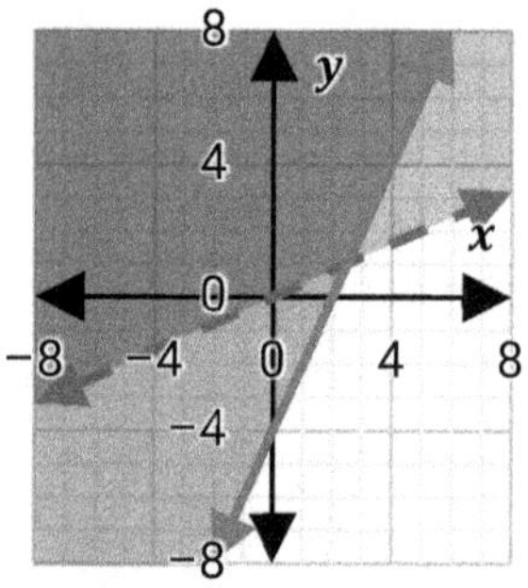

d. The given point is not in the solution.

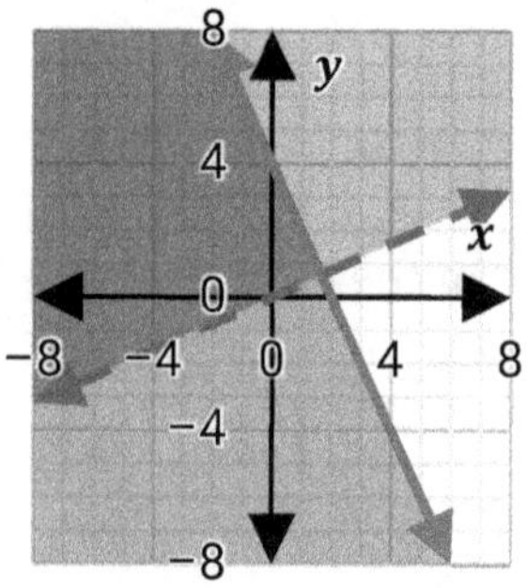

9. **Graph the system** $\begin{cases} y \leq 2x+3 \\ y > x-4 \end{cases}$**, and determine if the point** $(4, -3)$ **is in the solution set.**

a. The given point is not in the solution.

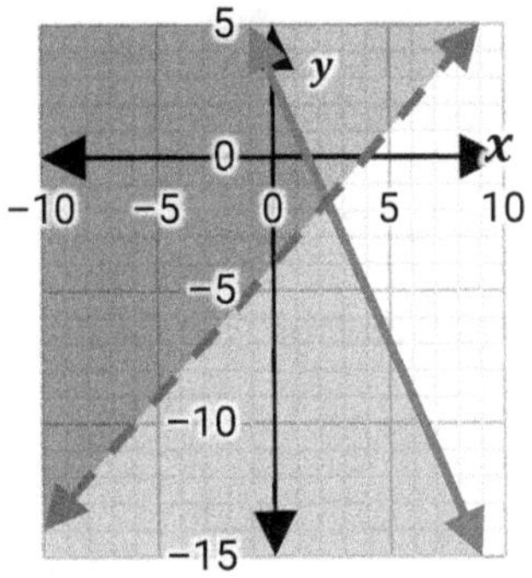

b. The given point is in the solution.

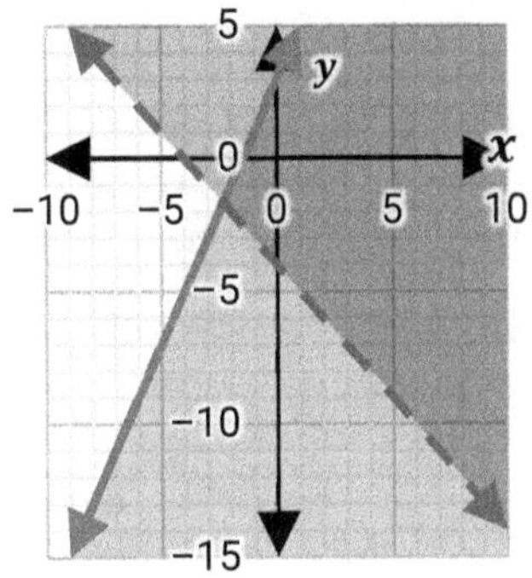

c. The given point is not in the solution.

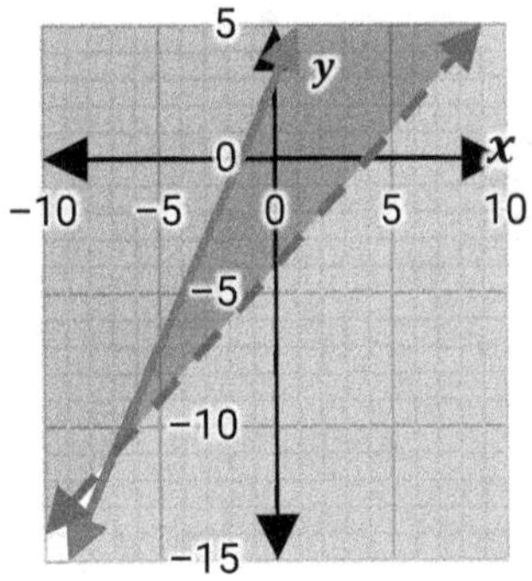

d. The given point is in the solution.

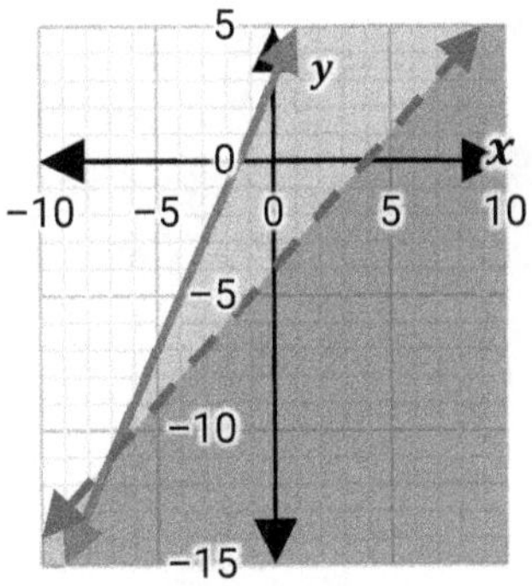

10. Graph the system $\begin{cases} y \leq -\frac{1}{3}x + 4 \\ y > \frac{3}{5}x \end{cases}$**, and determine if the point $(0, 2)$ is in the solution set.**

a. The given point is in the solution.

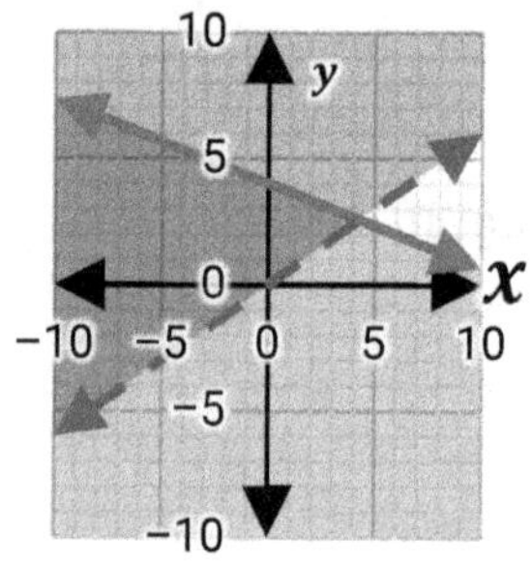

b. The given point is not in the solution.

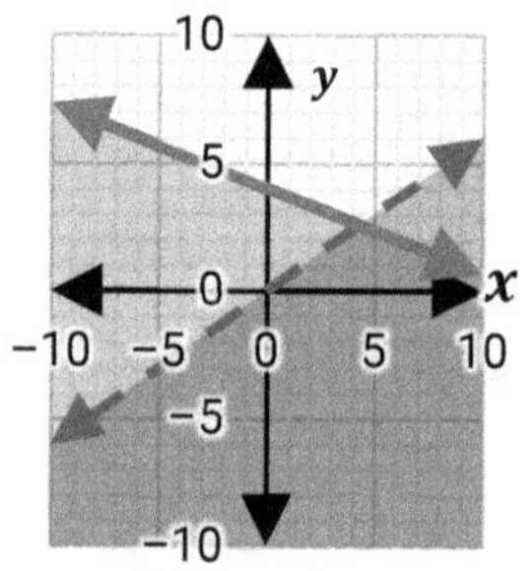

c. The given point is not in the solution.

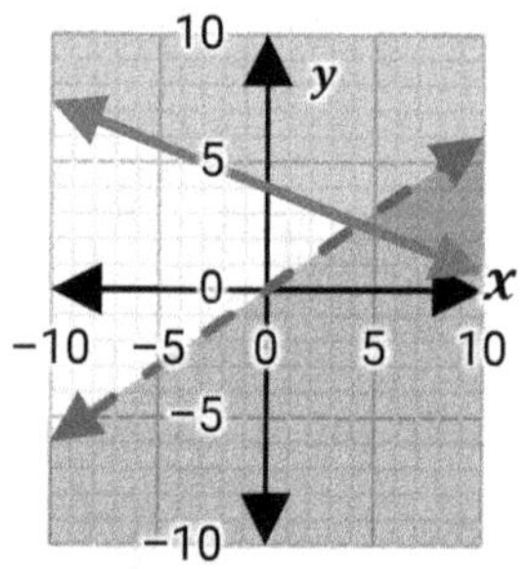

d. The given point is in the solution.

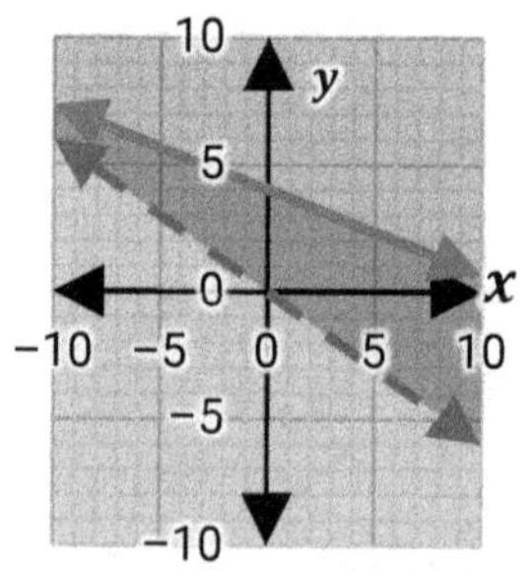

11. Graph the system $\begin{cases} y \leq -\frac{1}{2}x + 4 \\ y > \frac{3}{4}x \end{cases}$**, and determine if the point $(6, 3)$ is in the solution set.**

a. The given point is in the solution.

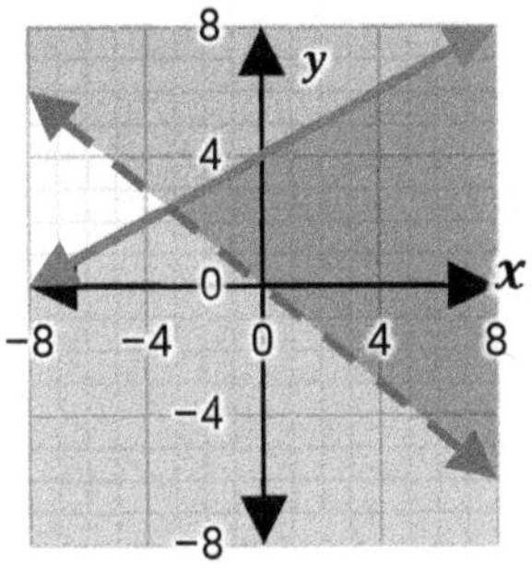

b. The given point is not in the solution.

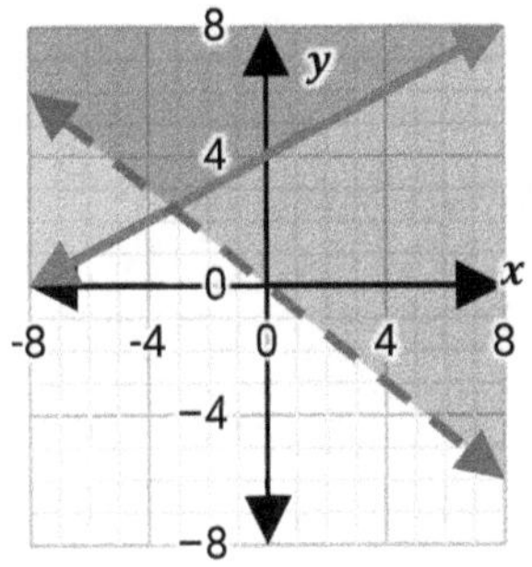

c. The given point is in the solution.

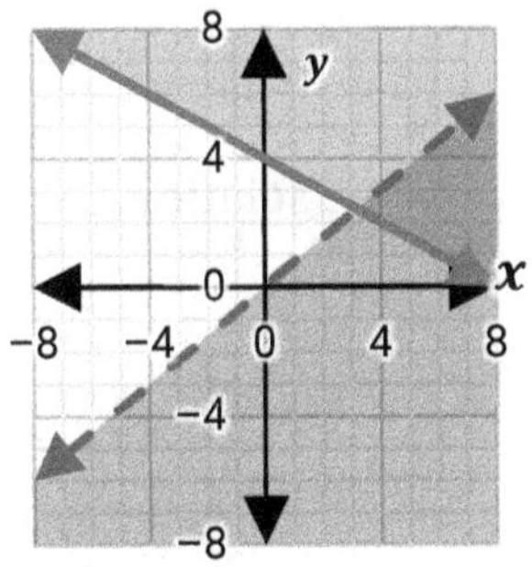

d. The given point is not in the solution.

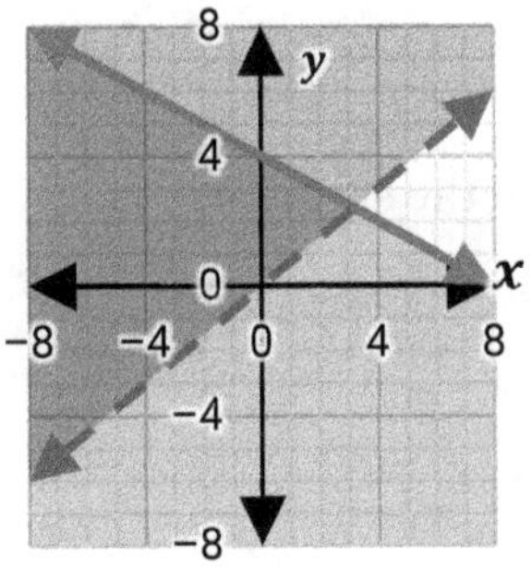

12. Graph the system $\begin{cases} y \leq 3 \\ x > -2 \end{cases}$, and determine if the point $(4, 2)$ is in the solution set.

a. The given point is in the solution.

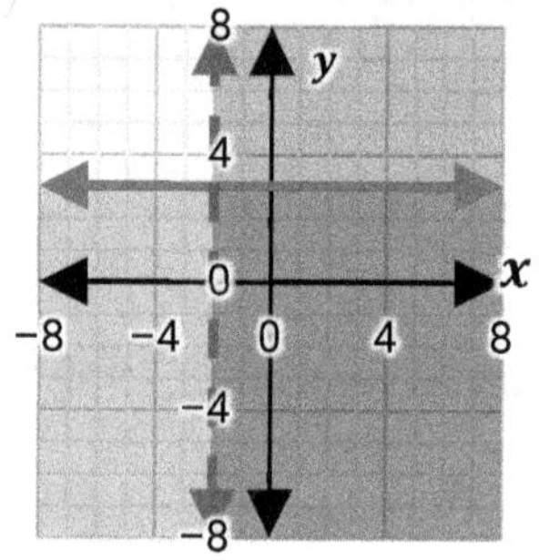

b. The given point is not in the solution.

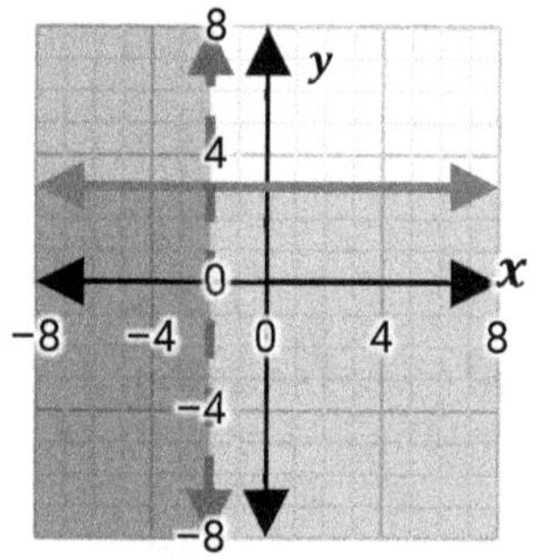

c. The given point is not in the solution.

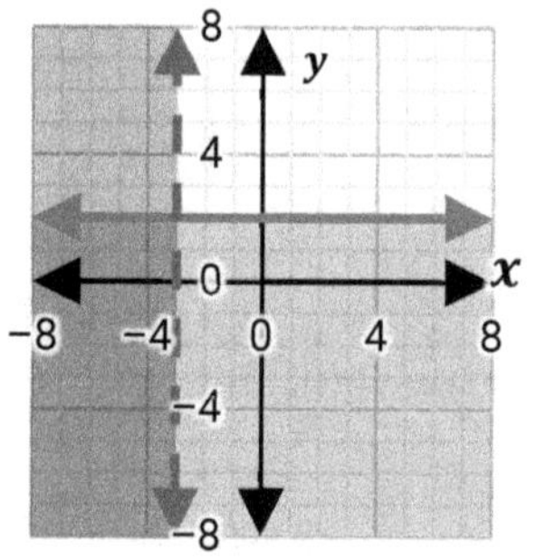

d. The given point is in the solution.

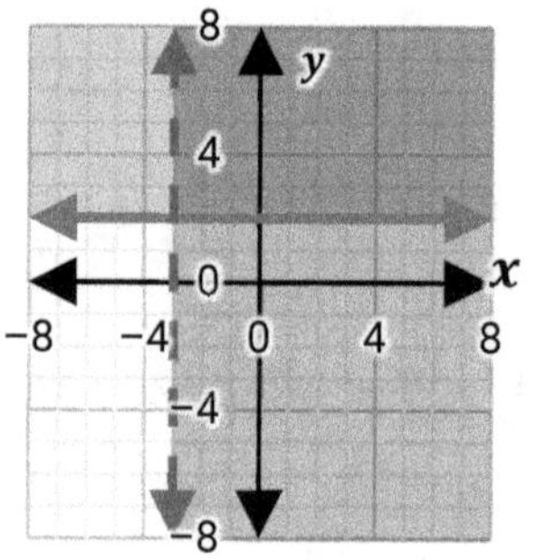

13. The sum of x and y is greater than or equal to twenty-five. The number y is greater than the sum of four and twice x. Which of the following graphs represents the possible values of x and y?

a. 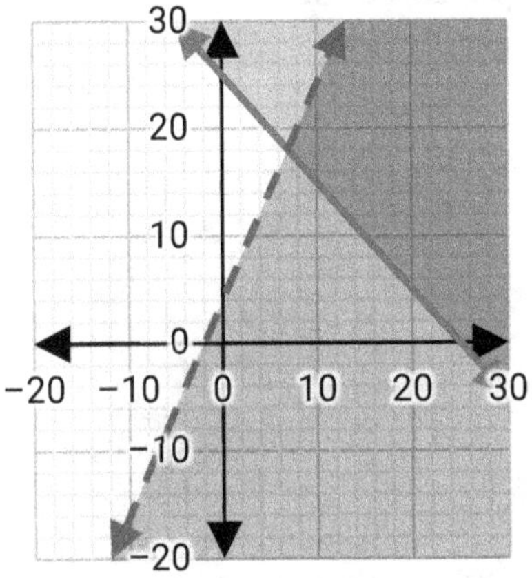

c. 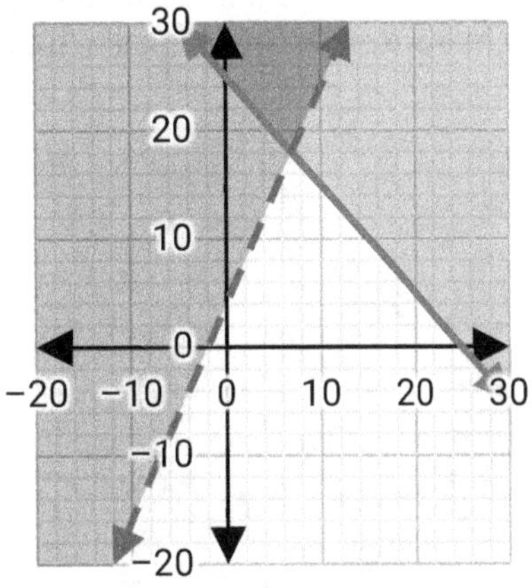

b. 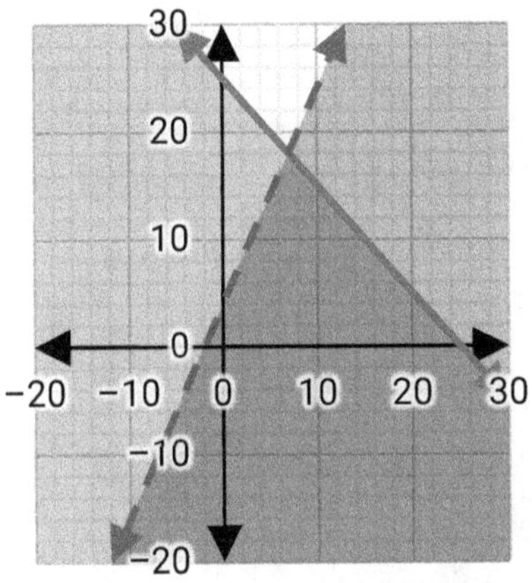

d. 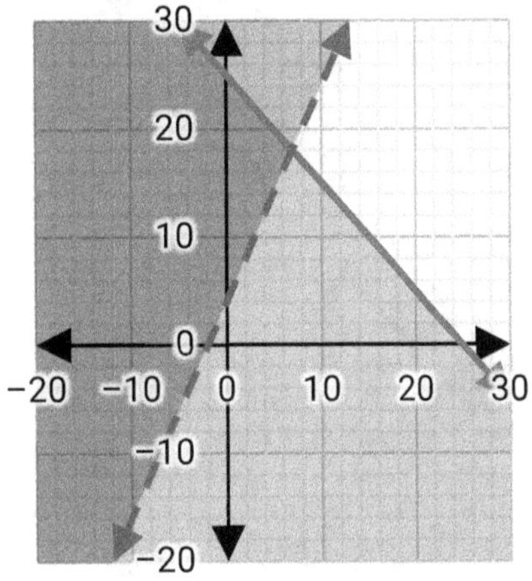

14. A department store never has more than thirty computer screens and keyboards on its shelf at any one time. If there are at least eight more keyboards than computer screens on the shelves, which of the following could be the graph of the shaded region for the various combinations of computer screens and keyboards the department store can have? Let x be the number of computer screens and y be the number of keyboards.

a. 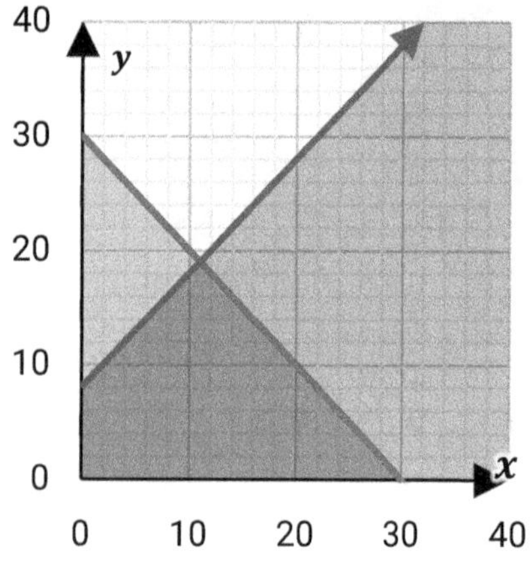

c. 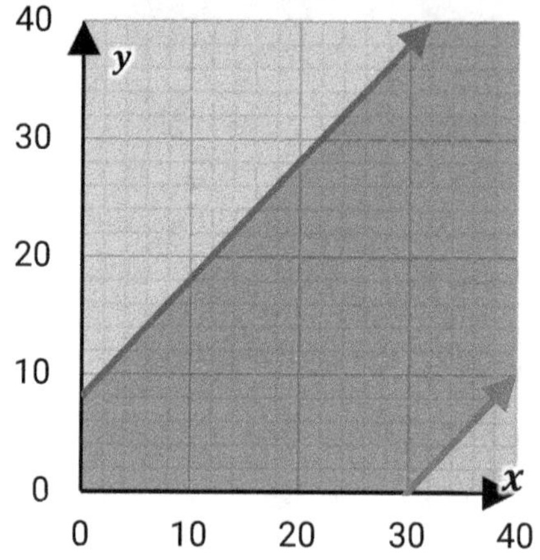

b. 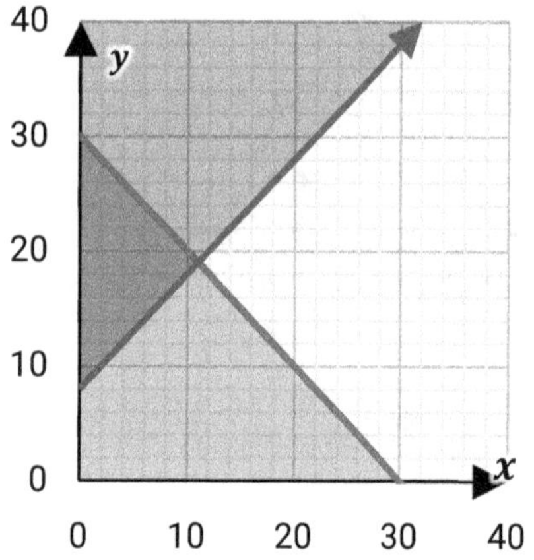

d. 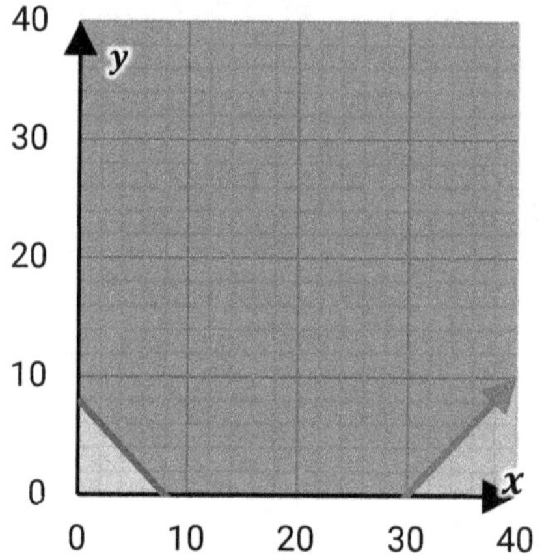

15. You have no more than fifteen hours to study for your upcoming tests in English and Biology. You spend less than twice as much time studying for your science test than your English test. If x is the number of hours spent studying for the English test, and y is the number of hours spent studying for the Biology test, which graph represents the region for the possible number of hours you can spend studying for the two tests?

a. 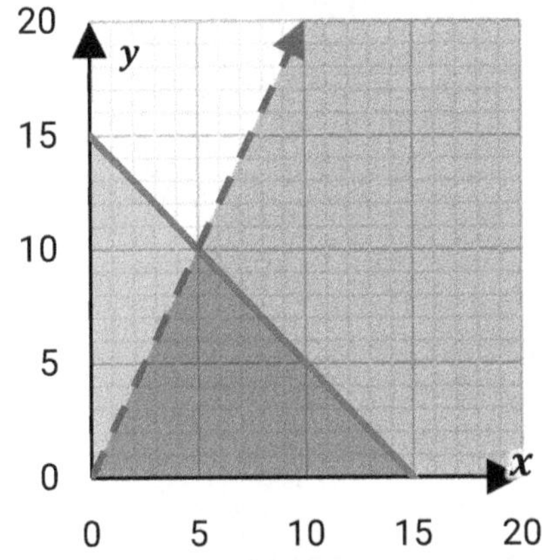

c. 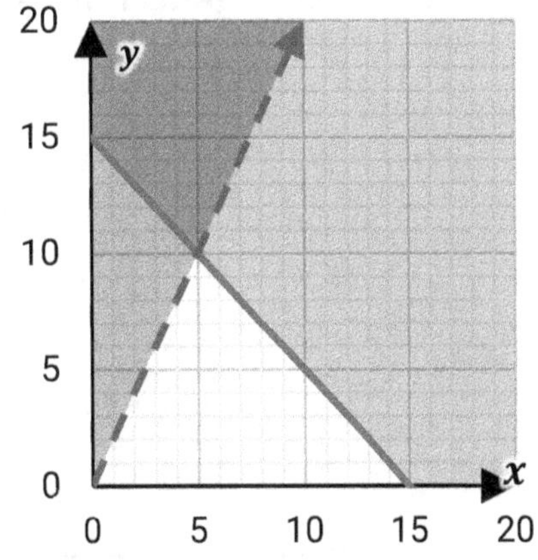

b. 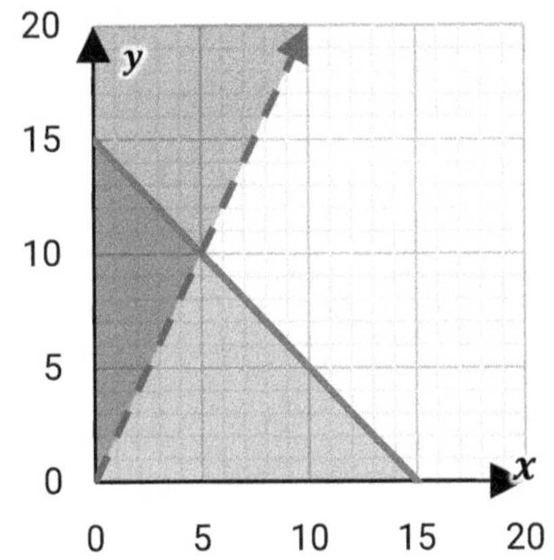

d. 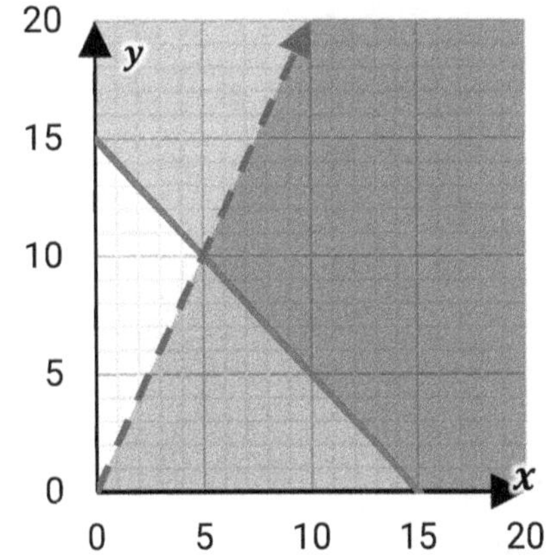

Chapter 5 Quiz

1. Graph the system of inequalities.

$$\begin{cases} 3x - 2y \geq -8 \\ x + 2y > 2 \end{cases}$$

Which of the four labeled regions represents the solutions to the system?

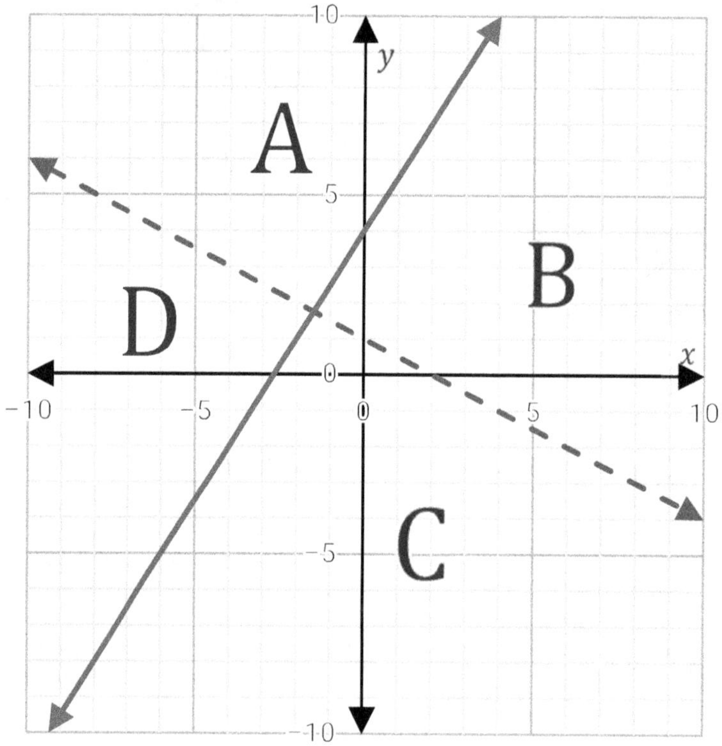

a. Region A
b. Region B
c. Region C
d. Region D

2. Solve the following system of equations using the elimination method.

$$\begin{cases} -3x + 6y = -9 \\ -3x - 2y = 7 \end{cases}$$

a. $(-3, -3)$
b. $(-1, -2)$
c. $(-5, -4)$
d. $(-7, 7)$

3. Solve the following system of equations using the elimination method:

$$\begin{cases} 4x + 2y = -6 \\ 6x + 5y = -21 \end{cases}$$

a. $\left(\frac{13}{2}, -12\right)$
b. $\left(\frac{5}{2}, -8\right)$
c. $\left(\frac{3}{2}, -6\right)$
d. No solution

4. Solve the following system of equations using the substitution method:

$$\begin{cases} 2x + 4y = 12 \\ 3x + 2y = -2 \end{cases}$$

a. $(-4,5)$
b. $(-2,2)$
c. $(2,2)$
d. $(4,1)$

5. You keep spare change in a jar. One day you count your coins and find out that you have 15 coins in the jar, all of them either dimes or quarters. You find that their value adds up to 300 cents. How many dimes and quarters are in the jar? To answer this question, set up an appropriate system of linear equations and solve it using the graphing method.

a. There are 7 dimes and 8 quarters.
b. There are 10 dimes and 8 quarters.
c. There are 9 dimes and 6 quarters.
d. There are 5 dimes and 10 quarters.

6. Solve the following system of equations using the substitution method.

$$\begin{cases} 3x - 2y = 30 \\ 5x + 10y = -30 \end{cases}$$

a. $(4, -5)$
b. $(4, -9)$
c. $(8, -3)$
d. $(6, -6)$

7. Graph the system of inequalities.

$$\begin{cases} 2x - y \leq -5 \\ x + 2y > -10 \end{cases}$$

Which of the four labeled regions represents the solutions to the system?

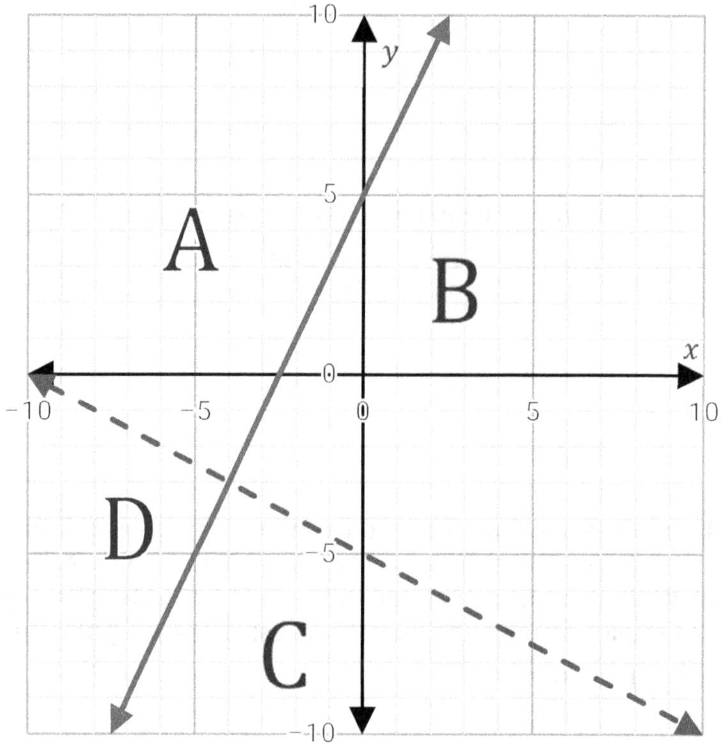

a. Region A
b. Region B
c. Region C
d. Region D

8. Solve the following system of equations using the elimination method:

$$\begin{cases} -2x + 3y = 11 \\ 4x - 6y = -20 \end{cases}$$

a. (2,5)
b. (−2,2)
c. (−1,3)
d. This system has no solution

9. Answer the following question by setting up an appropriate system of equations and solving it by the substitution method. John works two jobs. His day job pays $12 an hour, and his night job pays $18 an hour. Last week John worked 50 hours and earned $669. How many hours did he work at each job?

a. 37 hours at the day job, 13 hours at the night job
b. 37.5 hours at the day job, 12.5 hours at the night job
c. 38 hours at the day job, 12 hours at the night job
d. 38.5 hours at the day job, 11.5 hours at the night job

10. Your favorite barbecue restaurant sells small party packs for $30 that feed 6 people and large party packs for $50 that feed 12 people. If you spend $610 to buy party packs that feed 138 people, how many small and large packs did you buy? To answer this question, set up an appropriate system of linear equations and solve it using the elimination method.

a. You bought 13 small packs and 5 large packs.
b. You bought 12 small packs and 5 large packs.
c. You bought 7 small packs and 8 large packs.
d. You bought 5 small packs and 9 large packs.

11. Answer the following question by setting up an appropriate system of equations and solving it by the substitution method. Small candles cost $2 and burn for 8 hours. Large candles cost $6 and burn for 20 hours. You spend $120 buying candles that will burn for a total of 428 hours. How many candles of each size did you buy?

a. 41 small, 5 large
b. 33 small, 9 large
c. 21 small, 13 large
d. 15 small, 15 large

12. Graph the system of inequalities.

$$\begin{cases} x + 4y \leq -12 \\ x - 2y \leq 18 \end{cases}$$

Which of the following points is a solution to the system?

a. $(8.3, -8.3)$
b. $(7.3, -1.8)$
c. $(5.1, -5.4)$
d. $(12.2, -4.9)$

13. Solve the following system of equations using the graphing method.

$$\begin{cases} 2x + 2y = -4 \\ -3x + 3y = 12 \end{cases}$$

a. $(1,3)$
b. $(-6,2)$
c. $(2,-6)$
d. $(-3,1)$

14. Solve the following system of equations using the substitution method.

$$\begin{cases} -8x + 2y = 12 \\ 9x - 4y = 4 \end{cases}$$

a. $(-3, -6)$
b. $(4,8)$
c. $(-5, -14)$
d. $(-4, -10)$

15. Solve the following system of equations using the substitution method.

$$\begin{cases} 4x + y = 3 \\ -6x + 2y = -15 \end{cases}$$

a. $(3.5,3)$
b. $(4.5, -15)$
c. $(1.5, -3)$
d. $(2.5, -7)$

16. Solve the following system of equations using the elimination method.

$$\begin{cases} x - 2y = 9 \\ -3x - 5y = 28 \end{cases}$$

a. $(-1, -5)$
b. $(-11, -10)$
c. $(5, -2)$
d. $(-11,1)$

17. This is the graph of two lines k and l, whose equations form a system of two linear equations. Write the system of equations.

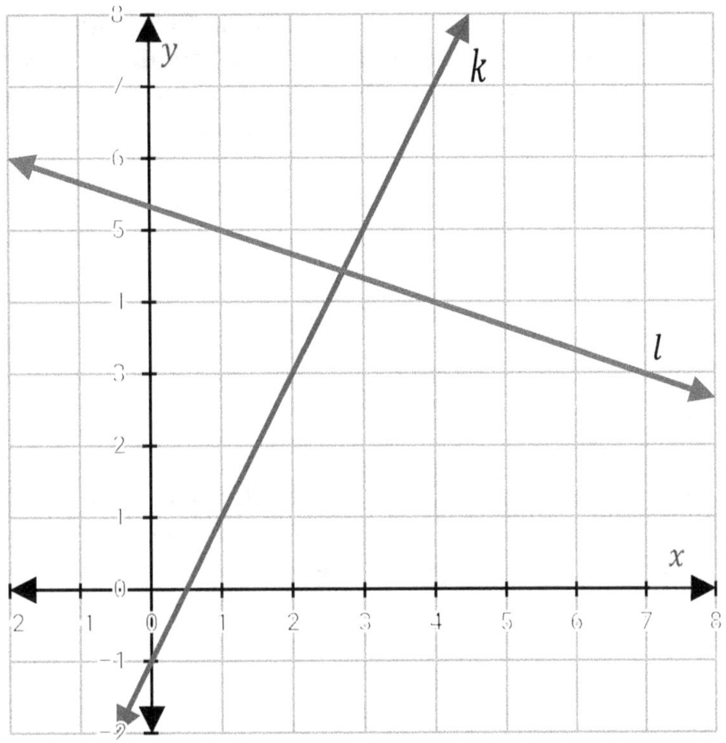

a. $\begin{cases} 2x - y = 1 \\ x + 3y = 16 \end{cases}$

b. $\begin{cases} 2x + y = 3 \\ x - 3y = -8 \end{cases}$

c. $\begin{cases} 3x - y = 2 \\ x + 2y = 12 \end{cases}$

d. $\begin{cases} x - 2y = -1 \\ 3x + y = 16 \end{cases}$

18. You have two sizes of boxes: small, heavy boxes and large, light boxes. You are loading them into a shipping container that has a fixed volume and that must weigh a specified amount. To meet the weight requirement, you find that you can pair 9 small boxes with 53 large ones, or else 24 small boxes with 3 large ones. To fill the volume exactly, you find that you can pair 12 small boxes with 16 large ones, or else 18 small boxes with 14 large ones. Let *x* be the number of small boxes and *y* be the number of large boxes. Write a system of linear equations that models your weight and volume requirements.

a. $\begin{cases} 3x + 10y = 557 \\ x + 3y = 60 \end{cases}$

b. $\begin{cases} 3x + 10y = 557 \\ 3x + y = 52 \end{cases}$

c. $\begin{cases} 10x + 3y = 249 \\ x + 3y = 60 \end{cases}$

d. $\begin{cases} 10x + 3y = 249 \\ 3x + y = 52 \end{cases}$

19. Solve the following system of equations using the elimination method:

$$\begin{cases} 4x + y = 27 \\ -3x + 3y = 6 \end{cases}$$

a. (4,11)
b. (7,9)
c. (5,7)
d. (6,3)

20. **A hardware store has a paint department and a tool department. Management predicts that next year the tool department's profit will be more than twice that of the paint department, and that together these two departments will produce a profit of over $50,000. Let x be the profit of the paint department (in thousands of dollars) and y be the profit of the tool department (in thousands of dollars). Set up a system of two inequalities that represents management's predictions. Solve the system graphically to determine which of these four regions shows the profits that meet management's predictions. Note that negative values of x and y represent losses, or negative profits.**

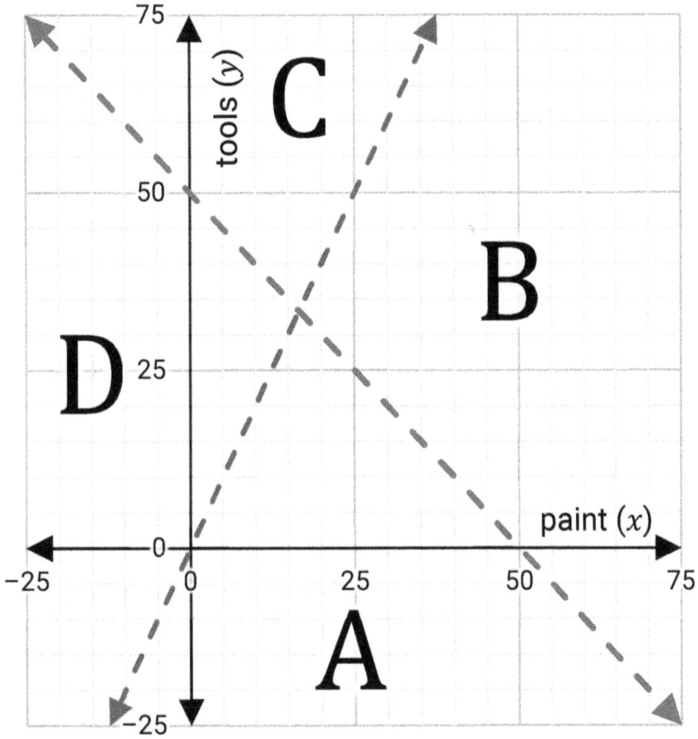

a. Region A
b. Region B
c. Region C
d. Region D

Chapter 6: Polynomials

6.1 Adding and Subtracting Polynomials

Polynomials can have constants, variables, and exponents. "Poly" means more than one, and "nomial" means name or term. In this case, **polynomial** refers to "many terms." Polynomials come in a wide variety, but they do not include negative exponents or division by a variable. For example, $\frac{5}{x}$ is not a polynomial because it divides by a variable. Similarly, $14x^{-3}$ is not a polynomial because it contains a negative exponent.

The following are examples of polynomials.

$5x$: This is a monomial, a polynomial that consists of only one term.

$2x - 4$: This is a binomial, a polynomial that consists of two terms.

$3x^4 - 5$: This is a binomial, a polynomial that consists of two terms.

$-8x^2 + 4x - 1$: This is a trinomial, a polynomial that consists of three terms.

The **degree** of a polynomial is the highest exponent that appears in the terms. For example, the polynomial $4x^2 + 3x - 9$ has a degree of 2 because the highest exponent is 2.

Polynomials can be added and subtracted by combining like terms. This can be accomplished using a vertical or horizontal format. For example, $(2x^2 + 3x + 4) + (3x^2 + 5x - 1)$ can be added vertically by stacking the polynomials and then combining the like terms.

$$\begin{array}{rrrr} & 2x^2 & +3x & +4 \\ + & 3x^2 & +5x & -1 \\ \hline & 5x^2 & +8x & +3 \end{array}$$

When subtracting polynomials vertically, make sure to distribute the negative, or subtraction, sign into the polynomial that is being subtracted. For example, when simplifying $(9x^2 - 5x + 8) - (5x^2 + 3x + 1)$, the subtraction symbol needs to be distributed to each term in $(5x^2 + 3x + 1)$. $(9x^2 - 5x + 8) - (5x^2 + 3x + 1)$ becomes $(9x^2 - 5x + 8) + (-5x^2 - 3x - 1)$. Now the polynomials can be set up vertically.

$$\begin{array}{rrrr} & 9x^2 & -5x & +8 \\ + & -5x^2 & -3x & -1 \\ \hline & 4x^2 & -8x & +7 \end{array}$$

Polynomials can also be added and subtracted horizontally. The same process of combining like terms will occur, but the polynomials will now be added or subtracted from left to right. For example, $(-8x^2 - 3x + 2) + (4x^2 - 2x + 7)$ can be added horizontally by placing the like terms next to each other and then combining.

$$(-8x^2 + 4x^2) + (-3x - 2x) + (2 + 7)$$

$$-4x^2 - 5x + 9$$

When subtracting polynomials horizontally, remember to distribute the negative sign into the polynomial that is being subtracted. For example, in the problem $\left(5x^2 - 3x + \frac{1}{2}\right) - (4x^2 + 3x - 2)$, the negative needs to be distributed to each term within $(4x^2 + 3x - 2)$.

$$\left(5x^2 - 3x + \frac{1}{2}\right) - (4x^2 + 3x - 2)$$

$$5x^2 - 3x + \frac{1}{2} - 4x^2 - 3x + 2$$

Now the polynomial can be rearranged so that the like terms are next to each other.

$$(5x^2 - 4x^2) + (-3x - 3x) + \left(\frac{1}{2} + 2\right)$$

Then, combine like terms.

$$x^2 - 6x + 2\frac{1}{2}$$

Review Video: Adding and Subtracting Polynomials
Visit mometrix.com/academy and enter code: 124088

Examples

Example 1:

$$(45ab + 3a + 2b) + (25ab + 3a + 8b)$$

Answer:

Group and combine terms with the same variables.

$$(45ab + 25ab) + (3a + 3a) + (2b + 8b)$$

$$70ab + 6a + 10b$$

Example 2:

$$(3xy + 8z) - (xy + z)$$

Answer:

$$\begin{array}{rrr} & 3xy & 8z \\ + & -xy & -z \\ \hline & 2xy & +7z \end{array}$$

Distribute the negative sign to the second set of parentheses and combine like terms.

Example 3:

$$(6x^2 + 5y) + (5x^2 + 8xy + 2y)$$

Answer:

Group and combine terms with the same variable raised to the same power.

$$(6x^2 + 5x^2) + (8xy) + (5y + 2y)$$

$$11x^2 + 8xy + 7y$$

EXAMPLE 4:

$$(8a^2 + 23b^2 - 5b) - (a^2 + 3b^2 + 2b)$$

Answer:

Distribute the negative through the second polynomial.

$$\begin{array}{rrrr} & 8a^2 & +23b^2 & -5b \\ - & (a^2 & +3b^2 & +2b) \\ \hline \end{array} \quad \rightarrow \quad \begin{array}{rrrr} & 8a^2 & +23b^2 & -5b \\ + & -a^2 & -3b^2 & -2b \\ \hline \end{array}$$

Combine terms with the same variable raised to the same power.

$$\begin{array}{rrrr} & 8a^2 & +23b^2 & -5b \\ + & -a^2 & -3b^2 & -2b \\ \hline & 7a^2 & +20b^2 & -7b \end{array}$$

EXAMPLE 5:

A rectangular poster has a length of $\frac{1}{2}x + 4y$ and a width of $\frac{1}{4}x + 1$. What is the perimeter of the poster?

Answer:

Calculate the perimeter, P, of the poster by adding the four sides of the poster.

$$P = \left(\frac{1}{2}x + 4y\right) + \left(\frac{1}{2}x + 4y\right) + \left(\frac{1}{4}x + 1\right) + \left(\frac{1}{4}x + 1\right)$$

Combine the like terms. Combine all the terms that contain x, and combine all the terms that contain y.

$$P = \frac{3}{2}x + 8y + 2$$

PRACTICE QUESTIONS

1. Find the sum: $(7x + 8) + (14x - 19)$.

a. $21x^2 - 11$
b. $21x + 27$
c. $21x - 11$
d. $10x$

2. Find the sum: $(14x^2 + 8x - 20) + (9x^2 - 10x + 32)$.

a. $23x^2 - 2x + 12$
b. $23x^4 - 2x^2 + 12$
c. $21x^3 + 12$
d. $23x^2 + 18x + 52$

3. Find the sum: $(11x + 15y - 9) + (7y - 24x - 35)$.

a. $27y^2 - 57$
b. $18x - 9y - 44$
c. $9xy - 44$
d. $-13x + 22y - 44$

4. Find the sum: $(5x^2 - 8y^2 - 14x - y + 16) + (x^2 + 3y^2 - 10x + 7y + 25)$.

a. $-18x^2 + y^2 + 41$
b. $6x^2 - 5y^2 - 24x + 6y + 41$
c. $6x^2 - 24x + 42$
d. $5x^2 - 5y^2 - 24x + 7y + 41$

5. Find the difference: $(-12x + 16) - (5x - 9)$.

a. $-17x + 7$
b. $-17x^2 + 25$
c. $8x$
d. $-17x + 25$

6. Find the difference: $(10x^2 + 14x + 18) - (9x^2 - x - 8)$.

a. $x^2 + 14x + 26$
b. $x^2 + 15x + 26$
c. $x^2 + 13x + 10$
d. $19x^2 + 13x + 10$

7. Find the difference: $(28x - 10y - 45) - (19x + 4y + 15)$.

a. -27
b. $9x - 14y - 60$
c. $9x - 6y - 60$
d. $9x - 6y - 30$

8. Find the difference $(4x^2 + 3xy - 2y^2 - 5x + 4y - 8) - (-2x^2 + 6xy + 7y^2 - 3x - 2y + 9)$.

a. $6x^2 - 3xy - 9y^2 - 2x + 6y + 1$
b. $6x^2 - 3xy - 9y^2 - 2x + 6y - 17$
c. $6x^2 + 9xy + 5y^2 - 8x + 2y + 1$
d. $2x^2 - 3xy - 9y^2 - 2x + 6y - 17$

9. Find the sum: $(2x^2 - 4x + 9) + (7x^2 - 9)$.

a. $9x^2 - 4x - 18$
b. $9x^2 - 4x$
c. $5x^2$
d. $9x^2 - 4x + 18$

10. Find the difference: $(-8x^2 + 5xy - 4y - 10) - (-8x^2 + 12y^2 + 5x - 6y - 10)$.

a. $5xy + 12y^2 + 5x - 10y - 20$
b. $5xy - 12y^2 - 5x + 2y - 20$
c. $5xy - 12y^2 - 5x + 2y$
d. $-16x^2 + 5xy - 12y^2 - 5x + 2y$

11. Find the following sum: $\left(\frac{1}{2}x + \frac{2}{9}y - \frac{2}{5}\right) + \left(\frac{1}{4}x + \frac{5}{9}y + \frac{3}{5}\right)$.

a. $\frac{3}{2}x + \frac{7}{9}y + \frac{1}{5}$
b. $\frac{3}{4}x + \frac{7}{9}y - 1$
c. $\frac{3}{4}x + \frac{1}{3}y + \frac{1}{5}$
d. $\frac{3}{4}x + \frac{7}{9}y + \frac{1}{5}$

12. Find the difference $(2.3x^2 - 4.9x + 7.8) - (4.1x^2 - 2.5x - 5.4)$.

a. $-1.8x^2 - 2.4x + 13.2$
b. $6.4x^2 - 2.4x + 13.2$
c. $-1.8x^2 - 7.4x + 2.4$
d. $-1.8x^2 - 2.4x - 2.4$

13. You have x nickels and 3 pennies in your left pocket. You have y dimes and 12 pennies in your right pocket. Represent the value, in cents, of the coins in each pocket as a polynomial, and then add the polynomials together to find an expression for the total value of coins you are carrying.

a. $15x + 120y$ cents
b. $5x + 10y + 15$ cents
c. $8x + 22y$ cents
d. $15xy + 15$ cents

14. You have a rectangular backyard with an area of $x^2 + 8x + 15$ square feet. In it you fence off a rectangular garden plot with an area of $x^2 - 6x + 8$ square feet. How much area is left in the backyard outside the garden plot?

a. $14x + 7$ square feet
b. $2x + 23$ square feet
c. $-14x - 7$ square feet
d. $2x^2 + 2x + 23$ square feet

15. You manufacture calculators that you sell for $30. If your factory's production rate is x calculators per day, then your daily production cost is $0.05x^2 + 12.45x + 500$ dollars. Assuming you sell every calculator that you produce, use polynomial subtraction to find an expression for your daily profit from producing x calculators per day.

a. $0.05x^2 + 42.45x + 500$ dollars
b. $17.5x - 500$ dollars
c. $0.05x^2 - 17.55x + 500$ dollars
d. $-0.05x^2 + 17.55x - 500$ dollars

6.2 Multiplying Monomials

Monomials are a subgroup of polynomials. **Monomials** consist of one term, and they come in a wide variety. For example, $18x$, $2x^2$, $-4xy^4$, and $24x^4y^2$ are all monomials. Monomial refers to "one term," binomial refers to "two terms," and trinomial refers to "three terms." **Terms** consist of numbers, variables, or both numbers and variables.

Polynomial Name	Number of Terms	Examples
Monomial	1	$4x, 7x^2, 3x^2y^3$
Binomial	2	$x + 4, x^2 - 2y, 4x^3 - 7$
Trinomial	3	$x^2 + 3y - 4, 17x^4 + 19x^2 + 2$

When multiplying monomials, it is important to remember the rules regarding multiplying exponents. Remember, when multiplying exponents, you *add* the exponents for terms with the same base. For example, $x^3 \times x^5 = x^8$. The bases, x, are the same, so the exponents are added.

Some monomials will have a coefficient, such as the 4 in $4x^3$. When multiplying monomials that have coefficients, multiply the coefficients and then add the exponents. For example, $4x^3 \times 9x^2$ is simplified by multiplying the coefficients, $4 \times 9 = 36$, and then adding the exponents for bases that are the same. $x^3 \times x^2 = x^5$, so $4x^3 \times 9x^2 = 36x^5$.

More advanced monomials will have coefficients and multiple variables. For example, the monomials $15x^3y$ and $3xy^5$ contain coefficients and multiple variables, x and y. When multiplying these two monomials, the coefficients need to be multiplied, and the exponents with the same base need to be added. The expression $(15x^3y)(3xy^5)$ becomes $45x^4y^6$.

Remember, when you see variables such as x and y, this is the same thing as x^1 and y^1.

Let's look at one more example where the monomials have three variables. The expression $(8a^4bc^2)(-2abc^6)$ has two coefficients and three variables with exponents. Multiply the coefficients, and then add the exponents for bases that are the same. The expression $(8a^4bc^2)(-2abc^6)$ becomes $-16a^5b^2c^8$.

Some monomials with exponents are raised to a power. For example, $(x^2)^6$. In this case, the Power of a Power Rule can be applied. When a power is raised to a power, and the bases are the same, multiply the exponents. The expression $(x^2)^6$ becomes x^{12}.

The same process applies to monomial expressions with multiple variables. For example, $(x^3y^2)^4$. To simplify, multiply the exponents. The expression $(x^3y^2)^4$ becomes $x^{3\times4}y^{2\times4}$ which simplifies to $x^{12}y^8$.

When multiplying monomials that are raised to a power, apply the Power to a Power Rule before multiplying. For example, when multiplying $(-2x^3y^3)^6 \times (5x^2y)$, start by raising the monomial $-2x^3y^3$ to the power of 6.

$$(64x^{18}y^{18}) \times (5x^2y)$$

Now, multiply the two monomials.

$$320x^{20}y^{19}$$

Review Video: Multiplying Monomials
Visit mometrix.com/academy and enter code: 567047

Examples

Example 1:

$$(4pr)(3pr)$$

Answer:

Multiply the coefficients and add powers of like bases.

$$(4pr)(3pr)$$

$$12p^{1+1}r^{1+1}$$

$$12p^2r^2$$

Example 2:

$$(2rt^3)(-3r^5)$$

Answer:

Multiply the coefficients, and add powers of like bases.

$$(2rt^3)(-3r^5)$$

$$-6r^{1+5}t^3$$

$$-6r^6t^3$$

Example 3:

Multiply the monomials. $(x^2y^2)^4 \times (4xy)$

Answer:

$(x^2y^2)^4$ has a power raised to a power. Multiply the exponents to get x^8y^8.

$$x^8y^8 \times 4xy$$

Now, multiply the monomials. Add the exponents if the bases are the same.

$$4x^9y^9$$

EXAMPLE 4:

The dimensions of a box are labeled below. Calculate the volume of the box by multiplying the monomials.

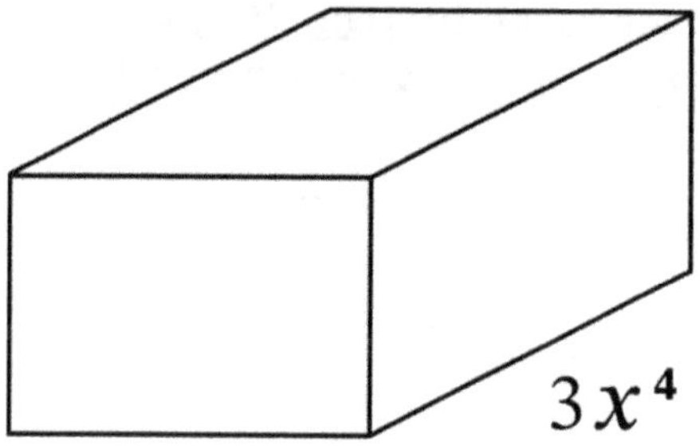

Answer:

When multiplying monomials with the same base, multiply the coefficients and add the exponents. $\frac{1}{4}x^{13}$, $3x^4$, and $\frac{1}{2}x^5$ all have a base of x.

$$V = \left(\frac{1}{4}x^{13}\right)\left(\frac{1}{2}x^5\right)(3x^4)$$
$$V = \left(\frac{1}{8}x^{18}\right)(3x^4)$$
$$V = \frac{3}{8}x^{22}$$

EXAMPLE 5:

Becca is cutting out rectangular paper to make postcards. The cards have a length of $4x^9y^3$ and a width of $2x^4y$. Determine the area of the postcard in terms of x and y.

Answer:

The area of the rectangular postcards can be calculated using the area formula $A = l \times w$. In this scenario, the length is expressed as $4x^9y^3$ and the width is expressed as $2x^4y$. When multiplying monomials with the same base, multiply the coefficients and add the exponents.

$$A = l \times w$$
$$A = (4x^9y^3)(2x^4y)$$
$$A = 8x^{9+4}y^{3+1}$$
$$A = 8x^{13}y^4$$

The area of each postcard can be described as $A = 8x^{13}y^4$.

Practice Questions

1. Multiply $(3xy^2)(2x^4y)$.

a. $6x^4y^2$
b. $6x^5y^3$
c. $5x^5y^3$
d. $5x^2y^2$

2. Multiply $(4ab^3)(3a^5)$.

a. $7a^6b^8$
b. $7a^5b^{15}$
c. $12a^6b^3$
d. $12a^5b^3$

3. Multiply $(-5x^3y^3)(4x^2y^2)$.

a. $-20x^5y^5$
b. $-20x^6y^6$
c. $-x^5y^5$
d. x^9y^9

4. Multiply $(-4p^5q^7)(-6p^3q^4)$.

a. $-24p^9q^{10}$
b. $24p^{15}q^{28}$
c. $24p^8q^{11}$
d. $-24p^8q^{11}$

5. Multiply $\left(\frac{2}{3}x^5y^{-3}\right)(-12x^6y^5)$.

a. $\frac{8x^{11}}{y^2}$
b. $-\frac{8x^{30}}{y^{15}}$
c. $8x^{11}y^2$
d. $-8x^{11}y^2$

6. Multiply $\left(\frac{3}{5}r^2s^4\right)\left(\frac{1}{2}r^5s^{-1}\right)$.

a. $\frac{3}{10}r^{10}s^3$
b. $\frac{3}{10}r^7s^3$
c. $\frac{2}{5}r^{10}s^3$
d. $\frac{3r^{10}}{10s^3}$

7. Multiply $(2x^4y)^3(4x^2y^6)$.

a. $32x^{14}y^9$
b. $24x^{14}y^9$
c. $32x^9y^{10}$
d. $24x^6y^7$

8. Multiply $(-5x^3y)(-7x^2y^3)^2$.

a. $-35x^5y^6$
b. $-35x^5y^4$
c. $-245x^7y^6$
d. $-245x^7y^7$

9. Multiply $(-5p^3q^2)^2(p^2q^4)^3$.

a. $25p^{10}q^9$
b. $-25p^9q^{14}$
c. $25p^{12}q^{16}$
d. $-25p^{12}q^{16}$

10. Multiply $(4x^5y)^2\left(\frac{3}{4}x^3y^4\right)^2$.

a. $9x^{16}y^{10}$
b. $6x^{16}y^{10}$
c. $9x^{13}y^9$
d. $6x^{16}y^{10}$

11. Multiply $\left(\frac{2}{5}p^3q\right)^2(-5p^2q^2)^3(p^4q^3)$.

a. $20p^{16}q^{11}$
b. $-20p^{16}q^{11}$
c. $6p^{14}q^8$
d. $-6p^9q^6$

12. Multiply $(7x^4y)^2\left(\frac{2}{7}x^3y^4\right)^2(-4x^2y)^2$.

a. $-32x^9y^6$
b. $32x^{16}y^{11}$
c. $-64x^{18}y^{12}$
d. $64x^{18}y^{12}$

13. Which of the following expressions represents the area of a rectangle that has a length of $3xy^2$ and a width of $2xy$?

a. $5x^2y^2$
b. $5x^2y^2$
c. $6x^2y^3$
d. $6xy^2$

14. Which of the following expressions represents the volume of a right circular cone that has a radius of $4mn^3$ and a height of $3m^2n$? Note: The formula for the volume, V, of a right circular cone is $V = \frac{\pi}{3}r^2h$.

a. $16\pi m^3n^4$
b. $16\pi m^4n^7$
c. $4\pi m^4n^3$
d. $4\pi m^3n^7$

15. Let z be the product of the two monomials a^2b^5 and $2a^4b$. Which of the following expressions represents z?

a. $2a^6b^6$
b. $2a^2b^5$
c. $4a^6b^6$
d. $4a^2b^5$

6.3 Dividing Monomials

The process of dividing monomials is very similar to the process of multiplying monomials. Remember, when multiplying monomials, exponents with the same base can be *added*. This changes when dividing monomials. When exponents have the same base, they can be *subtracted*. For example, $\frac{x^4}{x^2} = x^{4-2} = x^2$. Both monomials have a base of x, so their exponents can be subtracted.

The same process occurs when coefficients are involved. For example, $\frac{40x^3y^5}{8xy^2}$ can be simplified by dividing the coefficients and then subtracting the exponents with like bases.

$$40 \div 8 = 5$$
$$x^3 \div x = x^{3-1} = x^2$$
$$y^5 \div y^2 = y^{5-2} = y^3$$

The result is $5x^2y^3$.

Sometimes it is helpful to view this process in expanded form.

The expression $\frac{40x^3y^5}{8xy^2}$ becomes $\frac{8 \cdot 5 \cdot x \cdot x \cdot x \cdot y \cdot y \cdot y \cdot y \cdot y}{8 \cdot x \cdot y \cdot y}$. We can see that the 8's will cancel out, as well as one of the x's and two of the y's. The expression $\frac{8 \cdot 5 \cdot x \cdot x \cdot x \cdot y \cdot y \cdot y \cdot y \cdot y}{8 \cdot x \cdot y \cdot y}$ becomes $5 \cdot x \cdot x \cdot y \cdot y \cdot y$ or $5x^2y^3$.

Remember, the bases need to be the same to subtract the exponents. For example, the expression $\frac{14a^2b}{7cd^2}$ will simplify to $\frac{2a^2b}{cd^2}$. The coefficients can be simplified, but none of the bases are the same, so their exponents will remain the same.

It is also important to remember that when a variable has no exponent, it is raised to the power of 1. For example, x is the same thing as x^1.

Let's look at one more example that has coefficients and more than one variable. The expression $\frac{24a^7b^2}{12ab}$ has coefficients that can be simplified.

$$24 \div 12 = 2$$

Now, we look for exponents that have the same base.

$$a^7 \div a = a^{7-1} = a^6$$
$$b^2 \div b = b^{2-1} = b$$

When $\frac{24a^7b^2}{12ab}$ is simplified, it becomes $2a^6b$.

Review Video: Dividing Monomials
Visit mometrix.com/academy and enter code: 584409

EXAMPLES

EXAMPLE 1:

Simplify the expression.

$$\frac{8x^5}{2x^2}$$

Answer:

Divide the coefficients, and subtract the exponents with the same base.

$$\frac{8}{2} = 4$$
$$\frac{x^5}{x^2} = x^{5-2} = x^3$$
$$\frac{8x^5}{2x^2} = 4x^3$$

EXAMPLE 2:

Simplify the expression.

$$-\frac{12x^3y}{15xy^2}$$

Answer:

Divide, or in this case simplify, the coefficient.

$$-\frac{12}{15} = -\frac{12 \div 3}{15 \div 3} = -\frac{4}{5}$$

Now, subtract the exponents.

$$x^{3-1} = x^2$$
$$y^{1-2} = y^{-1}$$

Remember, a negative exponent can be made positive by moving the base and exponent to the other side of the fraction.

It can be helpful to view the monomials in expanded form so that you can see what is being canceled out.

For example, $-\frac{12x^3y}{15xy^2}$ is the same thing as $-\frac{12 \cdot x \cdot x \cdot x \cdot y}{15 \cdot x \cdot y \cdot y}$. In this form, we can see how the x's and y's will cancel out. The expression $-\frac{12 \cdot x \cdot x \cdot x \cdot y}{15 \cdot x \cdot y \cdot y}$ becomes $\frac{4x^2}{5y}$.

$$\frac{8x^5}{2x^2} = 4x^3$$

EXAMPLE 3:

Julia is collecting glass pieces for a large stained-glass window. She shapes one of the pieces of glass into a right triangle. The area of the glass triangle is $14x^2$. The base of the triangle is $4x$. Use the area formula for a triangle to calculate the height of the triangular piece of glass.

Triangle area formula: $A = \frac{1}{2}bh$

Answer:

Values for the area and the base are provided in the problem. The height is unknown. Substitute the known values into the triangle area formula.

$$A = \frac{1}{2}bh$$
$$14x^2 = \frac{1}{2}(4x)h$$

The value of h can be determined using inverse operations. Start by multiplying both sides of the equation by 2. Then divide both sides by $4x$.

$$28x^2 = (4x)h$$
$$\frac{28x^2}{4x} = h$$

When dividing monomials with the same base, divide the coefficients and subtract the exponents.

$$28 \div 4 = 7$$
$$\frac{x^2}{x} = x^{2-1} = x$$

Therefore, the equation simplifies to $7x = h$. The height of the triangular piece of glass is $7x$.

PRACTICE QUESTIONS

1. Divide $15x^6$ by $5x^2$ and simplify.

a. 9
b. $3x^3$
c. $3x^4$
d. $10x^4$

2. Simplify the quotient $\frac{19x^3}{19x}$.

a. 3
b. $19x^2$
c. x^3
d. x^2

3. Simplify the quotient $(100z^8) \div (25z^7)$.

a. $75z$

b. $4z^{15}$

c. $4z$

d. $\frac{32}{7}$

4. Simplify the following quotient.

$$\frac{6x^8}{18x^2}$$

a. $\frac{x^4}{3}$

b. $3x^6$

c. $\frac{4}{3}$

d. $\frac{x^6}{3}$

5. Divide $14x^3$ by $7x^6$ and simplify.

a. $\frac{2}{x^2}$

b. $2x^3$

c. $\frac{2}{x^3}$

d. $2x^2$

6. Simplify the quotient $\frac{22x^8}{33x^{12}}$.

a. $\frac{1}{11x^4}$

b. $\frac{2}{3x^4}$

c. $\frac{2x^2}{3x^3}$

d. $\frac{4}{9}$

7. Simplify the quotient $(35x^{10}y^8) \div (5x^2y^2)$.

a. $7x^5y^4$

b. $7x^8y^6$

c. 140

d. $30x^8y^6$

8. Simplify the following quotient.

$$\frac{24x^9y^6}{36x^3y^6}$$

a. $\frac{2x^3y}{3}$

b. $\frac{2x^6y}{3}$

c. $\frac{2x^3}{3}$

d. $\frac{2x^6}{3}$

9. Divide $32s^2t^3$ by $8st$ and simplify.

a. $4s^2t^3$

b. $4st^2$

c. $24st^2$

d. 24

10. Simplify the quotient $-\frac{6xy^3}{15x^3y}$.

a. $-\frac{2y^3}{5x^3}$

b. $-\frac{y^2}{9x^2}$

c. $-\frac{2y^2}{5x^2}$

d. $-2x^4y^4$

11. Simplify the quotient $(54x^2y^3) \div (81x^4y^5)$.

a. $2x^2y^2$

b. $\frac{1}{27x^2y^2}$

c. $\frac{1}{5}$

d. $\frac{2}{3x^2y^2}$

12. Simplify the following quotient.

$$\frac{26x^{10}y^{12}}{13x^{20}y^4}$$

a. 3

b. $\frac{2y^8}{x^{10}}$

c. $\frac{13y^8}{x^{10}}$

d. $\frac{2y^3}{x^2}$

13. A rectangle has a length of $15y$ meters and an area of $90xy$ square meters. Use division of monomials to find the width of the rectangle.

a. $6y$ meters

b. $6x$ meters

c. $1{,}350xy^2$ meters

d. $\frac{1}{6x}$ meters

14. A rectangular solid box is $44y$ inches high and has volume $66xy^2$ cubic inches. Use division of monomials to find the area of the base of the box. Hint: The volume of a rectangular solid can be found by multiplying the area of the base by the height.

a. $3x$ square inches

b. $\frac{3xy^2}{2}$ square inches

c. $\frac{2}{3xy}$ square inches

d. $\frac{3xy}{2}$ square inches

15. You are a hot chocolate vendor. On a normal day, working in your food truck, you sell x cups of hot chocolate for y dollars each. But working at the concessions stand on game days at the local stadium, you triple the normal price and yet sell many more cups. On a particularly cold day working concessions at the stadium, you bring in $51xy$ dollars from selling hot chocolate. Use division of monomials to calculate how many cups you sold.

a. $51y$ cups

b. $17x$ cups

c. $51x$ cups

d. $17y$ cups

6.4 Multiplying Polynomials by Monomials

Remember, a monomial is a specific term for a polynomial that only contains one term.

Examples of monomials: x^2, $13abc$, $-8y^7$

Example of polynomials: $2x - 1$, $-5x^2 + \frac{1}{2}$, $4ax^4 - x^3 + 1$

When multiplying polynomials and monomials, use the distributive property. For example, $3x + 6$ is a polynomial, and $4x$ is a monomial. The monomial $4x$ needs to be multiplied by, or distributed to, each piece of the polynomial. The expression $4x(3x + 6)$ becomes $12x^2 + 24x$. The result is already in simplest form because there are no like terms to combine. This means that $12x^2 + 24x$ is fully simplified.

Let's look at another example.

$$(2a^2)(5a^4b^7 + 3a^2)$$

This is a monomial multiplied by a polynomial. Start by distributing the $2a^2$ into both pieces of the polynomial. Remember, you can add the exponents that have like bases.

$$(2a^2)(5a^4b^7 + 3a^2)$$

$$10a^{2+4}b^7 + 6a^{2+2}$$

$$10a^6b^7 + 6a^4$$

Remember, when simplifying terms with exponents, they can only be combined if they have the same base raised to the same power. For example, $2a^3$ and $3a^3$ can be combined to $5a^3$ because they have the same base, a, to the same power, 3. However, $2a^2$ and $5a^4$ cannot be combined because they do not have the same base raised to the same power.

Let's look at an example with a negative coefficient.

$$(-3a^3b^2)(4ab^3 + 4a^2 + b)$$

This is a monomial multiplied by a polynomial.

Start by multiplying, or distributing, the term $-3a^3b^2$ by each term in the polynomial.

$$(-3a^3b^2)(4ab^3 + 4a^2 + b)$$
$$(-3a^3b^2)(4ab^3) = -12a^4b^5$$
$$(-3a^3b^2)(4a^2) = -12a^5b^2$$
$$(-3a^3b^2)(b) = -3a^3b^3$$

Now, write these three parts together: $-12a^4b^5 - 12a^5b^2 - 3a^3b^3$.

Review Video: Multiplying Polynomials by Monomials
Visit mometrix.com/academy and enter code: 234401

Examples

Example 1:

Simplify the expression.

$$2xy^2(4y + 2x)$$

Answer:

Multiply $2xy^2$ by $4y$. Multiply the coefficients, and then add the exponents that have the same bases.

$$8xy^{2+1}$$

$$8xy^3$$

Multiply $2xy^2$ by $2x$. Multiply the coefficients, and then add the exponents that have the same bases.

$$4x^{1+1}y^2$$

$$4x^2y^2$$

Write the two terms together.

$$8xy^3 + 4x^2y^2$$

Example 2:

Simplify the expression.

$$4x^2y(3x^2 - 15x + 2xy)$$

Answer:

Multiply $4x^2y$ by each term in the parentheses.

Multiply the coefficients, and then add the exponents that have the same base.

$$(4x^2y)(3x^2) = 12x^4y$$
$$(4x^2y)(-15x) = -60x^3y$$
$$(4x^2y)(2xy) = 8x^3y^2$$

Write the three terms together.

$$12x^4y - 60x^3y + 8x^3y^2$$

Example 3:

Emma draws a rectangle with a length that is 5 units longer than its width, w. Write and simplify an expression to represent the area of the Emma's rectangle.

Answer:

Emma's rectangle has a width of w and a length that is 5 units longer than w. This means that the width is w and the length is $(w + 5)$. The area of a rectangle is calculated as $A = l \times w$. When the dimensions of Emma's rectangle are substituted in for l and w, the formula becomes $A = (w + 5)(w) = w(w + 5)$. To simplify the expression, distribute.

$$A = w(w + 5)$$
$$A = w^2 + 5w$$

The area of Emma's rectangle can be expressed as $w^2 + 5w$.

Practice Questions

1. Simplify the product $\frac{1}{3}x(6x - 12)$.

a. $2x - 12$

b. $2x^2 - 12x$

c. $2x^2 - 12$

d. $2x^2 - 4x$

2. Simplify the product $11x(2x^2 + 3x + 5)$.

a. $13x^3 + 14x^2 + 16x$

b. $22x^3 + 33x^2 + 55x$

c. $22x^2 + 88x$

d. $22x^3 + 3x + 5$

3. Simplify the product $-4x^2(7x^2 - x - 9)$.

a. $-28x^4 - 4x^3 - 36x^2$

b. $-28x^4 + 4x^3 + 36x^2$

c. $-28x^4 - x - 9$

d. $-28x^4 + x^3 + 9x^2$

4. Simplify the product $2t^3(-6t^3 + 8t^2 - 3t + 4)$.

a. $-12t^6 + 16t^5 - 6t^4 + 8t^3$

b. $-12t^9 + 16t^6 - 6t^3 + 8$

c. $-12t^6 + 8t^2 - 3t + 4$

d. $-12t^6 + 8t^5 - 3t^4 + 4t^3$

5. Simplify the product $-3x^4(5x^6 - 7x^4 - 6x + 6)$.

a. $-15x^{10} - 7x^8 - 6x^5 + 6x^4$

b. $-15x^{10} + 21x^8 + 18x^5 - 18x^4$

c. $-15x^{10} - 7x^4 - 6x + 6$

d. $-15x^{10} - 21x^8 - 18x^5 + 18x^4$

6. Simplify the product $x^2y(2x^2 + 3x - 10)$.

a. $2x^4y + 3x - 10$

b. $-5x^2y$

c. $2x^4 + 3x - 10y$

d. $2x^4y + 3x^3y - 10x^2y$

7. Simplify the product $8x^5(-3x^2 - 2xy + 12y^2)$.

a. $-24x^7 - 2xy + 12y^2$

b. $-24x^7 - 2x^6y + 12x^5y^2$

c. $-24x^{10} - 16x^5y + 96x^5y^2$

d. $-24x^7 - 16x^6y + 96x^5y^2$

8. Simplify the product $\frac{3}{2}xy(6x^2 - 8xy + 3y^2)$.

a. $9x^2y - 12xy + \frac{9}{2}xy^2$

b. $9x^2y - 8xy + 3y^2$

c. $\frac{15}{2}x^3y - 9.5x^2y^2 + \frac{9}{2}xy^3$

d. $9x^3y - 12x^2y^2 + \frac{9}{2}xy^3$

9. Simplify the product $2st^2(27s^3 + 9s^2t + 3st^2 + t^3)$.

a. $54s^4t^2 + 18s^3t^3 + 6s^2t^4 + 2st^5$

b. $54s^4t^2 + 9s^3t^3 + 3s^2t^4 + st^5$

c. $54s^3t^2 + 18s^2t^2 + 6st^2 + 2st^3$

d. $54s^4t^2 + 9s^2t + 3st^2 + t^3$

10. Simplify the product $-10x^3y(4x^5 - 2y^5 + 9x^2y^2 - 3)$.

a. $-40x^{15}y + 20x^3y^5 - 90x^6y^2 + 30x^3y$

b. $-40x^8y + 20x^3y^6 - 90x^5y^3 + 30x^3y$

c. $-40x^8y - 20x^3y^6 + 90x^5y^3 - 30x^3y$

d. $-40x^8y - 2y^5 + 9x^2y^2 - 3$

11. Simplify the product $xy^2(-7x + 6y + 5xy)$.

a. $4xy^2$

b. $-7x^2y^2 + 6y + 5xy$

c. $-7x^2y^2 + 6xy^3 + 5x^2y^3$

d. $-7x^2 + 6y + 5xy^3$

12. Simplify the product $3x^2y^3(2x^2 + 6y^2 + 4x - 9)$.

a. $6x^4y^3 + 18x^2y^5 + 12x^3y^3 - 27x^2y^3$

b. $6x^4y^3 + 6x^2y^5 + 4y^3 - 9x^4y^3$

c. $6x^4y^3 + 6y^2 + 4x - 9$

d. $6x^4y^3 + 18x^2y^6 + 12x^2y^3 - 27x^2y^3$

13. Your company manufactures collectible plush animal toys. Last year you sold x toy penguins at a price of y dollars each. This year penguins have become wildly popular, and you expect to sell three times as many, even though you are going to raise the price by two dollars. Write an expression for the money you expect to receive from selling plush penguins next year in terms of x and y.

a. $3xy + 2$ dollars

b. $xy + 2x + 3y + 6$ dollars

c. $3xy + 6x$ dollars

d. $3xy + 6y$ dollars

14. You have a rectangular garden plot x feet long and y feet wide. You want to expand it by making it ten feet wider and four times as long. Write an expression for the area of the expanded plot in terms of x and y.

a. $4xy + 40$ square feet

b. $4xy + 40y$ square feet

c. $xy + 10x + 4y + 40$ square feet

d. $4xy + 40x$ square feet

15. You have a rectangular pen built out of fencing panels. Each panel is y meters long. The pen is twice as long as it is wide, with $2x$ panels along the length and x panels along the width. You are expanding the pen by adding a gate 2 meters long to each of the long sides of the rectangle. Write an expression for the area inside the expanded pen, measured in square meters.

a. $2x^2y^2 + 2$ square meters

b. $2x^2y^2 + 2xy$ square meters

c. $4x^2y^2$ square meters

d. $4xy + 4x + 4y + 4$ square meters

6.5 Multiplying Binomials

The **FOIL method** is a helpful way to remember the four steps for multiplying binomials. FOIL is an acronym that stands for First, Outer, Inner, Last. For example, when multiplying the binomials

$(x + 2)(x + 3)$, multiply the first terms, the outer terms, the inner terms, and the last terms.

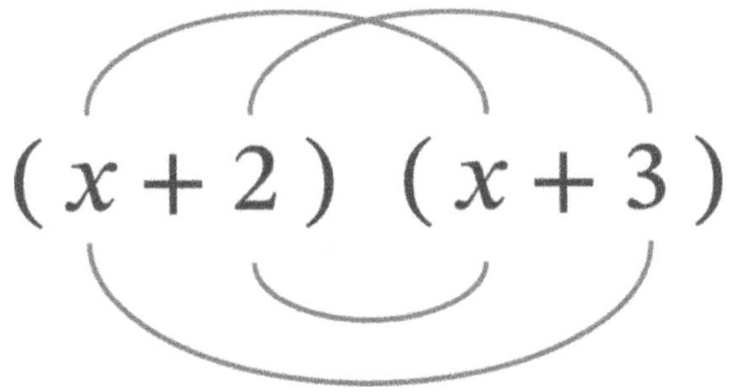

F: $x \cdot x = x^2$

O: $x \cdot 3 = 3x$

I: $2 \cdot x = 2x$

L: $2 \cdot 3 = 6$

Then, add all the terms together. $(x + 2)(x + 3)$ becomes $x^2 + 3x + 2x + 6$, which simplifies to $x^2 + 5x + 6$.

When using the FOIL strategy to multiply binomials, it is important to be aware of the signs that are attached to each term. For example, the rules for multiplying negatives will apply in the following example.

$$(m - 7)(-n + 5)$$

F: $m \cdot (-n) = -mn$

O: $m \cdot 5 = 5m$

I: $-7 \cdot (-n) = 7n$

L: $-7 \cdot 5 = -35$

$(m - 7)(-n + 5)$ becomes $-mn + 5m + 7n - 35$, which is in simplest form.

Some binomials will have multiple variables raised to a power. Remember, add the exponents when the bases are the same. For example, $(x + y^2)(3x - 2y)$ has multiple variables and some are raised to a power. The same rules for FOIL will apply.

$$(x + y^2)(3x - 2y)$$

F: $x \cdot 3x = 3x^2$
O: $x \cdot (-2y) = -2xy$

I: $y^2 \cdot 3x = 3xy^2$

L: $y^2 \cdot (-2y) = -2y^3$

$(x + y^2)(3x - 2y)$ becomes $3x^2 - 2xy + 3xy^2 - 2y^3$, which is in simplest form.

Review Video: Multiplying Terms Using the FOIL Method
Visit mometrix.com/academy and enter code: 854792

EXAMPLES

EXAMPLE 1:

Simplify the expression.

$$(5x + 3)(x + 2)$$

Answer:

F: First

O: Outer

I: Inner

L: Last

$$(5x + 3)(x + 2)$$

F: $5x \cdot x = 5x^2$

O: $5x \cdot 2 = 10x$

I: $3 \cdot x = 3x$

L: $3 \cdot 2 = 6$

The expression $(5x + 3)(x + 2)$ becomes $5x^2 + 10x + 3x + 6$, which simplifies to $5x^2 + 13x + 6$.

EXAMPLE 2:

Simplify the expression.

$$(-3x + 4y)(5x - y)$$

Answer:

F: First

O: Outer

I: Inner

L: Last

$$(-3x + 4y)(5x - y)$$

F: $-3x \cdot 5x = -15x^2$

O: $-3x \cdot (-y) = 3xy$

I: $4y \cdot 5x = 20xy$

L: $4y \cdot (-y) = -4y^2$

The expression $(-3x + 4y)(5x - y)$ becomes $15x^2 + 3xy + 20xy - 4y^2$, which simplifies to $-15x^2 + 23xy - 4y^2$.

EXAMPLE 3:

Joan is building a rectangular garden in her backyard. She wants to divide the garden into 4 sections so that she can contain different vegetables within separate regions. A sketch of her garden is shown below. What will be the total area of Joan's garden?

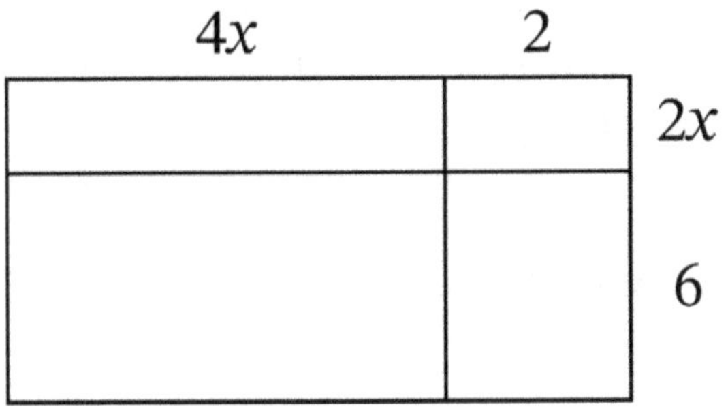

Answer:

The FOIL method can be used to determine the area of the rectangular garden. Start by identifying the area formula for a rectangle.

$$A = l \times w$$

The length of the garden can be described as $(4x + 2)$. The width of the garden can be described as $(2x + 6)$. Substitute these values into the area formula.

$$A = l \times w$$
$$A = (4x + 2)(2x + 6)$$

Use the FOIL method to multiply the two binomials.

"FOIL" is an acronym that indicates the order to multiply terms of polynomials.

F: First

O: Outer

I: Inner

L: Last

$$A = (4x + 2)(2x + 6)$$

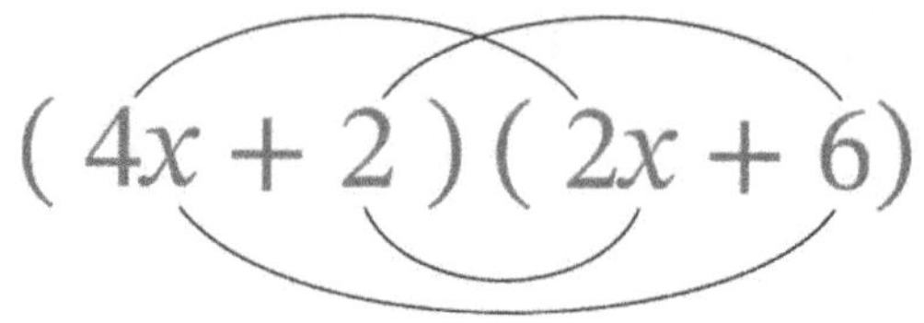

F: $4x \cdot 2x = 8x^2$

O: $4x \cdot 6 = 24x$

I: $2 \cdot 2x = 4x$

L: $2 \cdot 6 = 12$

$$A = 8x^2 + 24x + 4x + 12$$
$$A = 8x^2 + 28x + 12$$

The area of Joan's garden is $8x^2 + 28x + 12$.

Practice Questions

1. Multiply the following binomial: $(3x + 1)(x + 2)$.

a. $3x^2 + 5x + 2$
b. $3x^2 + 7x + 2$
c. $3x^2 + 2$
d. $3x^2 + 4$

2. Multiply the following binomial: $(4m - 1)(2m + 3)$.

a. $8m^2 + 3$
b. $8m^2 - 3$
c. $8m^2 + 10m - 3$
d. $8m^2 + 5m - 3$

3. Multiply the following binomials: $(4x + 9)(4x - 9)$.

a. $16x^2 - 72x - 81$
b. $16x^2 + 72x + 81$
c. $16x^2 + 81$
d. $16x^2 - 81$

4. Multiply the following binomials: $(3p - 2)(3p - 2)$.

a. $9p^2 + 4$
b. $9p^2 - 4$
c. $9p^2 + 12p - 4$
d. $9p^2 - 12p + 4$

5. Multiply the following binomials: $(-7r - 2)(3r + 4)$.

a. $21r^2 - 22r + 8$
b. $-21r^2 - 34r - 8$
c. $12r^2 - 8$
d. $-28r^2 - 34r - 6$

6. Multiply the following binomials: $(11m + 12p)(11m - 12p)$.

a. $22m^2 - 264mp - 24p^2$
b. $121m^2 - 46mp + 144p^2$
c. $121m^2 - 144p^2$
d. $22m^2 + 24p^2$

7. Multiply the following binomials: $(9x - 10y)(8x - 13y)$.

a. $72x^2 - 40xy + 130y^2$
b. $72x^2 - 197xy + 130y^2$
c. $72x^2 + 130y^2$
d. $72x^2 - 130y^2$

8. Multiply the following binomials: $(-9r + 4s)(-12r + 7s)$.

a. $-3r^2 - 111rs + 28s^2$
b. $108r^2 - 10rs + 11s^2$
c. $108r^2 - 111rs + 28s^2$
d. $-108r^2 + 111rs + 28s^2$

9. Multiply the following binomials: $(18x + 13y)(18x - 13y)$.

a. $36x^2 - 62xy + 26y^2$
b. $36x^2 + 72xy + 26y^2$
c. $324x^2 + 169y^2$
d. $324x^2 - 169y^2$

10. Multiply the following binomials: $(8c + 12d)(8c + 12d)$.

a. $96c^2 + 192cd + 96d^2$
b. $16c^2 + 40cd + 24d^2$
c. $64c^2 + 40cd + 144d^2$
d. $64c^2 + 192cd + 144d^2$

11. Multiply the following binomials: $(4m - 10n)(5m + 10n)$.

a. $20m^2 + 10mn - 100n^2$
b. $20m^2 - 10mn - 100n^2$
c. $9m^2 - 10mn - 120n^2$
d. $10m^2 + 10nm - 50n^2$

12. Multiply the following binomials: $(12x + 18y)(12x - 18y)$.

a. $144x^2 - 324y^2$

b. $144x^2 + 324y^2$

c. $144x^2 + 432xy + 324y^2$

d. $144x^2 + 60xy - 324y^2$

13. Which of the following expressions represents the area of a rectangle that has a length of $4m + 2$ and a height of $7m - 3$?

a. $28m^2 + 6$

b. $28m^2 - 6$

c. $28m^2 + 2m - 6$

d. $28m^2 - 2m - 6$

14. A square garden is bordered by a walkway of uniform width. The sides of the garden are x feet in length and the walkway is y feet wide. What is the total area of the garden and the walkway that surrounds it?

a. $x^2 + 4xy + y^2$

b. $x^2 + 2xy + 4y^2$

c. $x^2 + 4xy + 4y^2$

d. $x^2 + 4y^2$

15. Let $(3p + 5q)$ and $(4p - 9q)$ represent two integers. Which of the following represents their product?

a. $12p^2 - 45p^2$

b. $12p^2 - 45p^2$

c. $12p^2 + 7pq - 45p^2$

d. $12p^2 - 7pq - 45p^2$

6.6 Dividing Polynomials by Monomials

The process of dividing polynomials by monomials is very similar to the process of simplifying fractions. For example, the fraction $\frac{40}{8}$ can be thought of as $\frac{8(5)}{8}$. We can see how the 8's will cancel out. When the numerator and denominator are both divided by 8, the result is 5. The same process works for polynomials divided by monomials. Similar factors will cancel out until the fraction is in simplest form.

Remember, when dividing terms with powers, the exponents can be subtracted if they have the *same base.*

For example, $\frac{24x^5}{6x^2}$ can be reduced by simplifying the coefficients, and then subtracting the exponents because the base is the same, x.

The coefficients $\frac{24}{6}$ simplify to 4. The expression $\frac{x^5}{x^2}$ becomes x^{5-2}, which simplifies to x^3. The result is $4x^3$.

When the numerator is a polynomial with multiple terms, simply divide each term by the monomial in the denominator. For example, when dividing $18x^4 + 24x^3 - 15x^2$ by the monomial $3x$, divide each term by $3x$.

$$\frac{18x^4}{3x} = 6x^{4-1} = 6x^3$$
$$\frac{24x^3}{3x} = 8x^{3-1} = 8x^2$$
$$-\frac{15x^2}{3x} = -5x^{2-1} = -5x$$

Therefore, $\frac{18x^4+24x^3-15x^2}{3x}$ simplifies to $6x^3 + 8x^2 - 5x$.

The same process works when you are dividing a polynomial by a monomial that is negative. For example, the expression $\frac{15x^4+6x^3-9x^2}{-3x^2}$ can be split up into three separate division problems.

$$\frac{15x^4}{-3x^2} = -5x^{4-2} = -5x^2$$
$$\frac{6x^3}{-3x^2} = -2x^{3-2} = -2x$$
$$\frac{-9x^2}{-3x^2} = 3x^{2-2} = 3x^0 = 3$$

Therefore, $\frac{15x^4+6x^3-9x^2}{-3x^2}$ becomes $-5x^2 - 2x + 3$.

Sometimes more than one variable will be included. Remember, simplify the coefficients, and then subtract the powers that have the same base.

The expression $\frac{35x^5y^3-20x^4y+15x^2y^5}{5x^2y}$ can be split into three fractions.

$$\frac{35x^5y^3}{5x^2y} = 7x^{5-2}y^{3-1} = 7x^3y^2$$
$$\frac{-20x^4y}{5x^2y} = -4x^{4-2}y^{1-1} = -4x^2y^0 = -4x^2$$
$$\frac{15x^2y^5}{5x^2y} = 3x^{2-2}y^{5-1} = 3x^0y^4 = 3y^4$$

Therefore, $\frac{35x^5y^3-20x^4y+15x^2y^5}{5x^2y}$ simplifies to $7x^3y^2 - 4x^2 + 3y^4$.

Review Video: Dividing Polynomials by Monomials
Visit mometrix.com/academy and enter code: 253551

Examples

Example 1:

Simplify the expression.

$$\frac{4x^3 + 6x^2}{2x}$$

Answer:

Divide both terms of the binomial by $2x$. Divide the coefficients, and subtract the exponents with the same base.

$$\frac{4x^3}{2x} = 2x^{3-1} = 2x^2$$
$$\frac{6x^2}{2x} = 3x^{2-1} = 3x$$

Therefore, $\frac{4x^3+6x^2}{2x}$ simplifies to $2x^2 + 3x$.

Example 2:

Simplify the expression.

$$\frac{12x^3 + 8x^2 + 16x}{4x}$$

Answer:

Divide the coefficients, and subtract the exponents with the same base. All three terms in the trinomial will be divided by $4x$.

$$\frac{12x^3}{4x} = 3x^{3-1} = 3x^2$$
$$\frac{8x^2}{4x} = 2x^{2-1} = 2x$$

$$\frac{16x}{4x} = 4x^{1-1} = 4x^0 = 4$$

Therefore, $\frac{12x^3+8x^2+16x}{4x}$ simplifies to $3x^2 + 2x + 4$.

Or, think about the problem in expanded form.

The expression $\frac{12x^3+8x^2+16x}{4x}$ is the same thing as $\frac{12 \cdot x \cdot x \cdot x + 8 \cdot x \cdot x + 16 \cdot x}{4 \cdot x}$. In expanded form, it can be easier to see how the exponents cancel out.

$$\frac{12 \cdot x \cdot x \cdot x}{4 \cdot x} = 3 \cdot x \cdot x = 3x^2$$
$$\frac{8 \cdot x \cdot x}{4 \cdot x} = 2x$$
$$\frac{16 \cdot x}{4 \cdot x} = 4$$

Now, put the terms together: $3x^2 + 2x + 4$.

Practice Questions

1. Divide $x^3 - 3x^2 + 6x$ by x.

a. $x^2 - x + 2$

b. $x^2 - 3x + 6$

c. $x^4 - 3x^3 + 6x^2$

d. $x^5 - x^4 + 2x^2$

2. Divide $4y^3 + y^2 - 8y$ by y.

a. $y^3 + y^2 - 2y^2$

b. $4y^4 + y^3 - 8y^2$

c. $y^2 + y - 2$

d. $4y^2 + y - 8$

3. Divide $3m^4 - 6m^3 + 9m^2$ by $3m$.

a. $m^3 - 2m^2 + 3m$

b. $m^3 - 3m^2 + 6m$

c. $m^4 - 2m^3 + 3m^2$

d. $m^4 - 3m^3 + 6m^2$

4. Divide $2a^4 + 5a^3 - 6a^2$ by a^2.

a. $2a^6 + 5a^5 - 6a^4$

b. $2a^2 + 5a - 6$

c. $2a^8 + 5a^6 - 6a^4$

d. $2a^2 + 5a^3 - 6$

5. Divide $10x^5 - 25x^3 + 5x^2$ by $5x^2$.

a. $5x^7 - 20x^5 + x^4$

b. $2x^{10} - 5x^6 + x^4$

c. $2x^3 - 5x + 1$

d. $5x^3 - 20x + 1$

6. Divide $4x^4y^3 + 6x^3y^2 + 2xy$ by $2xy$.

a. $2x^3y^2 + 3x^2y + 1$

b. $2x^3y^2 + 4x^2y + 1$

c. $2x^5y^4 + 3x^4y^3 + x^2y^2$

d. $2x^4y^3 + 3x^3y^2 + xy$

7. Divide $8x^2y^3 - 16x^3y - 24x^3y^2$ by $-4x^2y$.

a. $2x^4y^3 - 4x^6y^2 - 6x^6y^2$

b. $-2x^4y^4 + 6x^5y^2 + 4x^5y^3$

c. $-2y^2 + 6xy - 4x$

d. $-2y^2 + 6xy + 4x$

8. Divide $m^3n^4 - 9m^4n^3 + 6m^2n^2 - 15mn^2$ by $3mn^2$.

a. $\frac{1}{3}m^2n^2 - 3m^3n + 2m - 5$

b. $\frac{1}{3}m^3n^8 - 3m^4n^6 + 2m^2n^4 - 5m^2n^4$

c. $-2m^3n^8 - 12m^4n^6 + 3m^2n^4 - 18m^2n^4$

d. $-2m^2n^2 - 12m^3n + 3m - 18$

9. Divide $3p^5q^2 + 6p^6q^3 + 3p^2q^3 + 12p^2q^2$ by $3p^2q$.

a. $p^2q + 23q^2 + pq^2 + 9pq^3$

b. $p^7q^3 + 3p^8q^4 + p^4q^3 + 9p^4q^3$

c. $p^3q + 2p^4q^2 + q^2 + 4q$

d. $p^2q + 2p^2q^2 + pq^2 + 4pq^3$

10. Divide $-x^7y^5 + 9x^4y^4 - 36x^3y^3$ by $-9x^3y^2$.

a. $-8x^4y^3 - 18xy^2 - 27y$

b. $\frac{1}{9}x^4y^3 - xy^2 + 4y$

c. $-2y^2 - 6xy - 4x$

d. $\frac{1}{9}x^{10}y^7 - x^7y^6 + 4x^6y^5$

11. Divide $m^{11}n^8 - 4m^8n^6 + 5m^6n^4$ by m^5n^3.

a. $m^{16}n^{11} - 3m^{13}n^9 + 4m^{11}n^7$
b. $m^6n^5 - 3m^3n^3 + 4mn$
c. $m^6n^5 - 4m^3n^3 + 5mn$
d. $m^{16}n^{11} - 4m^{13}n^9 + 5m^{11}n^7$

12. Divide $-5x^9y^5 + 10x^7y^3 - 15x^6y^3 + 20x^5y$ by $-10x^5y$.

a. $\frac{1}{2}x^{14}y^5 - x^{12}y^4 + \frac{3}{2}x^{11}y^4 - 2x^{10}y^2$
b. $-15x^{14}y^5 - x^{12}y^4 + 5x^{11}y^4 - 10x^{10}y^2$
c. $\frac{1}{2}x^4y^4 - x^2y^2 - \frac{3}{2}xy^2 - 2$
d. $\frac{1}{2}x^4y^4 - x^2y^2 + \frac{3}{2}xy^2 - 2$

13. The area of a rectangular swimming pool is given by the polynomial $3n^4 + 6n^3 + n^2$. If the width of the pool is $3n^2$, what is the length of the pool?

a. $n^2 + 2n + \frac{1}{3}$
b. $n^6 + 2n^5 + \frac{1}{3}n^4$
c. $n^6 + 3n^5 - 2n^4$
d. $n^2 + 3n - 2$

14. The revenue, in dollars, generated by selling computer laptop batteries at a price of $3xy$ dollars each can be represented by the expression $24xy^2 + 9x^3y - 12xy$. If the revenue of the batteries can be calculated by the product of the price and the number of units sold, what is the number of batteries sold in terms of x and y?

a. $21x^2y^3 + 6x^4y^2 - 9x^2y^2$
b. $21x^2 + 6y - 9$
c. $3x^2 + 8y - 4$
d. $3x^4y^2 + 8x^2y^3 - 4x^2y^2$

15. The volume of a rectangular box is $2m^2n^3 + 8mn^2 + 10m^2n$. If the product of the length and width is $2mn$, what is the height of the box in terms of m and n?

a. $m^3n^4 + 6m^2n^3 + 8m^3n^2$
b. $mn^2 + 6n + 8m$
c. $m^3n^4 + 4m^2n^3 + 5m^3n^2$
d. $mn^2 + 4n + 5m$

6.7 Dividing Polynomials by Binomials

When dividing a trinomial by a binomial, the long division algorithm can be used. Long division follows a 4-step process:

Step 1: Divide the first term by the first term.

Step 2: Multiply.

Step 3: Subtract.

Step 4: Bring down the next term.

Repeat.

For example, when dividing a trinomial like $x^2 + 5x + 6$ by $x + 2$, start by placing the divisor to the left of the long division symbol, and the dividend under the long division symbol.

$$x+2\overline{)x^2+5x+6}$$

Now follow the 4-step process:

Step 1: Divide the first term by the first term.

This means x^2 divided by x. Remember, subtract the exponents when the bases are the same: $x^2 \div x = x$.

$$\begin{array}{r} x \\ x+2\overline{)\,x^2 +5x +6} \end{array}$$

Step 2: Multiply: $x \cdot (x + 2) = x^2 + 2x$.

$$\begin{array}{r} x \\ x+2\overline{)\,x^2 +5x +6} \\ x^2 +2x \end{array}$$

Step 3: Subtract: $x^2 - x^2 = 0$ and $5x - 2x = 3x$.

$$\begin{array}{r} x \\ x+2\overline{)\,x^2 +5x +6} \\ \underline{-(x^2 +2x)} \\ 3x \end{array}$$

Step 4: Bring down the next term.

$$\begin{array}{r} x \\ x+2\overline{)\,x^2 +5x +6} \\ \underline{-(x^2 +2x)} \\ 3x +6 \end{array}$$

Repeat steps 1–4.

$$3x \div x = 3$$

$$3 \cdot (x + 2) = 3x + 6$$

$$\begin{array}{r|rrr} & x & +3 & \\ \hline x + 2 & x^2 & +5x & +6 \\ & -(x^2 & +2x) & \\ \hline & & 3x & +6 \end{array}$$

Subtract.

$$\begin{array}{r|rrr} & x & +3 & \\ \hline x + 2 & x^2 & +5x & +6 \\ & -(x^2 & +2x) & \\ \hline & & 3x & +6 \\ & - & (3x & +6) \\ \hline & & & 0 \end{array}$$

Review Video: Dividing Trinomials by Binomials
Visit mometrix.com/academy and enter code: 651465

Examples

Example 1:

Divide the expression.

$$\frac{4x^2 - 5x - 21}{x - 3}$$

Answer:

Use long division to solve. Place the denominator on the outside and the numerator on the inside.

$$x - 3 \overline{\smash{)}\, 4x^2 \quad -5x \quad -21}$$

Step 1: Divide the first term of the dividend by the first term of the divisor. This means $4x^2$ divided by x is $4x$.

$$\begin{array}{r|rrr} & & 4x & +7 \\ \hline x - 3 & 4x^2 & -5x & -21 \end{array}$$

Step 2: Multiply $4x$ by $x - 3$ to get $4x^2 - 12x$.

$$\begin{array}{r|rrr} & & 4x & +7 \\ \hline x - 3 & 4x^2 & -5x & -21 \\ & -(4x^2 & -12x) & \end{array}$$

Step 3: Subtract to form a new polynomial. Bring down the –21.

$$\begin{array}{r|rrr} & & 4x & +7 \\ \hline x - 3 & 4x^2 & -5x & -21 \\ & -(4x^2 & -12x) & \\ \hline & & 7x & -21 \end{array}$$

Step 4: Repeat the same process (steps 1–3) for the first term of the new binomial.

Divide the first term of the new binomial by first term of the divisor. This means $7x$ divided by x, which is 7. Multiply 7 by $x - 3$, which is $7x - 21$. Finally, subtract.

$$\begin{array}{r|rrr} & 4x & +7 & \\ x - 3 & 4x^2 & -5x & -21 \\ & -(4x^2 & -12x) & \\ \hline & & 7x & -21 \\ & & -(7x & -21) \\ \hline & & & 0 \end{array}$$

The quotient is $4x + 7$.

EXAMPLE 2:

Divide the expression.

$$\frac{-6x^2 + 29x - 28}{3x - 4}$$

Answer:

Use long division to solve. Place the denominator on the outside and the numerator on the inside.

$$3x - 4 \overline{\big) -6x^2 \quad +29x \quad -28}$$

Step 1: Divide the first term of the dividend by the first term of the divisor. This means $-6x^2$ divided by $3x$, which equals $-2x$.

$$\begin{array}{r|rrr} & & -2x & \\ 3x - 4 & -6x^2 & +29x & -28 \end{array}$$

Step 2: Multiply $-2x$ by $3x - 4$, which is $-6x^2 + 8x$.

$$\begin{array}{r|rrr} & & -2x & +7 \\ 3x - 4 & -6x^2 & +29x & -28 \\ & -(-6x^2 & +8x) & \\ \hline \end{array}$$

Step 3: Subtract and bring down the -28.

$$\begin{array}{r|rrr} & & -2x & +7 \\ 3x - 4 & -6x^2 & +29x & -28 \\ & -(-6x^2 & +8x) & \\ \hline & & 21x & -28 \end{array}$$

Step 4: Repeat steps 1–3.

Divide the first term of the new binomial by the first term of the divisor. This means $21x$ divided by $3x$, which equals 7. Multiply 7 by $3x - 4$, which equals $21x - 28$. Finally, subtract.

$$\begin{array}{r} -2x \quad +7 \\ 3x - 4\,\overline{)\,-6x^2 \; +29x \; -28} \\ -(-6x^2 \; +8x) \qquad \\ \hline 21x \; -28 \\ -\;(21x \; -28) \\ \hline 0 \end{array}$$

The quotient is $-2x + 7$.

Example 3:

Use long division to evaluate: $(2x^3 + 3x^2 - 30x + 7) \div (x - 3)$

Answer:

The polynomial $(2x^3 + 3x^2 - 30x + 7)$ can be divided by the binomial $(x - 3)$ using long division. The standard 4-step process for long division applies when dividing polynomials: divide, multiply, subtract, bring down.

$$\begin{array}{r} x - 3\,\overline{)\,2x^3 \; +3x^2 \; -30x \; +7} \end{array}$$

Step 1: Divide: $2x^3 \div x = 2x^2$

$$\begin{array}{r} 2x^2 \qquad\qquad\qquad \\ x - 3\,\overline{)\,2x^3 \; +3x^2 \; -30x \; +7} \end{array}$$

Step 2: Multiply: $2x^2 \cdot x = 2x^3$ and $2x^2 \cdot (-3) = -6x^2$

$$\begin{array}{r} 2x^2 \qquad\qquad\qquad \\ x - 3\,\overline{)\,2x^3 \; +3x^2 \; -30x \; +7} \\ -\;(2x^3 \; -6x^2) \qquad\qquad \end{array}$$

Step 3: Subtract: Start by distributing the negative within the parentheses. $-(2x^3 - 6x^2)$ becomes $(-2x^3 + 6x^2)$. Now, add $(2x^3 + 3x^2) + (-2x^3 + 6x^2) = 9x^2$.

$$\begin{array}{r} 2x^2 \qquad\qquad\qquad \\ x - 3\,\overline{)\,2x^3 \; +3x^2 \; -30x \; +7} \\ -\;(2x^3 \; -6x^2) \qquad\qquad \\ 9x^2 \qquad\qquad\qquad \end{array}$$

Step 4: Bring down: Bring down the next term, which is $-30x$. Repeat the process.

$$\begin{array}{r} 2x^2 \; +9x \; -3 \qquad \\ x - 3\,\overline{)\,2x^3 \; +3x^2 \; -30x \; +7} \\ -\;(2x^3 \; -6x^2) \qquad\qquad \end{array}$$

$$\begin{array}{r} 9x^2 \quad -30x \\ - \quad (9x^2 \quad -27x) \\ \hline -3x \quad +7 \\ - \quad (-3x \quad +9) \\ \hline -2 \end{array}$$

The value -2 represents the remainder. Divide the remainder by the original divisor $x - 3$. The result is the fraction $-\frac{2}{x-3}$.

$$(2x^3 + 3x^2 - 30x + 7) \div (x - 3) = 2x^2 + 9x - 3 - \frac{2}{x - 3}$$

Practice Questions

1. Use polynomial long division to calculate the quotient $(4x^2 + 27x + 35) \div (x + 5)$.

a. $4x + 47$

b. $4x + 7 + \frac{70}{x+5}$

c. $4x + 7$

d. $4x + 47 + \frac{87}{x+5}$

2. Use polynomial long division to calculate the quotient $(6x^2 + 53x + 59) \div (x + 8)$.

a. $6x + 101 + \frac{747}{x+8}$

b. $6x + 101 + \frac{867}{x+8}$

c. $6x + 5 + \frac{19}{x+8}$

d. $6x + 5 + \frac{99}{x+8}$

3. Use polynomial long division to calculate the quotient $(-3x^2 + 8x + 9) \div (x + 4)$.

a. $-3x - 4 + \frac{25}{x+4}$

b. $-3x + 20 - \frac{71}{x+4}$

c. $-3x + 20 + \frac{89}{x+4}$

d. $-3x - 4$

4. Use polynomial long division to calculate the quotient $(10x^2 + 3x - 13) \div (x - 1)$.

a. $10x + 13$

b. $10x + 13 - \frac{26}{x-1}$

c. $10x - 7$

d. $10x - 7 - \frac{20}{x-1}$

5. Use polynomial long division to calculate the quotient $(-6x^2 + 41x - 29) \div (x - 6)$.

a. $-6x + 77 - \frac{491}{x-6}$

b. $-6x + 5 - \frac{59}{x-6}$

c. $-6x + 5 + \frac{1}{x-6}$

d. $-6x + 77 + \frac{433}{x-6}$

6. Use polynomial long division to calculate the quotient $(5x^2 + 58x - 29) \div (x + 12)$.

a. $5x + 118$

b. $5x + 118 + \frac{1,387}{x+12}$

c. $5x - 2 - \frac{53}{x+12}$

d. $5x - 2 - \frac{5}{x+12}$

7. Use polynomial long division to calculate the quotient $(11x^2 - 120x + 189) \div (x - 9)$.

a. $11x - 21$

b. $11x - 219 + \frac{2,160}{x-9}$

c. $11x - 219$

d. $11x - 21 + \frac{378}{x-9}$

8. Use polynomial long division to calculate the quotient $(8x^2 - 19x + 39) \div (x - 2)$.

a. $8x - 35 + \frac{109}{x-2}$

b. $8x - 35 - \frac{31}{x-2}$

c. $8x - 3 + \frac{45}{x-2}$

d. $8x - 3 + \frac{33}{x-2}$

9. Use polynomial long division to calculate the quotient $(6x^2 - 23x + 3) \div (2x - 5)$.

a. $3x - 19 - \frac{92}{2x-5}$

b. $3x - 4 - \frac{17}{2x-5}$

c. $3x - 19 + \frac{98}{2x-5}$

d. $3x - 4 + \frac{23}{2x-5}$

10. Use polynomial long division to calculate the quotient $(3x^2 - 5x - 38) \div (x + 3)$.

a. $3x - 14 - \frac{80}{x+3}$

b. $3x + 4 - \frac{50}{x+3}$

c. $3x - 14 + \frac{4}{x+3}$

d. $3x + 4 - \frac{26}{x+3}$

11. Use polynomial long division to calculate the quotient $(7x^2 - 61x - 1{,}987) \div (x - 13)$.

a. $7x - 152 - \frac{11}{x-13}$

b. $7x - 152$

c. $7x + 30 - \frac{1{,}597}{x-13}$

d. $7x + 30 - \frac{2{,}377}{x-13}$

12. Use polynomial long division to calculate the quotient $(12x^2 - x - 88) \div (3x + 8)$.

a. $4x + 11 + \frac{176}{3x+8}$

b. $4x - 11 - \frac{176}{3x+8}$

c. $4x - 11$

d. $4x + 11$

13. A rectangle has an area of $4x^2 + 29x + 50$ square centimeters. Its width is $x + 5$ centimeters. Use polynomial long division to find the length of the rectangle.

a. $4x + 10$ centimeters

b. $4x + 12 - \frac{10}{x+5}$ centimeters

c. $4x + 11 - \frac{5}{x+5}$ centimeters

d. $4x + 9 + \frac{5}{x+5}$ centimeters

14. A business has a square parking lot with an area of x^2 square meters. Because business is booming, the owner wants to expand it into a rectangular lot 10 meters wider than the current lot with a new area of $2x^2 + 19x - 10$ square meters. Use polynomial long division to find the length of the new lot.

a. $2x - 2$ meters

b. $2x + 2$ meters

c. $2x + 1$ meters

d. $2x - 1$ meters

15. You manufacture widgets. Each one costs you $12 to make. As you raise the selling price, your profit per widget increases, but you sell fewer widgets per week. When you set the selling price at x dollars, your total profit for all widget sales during the week is $-3x^2 + 536x - 6,000$ dollars. Use polynomial long division to find the number of widgets you sell per week.

a. $400 - 3x$ widgets

b. $500 - 3x$ widgets

c. $600 - 3x$ widgets

d. $750 - 3x$ widgets

6.8 Factoring Using Common Monomial Factors

One way to simplify a polynomial is to factor out a monomial. Some polynomials will contain terms that all share a common factor. For example, consider the polynomial $9x^3 + 27x^2 + 24x$. Take a close look at each separate term. Is there a factor that is shared by all terms? If so, this can be factored out of the polynomial.

In this example, the greatest common factor of 9, 27, and 24, is 3. Factor this out. Now look at the variables. The greatest common factor of x^3, x^2, and x, is x. Factor this out. The monomial $3x$ can be factored out of the polynomial $9x^3 + 27x^2 + 24x$. When $3x$ is factored out, $3x^2 + 9x + 8$ remains. So, $3x(3x^2 + 9x + 8)$ is the factored polynomial.

Some polynomials will contain more than one variable. For example, the polynomial $55x^3y^2 + 20xy^2$. The polynomial can be simplified by factoring out the GCF (greatest common factor).

The GCF of 55 and 20 is 5. Factor this out.

The GCF of x^3y^2 and xy^2 is xy^2. Factor this out.

When the GCFs are factored out, the result is $5xy^2(11x^2 + 4)$.

Check the answer by multiplying $5xy^2$ and $11x^2 + 4$. The result will be the original polynomial.

Not every polynomial will have a monomial to factor out. Factoring out a monomial only works when there is a common factor for all terms within the polynomial.

For example, the trinomial $x^2 + 7x + 3$ consists of three terms: x^2, $7x$, and 3. There is no factor that all three terms share, so a monomial cannot be factored out.

Review Video: Factoring Out Common Monomial Factors
Visit mometrix.com/academy and enter code: 398578

EXAMPLES

EXAMPLE 1:

Factor the expression.

$$12x^3 - 16x^6$$

Answer:

Factor out the greatest common factor of the two terms in the polynomial. Both 12 and 16 have a factor of 4. Both x^3 and x^6 have a factor of x^3. The greatest common factor of $12x^3$ and $-16x^6$ is $4x^3$.

Remove this factor, and divide each term of the polynomial by $4x^3$ to get the expression inside parentheses.

$$4x^3(3 - 4x^3)$$

Check your work by multiplying the monomial by the polynomial. If the result is the original polynomial, then the factoring was done correctly.

$$4x^3(3 - 4x^3) = 12x^3 - 16x^9$$

$12x^3 - 16x^6$ was the original polynomial, so $4x^3(3 - 4x^3)$ is the correct factored form of the polyonomial.

Example 2:

Factor the expression.

$$9x^2y^4 - 27xy^2$$

Answer:

Factor out the greatest common factor of the two terms in the polynomial. The greatest common factor of 9 and 27 is 9. The greatest common factor of x^2y^4 and xy^2 is xy^2. This means that we can pull out the factor $9xy^2$ from the polynomial $9x^2y^4 - 27xy^2$. When you pull out the factor $9xy^2$ from the polynomial, you are essentially dividing by $9xy^2$.

When $9xy^2$ is removed, $xy^2 - 3$ is left inside the parentheses.

$$9xy^2(xy^2 - 3)$$

Check your work by multiplying the monomial by the binomial. If the result is the original polynomial, then the factoring was done correctly.

$$9xy^2(xy^2 - 3) = 9x^2y^4 - 27xy^2$$

$9x^2y^4 - 27xy^2$ was the original polynomial, so $9xy^2(xy^2 - 3)$ is the correct factored form of the polynomial.

Example 3:

Factor the polynomial completely: $-x^2 - 3x + 54$

Answer:

When a trinomial has a negative leading coefficient, it is often helpful to factor out the negative first. Factoring out the negative first creates an easier breakdown process from trinomial to binomial. Start by factoring out the -1 from $-x^2 - 3x + 54$.

$$-(x^2 + 3x - 54)$$

Now, factor the trinomial. Since $a = 1$, we only need to focus on the middle term and the last term. List factor pairs that multiply to -54 and combine to 3. The factor pair -6 and 9 multiply to -54 and combine to 3.

$$-(x - 6)(x + 9)$$

Practice Questions

1. Factor the polynomial $12x + 18$ by factoring out the greatest common factor.

a. $3(4x + 6)$

b. $6(2x + 3)$

c. $6x(2x + 3)$

d. $12\left(x + \frac{3}{2}\right)$

2. Factor the polynomial $15x^4 + 12x^2$ by factoring out the greatest common factor.

a. $3x^2(5x^2 + 4)$
b. $5x^2(3x^2 + 12x)$
c. $3x^2(15x^2 + 12)$
d. $3(5x^2 + 4)$

3. Factor the polynomial $10x^3 - 20x^2 - 35x$ by factoring out the greatest common factor.

a. $10(2x^2 - 4x - 7)$
b. $5x(2x^2 - 4x - 7)$
c. $x(10x^2 - 20x - 35)$
d. $5x(x^2 - 2x - 3.5)$

4. Completely factor the polynomial $-6x^5 + 22x^4 - 18x^3$ by factoring out the greatest common factor.

a. $-2x^3(3x^2 - 11x + 9)$
b. $x^3(-6x^2 + 22x - 18)$
c. $-2x^2(3x^3 - 11x^2 + 9x)$
d. $-6x^3(x^2 - 4x + 3)$

5. Factor the polynomial $27x^{12} + 72x^8 - 36x^4$ by factoring out the greatest common factor.

a. $9x^4(3x^8 + 8x^4 - 4)$
b. $9x^4(3x^3 + 8x^2 - 4x)$
c. $3x^4(9x^8 + 24x^4 - 12)$
d. $12x^4(2x^8 + 6x^4 - 3)$

6. Factor the polynomial $6x^4 + 15x^3 + 10x^2$ by factoring out the greatest common factor.

a. $x^2(6x^2 + 15x + 10)$
b. $3x^2(2x^2 + 5x + 3)$
c. $5x^2(x^2 + 3x + 2)$
d. $2x^2(3x^2 + 7x + 5)$

7. Factor the polynomial $8x^2y - 12xy^2$ by factoring out the greatest common factor.

a. $4x(2xy - 3y^2)$
b. $xy(8x - 12y)$
c. $4y(2x^2 - 3xy)$
d. $4xy(2x - 3y)$

8. Completely factor the polynomial $2x^2y^3 + 26y^5$ by factoring out the greatest common factor.

a. $2x^2(y^3 + 13y^5)$
b. $y^3(2x^2 + 26y^2)$
c. $2y^3(x^2 + 13y^2)$
d. $2y^2(x^2y + 13y^3)$

9. Factor the polynomial $30x^6y^3 - 40x^5y^4 + 10x^4y^5$ by factoring out the greatest common factor.

a. $5x^4y^3(6x^2 - 8xy + 2y^2)$
b. $10x^4y^3(3x^2 - 4xy + y^2)$
c. $30x^4y^3(x^2 - 40xy + 10y^2)$
d. $xy(30x^5y^2 - 40x^4y^3 + 10x^3y^4)$

10. Completely factor the polynomial $-15x^6y^2 + 9x^2y^3 - 12x^3y^4$ by factoring out the greatest common factor.

a. $-3x^2y^2(5x^4 - 3y + 4xy^2)$
b. $-x^2y^2(15x^4 - 9y + 12xy^2)$
c. $-3xy(5x^5y - 3xy^2 + 4x^2y^3)$
d. $-15x^2y^2(x^4 - 9y + 12xy^2)$

11. Factor the polynomial $10x^2y^3 + 20xy^4 - 15x^2y^4 + 40xy^5$ by factoring out the greatest common factor.

a. $xy^3(10x + 20y - 15xy + 40y^2)$
b. $5xy^3(2x + 4y - 3xy + 8y^2)$
c. $10xy(xy^2 + 20y^3 - 15xy^3 + 40y^4)$
d. $5xy(2xy^2 + 4y^3 - 3xy^3 + 8y^4)$

12. Factor the polynomial $54z^5w - 36z^3w^3 - 24z^2w^5$ by factoring out the greatest common factor.

a. $2z^2w(27z^3 - 18zw^2 - 12w^4)$
b. $3z^2w(18z^3 - 12zw^2 - 8w^4)$
c. $12z^2w(5z^3 - 3zw^2 - 2w^4)$
d. $6z^2w(9z^3 - 6zw^2 - 4w^4)$

13. A rectangle has length $5x$ cm and area $15x^2 + 20x$ cm^2. Factor the length out of the area to rewrite the area in the form Area = length × width.

a. Area $= 5x(7x)$ cm^2
b. Area $= 5x(3x + 4)$ cm^2
c. Area $= 5x(3x + 20)$ cm^2
d. Area $= 5x(23x)$ cm^2

14. A rectangular solid (a box) has a square base with an area of x^2 ft^2 and a volume of $2x^3 + x^2y^2$ ft^3. Factor the area of the base out of the volume to rewrite the volume in the form Volume = (area of base) × (height).

a. Volume $= x^2(2x + y^2)$ ft^3
b. Volume $= x^2(2x^3 + y^2)$ ft^3
c. Volume $= x^2(2x + x^2y^2)$ ft^3
d. Volume $= x^2(1 + y^2)$ ft^3

15. You sell widgets for a price of $3x$ dollars per widget. Last month you had a total income of $3x^2 + 3xy + 6x^2y^2$ dollars from widget sales. Factor the price out of the total income to rewrite your total income in the form: Total Income = (price per item) × (number of items sold).

a. Total Income $= 3x(x + 3xy + 6x^2y^2)$ dollars
b. Total Income $= 3x(x + y + 2xy^2)$ dollars
c. Total Income $= 3x(x + 3y + 6xy^2)$ dollars
d. Total Income $= 3x(x^2 + y + 2x^2y^2)$ dollars

6.9 Factoring Trinomials of the Form $x^2 + bx + c$

A **trinomial** is a polynomial that contains three terms that are separated by addition or subtraction. Trinomials usually take on the form $ax^2 + bx + c$. For example, $x^2 + 6x + 9$ is a trinomial consisting of the terms x^2 (meaning $a = 1$), $6x$, and 9. Many trinomials can be written as a product using the process of factoring. One way to factor a trinomial is to split it into the product of two binomials. This means rewriting the trinomial in the form $(x + a)(x + b)$. In other words, the process of factoring a trinomial is the opposite process of FOILing.

Let's consider the perfect square trinomial $x^2 + 6x + 9$. This trinomial is considered a **perfect square trinomial** because it can be factored into the product of two identical binomials. This process is the reverse process of FOIL, so start by writing two sets of parentheses.

$$(\quad)(\quad)$$

Look at the first term, x^2. The term x^2 can be expressed as x times x.

$$(x \quad)(x \quad)$$

Now look at the third term, 9. What two numbers will multiply to 9, but combine to form a middle term of 6? The answer is 3 and 3.

The last term, $3 \times 3 = 9$ and the middle term, $3 + 3 = 6$.

$(x \quad)(x \quad)$ becomes $(x + 3)(x + 3)$.

Check the answer by FOILing. When the binomials are FOILed, they will produce the original perfect square trinomial.

An important thing to notice about this trinomial is that it is a perfect square trinomial, meaning it can be written as the square of a binomial, $(x + 3)^2$. These trinomials can be identified by noticing the pattern $x^2 + 2ax + a^2$, which will factor to $(x + a)^2$.

Review Video: Factoring Trinomials of the Form $x^2 + bx + c$
Visit mometrix.com/academy and enter code: 270556

Examples

Example 1:

Factor the expression.

$$x^2 + 2x - 24$$

Answer:

The trinomial $x^2 + 2x - 24$ can be factored into two binomials.

x^2 is the product of x times x.

$$(x \quad)(x \quad)$$

-24 is the product of -4 times 6. The sum of -4 and 6 is 2, which combines to the middle term, $2x$.

$$(x + 6)(x - 4)$$

The answer can be checked by using FOIL. If the original trinomial is the result, then factoring was done correctly.

EXAMPLE 2:

Factor the expression.

$$x^2 + 9x + 18$$

Answer:

The trinomial $x^2 + 9x + 18$ can be factored into two binomials.

x^2 is the product of x times x.

$$(x\quad)(x\quad)$$

18 is the product of 6 and 3. The sum of 6 and 3 is 9, which creates the middle term, $9x$.

$$(x + 6)(x + 3)$$

The answer can be checked by using FOIL. If the original trinomial is the result, then factoring was done correctly.

EXAMPLE 3:

Factor the expression.

$$x^2 + 8x + 16$$

Answer:

The trinomial $x^2 + 8x + 16$ is a perfect square trinomial because it can be written as the square of a binomial. The trinomial can be factored into $(x + 4)^2$. When $(x + 4)$ is multiplied by $(x + 4)$, the result is $x^2 + 4x + 4x + 16$, or $x^2 + 8x + 16$. This can be checked using the FOIL technique.

EXAMPLE 4:

Amy is planting a garden with an area of $x^2 - 3x - 40$ square meters. If the garden is longer than it is wide, write its area in the factored form, Area = length × width.

Answer:

Factor the trinomial into the product of two binomials. Since $a = 1$, ask yourself the question, "What multiplies to −40 and combines to −3?" The answer is 5 and −8. This means that $x^2 - 3x - 40$ factors to $(x + 5)(x - 8)$. So, the area of her garden is $(x + 5)(x - 8)$ square meters.

PRACTICE QUESTIONS

1. Factor the trinomial $x^2 + 12x + 11$ into a product of binomials.

a. $(x + 1)(x + 12)$
b. $(x + 1)(x + 10)$
c. $(x + 1)(x + 11)$
d. $(x + 5)(x + 6)$

2. Factor the trinomial $x^2 + 10x + 24$ into a product of binomials.

a. $(x + 4)(x + 6)$
b. $(x + 1)(x + 24)$
c. $(x + 2)(x + 12)$
d. $(x + 3)(x + 8)$

3. Factor the trinomial $x^2 + 24x + 144$ into a product of binomials.

a. $(x + 6)(x + 24)$
b. $(x + 9)(x + 16)$
c. $(x + 8)(x + 18)$
d. $(x + 12)^2$

4. Factor the trinomial $x^2 - 14x + 24$ into a product of binomials.

a. $(x - 3)(x - 8)$
b. $(x - 2)(x - 12)$
c. $(x - 1)(x - 24)$
d. $(x - 4)(x - 6)$

5. Factor the trinomial $x^2 - 18x + 45$ into a product of binomials.

a. $(x - 3)(x - 15)$
b. $(x - 1)(x - 45)$
c. $(x - 5)(x - 9)$
d. $(x - 9)^2$

6. Factor the trinomial $x^2 - 10x + 25$ into a product of binomials.

a. $(x - 4)(x - 6)$
b. $(x - 1)(x - 25)$
c. $(x + 5)(x - 5)$
d. $(x - 5)^2$

7. Factor the trinomial $x^2 + 5x - 24$ into a product of binomials.

a. $(x + 3)(x - 8)$
b. $(x + 9)(x - 4)$
c. $(x + 6)(x - 4)$
d. $(x + 8)(x - 3)$

8. Factor the trinomial $x^2 + 3x - 28$ into a product of binomials.

a. $(x + 14)(x - 2)$
b. $(x + 4)(x - 7)$
c. $(x + 7)(x - 4)$
d. $(x + 8)(x - 5)$

9. Factor the trinomial $x^2 + 2x - 80$ into a product of binomials.

a. $(x + 20)(x - 18)$
b. $(x + 10)(x - 8)$
c. $(x + 16)(x - 5)$
d. $(x + 8)(x - 10)$

10. Factor the trinomial $x^2 - 10x - 39$ into a product of binomials.

a. $(x+3)(x-13)$
b. $(x+13)(x-3)$
c. $(x+1)(x-39)$
d. $(x+4)(x-14)$

11. Factor the trinomial $x^2 - 15x - 16$ into a product of binomials.

a. $(x+16)(x-1)$
b. $(x+1)(x-16)$
c. $(x+2)(x-17)$
d. $(x+2)(x-8)$

12. Factor the trinomial $x^2 - x - 30$ into a product of binomials.

a. $(x+4)(x-5)$
b. $(x+3)(x-10)$
c. $(x+6)(x-5)$
d. $(x+5)(x-6)$

13. Last year you managed a square ice-skating rink with an area of x^2 square meters. This year you have expanded the rink into a rectangle with an area of $x^2 + 12x + 20$ square meters. If the rectangle is longer than it is wide, write its area in the factored form, Area = length × width.

a. $(x+20)(x+1)$ square meters
b. $(x+8)(x+4)$ square meters
c. $(x+5)(x+4)$ square meters
d. $(x+10)(x+2)$ square meters

14. You have a square piece of land with a total area of x^2 square feet. To prepare it for farming, you fence it in and dig a drainage ditch, reducing the farmable area to a rectangle with area $x^2 - 32x + 175$ square feet. If the rectangle is longer than it is wide, write its area in the factored form, Area = length × width.

a. $(x-1)(x-175)$ square feet
b. $(x-5)(x-35)$ square feet
c. $(x-7)(x-25)$ square feet
d. $(x-8)(x-24)$ square feet

15. You are replacing a square window in your kitchen with a rectangular window that is taller but with a narrower base. The original window has an area of x^2 square centimeters, and the new window has an area of $x^2 + x - 72$ square centimeters. Write the area of the new window in the factored form, Area = base × height.

a. $(x-4)(x+18)$ square centimeters
b. $(x-9)(x+8)$ square centimeters
c. $(x-8)(x+9)$ square centimeters
d. $(x-6)(x+12)$ square centimeters

6.10 Factoring the Difference of Two Squares

There is a special category of binomials referred to as the difference of two squares. This type of binomial consists of two perfect squares that are separated by subtraction, which follows the form $a^2 - b^2$. When a binomial has the form $a^2 - b^2$, it can be factored into two products. This is essentially the reverse process of FOILing. The expression $a^2 - b^2$ always becomes $(a + b)(a - b)$ when factored.

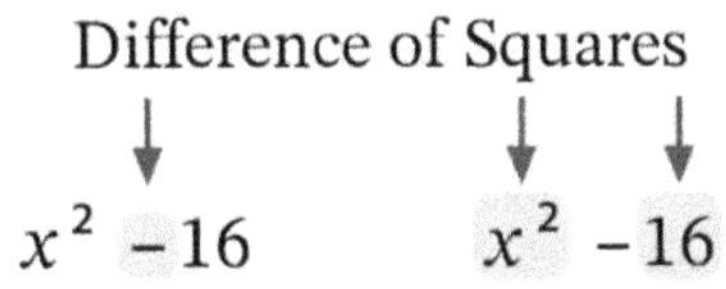

It is important to note that one sign will be addition and one sign will be subtraction when factoring. Having one of each sign is what allows the middle terms to cancel out. The expression $(a + b)(a - b)$ becomes $a^2 - ab + ab - b^2$ when FOILed, and the middle terms cancel out, leaving only $a^2 - b^2$.

To quickly identify whether a binomial can be written as the difference of two squares, two things need to be true:

1. Both terms are perfect squares.
2. The operation between the two terms is subtraction.

For example, the binomial $x^2 - 9$ follows both rules. x^2 and 9 are both perfect squares, and these terms are separated with the subtraction operation. Factor this binomial by following these steps:

1. Find the square root of both terms.

The square root of x^2 is x.

$$(x \quad)(x \quad)$$

The square root of 9 is 3.

$$(x \quad 3)(x \quad 3)$$

1. Remember to include addition and subtraction so that the middle terms cancel out when FOILing.

$$(x + 3)(x - 3)$$

Not all binomials represent the difference of two squares. The expression $x^2 + 36$ is a binomial with two perfect squares, but the terms are not separated by the subtraction operation. It cannot be factored using the difference of two squares method.

<u>Examples of binomials that can be written as the difference of two squares:</u>

$x^2 - 25$ becomes $(x + 5)(x - 5)$

$x^2 - 16y^2$ becomes $(x + 4y)(x - 4y)$

$36x^2 - 49$ becomes $(6x + 7)(6x - 7)$

> **Review Video: Factoring the Difference of Two Squares**
> Visit mometrix.com/academy and enter code: 128954

EXAMPLES

EXAMPLE 1:

Factor the expression.

$$x^2 - 25$$

Answer:

x^2 and 25 are both perfect squares, and the binomial $x^2 - 25$ shows the *difference* of these two squares (subtraction). This means that the binomial can be factored using the formula $a^2 - b^2 = (a + b)(a - b)$.

$$x^2 - 25 = (x)^2 - (5)^2$$
$$(x)^2 - (5)^2 = (x + 5)(x - 5)$$

The expression $(x + 5)(x - 5)$ is the factored version of the binomial $x^2 - 25$.

EXAMPLE 2:

Factor the expression.

$$x^2 - 36$$

Answer:

x^2 and 36 are both perfect squares, and the binomial $x^2 - 36$ shows the *difference* of these two squares (subtraction). This means that the binomial can be factored using the formula $a^2 - b^2 = (a + b)(a - b)$.

$$x^2 - 36 = (x)^2 - (6)^2$$
$$(x)^2 - (6)^2 = (x + 6)(x - 6)$$

The expression $(x + 6)(x - 6)$ is the factored version of the binomial $x^2 - 36$.

EXAMPLE 3:

Ben's house has a large rectangular window. The area of the window can be expressed as $x^2 - 36$. What are the dimensions of the window in terms of x?

Answer:

The area of a rectangle is calculated by multiplying the length and the width. The area of Ben's window is $x^2 - 36$. This means that ___ × ___ $= x^2 - 36$. To break apart $x^2 - 36$ into the product of two "side lengths," use the difference of two squares method.

The difference of two squares method states that $(a^2 - b^2) = (a + b)(a - b)$. In the case of $x^2 - 36$, identify the square root of x^2 and the square root of 36, and separate the squares with the symbols $-$ and $+$. $\sqrt{x^2} = x$ and $\sqrt{36} = 6$. The difference of two squares for $x^2 - 36$ becomes $(x +$

$6)(x - 6)$. This means that one dimension of the window is $(x + 6)$ and the other dimension is $(x - 6)$.

PRACTICE QUESTIONS

1. Determine whether the binomial $-25 + m^2$ can be expressed as a difference of perfect squares of variables and integers. If so, factor it into a product of linear binomials.

a. $(m - 5)^2$
b. $(m + 25)(m - 25)$
c. $(m + 5)(m - 5)$
d. The binomial is not the difference of perfect squares of variables and integers.

2. Determine whether the binomial $100 - x^2$ can be expressed as a difference of perfect squares of variables and integers. If so, factor it into a product of linear binomials.

a. $(10 - x)^2$
b. $(10 + x)(10 - x)$
c. $(25 + x)(4 - x)$
d. The binomial is not the difference of perfect squares of variables and integers.

3. Determine whether the binomial $x^2 + 16$ can be expressed as a difference of perfect squares of variables and integers. If so, factor it into a product of linear binomials.

a. $(x + 4)^2$
b. $(x + 16)^2$
c. $(x + 4)(x - 4)$
d. The binomial is not the difference of perfect squares of variables and integers.

4. Determine whether the binomial $t^2 - 49$ can be expressed as a difference of perfect squares of variables and integers. If so, factor it into a product of linear binomials.

a. $(t + 49)(t - 49)$
b. $(t + 7)(t - 7)$
c. $(t - 7)^2$
d. The binomial is not the difference of perfect squares of variables and integers.

5. Determine whether the binomial $x^2 - 1$ can be expressed as a difference of perfect squares of variables and integers. If so, factor it into a product of linear binomials.

a. $(x - 1)^2$
b. $(x + 1)^2$
c. $(x + 1)(x - 1)$
d. The binomial is not the difference of perfect squares of variables and integers.

6. Determine whether the binomial $x^2 - 6$ can be expressed as a difference of perfect squares of variables and integers. If so, factor it into a product of linear binomials.

a. $(x + 2)(x - 3)$
b. $(x + 6)(x - 6)$
c. $(x - 3)^2$
d. The binomial is not the difference of perfect squares of variables and integers.

7. Determine whether the binomial $25x^2 - 64$ can be expressed as a difference of perfect squares of variables and integers. If so, factor it into a product of linear binomials.

a. $(5x + 32)(5x - 32)$
b. $(5x - 8)^2$
c. $(5x + 8)(5x - 8)$
d. The binomial is not the difference of perfect squares of variables and integers.

8. Determine whether the binomial $x^2 - 36y^2$ can be expressed as a difference of perfect squares of variables and integers. If so, factor it into a product of linear binomials.

a. $(x + 6y)(x - 6y)$
b. $(x - 6y)^2$
c. $(x + 4y)(x - 9y)$
d. The binomial is not the difference of perfect squares of variables and integers.

9. Determine whether the binomial $4x^2 - 81$ can be expressed as a difference of perfect squares of variables and integers. If so, factor it into a product of linear binomials.

a. $(2x - 9)^2$
b. $(x + 3)(4x - 27)$
c. $(2x + 9)(2x - 9)$
d. The binomial is not the difference of perfect squares of variables and integers.

10. Determine whether the binomial $16p^2 - 25q^2$ can be expressed as a difference of perfect squares of variables and integers. If so, factor it into a product of linear binomials.

a. $(4p - 5q)^2$
b. $(4p + 5q)(4p - 5q)$
c. $(16p + q)(p - 25q)$
d. The binomial is not the difference of perfect squares of variables and integers.

11. Determine whether the binomial $9x^2 - 144$ can be expressed as a difference of perfect squares of variables and integers. If so, factor it into a product of linear binomials.

a. $(3x - 12)^2$
b. $(3x + 12)(3x - 12)$
c. $(x + 9)(9x - 16)$
d. The binomial is not the difference of perfect squares of variables and integers.

12. Determine whether the binomial $x^2 - 35y^2$ can be expressed as a difference of perfect squares of variables and integers. If so, factor it into a product of linear binomials.

a. $(x + 5y)(x - 7y)$
b. $(x - 5y)^2$
c. $(x + 35y)(x - 35y)$
d. The binomial is not the difference of perfect squares of variables and integers.

13. You have a collection of x^2 small square tiles all the same size, arranged in a large square. A rock falls on the square and breaks 9 of the tiles. You want to rearrange all the remaining tiles into a smaller square or rectangle. Write the number of undamaged tiles as a difference of perfect squares then factor it to get possible dimensions of the new square or rectangle.

a. $(x + 9)(x - 9)$
b. $(x - 3)(x - 3)$
c. $(x + 3)(x - 3)$
d. It is impossible to rearrange the remaining tiles into a square or rectangle.

14. You have a large, square recreation room, the floor of which is carpeted by $4x^2$ old square carpet tiles, each measuring 1 square foot. You decide to pull up these old tiles and to reuse them to replace the even older flooring in your rectangular den. When you finish, you have 100 of the old carpet tiles left over. Write the number of old carpet tiles used as a difference of perfect squares then factor it to get possible dimensions of the floor of the recreation room.

a. $(2x + 4)(2x - 25)$
b. $(4x + 100)(x - 1)$
c. $(2x - 10)(2x - 10)$
d. $(2x + 10)(2x - 10)$

15. A large baking sheet holds x^2 freshly baked chocolate chip cookies laid out in a square. After you eat 4 of them, you rearrange the remaining cookies in a smaller square or rectangle. Write the number of uneaten cookies as a difference of perfect squares then factor it to get possible dimensions of the new square or rectangle.

a. $(x + 4)(x - 4)$
b. $(x - 4)(x - 4)$
c. $(x + 2)(x - 2)$
d. $(x - 2)(x - 2)$

6.11 AC Method

A trinomial is a polynomial that has three terms. Trinomials often follow the form $ax^2 + bx + c$. It is common for trinomials to have an a-value of 1. For example, the trinomial $x^2 - 4x + 4$ has an a-value of 1. These kinds of trinomials can often be factored into the product of two binomials. For example, $x^2 - 4x + 4$ can be factored into $(x-2)(x-2)$.

More difficult trinomials will have values other than 1 for a. For example, the polynomial $15x^2 - 4x - 3$ has an a-value of 15. When factoring these trickier kinds of polynomials, we need to use the AC method. When using the AC method, we want to separate the middle term of the polynomial so we can factor out a common binomial from the expression.

The first step of the AC method is to multiply the a-value and the c-value together.

$$15 \cdot (-3) = -45$$

Now, we want to find two factors of –45 that combine to the value of b, which is –4.

$$\begin{aligned} 1 \cdot (-45) &= 45 \\ (-1) \cdot 45 &= 45 \\ 3 \cdot (-13) &= 45 \\ (-3) \cdot 13 &= 45 \\ 5 \cdot (-9) &= 45 \\ (-5) \cdot 9 &= 45 \end{aligned}$$

Those two factors are 5 and –9. Next, we rewrite the expression but replace the middle term with $5x - 9x$. This simplifies to the original $-4x$, so we know that we chose the correct factors.

$$15x^2 + 5x - 9x - 3$$

From here, pull out a common factor of $5x$ from the first two terms and a common factor of –3 from the last two terms.

$$5x(3x+1) - 3(3x+1)$$

Notice there is now a common binomial of $3x + 1$ in the first part of the expression and the second part of the expression. Factor this binomial out, which leaves the factor $5x - 3$. Therefore, the following expression is the factored form of the original expression $15x^2 - 4x - 3$.

$$(3x+1)(5x-3)$$

To check the answer, use the FOIL method to expand the binomial multiplication.

$$(3x+1)(5x-3) = 15x^2 - 9x + 5x - 3 = 15x^2 - 4x - 3$$

Since the FOIL method resulted in the original expression, the expression was factored correctly.

Here's another example of the AC method.

$$5x^2 - 28x - 12$$

First, multiply the a-term and the c-term.

$$5 \cdot (-12) = -60$$

Next, find two numbers that multiply to -60 and combine to -28. Those two factors are -30 and 2.

$$-30 \cdot 2 = -60$$
$$-30 + 2 = -28$$

Now, rewrite the original expression with the middle term replaced with either $-30x + 2x$ or $2x - 30x$. The order is not important, but you do need to make sure the signs are correct for each number.

$$5x^2 - 30x + 2x - 12$$

Next, pull out a common factor of $5x$ from the first two terms and a common factor of 2 from the last two terms.

$$5x(x - 6) + 2(x - 6)$$

Finally, factor out the common binomial $x - 6$ and write the expression in factored form.

$$(x - 6)(5x + 2)$$

Again, this expression czan always be multiplied to check that it is correct.

$$(x - 6)(5x + 2) = 5x^2 + 2x - 30x - 12 = 5x^2 - 28x - 12$$

Since this results in the original expression, the factored form is correct. The AC method can be used to factor any factorable quadratic equation.

EXAMPLES

EXAMPLE 1:

Factor the polynomial $5x^2 - x - 4$.

Answer:

First, multiply the a-value and the c-value together.

$$5 \cdot (-4) = -20$$

Then, find two numbers that multiply to -20 and combine to -1. These two numbers are -5 and 4. Rewrite the expression and replace the middle term with $-5x + 4x$.

$$5x^2 - 5x + 4x - 4$$

Now, factor a $5x$ from the first two terms of the expression and a 4 from the last two terms of the expression.

$$5x(x - 1) + 4(x - 1)$$

Finally, factor out the binomial $x - 1$ and rewrite the expression in factored form.

$$(x - 1)(5x + 4)$$

EXAMPLE 2:

Factor the polynomial $3x^2 + 5x + 2$.

Answer:

First, multiply the a- and c-terms together.

$$3 \cdot 2 = 6$$

Next, find two numbers that multiply to 6 and add to 5. These numbers are 3 and 2. Rewrite the middle term of the expression as $3x + 2x$.

$$3x^2 + 3x + 2x + 2$$

Then, factor out a $3x$ from the first two terms of the expression and a 2 from the last two terms of the expression.

$$3x(x + 1) + 2(x + 1)$$

Finally, factor out the $x + 1$ binomial.

$$(x + 1)(3x + 2)$$

Example 3:

Find the dimensions, in terms of x, of a rectangle that has the given area.

$$A = 6x^2 - 19x + 15$$

Answer

First, multiply the a-value and c-values together.

$$6 \cdot 15 = 90$$

Now, find two numbers that multiply to 90 and combine to −19. These two numbers are −9 and −10, so replace the middle term of the expression with $-9x - 10x$.

$$A = 6x^2 - 9x - 10x + 15$$

From here, factor a $3x$ from the first two terms and a −5 from the last two terms.

$$A = 3x(2x - 3) - 5(2x - 3)$$

Then, factor out the common binomial $2x - 3$.

$$A = (2x - 3)(3x - 5)$$

Therefore, the dimensions of a rectangle whose area is $6x^2 - 19x + 15$ are $2x - 3$ and $3x - 5$.

Example 4:

The garden in your backyard has an area of $9x^2 - 52x - 12$ square feet. Factor this trinomial into a product of binomials to get a possible way of writing the area as length times width.

Answer:

First, multiply the a-term and the c-term together.

$$9 \cdot (-12) = -108$$

Next, find two numbers that multiply to –108 and combine to –52. These numbers are –54 and 2. Rewrite the middle term of the expression as $-54x + 2x$.

$$9x^2 - 54x + 2x - 12$$

Then, factor out common factors of $9x$ from the first two terms of the expression and 2 from the last two terms of the expression.

$$9x(x - 6) + 2(x - 6)$$

Finally, factor out the common binomial $x - 6$.

$$(x - 6)(9x + 2)$$

Practice Questions

1. Factor the trinomial $6x^2 + 5x + 1$ into a product of binomials.

a. $(6x + 1)(x + 1)$
b. $(3x + 1)(x + 2)$
c. $(2x + 3)(x + 1)$
d. $(2x + 1)(3x + 1)$

2. Factor the trinomial $5x^2 + 36x + 7$ into a product of binomials.

a. $(x + 7)(5x + 1)$
b. $(7x + 1)(5x + 1)$
c. $(5x + 7)(x + 1)$
d. $(7x + 1)(x + 5)$

3. Factor the trinomial $4x^2 + 16x + 15$ into a product of binomials.

a. $(2x + 15)(2x + 1)$
b. $(4x + 15)(x + 1)$
c. $(4x + 3)(x + 5)$
d. $(2x + 3)(2x + 5)$

4. Factor the trinomial $10x^2 - 17x + 3$ into a product of binomials.

a. $(10x - 3)(x - 1)$
b. $(10x - 1)(x - 3)$
c. $(5x - 3)(2x - 1)$
d. $(2x - 3)(5x - 1)$

5. Factor the trinomial $7x^2 - 18x + 8$ into a product of binomials.

a. $(x - 2)(7x - 4)$
b. $(x - 4)(7x - 2)$
c. $(x - 1)(7x - 8)$
d. $(x - 8)(7x - 1)$

6. Factor the trinomial $10x^2 - 29x + 10$ into a product of binomials.

a. $(10x - 5)(x - 2)$
b. $(5x - 5)(2x - 2)$
c. $(2x - 5)(5x - 2)$
d. $(5x - 1)(2x - 10)$

7. Factor the trinomial $14x^2 - 5x - 1$ into a product of binomials.

a. $(2x - 1)(7x + 1)$
b. $(x - 1)(14x + 1)$
c. $(14x - 1)(x + 1)$
d. $(7x - 1)(2x + 1)$

8. Factor the trinomial $5x^2 - 54x - 11$ into a product of binomials.

a. $(5x - 11)(x + 1)$
b. $(x - 11)(5x + 1)$
c. $(5x - 1)(x + 11)$
d. $(x - 1)(5x + 11)$

9. Factor the trinomial $4x^2 + 8x - 21$ into a product of binomials.

a. $(x - 3)(4x + 7)$
b. $(2x - 7)(2x + 3)$
c. $(2x - 1)(2x + 21)$
d. $(2x - 3)(2x + 7)$

10. Factor the trinomial $30x^2 + 7x - 2$ into a product of binomials.

a. $(10x - 2)(3x + 1)$
b. $(15x - 2)(2x + 1)$
c. $(6x - 1)(5x + 2)$
d. $(2x - 1)(15x + 2)$

11. Factor the trinomial $6x^2 - 11x - 10$ into a product of binomials.

a. $(3x - 2)(2x + 5)$
b. $(3x + 2)(2x - 5)$
c. $(6x - 5)(x + 2)$
d. $(2x - 1)(3x + 10)$

12. Factor the trinomial $9x^2 - 12x - 5$ into a product of binomials.

a. $(3x + 1)(3x - 5)$
b. $(x - 1)(9x + 5)$
c. $(9x - 1)(x + 5)$
d. $(3x - 1)(3x + 5)$

13. Last year you constructed a square corn maze with an area of x^2 square meters. This year you have expanded the maze into a rectangle that is longer than it is wide and whose area is $5x^2 + 11x + 6$ square meters. Factor this trinomial into a product of binomials to get a possible way of writing the area as length times width.

a. $(5x + 2)(x + 3)$
b. $(5x + 3)(x + 2)$
c. $(5x + 6)(x + 1)$
d. $(5x + 1)(x + 6)$

14. You are building a rectangular pen out of fencing panels x feet long ($x > 1$). Each side of the pen consists of one or more panels plus a small gap to accommodate a gate, and the length of the pen exceeds its width. The total area of the pen is $42x^2 + 19x + 2$ square feet. Factor this trinomial into a product of binomials to find a possible way of writing the area as length times width.

a. $(7x + 1)(6x + 2)$
b. $(7x + 2)(6x + 1)$
c. $(14x + 1)(3x + 2)$
d. $(14x + 2)(3x + 1)$

15. You are replacing a square window with area x^2 square inches ($x \geq 4$) by a larger rectangular window with area $6x^2 + 5x - 21$ square inches. The base of the new window is larger than its height. Factor this trinomial into a product of binomials to get a possible way of writing its area as base times height.

a. $(3x + 3)(2x - 7)$
b. $(3x + 21)(2x - 1)$
c. $(3x + 7)(2x - 3)$
d. $(6x + 1)(x - 7)$

6.12 Multistep Factoring

The purpose of factoring polynomials is to write a polynomial as multiplication. Factoring requires you to look for a common factor, specifically the greatest common factor that all terms share. For example, consider the binomial $36x^4 + 20x^6$. Both terms consist of a coefficient and then the variable x, raised to a power. When determining the greatest common factor, determine the GCF of the coefficient and then the variable.

GCF of 36 and 20 is 4.

GCF of x^4 and x^6 is x^4.

Therefore, the GCF of $36x^4$ and $20x^6$ is $4x^4$. This can be factored out. When $4x^4$ is factored out, the result is $4x^4(9 + 5x^2)$. The factored version is equivalent to the original polynomial.

Many polynomials are factored into the product of two binomials. For example, $x^2 - 81$ is a binomial that can be factored into $(x + 9)(x - 9)$. Some polynomials cannot be factored until a monomial is factored out first.

For example, consider the trinomial $x^3 - 7x^2 + 10x$. At first glance, there does not appear to be a way to factor this into the product of two binomials. The polynomial is not a difference of two squares, and there does not seem to be any common factors within the coefficients. However, when we take a closer look, we see that all three terms do have something in common, an x.

When we factor this x out, we are left with $x(x^2 - 7x + 10)$. Take a closer look at $(x^2 - 7x + 10)$. This part can be factored even further.

The expression $(x^2 - 7x + 10)$ is equivalent to $(x - 2)(x - 5)$. We know this because when $(x - 2)(x - 5)$ is FOILed, the result is $x^2 - 5x - 2x + 10$, which simplifies to $x^2 - 7x + 10$. This means that $x(x^2 - 7x + 10)$ can be factored further to $x(x - 2)(x - 5)$.

Let's consider another tricky polynomial, such as $8x^5 - 16x^4 + 8x^3$. At this point, the trinomial can't be factored into a product of two binomials. However, we shouldn't give up quite yet. Start by factoring out a common monomial. In the trinomial $8x^5 - 16x^4 + 8x^3$, all three terms have a common factor, $8x^3$. Let's factor this monomial out of the polynomial.

$$8x^3(x^2 - 2x + 1)$$

Now look closely at $(x^2 - 2x + 1)$. The expression $(x^2 - 2x + 1)$ can be factored into the product of two binomials, $(x - 1)(x - 1)$. This means that the original polynomial $8x^5 - 16x^4 + 8x^3$ can be fully factored into $8x^3(x - 1)(x - 1)$, or in a more condensed form, $8x^3(x - 1)^2$.

Again, the thing to remember is that not all polynomials will break down and factor in a clear and obvious way. Sometimes multiple strategies need to be used, such as factoring out a monomial and then factoring what is remaining. The more you work with polynomials, the easier the factoring process will become!

Examples

Example 1:

Factor the expression.

$$x^3 - 7x^2 + 10x$$

Answer:

Each term in the trinomial $x^3 - 7x^2 + 10x$ has a common factor of x.

Factor out the x, and the result is $x(x^2 - 7x + 10)$.

Now factor the trinomial $(x^2 - 7x + 10)$ into $(x - 5)(x - 2)$.

$x(x^2 - 7x + 10)$ becomes $x(x - 5)(x - 2)$.

Example 2:

Factor the expression.

$$x^3 + 2x^2 - 15x$$

Answer:

Each term in the trinomial $x^3 + 2x^2 - 15x$ has a common factor of x. When x is factored out, the result is $x(x^2 + 2x - 15)$.

Now, factor the terms inside of the parentheses. For the first part of the polynomial, x^2 factors to $x \cdot x$. Then, find terms that multiply to -15 and will create middle terms that add to $2x$. Those terms are 5 and -3.

$x(x^2 + 2x - 15)$ becomes $x(x + 5)(x - 3)$.

Practice Questions

1. Factor the polynomial $-2x^5 - 24x^4 - 72x^3$ completely.

a. $x^3(2x - 6)^2$
b. $-2x^3(x + 6)^2$
c. $-2x^3(x + 4)(x + 9)$
d. $2x^5(x - 4)(x - 9)$

2. Factor the polynomial $4t^2 + 32t - 192$ completely.

a. $4(t + 4)(t - 12)$
b. $4(t + 12)(t - 4)$
c. $4(t + 8)(t - 6)$
d. $2(t + 24)(2t - 4)$

3. Factor the polynomial $20x^3 - 22x^2 + 6x$ completely.

a. $2x(2x - 1)(5x - 3)$
b. $2x(2x - 3)(5x - 1)$
c. $2x(10x - 3)(x - 1)$
d. $2x(10x - 1)(x - 3)$

4. Factor the polynomial $-21x^5 + 90x^4 - 24x^3$ completely.

a. $-3x^3(x-2)(7x-4)$
b. $-3x^3(x+2)(7x-4)$
c. $-3x^3(x+4)(7x-2)$
d. $-3x^3(x-4)(7x-2)$

5. Factor the polynomial $8x^6 - 96x^5 + 160x^4$ completely.

a. $4x^4(2x-2)(x-20)$
b. $8x^4(x-4)(x-5)$
c. $8x^4(x-2)(x-10)$
d. $4x^4(x-4)(2x-10)$

6. Factor the polynomial $40x^7 - 490x^5y^2$ completely.

a. $40x^5y^2(x+7)(x-7)$
b. $10x^5(2x-7y)^2$
c. $10x^5(2x+7y)(2x-7y)$
d. $10x^5(4x+y)(x-49y)$

7. Factor the polynomial $20x^{10} + 125x^9 - 105x^8$ completely.

a. $5x^8(2x-7)(2x+3)$
b. $5x^8(x+7)(4x-3)$
c. $5x^8(x-3)(4x+7)$
d. $5x^8(2x-3)(2x+7)$

8. Factor the polynomial $5x^3 + 45x^2 + 90x$ completely.

a. $x(5x+3)(x+30)$
b. $5x(x+1)(x+18)$
c. $5x(x+3)(x+6)$
d. $5x(x+2)(x+9)$

9. Factor the polynomial $90x^4 - 51x^3 - 6x^2$ completely.

a. $3x^2(6x+1)(5x-2)$
b. $3x^2(10x+1)(3x-2)$
c. $3x^2(5x+2)(6x-1)$
d. $6x^2(5x+1)(3x-1)$

10. Factor the polynomial $x^4 - 4x^3 - 45x^2$ completely.

a. $x^2(x+5)(x-9)$
b. $x^3(x+5)(x-9)$
c. $x^2(x+9)(x-5)$
d. $x^2(x+15)(x-3)$

11. Factor the polynomial $15x^4 + 36x^3 + 21x^2$ completely.

a. $3x^2(x + 7)(5x + 1)$
b. $3x^2(x + 1)(5x + 7)$
c. $x^2(3x + 1)(5x + 21)$
d. $x^2(x + 3)(15x + 7)$

12. Factor the polynomial $6x^2 - 486$ completely.

a. $3(x + 18)(2x - 9)$
b. $6(x - 9)^2$
c. $6(x + 9)(x - 9)$
d. $3(2x + 9)(x - 18)$

13. Your ice cream company used to sell ice cream in a cube-shaped package with volume x^3 cubic millimeters. To increase profits, you reduce the length and width of your package while leaving the height (and the price) the same. Your new package is a rectangular solid with a volume of $x^3 - 18x^2 + 72x$ cubic millimeters. Assume the length is greater than the width and you can find the dimensions of the new package by factoring the volume completely as a polynomial, rewrite the volume of the package in the form $V =$ length $\times$ width $\times$ height.

a. $V = (x - 4)(x - 18)x$ cubic millimeters
b. $V = (x - 2)(x - 36)x$ cubic millimeters
c. $V = (x - 6)(x - 12)x$ cubic millimeters
d. $V = (x - 8)(x - 9)x$ cubic millimeters

14. Your old storage tank for rainwater runoff consists of a cube-shaped hole in the ground with volume x^3 cubic meters. You expand it into a larger rectangular solid by increasing the length and width and doubling the depth, getting a new storage tank with volume $2x^3 + 58x^2 + 200x$ cubic meters. Assume the length is at least as large as the width and you can find the dimensions of the new tank by factoring the volume completely as a polynomial, rewrite the volume of the new tank in the form $V =$ length $\times$ width $\times$ depth.

a. $V = (x + 10)(x + 10)(2x)$ cubic meters
b. $V = (2x + 25)(x + 8)(x)$ cubic meters
c. $V = (x + 20)(x + 5)(2x)$ cubic meters
d. $V = (x + 25)(x + 4)(2x)$ cubic meters

15. A solar farm has a field entirely covered with solar cells. The field used to be square with an area of x^2 square feet, but recently the farm expanded the field into a larger rectangle. Each solar cell is the same size, and each square foot of the field holds the same number of cells. The total number of solar cells in the expanded field is $4x^2 + 96x + 512$. On this farm, as it happens, if you factor this polynomial completely, the factorization will express the total number of solar cells as (number of cells per square foot) $\times$ (length of the field in feet) $\times$ (width of the field in feet). Find this factorization.

a. $4(x + 16)(x + 8)$
b. $4(x + 4)(x + 32)$
c. $4(x + 2)(x + 32)$
d. $2(2x + 8)(x + 32)$

Chapter 6 Quiz

1. Factor completely: $6x^2 - 10x^4$.

a. $2x^2(3 - 5x^2)$
b. $x^2(6 - 10x^2)$
c. $2x(3x - 5x^3)$
d. $x(6x - 10x^3)$

2. The height of a triangle is three units longer than twice the square of its base, b. Which of the following represents the area of the triangle in terms of its base?

a. $b^3 + \frac{3}{2}b$
b. $b^2 + \frac{3}{2}b$
c. $\frac{5}{2}b^3 + \frac{7}{2}b$
d. $\frac{5}{2}b^3 + \frac{3}{2}b$

3. You are building a rectangular pen out of fencing panels. You want the pen to be twice as long as it is wide, so you are going to use x panels along the width and $2x$ panels along the length. Each panel is y meters long. Write an expression for the area inside the pen, measured in square meters, in terms of x and y.

a. $3x^2y^2$ square meters
b. $2x^2y^2$ square meters
c. $2xy^2$ square meters
d. $2x^2y$ square meters

4. You want to construct a rectangular pen using premade fence panels that are x feet long. The pen will be 10 panels long plus an extra 5 feet for a large gate to drive a yard tractor through. It will also be 8 panels wide plus an extra 2 feet for a small gate for people to enter and exit by. What will the area of the pen be in square feet?

a. $80x^2 + 10$ square feet
b. $80x^2 + 60x + 10$ square feet
c. $18x^2 + 25x + 7$ square feet
d. $80x^2 + 25x + 10$ square feet

5. Evaluate the product: $(x + 5)(x + 8)$.

a. $x^2 + 40x + 13$
b. $2x + 13$
c. $x^2 + 40$
d. $x^2 + 13x + 40$

6. Simplify: $(4x^2 - x + 6xy + 7) + (3xy - 5x^2 + 7x)$.

a. $-x^2 + 6x + 9xy + 7$
b. $9x^2 + 6x + 9xy + 7$
c. $4x^2 + 13x + 9xy$
d. $-x^2 + 14x + 9xy$

7. Completely factor $3x^3 + 18x^2 + 24x$.

a. $3(x + 4)(x + 6)$
b. $3x(x + 2)(x + 4)$
c. $(3x + 6)(x + 4)$
d. $(3x + 8)(x + 3)$

8. Simplify: $(13m^2 - 10n^2 + 3m - 8n) - (5m^2 + n^2 - 6m + 3)$.

a. $8m^2 - 9n^2 + 6m - 8n$
b. $8m^2 - 11n^2 - 8n$
c. $8m^2 - 11n^2 + 9m - 8n - 3$
d. $8m^2 - 9n^2 - 3m - 8n + 3$

9. The revenue, in dollars, generated by selling $n - 6$ units of a notebook binder can be represented by the expression $n^2 - 5n - 6$. If the revenue of the notebook binder can be calculated by the product of its price and the number of units sold, what is the price, in terms of n?

a. n
b. $2n$
c. $n - 1$
d. $n + 1$

10. The height of an object that is thrown from the top of a fifteen-story building can be modeled by the equation $h(t) = -16t^2 + 8t + 15$, where h is the height of the object in feet and t is time in seconds. Which of the following also represents the height of the object at time t?

a. $h(t) = -(4t - 5)(4t + 3)$
b. $h(t) = -(4t + 5)(4t - 3)$
c. $h(t) = -(8t - 5)(2t + 3)$
d. $h(t) = -(8t + 3)(4t - 5)$

11. Factor $x^2 + 11x + 24$.

a. $(x - 2)(x - 12)$
b. $(x - 3)(x - 8)$
c. $(x + 2)(x + 12)$
d. $(x + 3)(x + 8)$

12. Factor $x^2 - 25$.

a. $(x-5)^2$
b. $(x+5)^2$
c. $(x-5)(x+5)$
d. Cannot be factored

13. A rectangular box has a volume of $24x^2y + 18xy^2 + 2xy$ cubic meters. The area of the base of the box is $2xy$ meters. Find the height of the box if the volume of a rectangular solid can be found by multiplying the area of the base by the height.

a. 43 meters
b. $12x + 18xy^2 + 2xy$ meters
c. $48x^3y^2 + 36x^2y^3 + 4x^2y^2$ meters
d. $12x + 9y + 1$ meters

14. Divide $x^2 + x - 6$ by $x - 2$.

a. $x - 1$
b. $x + 2$
c. $x - 3$
d. $x + 3$

15. The revenue, r, generated by selling x number of laptop computers is $r(x) = 10x^3 + 16x^2 + 6x$. Which of the following is also a representation of the revenue generated by selling the laptops in terms of x?

a. $2x(x+3)(5x+1)$
b. $2x(x+1)(5x+3)$
c. $2(5x^2+3x)(x+1)$
d. $x(10x+2)(x+3)$

16. The profit that a department store makes for selling x number of printers can be represented by the expression $8x^3 - 12x^2 + 2x$. Which of the following is the complete factored form of the profit for selling x number of computers?

a. $2x(4x^2 - 6x + 1)$
b. $2x(4x^2 - 6x - 1)$
c. $4x^2(2x - 3)$
d. $2x(4x^2 - 1)$

17. The area of a rectangle is given by the expression $144x^2 - 225y^2$. Which of the following represents the dimensions of the rectangle in terms of x and y?

a. length = $(36x - 25y)$, width = $(4x - 9y)$
b. length = $(36x + 25y)$, width = $(4x - 9y)$
c. length = $(12x + 15y)$, width = $(12x - 15y)$
d. length = $(12x - 15y)$, width = $(12x - 15y)$

18. Simplify $\frac{3x^4y^2}{6x^3y^4}$.

a. $\frac{y^2}{2x}$
b. $\frac{x}{2y^2}$
c. $\frac{x^2}{2y^2}$
d. $\frac{x^7}{2y^6}$

19. The area of a rectangular garden is $x^2 + 10x - 600$ square feet. If the length of the garden is $x + 30$ feet, what is its width in terms of x?

a. $x - 60$
b. $x + 10$
c. $x - 20$
d. $x + 20$

20. Find the quotient:

$$\frac{6x^3 - 10x^2 + 16x}{2x}$$

a. $3x^2 - 10x + 16$
b. $3x^3 - 10x^2 + 16$
c. $3x^2 - 5x + 8$
d. $-7x^2 + 16x$

Chapter 7: Algebraic Fractions

7.1 Simplifying Algebraic Fractions

Simplifying algebraic fractions is similar to the process of simplifying regular fractions. For example, the fraction $\frac{50}{100}$ can be simplified to $\frac{1}{2}$. When simplifying a fraction, a factor needs to cancel out from the numerator and the denominator. $\frac{50}{100}$ can be written as $\frac{50\times1}{50\times2}$. We can see a common factor of 50 in the numerator and denominator, so they can be canceled out. So, $\frac{50\times1}{50\times2}$ becomes $\frac{1}{2}$. The same process applies when simplifying algebraic fractions.

Consider the fraction $\frac{5xy^3}{15x^3y^2}$. This fraction consists of constants, 5 and 15, as well as variables raised to a power. Do the constants have any factors in common? Yes. Both can be divided by 5. When the numerator and denominator are divided by 5, $\frac{5xy^3}{15x^3y^2}$ becomes $\frac{xy^3}{3x^3y^2}$. It can be helpful to think about the fraction in expanded form so that we can easily see what will cancel out.

The expression $\frac{xy^3}{3x^3y^2}$ in expanded form is $\frac{x\cdot y\cdot y\cdot y}{3\cdot x\cdot x\cdot x\cdot y\cdot y}$.

When x's and y's are canceled out from the numerator and denominator, we are left with $\frac{y}{3\cdot x\cdot x}$, or $\frac{y}{3x^2}$.

Some algebraic fractions need to be factored before they can be simplified. For example, the fraction $\frac{xy+y^2}{x^2-y^2}$ has a numerator and denominator that has addition and subtraction, instead of multiplication. So, we cannot divide, or "cancel" anything on top and bottom of the fraction. When the numerator is factored, $xy + y^2$ becomes $y(x + y)$. When the denominator is factored, $x^2 - y^2$ becomes $(x + y)(x - y)$.

$$\frac{xy + y^2}{x^2 - y^2} = \frac{y(x + y)}{(x + y)(x - y)}$$

Now it is more obvious that the numerator and denominator have something in common. There is an $(x + y)$ in both the numerator and denominator, so these can cancel out. The expression $\frac{y(x+y)}{(x+y)(x-y)}$ simplifies to $\frac{y}{x-y}$.

Another way to simplify an algebraic fraction is to start by factoring out a negative. Sometimes factoring out a negative allows us to see new ways to simplify a fraction. For example, let's consider the fraction $\frac{-x-y}{x+y}$. At first glance it does not look like the terms will cancel out because the signs do not match. However, when the negative is factored out of the numerator, the fraction $\frac{-x-y}{x+y}$ becomes $\frac{-1(x+y)}{x+y}$. Now that the negative is factored out, it is clear the $x + y$ terms will cancel out, leaving only -1.

Review Video: Simplifying Rational Polynomial Functions
Visit mometrix.com/academy and enter code: 351038

Review Video: Reducing Rational Expressions
Visit mometrix.com/academy and enter code: 788868

EXAMPLES

EXAMPLE 1:

Simplify the algebraic fraction $\frac{5x^3y}{10xy^2}$.

Answer:

The greatest common factor (GCF) of $5x^3y$ and $10xy^2$ is $5xy$. Divide the numerator and denominator by the GCF. Remember the exponent rule $\frac{x^m}{x^n} = x^{m-n}$.

$$\frac{5x^3y \div 5xy}{10xy^2 \div 5xy} = \frac{1x^{3-1}y^{1-1}}{2x^{1-1}y^{2-1}} = \frac{x^2}{2y}$$

It can be helpful to see this in expanded form.

$$\frac{5x^3y}{10xy^2} = \frac{5 \cdot x \cdot x \cdot x \cdot y}{5 \cdot 2 \cdot x \cdot y \cdot y} = \frac{x^2}{2y}$$

EXAMPLE 2:

$$\frac{x^2 + 6x}{9x}$$

Answer:

Factor the numerator and the denominator.

$$\frac{x^2 + 6x}{9x} = \frac{x(x + 6)}{x(9)}$$

Now we can cancel out the x's.

$$\frac{\cancel{x}(x + 6)}{\cancel{x}(9)} = \frac{x + 6}{9}$$

EXAMPLE 3:

$$\frac{-x - 2}{x + 2}$$

Answer:

Factor out –1 from the numerator.

$$\frac{-x - 2}{x + 2} = \frac{-1(x + 2)}{x + 2}$$

The factor $(x + 2)$ appears in the numerator and denominator. These two binomials cancel out.

$$\frac{-1\cancel{(x + 2)}}{\cancel{x + 2}} = -1$$

EXAMPLE 4:

Simplify the following algebraic fraction.

$$\frac{12x^3 - 12x^2 - 24x}{4x^3 - 24x^2 + 32x}$$

Answer:

When simplifying an algebraic fraction, first check to see if there's a greatest common factor (GCF) that can be factored out in the numerator and denominator. The terms in the numerator have a GCF of $12x$ that can be factored out, and the terms in the denominator have a GCF of $4x$ that can be factored out.

$$\frac{12x(x^2 - x - 2)}{4x(x^2 - 6x + 8)}$$

Next, factor the two quadratic expressions.

$$\frac{12x(x-2)(x+1)}{4x(x-2)(x-4)}$$

Finally, cancel out common factors from the numerator and denominator. There is a factor of 4 that can be canceled out, leaving a 3 in the numerator and a 1 in the denominator. The 1 doesn't need to be written since we are multiplying the terms in the denominator. There are also common factors of x and $x - 2$ that can be canceled out. This leaves the following simplified expression.

$$\frac{3(x+1)}{x-4}$$

EXAMPLE 5:

A circle has an area of $2x^3 + 8x^2$ and a circumference of $x^2 + x - 12$. Find the ratio of the circle's area to its circumference and simplify the ratio.

Answer:

First, create the ratio by creating a fraction with the area expression in the numerator and the circumference expression in the denominator.

$$\frac{\text{area}}{\text{circumference}} = \frac{2x^3 + 8x^2}{x^2 + x - 12}$$

Now, we need to simplify the ratio. For this ratio, start by factoring out the GCF in the numerator.

$$\frac{2x^2(x+4)}{x^2 + x - 12}$$

Next, factor the quadratic denominator.

$$\frac{2x^2(x+4)}{(x+4)(x-3)}$$

Finally, to simplify the ratio, cancel out the common factor of $x + 4$ from the numerator and denominator.

$$\frac{2x^2}{x - 3}$$

Therefore, the ratio of the circle's area to its circumference is $\frac{2x^2}{x-3}$.

Practice Questions

1. Simplify the following algebraic fraction.

$$\frac{6x^6}{30x^3}$$

a. $\frac{3x^3}{15}$
b. $\frac{x^2}{5}$
c. $\frac{3x^2}{15}$
d. $\frac{x^3}{5}$

2. Simplify the following algebraic fraction.

$$\frac{48x^4y^2}{12xy^5}$$

a. $\frac{4x^3}{y^3}$
b. $\frac{32}{5}$
c. $\frac{4x^4}{y^3}$
d. $\frac{4x^4}{y^5}$

3. Simplify the following algebraic fraction.

$$\frac{15x^3}{20x^{12}}$$

a. $\frac{1}{5x^9}$
b. $\frac{3}{4x^9}$
c. $\frac{9}{40}$
d. $\frac{3}{4x^4}$

4. Simplify the following algebraic fraction.

$$\frac{35x^2y^3}{14x^6y^6}$$

a. $\frac{21}{x^4y^3}$
b. $\frac{5}{12}$
c. $\frac{5}{2x^4y^3}$
d. $\frac{5}{2x^3y^2}$

5. Simplify the following algebraic fraction.

$$\frac{3x^3 + 12x^2}{5x^4 - 35x^3}$$

a. $-\frac{1}{2x}$
b. $\frac{3(x+4)}{5(x-7)}$
c. $-\frac{1}{2x^2}$
d. $\frac{3(x+4)}{5x(x-7)}$

6. Simplify the following algebraic fraction.

$$\frac{10x^4y + 40x^3y}{10x^2y^2 - 5xy^3}$$

a. $\frac{x^2(x+8)}{y(x-y)}$
b. $\frac{9x^2}{y-y^2}$
c. $\frac{6x^2}{y}$
d. $\frac{2x^2(x+4)}{y(2x-y)}$

7. Simplify the following algebraic fraction.

$$\frac{6a^5b - 18ab}{18a^2b^3 + 180a^2b^2}$$

a. $\frac{a^4-3}{3ab(b+10)}$
b. $\frac{a^3-1}{b(3b+10a)}$
c. $\frac{a^3-3}{3b(b+10)}$
d. $\frac{a^4-3}{12(b+10)}$

8. Simplify the following algebraic fraction.

$$\frac{x^2 + 3x}{3x + 9}$$

a. $\frac{x^2}{3}$

b. $\frac{x^2}{9}$

c. $\frac{x^2+3}{12}$

d. $\frac{x}{3}$

9. Simplify the following algebraic fraction.

$$\frac{2x^2 - 50}{6x^2 + 60x + 150}$$

a. $\frac{(x-5)^2}{3(x+5)^2}$

b. $\frac{-1}{3(10x+1)}$

c. $\frac{x-5}{3(x+5)}$

d. $\frac{2x}{3(x+5)}$

10. Simplify the following algebraic fraction.

$$\frac{16x^3 + 144x^2 + 320x}{20x^3 + 20x^2 - 240x}$$

a. $\frac{4(x+10)}{5(x-6)}$

b. $\frac{4(x+4)(x+5)}{5(x+3)(x-4)}$

c. $\frac{44}{7}$

d. $\frac{4(x+5)}{5(x-3)}$

11. Simplify the following algebraic fraction.

$$\frac{4x^3 - 20x^2}{30x - 6x^2}$$

a. $-\frac{2x}{3}$

b. $\frac{2x}{3}$

c. $\frac{(x^2-5)}{6}$

d. $\frac{2x(x-5)}{3(5-x)}$

12. Simplify the following algebraic fraction.

$$\frac{42x^2y^2 - 21x^3y}{14x^2y^3 - 28xy^4}$$

a. $\frac{3x(2y-x)}{2y^2(x-2y)}$
b. $-\frac{3x}{2y^2}$
c. $\frac{3(1+x)(1-x)}{y(1+2y)(1-2y)}$
d. $\frac{3x}{2y^2}$

13. A rectangular solid has a volume of $2x^3 + 20x^2 + 48x$, and its base has an area of $2x^2 + 8x$. Find the ratio of the solid's base area to its volume and simplify the ratio.

a. $\frac{1}{x+6}$
b. $\frac{2}{20x^2+x+6}$
c. $\frac{x+4}{(x+2)(x+12)}$
d. $\frac{x+8}{(x+3)(x+16)}$

14. A circle with radius r has a circumference of $2\pi r$ and an area of πr^2. Find the ratio of its circumference to its area and simplify the ratio as much as possible.

a. $\frac{2\pi}{r}$
b. $\frac{2}{r}$
c. $\frac{2}{\pi r}$
d. $\frac{1}{2}$

15. A pyramid has a volume of $2x^4 + 12x^3 + 16x^2$, and its base has an area of $6x^3 + 24x^2$. Find the ratio of the pyramid's base area to its volume and simplify the ratio.

a. $\frac{3(x+24)}{(x+2)(x+4)}$
b. $\frac{5}{x(17x+1)}$
c. $\frac{3}{x+2}$
d. $\frac{3(x+4)}{(x+1)(x+8)}$

7.2 Solving Proportions Containing Algebraic Fractions

A **ratio** shows the relationship between two quantities. A **proportion** is a mathematical comparison between two ratios. For example, $\frac{3}{4} = \frac{6}{8}$ shows that $\frac{3}{4}$ is proportional to $\frac{6}{8}$.

Ratios can be expressed in three ways: word form, fraction form, or with a colon. For example, the ratio "one out of three", $\frac{1}{3}$, and $1 : 3$ all represent the same ratio.

In algebra, we are often asked to solve proportions using the process of cross multiplication. Cross multiplication is a quick and efficient way to identify missing pieces within a proportion. For example, consider the proportion $\frac{4}{1} = \frac{12}{x}$. We can calculate the value of x by using cross multiplication. Multiply the numerator of the first fraction by the denominator of the second fraction. Then multiply the denominator of the first fraction by the numerator of the second fraction. Set the two values equal to each other and then solve for the missing variable.

$$\frac{4}{1} = \frac{12}{x}$$

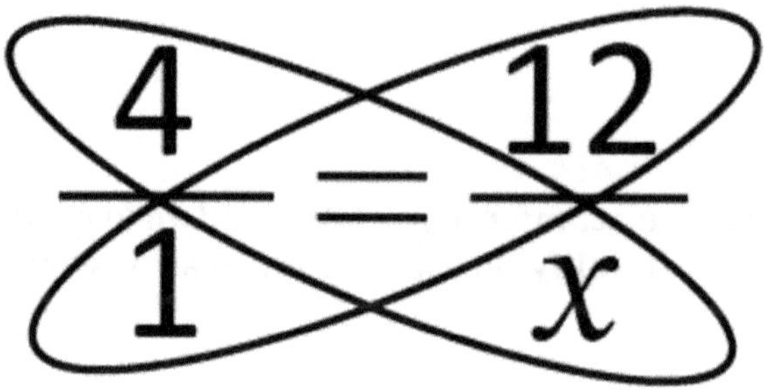

$$4x = 12$$

Now solve for x using inverse operations. Divide both sides by 4.

$$x = 3$$

The same process works for more complex algebraic proportions. For example, let's consider the proportion $\frac{2}{1} = \frac{2-6x}{3x}$. We have two ratios that are set equal to each other in a proportion, which means we can use cross multiplication to solve for the variable x. Multiply the left numerator by the right denominator, and multiply the left denominator by the right numerator.

$$2(3x) = 1(2 - 6x)$$
$$6x = -6x + 2$$

Solve for x. Add $6x$ to both sides.

$$12x = 2$$

Divide both sides by 12.

$$x = \frac{1}{6}$$

Some proportions will have variables on both sides. For example, let's consider the proportion $\frac{x+1}{5} = \frac{x-3}{3}$. In order to solve the proportion, we need to isolate the variable x. This can be achieved with cross multiplication.

$$3(x+1) = 5(x-3)$$

Now, multiply using the distributive property.

$$3x + 3 = 5x - 15$$

Now use inverse operations and combine like terms.

$$-2x + 3 = -15$$
$$-2x = -18$$

Divide both sides by –2.

$$x = 9$$

Review Video: Solving Proportions Containing Algebraic Fractions
Visit mometrix.com/academy and enter code: ???

Examples

Example 1:

Solve for x:

$$\frac{1}{x+4} = \frac{3}{x-2}$$

Answer:

Use cross multiplication to solve for x.

Multiply the numerator of the first fraction by the denominator of the second fraction. Multiply the denominator of the first fraction by the numerator of the second fraction. Set these values equal to each other, and then solve for x using inverse operations.

$$\frac{1}{x+4} = \frac{3}{x-2}$$
$$1(x-2) = 3(x+4)$$
$$x - 2 = 3x + 12$$
$$-2 = 2x + 12$$
$$-14 = 2x$$
$$-7 = x$$

Example 2:

Solve for x:

$$\frac{x}{72-x} = \frac{2}{7}$$

Answer:

Solve for x using cross multiplication. Multiply the numerator of the first fraction by the denominator of the second fraction. Multiply the denominator of the first fraction by the numerator of the second fraction. Set these values equal to each other, and then solve for x using inverse operations.

$$\begin{aligned}\frac{x}{72-x} &= \frac{2}{7} \\ x(7) &= 2(72-x) \\ 7x &= 144-2x \\ 9x &= 144 \\ x &= 16\end{aligned}$$

EXAMPLE 3:

Solve the following proportion:

$$\frac{x+5}{x-10} = \frac{x+2}{2x-4}$$

Answer:

Cross multiply, and then solve the quadratic equation.

$$(x+5)(2x-4) = (x-10)(x+2)$$

Multiply the binomials on both sides of the equation using the FOIL method.

$$\begin{aligned}2x^2 - 4x + 10x - 20 &= x^2 + 2x - 10x - 20 \\ 2x^2 + 6x - 20 &= x^2 - 8x - 20\end{aligned}$$

Move all terms to the left side of the equation.

$$x^2 + 14x = 0$$

Factor the left side of the equation.

$$x(x+14) = 0$$

Split apart the equation and solve both parts for x.

$$x = 0 \text{ and } x + 14 = 0$$

$$x = 0 \text{ and } x = -14$$

EXAMPLE 4:

A movie theater has x adults and x children in a room. After 4 adults and 6 children leave, the ratio of adults to children is $\frac{11}{9}$. Set up and solve a proportion to find the value of x.

Answer:

After 4 adults and 6 children leave, there are $x-4$ adults in the room and $x-6$ children in the room. The new ratio is equal to $\frac{11}{9}$. The following proportion represents this scenario.

$$\frac{x-4}{x-6}=\frac{11}{9}$$

Solve this proportion for x by cross multiplying.

$$9(x-4)=11(x-6)$$

Distribute on both sides of the equation.

$$9x-36=11x-66$$

Subtract $9x$ from both sides.

$$-36=2x-66$$

Add 66 to both sides of the equation.

$$30=2x$$

Finally, divide both sides by 2.

$$15=x$$

Practice Questions

1. Solve the following proportion.

$$\frac{6}{x+6}=\frac{2}{3}$$

a. $x=-3$
b. $x=15$
c. $x=3$
d. $x=-\frac{9}{2}$

2. Solve the following proportion.

$$\frac{1}{2}=\frac{5}{7-5x}$$

a. $x=5$
b. $x=-\frac{3}{5}$
c. $x=1$
d. $x=\frac{3}{5}$

3. Solve the following proportion.

$$\frac{7}{3x-5} = \frac{4}{x+5}$$

a. $x = \frac{1}{11}$
b. $x = 2$
c. $x = 3$
d. $x = 11$

4. Solve the following proportion.

$$\frac{3}{x+2} = \frac{5}{x-10}$$

a. $x = -20$
b. $x = 20$
c. $x = -5$
d. $x = -10$

5. Solve the following proportion.

$$\frac{6-9x}{3-4x} = \frac{3}{2}$$

a. $x = -\frac{1}{2}$
b. $x = \frac{1}{2}$
c. $x = -\frac{1}{10}$
d. $x = 2$

6. Solve the following proportion.

$$\frac{2x+3}{2} = \frac{x+3}{3}$$

a. $x = -\frac{3}{4}$
b. $x = \frac{3}{4}$
c. $x = -\frac{15}{4}$
d. $x = -\frac{4}{3}$

7. Solve the following proportion.

$$\frac{1}{x+2} = \frac{x+3}{x^2+16}$$

a. $x = -3$
b. $x = 2$
c. $x = \frac{22}{5}$
d. $x = 5$

8. Solve the following proportion.

$$\frac{x+2}{x+6} = \frac{x-6}{x-4}$$

a. $x = \frac{1}{14}$
b. $x = -14$
c. $x = 22$
d. $x = 14$

9. Solve the following proportion.

$$\frac{2x-4}{x-3} = \frac{x+7}{x+5}$$

a. $x = -1$
b. $x = 1$
c. $x = 4$
d. $x = -2$

10. Solve the following proportion.

$$\frac{x+6}{2x+6} = \frac{x-2}{x+10}$$

a. $x = -8$ or $x = 9$
b. $x = -3$ or $x = 24$
c. $x = -6$ or $x = 12$
d. $x = -4$ or $x = 18$

11. Solve the following proportion.

$$\frac{x+3}{x+8} = \frac{x+5}{2x+4}$$

a. $x = -14$ or $x = 2$
b. $x = -4$ or $x = 7$
c. $x = -2$ or $x = 14$
d. $x = -7$ or $x = 4$

12. Solve the following proportion.

$$\frac{-x-16}{x+4} = \frac{x-8}{x+2}$$

a. $x = -8$ or $x = 16$
b. $x = 2$ or $x = 4$
c. $x = 0$ or $x = 7$
d. $x = -7$ or $x = 0$

13. A pitcher of lemonade is diluted with 2 cups of water to get 5 cups of lemonade. You want to dilute the mixture with additional water until water makes up $\frac{7}{13}$ of the lemonade. Let x be the number of cups of water you need to add. Set up and solve a proportion of algebraic fractions to find x.

a. $3\frac{1}{2}$ cups

b. $\frac{1}{2}$ cup

c. $1\frac{1}{2}$ cups

d. $2\frac{1}{2}$ cups

14. A school assembly has x boys and x girls in the audience. After 4 boys and 8 girls leave to go on a field trip, the ratio of boys to girls becomes $\frac{16}{15}$. Set up and solve a proportion of algebraic fractions to find x.

a. $x = 17$

b. $x = 56$

c. $x = 68$

d. $x = 46$

15. You are trying to mix up an oil-and-vinegar salad dressing. You begin by mixing 17 cups of olive oil with 4 cups of vinegar, but the dressing is not tart enough. To fix this, you let the dressing separate, pour off x cups of olive oil from the top, and then add x cups of vinegar to replace it. In your final mixture, which tastes perfect, the ratio of oil to vinegar is $\frac{5}{2}$. Set up and solve a proportion of algebraic fractions to find x.

a. $3\frac{1}{2}$ cups

b. 2 cups

c. 3 cups

d. $2\frac{1}{2}$ cups

7.3 Multiplying Algebraic Fractions

In general, a three-step process can be used to multiply fractions.

1. Simplify, if possible.
2. Factor, if possible.
3. Multiply the numerators.
4. Multiply the denominators.

Simplifying before multiplying is a convenient strategy because it creates smaller numbers that are easier to work with. Simplifying first saves time in the long run. For example, when multiplying the fractions $\frac{3}{5} \cdot \frac{4}{15}$, we can simplify before multiplying. Notice that the 3 and the 15 are on opposite sides of the fraction bar, and they have a common factor. Their common factor is 3, so this can be divided out of both values. $\frac{3}{5} \cdot \frac{4}{15}$ becomes $\frac{1}{5} \cdot \frac{4}{5}$. The fractions cannot be simplified further, so the multiplication process can begin. Multiply the numerators and multiply the denominators. $\frac{1}{5} \cdot \frac{4}{5}$ becomes $\frac{4}{25}$. The same process applies to multiplying algebraic fractions that have variables and exponents.

For example, when multiplying algebraic fractions such as $\frac{4x}{3} \cdot \frac{x^2}{4}$, start by simplifying, if possible. First, look for any factors that appear on opposite sides of the fraction bar that can cancel out. In this case, a 4 can be canceled out. Divide the numerator of the first fraction and the denominator of the second fraction by 4.

$$\frac{x}{3} \cdot \frac{x^2}{1}$$

Now we can begin multiplying. Multiply numerator by numerator and denominator by denominator.

$$\frac{x^3}{3}$$

Remember, when multiplying exponents, add the exponents when the bases are the same.

Let's take a look at a more complex example that highlights the importance of simplifying first.

$$\frac{4x+8}{x^2-25} \cdot \frac{x-5}{5x+10}$$

At first glance, multiplying these algebraic fractions looks quite daunting! However, simplifying first will create a more manageable problem.

If there are no obvious factors that cancel out, try factoring the binomials.

$\frac{4x+8}{x^2-25} \cdot \frac{x-5}{5x+10}$	Factor the binomials.
$\frac{4(x+2)}{(x+5)(x-5)} \cdot \frac{x-5}{5(x+2)}$	Now it becomes more obvious.

Now that the fractions are all factored, there are obvious pieces that will cancel out. Simplify the fractions by canceling out terms that appear on opposite sides of the fraction bar.

$$\frac{\cancel{4(x+2)}}{(x+5)\cancel{(x-5)}} \cdot \frac{\cancel{x-5}}{5\cancel{(x+2)}} = \frac{4}{x+5} \cdot \frac{1}{5}$$

Now that the fractions are simplified, multiply the numerators and denominators.

$$\frac{4}{x+5} \cdot \frac{1}{5} = \frac{4}{5(x+5)} = \frac{4}{5x+25}$$

Review Video: Multiplying Algebraic Fractions
Visit mometrix.com/academy and enter code: ???

Examples

Example 1:

$$\frac{6x^2}{5y} \cdot \frac{10y^2}{3x}$$

Answer:

If possible, simplify first.

$$\frac{6x^2}{5y} \cdot \frac{10y^2}{3x}$$

Simplify the coefficients and the variables.

$$\frac{2x}{1} \cdot \frac{2y}{1}$$

Now, multiply straight across.

$$4xy$$

Example 2:

$$\frac{4x-20}{10} \cdot \frac{5x}{x-5}$$

Answer:

Factor, if possible.

$$\frac{4(x-5)}{10} \cdot \frac{5x}{x-5}$$

Simplify before multiplying by canceling out the binomial $(x-5)$.

$$\frac{4}{10} \cdot \frac{5x}{1}$$

Cancel out a factor of 5.

$$\frac{4}{2} \cdot \frac{x}{1}$$

Cancel out a factor of 2.

$$\frac{2}{1} \cdot \frac{x}{1}$$

Multiply straight across.

$$2x$$

EXAMPLE 3:

$$\frac{x-2}{4} \cdot \frac{12}{2x-4}$$

Answer:

Factor, if possible.

$$\frac{x-2}{4} \cdot \frac{12}{2(x-2)}$$

Now, simplify before multiplying. The binomial $(x - 2)$ cancels out in the numerator and the denominator.

$$\frac{1}{4} \cdot \frac{12}{2}$$

Cancel out a common factor of 4.

$$\frac{1}{1} \cdot \frac{3}{2}$$

Now, multiply straight across.

$$\frac{3}{2}$$

EXAMPLE 4:

Calculate the following product.

$$\frac{7x}{x^2 + 7x - 60} \cdot \frac{-4x^2 + 28x - 40}{35x^2 - 77x}$$

Answer:

When multiplying complicated algebraic fractions like these, it's often best to simplify before multiplying through. Start by pulling out any common factors in the numerators and denominators of the fractions.

$$\frac{7x}{x^2 + 7x - 60} \cdot \frac{-4(x^2 - 7x + 10)}{7x(5x - 11)}$$

Notice that a negative common factor was pulled out of the right fraction's numerator since the value in front of x^2 was negative. Next, factor the quadratic expressions.

$$\frac{7x}{(x - 5)(x + 12)} \cdot \frac{-4(x - 5)(x - 2)}{7x(5x - 11)}$$

Now, cancel common factors from the fractions.

$$\frac{\cancel{7x}}{\cancel{(x - 5)}(x + 12)} \cdot \frac{-4\cancel{(x - 5)}(x - 2)}{\cancel{7x}(5x - 11)}$$

$$\frac{1}{x + 12} \cdot \frac{-4(x - 2)}{5x - 11}$$

Then, distribute to simplify the right numerator.

$$\frac{1}{x + 12} \cdot \frac{-4x + 8}{5x - 11}$$

Finally, multiply the fractions.

$$\frac{-4x + 8}{5x^2 + 49x - 132}$$

EXAMPLE 5:

Calculate the following product. Simplify your answer as much as possible.

$$\frac{x^3 + 3x^2 - 28x}{x^2 + x - 6} \cdot \frac{x^2 + 4x - 12}{x^3 - x^2 - 56x}$$

Answer:

First, pull out any factors common to all three terms in either of the numerators or denominators.

$$\frac{x(x^2 + 3x - 28)}{x^2 + x - 6} \cdot \frac{x^2 + 4x - 12}{x(x^2 - x - 56)}$$

Then, factor all the trinomials as a product of two binomials.

$$\frac{x(x + 7)(x - 4)}{(x + 3)(x - 2)} \cdot \frac{(x - 2)(x + 6)}{x(x + 7)(x - 8)}$$

Now, cancel out the common factors of x and $x + 7$ from the left numerator and the right denominator and a common factor of $x - 2$ from the left denominator and right numerator.

$$\frac{x-4}{x+3} \cdot \frac{x+6}{x-8}$$

Finally, multiply the fractions together.

$$\frac{x^2 + 2x - 24}{x^2 - 5x - 24}$$

EXAMPLE 6:

A store is having a two-day sale where you receive a discount if you show a coupon at checkout. A person can use the coupon once per day during the sale. The first day, $\frac{7x-21}{4x+24}$ of the people who received the coupon used it. The second day, $\frac{16}{12x-36}$ of the people who used the coupon the first day used it again. What fraction of people who received the coupon used it both days? Write your answer as a reduced algebraic fraction.

Answer:

To find this fraction, multiply the fraction of people who used the coupon the first day by the fraction of people who used it again the second day.

$$\frac{7x-21}{4x+24} \cdot \frac{16}{12x-36}$$

Factor the numerators and denominators where possible.

$$\frac{7(x-3)}{4(x+6)} \cdot \frac{16}{12(x-3)}$$

Cancel common factors.

$$\frac{7 \cdot 1}{1(x+6)} \cdot \frac{4}{12 \cdot 1}$$

$$\frac{7 \cdot 1}{1(x+6)} \cdot \frac{1}{3 \cdot 1}$$

Simplify the fractions.

$$\frac{7}{x+6} \cdot \frac{1}{3}$$

Multiply the fractions.

$$\frac{7}{3(x+6)} = \frac{7}{3x+18}$$

Therefore, $\frac{7}{3x+18}$ of the people who received the coupon used it on both days.

PRACTICE QUESTIONS

1. Calculate the following product. Simplify your answer.

$$\frac{x+5}{x+10} \cdot \frac{x+5}{x-10}$$

a. $\frac{x^2+10x+25}{x^2-100}$

b. $\frac{x^2+10x+25}{x^2-20x-100}$

c. $\frac{x^2+25}{x^2-100}$

d. $\frac{x^2+25}{x^2-20x-100}$

2. Calculate the following product. Simplify your answer.

$$\frac{3}{5x} \cdot \frac{10x^2+15x}{6x+3}$$

a. $\frac{x+3}{x+1}$

b. $\frac{17x}{2x+3}$

c. $\frac{2x+3}{2x+1}$

d. 2

3. Calculate the following product. Simplify your answer.

$$\frac{3x}{4x-8} \cdot \frac{14x-28}{12x+15}$$

a. $\frac{7x^2-14x}{8x^2-6x-20}$

b. $\frac{42x^2-84x}{48x^2-36x-120}$

c. $\frac{7x-8}{-8x-10}$

d. $\frac{7x}{8x+10}$

4. Calculate the following product. Simplify your answer.

$$\frac{4x+1}{30} \cdot \frac{20x-30}{2x+11}$$

a. $\frac{80x^2+20x}{2x+11}$

b. $\frac{40x^2-130x-30}{30x+330}$

c. $\frac{8x^2-10x-3}{6x+33}$

d. $\frac{4x^2-4x-3}{6x+33}$

5. Calculate the following product. Simplify your answer.

$$\frac{20-5x}{10x+11}\cdot\frac{10}{7x-28}$$

a. $\frac{190}{91x-234}$
b. $\frac{-50x+20}{7x^2+49x-308}$
c. $\frac{-50}{70x+77}$
d. $\frac{-5x+20}{7x^2+49x-308}$

6. Calculate the following product. Simplify your answer.

$$\frac{9x+6}{8x-40}\cdot\frac{4x-20}{15x+10}$$

a. $\frac{9x+6}{30x+20}$
b. $\frac{3}{10}$
c. $\frac{9x^2+39x-30}{30x^2-130x-100}$
d. $\frac{18x^2-3x-3}{60x^2-146x-10}$

7. Calculate the following product. Simplify your answer.

$$\frac{7x+3}{-2x^2+16x-30}\cdot\frac{15-3x}{4}$$

a. $\frac{21x+9}{8x-24}$
b. $\frac{9+21x}{24-8x}$
c. $\frac{-21x^2+96x+45}{-8x^2+64x-120}$
d. $\frac{21x^2+3x+3}{8x^2+2x-8}$

8. Calculate the following product. Simplify your answer.

$$\frac{2x^2+20x+18}{x^2-12x+11}\cdot\frac{x-11}{4x+36}$$

a. $\frac{x+1}{2x-2}$
b. $\frac{x^2-10x-11}{2x^2-24x+22}$
c. $\frac{x^2+10x+9}{2x^2+16x-18}$
d. $\frac{2x+2}{4x-4}$

9. Calculate the following product. Simplify your answer.

$$\frac{9x}{5x^2+10x-40} \cdot \frac{-2x^2-4x+16}{6x^2-39x}$$

a. $\frac{-6}{10x-65}$

b. $\frac{3x}{2x-13}$

c. $\frac{-6x-24}{10x^2-25x-260}$

d. $\frac{12-6x}{10x^2-85x+130}$

10. Calculate the following product. Simplify your answer.

$$\frac{12x^2+36x}{x^2+7x+10} \cdot \frac{x^2+9x+20}{8x^2+24x}$$

a. $\frac{3x^2+21x+36}{2x^2+10x+12}$

b. $\frac{3x+12}{2x+4}$

c. $\frac{3x+3}{2x+1}$

d. $\frac{3x^2+27x+60}{2x^2+19x+20}$

11. Calculate the following product. Simplify your answer.

$$\frac{x^2+6x-7}{x^2-x-20} \cdot \frac{x^2-2x-15}{x^2+5x-14}$$

a. $\frac{x^2+2x-3}{x^2+2x-8}$

b. $\frac{x^2+2x-3}{x^2-6x+8}$

c. $\frac{x^2-4x+3}{x^2+2x-8}$

d. $\frac{x^2-4x+3}{x^2-6x+8}$

12. Calculate the following product. Simplify your answer.

$$\frac{x^3-7x^2-8x}{x^2-8x+12} \cdot \frac{x^2-3x+2}{x^3-2x^2-3x}$$

a. $\frac{x^3-9x^2+8x}{x^3-9x^2+18x}$

b. $\frac{x^2-x+4}{x^2-x+9}$

c. $\frac{x^2-9x+8}{x^2-9x+18}$

d. $\frac{x^2-5x+4}{x^2-6x+9}$

13. On a particular day, a clothing manufacturer produces a large number of socks. The socks are inspected for quality, and the fraction with noticeable flaws is $\frac{x-4}{3x^2+6x}$. Of these socks with flaws, the fraction good enough to be sold at a discount as "factory seconds" is $\frac{3x}{x^2-16}$, with the rest being discarded. What fraction of the day's production will be sold as factory seconds? Write your answer as a reduced algebraic fraction.

a. $\frac{1}{x^2+6x+8}$
b. $\frac{12}{9x^2-144x}$
c. $\frac{3x^2-12x}{4x^2+6x-16}$
d. $\frac{6x}{x^2-16}$

14. One day on the stock market, the fraction of stocks that increased in price was $\frac{9x+36}{16x+32}$. Of these, the fraction that increased in price again the next day was $\frac{8}{5x+20}$. What fraction of stocks increased on both days? Write your answer as a reduced algebraic fraction.

a. $\frac{72x+288}{80x^2+480x+640}$
b. $\frac{6}{5x+20}$
c. $\frac{9}{10x+20}$
d. $\frac{72x+288}{80x^2+640}$

15. In a clinical trial of vaccines, the fraction of volunteers receiving Vaccine A was $\frac{5x}{4x^2+20x+16}$. Of these, the fraction developing therapeutic levels of antibodies after three weeks was $\frac{2x+8}{3x^2+6x}$. What fraction of the original group of volunteers both received Vaccine A and developed therapeutic levels of antibodies after three weeks? Write your answer as a reduced algebraic fraction.

a. $\frac{5}{24x^2+12}$
b. $\frac{5}{6x^2+12}$
c. $\frac{5}{6x^2+18x+12}$
d. $\frac{5}{27x^2+18x}$

7.4 Dividing Algebraic Fractions

When dividing fractions, we multiply the first fraction by the reciprocal of the second fraction. This is often referred to as "flip and multiply." For example, $\frac{1}{5} \div \frac{3}{4}$ is solved by "flipping" (finding the reciprocal of) the second fraction, and then multiplying the two fractions. For example, $\frac{1}{5} \div \frac{3}{4}$ becomes $\frac{1}{5} \cdot \frac{4}{3}$. Now, multiply numerator by numerator and denominator by denominator to get $\frac{1}{5} \cdot \frac{4}{3} = \frac{4}{15}$.

The same process works when dividing algebraic fractions.

For example, consider the following: $\frac{x-3}{4} \div \frac{x}{8}$.

This problem can be solved by multiplying $\frac{x-3}{4}$ by the reciprocal of $\frac{x}{8}$.

$$\frac{x-3}{4} \cdot \frac{8}{x}$$

Before we multiply, remember to look for ways to simplify first. Remember, simplifying before multiplying is a great way to save time and eliminate extra work in the long run. $\frac{x-3}{4} \cdot \frac{8}{x}$ can be simplified because the denominator of the first fraction and the numerator of the second fraction can both be divided by 4.

$$\frac{x-3}{1} \cdot \frac{2}{x}$$

Now, multiply numerator by numerator and denominator by denominator.

$$\frac{2x-6}{x}$$

Some division problems will result in binomials being canceled out. For example, consider the following problem.

$$\frac{5(x-2)}{x+3} \div \frac{x-2}{2}$$

Rewrite the problem by flipping the second fraction and changing the division operation to multiplication.

$$\frac{5(x-2)}{x+3} \cdot \frac{2}{x-2}$$

Before multiplying, check to see if anything cancels out.

In this case, a binomial cancels out. The binomial $(x-2)$ appears on both sides of the fraction bar, so cancel it out.

$$\frac{5}{x+3} \cdot \frac{2}{1}$$

Now, multiply straight across.

$$\frac{10}{x+3}$$

The important thing to remember when dividing algebraic fractions is to multiply by the reciprocal of the second fraction ("flip and multiply"). Don't forget to simplify before you multiply in order to save time and keep the numbers small and easy to work with.

Review Video: Dividing Algebraic Fractions
Visit mometrix.com/academy and enter code: ???

Examples

Example 1:

$$\frac{20x}{16} \div \frac{2x^3}{3x}$$

Answer:

When dividing fractions, multiply the first fraction by the reciprocal of the second fraction.

$$\frac{20x}{16} \cdot \frac{3x}{2x^3}$$

Simplify before multiplying.

$$\frac{5}{4} \cdot \frac{3}{2x}$$

Now, multiply straight across.

$$\frac{15}{8x}$$

Example 2:

$$\frac{10}{5x} \div \frac{18x}{5x}$$

Answer:

When dividing fractions, multiply the first fraction by the reciprocal of the second fraction.

$$\frac{10}{5x} \cdot \frac{5x}{18x}$$

Simplify before multiplying.

$$\frac{5}{1} \cdot \frac{1}{9x}$$

Now, multiply straight across.

$$\frac{5}{9x}$$

EXAMPLE 3:

$$\frac{2x^2}{y-3} \div \frac{x}{4}$$

Answer:

When dividing fractions, multiply the first fraction by the reciprocal of the second fraction.

$$\frac{2x^2}{y-3} \cdot \frac{4}{x}$$

Simplify before multiplying.

$$\frac{2x}{y-3} \cdot \frac{4}{1}$$

Now, multiply straight across.

$$\frac{8x}{y-3}$$

EXAMPLE 4:

Calculate the following quotient. Simplify your answer.

$$\frac{6x}{-2x^2 - 10x + 22} \div \frac{2x^2 - 2x}{7x^2 + 35x - 77}$$

Answer:

First, change the division sign to multiplication and flip the second fraction.

$$\frac{6x}{-2x^2 - 10x + 22} \cdot \frac{7x^2 + 35x - 77}{2x^2 - 2x}$$

Now, factor out a GCF from each numerator or denominator possible.

$$\frac{6x}{-2(x^2 + 5x - 11)} \cdot \frac{7(x^2 + 5x - 11)}{2x(x-1)}$$

From here, cancel out the common factors of $x^2 + 5x - 11$ and $2x$ from the fractions.

$$\frac{3}{-2} \cdot \frac{7}{x-1}$$

Finally, multiply the fractions.

$$\frac{21}{-2x+2}$$

Example 5:

Happy Harvest has a revenue of $6x^2 + 24x$ dollars. $x^2 + 7x - 18$ of these dollars goes toward charitable giving. Goodwill Gardens has a revenue of $7x^2 + 28x$ dollars. $x^2 + x - 6$ of these dollars goes toward charitable giving. What is the ratio of Happy Harvest's giving to Goodwill Gardens' giving? Simplify your answer.

Answer:

Start by writing out a fraction that represents Happy Harvest's giving to its total revenue.

$$\frac{x^2 + 7x - 18}{6x^2 + 24x}$$

Then, do the same for Goodwill Gardens.

$$\frac{x^2 + x - 6}{7x^2 + 28x}$$

To find a ratio of the two companies' giving amounts, the fractions need to be divided.

$$\frac{x^2 + 7x - 18}{6x^2 + 24x} \div \frac{x^2 + x - 6}{7x^2 + 28x}$$

When dividing fractions, flip the second fraction and then multiply the two fractions.

$$\frac{x^2 + 7x - 18}{6x^2 + 24x} \cdot \frac{7x^2 + 28x}{x^2 + x - 6}$$

Now, factor each quadratic expression.

$$\frac{(x+9)(x-2)}{6x(x+4)} \cdot \frac{7x(x+4)}{(x+3)(x-2)}$$

From here, cancel any common factors.

$$\frac{x+9}{6} \cdot \frac{7}{x+3}$$

Finally, multiply the two fractions.

$$\frac{7(x+9)}{6(x+3)} = \frac{7x+63}{6x+18}$$

This fraction represents the ratio between Happy Harvest's giving and Goodwill Gardens' giving.

Practice Questions

1. Calculate the following quotient. Simplify your answer.

$$\frac{x+4}{x+8} \div \frac{x-8}{x+4}$$

a. $\frac{x^2+16}{x^2-64}$

b. $\frac{x^2+8x+1}{x^2-4}$

c. $\frac{x^2+8x+16}{x^2-16x-64}$

d. $\frac{x^2+8x+16}{x^2-64}$

2. Calculate the following quotient. Simplify your answer.

$$\frac{2}{7x} \div \frac{6x+8}{21x^2+35x}$$

a. $\frac{3x+5}{3x+4}$

b. $\frac{x^2+5x}{x^2+4x}$

c. $\frac{42x^2+70x}{42x^2+56x}$

d. $\frac{x+5}{x+4}$

3. Calculate the following quotient. Simplify your answer.

$$\frac{5x}{6x-42} \div \frac{10x+15}{21x-147}$$

a. $\frac{7x}{4x+6}$

b. $\frac{7}{10}$

c. $\frac{-42x}{70x-490}$

d. $\frac{21x^2-147x}{12x^2-66x-126}$

4. Calculate the following quotient. Simplify your answer.

$$\frac{6x+1}{15} \div \frac{3x+10}{15x-10}$$

a. $\frac{10x^2+3x-1}{5x+5}$

b. $\frac{2x^2-9x-2}{x+30}$

c. $\frac{18x^2-9x-2}{9x+30}$

d. $\frac{18x^2-x-2}{x+30}$

5. Calculate the following quotient. Simplify your answer.

$$\frac{18-6x}{7x+2} \div \frac{5x-15}{7}$$

a. $\frac{-42}{35x+10}$
b. $\frac{126-42x}{x^2-19x-6}$
c. $\frac{-6}{5x+10}$
d. $\frac{126+7x}{7x^2-19x+1}$

6. Calculate the following quotient. Simplify your answer.

$$\frac{4x+6}{9x-36} \div \frac{14x+21}{3x-12}$$

a. $\frac{2x+2}{7x+7}$
b. $\frac{2}{21}$
c. $\frac{12x+18}{42x+63}$
d. $\frac{2x-8}{21x-84}$

7. Calculate the following quotient. Simplify your answer.

$$\frac{5x+4}{-3x^2+18x-24} \div \frac{5}{8-2x}$$

a. $\frac{8-2x}{6-3x}$
b. $\frac{2x+4}{3x-15}$
c. $\frac{10x+8}{15x-30}$
d. $\frac{2x+8}{3x-6}$

8. Calculate the following quotient. Simplify your answer.

$$\frac{3x^2+15x+18}{x^2-9x+14} \div \frac{6x+18}{x-7}$$

a. $\frac{x^2+5x+6}{2x^2+2x-12}$
b. $\frac{x^2-5x-14}{2x^2-18x+28}$
c. $\frac{x+2}{2x-4}$
d. $\frac{x+1}{2x-2}$

9. Calculate the following quotient. Simplify your answer.

$$\frac{8x}{3x^2 + 6x - 45} \div \frac{12x^2 - 44x}{-5x^2 - 10x + 75}$$

a. $\frac{-10x}{9x^2-33x}$

b. $\frac{30-10x}{9x^2-60x+99}$

c. $\frac{-10x-50}{9x^2+12x-165}$

d. $\frac{-10}{9x-33}$

10. Calculate the following quotient. Simplify your answer.

$$\frac{9x^2 + 9x}{x^2 + 6x + 8} \div \frac{6x^2 + 6x}{x^2 + 5x + 6}$$

a. $\frac{3x^2+24x+45}{2x^2+18x+40}$

b. $\frac{3x+9}{2x+8}$

c. $\frac{3x^2+18x+27}{2x^2+14x+24}$

d. $\frac{9x+27}{6x+24}$

11. Calculate the following quotient. Simplify your answer.

$$\frac{x^2 + 4x - 12}{x^2 - x - 12} \div \frac{x^2 + 5x - 6}{x^2 - 2x - 8}$$

a. $\frac{x^2+8x+12}{x^2+6x+9}$

b. $\frac{x^2-4x+4}{x^2+2x-3}$

c. $\frac{x^2-4}{x^2+2x-3}$

d. $\frac{-3}{2x-2}$

12. Calculate the following quotient. Simplify your answer.

$$\frac{x^3 - 6x^2 - 7x}{x^2 - 9x + 20} \div \frac{x^3 - 2x^2 - 3x}{x^2 - 7x + 10}$$

a. $\frac{x^2-9x+7}{x^2-7x+6}$

b. $\frac{x^2-12x+35}{x^2-13x+30}$

c. $\frac{x^3-9x^2+14x}{x^3-7x^2+12x}$

d. $\frac{x^2-9x+14}{x^2-7x+12}$

13. **A red rectangle has an area of $x^2 + 7x + 10$ and a length of $x + 1$. A blue rectangle has an area of $x^2 + 9x + 20$ and a length of $x + 3$. Write an algebraic expression for the width of each rectangle. Then find an expression for the ratio of the width of the red rectangle to the width of the blue rectangle. Simplify your answer.**

a. $\frac{x^2+5x+6}{x^2+5x+4}$
b. $\frac{8x+10}{5x+10}$
c. $\frac{x^2+x+3}{x^2+x+2}$
d. $\frac{x+3}{x+2}$

14. **You sell $x + 4$ widgets and make a total gross income of \$100. You sell $x + 8$ gadgets and make a total gross income of \$200. Write an algebraic expression for the price of a widget and one for the price of a gadget. Then, find an expression for the ratio of the price of a widget to the price of a gadget. Simplify your answer.**

a. $\frac{x+8}{2x+8}$
b. $\frac{20{,}000}{x^2+12x+32}$
c. $\frac{2x+8}{x+8}$
d. $\frac{x+1}{2x+1}$

15. **A Christmas present is in a box, a rectangular solid, with a volume of $10x^2 + 50x$. Its base has an area of $x^2 + 11x + 10$. A birthday present is in a box, also a rectangular solid, with a volume of $15x^2 + 75x$. Its base has an area of $x^2 + 13x + 30$. Find an algebraic expression for the height of each box. Then, find an expression for the ratio of the height of the Christmas box to the height of the birthday box. Simplify your answer. (Hint: Volume of rectangular solid = area of base × height)**

a. $\frac{2x+2}{3x+1}$
b. $\frac{2}{3}$
c. $\frac{2x+6}{3x+3}$
d. $\frac{10x+30}{15x+15}$

7.5 Adding and Subtracting Algebraic Fractions

When adding and subtracting fractions, it is crucial to set up a common denominator. Common denominators can be set up by identifying the least common multiple (LCM) between both denominators. For example, $\frac{1}{2}+\frac{1}{3}$ can be solved by setting up a common denominator of 6 because 6 is the least common multiple of 2 and 3.

$$\frac{1}{2}=\frac{?}{6}$$

We multiplied the denominator by 3 to get a denominator of 6, so multiply the numerator by 3.

$$\frac{1}{2}\cdot\frac{3}{3}=\frac{3}{6}$$
$$\frac{1}{3}=\frac{?}{6}$$

We multiplied the denominator by 2 to get a denominator of 6, so multiply the numerator by 2.

$$\frac{1}{3}\cdot\frac{2}{2}=\frac{2}{6}$$

Now that we have a common denominator, we can add the fractions:

$$\frac{3}{6}+\frac{2}{6}=\frac{5}{6}$$

The same process works when adding or subtracting algebraic fractions. Start by determining the LCM of the denominators, and then add or subtract.

For example, consider the following: $\frac{3x}{2}+\frac{4x}{5}$.

The LCM of 2 and 5 is 10, so convert the fractions to each have a denominator of 10 by multiplying the first fraction by $\frac{5}{5}$ and the second fraction by $\frac{2}{2}$.

$$\frac{5}{5}\left(\frac{3x}{2}\right)+\frac{2}{2}\left(\frac{4x}{5}\right)=\frac{15x}{10}+\frac{8x}{10}$$

Now, add the fractions.

$$\frac{15x}{10}+\frac{8x}{10}=\frac{23x}{10}$$

Some algebraic fractions will have variables in the denominator. The same LCM strategy will work for these types of addition and subtraction problems. For example, consider the following:

$$\frac{5}{8x}-\frac{1}{2x}$$

The denominators need to be the same. If we multiply $\frac{1}{2x}$ by $\frac{4}{4}$, it becomes $\frac{4}{8x}$. Now the denominators are the same and we can subtract.

$$\frac{5}{8x} - \frac{1}{2x}$$

$$\frac{5}{8x} - \frac{1}{2x} \cdot \frac{4}{4}$$

$$\frac{5}{8x} - \frac{4}{8x} = \frac{1}{8x}$$

The same LCM strategy works for fractions with multiple variables.

$$\frac{6}{2xy} + \frac{2}{x}$$

Again, the denominators need to be the same. Multiply the second fraction by $\frac{2y}{2y}$.

$$\frac{(2y)(2)}{(2y)(x)} = \frac{4y}{2xy}$$

Now, both fractions have the same denominator and they can be added. Then, simplify the fraction.

$$\frac{6}{2xy} + \frac{2}{x}$$

$$\frac{6}{2xy} + \frac{4y}{2xy} = \frac{4y + 6}{2xy} = \frac{2(2y + 3)}{2xy} = \frac{2y + 3}{xy}$$

The important thing to remember is that denominators need to be the same before adding or subtracting algebraic fractions, and fractions should always be simplified if possible.

Review Video: Adding and Subtracting Algebraic Fractions
Visit mometrix.com/academy and enter code: ???

EXAMPLES

EXAMPLE 1:

$$\frac{3x - 4}{x - 4} + \frac{x - 5}{x - 4}$$

Answer:

When denominators are the same, simply add the numerators.

$$\frac{(3x - 4) + (x - 5)}{x - 4}$$

From here, combine like terms.

$$\frac{4x - 9}{x - 4}$$

EXAMPLE 2:

$$\frac{5}{ab} + \frac{6}{ac}$$

Answer:

When adding fractions, the denominators need to be the same. Multiply the first fraction by $\frac{c}{c}$ and the second fraction by $\frac{b}{b}$. Now both denominators are abc.

$$\frac{c}{c}\left(\frac{5}{ab}\right) + \frac{b}{b}\left(\frac{6}{ac}\right) = \frac{5c}{abc} + \frac{6b}{abc}$$

Now that the denominators are the same, simply add the numerators.

$$\frac{5c + 6b}{abc}$$

EXAMPLE 3:

$$\frac{2y}{x} - \frac{3}{y}$$

Answer:

When subtracting fractions, the denominators need to be the same. Multiply the first fraction by $\frac{y}{y}$, and multiply the second fraction by $\frac{x}{x}$. Now, both fractions have a denominator of xy.

$$\frac{y}{y}\left(\frac{2y}{x}\right) - \frac{x}{x}\left(\frac{3}{y}\right) = \frac{2y^2}{xy} - \frac{3x}{xy}$$

Now, subtract the numerators.

$$\frac{2y^2 - 3x}{xy}$$

EXAMPLE 4:

$$\frac{7x^2 - 5x + 3}{x - 1} - \frac{4x^2 + 9x - 6}{x - 1}$$

Answer:

To subtract fractions, first check to make sure the denominators are the same. Since they are, start by subtracting the numerators.

$$\frac{7x^2 - 5x + 3 - (4x^2 + 9x - 6)}{x - 1}$$

Since the second fraction is being subtracted, the subtraction sign, or negative, needs to be distributed to each part of the second trinomial.

$$\frac{7x^2 - 5x + 3 - 4x^2 - 9x + 6}{x - 1}$$

Now, combine like terms.

$$\frac{3x^2 - 14x + 9}{x - 1}$$

Since the trinomial in the numerator can't be factored at all, the fraction can't be simplified any further, and this is the final answer.

EXAMPLE 5:

You own a movie theater and want to find out how many tickets you sell over the course of a month. The first two weeks of the month you sell $\frac{4x^2-3x+1}{2x+2}$ tickets, and the last two weeks of the month you sell $\frac{6x^2+7x-9}{2x+2}$ tickets. Write a simplified expression that represents the total number of tickets you sold over the month.

Answer:

First, write an addition problem to represent the situation.

$$\frac{4x^2 - 3x + 1}{2x + 2} + \frac{6x^2 + 7x - 9}{2x + 2}$$

In order to add fractions, the denominators need to be the same. Since they already are, add the numerators and keep the denominator the same.

$$\frac{10x^2 + 4x - 8}{2x + 2}$$

The final step is to simplify the fraction. Factor a 2 out in both the numerator and denominator.

$$\frac{2(5x^2 + 2x - 4)}{2(x + 1)}$$

Cancel out the common factor of 2 to find the simplified expression that represents this scenario.

$$\frac{5x^2 + 2x - 4}{x + 1}$$

PRACTICE QUESTIONS

1. Evaluate the following sum. Simplify your answer.

$$\frac{x^2 + 3x - 8}{5x} + \frac{7x - 12}{5x}$$

a. $\frac{x^2+10x-20}{10x}$

b. $\frac{x^2+10x-20}{5x}$

c. $\frac{x^2+10x-4}{x}$

d. $x^2 - 18$

2. Evaluate the following difference. Simplify your answer.

$$\frac{3x^2 - 7x + 12}{2x^2} - \frac{x^2 + 3x + 12}{2x^2}$$

a. $\frac{x^2-2x+12}{x^2}$
b. $\frac{x-5}{x}$
c. $1 - 10x$
d. $\frac{2x^2-10x}{2x^2}$

3. Evaluate the following sum. Simplify your answer.

$$\frac{3x^2 + 30x - 19}{4x^2 + 12x} + \frac{5x^2 - 2x + 19}{4x^2 + 12x}$$

a. $\frac{8x^2+28x}{4x^2+12x}$
b. $\frac{2x+7}{2x+6}$
c. $\frac{2x+7}{x+3}$
d. $\frac{2x^2+7x}{x^2+3x}$

4. Evaluate the following difference. Simplify your answer.

$$\frac{4x^2 - 3x - 17}{6x^2 - 24x} - \frac{x^2 - 9x + 7}{6x^2 - 24x}$$

a. $\frac{x^2+2x-8}{2x^2-8x}$
b. $\frac{x^2+x-1}{x^2-x}$
c. $\frac{3x^2+x-1}{x^2-x}$
d. $\frac{x-2}{2x}$

5. Evaluate the following sum. Simplify your answer.

$$\frac{8x^2 - 30x - 12}{x^3 - 4x^2} + \frac{-2x^2 + 6x + 12}{x^3 - 4x^2}$$

a. $\frac{6x-6}{x^2-x}$
b. $\frac{6x-24}{x^2-4x}$
c. $\frac{6x^2-24x}{x^3-4x^2}$
d. $\frac{6}{x}$

6. Evaluate the following difference. Simplify your answer.

$$\frac{5x^2 + 10x + 20}{x^2 + 3x - 4} - \frac{3x^2 - 8x - 20}{x^2 + 3x - 4}$$

a. $\frac{2x^2+2x}{x^2+3x-4}$
b. $\frac{2x}{x-4}$
c. $\frac{2x+10}{x-1}$
d. $\frac{2x^2+6x+10}{x^2+x-4}$

7. Evaluate the following sum. Simplify your answer.

$$\frac{x + 11}{5} + \frac{14}{x}$$

a. $\frac{x^2+11x+14}{x}$
b. $\frac{x+25}{x+5}$
c. $\frac{x^2+11x+70}{5x}$
d. $\frac{x+25}{5x}$

8. Evaluate the following difference. Simplify your answer.

$$\frac{x}{3} - \frac{5x^2 - 4}{3x}$$

a. $\frac{-4x^2+4}{3x}$
b. $\frac{-4x+4}{3}$
c. $\frac{-4x^2-4}{3x}$
d. $\frac{x-5x^2+4}{3-3x}$

9. Evaluate the following sum. Simplify your answer.

$$\frac{5x - 2}{4x} + \frac{7x + 3}{6x}$$

a. $\frac{29x^2}{12x^2}$
b. $\frac{29x-1}{6x}$
c. $\frac{12x+1}{10x}$
d. $\frac{29}{12}$

10. Evaluate the following difference. Simplify your answer.

$$\frac{10x^2 + 5x - 9}{8x^2} - \frac{10x - 4}{8x}$$

a. $\frac{9x-9}{4x^2}$
b. $\frac{x-9}{8x^2}$
c. $\frac{10x^2-5x-5}{8x^2-8x}$
d. $\frac{9x-9}{8x^2}$

11. Evaluate the following sum. Simplify your answer.

$$\frac{7x - 9}{5x^2} + \frac{2x + 4}{3x^3}$$

a. $\frac{9x-5}{5x^2+3x^3}$
b. $\frac{21x^2-17x+20}{15x^3}$
c. $\frac{21x^4-17x^3+20x^2}{15x^5}$
d. $\frac{21x^2-17x+40}{30x^3}$

12. Evaluate the following difference. Simplify your answer.

$$\frac{x^2 - 9x + 3}{6x^4} - \frac{x^2 - 12x + 4}{8x^4}$$

a. $\frac{x^2-3x+1}{x^2}$
b. $\frac{1-3x}{2x^4}$
c. $\frac{1}{24x^2}$
d. $\frac{x^2-72x+24}{24x^2}$

13. A rectangle has a length of $\frac{x}{5}$ and a width of $\frac{5}{x}$. Find the perimeter of the rectangle. Simplify your answer.

a. $\frac{2x+10}{5+x}$
b. $\frac{2x+10}{x}$
c. $\frac{x^2+25}{5x}$
d. $\frac{2x^2+50}{5x}$

14. One week your company manufactures a large batch of widgets at an average cost of $\frac{x^2+4x-21}{x+10}$ dollars per widget and sells them at an average price of $\frac{x^2+11x+24}{x+10}$ dollars per widget. Find an expression for your average dollars of profit per widget. Simplify your answer. Hint: When the number of widgets produced equals the number of widgets sold, as it does here, the average profit equals the average price minus the average cost.

a. $\frac{7x+45}{x+10}$

b. $\frac{x^2+15x+3}{x+10}$

c. 8

d. $\frac{7x+9}{x+2}$

15. Two boxes (rectangular solids) both have bases with an area of $4x^2$. The taller box has a volume of $9x^2 + 11x - 10$, and the shorter box has a volume of $7x^2 + 7x - 10$. Find the difference in height between the taller box and the shorter one. Simplify your answer. Hint: Volume of rectangular solid = area of base × height

a. $\frac{x+1}{x}$

b. $2 + x$

c. $8x^4 - 16x^3$

d. $\frac{x+2}{2x}$

7.6 Adding and Subtracting Algebraic Fractions with Binomial Denominators

When adding or subtracting fractions, a common denominator is required. As we saw in the previous section, this rule applies to addition and subtraction of algebraic fractions as well. Some problems will already have a common denominator, such as the expression $\frac{9x}{y-3} - \frac{5x}{y-3}$. Both fractions have the same denominator, so simply subtract the numerators.

$$\frac{9x}{y-3} - \frac{5x}{y-3} = \frac{9x-5x}{y-3} = \frac{4x}{y-3}$$

When fractions have unlike denominators, a few more steps are required in order to set up a common denominator. For example, when adding the fractions $\frac{7}{x+4} + \frac{3}{x+6}$, we first need to address the denominators. One denominator is $(x+4)$ and the other is $(x+6)$. If we multiply the first fraction by $\frac{x+6}{x+6}$ and the second fraction by $\frac{x+4}{x+4}$, we create a common denominator.

$$\frac{7(x+6)}{(x+4)(x+6)} + \frac{3(x+4)}{(x+6)(x+4)}$$

This looks like a much longer problem, but once we combine the numerators, we have a more manageable problem.

$$\frac{7(x+6)+3(x+4)}{(x+4)(x+6)}$$

Now, start to simplify the numerator by distributing.

$$\frac{7x+42+3x+12}{(x+4)(x+6)}$$

Next, combine like terms.

$$\frac{10x+54}{(x+4)(x+6)}$$

The expressions $\frac{10x+54}{(x+4)(x+6)}$ and $\frac{10x+54}{x^2+10x+24}$ are both acceptable answers.

Some algebraic fractions will need to be factored before adding or subtracting. For example, $\frac{x}{x^2-25} - \frac{6}{x-5}$. Notice the difference of two squares in the denominator. In order to create a common denominator, start by factoring (x^2-25) into $(x+5)(x-5)$.

$$\frac{x}{(x+5)(x-5)} - \frac{6}{x-5}$$

The problem appears longer, but now it becomes more obvious what we need to do to create a common denominator. The common denominator will be $(x + 5)(x - 5)$. Multiply the second fraction by $\frac{x+5}{x+5}$.

$$\frac{x}{(x+5)(x-5)} - \frac{6(x+5)}{(x-5)(x+5)}$$

Now, combine the numerators. Be sure to distribute the minus sign.

$$\frac{x - 6x - 30}{(x+5)(x-5)}$$

Then, combine like terms in the numerator.

$$\frac{-5x - 30}{(x+5)(x-5)}$$

If we FOIL the denominator, we get the following expression.

$$\frac{-5x - 30}{x^2 - 25}$$

As you can see, adding and subtracting algebraic fractions that contain binomials and unlike denominators usually results in complex answers. The important thing to remember is to create a common denominator before adding or subtracting.

Review Video: Rational Expression
Visit mometrix.com/academy and enter code: 415183

Review Video: Adding and Subtracting Algebraic Fractions with Binomial Denominators
Visit mometrix.com/academy and enter code: ???

Examples

Example 1:

$$\frac{9y}{x+3} - \frac{4y}{x+3}$$

Answer:

When subtracting fractions with like denominators, simply subtract the numerators. The denominator will remain the same.

$$\frac{9y}{x+3} - \frac{4y}{x+3} = \frac{5y}{x+3}$$

Example 2:

$$-\frac{5x}{8x+7} - \frac{6x^3}{3x+1}$$

Answer:

When adding and subtracting fractions, we need a common denominator.

Right now, we have two binomial denominators $(8x + 7)$ and $(3x + 1)$. Multiply the first fraction by $\frac{3x+1}{3x+1}$ and the second fraction by $\frac{8x+7}{8x+7}$ to get a common denominator.

$$-\frac{5x(3x+1)}{(3x+1)(8x+7)} - \frac{6x^3(8x+7)}{(8x+7)(3x+1)}$$

Simplify the numerator and denominator.

$$-\frac{5x(3x+1)}{(3x+1)(8x+7)} - \frac{6x^3(8x+7)}{(8x+7)(3x+1)} = \frac{-15x^2 - 5x - 48x^4 - 42x^3}{24x^2 + 29x + 7}$$

Arrange the terms in the numerator from highest exponent to lowest exponent.

$$\frac{-48x^4 - 42x^3 - 15x^2 - 5x}{24x^2 + 29x + 7}$$

Example 3:

$$\frac{4}{x+1} + \frac{5}{y+2}$$

Answer:

When adding and subtracting fractions, we need a common denominator.

Right now, we have two binomial denominators $(x + 1)$ and $(y + 2)$.

Multiply the first fraction by $\frac{y+2}{y+2}$ and the second fraction by $\frac{x+1}{x+1}$ to get fractions with a common denominator.

$$\frac{4(y+2)}{(y+2)(x+1)} + \frac{5(x+1)}{(x+1)(y+2)}$$

Add the numerators.

$$\frac{4(y+2) + 5(x+1)}{(x+1)(y+2)}$$

Simplify the numerator and denominator.

$$\frac{4y + 8 + 5x + 5}{xy + 2x + y + 2}$$

Combine like terms.

$$\frac{5x + 4y + 13}{xy + 2x + y + 2}$$

EXAMPLE 4:

$$\frac{x}{x^2 - 14x + 49} - \frac{1}{x - 7}$$

Answer:

Factor the denominator of the first fraction.

$$\frac{x}{(x-7)(x-7)} - \frac{1}{x-7}$$

Create a common denominator by multiplying the second fraction by $\frac{x-7}{x-7}$.

$$\frac{x}{(x-7)(x-7)} - \frac{1}{x-7} = \frac{x}{(x-7)(x-7)} - \frac{1(x-7)}{(x-7)(x-7)}$$

Now that we have a common denominator, we can simplify the numerators and denominator, and then subtract. Be sure to distribute the minus sign.

$$\frac{x}{(x-7)(x-7)} - \frac{1(x-7)}{(x-7)(x-7)}$$

$$\frac{x - 1(x-7)}{(x-7)(x-7)}$$

$$\frac{x - x + 7}{x^2 - 14x + 49}$$

$$\frac{7}{x^2 - 14x + 49}$$

PRACTICE QUESTIONS

1. Simplify: $\frac{5}{x-1} + \frac{2}{x-1}$

a. $\frac{3}{x-1}$
b. $\frac{7}{x-1}$
c. $\frac{7}{2x-2}$
d. $\frac{7}{(x-1)^2}$

2. Simplify: $\frac{3}{y+1} + \frac{5}{y-1}$

a. $\frac{8y+2}{2y}$
b. $\frac{8}{2y}$
c. $\frac{8}{(y+1)(y-1)}$
d. $\frac{8y+2}{(y+1)(y-1)}$

3. Simplify: $\frac{6}{x+4} - \frac{x}{x+4}$

a. $\frac{6-x}{(x+4)-(x+4)}$

b. $\frac{6-x}{x+4}$

c. $\frac{x-6}{x+4}$

d. $\frac{6-x}{(x+4)^2}$

4. Simplify: $\frac{x}{x^2-9} - \frac{1}{x+3}$

a. $\frac{3}{(x+3)(x-3)}$

b. $\frac{x-1}{x^2-9}$

c. $\frac{3}{x^2-x-12}$

d. $\frac{-1}{-x^2+12}$

5. Simplify: $\frac{x}{x-2} + \frac{3x}{x-2}$

a. $\frac{4x}{2x-4}$

b. $\frac{4x}{x^2-4}$

c. $\frac{4x}{x-2}$

d. $\frac{x+3x}{(x-2)^2}$

6. Simplify: $\frac{4}{x+2} + \frac{8}{x-3}$

a. $\frac{12}{2x-1}$

b. $\frac{12x+4}{2x-1}$

c. $\frac{12x+4}{(x+2)(x-3)}$

d. $\frac{12}{(x+2)(x-3)}$

7. Simplify: $\frac{x-7}{6x+3} - \frac{x}{6x+3}$

a. $\frac{-7}{1}$

b. $\frac{-7}{(6x+3)^2}$

c. $\frac{2x-7}{6x+3}$

d. $\frac{-7}{6x+3}$

8. Simplify: $\frac{6}{x+4} - \frac{2}{x+1}$

a. $\frac{4}{(x+4)(x+1)}$

b. $\frac{4x-2}{(x+4)(x+1)}$

c. $\frac{4}{3}$

d. $\frac{4x-2}{3}$

9. Simplify: $\frac{2x+6}{x^2-5} + \frac{x+2}{x^2-5}$

a. $\frac{3x+8}{x^2-5}$

b. $\frac{3x+8}{2x^2-10}$

c. $\frac{3x+8}{2(x^2-5)}$

d. $\frac{8}{2x-10}$

10. Simplify: $\frac{6}{k^2-4} + \frac{k}{k+2}$

a. $\frac{k^2-2k+6}{k^2-4}$

b. $\frac{6}{k^2-4}$

c. $\frac{2+k}{6}$

d. $\frac{6+k}{k-2}$

11. Simplify: $\frac{6x-8}{4x^3+2} - \frac{7x+3}{4x^3+2}$

a. $\frac{-x-11}{4x^3+2}$

b. $\frac{-x-5}{4x^3+2}$

c. $\frac{-x-11}{(4x^3+2)-(4x^3-2)}$

d. $\frac{x+11}{4x^3+2}$

12. Simplify: $\frac{y}{y+1} - \frac{y}{y+3}$

a. $\frac{2y}{2y+4}$

b. $-y$

c. $\frac{2y}{y^2+4y+3}$

d. $\frac{2y}{y^2+4}$

13. **Adam and Joey are filling a garden bed with soil. Adam shoveled in $\frac{3m}{m+4}$ ft^3, and Joey shoveled in $\frac{4m}{m+5}$ ft^3. How much soil did Adam and Joey shovel into the garden bed altogether?**

a. $\frac{-1m}{(m+4)(m+5)}$ ft^3
b. $\frac{7m^2+31m}{(m+4)(m+5)}$ ft^3
c. $\frac{1m}{(m+4)(m+5)}$ ft^3
d. $\frac{7m^2+31m}{2m+9}$ ft^3

14. **Monica is reupholstering a bench. She has $\frac{a^2}{a+8}$ yards of fabric. She uses $\frac{4}{a+8}$ yards of her fabric to cover the bench. How many yards of fabric is she left with?**

a. $\frac{a^2-4}{2a+16}$ yd
b. $\frac{a^2+4}{2a+16}$ yd
c. $\frac{a^2+4}{a+8}$ yd
d. $\frac{a^2-4}{a+8}$ yd

15. **Martin's bookstore received a shipment of two boxes. The first box weighed $\frac{5}{x+1}$ lb, and the second box weighed $\frac{2}{x^2-1}$ lb. How much did the shipment weigh in all?**

a. $\frac{3}{(x+1)(x-1)}$ lb
b. $\frac{7}{(x+1)(x-1)}$ lb
c. $\frac{-3x+5}{(x+1)(x-1)}$ lb
d. $\frac{5x-3}{(x+1)(x-1)}$ lb

7.7 Solving Equations Involving Algebraic Fractions

When solving an equation, the goal is to isolate the variable. This often requires skills such as combining like terms, creating common denominators, and using inverse operations. When fractions are involved, it is essential to create a common denominator before adding or subtracting. For example, before adding $\frac{x}{3}+\frac{x}{5}$, we need to create a common denominator. 15 is the least common multiple of 3 and 5, so both denominators will become 15. The expression $\frac{x}{3}+\frac{x}{5}$ becomes $\frac{5x}{15}+\frac{3x}{15}$, which combines to $\frac{8x}{15}$.

Equations can contain fractions on one side or fractions on both sides.

Equations with Fractions on One Side

The equation $\frac{x}{9}-\frac{x}{3}=-1$ has fractions on one side of the equation. To simplify the left side of the equation, start by creating a common denominator. The expression $\frac{x}{9}-\frac{x}{3}$ has a common denominator of 9, because 9 is the least common multiple of 9 and 3. Multiply $\frac{x}{3}$ by $\frac{3}{3}$.

$$\frac{x}{9}-\frac{3x}{9}=-1$$

Now that there is a common denominator, the terms can be subtracted.

$$-\frac{2x}{9}=-1$$

Now the equation becomes solvable in two steps. Multiply both sides by 9.

$$-2x=-9$$

Then, divide both sides by –2.

$$x=\frac{9}{2}$$

Equations with Fractions on Both Sides

A few more steps are required when solving equations with fractions on both sides. For example, $\frac{1-x}{4}-\frac{2x+2}{6}=\frac{x}{8}$ has two fractions on the left side and one fraction on the right side. Creating a common denominator is a great place to start, especially because we are subtracting fractions on the left side.

All three denominators have a common multiple of 24. Multiply the entire equation by 24 to eliminate the fractions.

$$24\left(\frac{1-x}{4}-\frac{2x+2}{6}=\frac{x}{8}\right)$$
$$6(1-x)-4(2x+2)=3x$$

Now we can distribute and combine like terms.

$$6-6x-8x-8=3x$$
$$-14x-2=3x$$

From here, add $14x$ to both sides.

$$-2 = 17x$$

Finally, divide both sides of the equation by 17.

$$x = -\frac{2}{17}$$

Whether you are working with equations with one fraction or multiple fractions, it is important to create a common denominator when addition and subtraction are involved. After a common denominator is found, multiply both sides of the equation by the common denominator to eliminate the fractions. Then work to isolate the variable by combining like terms and using inverse operations.

Review Video: Solving Equations Involving Algebraic Fractions
Visit mometrix.com/academy and enter code: 237770

Examples

Example 1:

$$4x + 3 = \frac{2x}{7}$$

Answer:

Multiply both sides of the equation by 7 to eliminate the fraction.

$$7(4x + 3) = 2x$$

Distribute on the left side.

$$28x + 21 = 2x$$

Subtract $28x$ from both sides of the equation.

$$21 = -26x$$

Divide both sides of the equation by 26.

$$-\frac{21}{26} = x$$

Example 2:

$$\frac{3x}{10} = \frac{x}{2} + 3$$

Answer:

Subtract the term $\frac{x}{2}$ from both sides of the equation.

$$\frac{3x}{10} - \frac{x}{2} = 3$$

Create a common denominator by multiplying the second fraction by $\frac{5}{5}$.

$$\frac{3x}{10} - \frac{5x}{10} = 3$$

Subtract the numerators.

$$\frac{3x - 5x}{10} = 3$$
$$-\frac{2x}{10} = 3$$

Multiply both sides of the equation by 10.

$$-2x = 30$$

Divide both sides by -2.

$$x = -15$$

EXAMPLE 3:

Jason and Gabrielle are buying a $10 gift for their mom. Jason contributes $\frac{x-6}{4}$ dollars to the gift, and Gabrielle contributes $\frac{x+3}{3}$ dollars to the gift. How much money does each sibling contribute?

Answer:

To find how much each sibling contributes, start by adding the fractions together and setting it equal to the total amount of the gift.

$$\frac{x-6}{4} + \frac{x+3}{3} = 10$$

Now, solve for x. First, convert the fractions to have a common denominator.

$$\frac{3}{3}\left(\frac{x-6}{4}\right) + \frac{4}{4}\left(\frac{4x+12}{3}\right) = 10$$
$$\frac{3x-18}{12} + \frac{4x+12}{12} = 10$$

Next, add the fractions, and solve the equation by normal algebra methods.

$$\frac{7x-6}{12} = 10$$
$$7x = 126$$
$$x = 18$$

Now that we know the value of x, substitute this value into the expression for each sibling to find out how much they contributed. Start with Jason.

$$\frac{(18)-6}{4} = \frac{12}{4} = 3$$

Jason contributed \$3 toward the gift. Now, solve for Gabrielle's contribution.

$$\frac{(18) + 3}{3} = \frac{21}{3} = 7$$

Gabrielle contributed \$7 toward the gift.

Practice Questions

1. Solve for x.

$$\frac{x + 1}{3} + \frac{x - 4}{2} = 10$$

a. $x = 8$
b. $x = 10$
c. $x = 12$
d. $x = 14$

2. Solve for x.

$$\frac{x}{2} = \frac{x}{4} + 2$$

a. $x = 2$
b. $x = 8$
c. $x = 4$
d. $x = 6$

3. Solve for x.

$$\frac{2x}{5} - \frac{x}{4} = 3$$

a. $x = 20$
b. $x = 30$
c. $x = 3$
d. $x = 2$

4. Solve for x.

$$\frac{x + 1}{2} - 6 = \frac{x - 3}{6}$$

a. $x = 12$
b. $x = 9$
c. $x = 15$
d. $x = 6$

5. Solve for x.

$$\frac{x-1}{3}+\frac{x+1}{2}=1$$

a. $x = 5$
b. $x = -1$
c. $x = 25$
d. $x = 1$

6. Solve for x.

$$\frac{2x-2}{6}=9+\frac{2x-4}{8}$$

a. $x = 1$
b. $x = 2.5$
c. $x = 106$
d. $x = 105$

7. Solve for x.

$$\frac{x+8}{5}-\frac{x-6}{3}=4$$

a. $x = -57$
b. $x = 57$
c. $x = -3$
d. $x = 3$

8. Solve for x.

$$\frac{x+2}{2}=4-\frac{x+4}{3}$$

a. $x = 2$
b. $x = 10$
c. $x = 38$
d. $x = \frac{17}{5}$

9. Solve for x.

$$\frac{x-1}{2}+\frac{2x+3}{4}=1$$

a. $x = 2$
b. $x = \frac{3}{4}$
c. $x = \frac{1}{2}$
d. $x = 4$

10. Solve for x.

$$\frac{4x-2}{6} = 9 + \frac{4x-4}{8}$$

a. $x = 208$
b. $x = 59$
c. $x = 49$
d. $x = 53$

11. Solve for x.

$$\frac{x}{6} - \frac{2x}{3} = 3$$

a. $x = 6$
b. $x = -6$
c. $x = \frac{18}{5}$
d. $x = -\frac{18}{5}$

12. Solve for x.

$$\frac{2x-1}{4} = 2 - \frac{x-1}{5}$$

a. $x = \frac{49}{6}$
b. $x = \frac{7}{2}$
c. $x = 3$
d. $x = -3$

13. Sarah and Lucan are putting marbles into a bag. Sarah put $\frac{x+2}{2}$ marbles in the bag, and Lucan put $\frac{x+4}{8}$ marbles in the bag. Altogether they had 9 marbles. Solve for x to determine how many marbles Sarah and Lucan each put into the bag.

a. Sarah put 5 marbles into the bag, and Lucan put 4 marbles into the bag.
b. Sarah put 8 marbles into the bag, and Lucan put 1 marble into the bag.
c. Sarah put 7 marbles into the bag, and Lucan put 2 marbles into the bag.
d. Sarah put 3 marbles into the bag, and Lucan put 6 marbles into the bag.

14. Alex had $\frac{6x-1}{4}$ pencils in his backpack. After lending $\frac{5-2x}{2}$ pencils to Jasmine, Alex had 1 pencil remaining. How many pencils did Alex start with? How many did he lend to Jasmine?

a. Alex started with 10 pencils and lent 9 pencils to Jasmine.
b. Alex started with 3 pencils and lent 2 pencils to Jasmine.
c. Alex started with 5 pencils and lent 4 pencils to Jasmine.
d. Alex started with 2 pencils and lent 1 pencil to Jasmine.

15. There are $\frac{2x}{5}$ boys in Ms. Smith's class and $\frac{3x}{10}$ boys in Ms. Cole's class. Both classes combined have a total of 14 boys. Solve for x to identify the number of boys in each class.

a. There are 5 boys in Ms. Smith's class and 9 boys in Ms. Cole's class.
b. There are 10 boys in Ms. Smith's class and 4 boys in Ms. Cole's class.
c. There are 8 boys in Ms. Smith's class and 6 boys in Ms. Cole's class.
d. There is 1 boy in Ms. Smith's class and 13 boys in Ms. Cole's class.

Chapter 7 Quiz

1. Divide the following algebraic fractions.

$$\frac{3x^2 - x - 2}{x + 2} \div \frac{6x + 4}{x^2 - 4}$$

a. $\frac{3x^2-x-2}{6}$

b. $\frac{3x^2-x}{6}$

c. $\frac{x^2-3x-2}{2}$

d. $\frac{x^2-3x+2}{2}$

2. Solve the equation.

$$\frac{m + 3}{4} - \frac{m - 2}{10} = 1$$

a. $m = -2$

b. $m = 3$

c. $m = -\frac{2}{3}$

d. $m = \frac{1}{3}$

3. The area of a rectangle can be expressed by the expression $\frac{4t^2-5t-6}{2t^2+5t+2}$. If the rectangle's width can be expressed by $\frac{4t-8}{6t+3}$, which of the following expressions represents the length of the rectangle in terms of t?

a. $\frac{12t+9}{4t+8}$

b. $\frac{3t+9}{t+4}$

c. $\frac{4t+4}{12t+9}$

d. $\frac{t+4}{3t+9}$

4. Solve the equation.

$$\frac{2 + t}{t} = \frac{3 + t}{t - 3}$$

a. $t = -3$

b. $t = -\frac{2}{3}$

c. $t = -\frac{3}{2}$

d. No solution

5. The length of a rectangle is represented by the expression $\frac{n+6}{2n+8}$ and the width by the expression $\frac{2n+5}{3n+12}$. Which of the following expressions represents the perimeter of the rectangle?

a. $\frac{3n+11}{5n+17}$
b. $\frac{3n+11}{6n+24}$
c. $\frac{7}{12}$
d. $\frac{7}{3}$

6. Simplify the following algebraic expression. Leave your answer in factored form.

$$\frac{6y^2 + 20y - 16}{18 - 27y}$$

a. $-\frac{2(y+4)}{9}$
b. $\frac{2(y+4)}{9}$
c. $-\frac{2(y-4)}{3}$
d. $\frac{2(3y-2)(y+4))}{9(3y+2)}$

7. Solve the equation.

$$\frac{a-2}{4a-8} = \frac{1}{a+2}$$

a. $a = 2$
b. $a = -2$
c. $a = 4$
d. No solution

8. Multiply the following algebraic fractions.

$$\frac{x^2 - x - 6}{2x + 5} \cdot \frac{4x^2 + 20x + 25}{2x^2 - 4x - 6}$$

a. $\frac{x^2+9x+7}{2x+2}$
b. $\frac{2x^2+9x+10}{2x+2}$
c. $\frac{x^2+7x+10}{x+1}$
d. $\frac{2x^2+7x+10}{(x-3)^2}$

9. Combine the following algebraic fractions, and simplify the answer if possible.

$$\frac{2}{x-4} - \frac{x+2}{x-4}$$

a. $\frac{1}{x-4}$
b. $\frac{-x}{x-4}$
c. $\frac{-1}{x-4}$
d. $\frac{x}{x-4}$

10. Combine the following algebraic expressions, and simplify the answer if possible:

$$\frac{5x+2}{3x^2} + \frac{x-2}{3x^2}$$

a. $\frac{3}{x}$
b. $\frac{2}{x}$
c. $\frac{2}{x^2}$
d. $\frac{6}{x^2}$

11. A rectangular field is enclosed by a metal fence on three sides and the side of a barn by the fourth side. The length of the field, in feet, is given by the expression $\frac{n+3}{2}$ and the width, in feet, is given by the expression $\frac{2n-1}{3}$ feet. If 109 feet of feet are used to bound the three sides of the field requiring fencing, both widths and one of the lengths of the field, what must be the value of n?

a. $n = 22$
b. $n = 16$
c. $n = 59$
d. $n = 29$

12. Combine the following algebraic fractions and simplify the answer if possible.

$$\frac{2a-7}{a^2-4} + \frac{5-a}{a^2-4}$$

a. $\frac{1}{a-2}$
b. $\frac{1}{a+2}$
c. $\frac{3a-12}{a^2-4}$
d. $\frac{a}{a^2-4}$

13. Simplify the following algebraic expression. Leave your answer in factored form.

$$\frac{3a^3 - 6a^2 - 24a}{5a^2 - 20a}$$

a. $\frac{3(a+2)}{5a}$
b. $\frac{3a(a-4)(a+2)}{5a(a-4)}$
c. $\frac{3(a+2)}{5}$
d. $\frac{3(a+4)(a-2)}{5(a-4)}$

14. Combine the following algebraic expressions, and simplify the answer if possible.

$$\frac{13y + 4}{5y} - \frac{3y - 1}{5y}$$

a. $2 + \frac{3}{5y}$
b. $2 + y$
c. $\frac{2y+1}{y}$
d. $\frac{10y+3}{5y}$

15. The total cost, in dollars, for producing generators can be represented by the expression $\frac{3t^2+14t+8}{5t+10}$. The number of generators produced can be represented by the expression $3t + 2$. If the unit cost of the generators is determined by taking the product of the total cost and the reciprocal of the number of generators produced, which of the following expressions represents the unit cost of the generators in terms of t?

a. $\frac{t+4}{5t+10}$
b. $\frac{t+1}{5t+2}$
c. $\frac{3t^2+14t+8}{15t^2+40t+20}$
d. $\frac{3t^2+7t+2}{15t^2+25t+10}$

16. Divide the following algebraic fractions:

$$\frac{x^2 + 3x - 4}{x^2 - 9} \div \frac{x^2 + 4x - 5}{x^2 - 2x - 3}$$

a. $\frac{x^2+3x-4}{x^2+2x-15}$
b. $\frac{x^2-3x-4}{x^2-2x-15}$
c. $\frac{x^2+5x+4}{x^2+8x+15}$
d. $\frac{x^2+5x-4}{x^2+8x+15}$

17. A car and a bus are traveling towards the same bus station. When the car is 20 miles from the bus station, the bus is 15 miles away from the station. The car is traveling 10 miles per hour faster than the bus. At these distances from the bus station, what will the speed of the bus need to be for the car and the bus to meet at the station at the same time? (Note that distance traveled, D, can be found using the formula $D = rt$, where r is the speed in miles per hour, and t is time in hours.)

a. 35 mph
b. 45 mph
c. 30 mph
d. The car and bus cannot meet at the bus station at the same time.

18. Solve the equation:

$$\frac{2y}{3} = \frac{y-3}{6} + 9$$

a. $y = 18$
b. $y = 17$
c. $y = -24$
d. $y = -21$

19. In the study of Calculus, the equation $m = \frac{(x+h)^2+(x+h)-(x^2+x)}{h}$ is frequently used. Which of the following expressions gives the value for m in simplified form?

a. $2x^2 + 1$
b. $2x + h + 1$
c. $2x + h$
d. $xh + h^2$

20. Multiply the following algebraic fractions:

$$\frac{x^2 - 1}{x^2 + 2x + 1} \cdot \frac{x + 1}{x^2 - 3x - 4}$$

a. $\frac{x-1}{x^2-3x-4}$
b. $\frac{x+1}{x^2+5x+4}$
c. $\frac{(x+1)^2}{x^2+5x+4}$
d. $\frac{(x+1)^2}{x^2-3x-4}$

Chapter 8: Quadratic Functions

8.1 Domain and Range of Quadratic Functions

A **quadratic function** is a function that creates a curve, known as a parabola, when graphed. Quadratic functions can be written in the form $ax^2 + bx + c = 0$, where $a \neq 0$. For example, $2x^2 + 4x + 6 = 0$ is a quadratic function.

The **domain** of a quadratic function represents all x-values of the function, and the **range** represents all the y-values of the function.

When determining the domain and range from a graph, we are looking for the minimum and maximum x- and y-values. For example, the graph below shows a quadratic function.

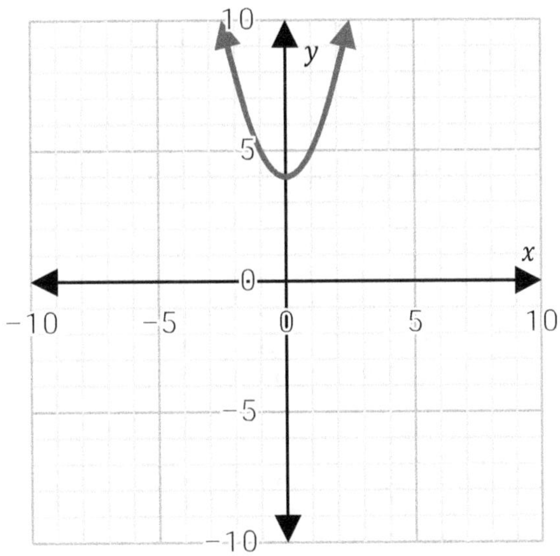

The domain, x-values, extend to the left and to the right indefinitely. The arrows on the graph imply that the function extends indefinitely. Since the x-values can be anything, we say that the domain of the function is all real numbers. This can be written as D: $-\infty < x < \infty$.

The range, y-values, do not extend indefinitely in the positive and negative direction. The vertex of the parabola is at the value $y = 4$. The arrows extend in the positive direction up from $y = 4$. This means that the range of the graph is all real numbers greater than or equal to 4. This can be written as R: $y \geq 4$.

When determining the domain and range from an equation, we are again looking for all the x-values (domain), and all the y-values (range), that satisfy the function. For example, let's consider the quadratic function $f(x) = 3x^2 + 6x - 2$.

The domain represents all x-values that can be input values for the function. In this case, any real number can be inserted into the function for x. Therefore, the domain of the function is all real numbers, or D: $-\infty < x < \infty$.

The range of the function represents all the output values for the function. One way to find the range is by graphing. When graphed, we see that the vertex of the parabola is at $(-1, -5)$, and opens up. This shows us that the range includes y-values that are greater than or equal to -5. The range of the function is all real numbers greater than or equal to -5, or R: $y \geq -5$.

Another option for finding the range is to find the vertex of the parabola using a formula.

The formula $x = -\frac{b}{2a}$ can be used to find the x-coordinate of the vertex. For example, in the quadratic equation $y = 5x^2 - 20x + 15$, the value of a is 5, the value of b is -20, and the value of c is 15. Plug a and b are into the vertex formula.

$$x = -\frac{b}{2a}$$
$$x = -\frac{-20}{2(5)}$$
$$x = -\frac{-20}{10}$$
$$x = -(-2)$$
$$x = 2$$

This means that the x-value of the vertex is 2. To find the y-value of the vertex, plug in 2 for x in the original equation.

$$y = 5x^2 - 20x + 15$$
$$y = 5(2)^2 - 20(2) + 15$$
$$y = 5(4) - 20(2) + 15$$
$$y = 20 - 40 + 15$$
$$y = -5$$

This means that the y-value of the vertex is -5. The point $(2, -5)$ is the vertex of the parabola.

Now we can determine the domain and range.

Look at the original equation: $y = 5x^2 - 20x + 15$.

The domain is all real numbers because any value can be plugged in for x.

Now, determine the range. If the x^2-term is positive, the parabola opens up. If the x^2-term is negative, the parabola opens down. In this case, $5x^2$ is positive, so the parabola opens up. The vertex of the parabola is $(2, -5)$. This means that the range is any real number greater than or equal to -5. Note, if the parabola opened down, the range would be "less than or equal to -5."

Now, we know that the domain is all real numbers because any value can be plugged in for x. We also know that the range is all real numbers greater than or equal to -5 because the parabola opens up from the vertex $(2, -5)$.

$$\text{D: } -\infty < x < \infty$$

$$\text{R: } y > -5$$

Review Video: Domain and Range of Quadratic Functions
Visit mometrix.com/academy and enter code: 331768

EXAMPLES

EXAMPLE 1:

Determine the domain and range of the quadratic function based on the graph.

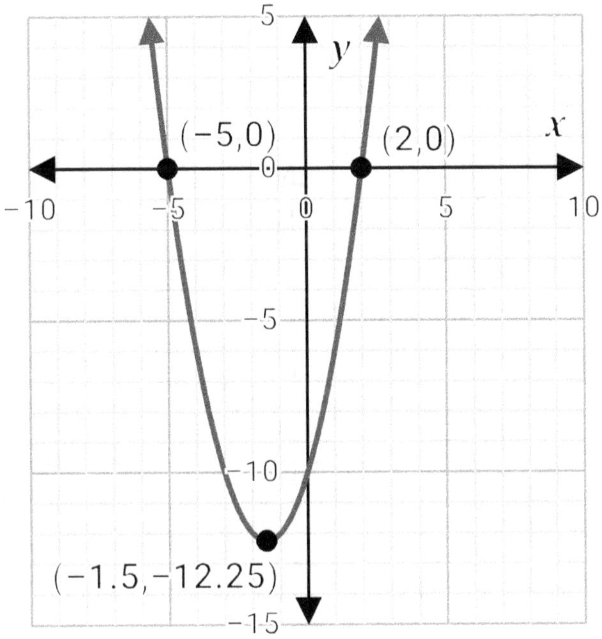

Answer:

The domain represents all possible x-values. On this graph, the x-values extend indefinitely in the left and right direction. This means that all x-values are possible. The domain is $-\infty < x < \infty$.

The range represents all possible y-values. On this graph, the y-values extend up from −12.25. This means that any y-value greater than or equal to −12.25 is possible. The range is $y \geq -12.25$.

EXAMPLE 2:

Determine the domain and range for the quadratic equation.

$$y = x^2 + 5x - 2$$

Answer:

The domain represents all possible x-values for the function. Since any value of x can be plugged into the formula, x can be all real numbers. The domain is $-\infty < x < \infty$.

The range can be determined using the vertex formula. The range of the quadratic will be all values above or below the vertex.

Vertex Formula: $x = -\frac{b}{2a}$

$x = -\frac{5}{2(1)} = -2.5$ is the x-coordinate of the vertex.

Now plug this in for x, and calculate the corresponding y-value.

$y = (-2.5)^2 + 5(-2.5) - 2 = -8.25$ is the y-coordinate of the vertex.

We know that the vertex of the parabola is at $(-2.5, -8.25)$. Now we need to determine if it opens up or down. If $a > 0$, the parabola opens up. If $a < 0$, the parabola opens down. In this case, $a = 1$ so the parabola opens up. The domain is all y-values above and including the vertex. The range is $y \geq -8.25$.

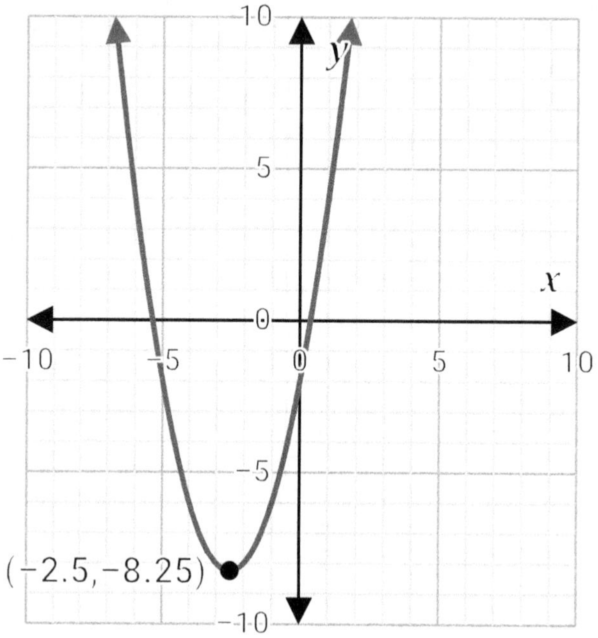

EXAMPLE 3:

When Sarah bounces a basketball, she notices that 3 seconds pass between the first and second bounce. Let $h(t)$ be the height, in inches, of the basketball above the ground t seconds after the first bounce and before the second bounce. The following graph of a quadratic equation shows the values of $h(t)$. What is the domain and range of $h(t)$?

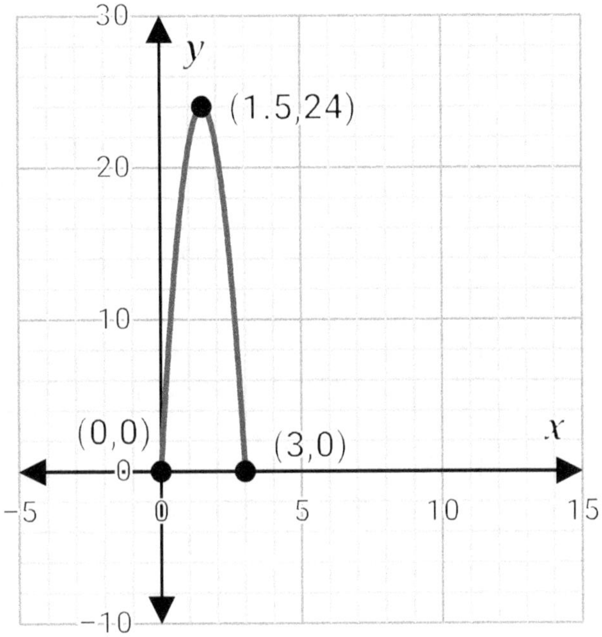

Answer:

Start by finding the domain of the graph. The domain represents all possible x-values. Since we are only looking at the time between bounces and the time cannot be negative, the domain is from 0 seconds to 3 seconds.

$$\text{Domain: } 0 \leq t \leq 3$$

Now find the range of the graph. The range represents all possible y-values. The bounce starts at the ground and reaches a height of 24 inches, so the range is from 0 inches to 24 inches.

$$\text{Range: } 0 \leq h(t) \leq 24$$

Practice Questions

1. Find the domain of the quadratic function *f* whose graph appears here.

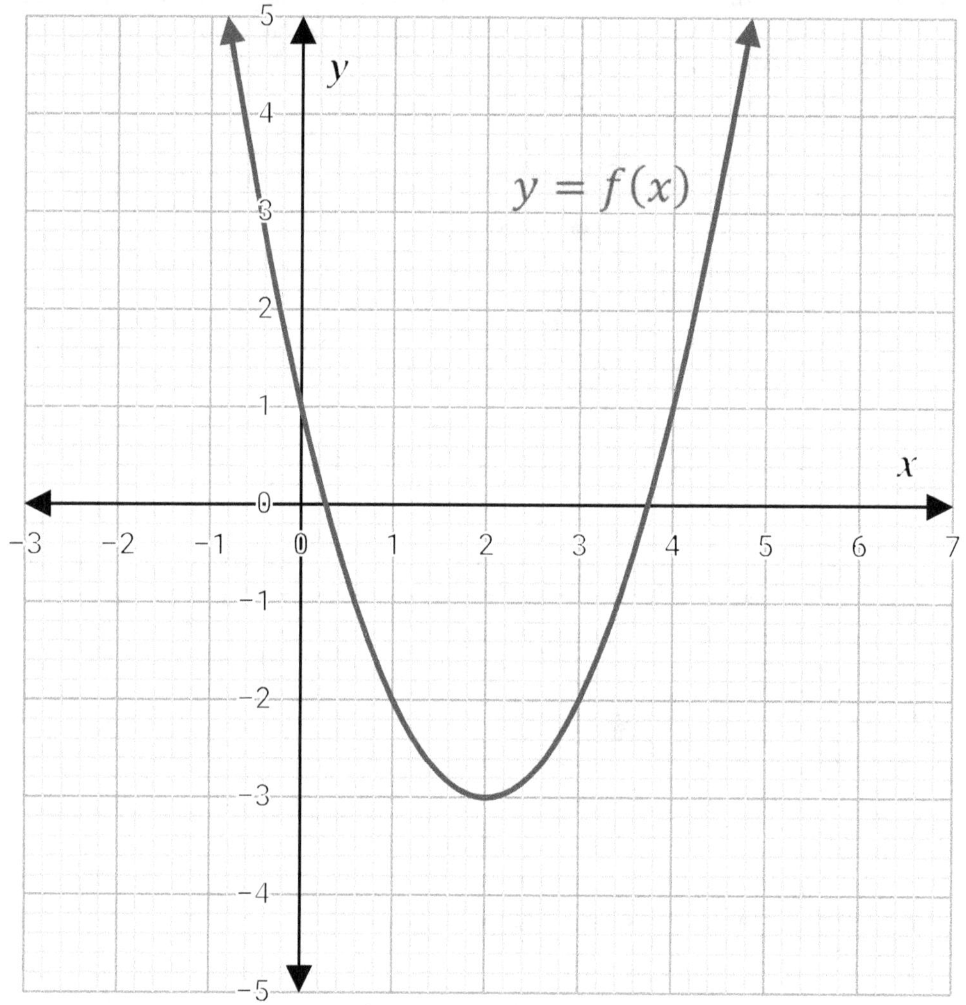

a. $-\infty \leq x \leq \infty$

b. $x \geq 1$

c. $x \leq 3$

d. $x \geq 3$

2. Find the domain and range of the quadratic function *f* whose graph appears here.

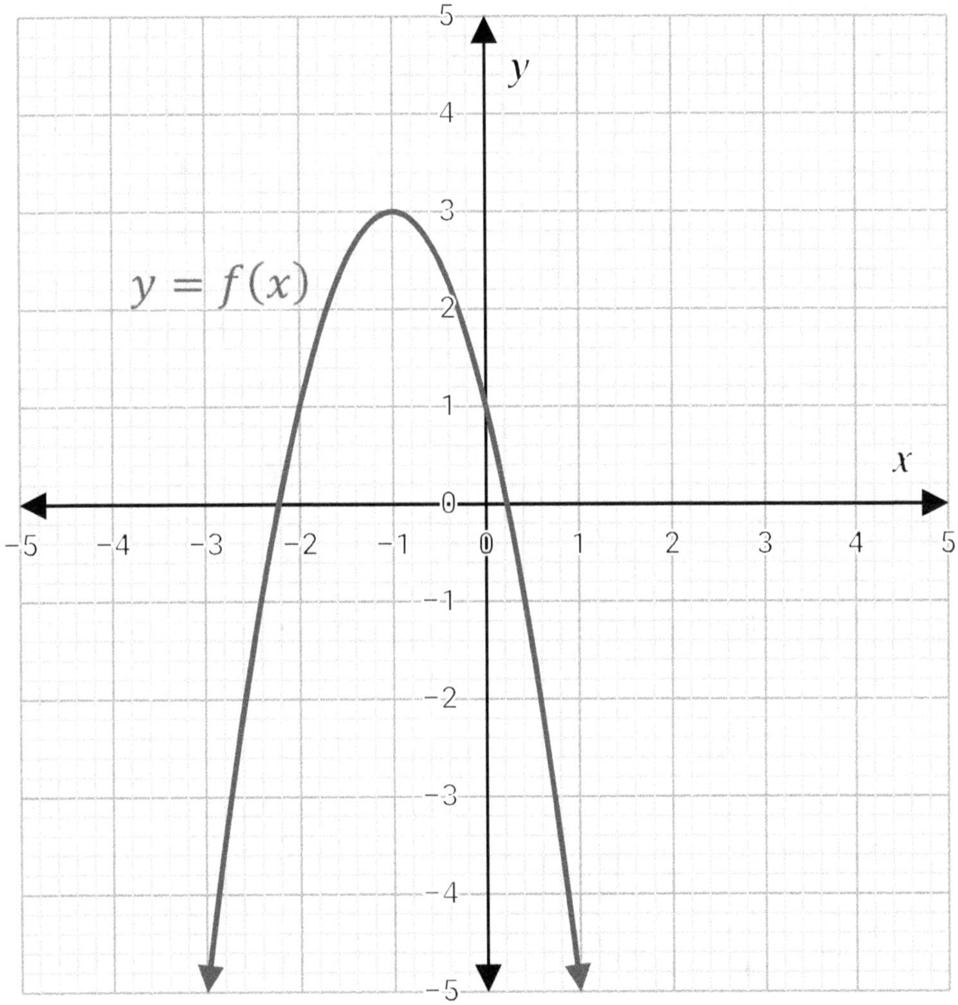

a. Domain: $-\infty \leq x \leq \infty$, Range: $y \leq 3$
b. Domain: $-\infty \leq x \leq \infty$, Range: $y \geq 3$
c. Domain: $-\infty \leq x \leq \infty$, Range: $-\infty \leq y \leq \infty$
d. Domain: $x \geq -1$, Range: $y \leq 3$

3. Find the range of the quadratic function f whose graph appears here.

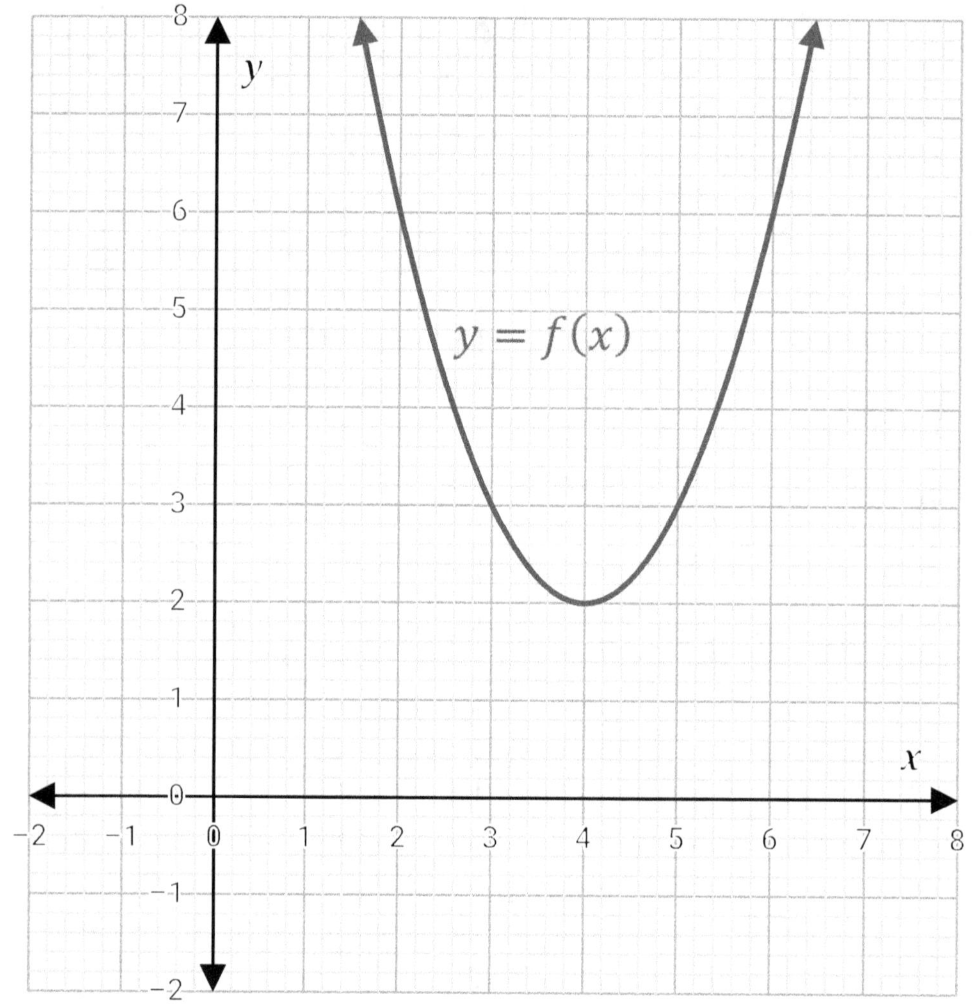

a. $y \geq 0$
b. $-\infty \leq y \leq \infty$
c. $y \geq 2$
d. $y \leq 2$

4. Find the range of the quadratic function *f* whose graph appears here.

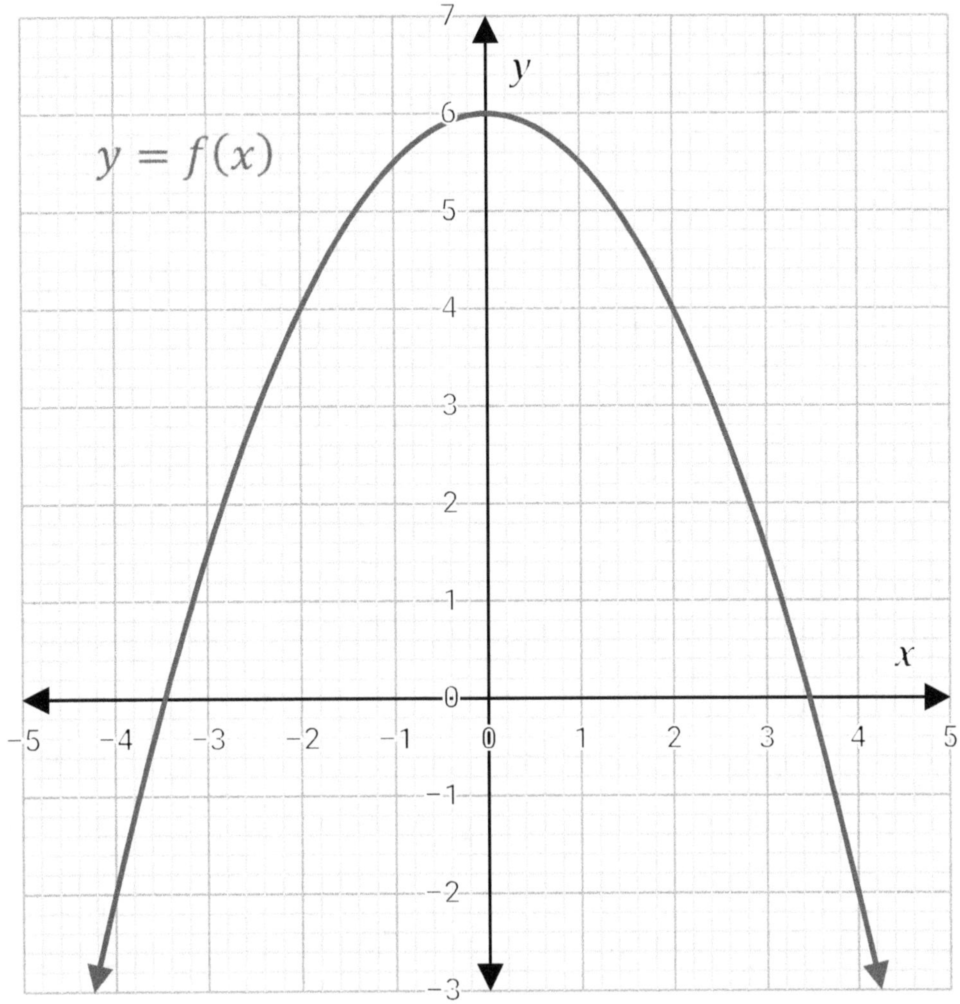

a. $y \geq 6$
b. $-\infty \leq y \leq \infty$
c. $y \leq 6$
d. $y \leq 0$

5. Find the range of the quadratic function *f* whose graph appears here.

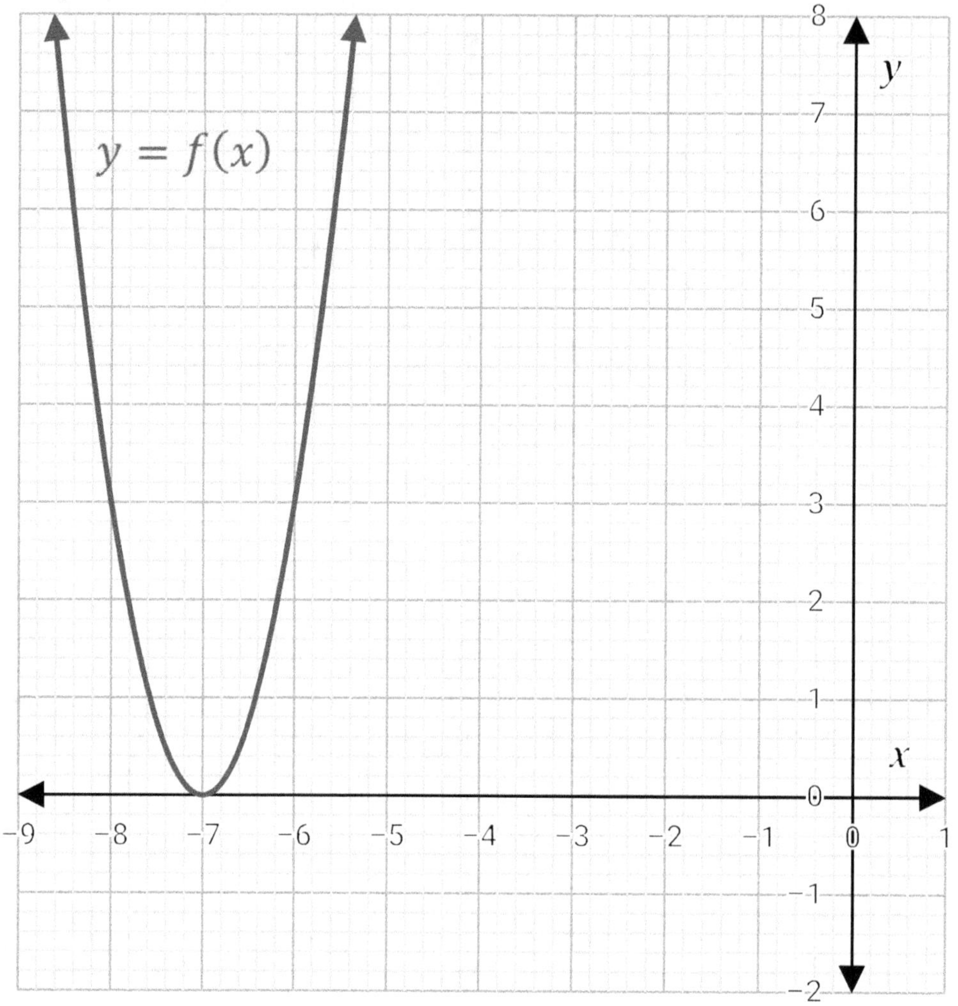

a. $-\infty \leq y \leq \infty$
b. $y \geq -7$
c. $y \geq 0$
d. $y \leq -7$

6. Find the range of the quadratic function f whose graph and vertex appear here.

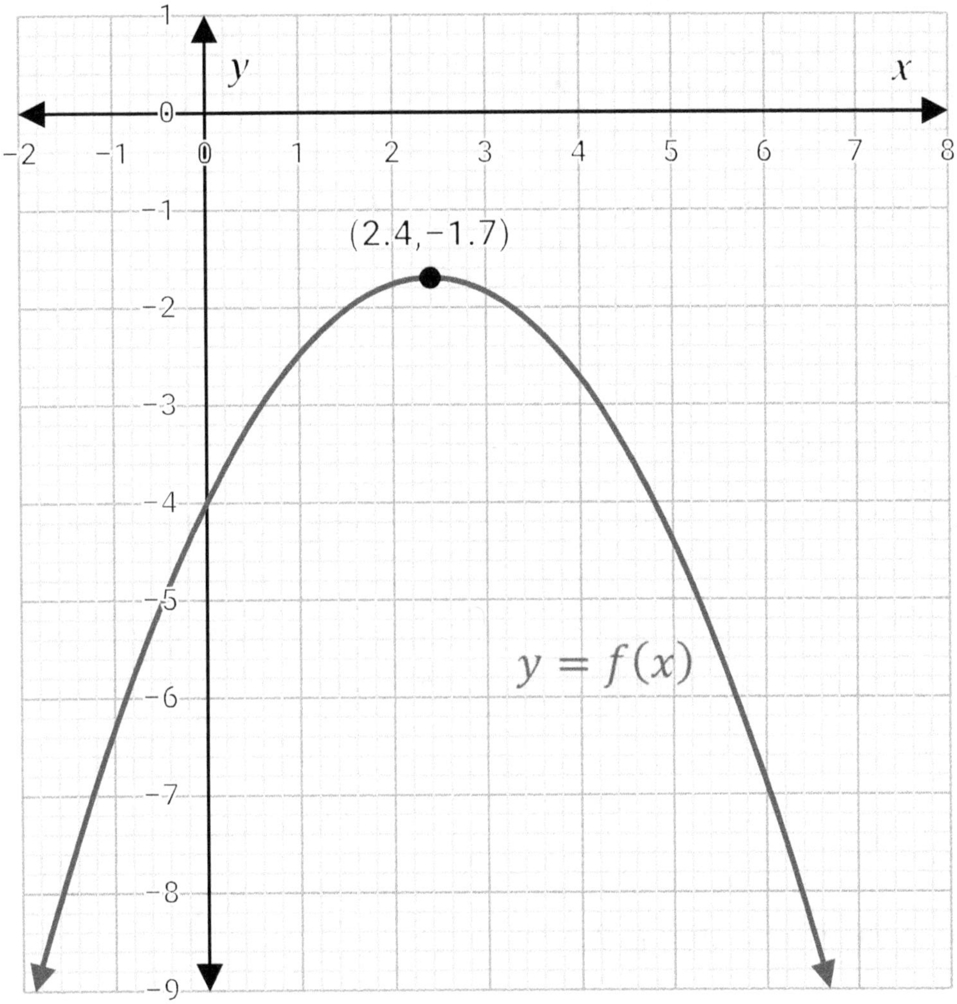

a. $y \leq 2.4$
b. $y \leq -1.7$
c. $y \geq -1.7$
d. $y \geq 2.4$

7. Find the domain of the function $f(x) = -2x^2 - 12x - 12$.

a. $x \leq 6$
b. $-\infty \leq x \leq \infty$
c. $x \leq -3$
d. $x \leq 0$

8. Find the domain and range of the function $f(x) = x^2 - 8x + 14$.

a. Domain: $-\infty \leq x \leq \infty$, Range: $y \leq -2$
b. Domain: $x \geq 4$, Range: $y \geq -2$
c. Domain: $-\infty \leq x \leq \infty$, Range: $y \geq 4$
d. Domain: $-\infty \leq x \leq \infty$, Range: $y \geq -2$

9. Find the range of the function $f(x) = -x^2 - 10x - 26$.

a. $y \leq -1$
b. $y \geq -1$
c. $y \leq -5$
d. $y \geq -5$

10. Find the range of the function $f(x) = 6x^2 - 11$.

a. $y \geq 6$
b. $y \geq 0$
c. $y \geq -11$
d. $y \leq -11$

11. Find the range of the function $f(x) = -3x^2 + 42x - 147$.

a. $y \leq 49$
b. $y \leq 7$
c. $y \leq -7$
d. $y \leq 0$

12. Find the range of the function $f(x) = 2x^2 + 6x + 7$.

a. $y \geq -\frac{3}{2}$
b. $y \leq -\frac{3}{2}$
c. $y \leq \frac{5}{2}$
d. $y \geq \frac{5}{2}$

13. You throw a rock straight upward, and 6 seconds later it lands at your feet. Let $s(t)$ be the height, in feet, of the rock above the ground t seconds after you throw it. Then s is a quadratic function. If the following diagram shows the graph of $s(t)$, including the vertex at $(3, 144)$, find the domain and range of the function.

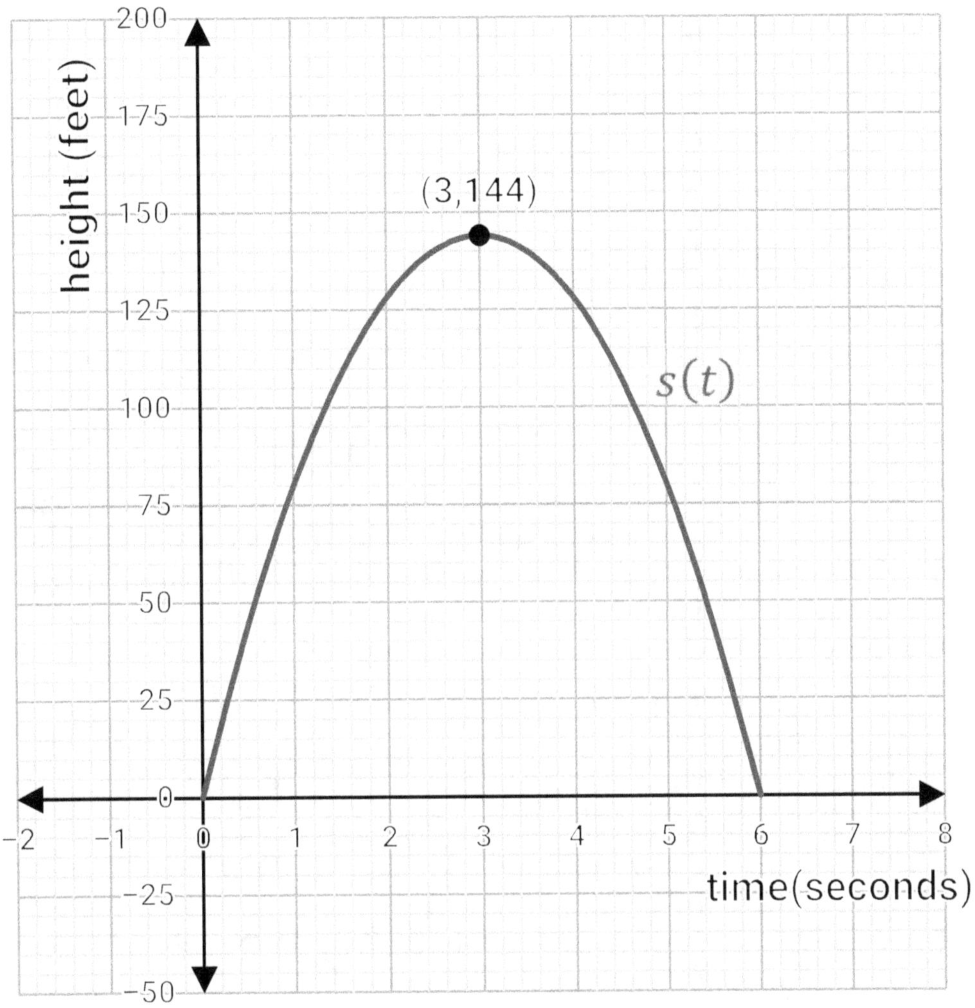

a. Domain: $0 \leq t \leq 3$, Range: $0 \leq s(t) \leq 144$
b. Domain: $0 \leq t \leq 6$, Range: $0 \leq s(t) \leq 144$
c. Domain: $0 \leq t \leq 6$, Range: $0 \leq s(t) \leq 3$
d. Domain: $0 \leq t \leq 144$, Range: $0 \leq s(t) \leq 6$

14. You manufacture widgets at a factory that can produce and sell up to 100 widgets per day. Careful research shows that a quadratic function $P(x)$, whose graph appears here with its endpoints and vertex labeled, is a good model for your profit, in dollars, when you produce x widgets per day. What is the range of $P(x)$, your range of possible daily profits?

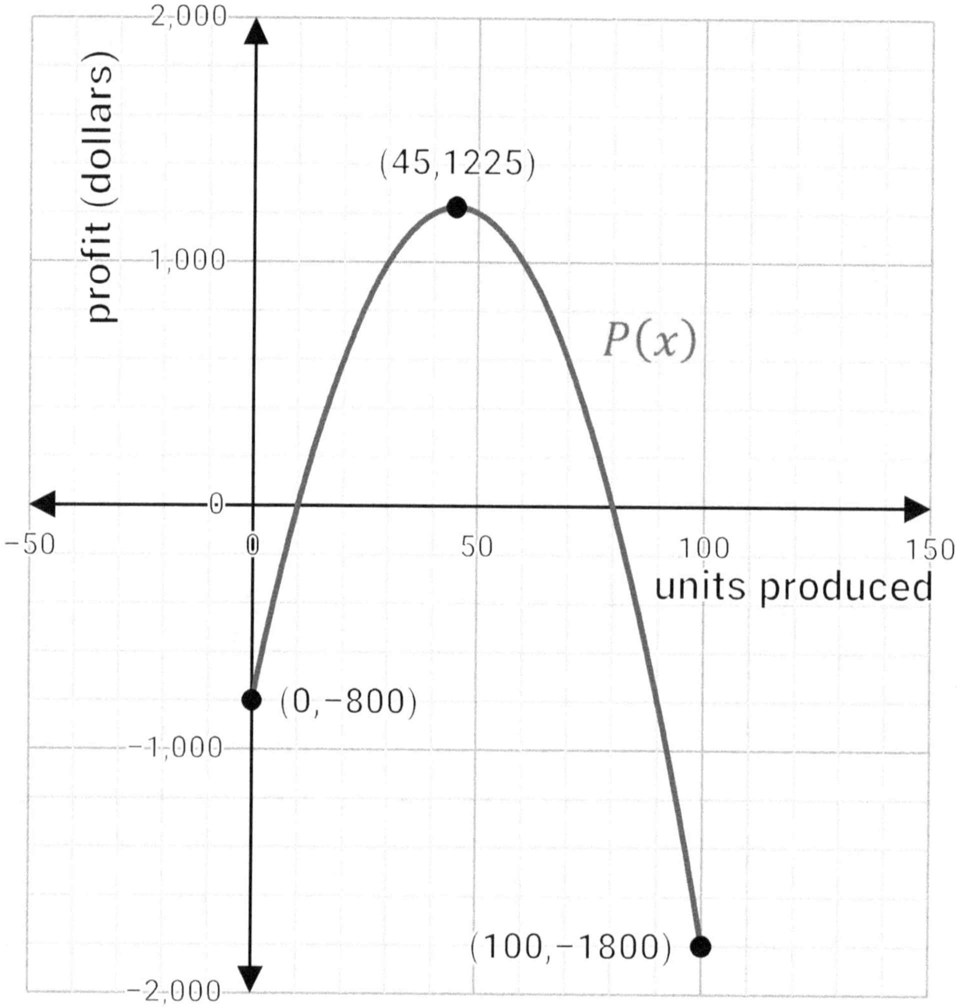

a. $-1{,}800 \leq P(x) \leq 1{,}225$
b. $-1{,}800 \leq P(x) \leq -800$
c. $10 \leq P(x) \leq 80$
d. $0 \leq P(x) \leq 100$

15. Standing on a tower, you extend your arm out and throw a rock straight upward. Four seconds later it lands on the ground below. Suppose that $s(t) = -16t^2 + 48t + 64$ is the height, in feet, of the rock above the ground t seconds after you throw it. Find the range of s.

a. $0 \leq s(t) \leq 4$
b. $0 \leq s(t) \leq 64$
c. $0 \leq s(t) \leq 100$
d. $1.5 \leq s(t) \leq 100$

8.2 Writing Equations Given Vertex and a Point

Many quadratic equations are expressed in **standard form**: $y = ax^2 + bx + c$, where $a \neq 0$. However, quadratic equations can also be expressed in **vertex form**: $y = a(x-h)^2 + k$, where $a \neq 0$.

Let's review what the constants and variables represent in vertex form.

$$y = a(x-h)^2 + k$$

In vertex form, a is a constant that indicates if the parabola opens up or down. A positive value for a means the parabola opens up. A negative value for a means the parabola opens down.

The variables x and y represent an ordered pair (x, y) that is on the parabola. There can be many values of x and y for a quadratic function because x and y can represent any ordered pair on the parabola.

The constants h and k represent the ordered pair of the vertex (h, k). The values for h and k are unique from other ordered pairs because they represent one specific point on the parabola.

When we know a point on the graph and the vertex of the parabola, we have enough information to express the equation of the parabola in vertex form. For example, if we know that $(0, -3)$ is a point on the graph and the vertex is $(2, -5)$, we can write the equation in vertex form by plugging in the values for the variables we are given.

$$y = a(x-h)^2 + k$$
$$-3 = a(0-2)^2 + (-5)$$

Now, solve for a. Remember to use the order of operations carefully on this step.

$$-3 = a(0-2)^2 + (-5)$$
$$-3 = a(-2)^2 - 5$$
$$-3 = 4a - 5$$
$$2 = 4a$$
$$\frac{1}{2} = a$$

Now that we know the value of a, plug this into the vertex form equation. Plug in values for a, h, and k. Leave x and y as variables. $y = a(x-h)^2 + k$ becomes $y = \frac{1}{2}(x-2)^2 - 5$. This is the equation in vertex form.

Examples

Example 1:

Write a quadratic equation in vertex form that has a vertex of $(1, -8)$ and passes through the point $(2, -6)$.

Answer:

Vertex form: $y = a(x-h)^2 + k$

Plug in the point $(2, -6)$ for x and y.

$$-6 = a(2-h)^2 + k$$

Plug in the vertex $(1,-8)$ for h and k.

$$-6 = a(2-1)^2 - 8$$

Solve for a.

$$\begin{aligned} -6 &= a(1)^2 - 8 \\ -6 &= a - 8 \\ 2 &= a \end{aligned}$$

Plug in values for a, h, and k into vertex form.

$$y = 2(x-1)^2 - 8$$

Example 2:

Write an equation in vertex form that has a vertex of $(-2,5)$ and passes through the point $(-4,-3)$.

Answer:

Vertex form: $y = a(x-h)^2 + k$

Plug in the point $(-4,-3)$ for x and y.

$$-3 = a(-4-h)^2 + k$$

Plug in the vertex $(-2,5)$ for h and k.

$$\begin{aligned} -3 &= a(-4-(-2))^2 + 5 \\ -3 &= a(-4+2)^2 + 5 \end{aligned}$$

Solve for a.

$$\begin{aligned} -3 &= a(-2)^2 + 5 \\ -3 &= 4a + 5 \\ -8 &= 4a \\ -2 &= a \end{aligned}$$

Plug in values for a, h, and k into vertex form.

$$y = -2(x+2)^2 + 5$$

Example 3:

An interior designer is designing an archway in the living room of a home. While explaining her idea to her contractor, they sketch out the design on a coordinate plane. The top of the arch is at the point $(-4,3)$, and the arch also passes through the point $(2,-1)$. What quadratic equation best represents the archway?

Answer:

Use the vertex form of a quadratic equation, $y = a(x-h)^2 + k$, where (h,k) is the vertex point and (x,y) is another point on the curve. Substitute the values for the vertex point.

$$y = a\left(x-(-4)\right)^2 + 3$$

$$y = a(x+4)^2 + 3$$

Now, substitute the values for the other point given.

$$-1 = a(2+4)^2 + 3$$

From here, use inverse operations to solve for the value of a.

$$\begin{aligned} -1 &= a(6)^2 + 3 \\ -1 &= 36a + 3 \\ -4 &= 36a \\ -\frac{1}{9} &= a \end{aligned}$$

Now, substitute this value for a and the values for h and k into the vertex form equation.

$$y = -\frac{1}{9}(x+4)^2 + 3$$

Therefore, the equation that best represents the archway is $y = -\frac{1}{9}(x+4)^2 + 3$.

PRACTICE QUESTIONS

1. Write the equation of a quadratic function in vertex form that has a vertex at $(2,3)$ and passes through the point $(3,8)$.

a. $y = 11(x-2)^2 + 3$
b. $y = 5(x-2)^2 + 3$
c. $y = 5(x^2 - 2) + 3$
d. $y = 5(x+2)^2 - 3$

2. Write the equation of a quadratic function in vertex form that has a vertex at $(0,0)$ and passes through the point $(1,3)$.

a. $y = 2x^2$
b. $y = \frac{1}{3}x^2$
c. $y = x^2$
d. $y = 3x^2$

3. Write the equation of a quadratic function in vertex form that has a vertex at $(3,-2)$ and passes through the point $(5,2)$.

a. $y = (x-3)^2 - 2$
b. $y = 4(x-3)^2 - 2$
c. $y = (x+3)^2 + 2$
d. $y = 4(x+3)^2 + 2$

4. Write the equation of a quadratic function in vertex form that has a vertex at $(-4, 1)$ and passes through the point $(-2, 5)$.

a. $y = 6(x + 4)^2 - 1$

b. $y = (x^2 + 4) + 1$

c. $y = (x + 4)^2 + 1$

d. $y = 6(x + 4)^2 + 1$

5. Write the equation of a quadratic function in vertex form that has a vertex at $(4, 2)$ and passes through the point $(3, 3)$.

a. $y = (x + 4)^2 - 2$

b. $y = (x - 4)^2 + 2$

c. $y = -(x - 4)^2 + 2$

d. $y = 5(x - 4)^2 + 2$

6. Write the equation of a quadratic function in vertex form that has a vertex at $(4, -1)$ and passes through the point $(3, 3)$.

a. $y = 4(x + 4)^2 - 1$

b. $y = 4(x - 4)^2 - 1$

c. $y = 2(x - 4)^2 - 1$

d. $y = 4(x - 4)^2 + 1$

7. Write the equation of a quadratic function in vertex form that has a vertex at $(4, -9)$ and passes through the point $(2, -1)$.

a. $y = 2(x^2 - 4) - 9$

b. $y = 2(x - 4)^2 + 9$

c. $y = 4(x - 4)^2 - 9$

d. $y = 2(x - 4)^2 - 9$

8. Write the equation of a quadratic function in vertex form that has a vertex at $(-3, -1)$ and passes through the point $(-2, 0)$.

a. $y = (x - 3)^2 - 1$

b. $y = (x - 3)^2 + 1$

c. $y = (x + 3)^2 - 1$

d. $y = (x + 3)^2 + 1$

9. Write the equation of a quadratic function in vertex form that has a vertex at $(-1, -4)$ and passes through the point $(2, -1)$.

a. $y = 3(x + 1)^2 + 4$

b. $y = \frac{1}{3}(x + 1)^2 + 4$

c. $y = 3(x + 1)^2 - 4$

d. $y = \frac{1}{3}(x + 1)^2 - 4$

10. Write the equation of a quadratic function in vertex form that has a vertex at $(-6, 8)$ and passes through the point $(-4, 10)$.

a. $y = \frac{1}{2}(x + 6)^2 + 8$
b. $y = 2(x + 6)^2 + 8$
c. $y = \frac{1}{2}x^2 + 14$
d. $y = 2(x - 6)^2 - 8$

11. Write the equation of a quadratic function in vertex form that has a vertex at $(0, 4)$ and passes through the point $(8, -12)$.

a. $y = -\frac{1}{4}x^2 + 4$
b. $y = \frac{1}{4}x^2 + 4$
c. $y = -\frac{1}{4}x^2 - 4$
d. $y = -4x^2 + 4$

12. Write the equation of a quadratic function in vertex form that has a vertex at $(-4, 6)$ and passes through the point $(-1, 9)$.

a. $y = 3(x + 4)^2 + 6$
b. $y = \frac{1}{3}(x + 4)^2 + 6$
c. $y = \frac{1}{3}(x^2 + 4) + 6$
d. $y = \frac{1}{3}(x - 4)^2 - 6$

13. Marco is a civil engineer who is examining a model of an archway on the coordinate plane. If the highest point of the archway's graph is at the point $(0, 0)$ and it also passes through the point $(-2, -12)$, which quadratic equation best represents the archway?

a. $y = 3x^2$
b. $y = -3x^2$
c. $y = -\frac{1}{3}x^2$
d. $y = \frac{1}{3}x^2$

14. The pathway of a water jet in an outdoor fountain follows a parabolic trajectory. As the fountain's jet sprays out water, gravity pulls it back down. When graphed on the coordinate plane, this parabola has a vertex of $(-3, 0)$ and passes through the point $(-5, -4)$. Based on this information, which quadratic equation best represents the scenario?

a. $y = -4(x + 3)^2$
b. $y = 4(x + 3)^2$
c. $y = (x + 3)^2$
d. $y = -(x + 3)^2$

15. Buzzy is a structural engineer who is designing a new roller coaster. He is working on a portion of the tracks that is in the shape of a parabola. When graphed on the coordinate plane, this parabola has a vertex of $(2,5)$ and passes through the point $(3,7)$. Buzzy needs to write a quadratic equation in vertex form that represents this portion of the tracks. Which equation should Buzzy write?

a. $y = 2(x-2)^2 + 5$
b. $y = 12(x-2)^2 + 5$
c. $y = -2(x-2)^2 + 5$
d. $y = -12(x-2)^2 + 5$

8.3 Factoring Quadratic Equations

One way to solve quadratic equations is by factoring the equation. To do this, the equation needs to be in standard form.

Standard Form: $y = ax^2 + bx + c$

$$0 = x^2 + 8x + 15$$

The values of x can be determined by factoring the equation. When factoring a quadratic equation where $a = 1$, start by separating the expression on the right into two sets of parentheses with an x in each one.

$$0 = (x \quad)(x \quad)$$

From here, determine what two numbers multiply to 15 and add to 8. Those numbers are +3 and +5, so write these in the parentheses after the x's.

$$0 = (x + 3)(x + 5)$$

We want to find the roots of the function—i.e., when y is equal to 0—so solve for x by setting each factor equal to zero.

$$x + 3 = 0 \qquad\qquad x + 5 = 0$$
$$x = -3 \qquad\qquad x = -5$$

This means that the solutions to the quadratic equation are $x = -3$ and $x = -5$. When the quadratic function is graphed, the parabola will pass through the x-axis at $x = -3$ and $x = -5$.

Review Video: Factoring Quadratic Equations
Visit mometrix.com/academy and enter code: 336566

Examples

Example 1:

Find the solutions for the quadratic equation by factoring.

$$2x^2 + 7x + 6 = 0$$

Answer:

Factor the quadratic.

$$(2x + 3)(x + 2) = 0$$

Set each factor equal to 0, and then solve for x.

$$2x + 3 = 0 \qquad\qquad x + 2 = 0$$
$$2x = -3 \qquad\qquad x = -2$$
$$x = -\frac{3}{2}$$

These x-values are the solutions for the quadratic function.

$$x = -\frac{3}{2} \text{ and } x = -2$$

Example 2:

Solve the equation $x^2 + x = 20$ by factoring.

Answer:

First, convert the quadratic equation to standard form by setting the right side equal to 0.

$$x^2 + x - 20 = 0$$

From here, factor the left side of the equation.

$$(x + 5)(x - 4) = 0$$

Then, set each factor equal to 0 and solve for x.

$$\begin{aligned} x + 5 &= 0 \\ x &= -5 \end{aligned} \qquad \begin{aligned} x - 4 &= 0 \\ x &= 4 \end{aligned}$$

The solutions to the quadratic equation are $x = -5$ and $x = 4$.

Example 3:

A rectangular garden has an area of 10 square meters. Its length is l and its width is $3l + 5$. What is the width of the garden?

Answer:

First, write an equation to represent the scenario. The area of a rectangle is found by multiplying length by width.

$$\begin{aligned} A &= lw \\ 10 &= l(3l + 5) \end{aligned}$$

From here, convert the equation to standard form of a quadratic equation.

$$\begin{aligned} 10 &= 3l^2 + 5l \\ 0 &= 3l^2 + 5l - 10 \end{aligned}$$

Now, factor the equation and solve for the zeroes.

$$\begin{aligned} 0 &= (3l + 5)(l - 2) \\ 3l + 5 &= 0 \\ 3l &= -5 \end{aligned} \qquad \begin{aligned} l &= -\frac{5}{3} \\ l - 2 &= 0 \\ l &= 2 \end{aligned}$$

Since you can't have a negative length, the value for l must be positive. Therefore, the length of the garden is 2 meters. To find the width, substitute 2 for l in $3l + 5$.

$$3(2) + 5 = 6 + 5 = 11$$

The width of the garden is 11 square meters.

Practice Questions

1. Solve the equation $x^2 + 9x = -8$ by factoring.

a. $x = 9$ and $x = -8$

b. $x = -8$ and $x = -1$

c. $x = -9$ and $x = 1$

d. $x = 1$ and $x = -8$

2. Solve the equation $x^2 - 6x + 8 = 0$ by factoring.

a. $x = -2$ and $x = -4$

b. $x = 3$ and $x = -1$

c. $x = 2$ and $x = 4$

d. $x = -3$ and $x = -1$

3. Solve the equation $x^2 + 2x - 15 = 0$ by factoring.

a. $x = -3$ and $x = 5$

b. $x = -5$ and $x = 3$

c. $x = 2$ and $x = 8$

d. $x = -2$ and $x = -8$

4. Solve the equation $x^2 = x + 12$ by factoring.

a. $x = 12$ and $x = 1$

b. $x = -4$ and $x = 3$

c. $x = -6$ and $x = 2$

d. $x = -3$ and $x = 4$

5. Solve the equation $2x^2 + 5x = -3$ by factoring.

a. $x = \frac{2}{3}$ and $x = -3$

b. $x = -\frac{3}{2}$ and $x = -1$

c. $x = \frac{2}{3}$ and $x = 1$

d. $x = 5$ and $x = 3$

6. Solve the equation $x^2 = 15x - 56$ by factoring.

a. $x = 7$ and $x = 8$

b. $x = -3$ and $x = -5$

c. $x = -7$ and $x = -8$

d. $x = 3$ and $x = 5$

7. Solve the equation $2x^2 + x = 21$ by factoring.

a. $x = -\frac{7}{2}$ and $x = 3$

b. $x = -14$ and $x = 3$

c. $x = 2$ and $x = 21$

d. $x = -3$ and $x = 7$

8. Solve the equation $x^2 = 7x + 18$ by factoring.

a. $x = -3$ and $x = 6$

b. $x = -9$ and $x = 2$

c. $x = -2$ and $x = 9$

d. $x = -6$ and $x = 3$

9. Solve the equation $2(x^2 - 6x + 8) = 0$ by factoring.

a. $x = 2$ and $x = 4$

b. $x = \frac{1}{2}$ and $x = 4$

c. $x = 8$ and $x = -4$

d. $x = 2$ and $x = -8$

10. Solve the equation $x^2 + 6x + 8 = 0$ by factoring.

a. $x = 3$

b. $x = 6$ and $x = 9$

c. $x = 2$ and $x = 4$

d. $x = -4$ and $x = -2$

11. Solve the equation $9x^2 = 64$ by factoring.

a. $x = -\frac{3}{8}$ and $x = \frac{3}{8}$

b. $x = -3$ and $x = 8$

c. $x = -8$ and $x = 3$

d. $x = -\frac{8}{3}$ and $x = \frac{8}{3}$

12. Solve the equation $x^2 - 49 = 0$ by factoring.

a. $x = 1$ and $x = -49$

b. $x = -7$ and $x = 7$

c. $x = 0$

d. $x = -7$ and $x = 49$

13. A plot of land has a width of w feet and a length of $2w + 3$ feet. If the plot of land has an area of 90 square feet, what is the width of the land?

a. 5 feet
b. 15 feet
c. 6 feet
d. $\frac{15}{2}$ feet

14. Maddie is creating a garden in her backyard. The area of her garden is 24 square meters. The length of the garden is 2 meters more than its width. What is the length of the garden?

a. 10 meters
b. 8 meters
c. 6 meters
d. 4 meters

15. The product of two numbers is 45, and their sum is 14. Which quadratic equation represents this scenario where s is one of the numbers?

a. $s(45 + s) = 14$
b. $s(45 - s) = 14$
c. $s(14 + s) = 45$
d. $s(14 - s) = 45$

8.4 Writing Quadratic Functions Given Solutions and Graphs

When you know the solutions of a quadratic function, you can work backwards to find the original equation. When the solutions are known, they can be substituted along with a known point into the **intercept form** of a quadratic equation, $y = a(x - p)(x - q)$, where a is a constant, and p and q are solutions to the quadratic function.

Let's consider a quadratic that has the solutions $x = -1$ and $x = 3$ and that passes through the point $(0, -5)$. To find the original quadratic function, start by substituting each solution into the intercept form equation.

$$y = a(x + 1)(x - 3)$$

From here, substitute the given point in for x and y, then solve for a.

$$\begin{aligned} -5 &= a(0 + 1)(0 - 3) \\ -5 &= a(1)(-3) \\ -5 &= -3a \\ \frac{5}{3} &= a \end{aligned}$$

Finally, substitute this value in for a in the previous intercept form equation and use the FOIL method to convert the equation to standard form.

$$\begin{aligned} y &= \frac{5}{3}(x + 1)(x - 3) \\ y &= \frac{5}{3}(x^2 - 3x + x - 3) \\ y &= \frac{5}{3}(x^2 - 2x - 3) \\ y &= \frac{5}{3}x^2 - \frac{10}{3}x - 5 \end{aligned}$$

Similarly, when you know what the graph of a quadratic function looks like, you can write the original equation.

For example, a quadratic function is graphed below. The **solutions** are the x-intercepts, or roots. We can see that the x-intercepts are located at $x = -1$ and $x = 2$, and we can see that the graph passes through the point $(0, -2)$.

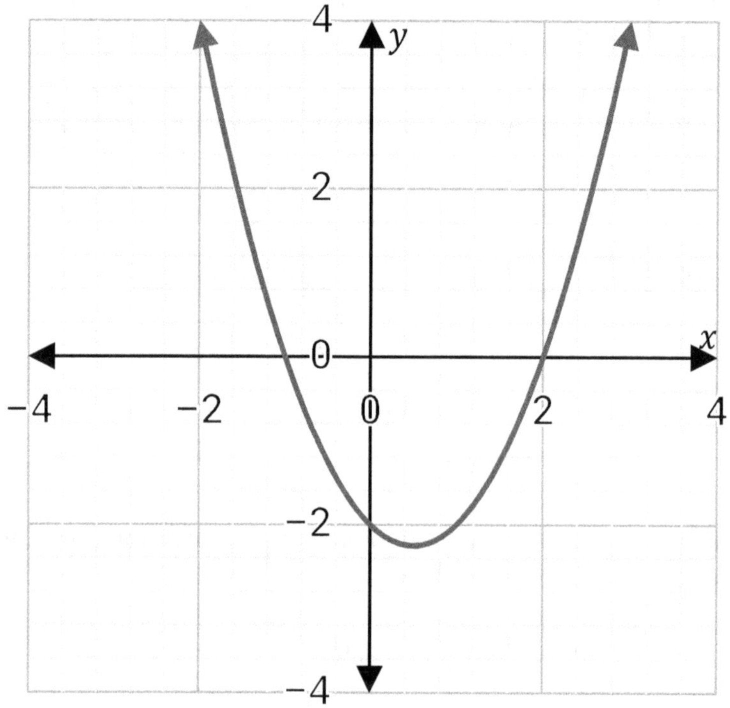

Start by substituting the solutions into the intercept form equation.

$$y = a(x + 1)(x - 2)$$

From here, substitute the additional point, $(0, -2)$, for x and y, then solve for a.

$$\begin{aligned} -2 &= a(0 + 1)(0 - 2) \\ -2 &= a(1)(-2) \\ -2 &= -2a \\ 1 &= a \end{aligned}$$

Now, substitute this into the previous intercept form equation.

$$y = 1(x + 1)(x - 2)$$

Finally, convert the equation to standard form using the FOIL method.

$$\begin{aligned} y &= (x + 1)(x - 2) \\ y &= x^2 - 2x + x - 2 \\ y &= x^2 - x - 2 \end{aligned}$$

Examples

Example 1:

Write a quadratic function in standard form that has the solutions $x = -3$ and $x = 5$ and that passes through the point $(2, -10)$.

Answer:

Start by substituting the solutions into the intercept form equation.

$$y = a(x + 3)(x - 5)$$

Then, substitute the given point, $(2, -10)$, for x and y, and solve for a.

$$\begin{aligned} -10 &= a(2 + 3)(2 - 5) \\ -10 &= a(5)(-3) \\ -10 &= -15a \\ \frac{2}{3} &= a \end{aligned}$$

Next, substitute this back into the intercept form equation from before.

$$y = \frac{2}{3}(x + 3)(x - 5)$$

Finally, convert the equation to standard form

$$\begin{aligned} y &= \frac{2}{3}(x^2 - 5x + 3x - 15) \\ y &= \frac{2}{3}(x^2 - 2x - 15) \\ y &= \frac{2}{3}x^2 - \frac{4}{3}x - 10 \end{aligned}$$

EXAMPLE 2:

Use the graph to write a quadratic function in standard form.

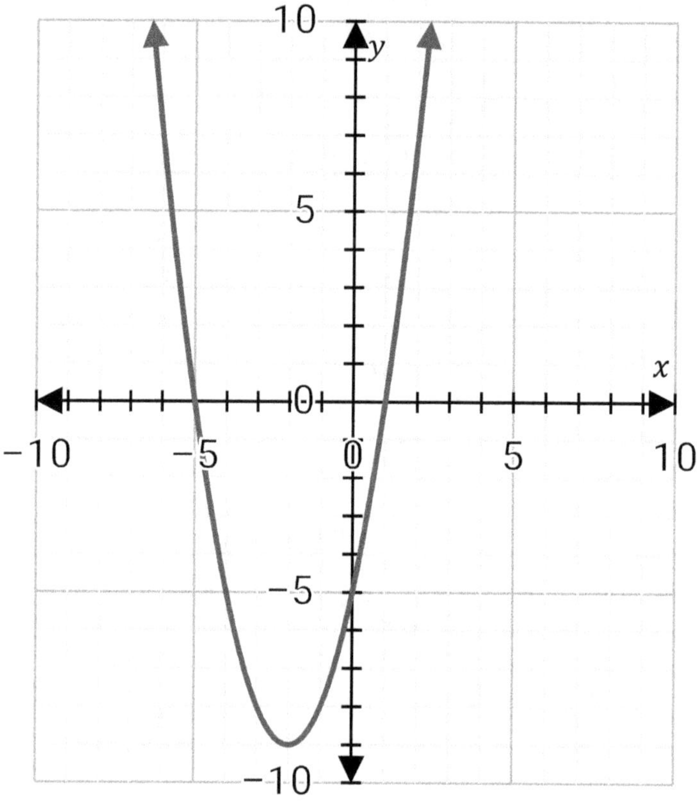

Answer:

Use the graph to identify the solutions. The solutions are also referred to as the "roots" or the "zeros." The solutions are found at the x-intercepts. In this case, $(-5,0)$ and $(1,0)$.

First, substitute the solutions into the intercept form equation.

$$y = a(x + 5)(x - 1)$$

From here, identify another point on the graph. The vertex is at $(-2, -9)$, so this point can be used for x and y. Substitute this into the intercept form equation and solve for a.

$$\begin{aligned} -9 &= a(-2 + 5)(-2 - 1) \\ -9 &= a(3)(-3) \\ -9 &= -9a \\ 1 &= a \end{aligned}$$

Substitute this back into the intercept form equation from earlier.

$$y = 1(x + 5)(x - 1)$$

Finally, convert the equation to standard form.

$$y = (x + 5)(x - 1)$$
$$y = x^2 - x + 5x - 5$$
$$y = x^2 + 4x - 5$$

The graph represents the quadratic function $y = x^2 + 4x - 5$.

EXAMPLE 3:

Write an equation in standard form for the quadratic function shown below.

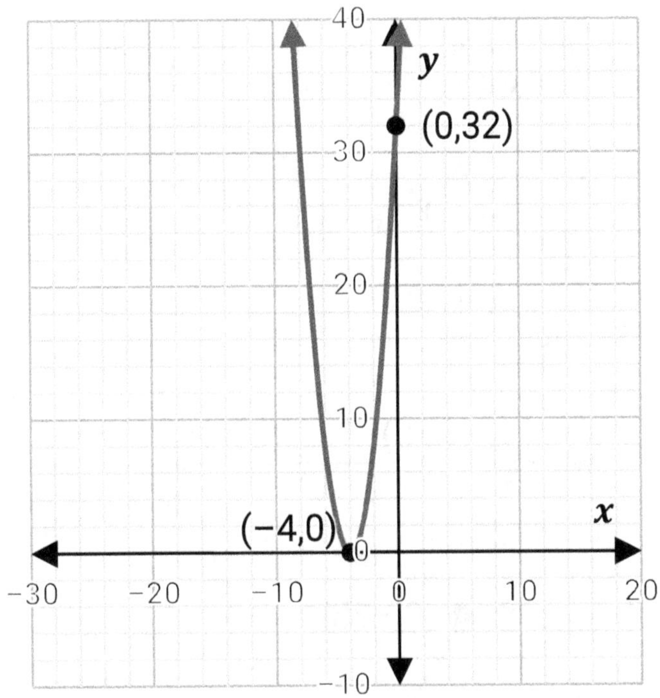

Answer:

The intercept form of a quadratic equation is $y = a(x - p)(x - q)$, where p and q are the roots of the equation. Since –4 is the only real root for the equation, it is a repeated root, which means we need to substitute –4 in for both p and q.

$$y = a(x - (-4))(x - (-4))$$
$$y = a(x + 4)(x + 4)$$

Next, substitute the point (0,32) in for x and y, and solve for a.

$$32 = a(0 + 4)(0 + 4)$$
$$32 = a(4)(4)$$
$$32 = 16a$$
$$2 = a$$

Now that the value of a is known, substitute it into the intercept form equation.

$$y = 2(x + 4)(x + 4)$$

Finally, convert the equation to standard form by multiplying the binomials and then distributing the 2.

$$y = 2(x^2 + 4x + 4x + 16)$$
$$y = 2(x^2 + 8x + 16)$$
$$y = 2x^2 + 16x + 32$$

EXAMPLE 4:

Jacob's teacher asks him to find an object in real life that is in the shape of a parabola. He chooses a jump rope that he holds on either handle and sketches it on a coordinate plane, as shown below. Based on the graph, what is the equation of this function?

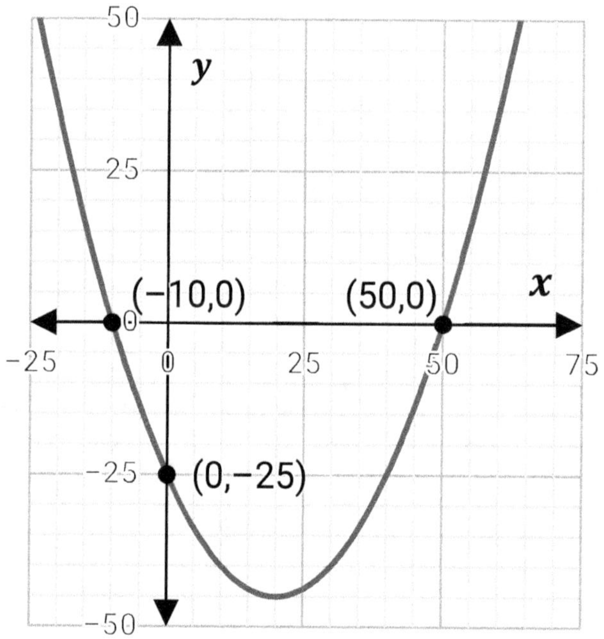

Answer:

To find this equation, use the intercept form of a quadratic equation, $y = a(x - p)(x - q)$, where p and q are the roots of the function. Substitute the roots from the graph into the equation.

$$y = a(x - (-10))(x - 50)$$
$$y = a(x + 10)(x - 50)$$

Next, multiply the binomials together.

$$y = a(x^2 - 40x - 500)$$

Then, substitute the point $(0, -25)$ in for x and y.

$$-25 = a[(0)^2 - 40(0) - 500]$$

From here, solve for a.

$$-25 = a(0 - 0 - 500)$$
$$-25 = -500a$$
$$\frac{1}{20} = a$$

Finally, substitute this value for a into the equation $y = a(x^2 - 40x - 500)$ from before and simplify.

$$y = \frac{1}{20}(x^2 - 40x - 500)$$
$$y = \frac{1}{20}x^2 - 2x - 25$$

PRACTICE QUESTIONS

1. Write an equation in standard form for the quadratic function shown below, with roots at $x = -3$ and $x = 4$.

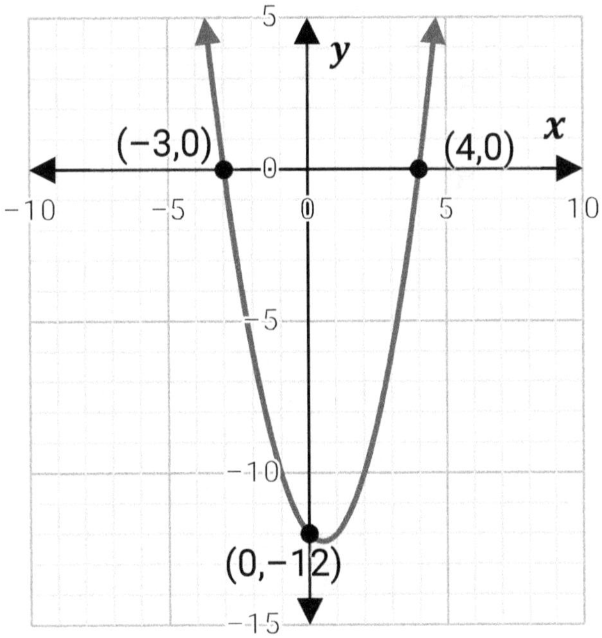

a. $y = x^2 - 7x - 7$
b. $y = x^2 + 7x + 7$
c. $y = -x^2 + x + 12$
d. $y = x^2 - x - 12$

2. Write an equation in standard form for the quadratic function shown below, with roots at $x = 2$ and $x = 6$.

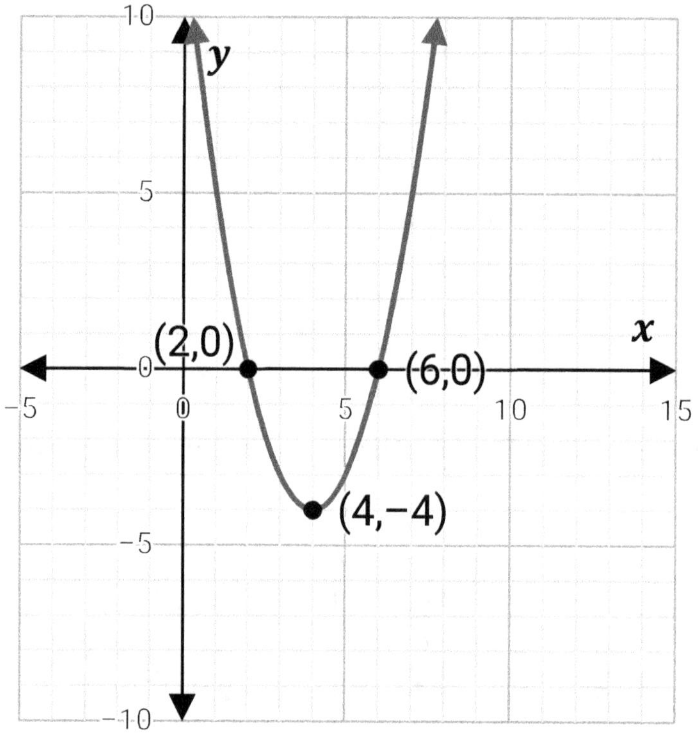

a. $y = x^2 - 8x + 12$
b. $y = x^2 + 8x + 12$
c. $y = x^2 + 8x - 8$
d. $y = x^2 - 12x + 8$

3. Write an equation in standard form for the quadratic function shown below, with roots at $x = -1$ and $x = 5$.

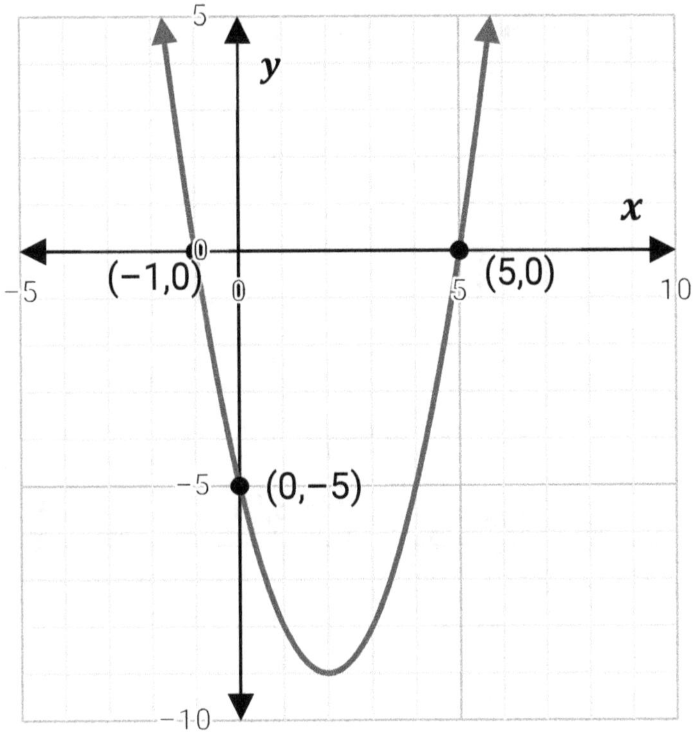

a. $y = x^2 - 6x + 5$
b. $y = x^2 + x - 5$
c. $y = x^2 - 5x - 5$
d. $y = x^2 - 4x - 5$

4. Write an equation in standard form for the quadratic function shown below.

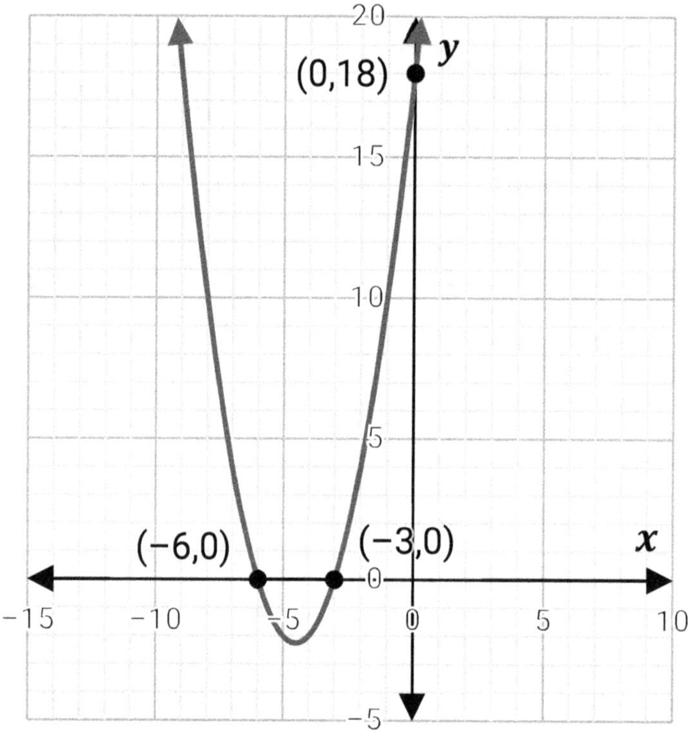

a. $y = x^2 + 9x + 18$

b. $y = x^2 - 9x + 18$

c. $y = x^2 + 9x - 9$

d. $y = x^2 - 9x + 9$

5. Write an equation in standard form for the quadratic function shown below.

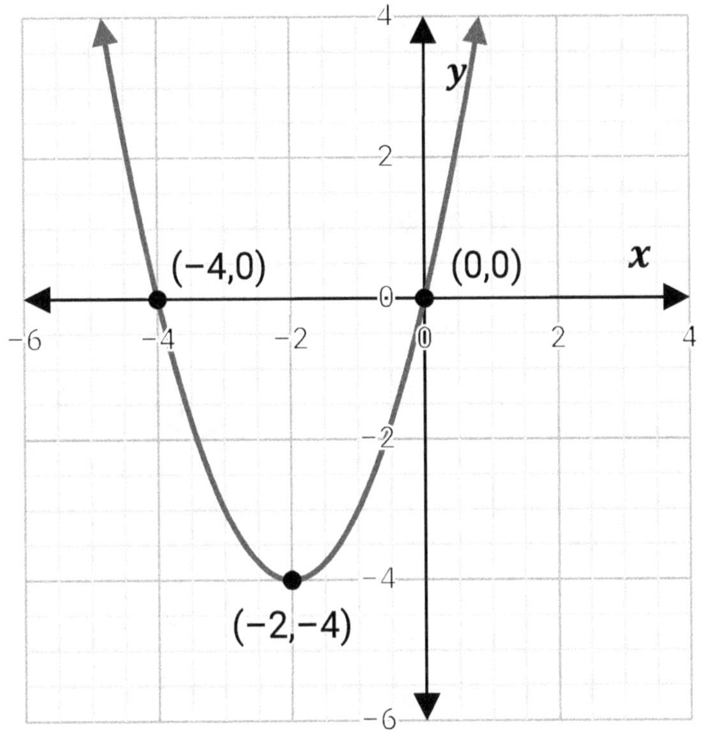

a. $y = x^2 + 4$
b. $y = x^2 - 4$
c. $y = x^2 + 4x$
d. $y = x^2 - 4x$

6. Write an equation in standard form for the quadratic function shown below.

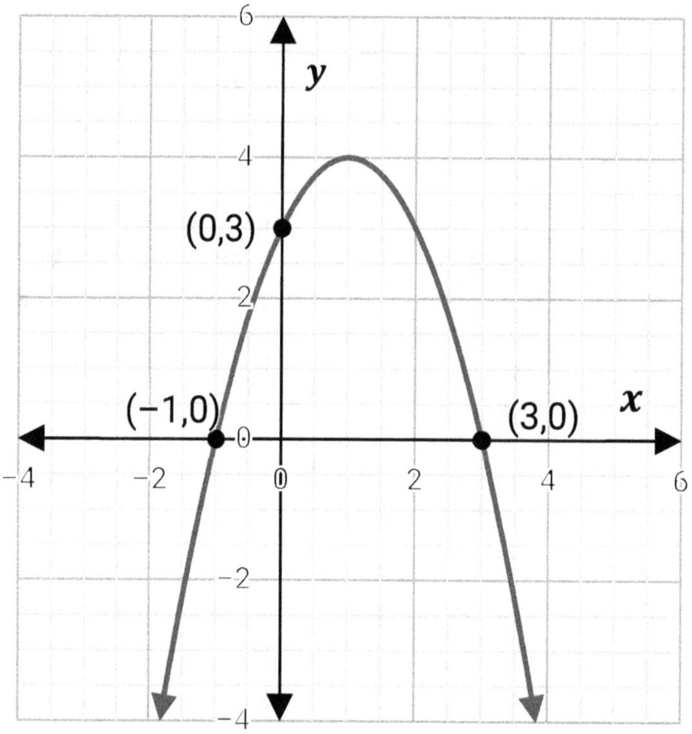

a. $y = x^2 - 2x - 3$
b. $y = -x^2 + 2x + 3$
c. $y = -x^2 - 2x - 3$
d. $y = -x^2 - 4x + 3$

7. Write an equation in standard form for the quadratic function shown below.

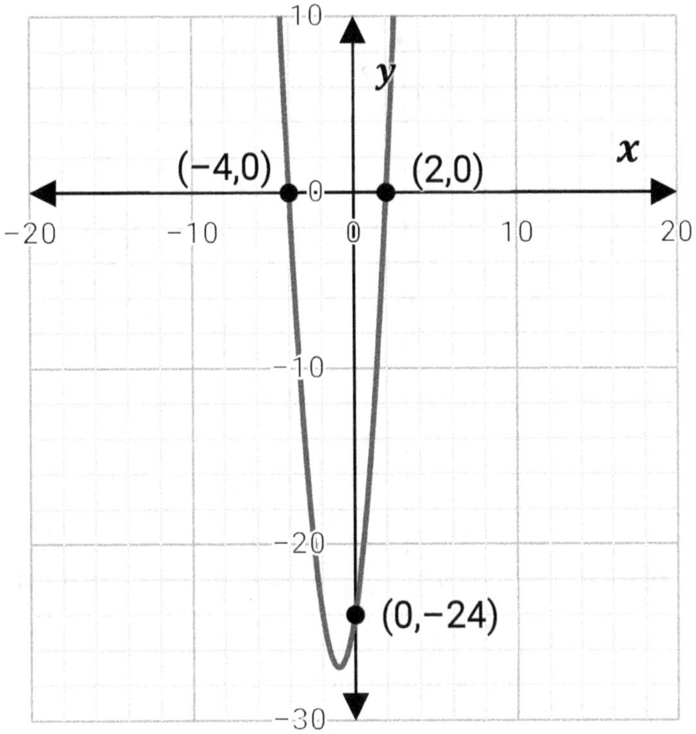

a. $y = x^2 + 2x - 8$
b. $y = 3x^2 + 6x - 24$
c. $y = 3x^2 + 2x - 8$
d. $y = 3x^2 + 5x - 5$

8. Write an equation in standard form for the quadratic function shown below.

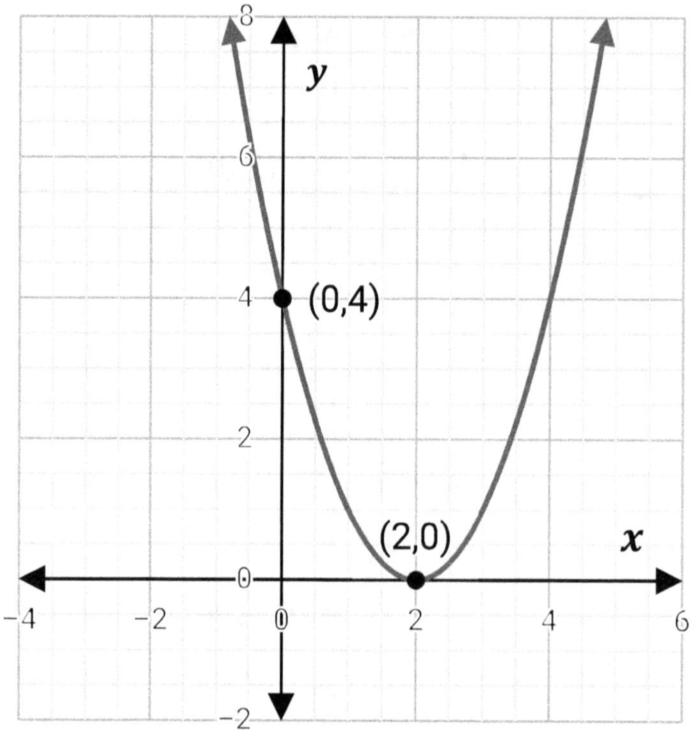

a. $y = x^2 - 4x + 4$
b. $y = 4x^2 - 16x + 16$
c. $y = x^2 + 4x - 4$
d. $y = 4x^2 + 16x - 16$

9. Write an equation in standard form for the quadratic function shown below.

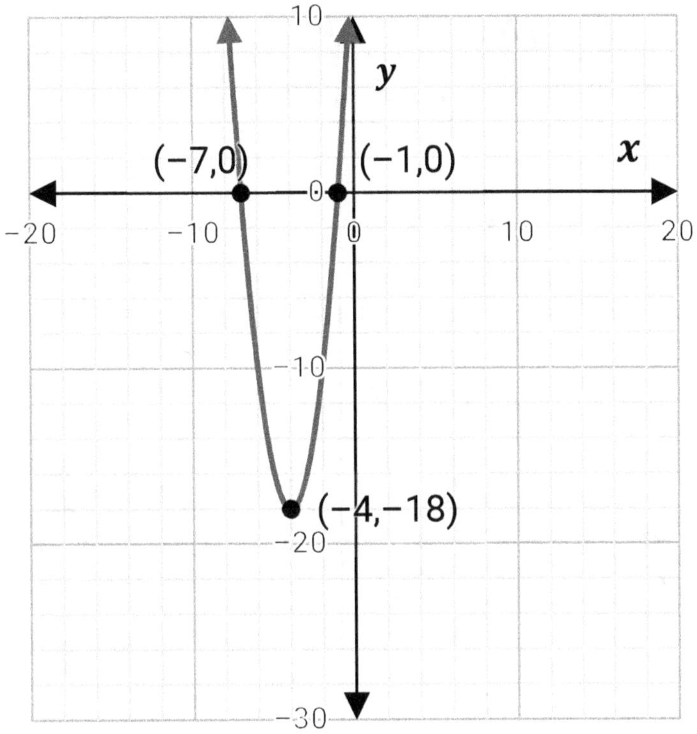

a. $y = 2x^2 - 16x + 14$
b. $y = x^2 - 8x + 7$
c. $y = x^2 + 8x + 7$
d. $y = 2x^2 + 16x + 14$

10. Write an equation in standard form for the quadratic function shown below.

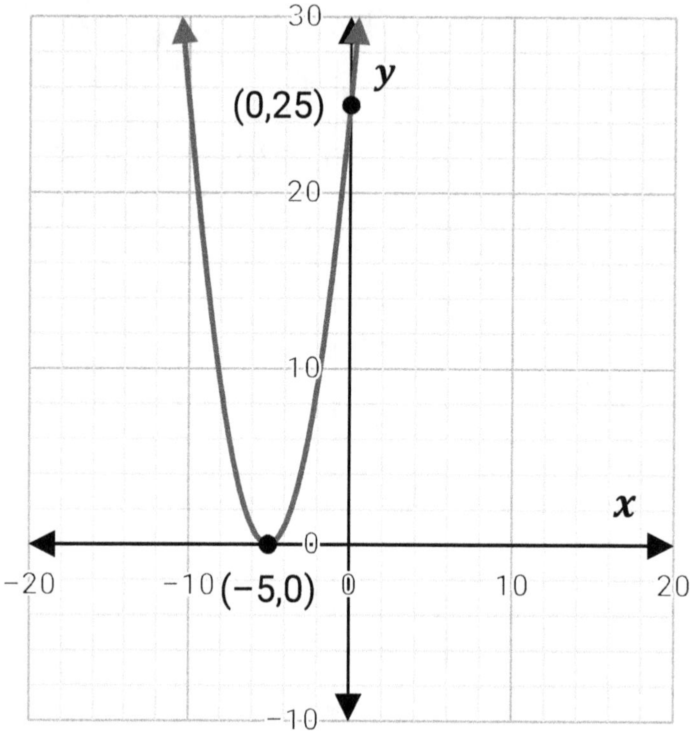

a. $y = x^2 - 10x - 10$
b. $y = x^2 + 10x + 10$
c. $y = x^2 + 10x + 25$
d. $y = x^2 - 10x + 25$

11. Write an equation in standard form for the quadratic function shown below.

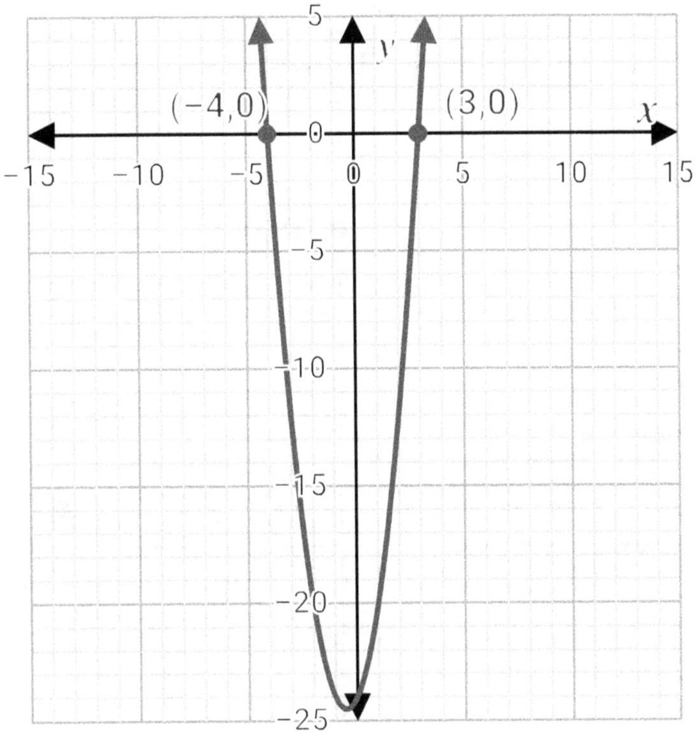

a. $y = -x^2 + 2x + 24$
b. $y = x^2 + x - 12$
c. $y = 2x^2 + 2x - 24$
d. $y = 2x^2 + 2x - 12$

12. Write an equation in standard form for the quadratic function shown below.

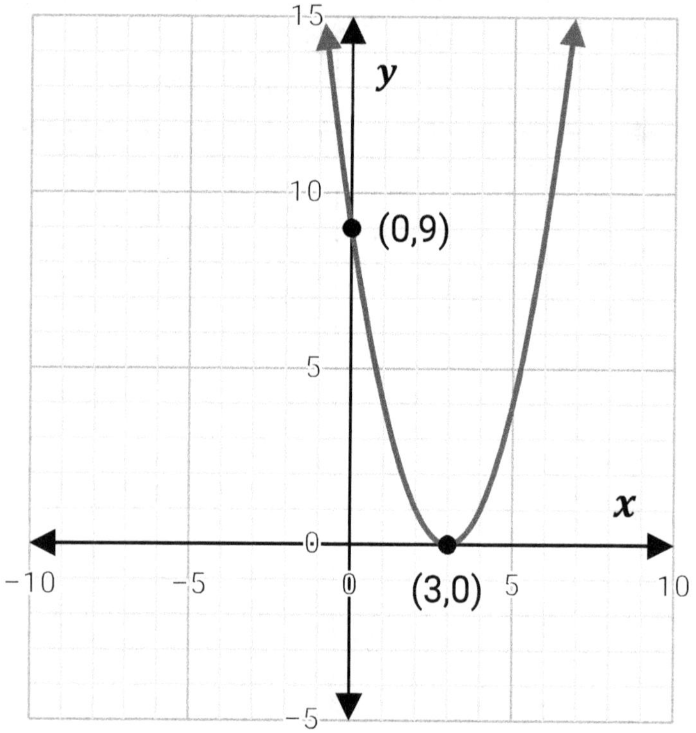

a. $y = -x^2 - 3x + 6$
b. $y = x^2 + 3x - 6$
c. $y = x^2 - 6x + 9$
d. $y = -x^2 + 6x - 9$

13. Marc bounces a basketball and watches as it rebounds back up into the air. As he observes the basketball, he notices that its path takes the shape of a parabola. Marc sketches the basketball's path on a coordinate plane, as shown below. Based on the graph, what is the equation of this function?

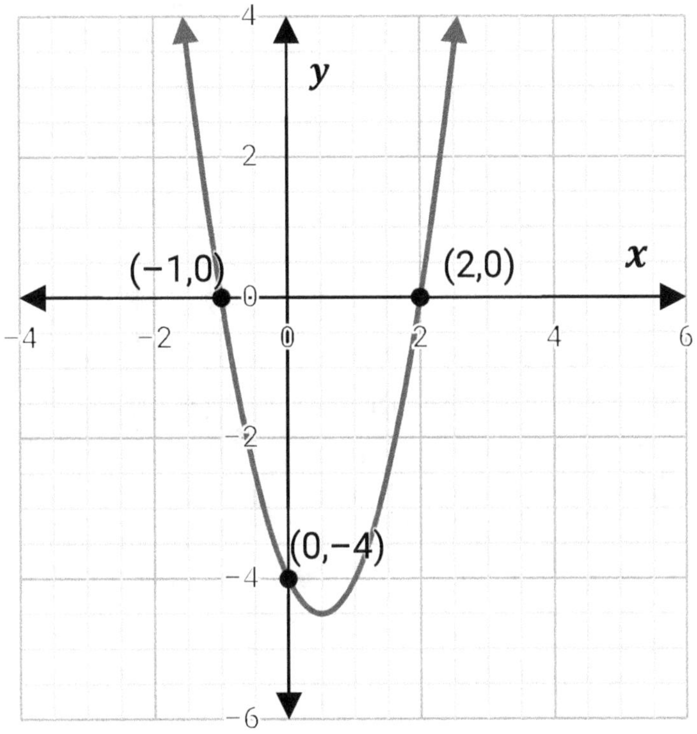

a. $y = -x^2 - x - 2$
b. $y = -2x^2 + 2x + 4$
c. $y = x^2 + x + 1$
d. $y = 2x^2 - 2x - 2$

14. Alen is a graphic designer who is overseeing the creation of a new logo. Part of the logo is in the shape of a parabola, as shown on the coordinate plane below. Alen needs to write an equation for this portion of the logo. Which quadratic equation, in standard form, represents Alen's logo?

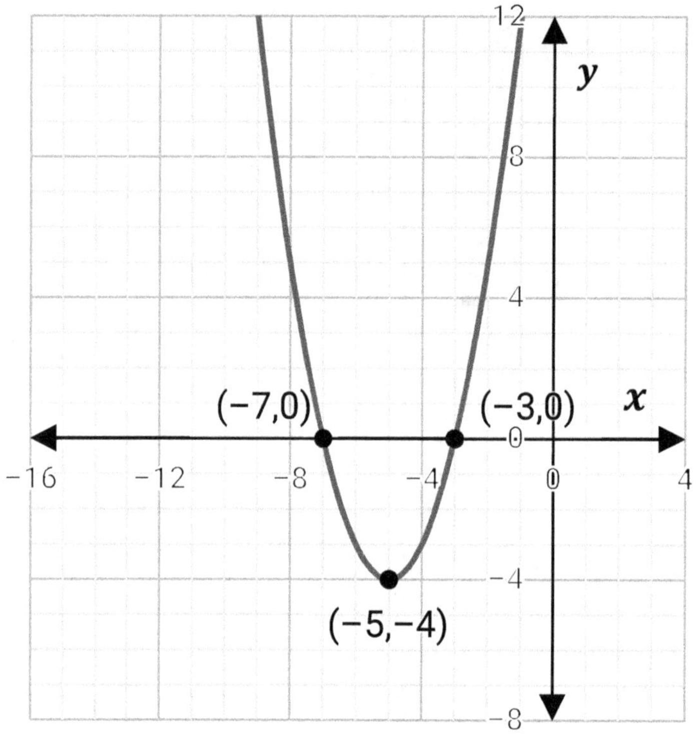

a. $y = x^2 + 10x + 21$
b. $y = -x^2 - 10x - 21$
c. $y = x^2 + 10x + 10$
d. $y = -x^2 - 4x - 10$

15. A logo is made up of two parabolic curves, one of which is represented on the coordinate plane below. What is the quadratic equation, written in standard form, of this function?

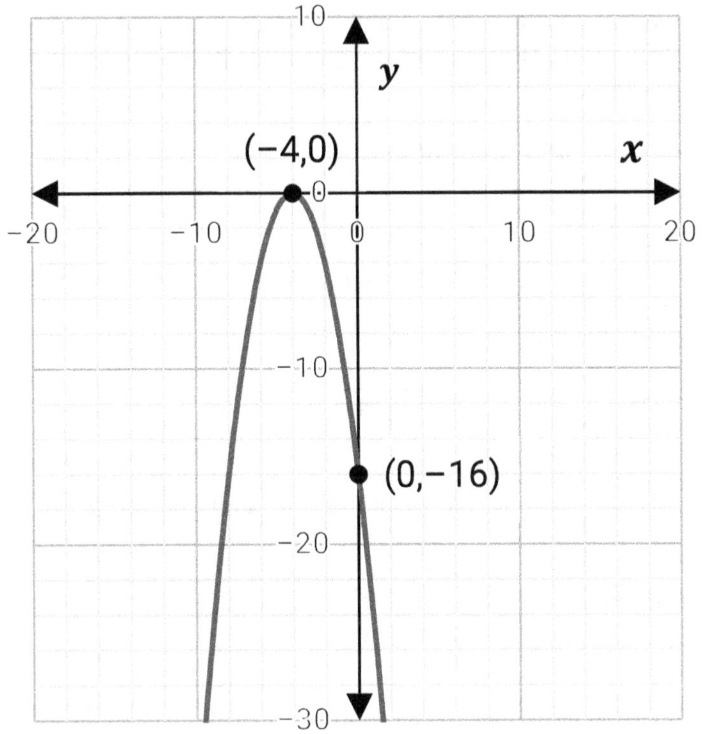

a. $y = x^2 + 16x + 8$
b. $y = -x^2 - 16x - 8$
c. $y = x^2 + 8x + 16$
d. $y = -x^2 - 8x - 16$

8.5 Solving Quadratic Equations Using Square Roots

When we solve quadratic equations, we are looking for the x-values, called the "roots," that tell us where the parabola will intersect the x-axis. Quadratic equations can be solved by factoring, completing the square, the quadratic formula, or by using square roots. Let's take a closer look at the square root method.

For example, consider the quadratic equation $(x + 4)^2 - 15 = 49$. We can solve for x by solving the multi-step equation using square roots. Start by adding 15 to both sides. This isolates the squared quantity.

$$(x + 4)^2 = 64$$

Now, find the square root of both sides.

$$\sqrt{(x + 4)^2} = \sqrt{64}$$
$$(x + 4) = \pm 8$$

Remember, $(8)^2$ and $(-8)^2$ are both 64, so use a $\pm$ symbol when taking a square root when solving an equation.

Now, subtract 4 from both sides.

$$x + 4 = 8 \qquad\qquad x + 4 = -8$$
$$x = 4 \qquad\qquad x = -12$$

This means when the quadratic equation is graphed, the parabola will pass through the x-axis at $(-12,0)$ and $(4,0)$.

Sometimes the square root method will not simplify to a whole number. For example, you might be left with an answer like $x = \pm\sqrt{13}$.

Consider the quadratic equation $6x^2 - 15 = 27$.

Start by isolating the x^2 term by adding 15 to both sides.

$$6x^2 = 42$$

Now, divide both sides by 6.

$$x^2 = 7$$

Take the square root of both sides.

$$x = \pm\sqrt{7}$$

7 is not a perfect square, so we can't simplify it. The answer becomes $x = \sqrt{7}$ and $x = -\sqrt{7}$.

EXAMPLES

EXAMPLE 1:

Determine the solutions for the quadratic equation.

$$2x^2 + 3 = 131$$

Answer:

Isolate the x^2 term. First, subtract 3 from both sides.

$$2x^2 = 128$$

Divide both sides by 2.

$$x^2 = 64$$

Take the square root of both sides.

$$\sqrt{x^2} = \sqrt{64}$$

Remember, the square root will have 2 solutions, +8 and –8.

$$x = 8 \text{ and } x = -8$$

EXAMPLE 2:

Determine the solutions for the quadratic equation.

$$(x-9)^2 + 3 = 19$$

Answer:

First, isolate the squared term on one side by subtracting 3 from both sides of the equation.

$$(x-9)^2 = 16$$

Now, take the square root of both sides.

$$x - 9 = \pm 4$$

Separate the equation into two separate equations and solve them both for x.

$$x - 9 = 4 \qquad\qquad x - 9 = -4$$
$$x = 13 \qquad\qquad x = 5$$

The solutions to the equation are $x = 5$ and $x = 13$.

EXAMPLE 3:

Sarah is painting a rectangular wall in her bedroom. She knows the area of the wall is 432 square feet. She also knows that the width is three times the length. Write a quadratic equation that represents this scenario and solve it to find the length and width of the wall.

Answer:

First, write an equation that represents this scenario. The area of a rectangle is found by multiplying the length by the width.

$$A = lw$$

Since the width is three times the length, substitute $3l$ for w in the equation. We also know the area is 432 square feet, so substitute that information in as well.

$$432 = l(3l)$$

Now, solve for l. Start by multiplying on the right side.

$$432 = 3l^2$$

Then, divide both sides by 3.

$$144 = l^2$$

Next, take the square root of both sides.

$$\pm 12 = l$$

Since the length of the wall cannot be a negative number, it is 12 feet long. Now, find the width of the wall by multiplying this by 3.

$$12 \times 3 = 36$$

The width of the wall is 36 feet.

Practice Questions

1. Solve the quadratic equation by taking square roots.

$$x^2 - 9 = 7$$

a. $x = 4$ or $x = -4$
b. $x = 2$ or $x = -2$
c. $x = 4$
d. $x = 2$

2. Solve the quadratic equation by taking square roots.

$$(x - 3)^2 = 36$$

a. $x = 39$ or $x = -39$
b. $x = 9$ or $x = -3$
c. $x = 3$ or $x = -3$
d. $x = 9$ or $x = -9$

3. Solve the quadratic equation by taking square roots.

$$\frac{1}{2}x^2 = 50$$

a. $x = 5$ or $x = -5$
b. $x = 25$ or $x = -25$
c. $x = 10$ or $x = -10$
d. $x = 100$ or $x = -100$

4. Solve the quadratic equation by taking square roots.

$$(x + 7)^2 - 64 = 0$$

a. $x = 15$ or $x = -15$
b. $x = 1$ or $x = -1$
c. $x = 1$ or $x = -15$
d. $x = 57$ or $x = -57$

5. Solve the quadratic equation by taking square roots.

$$8x^2 - 8 = 192$$

a. $x = 20$ or $x = -20$
b. $x = 5$
c. $x = -20$
d. $x = 5$ or $x = -5$

6. Solve the quadratic equation by taking square roots:

$$(x - 8)^2 + 9 = 130$$

a. $x = 19$
b. $x = 11$
c. $x = 19$ or $x = -3$
d. $x = 11$ or $x = -11$

7. Solve the quadratic equation by taking square roots.

$$4x^2 - 2 = 14$$

a. $x = 4$ or $x = -4$
b. $x = 2$ or $x = -2$
c. $x = 2$ or $x = -4$
d. $x = -2$ or $x = 4$

8. Solve the quadratic equation by taking square roots.

$$2x^2 + 4 = 22$$

a. $x = 3$ or $x = -9$
b. $x = 18$ or $x = -18$
c. $x = 9$ or $x = -9$
d. $x = 3$ or $x = -3$

9. Solve the quadratic equation by taking square roots.

$$2(x - 1)^2 - 3 = 47$$

a. $x = 6$ or $x = -4$
b. $x = 4$ or $x = -6$
c. $x = 4$ or $x = -4$
d. $x = 5$ or $x = -5$

10. Solve the quadratic equation by taking square roots.

$$4x^2 + 24 = 600$$

a. $x = 12$ or $x = -12$
b. $x = 144$ or $x = -144$
c. $x = 12$ or $x = -144$
d. $x = 144$ or $x = -12$

11. Solve the quadratic equation by taking square roots.

$$9(2x - 3)^2 + 6 = 447$$

a. $x = 5$ or $x = -5$
b. $x = 2$ or $x = -2$
c. $x = 5$ or $x = -2$
d. $x = 2$ or $x = -5$

12. Solve the quadratic equation by taking square roots.

$$x^2 + 6 = 26$$

a. $x = 2$ or $x = -2$
b. $x = 5$ or $x = -5$
c. $x = 20$ or $x = -20$
d. $x = 2\sqrt{5}$ or $x = -2\sqrt{5}$

13. The length of a rectangle equals twice its width, and the area is 162 square inches. What is the length of the rectangle?

a. 2 inches
b. 81 inches
c. 18 inches
d. 9 inches

14. Tim and Mary Kate are building a square patio. They have enough building materials for the patio to have an area of 169 square feet. The length of one of the sides of the patio is represented by the expression $x + 5$. Write a quadratic equation that represents this scenario and solve it by taking square roots. What is the perimeter of the patio?

a. 13 feet
b. 52 feet
c. 169 feet
d. 8 feet

15. A pen is dropped out of a window that is 400 feet above the ground. The speed at which the pen will hit the ground can be calculated using the formula $s^2 = 64h$, where s stands for the speed of the falling object, and h stands for the distance the object falls. Based on this information, solve the equation by taking square roots to find the speed of the pen when it hit the ground.

a. 25,600 feet per second
b. 160 feet per second
c. -160 feet per second
d. 400 feet per second

8.6 Completing the Square

Quadratic equations can be solved by completing the square. Remember, a quadratic equation in standard form is $ax^2 + bx + c = 0$. Completing the square is used to find the roots of quadratic equations. **Roots** are the values of x where the graph crosses the x-axis. When completing the square, we are attempting to identify values of x where the parabola intersects the x-axis—if such values exist. This process is especially useful if the quadratic equation doesn't factor and you don't want to use the quadratic formula.

Let's solve for x in the following quadratic equation by completing the square: $x^2 + 6x - 7 = 0$.

The process for completing the square can be done in 6 steps:

Step 1: Move the constant to the right side of the equation.

$$x^2 + 6x = 7$$

Step 2: Make sure the x^2-term is 1. If it isn't 1, factor it out so it becomes 1.

In this case, x is 1.

$$x^2 + 6x = 7$$

Step 3: Take the coefficient of the x-term, divide it by 2, then square it.

$$\left(\frac{6}{2}\right)^2 = (3)^2 = 9$$

Step 4: Add this number to both sides of the equation.

$$x^2 + 6x + 9 = 7 + 9$$

Step 5: Factor the left side of the equation as the square of a binomial, and simplify the right side of the equation.

$$(x + 3)^2 = 16$$

Step 6: Solve for x.

Start by taking the square root of both sides.

$$(x + 3)^2 = 16$$
$$x + 3 = \pm 4$$

Remember, when you find the square root of a constant, like 16, you will have a positive and negative answer.

$x + 3 = 4$	$x + 3 = -4$
$x = 1$	$x = -7$

Therefore, the roots of this quadratic are $x = 1$ and $x = -7$.

The same process will work for quadratic equations that do not have an x^2-term of 1. For example, $-2x^2 + 36x - 10 = 24$. In this case, the coefficient of the x^2-term is –2.

Let's apply the 6-step process for completing the square to solve for the roots of this equation.

Step 1: Move the constant to the right side of the equation.

$$-2x^2 + 36x = 34$$

Step 2: Make sure the x^2-term is 1. If it isn't 1, factor it out so it becomes 1.

$$-2(x^2 - 18x) = 34$$

Divide both sides by –2.

$$x^2 - 18x = -17$$

Step 3: Take the coefficient of the x-term, divide it by 2, then square it.

$$-\frac{18}{2} = -9$$
$$(-9)^2 = 81$$

Step 4: Add this number to both sides of the equation.

$$x^2 - 18x + 81 = -17 + 81$$

Step 5: Factor the left side of the equation as the square of a binomial, and simplify the right side of the equation.

Hint: The constant inside the binomial that is being squared is always half the coefficient of the x-term.

$$(x - 9)^2 = 64$$

Step 6: Solve for x.

Start by taking the square root of both sides.

$$(x - 9)^2 = 64$$
$$x - 9 = \pm 8$$

Remember, when you find the square root of a constant, like 64, you will have a positive and negative answer.

$$x - 9 = 8 \qquad\qquad x - 9 = -8$$
$$x = 17 \qquad\qquad x = 1$$

Therefore, the roots of this quadratic are $x = 17$ and $x = 1$.

Review Video: Completing the Square
Visit mometrix.com/academy and enter code: 982479

Examples

Example 1:

Solve the quadratic by completing the square.

$$x^2 + 4x - 12 = 0$$

Answer:

First, move the constant to the right side.

$$x^2 + 4x = 12$$

From here, make sure the coefficient of the x^2-term is 1. In this case, it is already 1.

Then, identify the x-term, which is 4. Divide this number by 2, and then square it.

$$\left(\frac{4}{2}\right)^2 = (2)^2 = 4$$

Add this number to both sides.

$$x^2 + 4x + 4 = 12 + 4$$

Write the left side as a binomial squared and add the constants on the right side.

$$(x + 2)^2 = 16$$

Take the square root of both sides.

$$\sqrt{(x + 2)^2} = \sqrt{16}$$
$$x + 2 = \pm 4$$

Remember, when using a square root to solve an equation, there is a positive and negative case. Here, 4 and –4. Solve both equations for x.

$$x + 2 = 4 \qquad x + 2 = -4$$
$$x = 2 \qquad x = -6$$

Therefore, the solutions to this quadratic equation are $x = 2$ and $x = -6$.

Example 2:

Solve the quadratic by completing the square.

$$x^2 - 6x + 5 = 0$$

Answer:

First, move the constant to the right side.

$$x^2 - 6x = -5$$

From here, make sure the coefficient of the x^2-term is 1. In this case, it is already 1.

Then, identify the x-term, which is –6. Divide this number by 2, and then square it.

$$\left(\frac{-6}{2}\right)^2 = (-3)^2 = 9$$

Add this number to both sides.

$$x^2 - 6x + 9 = -5 + 9$$

Write the left side as a binomial squared and add the constants on the right side.

$$(x-3)^2 = 4$$

Take the square root of both sides.

$$\sqrt{(x-3)^2} = \sqrt{4}$$
$$x - 3 = \pm 2$$

Solve each equation for x.

$$x - 3 = 2 \qquad x - 3 = -2$$
$$x = 5 \qquad x = 1$$

Therefore, the solutions to this quadratic are $x = 5$ and $x = 1$.

Example 3:

Solve the quadratic equation by completing the square: $4x^2 + 13x + 5 = 0$.

Answer:

To complete the square, start by dividing all terms by 4 to make the value of a be 1.

$$\frac{4x^2 + 13x + 5 = 0}{4}$$
$$x^2 + \frac{13}{4}x + \frac{5}{4} = 0$$

From here, move the constant to the right side of the equation.

$$x^2 + \frac{13}{4}x = -\frac{5}{4}$$

Then, complete the square on the left side of the equation by dividing b by 2, squaring it, and adding the value to both sides.

$$\left(\frac{\frac{13}{4}}{2}\right)^2 = \left(\frac{13}{8}\right)^2 = \frac{169}{64}$$

$$x^2 + \frac{13}{4}x + \frac{169}{64} = -\frac{5}{4} + \frac{169}{64}$$

$$x^2 + \frac{13}{4}x + \frac{169}{64} = \frac{89}{64}$$

Next, create a squared binomial on the left side, and combine like terms on the right side.

$$\left(x+\frac{13}{8}\right)^2=\frac{89}{64}$$

From here, square root both sides of the equation.

$$x+\frac{13}{8}=\pm\frac{\sqrt{89}}{8}$$

Separate the equation into two separate equations, and solve for x.

$$x+\frac{13}{8}=\frac{\sqrt{89}}{8}$$
$$x=\frac{-13+\sqrt{89}}{8}$$

$$x+\frac{13}{8}=-\frac{\sqrt{89}}{8}$$
$$x=\frac{-13-\sqrt{89}}{8}$$

Therefore, the values of x are $x=\frac{-13+\sqrt{89}}{8}$ and $x=\frac{-13-\sqrt{89}}{8}$.

EXAMPLE 4:

A lighting engineer is setting up lights for a performer. To understand how a particular spotlight moves, he graphs it on a coordinate plane. The quadratic equation that represents the light's path is $x^2-4x-14=0$. Solve the equation for x by completing the square to figure out at which point(s) the light will be off the stage.

Answer:

This is a quadratic equation that needs to be solved for x by completing the square. Start by moving the constant to the right side of the equation.

$$x^2-4x=14$$

Next, complete the square on the left side of the equation by dividing b by 2, squaring it and adding it to both sides.

$$\left(-\frac{4}{2}\right)^2=(-2)^2=4$$
$$x^2-4x+4=14+4$$

From here, create a squared binomial on the left side and combine like terms on the right side.

$$(x-2)^2=18$$

Next, take the square root of both sides.

$$x-2=\pm\sqrt{18}$$

Simplify the square root.

$$x-2=\pm\sqrt{9\times 2}$$
$$x-2=\pm 3\sqrt{2}$$

Finally, add 2 to both sides of the equation.

$$x = 2 \pm 3\sqrt{2}$$

Therefore, the light will be off the stage when $x = 2 + 3\sqrt{2}$ and when $x = 2 - 3\sqrt{2}$.

PRACTICE QUESTIONS

1. Solve the quadratic equation by completing the square: $x^2 - 12x + 11 = 0$.

a. $x = 6$ or $x = 5$
b. $x = 11$ or $x = 1$
c. $x = -1$ or $x = 11$
d. $x = 36$ or $x = -6$

2. Solve the quadratic equation by completing the square: $x^2 + 6x + 8 = 0$.

a. $x = -2$ or $x = -4$
b. $x = 2$ or $x = 4$
c. $x = 3$ or $x = -8$
d. $x = 3$ or $x = 9$

3. Solve the quadratic equation by completing the square: $x^2 + 8x + 7 = 0$.

a. $x = 12$ or $x = -12$
b. $x = 4$ or $x = 16$
c. $x = 1$ or $x = -8$
d. $x = -1$ or $x = -7$

4. Solve the quadratic equation by completing the square: $12x^2 - 24x - 36 = 0$.

a. $x = -3$ or $x = -1$
b. $x = 2$ or $x = 3$
c. $x = 1$ or $x = -1$
d. $x = 3$ or $x = -1$

5. Solve the quadratic equation by completing the square: $2x^2 - 20x + 32 = 0$.

a. $x = 5$ or $x = 3$
b. $x = 8$ or $x = 2$
c. $x = -5$ or $x = 5$
d. $x = 4$ or $x = -2$

6. Solve the quadratic equation by completing the square: $6x^2 + 12x - 48 = 0$.

a. $x = 1$ or $x = -3$
b. $x = 2$ or $x = -4$
c. $x = 1$ or $x = 4$
d. $x = 2$ or $x = -3$

7. Solve the quadratic equation by completing the square: $5x^2 + 20x - 60 = 0$.

a. $x = 2$ or $x = -4$

b. $x = 4$ or $x = 6$

c. $x = 2$ or $x = -6$

d. $x = -2$ or $x = -4$

8. Solve the quadratic equation by completing the square: $4x^2 - 12x - 16 = 0$.

a. $x = 4$ or $x = -1$

b. $x = 25$ or $x = 4$

c. $x = 2$ or $x = 5$

d. $x = 4$ or $x = 2$

9. Solve the quadratic equation by completing the square: $4x^2 + 20x + 25 = 0$.

a. $x = -\frac{5}{2}$ or $x = 0$

b. $x = \frac{5}{2}$ or $x = 0$

c. $x = -\frac{5}{2}$

d. $x = \frac{5}{2}$

10. Solve the quadratic equation by completing the square: $3x^2 + 8x - 3 = 0$.

a. $x = \frac{1}{6}$ or $x = 3$

b. $x = 10$ or $x = 6$

c. $x = 6$ or $x = 3$

d. $x = -3$ or $x = \frac{1}{3}$

11. Solve the quadratic equation by completing the square: $3x^2 - 5x + 2 = 0$.

a. $x = 6$ or $x = \frac{5}{6}$

b. $x = 1$ or $x = \frac{2}{3}$

c. $x = 3$ or $x = 1$

d. $x = \frac{2}{3}$ or $x = 36$

12. Solve the quadratic equation by completing the square: $x^2 + 6x - 3 = 0$.

a. $x = 2 \pm 3\sqrt{2}$

b. $x = 2$ or $x = -3$

c. $x = 5$ or $x = 1$

d. $x = -3 \pm 2\sqrt{3}$

13. A ride at an amusement park takes passengers to the top of a tower and then drops them 57 feet. The quadratic equation that represents this ride is $-16t^2 + 64t + 57 = 0$, where t stands for the time, in seconds, it takes for riders to reach ground level. Solve the equation for t by completing the square to figure out how many seconds it will take for riders to reach ground level.

a. $4\frac{3}{4}$ seconds
b. $2\frac{3}{4}$ seconds
c. 11 seconds
d. 4 seconds

14. Jill is framing a picture. The length of the frame is 6 inches longer than its width, and the picture and frame has a total area of 520 square inches. Given this information, write a quadratic equation to represent this scenario and solve it by completing the square to identify the dimensions of the frame.

a. The width of the frame is 20 inches, and the length is 26 inches.
b. The width of the frame is 22 inches, and the length is 28 inches.
c. The width of the frame is 26 inches, and the length is 32 inches.
d. The width of the frame is 20 inches, and the length is 26 inches.

15. Miguel throws a football straight up into the air, from 3 feet above the ground, with a velocity of 14 feet per second. The quadratic equation that represents this scenario is $5t^2 - 14t - 3 = 0$, where t stands for time in seconds. Solve the equation for t by completing the square to figure out how many seconds it will take for the football to hit the ground.

a. 7 seconds
b. 15 seconds
c. 3 seconds
d. $\frac{1}{5}$ second

8.7 Converting Between Standard and Vertex Forms

The standard form of a quadratic equation is commonly used when graphing. Vertex form is helpful because it provides information about the vertex of the parabola.

Let's review both forms.

Standard form: $f(x) = ax^2 + bx + c$

For example, $f(x) = -3x^2 - 6x - 9$ is in standard form.

Vertex form: $f(x) = a(x - h)^2 + k$

For example, $f(x) = -2(x - 3)^2 + 4$ is in vertex form.

<u>Converting from standard form to vertex form:</u>

Converting from standard form to vertex form requires us to complete the square and follows a 7-step process. Let's convert the standard form equation $y = -3x^2 - 6x - 9$ into vertex form by completing the square.

Step 1: Make sure that the coefficient of x^2 is a 1. In this case, it is not a 1. Factor out the –3 so that x^2 is alone.

$$y = -3(x^2 + 2x + 3)$$

Step 2: Identify the coefficient of x. In this case it is a 2.

Step 3: Halve this value, and then square the result.

$$\frac{2}{2} = 1$$
$$(1)^2 = 1$$

Step 4: Add and subtract this number (1) right after the x-term.

$$y = -3(x^2 + 2x + 1 - 1 + 3)$$

Step 5: Identify the first three terms in the parentheses: $x^2 + 2x + 1$. Factor this. $x^2 + 2x + 1$ becomes $(x + 1)^2$. Plug this back into the equation.

$$y = -3[(x + 1)^2 - 1 + 3]$$

Step 6: Simplify the last two numbers.

$$y = -3[(x + 1)^2 + 2]$$

Step 7: Distribute the outside number.

$$y = -3(x + 1)^2 - 6$$

The equation is now in vertex form.

Converting from vertex form to standard form:

The process for converting from vertex form to standard form can be done in 4 steps. Let's convert the equation $y = 2(x + 1)^2 - 4$ from vertex form to standard form.

Step 1: Rewrite the equation using two binomials.

$$y = 2(x + 1)(x + 1) - 4$$

Step 2: FOIL the binomials.

$$y = 2(x^2 + 2x + 1) - 4$$

Step 3: Distribute.

$$y = 2x^2 + 4x + 2 - 4$$

Step 4: Combine the constants.

$$y = 2x^2 + 4x - 2$$

The equation is now in standard form.

> **Review Video: How to Convert Between Standard Form and Vertex Form**
> Visit mometrix.com/academy and enter code: 916601

Examples

Example 1:

Convert the equation from standard form to vertex form.

$$y = x^2 + 8x + 3$$

Answer:

Start by completing the square.

$$\left(\frac{8}{2}\right)^2 = (4)^2 = 16$$
$$y = (x^2 + 8x + 16) - 16 + 3$$

Create a perfect square trinomial.

$$y = (x + 4)^2 - 16 + 3$$

Combine the constants.

$$y = (x + 4)^2 - 13$$

Example 2:

Convert the equation from vertex form to standard form.

$$y = 2(x - 3)^2 + 4$$

Answer:

Expand the squared binomial.

$$y = 2(x - 3)(x - 3) + 4$$

Multiply the binomials by using the FOIL method.

$$y = 2(x^2 - 6x + 9) + 4$$

Distribute the 2.

$$y = 2x^2 - 12x + 18 + 4$$

Combine the constants.

$$y = 2x^2 - 12x + 22$$

Example 3:

The arc of a cannonball being shot out of a cannon can be represented by the quadratic equation $y = -\frac{1}{25}(x + 50)^2 + 150$. What is the standard form of this quadratic equation?

Answer:

To convert the equation to standard form, start by expanding the square $(x + 50)^2$.

$$y = -\frac{1}{25}(x + 50)(x + 50) + 150$$

Then, multiply the binomials.

$$y = -\frac{1}{25}(x^2 + 100x + 2{,}500) + 150$$

Now, distribute $-\frac{1}{25}$ to each term in parentheses.

$$y = -\frac{1}{25}x^2 - 4x - 100 + 150$$

Finally, combine like terms.

$$y = -\frac{1}{25}x^2 - 4x + 50$$

Practice Questions

1. Convert the quadratic equation from vertex form to standard form.

$$\boldsymbol{y = 3(x - 1)^2 + 2}$$

a. $y = 3x^2 - 6x + 3$
b. $y = 3x^2 - 3x - 6$
c. $y = 3x^2 - 2x + 3$
d. $y = 3x^2 - 6x + 5$

2. Convert the quadratic equation from standard form to vertex form.

$$y = 2x^2 - 4x + 5$$

a. $y = 2(x - 3)^2 + 1$
b. $y = 2(x - 1)^2 + 3$
c. $y = 2(x + 1)^2 - 3$
d. $y = 2(x + 3)^2 - 1$

3. Convert the quadratic equation from vertex form to standard form.

$$y = (x + 3)^2 - 1$$

a. $y = x^2 + 6x + 9$
b. $y = x^2 + 6x + 8$
c. $y = x^2 + 9x + 8$
d. $y = x + 8$

4. Convert the quadratic equation from standard form to vertex form.

$$y = 2x^2 - 8x + 9$$

a. $y = 2(x - 2)^2 + 1$
b. $y = 2(x + 2)^2 + 1$
c. $y = 2(x - 1)^2 + 2$
d. $y = 2(x - 4)^2 + 1$

5. Convert the quadratic equation from vertex form to standard form.

$$y = 2(x + 2)^2 + 7$$

a. $y = 2x^2 + 4x + 15$
b. $y = 2x^2 + 4x + 11$
c. $y = 2x^2 + 8x + 15$
d. $y = 2x^2 + 8x + 1$

6. Convert the quadratic equation from standard form to vertex form.

$$y = x^2 - 4x - 3$$

a. $y = (x - 2)^2 - 7$
b. $y = (x + 7)^2 - 2$
c. $y = (x + 2)^2 + 7$
d. $y = (x - 7)^2 + 2$

7. Convert the quadratic equation from vertex form to standard form.

$$y = 4(x + 5)^2 - 25$$

a. $y = 4x^2 + 10x$
b. $y = 4x^2 + 40x + 75$
c. $y = 4x^2 + 10x + 75$
d. $y = 4x^2 + 40x$

8. Convert the quadratic equation from standard form to vertex form.

$$y = x^2 + 8x + 20$$

a. $y = (x - 4)^2 - 4$
b. $y = (x + 4)^2 - 4$
c. $y = (x - 4)^2 + 4$
d. $y = (x + 4)^2 + 4$

9. Convert the quadratic equation from vertex form to standard form.

$$y = -(x - 6)^2$$

a. $y = -x^2 + 12x - 36$
b. $y = x^2 - 12x + 36$
c. $y = -x^2 + 12x - 12$
d. $y = x^2 - 12x + 12$

10. Convert the quadratic equation from standard form to vertex form.

$$y = -x^2 + 2x - 2$$

a. $y = -(x - 1)^2 - 1$
b. $y = -(x + 1)^2 - 1$
c. $y = (x - 1)^2 - 1$
d. $y = -(x - 1)^2 + 1$

11. Convert the quadratic equation from vertex form to standard form.

$$y = \frac{1}{2}(x - 4)^2 - 3$$

a. $y = \frac{1}{2}x^2 - 8x + 5$
b. $y = \frac{1}{2}x^2 - 8x + 13$
c. $y = \frac{1}{2}x^2 - 4x - 3$
d. $y = \frac{1}{2}x^2 - 4x + 5$

12. Convert the quadratic equation from standard form to vertex form.

$$y = x^2 + 2x - 6$$

a. $y = (x - 1)^2 + 8$
b. $y = (x + 1)^2 - 8$
c. $y = (x + 1)^2 - 7$
d. $y = (x - 1)^2 - 7$

13. The arc of a football for a particular throw can be modeled by the quadratic equation $y = -\frac{1}{100}(x - 20)^2 + 8$. What is the standard form of this quadratic equation?

a. $y = \frac{1}{100}x^2 - \frac{2}{5}x - 12$
b. $y = -\frac{1}{100}x^2 + \frac{2}{5}x + 12$
c. $y = -\frac{1}{100}x^2 + \frac{2}{5}x + 4$
d. $y = \frac{1}{100}x^2 + \frac{2}{5}x - 4$

14. The equation $y = x^2 - 12x + 45$ models the number of records, y, sold in a record store x days after a band played at a publicity event. What is the vertex form of this quadratic equation?

a. $y = (x - 6)^2 + 9$
b. $y = (x - 9)^2 + 6$
c. $y = (x + 6)^2 - 9$
d. $y = (x + 9)^2 - 6$

15. Mr. Dawson's physics class is launching a rocket from a platform. The height of the rocket in feet x seconds after the launch is represented by the equation $y = -5(x - 4)^2 + 200$. What is the standard form of this quadratic equation?

a. $y = -5x^2 - 8x + 216$
b. $y = -5x^2 + 40x + 120$
c. $y = x^2 - 8x + 16$
d. $y = -5x^2 - 40x - 120$

8.8 The Quadratic Formula

The **quadratic formula** is another method used to find the solutions of quadratic equations. The quadratic formula is a useful formula to have memorized because it can be used to solve any quadratic equation, even ones that are not factorable.

$$\text{Quadratic formula: } x = \frac{-b \pm \sqrt{b^2 - 4ac}}{2a}$$

To find the values of a, b, and c, the quadratic equation needs to be in **standard form**, $ax^2 + bx + c = 0$. For example, the quadratic equation $2x^2 - 7x - 4 = 0$ follows the form $ax^2 + bx + c = 0$, so a, b, and c can be identified. In this case, $a = 2$, $b = -7$, and $c = -4$. These values can be plugged into the quadratic formula to determine the solutions. Remember, the solutions are the x-intercepts. This means we are locating the point (or points) where the parabola crosses the x-axis.

Substitute the values of a, b, and c into the quadratic formula.

$$x = \frac{-b \pm \sqrt{b^2 - 4ac}}{2a}$$

$$x = \frac{-(-7) \pm \sqrt{(-7)^2 - 4(2)(-4)}}{2(2)}$$

Simplify the equation.

$$x = \frac{7 \pm \sqrt{49 + 32}}{4}$$

$$x = \frac{7 \pm \sqrt{81}}{4}$$

$$x = \frac{7 \pm 9}{4}$$

Write the equation as two separate equations.

$$x = \frac{7 + 9}{4}$$

$$x = \frac{16}{4}$$

$$x = 4$$

$$x = \frac{7 - 9}{4}$$

$$x = -\frac{2}{4}$$

$$x = -\frac{1}{2}$$

These two x-values represent the solutions, or **zeros**, of the quadratic equation. When the equation is graphed, the parabola will cross the x-axis at these two locations.

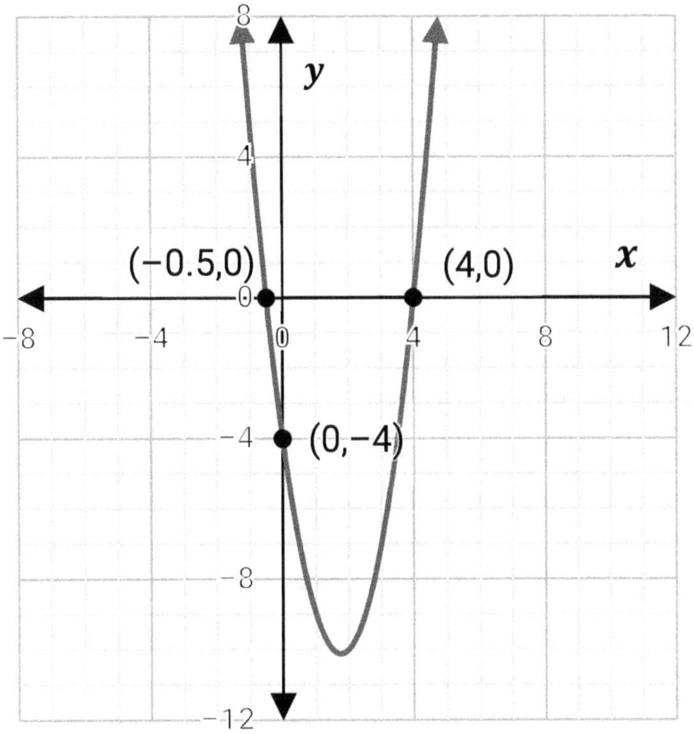

Review Video: Using the Quadratic Formula
Visit mometrix.com/academy and enter code: 163102

Examples

Example 1:

Use the quadratic formula to solve the quadratic equation.

$$x^2 + 4x + 3 = 0$$

Answer:

Substitute the values of a, b, and c into the quadratic formula.

$$a = 1, b = 4, c = 3$$

$$x = \frac{-b \pm \sqrt{b^2 - 4ac}}{2a}$$

$$x = \frac{-4 \pm \sqrt{(4)^2 - 4(1)(3)}}{2(1)}$$

Simplify the equation.

$$x = \frac{-4 \pm \sqrt{16 - 12}}{2}$$
$$x = \frac{-4 \pm \sqrt{4}}{2}$$
$$x = \frac{-4 \pm 2}{2}$$

Write the equation as two separate equations.

$$x = \frac{-4 + 2}{2} = \frac{-2}{2} = -1 \qquad x = \frac{-4 - 2}{2} = \frac{-6}{2} = -3$$

The solutions to this quadratic are $x = -1$ and $x = -3$.

EXAMPLE 2:

Use the quadratic formula to solve the quadratic equation.

$$2x^2 - 4x - 30 = 0$$

Answer:

Substitute the values of a, b, and c into the quadratic formula.

$$a = 2, b = -4, c = -30$$

$$x = \frac{-b \pm \sqrt{b^2 - 4ac}}{2a}$$
$$x = \frac{4 \pm \sqrt{(-4)^2 - 4(2)(-30)}}{2(2)}$$

Simplify the equation.

$$x = \frac{4 \pm \sqrt{16 + 240}}{4}$$
$$x = \frac{4 \pm \sqrt{256}}{4}$$
$$x = \frac{4 \pm 16}{4}$$

Write the equation as two separate equations.

$$x = \frac{4 + 16}{4} = \frac{20}{4} = 5 \qquad x = \frac{4 - 16}{4} = -\frac{12}{4} = -3$$

The solutions to this quadratic equation are $x = 5$ and $x = -3$.

EXAMPLE 3:

Solve the quadratic equation using the quadratic formula.

$$6x^2 + 2x - 5 = 0$$

Answer:

First, determine the values for a, b, and c.

$$a = 6, b = 2, c = -5$$

Next, substitute these values into the quadratic equation, $x = \frac{-b \pm \sqrt{b^2 - 4ac}}{2a}$.

$$x = \frac{-(2) \pm \sqrt{(2)^2 - 4(6)(-5)}}{2(6)}$$

Now, simplify the equation.

$$x = \frac{-2 \pm \sqrt{4 + 120}}{12}$$
$$x = \frac{-2 \pm \sqrt{124}}{12}$$

Then, simplify the square root.

$$x = \frac{-2 \pm \sqrt{4 \cdot 31}}{12}$$
$$x = \frac{-2 \pm 2\sqrt{31}}{12}$$

Finally, simplify the fraction by dividing the numerator and denominator by 2.

$$x = \frac{-1 \pm \sqrt{31}}{6}$$

Therefore, the solutions to the quadratic equation are $x = \frac{-1+\sqrt{31}}{6}$ and $x = \frac{-1-\sqrt{31}}{6}$.

EXAMPLE 4:

Solve the following quadratic equation using the quadratic formula.

$$7x^2 - 4x + 3 = 0$$

Answer:

Start by determining the values for a, b, and c in the equation.

$$a = 7, b = -4, c = 3$$

Now, substitute these values into the quadratic formula, $x = \frac{-b \pm \sqrt{b^2-4ac}}{2a}$.

$$x = \frac{-(-4) \pm \sqrt{(-4)^2 - 4(7)(3)}}{2(7)}$$

From here, simplify.

$$x = \frac{4 \pm \sqrt{16 - 84}}{14}$$
$$x = \frac{4 \pm \sqrt{-68}}{14}$$

Since there is a negative number under the radical, this quadratic equation has no real solutions.

Example 5:

A commercial kitchen space is 9 meters longer than it is wide and has an area of 52 square meters. Write a quadratic equation that represents this scenario, and solve it using the quadratic formula to find the length and width of the kitchen.

Answer:

First, write an equation that represents this scenario. Multiply the length and width of the flag and set it equal to the area. Use w to represent the width and $w + 9$ to represent the length.

$$lw = A$$
$$(w + 9)(w) = 52$$

Convert the equation into standard form of a quadratic equation.

$$w^2 + 9w = 52$$
$$w^2 + 9w - 52 = 0$$

Now, identify the values for a, b, and c.

$$a = 1, b = 9, c = -52$$

Substitute these values into the quadratic formula, $x = \frac{-b \pm \sqrt{b^2-4ac}}{2a}$.

$$w = \frac{-(9) \pm \sqrt{(9)^2 - 4(1)(-52)}}{2(1)}$$

Now, simplify.

$$w = \frac{-9 \pm \sqrt{81 + 208}}{2}$$
$$w = \frac{-9 \pm \sqrt{289}}{2}$$
$$w = \frac{-9 \pm 17}{2}$$

From here, split the equation into two separate equations to solve for x.

$$w = \frac{-9 + 17}{(2)} = \frac{8}{2} = 4$$
$$w = \frac{-9 - 17}{2} = \frac{-26}{2} = -13$$

Therefore, the values for w are $w = 4$ and $w = -13$. However, since width can't be negative, only the positive value applies to this scenario. Therefore, the width of the kitchen is 4 meters. Since the length is 9 more than the width, the length of the kitchen is 13 meters.

Practice Questions

1. Solve using the quadratic formula: $x^2 + x - 6 = 0$.

a. $x = 3$ and $x = -2$

b. $x = 2$ and $x = -3$

c. $x = 6$ and $x = 4$

d. $x = -6$ and $x = -4$

2. Solve using the quadratic formula: $x^2 + 4x + 3 = 0$.

a. $x = -2$ and $x = -6$

b. $x = 0$ and $x = -4$

c. $x = -1$ and $x = -3$

d. $x = 1$ and $x = 3$

3. Solve using the quadratic formula: $x^2 - 6x - 16 = 0$.

a. $x = 16$ and $x = -4$

b. $x = 6$ and $x = 0$

c. $x = 8$ and $x = -2$

d. $x = 8$ and $x = 2$

4. Solve using the quadratic formula: $x^2 - 8x + 12 = 0$.

a. $x = 12$ and $x = -8$

b. $x = -4$ and $x = -12$

c. $x = -6$ and $x = -2$

d. $x = 6$ and $x = 2$

5. Solve using the quadratic formula: $2x^2 + 2x - 12 = 0$.

a. $x = 2$ and $x = -3$

b. $x = -12$ and $x = 8$

c. $x = 24.5$ and $x = -25.5$

d. $x = -8$ and $x = 12$

6. Solve using the quadratic formula: $5x^2 - 80 = 0$.

a. $x = 0$ and $x = 40$
b. $x = 5$ and $x = 80$
c. $x = 40$ and $x = -40$
d. $x = 4$ and $x = -4$

7. Solve using the quadratic formula: $4x^2 + 2x - 42 = 0$.

a. $x = 2$ and $x = -42$
b. $x = 7$ and $x = 4$
c. $x = 3$ and $x = -4$
d. $x = 3$ and $x = -\frac{7}{2}$

8. Solve using the quadratic formula: $2x^2 - 14x + 24 = 0$.

a. $x = 4$ and $x = 3$
b. $x = -4$ and $x = -3$
c. $x = 9$ and $x = 5$
d. $x = -9$ and $x = -5$

9. Solve using the quadratic formula: $3x^2 - x - 4 = 0$.

a. $x = \frac{3}{4}$ and $x = \frac{4}{3}$
b. $x = \frac{4}{3}$ and $x = -1$
c. $x = -\frac{4}{3}$ and $x = -1$
d. $x = 6$ and $x = 8$

10. Solve using the quadratic formula: $8x^2 + 14x + 3 = 0$.

a. $x = \frac{1}{2}$ and $x = \frac{1}{4}$
b. $x = -12$ and $x = -2$
c. $x = -24$ and $x = -4$
d. $x = -\frac{1}{4}$ and $x = -\frac{3}{2}$

11. Solve using the quadratic formula: $4x^2 + 7x - 15 = 0$.

a. $x = 7$ and $x = 4$
b. $x = 5$ and $x = -12$
c. $x = \frac{5}{4}$ and $x = -3$
d. $x = \frac{4}{5}$ and $x = 3$

12. Solve using the quadratic formula: $5x^2 + 50x + 125 = 0$.

a. $x = 0$

b. $x = 0$ or $x = -5$

c. $x = -5$

d. $x = 5$ or $x = -5$

13. The length of a rectangular flag is 12 inches more than its width, and the area is 325 square inches. Write a quadratic equation that represents this scenario and solve it by using the quadratic formula to identify the length and width of the flag.

a. The width of the flag is 14 inches, and the length is 26 inches.

b. The width of the flag is 13 inches, and the length is 25 inches.

c. The width of the flag is 12 inches, and the length is 24 inches.

d. The width of the flag is 16 inches, and the length is 28 inches.

14. A pack of gum is thrown out of a window. The speed at which the pack of gum will hit the ground can be calculated using the formula $-2v^2 + 5v + 900 = 0$, where v stands for the velocity of the falling object. Based on this information, solve the equation by using the quadratic formula to find the speed of the pack of gum when it hits the ground in feet per second. (Hint: speed is the absolute value of velocity.)

a. -20 feet per second

b. 22.5 feet per second

c. 85 feet per second

d. 127.5 feet per second

15. Nicole is installing a concrete slab in the shape of a rectangle in her backyard with an area of 540 square feet. The length of the rectangle is represented by the expression $x - 4$. The width of the rectangle is represented by the expression $x + 8$. Write a quadratic equation that represents this scenario and solve it by using the quadratic formula to find the perimeter of the concrete slab.

a. 22 feet

b. 36 feet

c. 60 feet

d. 96 feet

8.9 The Discriminant

Real solutions for quadratic equations refer to the x-intercepts of the graphed parabola. When a quadratic is graphed, it will touch the x-axis either once, twice, or not at all. This means that quadratics can have either 1, 2, or 0 real solutions.

A parabola that never crosses the x-axis will have 0 real solutions.

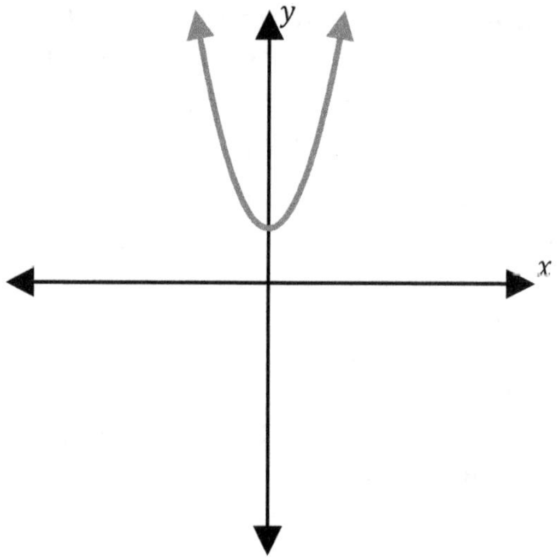

A parabola that has its vertex somewhere on the x-axis will have 1 real solution.

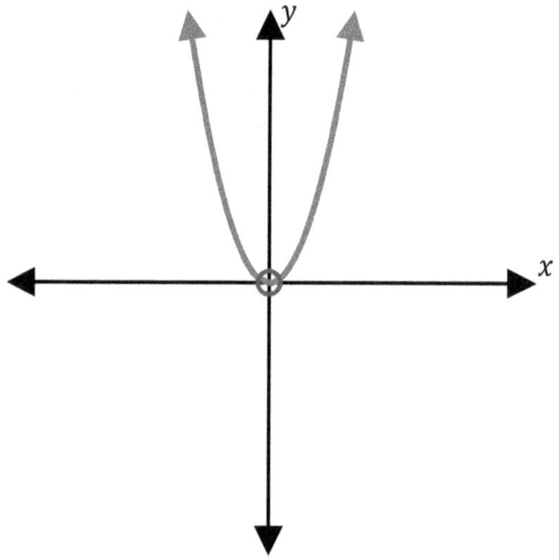

A parabola that dips above or below the x-axis will have 2 real solutions.

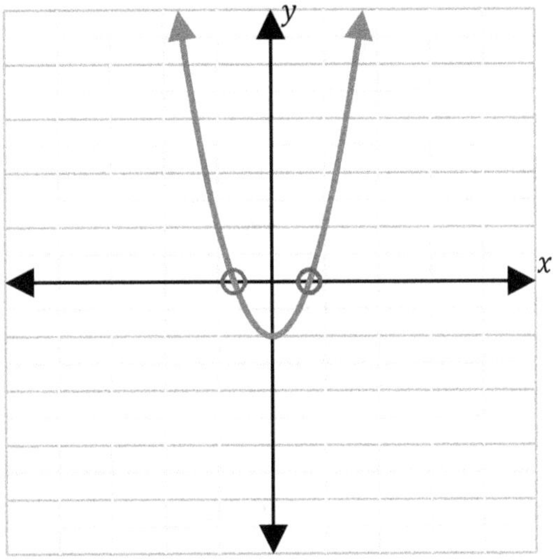

We can use the discriminant to determine the number of real solutions for a quadratic equation. The **discriminant** refers to the $b^2 - 4ac$ portion of the quadratic formula.

$$\text{Quadratic Formula: } x = \frac{-b \pm \sqrt{b^2 - 4ac}}{2a}$$

$$\text{Discriminant: } b^2 - 4ac$$

If the discriminant, D, is greater than zero, then there will be two real solutions to the quadratic equation.

$$D > 0 \text{ means 2 solutions}$$

If the discriminant is equal to zero, then there will be one real solution to the quadratic equation.

$$D = 0 \text{ means 1 real solution}$$

If the discriminant is less than zero, then there will be no real solutions to the quadratic.

$$D < 0 \text{ means no real solutions}$$

For example, the following quadratic equation has two real solutions.

$$9x^2 + 6x - 4 = 0$$

To determine the number of real solutions, substitute the values $a = 9$, $b = 6$, and $c = -4$ into the discriminant equation.

$$\begin{aligned} D &= b^2 - 4ac \\ D &= (6)^2 - 4(9)(-4) \\ D &= 36 - 4(9)(-4) \\ D &= 36 + 144 \\ D &= 180 \end{aligned}$$

The discriminant is greater than 0, $180 > 0$, so the quadratic equation has 2 real solutions.

The following quadratic equation has one real solution.

$$9x^2 + 6x + 1 = 0$$

To determine the number of real solutions, substitute the values $a = 9$, $b = 6$, and $c = 1$ into the discriminant equation.

$$D = b^2 - 4ac$$
$$D = (6)^2 - 4(9)(1)$$
$$D = 36 - 4(9)(1)$$
$$D = 36 - 36$$
$$D = 0$$

The discriminant is zero, so the quadratic has 1 real solution.

The following quadratic equation has no real solutions.

$$9x^2 + 6x + 5 = 0$$

To determine the number of real solutions, substitute the values $a = 9$, $b = 6$, and $c = 5$ into the discriminant formula.

$$D = b^2 - 4ac$$
$$D = (6)^2 - 4(9)(5)$$
$$D = 36 - 4(9)(5)$$
$$D = 36 - 180$$
$$D = -144$$

The discriminant is less than 0, $-144 < 0$, so the quadratic equation has no real solutions.

EXAMPLES

EXAMPLE 1:

Use the discriminant to determine how many times the graph will touch the x-axis.

$$x^2 - 5x + 6 = 0$$

Answer:

The discriminant formula: $D = b^2 - 4ac$

A parabola will touch the x-axis either once, twice, or not at all.

If the discriminant is zero, the graph touches the x-axis once. If the discriminant is positive, the graph touches the x-axis twice. If the discriminant is negative, the graph will not touch the x-axis.

Substitute the values $a = 1$, $b = -5$, and $c = 6$ into the discriminant formula.

$$\begin{aligned} D &= b^2 - 4ac \\ D &= (-5)^2 - 4(1)(6) \\ D &= 25 - 4(1)(6) \\ D &= 25 - 24 \\ D &= 1 \end{aligned}$$

The discriminant is positive, which means the graph will touch the x-axis twice.

EXAMPLE 2:

Use the discriminant to determine how many times the graph will touch the x-axis.

$$4x^2 - 8x + 4 = 0$$

Answer:

The discriminant formula: $D = b^2 - 4ac$

A parabola will touch the x-axis either once, twice, or not at all.

If the discriminant is zero, the graph touches the x-axis once. If the discriminant is positive, the graph touches the x-axis twice. If the discriminant is negative, the graph will not touch the x-axis.

Substitute the values $a = 4$, $b = -8$, and $c = 4$ into the discriminant formula.

$$\begin{aligned} D &= b^2 - 4ac \\ D &= (-8)^2 - 4(4)(4) \\ D &= 64 - 4(4)(4) \\ D &= 64 - 64 \\ D &= 0 \end{aligned}$$

This means that the graph will touch the x-axis once.

EXAMPLE 3:

Sandra runs her own company and created an equation to represent her profits for the year. The equation she comes up with is $P = 4x^2 - 2x - 21{,}600$, where x is the number of units sold Will Sandra's company turn be able to turn a profit of $50,000 this year?

Answer:

Use the discriminant to tell if Sandra's company will be able to meet the desired profit. Start by identifying the values of a, b, and c in the equation.

$$a = 4, b = -2, c = -21{,}600$$

Set the equation equal to zero and then substitute these values into the discriminant formula, $D = b^2 - 4ac$, and solve.

$$50{,}000 = 4x^2 - 2x - 21{,}600$$
$$0 = 4x^2 - 2x - 71{,}600$$
$$D = (-2)^2 - 4(4)(-71{,}600)$$
$$D = 4 + 1{,}145{,}600$$
$$D = 1{,}145{,}604$$

Since the discriminant is greater than 0, there are two solutions to the equation. Therefore, it is possible for Sandra's company to make the desired profit.

Practice Questions

1. Calculate the discriminant of the quadratic equation below. Use this information to determine how many times the graph of this function touches the x-axis.

$$x^2 + 9x + 14 = 0$$

a. $D = -25$. The graph touches the x-axis 0 times.
b. $D = 25$. The graph touches the x-axis 1 time.
c. $D = 25$. The graph touches the x-axis 2 times.
d. $D = -25$. The graph touches the x-axis 2 times.

2. Calculate the discriminant of the quadratic equation below. Use this information to determine how many times the graph of this function will touch the x-axis.

$$x^2 - 8x + 16 = 0$$

a. $D = 0$. The graph touches the x-axis 0 times.
b. $D = 0$. The graph touches the x-axis 1 time.
c. $D = 64$. The graph touches the x-axis 2 times.
d. $D = 64$. The graph touches the x-axis 1 time.

3. Calculate the discriminant of the quadratic equation below. Use this information to determine how many times the graph of this function will touch the x-axis.

$$x^2 - 4x + 8 = 0$$

a. $D = 16$. The graph touches the x-axis 2 times.
b. $D = -16$. The graph touches the x-axis 1 time.
c. $D = 16$. The graph touches the x-axis 0 times.
d. $D = -16$. The graph touches the x-axis 0 times.

4. Calculate the discriminant of the quadratic equation below. Use this information to determine how many times the graph of this function will touch the x-axis.

$$3x^2 - 12x + 12 = 0$$

a. $D = 0$. The graph touches the x-axis 0 times.
b. $D = 0$. The graph touches the x-axis 1 time.
c. $D = 144$. The graph touches the x-axis 1 time.
d. $D = 288$. The graph touches the x-axis 2 times.

5. Calculate the discriminant of the quadratic equation below. Use this information to determine how many times the graph of this function will touch the x-axis.

$$5x^2 - x - 1 = 0$$

a. $D = 19$. The graph touches the x-axis 2 times.
b. $D = -19$. The graph touches the x-axis 0 times.
c. $D = -21$. The graph touches the x-axis 0 times.
d. $D = 21$. The graph touches the x-axis 2 times.

6. Calculate the discriminant of the quadratic equation below. Use this information to determine how many times the graph of this function will touch the x-axis.

$$x^2 - 6x + 15 = 0$$

a. $D = -24$. The graph touches the x-axis 0 times.
b. $D = 24$. The graph touches the x-axis 2 times.
c. $D = -48$. The graph touches the x-axis 0 times.
d. $D = 72$. The graph touches the x-axis 2 times.

7. Calculate the discriminant of the quadratic equation below. Use this information to determine how many times the graph of this function will touch the x-axis.

$$2x^2 - 8x - 13 = 0$$

a. $D = 168$. The graph touches the x-axis 2 times.
b. $D = -40$. The graph touches the x-axis 0 times.
c. $D = 40$. The graph touches the x-axis 2 times.
d. $D = 120$. The graph touches the x-axis 2 times.

8. Calculate the discriminant of the quadratic equation below. Use this information to determine how many times the graph of this function will touch the x-axis.

$$-3x^2 - 10x - 9 = 0$$

a. $D = -8$. The graph touches the x-axis 0 times.
b. $D = 8$. The graph touches the x-axis 2 times.
c. $D = 208$. The graph touches the x-axis 2 times.
d. $D = -88$. The graph touches the x-axis 0 times.

9. Calculate the discriminant of the quadratic equation below. Use this information to determine how many times the graph of this function will touch the x-axis.

$$x^2 - 6x + 9 = 0$$

a. $D = -24$. The graph touches the x-axis 0 times.
b. $D = 72$. The graph touches the x-axis 2 times.
c. $D = 0$. The graph touches the x-axis 1 time.
d. $D = 0$. The graph touches the x-axis 0 times.

10. Calculate the discriminant of the quadratic equation below. Use this information to determine how many times the graph of this function will touch the x-axis.

$$-3x^2 + 5x - 11 = 0$$

a. $D = 107$. The graph touches the x-axis 2 times.
b. $D = -107$. The graph touches the x-axis 0 times.
c. $D = -122$. The graph touches the x-axis 0 times.
d. $D = 122$. The graph touches the x-axis 2 times.

11. Calculate the discriminant of the quadratic equation below. Use this information to determine how many times the graph of this function will touch the x-axis.

$$-5x^2 - 16x - 12 = 0$$

a. $D = 0$. The graph touches the x-axis 1 time.
b. $D = -16$. The graph touches the x-axis 0 times.
c. $D = 16$. The graph touches the x-axis 2 times.
d. $D = -208$. The graph touches the x-axis 2 times.

12. Calculate the discriminant of the quadratic equation below. Use this information to determine how many times the graph of this function will touch the x-axis.

$$x^2 - 12x + 36 = 0$$

a. $D = 120$. The graph touches the x-axis 2 times.
b. $D = -120$. The graph touches the x-axis 0 times.
c. $D = 0$. The graph touches the x-axis 1 time.
d. $D = 288$. The graph touches the x-axis 2 times.

13. Is it possible for the product of a number and 7 less than that number to be 256?

a. Yes, because the discriminant is positive.
b. Yes, because the discriminant is zero.
c. No, because the discriminant is negative.
d. Solving for the discriminant will not give enough information to answer this question.

14. Is it possible for the product of 3 times a number and 4 more than that number to equal -117?

a. Yes, because the discriminant is positive.
b. Yes, because the discriminant is zero.
c. No, because the discriminant is negative.
d. Solving for the discriminant will not give enough information to answer this question.

15. Elizabeth is in charge of finances for her company and created an equation to represent the company's profits for the year. The equation she comes up with is $f(x) = 7x^2 - 20x + 954$. Will the company be able to turn a profit of $10,000 this year?

a. Yes, because the discriminant is positive.
b. Yes, because the discriminant is zero.
c. No, because the discriminant is negative.
d. Solving for the discriminant will not give enough information to answer this question.

8.10 Graphing Quadratic Functions

Quadratic equations form an arc called a parabola when they are graphed. Parabolas can be graphed if two things are known:

- If the parabola opens up or down.
- The location of the vertex.

When a quadratic equation is in standard form, we can determine its vertex and whether the parabola will open up or down. Remember, standard form for a quadratic equation is $y = ax^2 + bx + c$.

Does the Parabola Open Up or Down?

When you look at a quadratic equation, if $a > 0$, then the parabola opens up. If $a < 0$, then the parabola opens down. For example, the quadratic equation $y = 2x^2 + 3x - 4$ has an a-value of 2. The value of a is positive, so the parabola will open up. However, the equation $y = -3x^2 - 2x + 5$ has an a-value of –3. Since a is negative, the parabola will open down.

Where Is the Vertex?

The vertex of a parabola can be found using the vertex formula.

Vertex formula: $h = -\frac{b}{2a}$, where h represents the x-value of the ordered pair of the vertex (h, k).

For example, to find the vertex for the quadratic equation $y = -3x^2 + 6x + 5$, start by determining the values for a and b.

$$a = -3$$
$$b = 6$$

Plug a and b into the vertex formula and simplify.

$$h = -\frac{b}{2a}$$
$$h = -\frac{6}{2(-3)}$$
$$h = -\frac{6}{-6}$$
$$h = 1$$

Now, find the corresponding y-value, k, by plugging in 1 for x in the original equation.

$$k = -3(1)^2 + 6(1) + 5$$
$$k = -3(1) + 6(1) + 5$$
$$k = -3 + 6 + 5$$
$$k = 8$$

This means that the vertex (h, k) is at the point (1,8).

In the equation $y = -3x^2 + 6x + 5$, a is negative so the parabola opens down from the vertex at (1,8). This is enough information to sketch the graph.

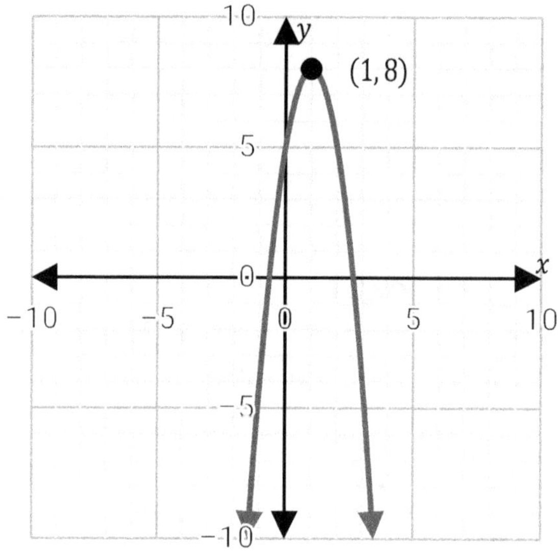

However, it is often helpful to graph a few more points on the parabola so that the graph becomes more accurate. Plug in a few values for x, and calculate the corresponding y-values. Then plot these points on the graph.

For example, plug in 0 for x and calculate the corresponding y-value.

$$y = -3(0)^2 + 6(0) + 5$$
$$y = 0 + 0 + 5$$
$$y = 5$$

This means that the point (0,5) is on the graphed parabola. Because of the symmetrical nature of parabolas, we can determine that the point (2,5) is also on the parabola. An infinite number of points can be plotted on the graph. The more points you plot, the more accurate the sketch of the graph becomes.

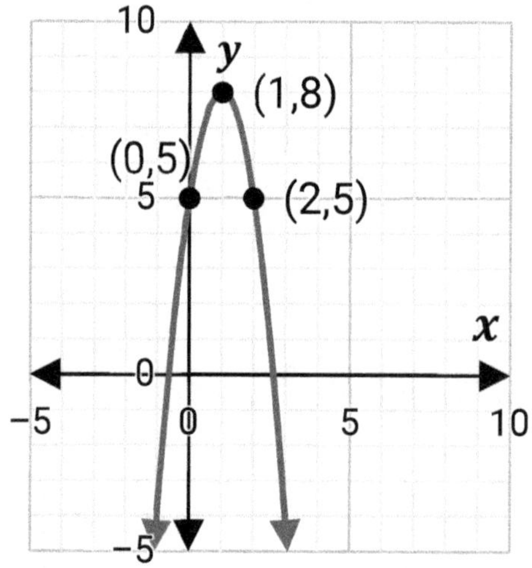

Examples

Example 1:

Graph the quadratic equation.

$$y = x^2 + 2x + 2$$

Answer:

If a is positive, the parabola will open up. If a is negative, the parabola will open down. In this case, a is 1, so the parabola will open up.

Now, find the vertex using the vertex formula: $h = -\frac{b}{2a}$.

$$h = \frac{-2}{2} = -1$$

Plug $x = -1$ into the original equation to find the corresponding y-value.

$$k = (-1)^2 + 2(-1) + 2 = 1 - 2 + 2 = 1$$

This means the vertex is $(-1,1)$.

Once we know the direction of the parabola (up), and the vertex $(-1,1)$, we can graph the function.

You can determine a few more ordered pairs to make the graph more accurate. Plug in any value for x, and calculate the corresponding y-value.

For example, plug in 0 for x, and solve for y.

$$y = (0)^2 + 2(0) + 2 = 2$$

This means that $(0,2)$ is a point on the parabola.

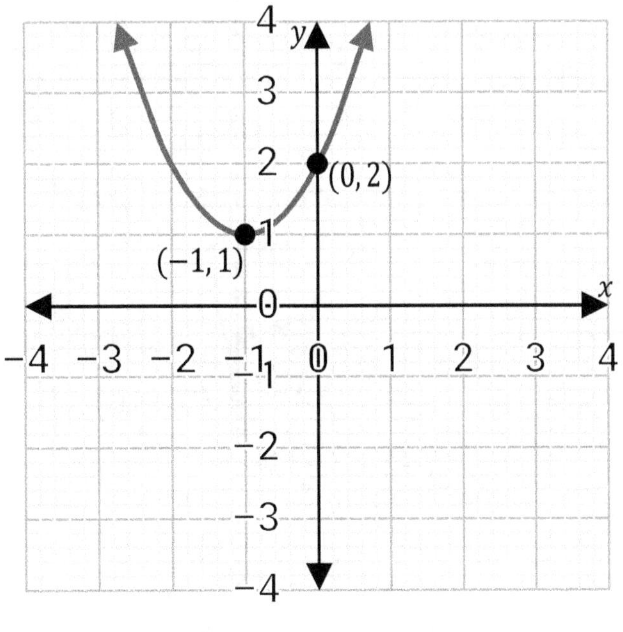

EXAMPLE 2:

Graph the quadratic equation.

$$y = 2x^2 + 4x - 10$$

Answer:

If a is positive, the parabola will open up. If a is negative, the parabola will open down. In this case, a is 2, so the parabola will open up.

Now, find the vertex using the vertex formula: $h = -\frac{b}{2a}$.

$$h = -\frac{4}{2(2)} = -\frac{4}{4} = -1$$

Now plug in $x = -1$ into the original equation to determine the value of k.

$$k = 2(-1)^2 + 4(-1) - 10 = 2 - 4 - 10 = -12$$

The vertex is $(-1, -12)$.

We have enough information to graph the function because we know the vertex and the direction of the parabola.

We can plot more points on the parabola to make the graph more accurate. Plug in values for x to calculate the corresponding y-values. For example, plug in 0 for x, and calculate the y-value.

$$y = 2(0)^2 + 4(0) - 10 = 0 + 0 - 10 = -10$$

This means that $(0, -10)$ is a point on the parabola.

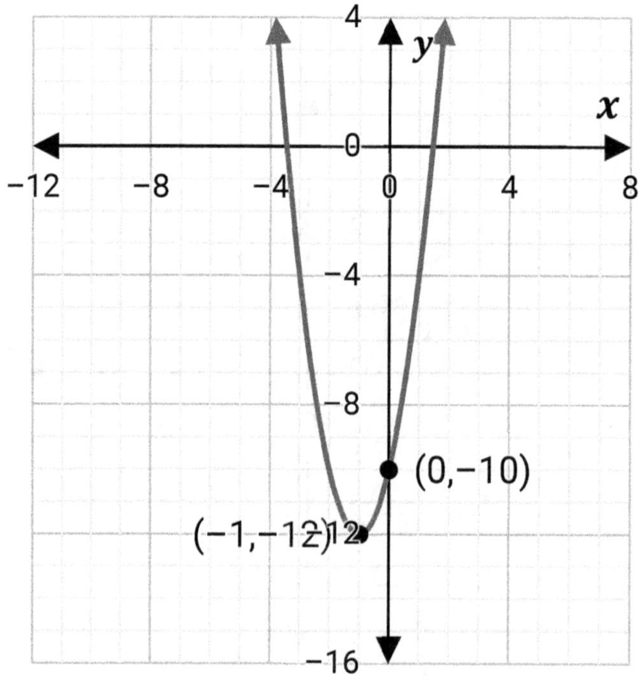

EXAMPLE 3:

Graph the quadratic equation.

$$-2x^2 + 4x - 10$$

Answer:

If a is positive, the parabola will open up. If a is negative, the parabola will open down. In this case, a is -2, so the parabola will open down.

Now, find the vertex using the vertex formula $h = -\frac{b}{2a}$.

$$h = -\frac{4}{2(-2)} = -\frac{4}{-4} = 1$$

Now plug in $x = 1$ into the original equation to find the value of k.

$$k = -2(1)^2 + 4(1) - 10 = -2(1) + 4(1) - 10 = -2 + 4 - 10 = -8$$

The vertex is $(1, -8)$.

We have enough information to graph the parabola because we know the vertex and the direction of the parabola.

We can plot more points on the parabola to make it more accurate. Plug in values for x to calculate the corresponding y-values. For example, use $x = 0$ and solve for y.

$$y = -2(0)^2 + 4(0) - 10 = 0 + 0 - 10 = -10$$

This means that $(0, -10)$ is a point on the parabola.

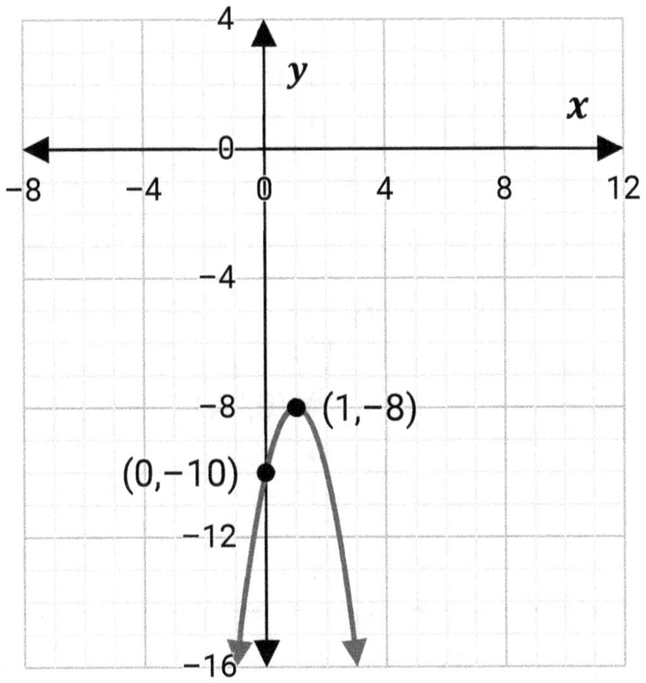

EXAMPLE 4:

What is the graph of the equation $y = 3x^2 + 14$?

Answer:

Start by identifying the values of a, b, and c in the equation.

$$a = 3, b = 0, c = 14$$

Since a is a positive integer, the parabola opens upward.

Next, find the vertex (h, k), starting with h. Substitute values for a and b into the formula $h = -\frac{b}{2a}$ and simplify.

$$h = -\frac{0}{2(3)} = -\frac{0}{6} = 0$$

Since k is the y-value of the vertex, substitute 0 into the original equation for x and simplify.

$$k = 3(0)^2 + 14 = 0 + 14 = 14$$

Therefore, the vertex of the parabola is at the point (0,14).

From here, find another point on the graph by substituting a value for x and solving for y. For example, solve for y when $x = 2$.

$$y = 3(2)^2 + 14 = 3(4) + 14 = 12 + 14 = 26$$

Therefore, the point (2,26) is on the graph. Given the parabola opens up, the vertex is at (0,14), and the graph also passes through the point (2,26), the following graph is the graph of the equation.

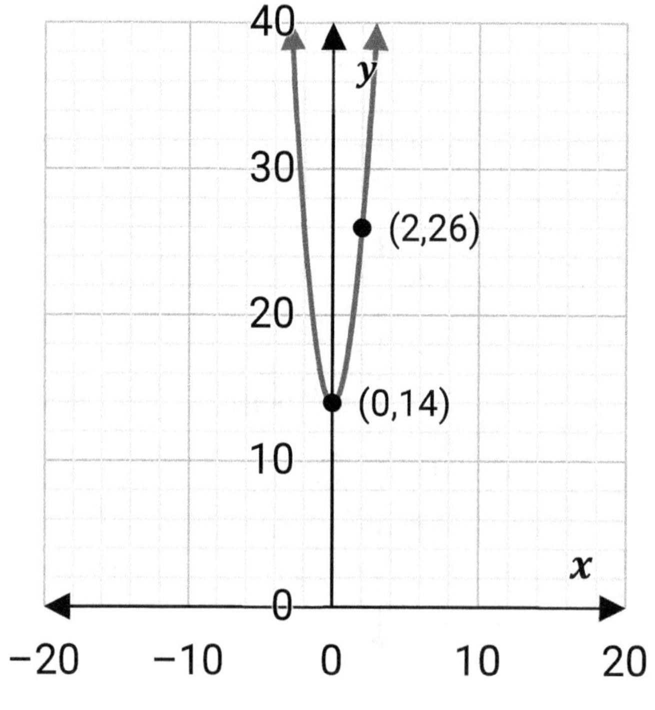

EXAMPLE 5:

Steven is designing a bridge. With the parameters he's been given, the equation for the bridge is $y = -\frac{1}{5}x^2 + 5x + 3$. He needs to sketch a graph of the bridge for the next meeting. What will the sketch look like?

Answer:

Start by identifying the values of a, b, and c in the equation.

$$a = -\frac{1}{5}, b = 5, c = 3$$

Since a is a negative number, the parabola opens downward, extending in a negative direction.

Next, find the vertex (h, k), starting with h. Substitute values for a and b into the formula $h = -\frac{b}{2a}$ and simplify.

$$h = -\frac{5}{2\left(-\frac{1}{5}\right)} = -\frac{5}{-\frac{2}{5}} = -\frac{5}{1} \div \left(-\frac{2}{5}\right) = -\frac{5}{1} \cdot \left(-\frac{5}{2}\right) = \frac{25}{2} = 12.5$$

Since k is the y-value of the vertex, substitute $\frac{25}{2}$ into the original equation for x and simplify.

$$k = -\frac{1}{5}\left(\frac{25}{2}\right)^2 + 5\left(\frac{25}{2}\right) + 3 = -\frac{1}{5}\left(\frac{625}{4}\right) + 5\left(\frac{25}{2}\right) + 3 = -\frac{625}{20} + \frac{125}{2} + 3 = \frac{137}{4} = 34.25$$

Therefore, the vertex of the equation is (12.5,34.25).

From here, find another point on the graph by substituting a value for x and solving for y. For example, solve for y when $x = 0$.

$$y = -\frac{1}{5}(0)^2 + 5(0) + 3 = 0 + 0 + 3 = 3$$

Therefore, the point (0,3) is on the graph. Given the parabola opens down, the vertex is at (12.5,34.25), and the graph also passes through the point (0,3), the following graph can be drawn.

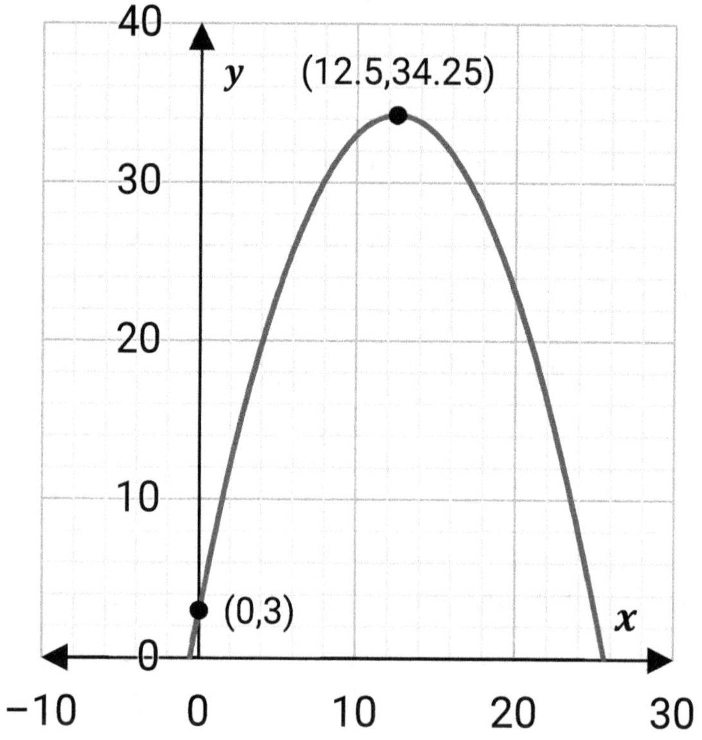

PRACTICE QUESTIONS

1. Which graph matches the equation $y = x^2 - 2x - 3$?

a. 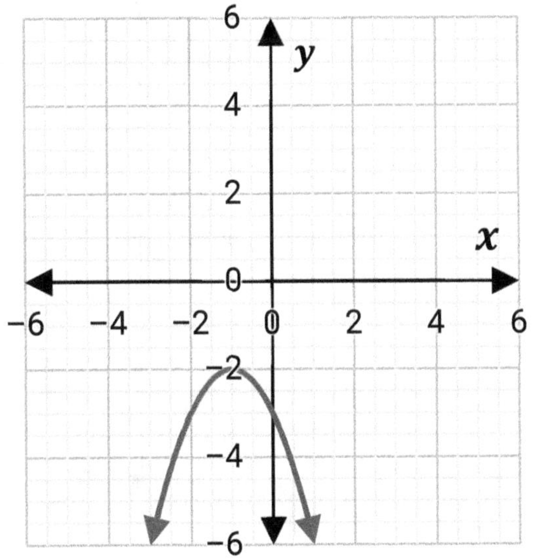

c. 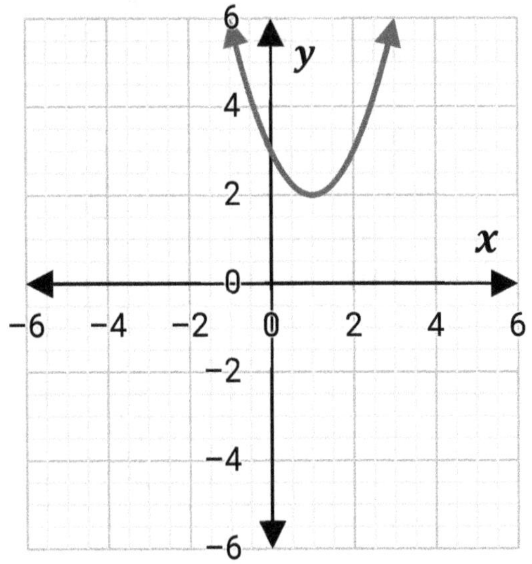

b. 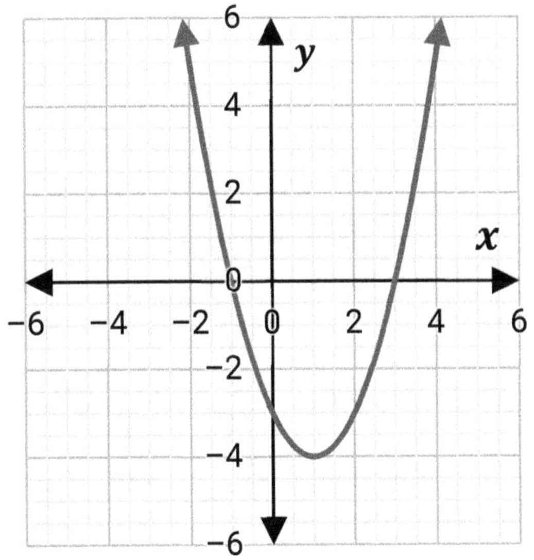

d. 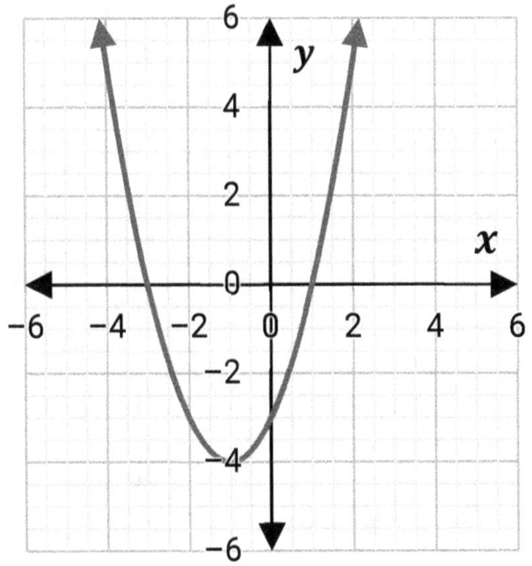

2. Which graph matches the equation $y = x^2 - 4x - 5$?

a. 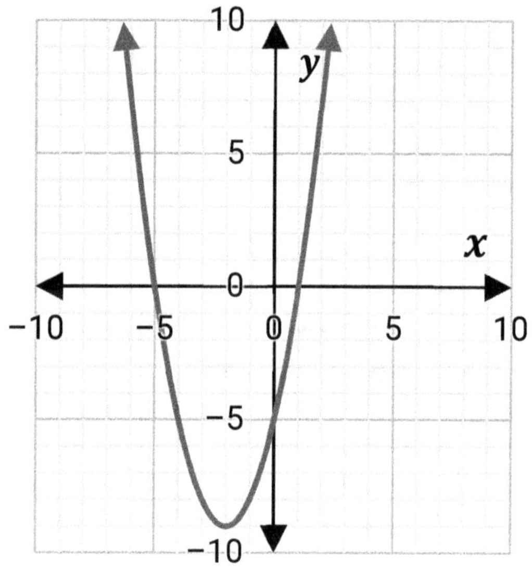

c. 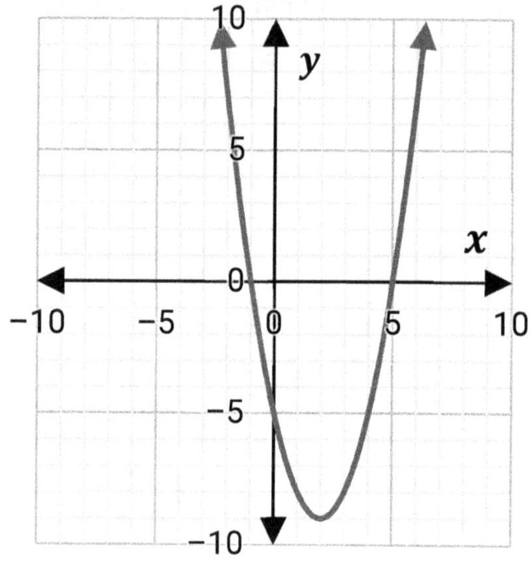

b. 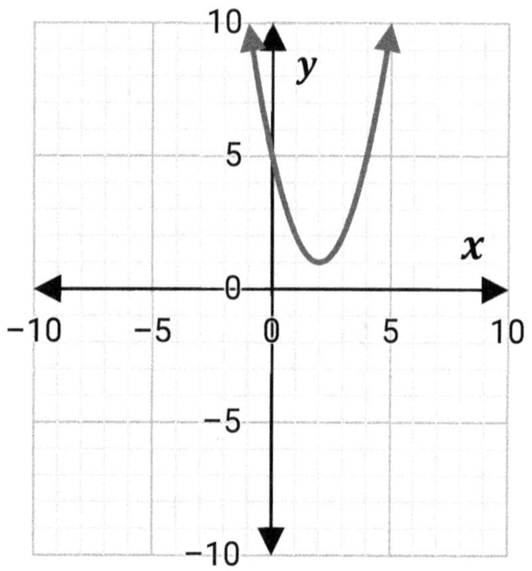

d. 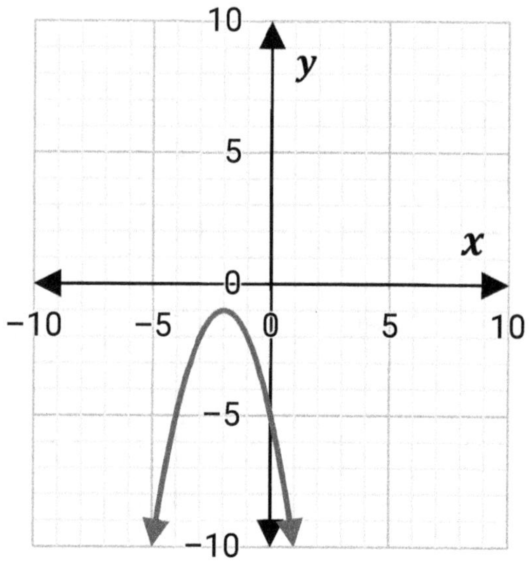

3. Which graph matches the equation $y = 3x^2 - 12x + 9$?

a. 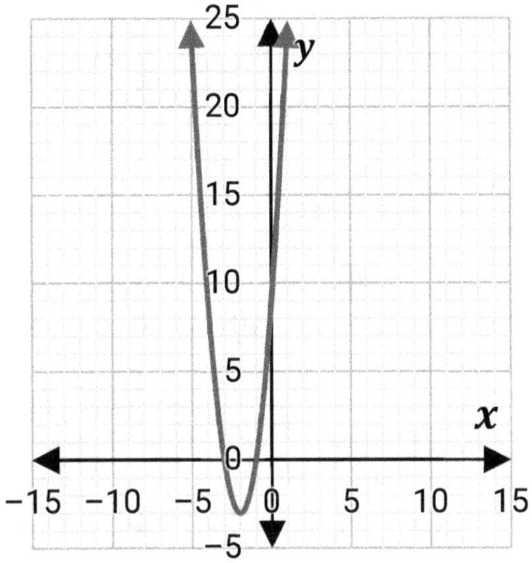

c. 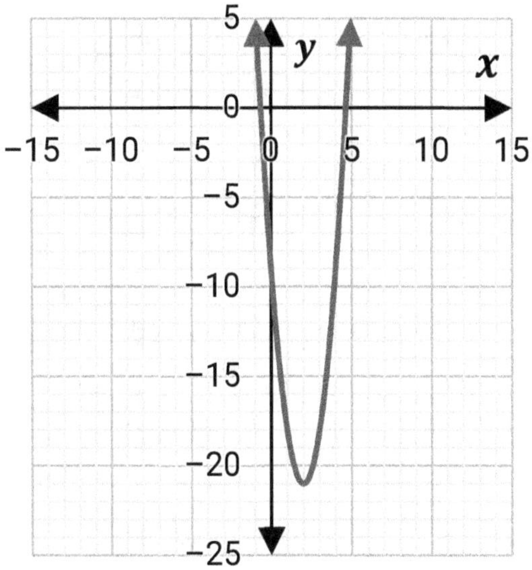

b. 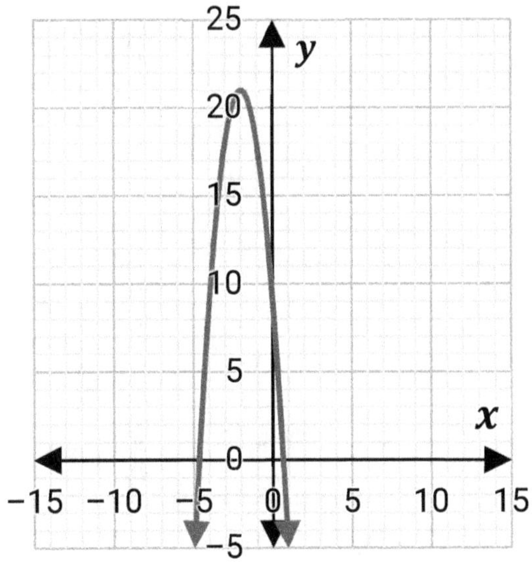

d. 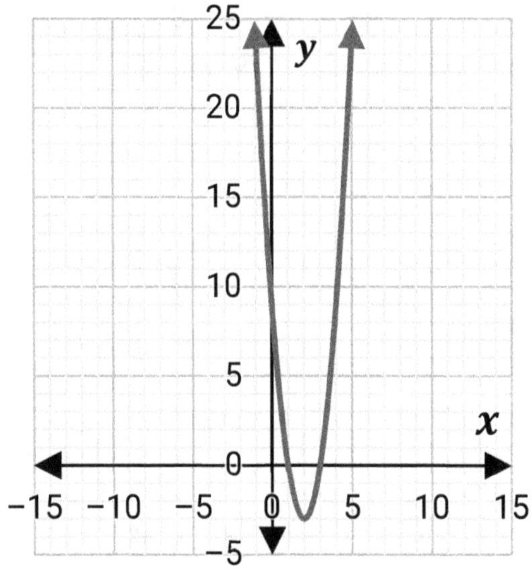

4. Which graph matches the equation $y = x^2 - 10x + 9$?

a. 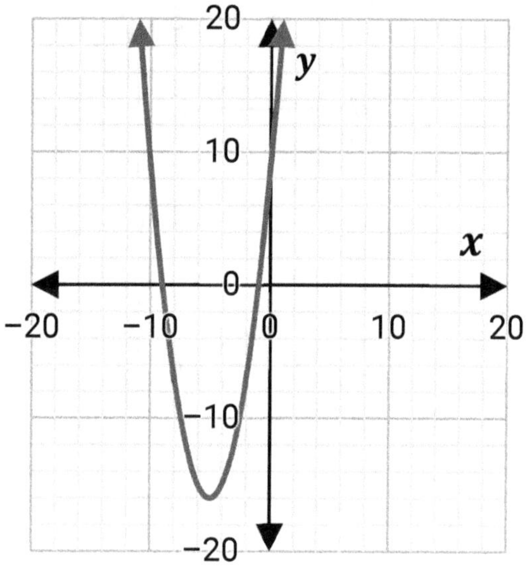

c. 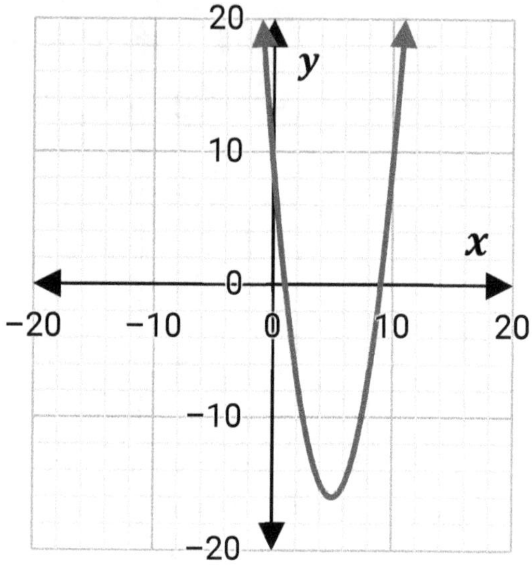

b. 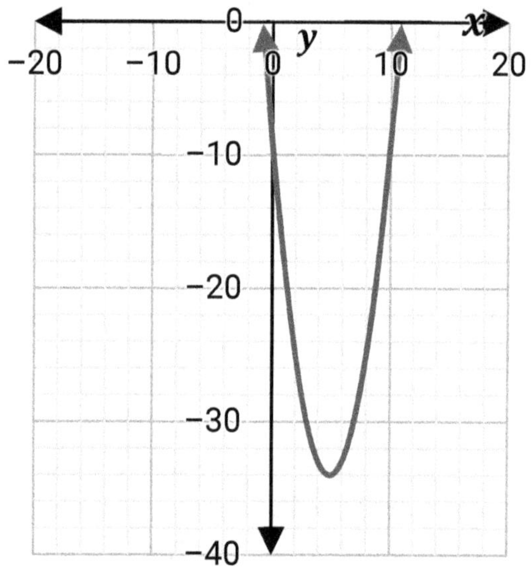

d. 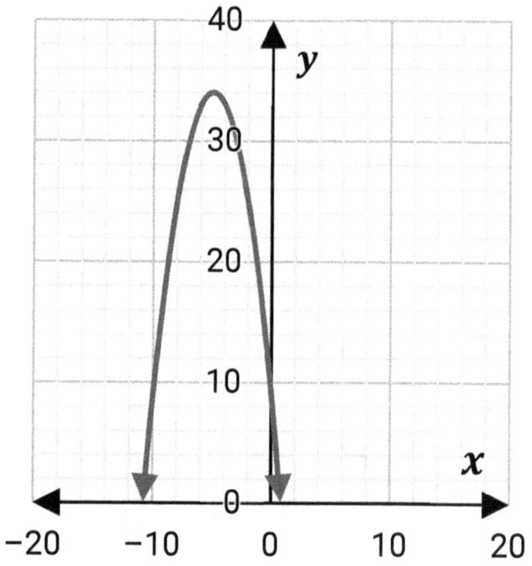

5. Which graph matches the equation $y = x^2 - 1$?

a.

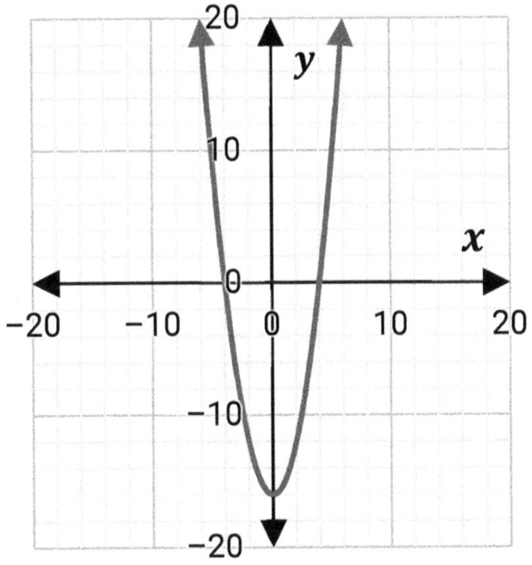

b.

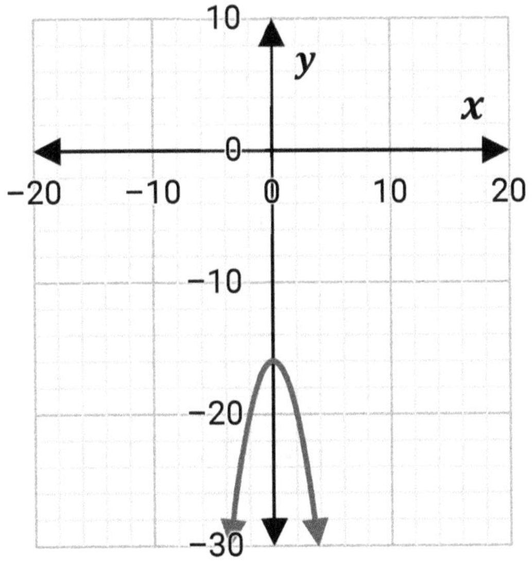

c.

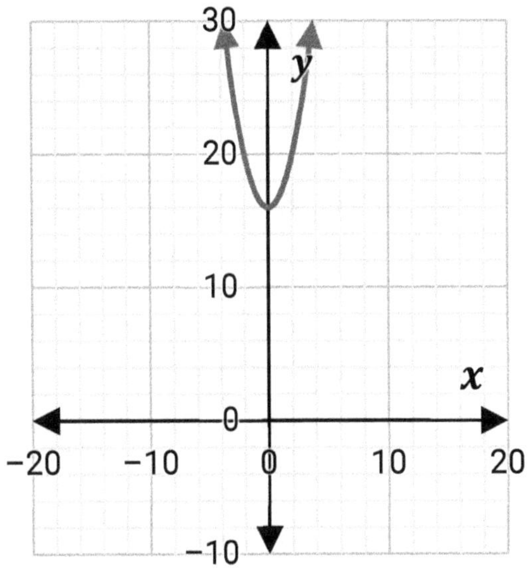

d.

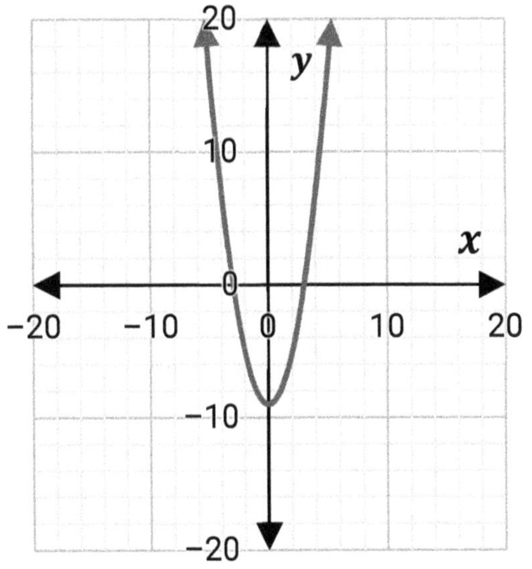

6. Which graph matches the equation $y = -x^2 - 2x - 1$?

a. 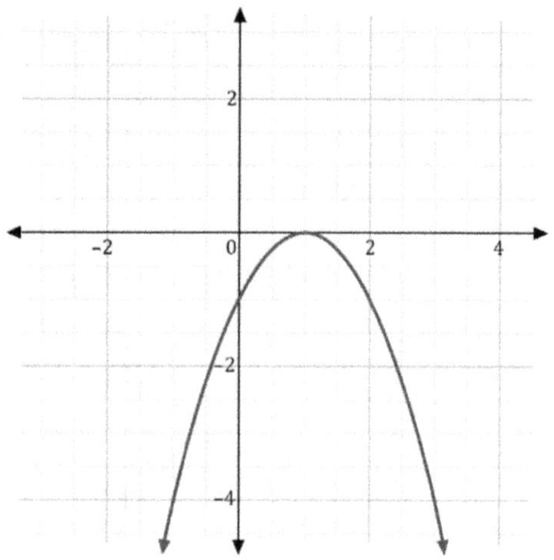

c. 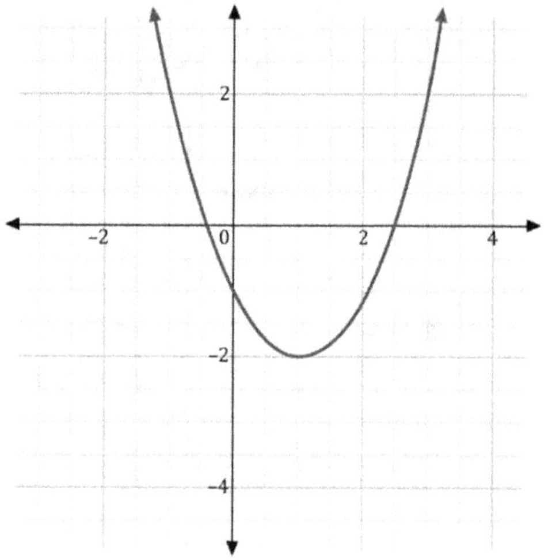

b. 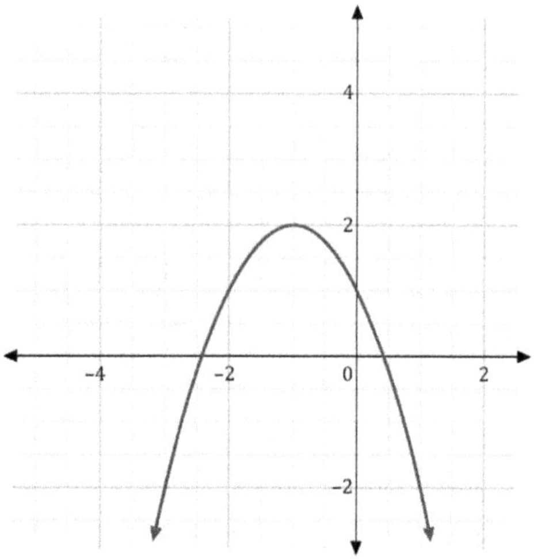

d. 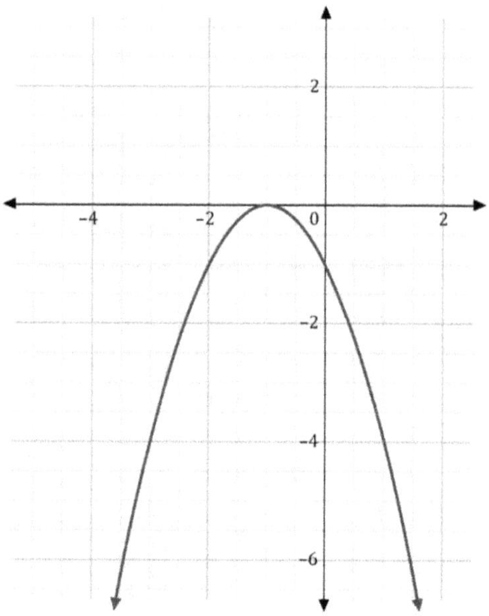

7. Which graph matches the equation $y = -x^2 + 6x + 5$?

a. 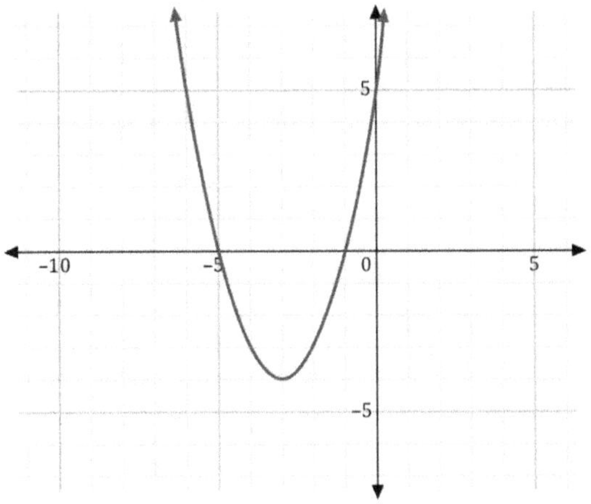

b. 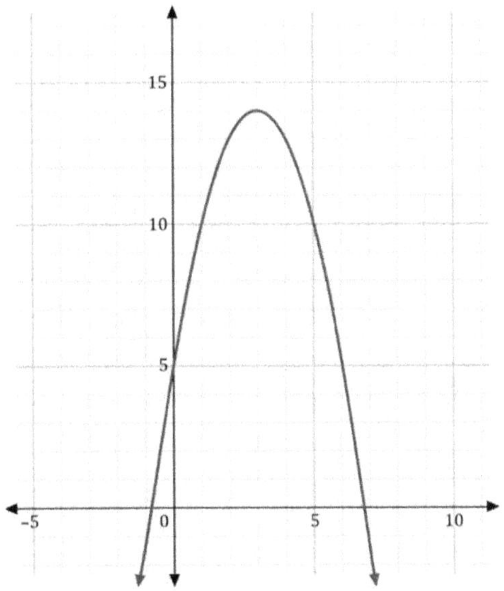

c. 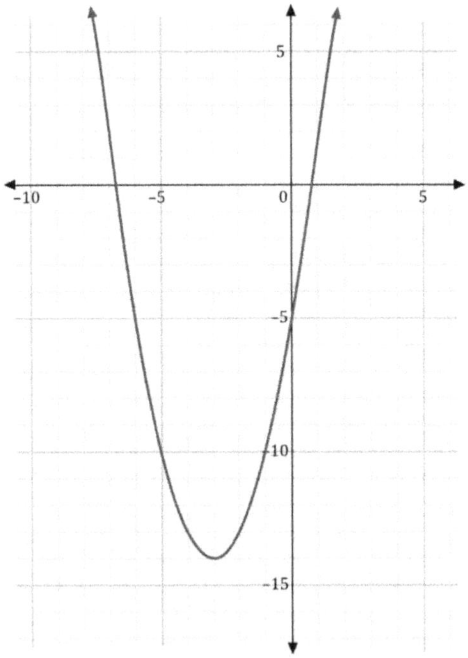

d. 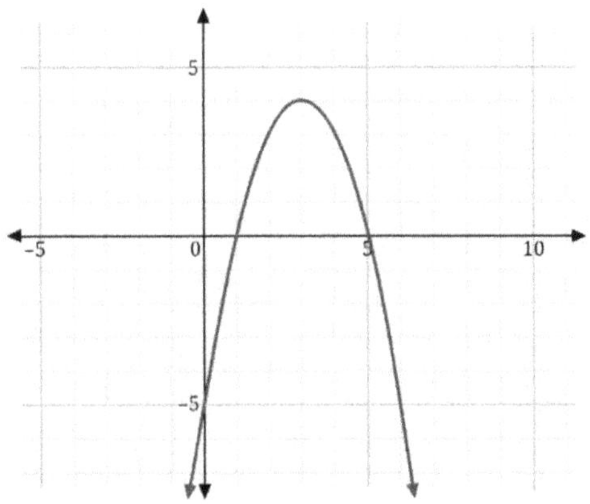

8. Which graph matches the equation $y = -2x^2 - 8x + 9$?

a.

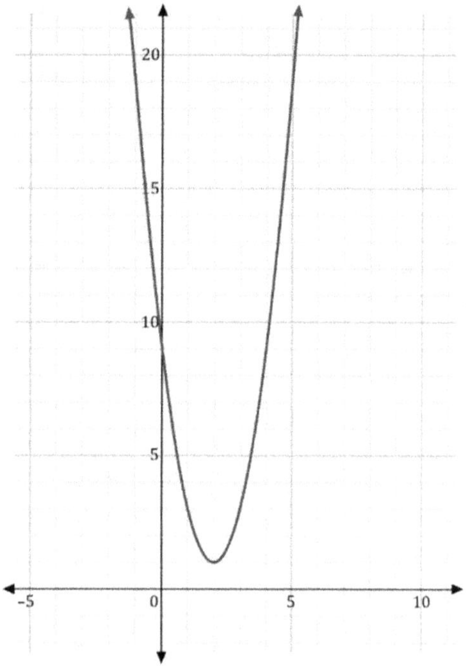

b.

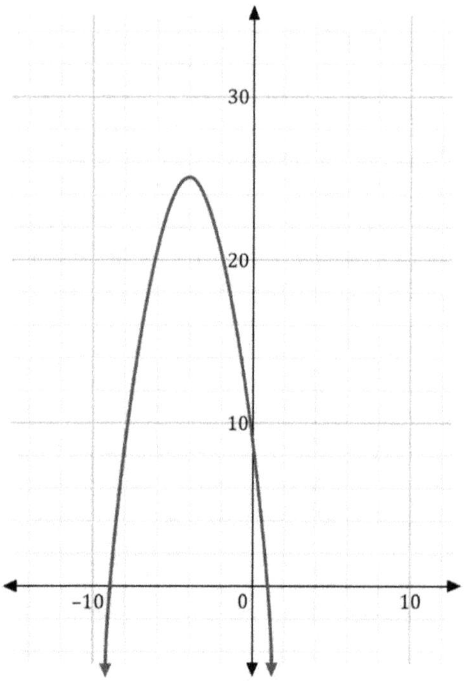

c.

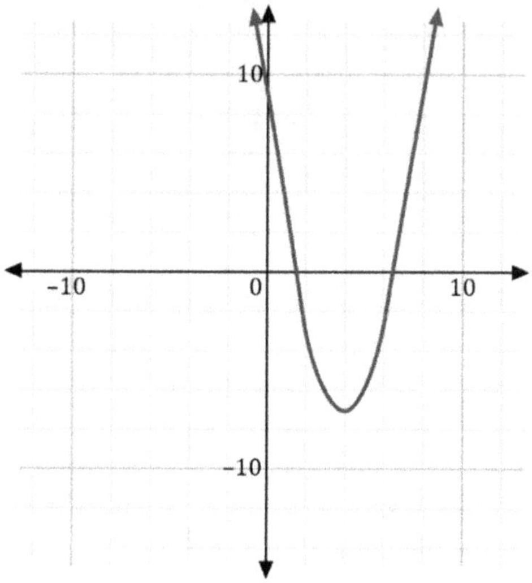

d.

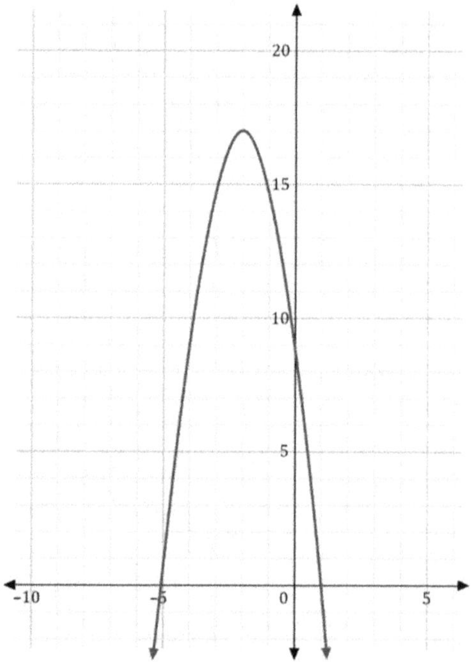

9. Which graph matches the equation $y = -\frac{1}{2}x^2 - 4x +?$

a.

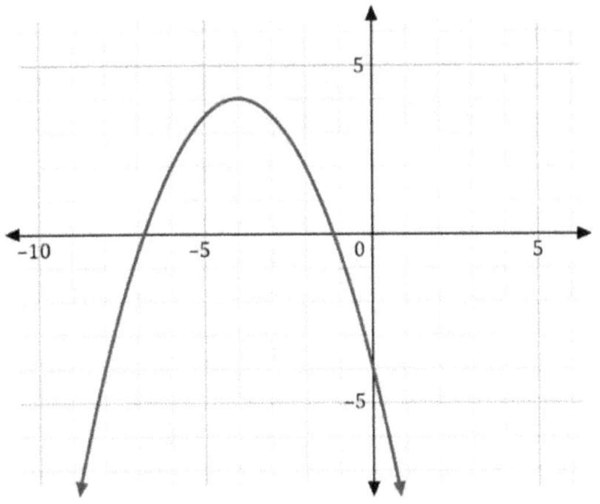

c.

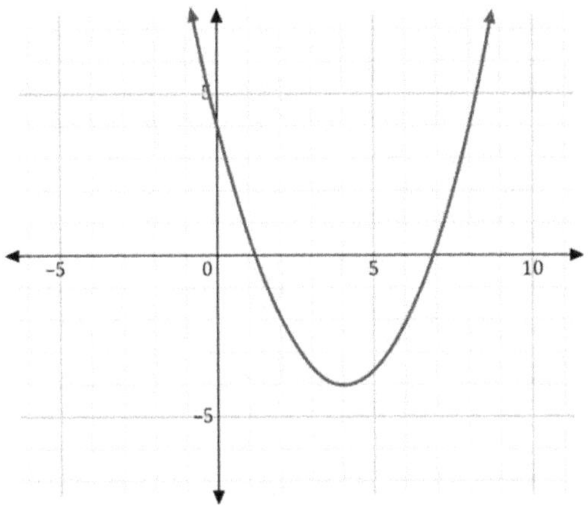

b.

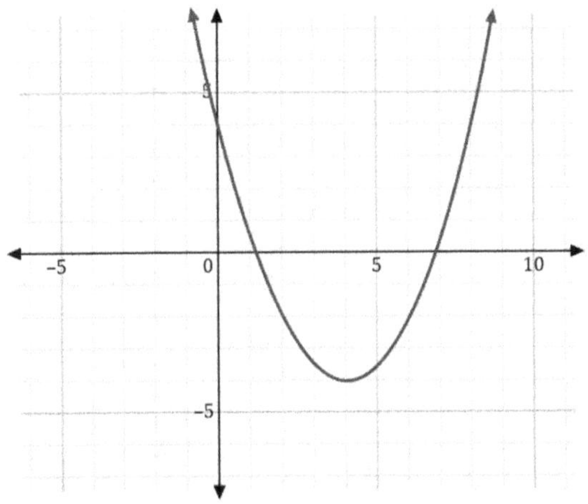

d.

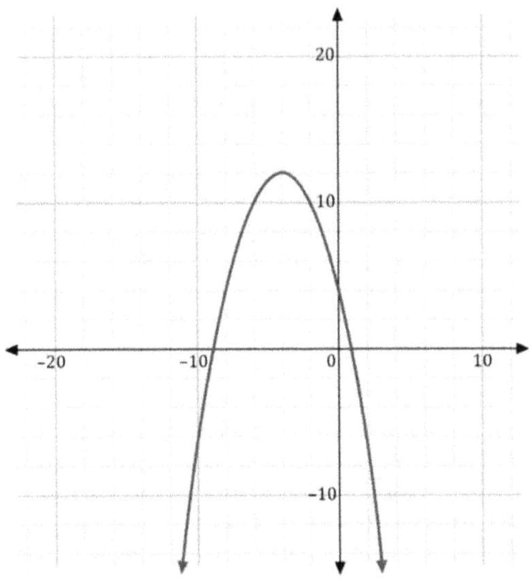

10. **Which graph matches the equation $y = -2x^2 + 1$?**

a. 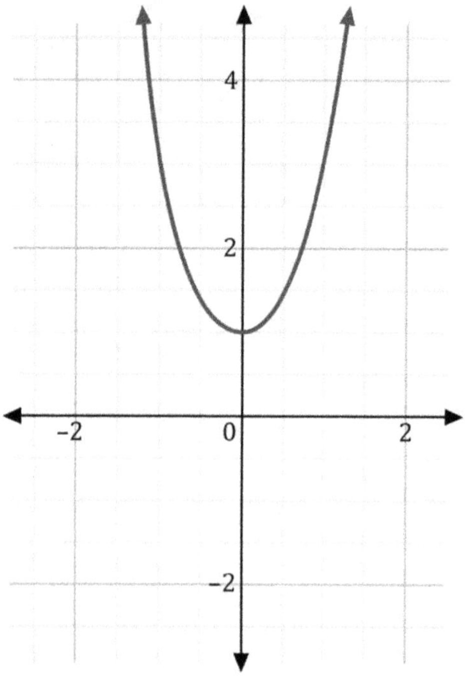

c. 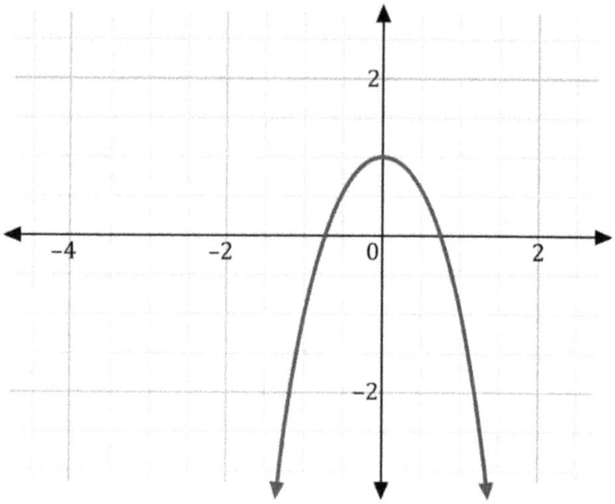

b. 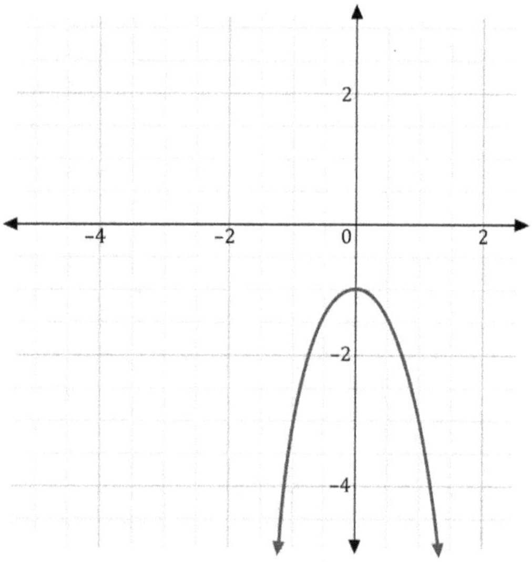

d. 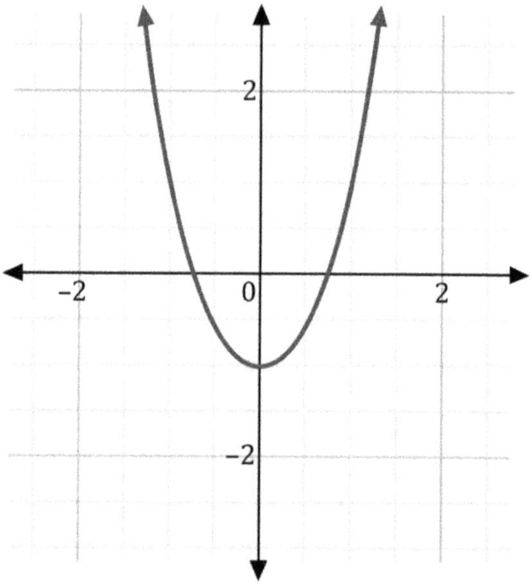

11. Which graph matches the equation $y = 3x^2 + 6x - 7$?

a.

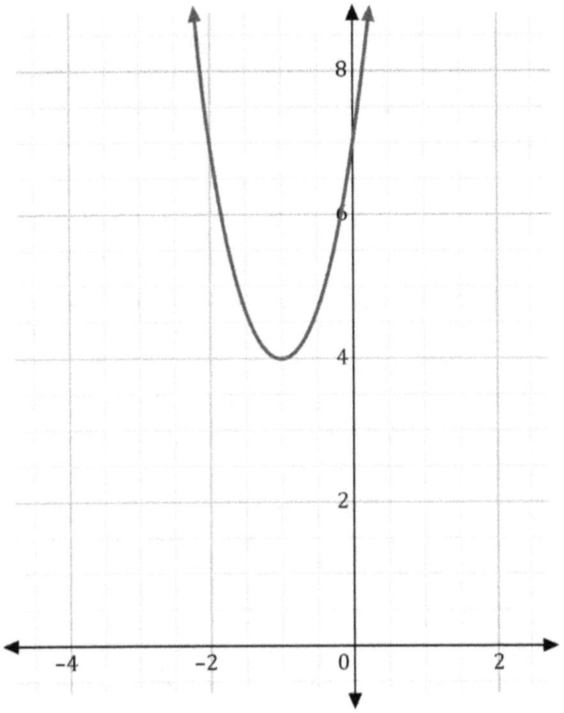

c.

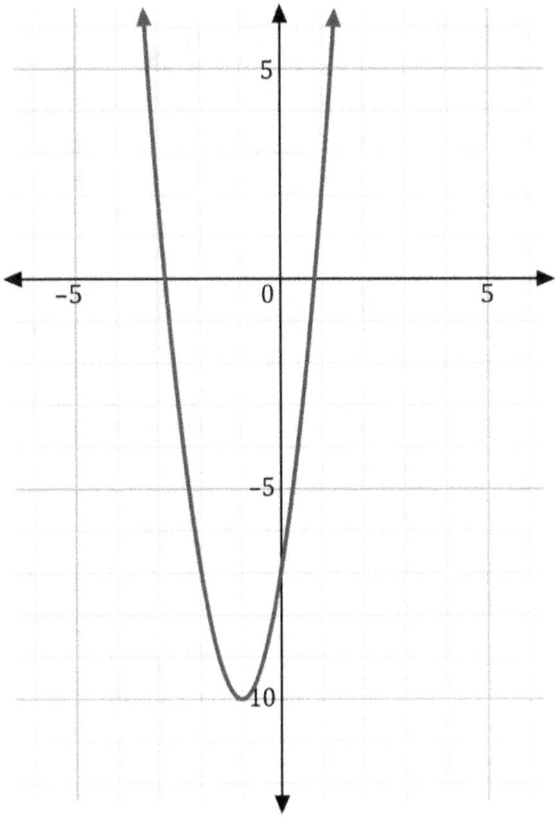

b.

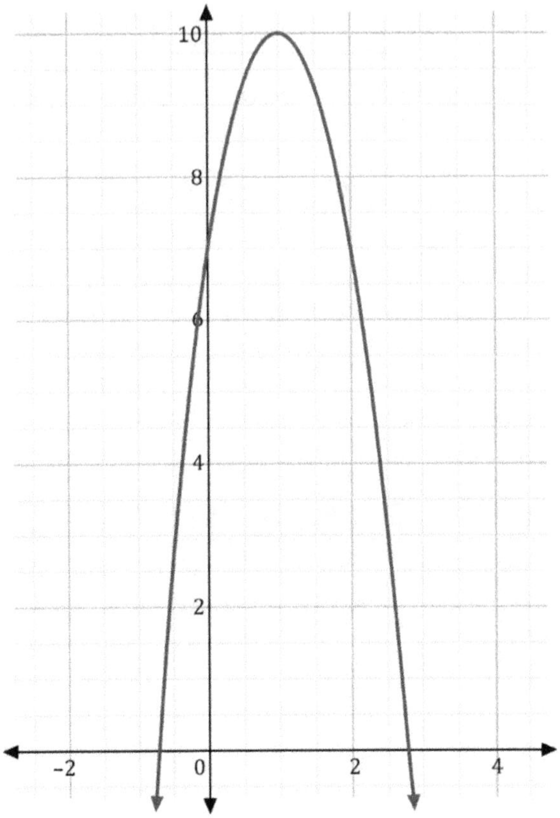

d.

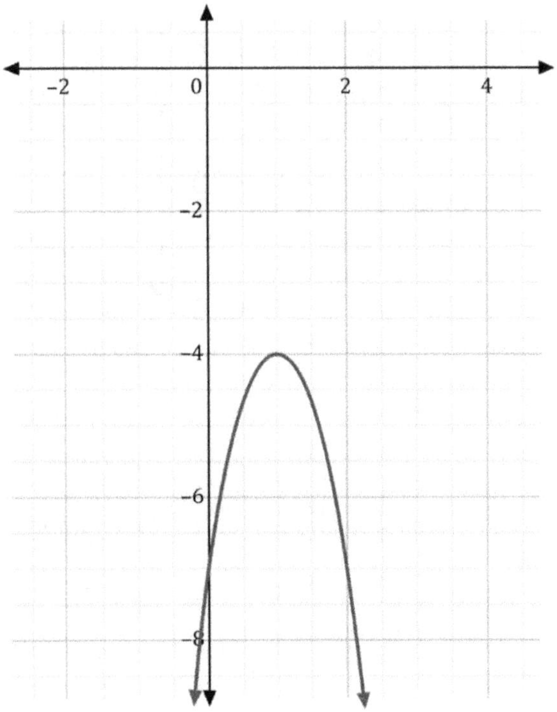

12. Which graph matches the equation $y = x^2 - 4x - 8$?

a. 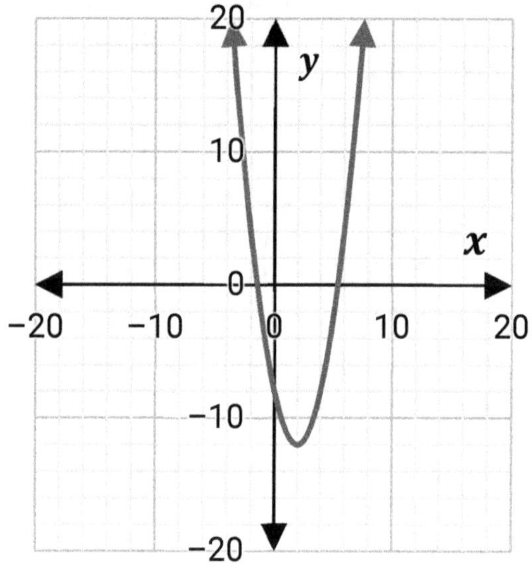

c. 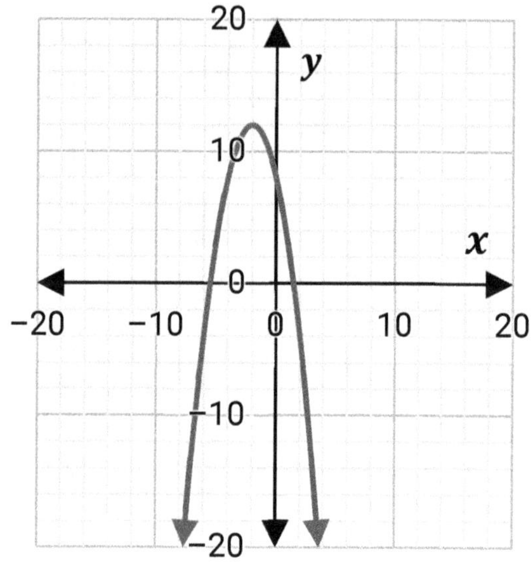

b. 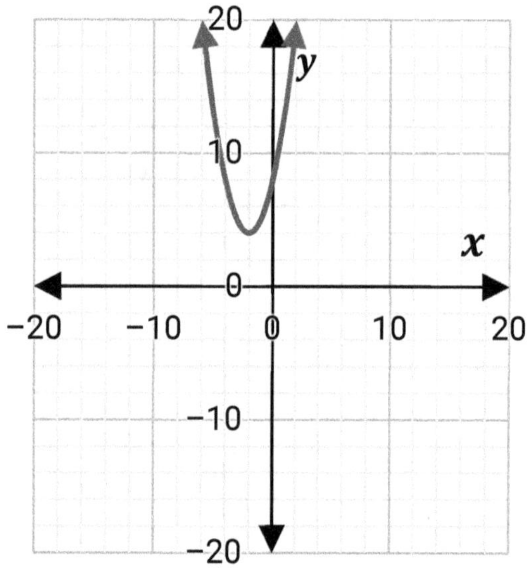

d. 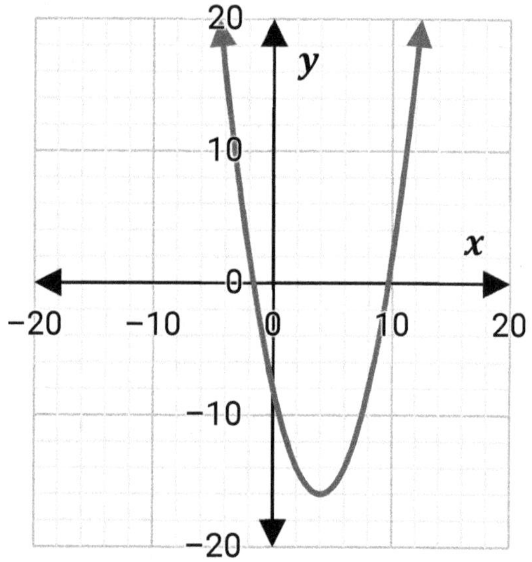

13. A physics class is experimenting with a model rocket. After s seconds, the rocket's height in feet, $h(s)$, is modeled by the equation $h(s) = -5s^2 + 20s + 3$. Graph the equation and identify the maximum height of the rocket.

a. 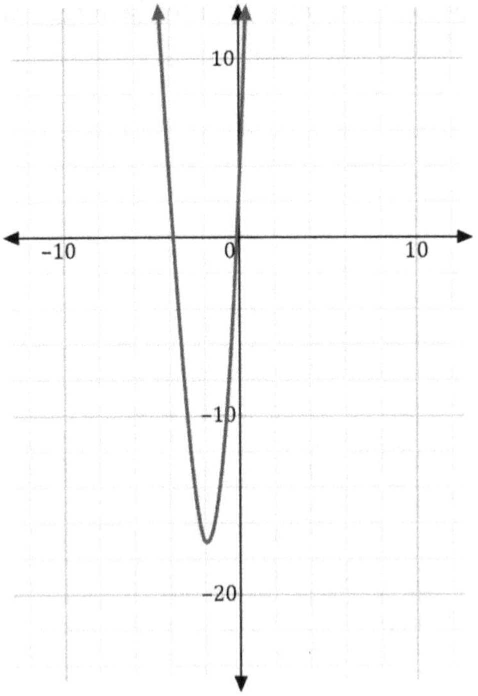

The maximum height is 3 feet.

b. 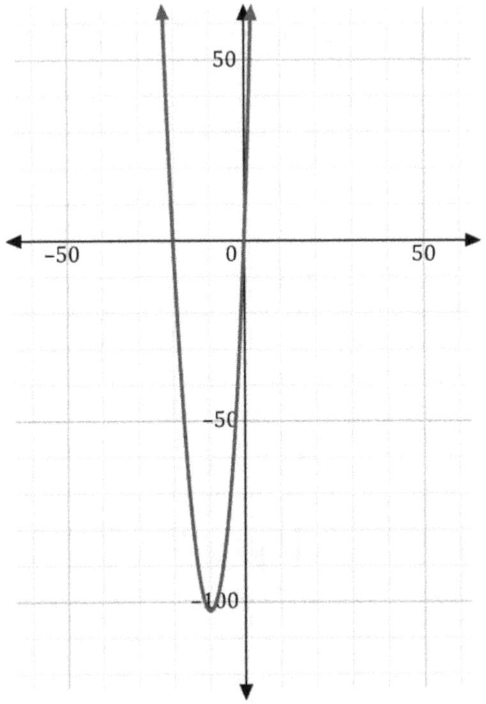

The maximum height is 23 feet.

c. 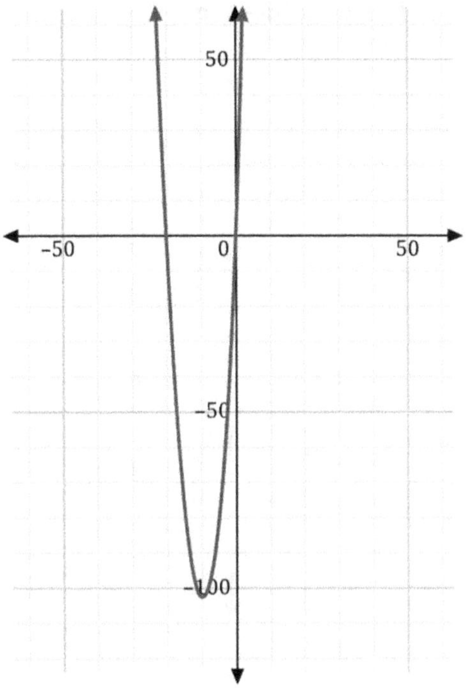

The maximum height is –103 feet.

d. 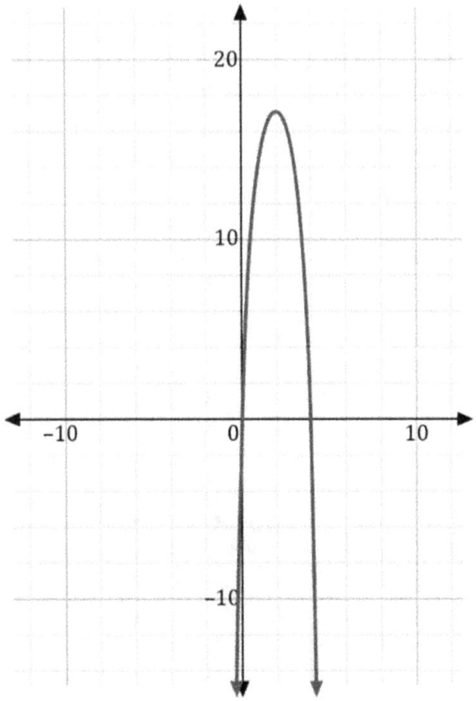

The maximum height is 17 feet.

14. Jack is at a carnival and wants to try the high striker, a game that tests your strength by determining how far in the air you can move a weight with a hammer. If the weight reaches 17 feet, you win the game. The equation $h(t) = -16t^2 + 32t + 1$ represents the height, $h(t)$, t seconds after Jack hits the weight. Graph the equation and find out whether Jack won the game.

a. 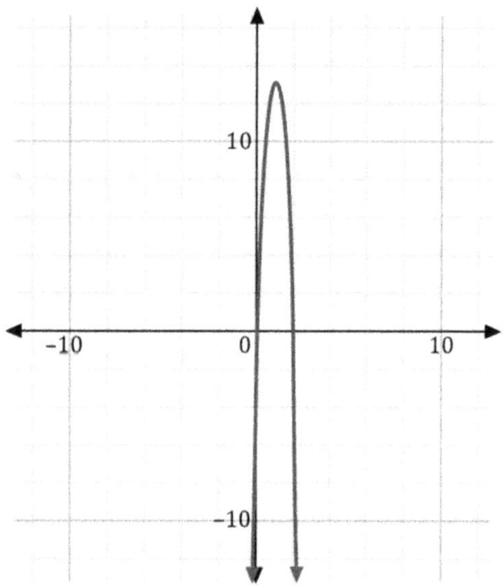

Jack's weight reached 13 feet, so he did not win the game.

b. 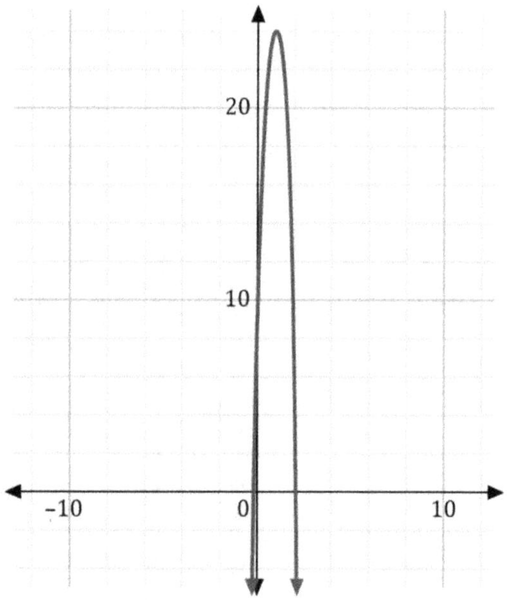

Jack's weight reached 24 feet, so he won the game.

c. 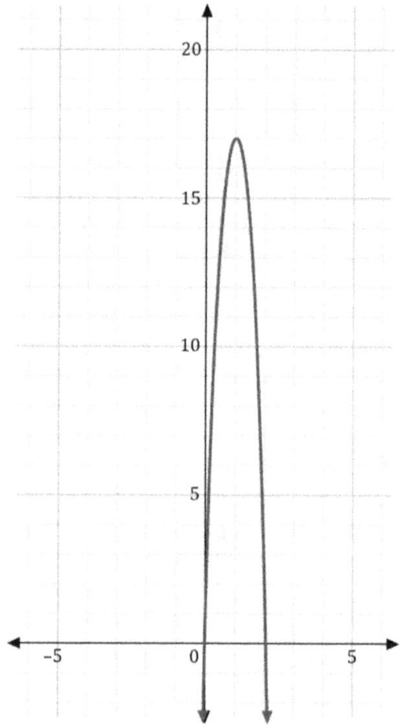

Jack's weight reached 17 feet, so he won the game.

d. 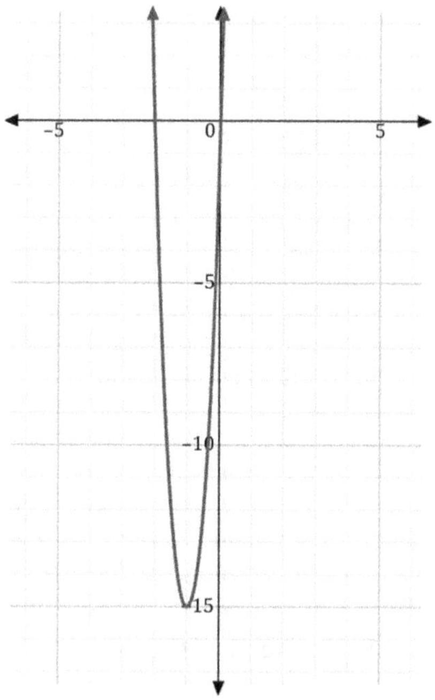

Jack's weight reached –15 feet, so he did not win the game.

15. A batter hits a baseball from ground level with an initial vertical velocity of 64 feet per second. The equation $y = -16x^2 + 64x$ represents the height of the baseball in the air with respect to the number of seconds that have passed. Based on this information, graph the equation to figure out the height of the baseball after it's been in the air for 3 seconds.

a. 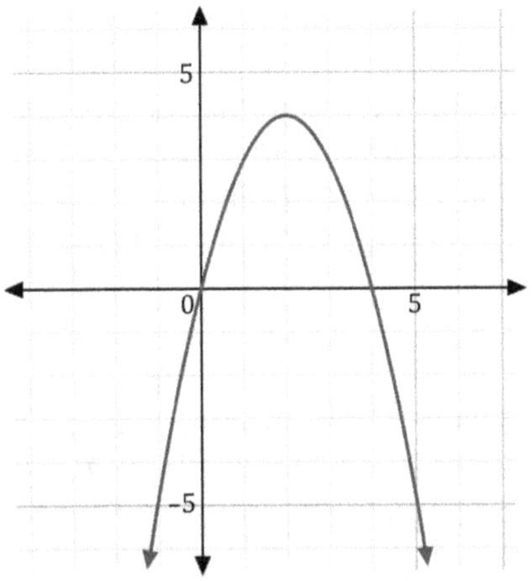

The height of the baseball at 3 seconds is 3 feet.

b. 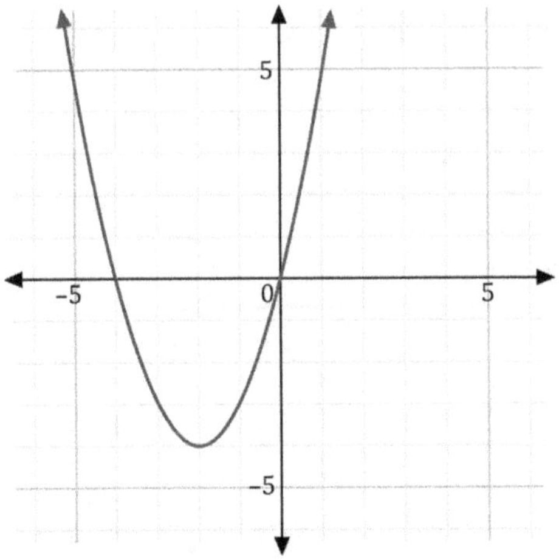

The height of the baseball at 3 seconds is –3 feet.

c. 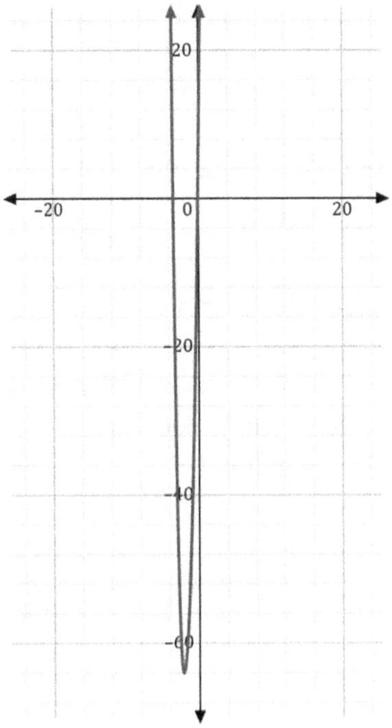

The height of the baseball at 3 seconds is –48 feet.

d. 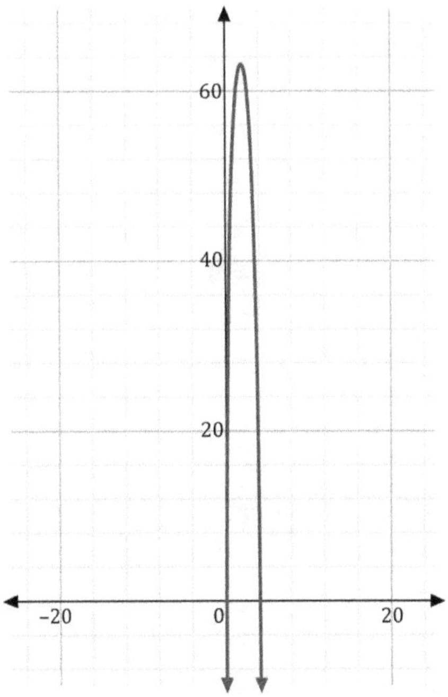

The height of the baseball at 3 seconds is 48 feet.

8.11 Manipulating Quadratic Functions

A **function manipulation**, also called a transformation, is when the graph of the original function, $f(x)$, gets moved on the coordinate plane or is changed in size. Two types of function manipulations are translations and dilations.

Translations

A **translation** is when the graph of a function slides in a specific direction on the coordinate plane. A translation can occur either vertically or horizontally. Vertical translations shift the graph of the function up or down along the y-axis. Horizontal translations shift the graph of the function left or right along the x-axis. When translations occur, the graph of the function does not change in shape or size, nor is it rotated or reflected.

Vertical Translations

Vertical translations happen when the graph of a function is shifted either up or down by k units. To shift the graph of the original function up, add k to $f(x)$ to get $g(x) = f(x) + k$. Doing so results in k getting added to all y-coordinates on the graph of the function. For instance, $g(x) = f(x) + 4$ moves the graph 4 units up.

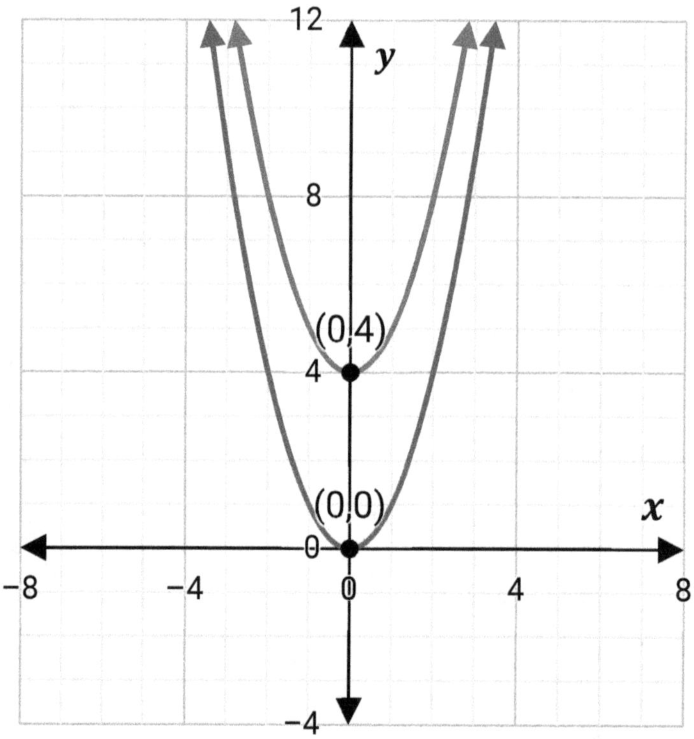

To shift the graph of the original function down, subtract k from $f(x)$ to get $g(x) = f(x) - k$. Doing so results in k getting subtracted from all y-coordinates on the graph of the function. For instance, $g(x) = f(x) - 4$ moves the graph 4 units down.

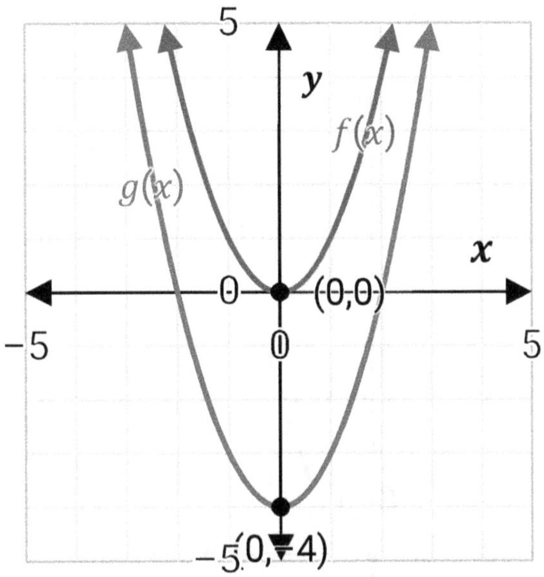

HORIZONTAL TRANSLATIONS

Horizontal translations occur when the graph of a function is shifted either left or right by h units. To shift the graph of the original function to the left, add h units to x inside of parentheses to get $g(x) = f(x + h)$. Adding h inside of parentheses shifts the graph in a negative direction, to the left. When a function manipulation occurs inside the parentheses, the graph moves in the opposite direction of the sign given in the function. For example, $g(x) = f(x + 3)$ moves the graph 3 units to the left.

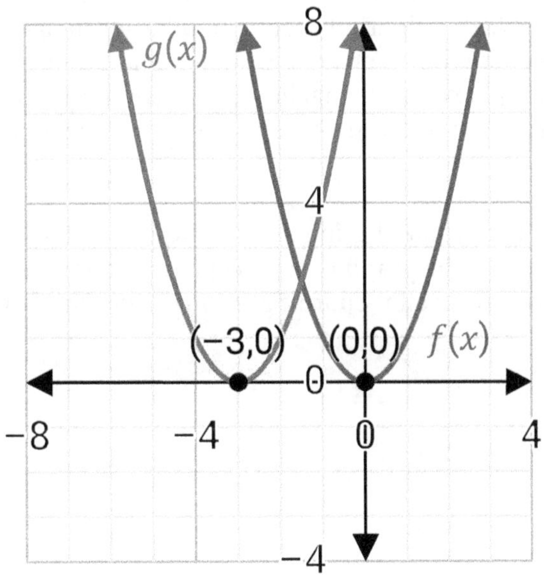

To shift the graph of the original function to the right, subtract h units from x inside the parentheses to get $g(x) = f(x - h)$. Subtracting h inside of parentheses shifts the graph in a positive direction, to the right. When a function manipulation occurs inside the parentheses, the graph moves in the opposite direction of the sign given in the function. For example, $g(x) = f(x - 3)$ moves the graph 3 units to the right.

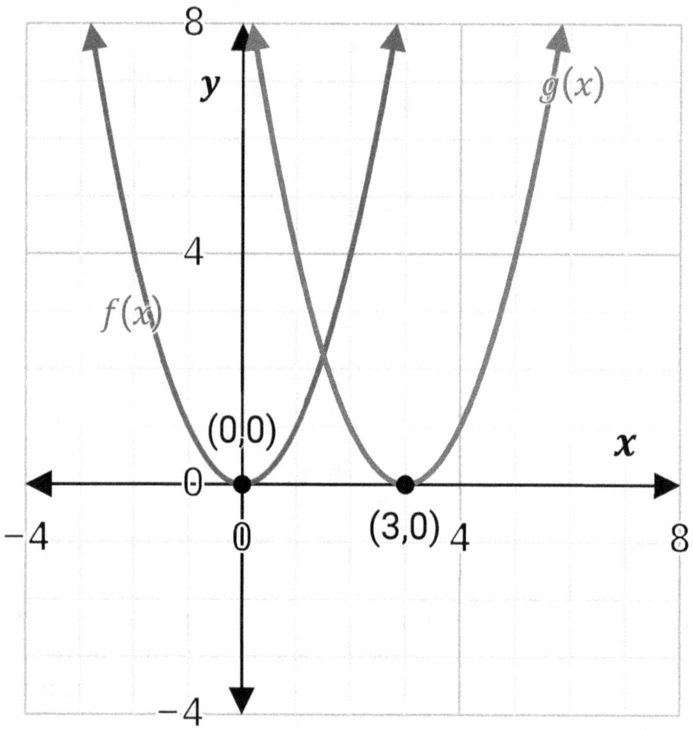

Dilations

Another way in which functions are manipulated is with dilations. **Dilation** is when the size of a function's graph is changed. When a dilation results in a larger image, it is called an **enlargement**, and the graph of the original function is stretched. When a dilation results in a smaller image, it is called a **reduction**, and the graph of the original function is compressed. The **scale factor** of a dilation determines how much larger or smaller the new graph will be in comparison to the parent function.

Vertical Dilations

A **vertical dilation** is when the graph of a function is either stretched or compressed in a vertical direction. To stretch or compress the graph of the original function vertically, multiply $f(x)$ by a to get $g(x) = af(x)$. Doing so results in all y-coordinates on the graph of the function getting multiplied by a.

If the scale factor, a, is greater than 1, then the graph of the parent function will stretch vertically, moving away from the x-axis. For example, $g(x) = 6f(x)$ stretches the graph of the parent function vertically by a factor of 6 because each y-coordinate gets multiplied by 6.

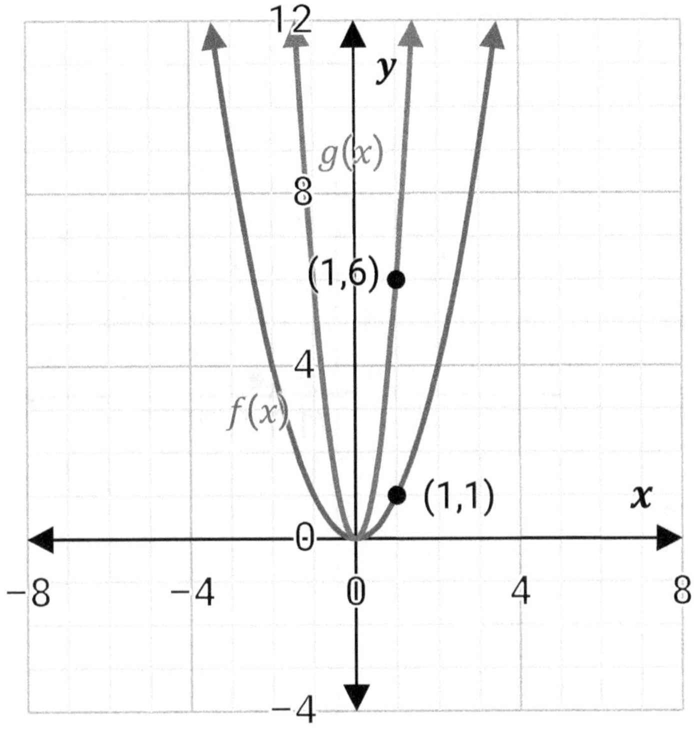

If a is a number between 0 and 1, then the graph of the parent function will compress vertically, moving toward the x-axis. For example, $g(x) = \frac{1}{6}f(x)$ compresses the graph of the parent function vertically by a factor of $\frac{1}{6}$ because each y-coordinate gets multiplied by $\frac{1}{6}$.

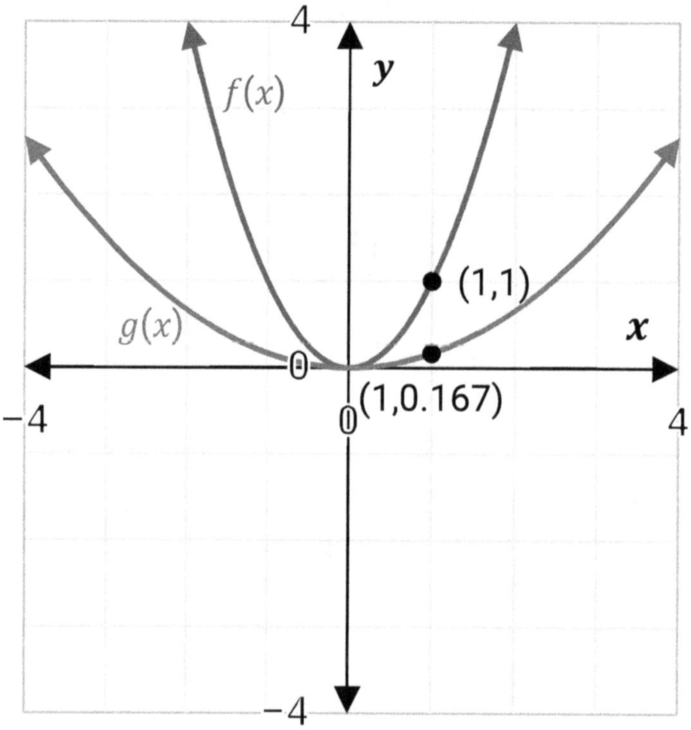

Horizontal Dilations

A **horizontal dilation** is when the graph of a function is either stretched or compressed in a horizontal direction. To stretch or compress the graph of the parent function $f(x)$ horizontally, multiply x by b inside the parentheses to get $g(x) = f(bx)$. Doing so results in all x-coordinates on the graph of the function getting divided by b. This is because horizontal dilations occur inside the parentheses, causing the graph of the function to do the inverse operation of the given operation. Since the inverse of multiplication is division, each x-coordinate gets divided by b.

If b is greater than 1, the graph of the parent function will compress horizontally, moving in toward the y-axis. For example, $g(x) = f(2x)$ compresses the graph of the parent function horizontally by a factor of $\frac{1}{2}$ because each x-coordinate is divided by 2.

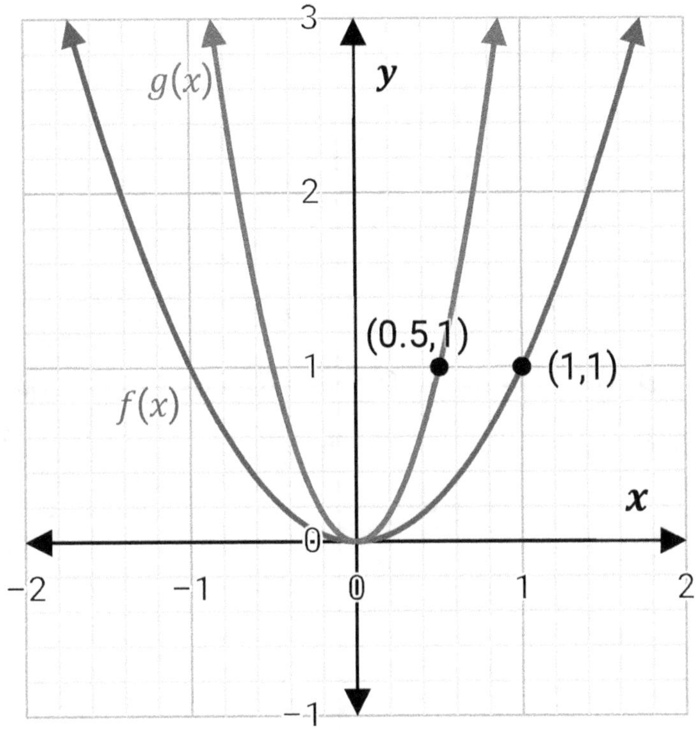

If b is a number between 0 and 1, the graph of the parent function will stretch horizontally, moving away from the y-axis. For example, $g(x) = f\left(\frac{1}{2}x\right)$ stretches the graph of the parent function horizontally by a factor of 2 because each x-coordinate is divided by $\frac{1}{2}$.

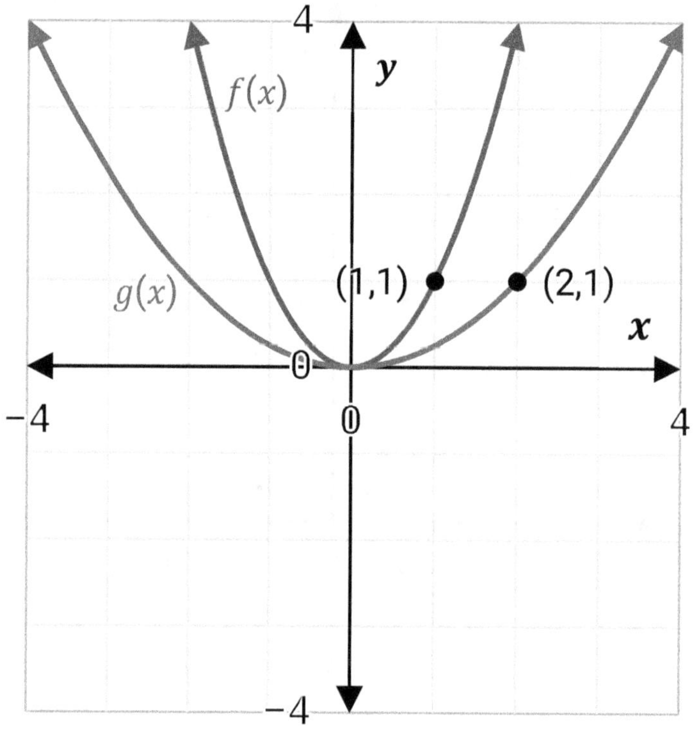

EXAMPLES

EXAMPLE 1:

What happens to the graph of $f(x) = x^2$ when $f(x)$ is replaced by $3f(x)$ to get $g(x) = 3x^2$?

Answer:

This function manipulation is a dilation, and the graph of $f(x) = x^2$ is stretched in a vertical direction by a factor of 3. Multiplying $f(x)$ by 3 results in all y-coordinates on the graph of the function getting multiplied by 3. On the coordinate plane shown, the point (1,1) on the parent function $f(x)$ becomes the point (1,3) on $g(x) = 3x^2$.

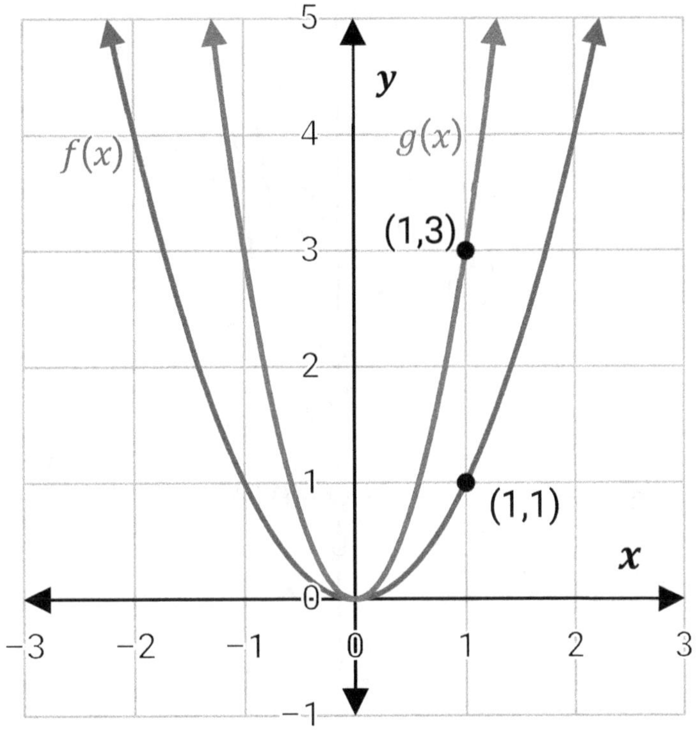

EXAMPLE 2:

What happens to the graph of $f(x) = x^2$ when $f(x)$ is replaced by $f(x) + 4$ to get $g(x) = x^2 + 4$?

Answer:

This function manipulation is a vertical translation, and the graph of $f(x) = x^2$ is shifted 4 units up. Adding 4 to x^2 results in 4 getting added to all y-coordinates on the graph of the function. On the coordinate plane shown, the point (0,0) on the parent function $f(x)$ becomes the point (0,4) on $g(x) = x^2 + 4$.

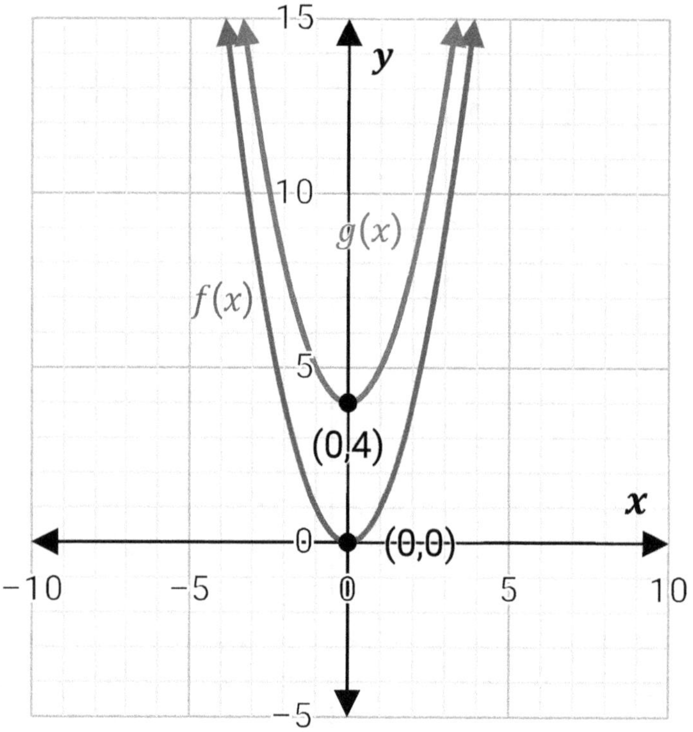

EXAMPLE 3:

What happens to the graph of $f(x) = x^2$ when $f(x)$ is replaced by $f(x-1)$ to get $g(x) = (x-1)^2$?

Answer:

This function manipulation is a horizontal translation, and the graph of $f(x) = x^2$ is shifted 1 unit to the right. Horizontal translations are represented within the parentheses of the equation. When a function manipulation occurs inside the parentheses, the graph moves in the opposite direction of the sign given in the function. Subtracting 1 inside parentheses results in 1 getting added to all x-coordinates on the graph of the function. On the coordinate plane shown, the point (0,0) on the parent function $f(x)$ becomes the point (1,0) on $g(x) = (x-1)^2$.

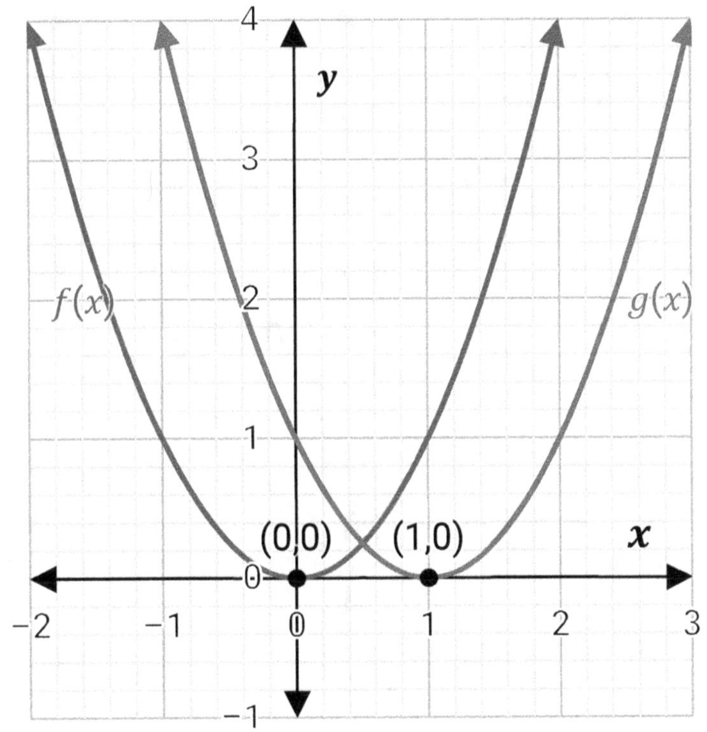

EXAMPLE 4:

What happens to the graph of $f(x) = x^2$ when $f(x)$ is replaced by $f(5x)$ to get $g(x) = (5x)^2$?

Answer:

The function manipulation is a dilation, and the graph of $f(x) = x^2$ is compressed in a horizontal direction by a factor of $\frac{1}{5}$. Multiplying x by 5 inside the parentheses results in all x-coordinates on the graph of the function getting multiplied by $\frac{1}{5}$. On the coordinate plane shown, the point (1,1) on the parent function $f(x)$ becomes the point $\left(\frac{1}{5}, 1\right)$ on $g(x) = (5x)^2$.

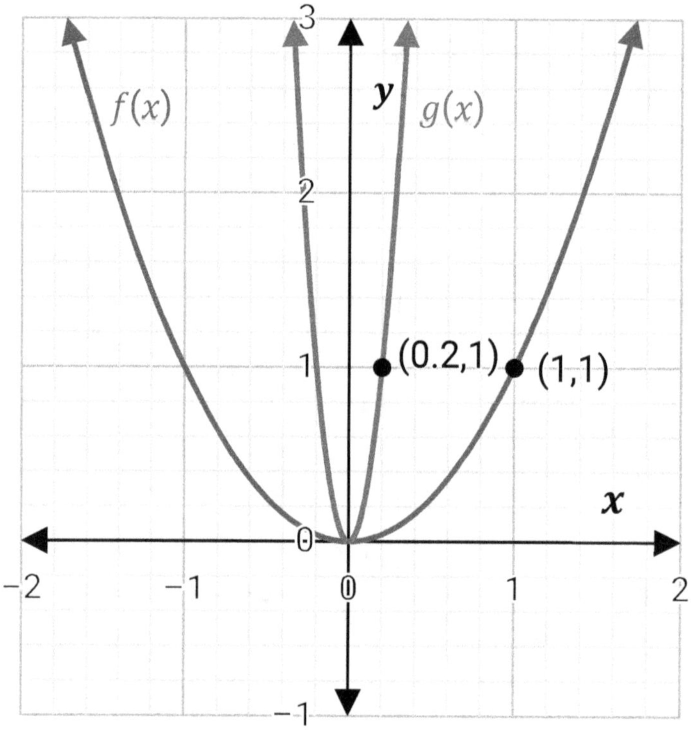

PRACTICE QUESTIONS

1. Which function represents the graph of $f(x) = x^2$ after it has been shifted 7 units down?

a. $g(x) = -7x^2$
b. $g(x) = x^2 - 7$
c. $g(x) = (x - 7)^2$
d. $g(x) = (-7x)^2$

2. Which function represents the graph of $f(x) = x^2$ after it is shifted 2 units to the right?

a. $g(x) = x^2 + 2$
b. $g(x) = x^2 - 2$
c. $g(x) = (x + 2)^2$
d. $g(x) = (x - 2)^2$

3. Which function represents the graph of $f(x) = x^2$ after it is stretched vertically by a factor of 6?

a. $g(x) = x^2 + 6$
b. $g(x) = 6x^2$
c. $g(x) = (x + 6)^2$
d. $g(x) = \frac{1}{6}x^2$

4. Which function represents the graph of $f(x) = x^2$ after it is compressed horizontally by a factor of $\frac{1}{2}$?

a. $g(x) = (2x)^2$
b. $g(x) = 2x^2$
c. $g(x) = x^2 + 2$
d. $g(x) = \left(\frac{x}{2}\right)^2$

5. Which function represents the graph of $f(x) = x^2$ after it is shifted 3 units down?

a. $g(x) = x^2 - 3$
b. $g(x) = -3x^2$
c. $g(x) = (x - 3)^2$
d. $g(x) = (x + 3)^2$

6. Which function represents the graph of $f(x) = x^2$ after it is shifted 5 units to the left?

a. $g(x) = -5x^2$
b. $g(x) = (x - 5)^2$
c. $g(x) = (x + 5)^2$
d. $g(x) = x^2 - 5$

7. Which function represents the graph of $f(x) = x^2$ after it is compressed vertically by a factor of $\frac{1}{2}$?

a. $g(x) = x^2 + \frac{1}{2}$
b. $g(x) = \frac{1}{2}x^2$
c. $g(x) = \left(\frac{x}{2}\right)^2$
d. $g(x) = \left(x + \frac{1}{2}\right)^2$

8. The function $f(x) = x^2$ is translated to create the function $g(x) = x^2 + 5$. Which statement describes the translation?

a. The graph of $f(x)$ shifted up 5 units to create $g(x)$.
b. The graph of $f(x)$ shifted down 5 units to create $g(x)$.
c. The graph of $f(x)$ shifted left 5 units to create $g(x)$.
d. The graph of $f(x)$ shifted right 5 units to create $g(x)$.

9. The function $f(x) = x^2$ is translated to create the function $g(x) = x^2 - 11$. Which statement describes the translation?

a. The graph of $f(x)$ shifted up 11 units to create $g(x)$.
b. The graph of $f(x)$ shifted down 11 units to create $g(x)$.
c. The graph of $f(x)$ shifted left 11 units to create $g(x)$.
d. The graph of $f(x)$ shifted right 11 units to create $g(x)$.

10. The function $f(x) = x^2$ is translated to create the function $g(x) = (x - 6)^2$. Which statement describes the translation?

a. The graph of $f(x)$ shifted up 6 units to create $g(x)$.
b. The graph of $f(x)$ shifted down 6 units to create $g(x)$.
c. The graph of $f(x)$ shifted left 6 units to create $g(x)$.
d. The graph of $f(x)$ shifted right 6 units to create $g(x)$.

11. The function $f(x) = x^2$ is translated to create the function $g(x) = (x + 9)^2$. Which statement describes the translation?

a. The graph of $f(x)$ shifted up 9 units to create $g(x)$.
b. The graph of $f(x)$ shifted down 9 units to create $g(x)$.
c. The graph of $f(x)$ shifted left 9 units to create $g(x)$.
d. The graph of $f(x)$ shifted right 9 units to create $g(x)$.

12. The graph of $f(x) = x^2$ is shown on Graph A. The graph of $f(x)$ is translated to create the graph of $g(x)$. What is the equation of Graph B?

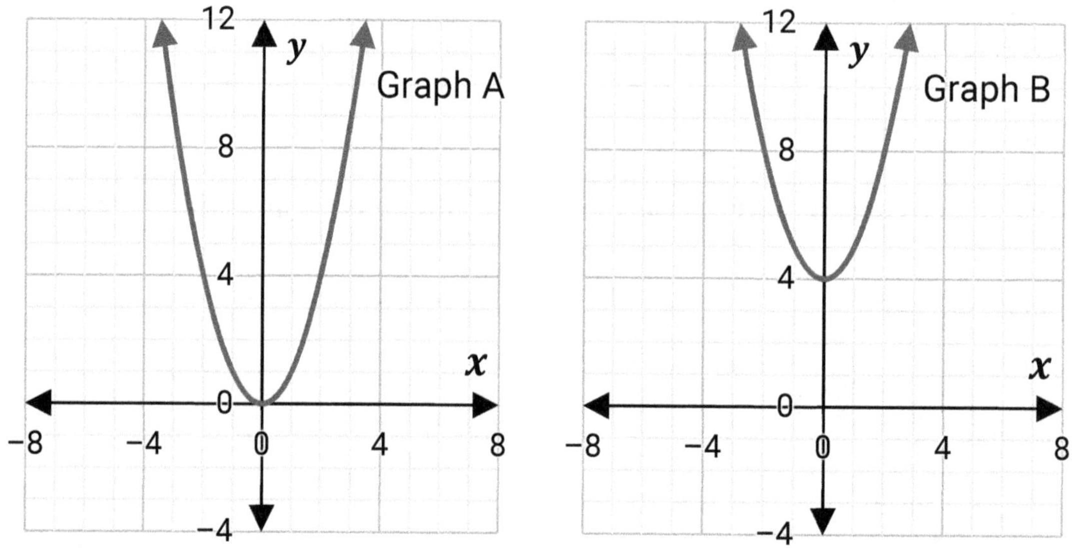

a. $g(x) = x^2 + 4$
b. $g(x) = x^2 - 4$
c. $g(x) = (x + 4)^2$
d. $g(x) = (x - 4)^2$

13. Jessica is a graphic designer. She needs to stretch an image vertically by a factor of 3. If the original image is represented by the function $f(x) = x^2$, which function represents this manipulation?

a. $g(x) = (x - 3)^2$
b. $g(x) = x^2 - 3$
c. $g(x) = -3x^2$
d. $g(x) = 3x^2$

14. An elementary school is ordering a new banner for their gymnasium. After seeing the design proof, they want to shift the logo's positioning. The original logo, represented by the function $f(x) = x^2$, needs to be moved 7 inches down. Which function represents this translation?

a. $g(x) = -7x^2$
b. $g(x) = (x - 7)^2$
c. $g(x) = x - 7$
d. $g(x) = x^2 - 7$

15. Jacob is a structural engineer. He is working on designing a new roller coaster for an amusement park. After creating an original design, he realizes he needs to change one of the roller coaster's dips. The dip, represented by the function $f(x) = x^2$, needs to be horizontally translated one foot to the right. Which function represents this translation?

a. $g(x) = (x - 1)^2$
b. $g(x) = (x + 1)^2$
c. $g(x) = x^2 + 1$
d. $g(x) = x^2 - 1$

Chapter 8 Quiz

1. The graph of the parabola on the coordinate shown below gives the profit, P in dollars, for a company that sells x number of computer hard drives. Which of the quadratic functions represents the profit in terms of the number of hard drives sold?

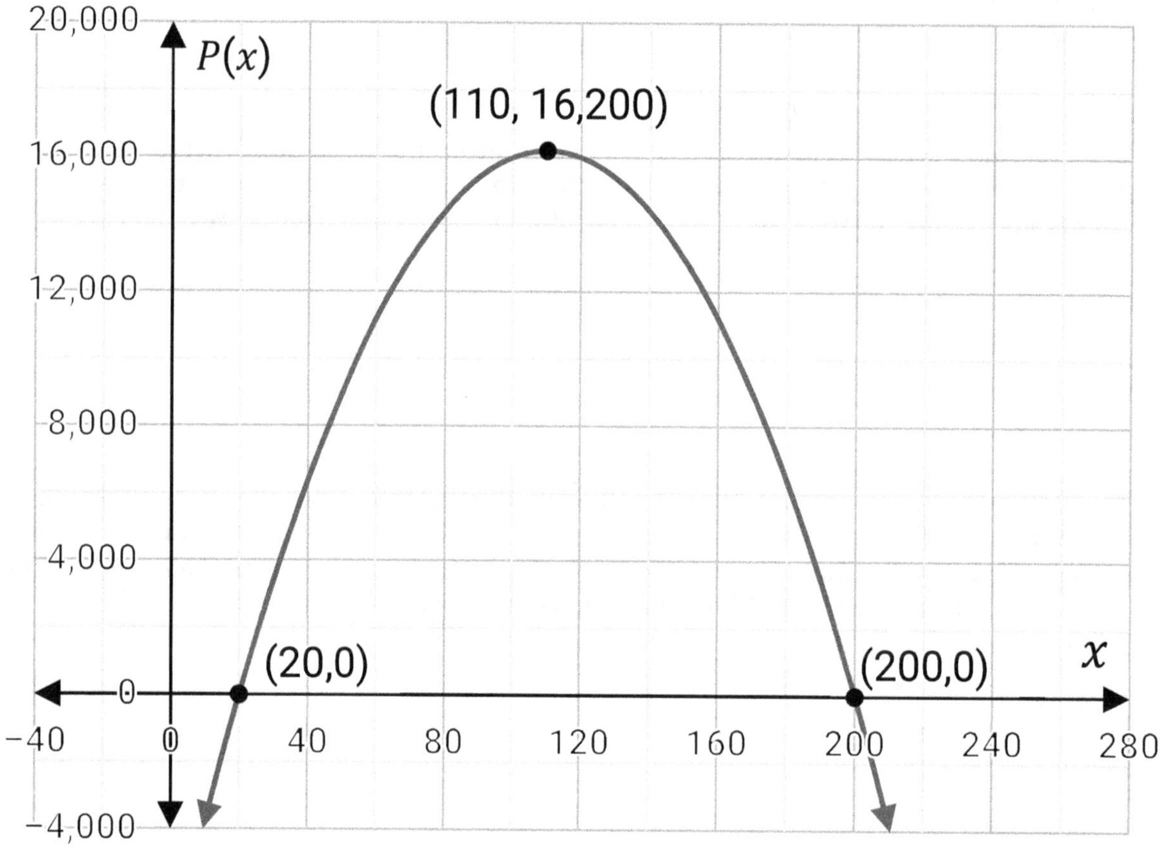

a. $P(x) = -x^2 + 220x - 4{,}000$
b. $P(x) = -x^2 + 220x + 16{,}200$
c. $P(x) = -2x^2 + 440x - 8{,}000$
d. $P(x) = 2x^2 - 440x + 8{,}000$

2. Solve the quadratic equation using the quadratic formula: $x^2 - 3x - 2 = 0$.

a. $x = \frac{3-\sqrt{17}}{2}$ or $x = \frac{3+\sqrt{17}}{2}$
b. $x = \frac{-3-\sqrt{17}}{2}$ or $x = \frac{-3+\sqrt{17}}{2}$
c. $x = \frac{-2-\sqrt{17}}{2}$ or $x = \frac{-2+\sqrt{17}}{2}$
d. $x = \frac{3-\sqrt{17}}{4}$ or $x = \frac{3+\sqrt{17}}{4}$

3. Write the equation of the parabola that has its vertex at $(3, -2)$ and contains the point $(4, 14)$.

a. $y = 16(x-2)^2 + 3$
b. $y = \frac{12}{49}(x+2)^2 - 3$
c. $y = \frac{17}{36}(x-3)^2 - 2$
d. $y = 16(x-3)^2 - 2$

4. Convert $y = x^2 - 6x + 5$ into vertex form.

a. $y = (x-3)(x-2)$
b. $y = (x-5)(x-1)$
c. $y = (x-3)^2 - 4$
d. $y = (x-3)^2 + 14$

5. You need to fence in your rectangular cabbage garden to protect the cabbage from rabbits. The length of the garden is 10 feet longer than the width and the area of the garden is 500 square feet. To the nearest tenth of a foot, what is the width of the garden?

a. 27.9 feet
b. 17.9 feet
c. 22.4 feet
d. 12.4 feet

6. Match the equation for the function $y = x^2 - 2x - 8$ with its appropriate graph.

a. 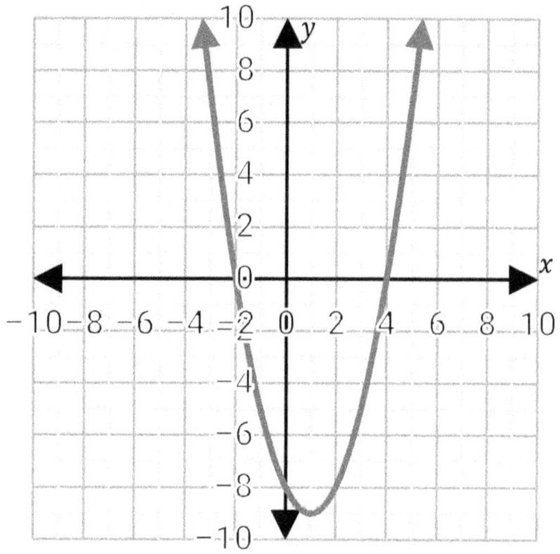

c. 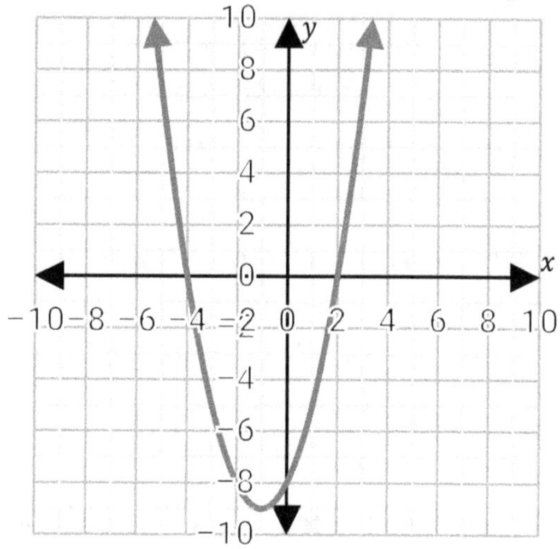

b. 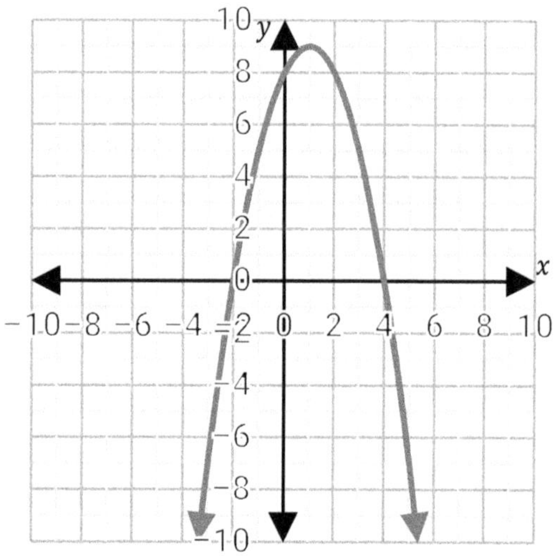

d. 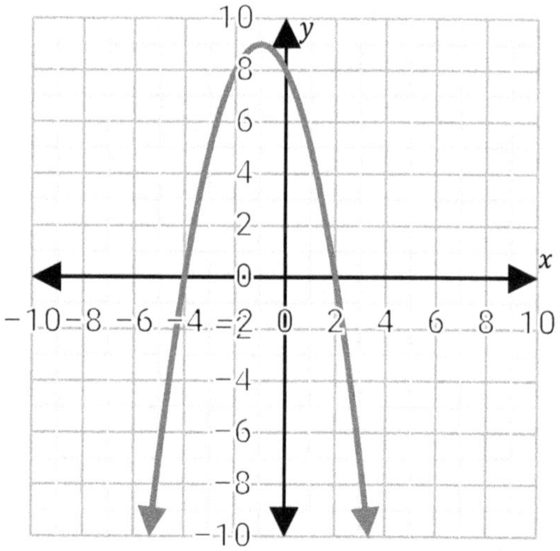

7. A creek runs underneath the opening of a stone bridge. The bridge's opening is in the shape of a parabola. The height of the opening is 20 feet and a width of 50 feet. An outline of the opening is shown below.

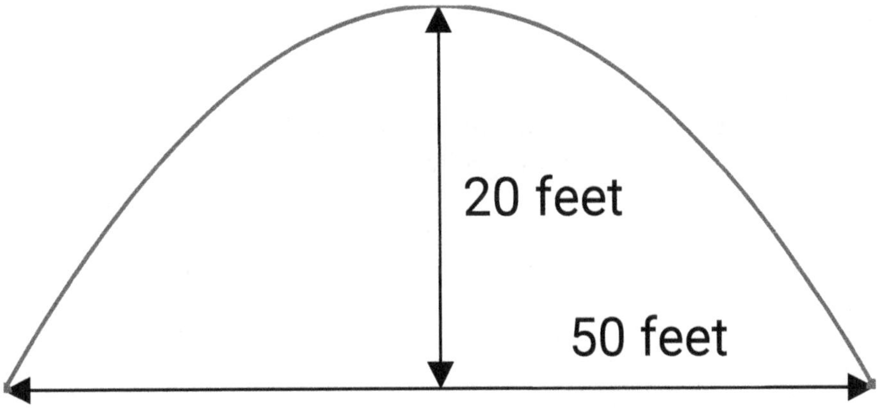

Which of the following equations could be used to represent the bridge's opening for the height of the bridge, *h* in feet, in terms of its width, *w* also in feet?

a. $h = -\frac{4}{125}(w - 20)^2$

b. $h = -\frac{4}{125}w^2 + 20$

c. $h = \frac{1}{25}w^2 + 25$

d. $h = \frac{1}{25}(w - 25)^2 + 20$

8. Twice the square of a number equals four times the number plus 30. Which of the following could be the values of the number?

a. -3 or 5

b. -5 or 3

c. -3 or 10

d. -2 or 15

9. A projectile is thrown downwards from the top of a building. In vertex form, the height, h in feet, of the projectile can be represented as $h = -16(t + 3)^2 + 208$, where t is time in seconds. In standard form, the height of an object can be represented by $h = -16t^2 + v_0t + h_0$, where v_0 is the initial velocity of the object in miles per hour and h_0 is the initial height of the object in feet. What is the initial height and velocity of the projectile?

a. $v_0 = 96$ mph and $h_0 = -64$ ft

b. $v_0 = -96$ mph and $h_0 = 64$ ft

c. $v_0 = -48$ mph and $h_0 = 13$ ft

d. $v_0 = -48$ mph and $h_0 = 208$ ft

10. Solve the quadratic equation by using square roots: $2x^2 = 72$.

a. $x = -9$ or $x = -4$

b. $x = -6$ or $x = 6$

c. $x = -6$ or $x = 12$

d. $x = -9$ or $x = 8$

11. The graph of $f(x) = x^2$ is shown on Graph A. The graph of $f(x)$ is translated to create the graph of $g(x)$. What is the equation of Graph B?

Graph A

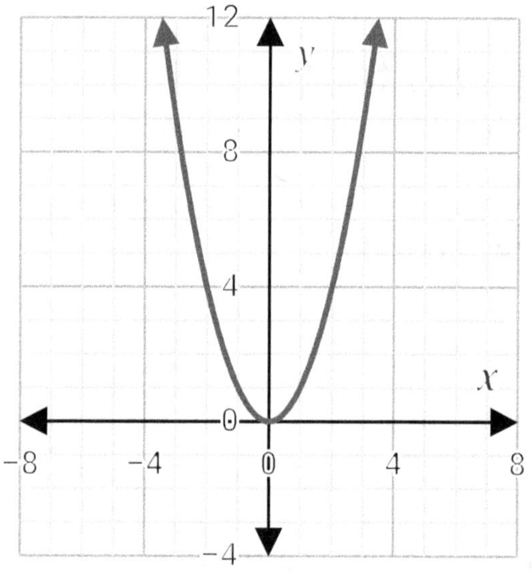

Graph B

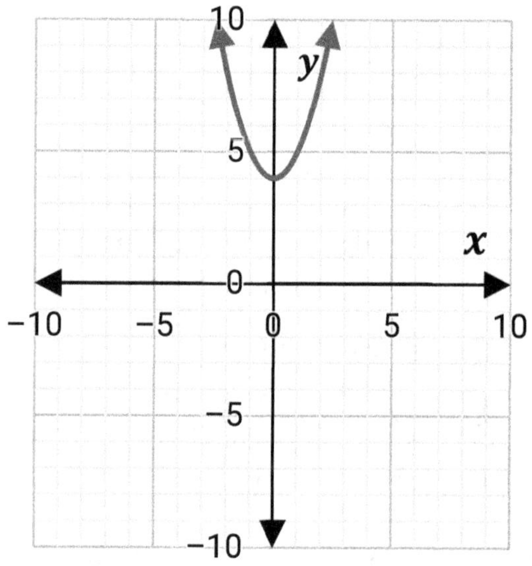

a. $g(x) = x^2 + 4$

b. $g(x) = x^2 - 4$

c. $g(x) = (x + 4)^2$

d. $g(x) = (x - 4)^2$

12. Write a function f in standard form, given that f is a quadratic function that has zeros of -2 and 1 and $f(2) = 12$.

a. $f(x) = 6x^2 + 6x - 12$

b. $f(x) = 3x^2 + 3x - 6$

c. $f(x) = \frac{1}{6}x^2 + \frac{1}{2}x - \frac{1}{6}$

d. $f(x) = \frac{1}{6}x^2 + 3x - 6$

13. The product, P, of twelve minus some integer and the integer itself can be expressed by the function $P = (12 - n)n$, where n is the integer. What is the range of values for the product?

a. $-\infty < P < \infty$
b. $P \leq 6$
c. $P \leq 36$
d. $P \geq 36$

14. From the ground, a rocket is launched in the air with an initial velocity of 64 feet per second. The height, h, of the object at time, t, in seconds can be found using the quadratic equation $h = -16t^2 + v_0 t$, where v_0 is the initial velocity of the object in feet per second. How many seconds after it is launched does the rocket return to the ground?

a. 8 sec
b. 4 sec
c. 2 sec
d. 16 sec

15. Solve this quadratic equation by factoring: $x^2 + 5x = -6$.

a. $x = -3$ and $x = -2$
b. $x = 2$ and $x = 3$
c. $x = -1$ and $x = 6$
d. $x = -6$ and $x = 1$

16. Find the discriminant for $3x^2 - x = -2$.

a. 23
b. -23
c. 25
d. -25

17. Is it possible for the product of 2 less than a number and 6 times the same number to be 312?

a. Yes, because the discriminant is positive.
b. Yes, because the discriminant is zero.
c. No, because the discriminant is negative.
d. Solving for the discriminant will not give enough information to answer this question.

18. Solve the following quadratic equation by completing the square: $x^2 + 10x + 24 = 0$.

a. $x = -6$ or $x = -4$
b. $x = 4$ or $x = 6$
c. $x = 3$ or $x = 8$
d. $x = -8$ or $x = -3$

19. State the domain and range for the graph of the parabola shown below.

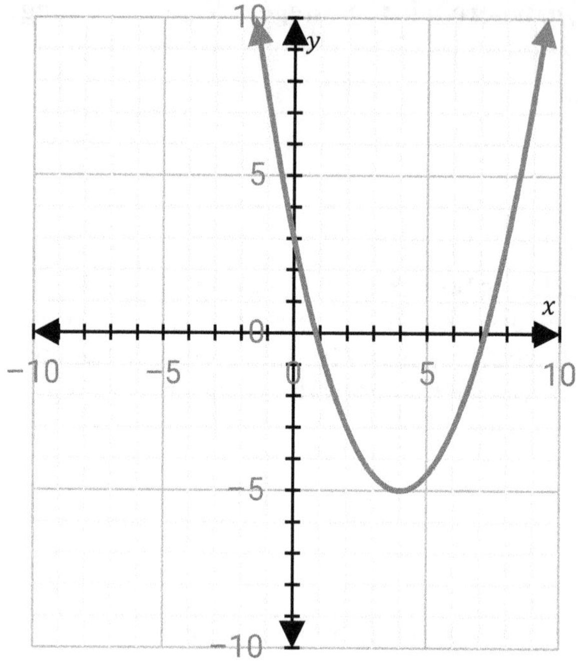

a. Domain: $x \geq -5$; Range: $-\infty < y < \infty$
b. Domain: $-\infty < x < \infty$; Range: $y \geq -5$
c. Domain: $x \geq -5$; Range: $y \geq 4$
d. Domain: $-\infty < x < \infty$; Range: $y \geq 4$

20. Which of the following is the graph of the quadratic function $y = x^2 + x - 12$?

a.

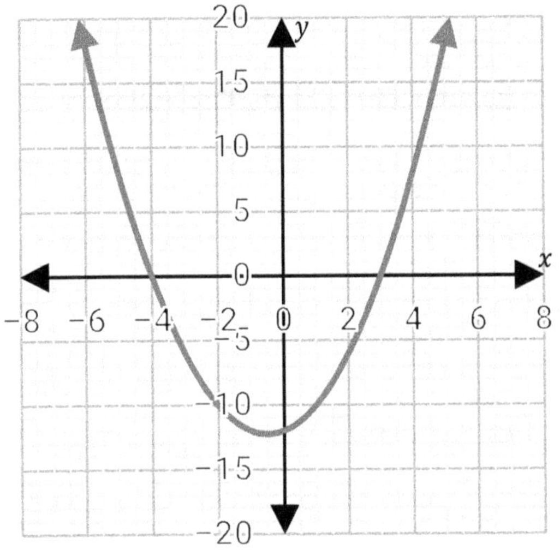

b.

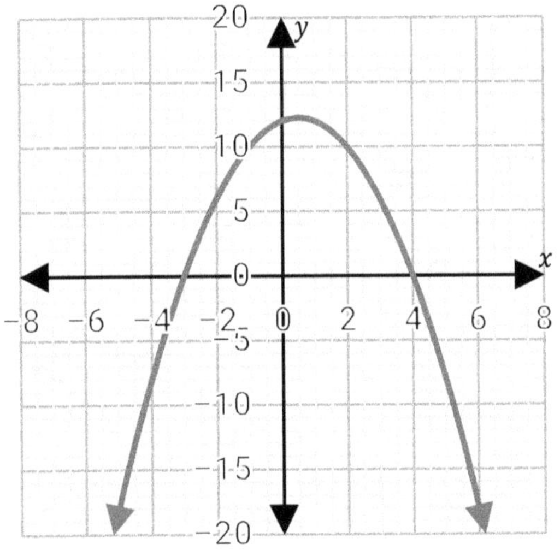

c.

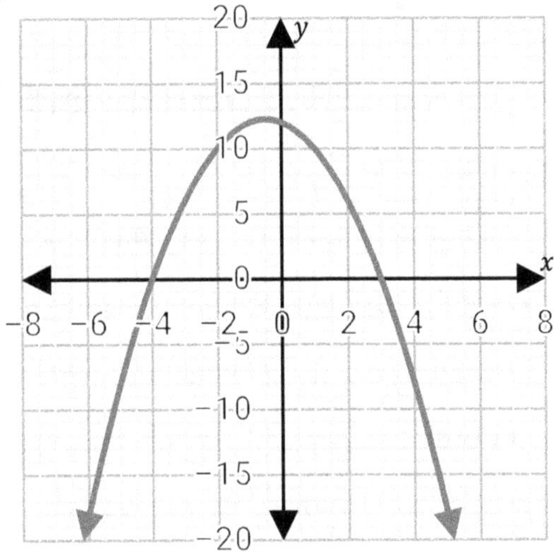

d.

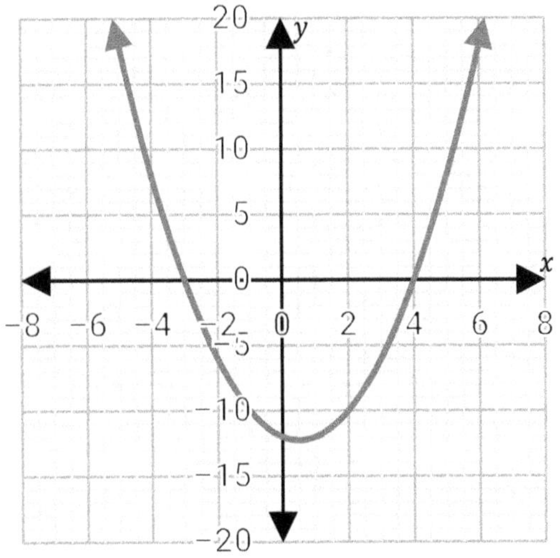

Chapter 9: Exponential Functions

9.1 Domain and Range of Exponential Functions

Exponential functions are often used to represent rapid growth or decay situations. Radioactive decay and population growth are examples of concepts that can be modeled with exponential functions. Exponential functions generally take the form $y = ab^x$.

When $b > 1$, the function represents exponential growth.

When $0 < b < 1$, the function represents exponential decay.

For example, $f(x) = 4^x$ represents exponential growth because $b > 1$.

$f(x) = 0.75^x$ represents exponential decay because $0 < b < 1$.

The domain and range of exponential functions can be expressed as inequalities. Remember, the **domain** represents the possible input (x) values for the function. The **range** represents the possible output (y) values for the function.

One way to determine the domain and range of exponential functions is to consider their graphs.

All exponential functions have a domain of all real numbers. That is, any x-value will satisfy the function. This is often expressed as $-\infty < x < \infty$. x-values will extend infinitely in the positive and negative directions.

However, the y-values in exponential functions will have some limitations. For example, the exponential function below only has y-values that are greater than 2. The dotted line is called an **asymptote**. The curve of the function will approach the asymptote but never touch the asymptote. The y-values extend above the dotted line, but y will never be equal to 2. This means that the range of this graph is $y > 2$ and the domain is $-\infty < x < \infty$.

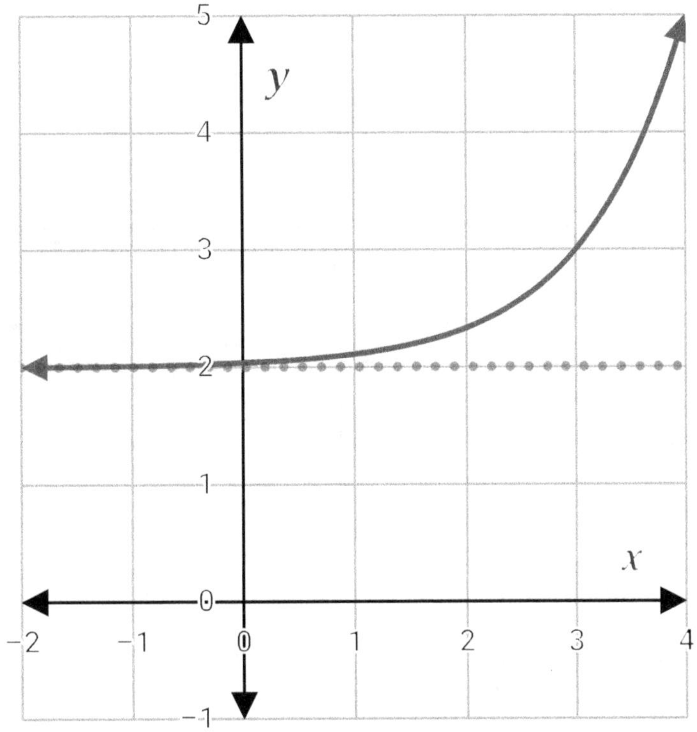

Let's look at another graphed exponential function.

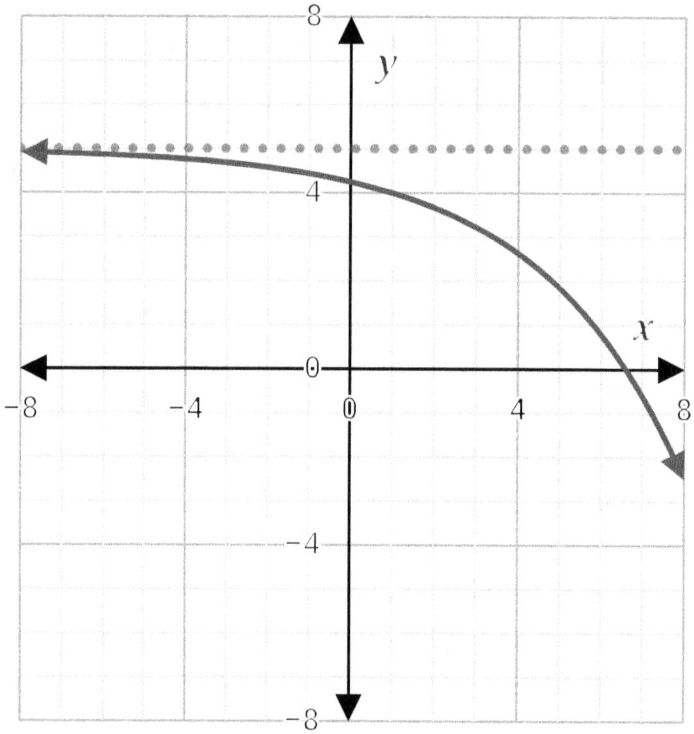

The domain is $-\infty < x < \infty$, and the range is any value less than 5. The y-values can get close to 5 but are never exactly 5. The range is $y < 5$.

Examples

Example 1:

Determine the domain and range of the exponential function graphed below, $f(x) = \frac{1}{2}(2)^x$.

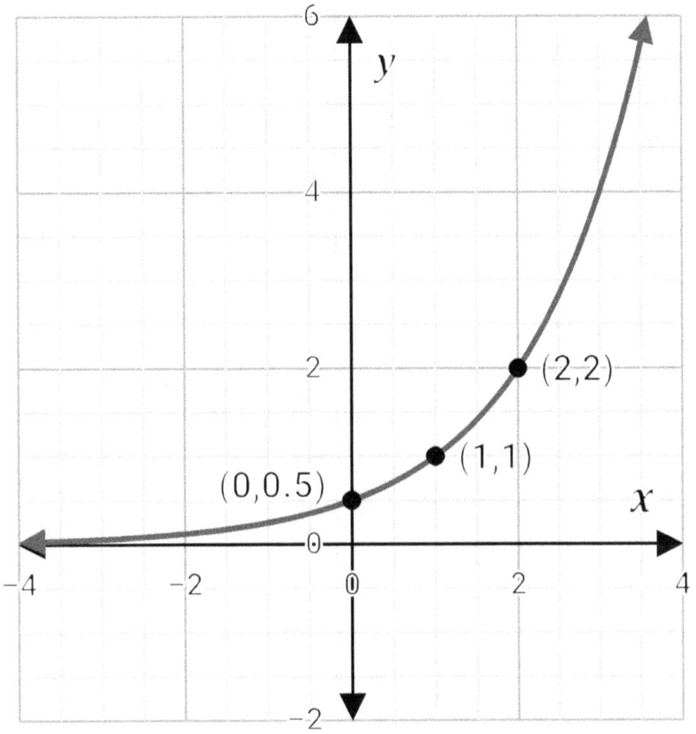

Answer:

The domain of a function is its input values, which is the independent variable x. On the coordinate plane, it is all possible x-values for which the function is defined. For any exponential function, the domain is the set of all real numbers, from negative to positive infinity.

$$-\infty < x < \infty$$

The range of a function is its output values, which is the dependent variable $f(x)$. On the coordinate plane, it is all y-values produced by the function. The range is limited by the horizontal asymptote and does not include the asymptote. The horizontal asymptote in the graph shown is 0, and the function exists above the asymptote. Therefore, the range of this exponential function is the set of all numbers greater than 0.

$$f(x) > 0$$

EXAMPLE 2:

Determine the domain and range of the exponential function graphed below, $f(x) = -2(3)^x + 4$.

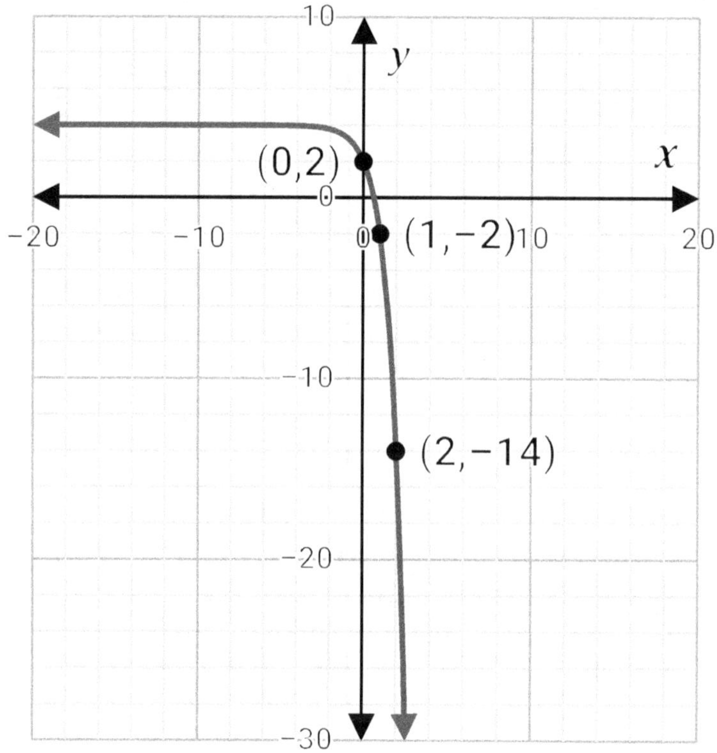

Answer:

The domain of a function is its input values, which is the independent variable x. On the coordinate plane, it is all possible x-values for which the function is defined. For any exponential function, the domain is the set of all real numbers, from negative to positive infinity.

$$-\infty < x < \infty$$

The range of a function is its output values, which is the dependent variable $f(x)$. On the coordinate plane, it is all y-values produced by the function. The range is limited by the horizontal asymptote and does not include the asymptote. The horizontal asymptote in the graph shown is 4, and the function exists below the asymptote. Therefore, the range of this exponential function is the set of all numbers less than 4.

$$f(x) < 4$$

If you are only given an equation, you can determine the horizontal asymptote by looking at the number added to, or subtracted from, the exponential term of the function. In this case, 4 is being added, so the horizontal asymptote will be at $y = 4$. To determine if the range values are above or below the asymptote, look as the sign of a. If a is positive, then the range is greater than the horizontal asymptote. If a is negative, then the range is less than the horizontal asymptote. In this case, $a = -2$, so the range values are all values less than 4.

EXAMPLE 3:

You go to the bank today and invest \$5,000 in an account that pays interest compounded continuously. You can estimate the amount of money in the account t years after today using the function $A(t) = 5{,}000 \cdot 1.04^t$ dollars. What is the range of this function?

Answer:

Since the investment starts today and t measures years (including fractional years) from today, the domain of this function is $t \geq 0$. When $b > 1$ (here $b = 1.04$), an exponential expression like b^t equals 1 when $t = 0$ and increases through all values larger than 1 as t increases. Thus, right now the money in the account is $A(0) = 5{,}000 \cdot 1.04^0 = 5{,}000 \cdot 1 = 5{,}000$ dollars, and as t grows, $A(t) = 5{,}000 \cdot 1.04^t$ will multiply 5,000 by all the numbers larger than 1, yielding every possible product larger than 5,000. So, the range of this function is $A(t) \geq 5{,}000$ dollars.

PRACTICE QUESTIONS

1. Find the domain of the function $f(x) = 5 \cdot 1.3^x$.

a. $x > 5$

b. $x > 6.5$

c. $-\infty < x < \infty$

d. $x > 0$

2. Find the range of the function $f(x) = 7 \cdot 2^x$.

a. $f(x) > 0$

b. $f(x) \geq 7$

c. $-\infty < f(x) < \infty$

d. $f(x) > 2$

3. Find the domain of the function $s(t) = -2.7(0.4)^t$.

a. $t < 0$

b. $t \leq -2.7$

c. $t > 0$

d. $-\infty < t < \infty$

4. Find the range of the function $f(x) = -9.3 \cdot 5^x$.

a. $-\infty < f(x) < \infty$

b. $f(x) \leq -46.5$

c. $f(x) \leq -9.3$

d. $f(x) < 0$

5. Find the domain and range of the function $f(x) = 11 \cdot 10^x$.

a. The domain is $-\infty < x < \infty$, and the range is $f(x) > 10$.

b. The domain is $-\infty < x < \infty$, and the range is $f(x) > 0$.

c. The domain is $-\infty < x < \infty$, and the range is $f(x) \geq 11$.

d. The domain is $x \geq 0$, and the range is $f(x) \geq 10$.

6. Find the domain and range of the function $f(x) = -21 \cdot 7^x$.

a. The domain is $-\infty < x < \infty$, and the range is $f(x) > 7$.
b. The domain is $-\infty < x < \infty$, and the range is $f(x) \leq -21$.
c. The domain is $-\infty < x < \infty$, and the range is $f(x) < 0$.
d. The domain is $x \geq 0$, and the range is $f(x) \geq 7$.

7. Find the range of the function $f(x) = 2\left(\frac{1}{3}\right)^x + 5$.

a. $0 < f(x) \leq 7$
b. $-\infty < f(x) < \infty$
c. $f(x) > 0$
d. $f(x) > 5$

8. Find the domain of the function $f(x) = 11.3(0.4)^x - 12$.

a. $x \geq 11.3$
b. $-\infty < x < \infty$
c. $x \geq -12$
d. $x \geq 0$

9. Find the range of the function $f(x) = -8 \cdot 3^x + 14$.

a. $f(x) > 14$
b. $f(x) < 14$
c. $f(x) < 0$
d. $f(x) > -8$

10. Find the domain and range of the function $f(x) = -4\left(\frac{3}{8}\right)^x - 20$.

a. The domain is $-\infty < x < \infty$, and the range is $f(x) > -20$.
b. The domain is $x \leq 0$, and the range is $f(x) \leq -24$.
c. The domain is $-\infty < x < \infty$, and the range is $f(x) \leq -26$.
d. The domain is $-\infty < x < \infty$, and the range is $f(x) < -20$.

11. Find the range of the function $f(x) = 1.1 \cdot 4^x - 9$.

a. $f(x) \geq -4.6$
b. $f(x) > 0$
c. $f(x) < -9$
d. $f(x) > -9$

12. Find the range of the function $f(x) = -2.7 \cdot 3^x + 4$.

a. $f(x) \leq -2.7$
b. $f(x) > 3$
c. $f(x) < 4$
d. $f(x) \geq 4$

13. **Today you invest \$1,000 in an account that pays interest compounded continuously. You estimate the amount of money in the account t years after today using the function $A(t) = 1,000 \cdot 1.03^t$ dollars. What is the domain of this function?**

a. $t \geq 1.03$

b. $t \geq 1{,}000$

c. $t \geq 0$

d. $-\infty < t < \infty$

14. **Today you invest \$2,500 in an account that pays interest compounded continuously. You estimate the amount of money in the account t years after today using the function $A(t) = 2,500 \cdot 1.02^t$ dollars. What is the range of this function?**

a. $A(t) \geq 2{,}500$

b. $A(t) \geq 0$

c. $-\infty < A(t) < \infty$

d. $A(t) \geq 1.02$

15. **At this moment you have a pure sample of 100 grams of a radioactive isotope. It undergoes radioactive decay at such a rate that in t days the amount of the isotope left is given by the function $A(t) = 100 \cdot 0.834^t$ grams. What is the range of this function?**

a. $0 < A(t) \leq 100$

b. $A(t) > 0$

c. $A(t) \leq 100$

d. $A(t) \geq 100$

9.2 Writing Exponential Functions

Exponential functions are used to describe situations involving rapid growth or decay. In general, exponential functions are expressed as $y = ab^x$, where a represents the initial amount, b represents the growth rate, and x represents the time period. When the growth rate is less than 0, it needs to be added to 1 for a growth function and subtracted from 1 for a decay function. Doing so makes sure to include the initial amount in the final value.

Let's look at an example of an exponential growth function.

If you own an antique book that increases its value 20% each year, how much will the book be worth after 15 years?

We can set up an exponential function to solve this.

The function $y = ab^x$ becomes $y = a(1 + r)^x$ when the situation involves a rate of increase. In this case, the initial amount is \$500, the growth rate is $20\% = 0.2$, and the time period is 15 years. Since the growth rate is less than 1, it needs to be added to 1.

$$y = 500(1 + 0.2)^{15}$$
$$y \approx 7{,}703.51$$

This means that after 20 years the book will be worth about \$7,703.51.

Now, let's look at an exponential decay function.

If a used car starts off with a value of \$1,024 and depreciates in value each year by 12%, what will the car be worth in 8 years?

We can solve this with an exponential function. In this case, the initial value is \$1,024, the rate of decay is $12\% = 0.12$, and the time period is 8 years. $y = ab^x$ becomes $y = a(1 - r)^x$. The decay rate, b, becomes $(r - 1)$ when we are talking about a rate of decrease less than 1.

$$y = 1{,}024(1 - 0.12)^8$$
$$y \approx 368.27$$

This means that after 8 years the car is worth only \$368.27.

If we were to create a table of values for this exponential function, it would look like this:

Year	1	2	3	4	5	6	7	8
Car Value	\$901.12	\$792.99	\$697.83	\$614.09	\$540.40	\$475.55	\$418.48	\$368.27

The big idea is that exponential functions are not increasing or decreasing at a consistent rate like linear functions. The rate of exponential functions has a "snowball effect," where the rate continues to build. This is what creates the sweeping look of the graphs.

EXAMPLES

EXAMPLE 1:

The Philadelphia 76ers sold 8,000 tickets to their most recent home game. For each consecutive win, the number of tickets sold increases by a factor of 1.2. This scenario is represented by the exponential function $f(x) = 8{,}000(1.2)^x$. Interpret the meaning of 8,000 and 1.2 in this scenario.

Answer:

Exponential functions are written in the form $f(x) = ab^x$, where a is the initial value, b is the common ratio, and x is the independent variable. In the exponential function given, $a = 8{,}000$ and $b = 1.2$. The number 8,000 is the initial value, a, which is the number of fans at the team's most recent home game. 1.2 is the common ratio, b, for the rate of exponential growth.

EXAMPLE 2:

Leah started a book club with her friends. In its first year, the book club had a total of 4 members. Each year after, the number of club members increased by a factor of 1.5. Based on this information, write an exponential function that represents this scenario.

Answer:

Exponential functions are written in the form $f(x) = ab^x$, where a is the initial value, b is the common ratio, and x is the independent variable. In the scenario given, the initial number of book club members was 4. The common ratio, which is the factor by which book club membership increased, is 1.5. Substitute these values into the exponential function for a and b to write the equation. The exponential function that represents this scenario is $f(x) = 4(1.5)^x$.

EXAMPLE 3:

Livonia, Michigan has a population of 95,000. The population is declining at a rate of 2% each year. Write an exponential function that represents the population of the city after x years.

Answer:

Exponential functions are written in the form $f(x) = ab^x$, where a is the initial value, b is the common ratio, and x is the independent variable. Since the initial population, a, of the city is 95,000 people, substitute 95,000 into the exponential function for a.

$$f(x) = ab^x$$
$$f(x) = 95{,}000b^x$$

Next, determine the value of b. Since the population is decreasing, this is an exponential decay problem. For exponential decay problems, $b = 1 - r$, where r stands for the decay rate expressed as a decimal. Since the population decreases at a rate of 2% each year, $r = 0.02$. Substitute this value into the exponential function.

$$f(x) = 95{,}000(1 - 0.02)^x$$

Simplify the function by subtracting.

$$f(x) = 95{,}000(0.98)^x$$

The exponential function that represents this scenario is $f(x) = 95{,}000(0.98)^x$.

EXAMPLE 4:

The table below shows the spread of a virus in a population of mosquitoes. Write an exponential function that represents the number of infected mosquitoes x days after the virus started spreading.

Mosquitoes	150	600	2,400	9,600	38,400
Day	0	1	2	3	4

Answer:

Exponential functions are written in the form $f(x) = ab^x$, where a is the initial value and b is the common ratio. Start by finding the initial value of the function. This is the number of mosquitoes infected on Day 0, which is 150.

$$a = 150$$

Now, determine the common ratio. Divide each number in the mosquitoes sequence by the previous number to determine the common ratio.

$$38{,}400 \div 9{,}600 = 4$$
$$9{,}600 \div 2{,}400 = 4$$
$$2{,}400 \div 600 = 4$$
$$600 \div 150 = 4$$

Therefore, the common ratio for this scenario is 4. Now that the values of a and b are known, substitute these into the equation.

$$f(x) = 150 \cdot 4^x$$

EXAMPLE 5:

Elizabeth and Henry buy a house for $275,000. They are told that the value of the house increases by 4% each year. How much will the house be worth in 10 years? Round your answer to the nearest hundredth.

Answer:

Exponential functions are written in the form $f(x) = ab^x$, where a is the initial value and b is the common ratio. Start by finding the initial value of the function. The initial value of this function is how much they paid initially.

$$a = 275{,}000$$

Now, determine the common ratio, b. Since this is an exponential growth problem, where the growth is represented by a percentage, $b = 1 + r$, where r stands for the growth rate represented by a decimal. Since the value of the house increases by 4% each year, $r = 0.04$.

$$b = 1 + 0.04 = 1.04$$

Now, substitute the values for a and b into the exponential function equation.

$$f(x) = 275{,}000 \cdot 1.04^x$$

Since the question asks for the value of the house after 10 years, substitute 10 for x and simplify.

$$f(10) = 275{,}000 \cdot 1.04^{10}$$
$$f(10) = 407{,}067.18$$

Therefore, the house will be worth $407,067.18 in 10 years.

PRACTICE QUESTIONS

1. Jamie opens a bank account with $20. She doubles the amount of money in the bank account each month. This scenario is represented by the exponential function $f(x) = 20 \cdot 2^x$. Interpret the meaning of 20 in this function.

a. 20 is the common ratio, which is the constant factor between consecutive terms in the sequence.
b. 20 is the initial value, which is the original amount of money in Jamie's bank account.
c. 20 is the common ratio, which is the original amount of money in Jamie's bank account.
d. 20 is the initial value, which is the constant factor between consecutive terms in the sequence.

2. A baseball team sells 300 tickets to their most recent game. For each consecutive game they win, the number of tickets sold increases by a factor of 1.5. Which function represents the number of tickets bought after the baseball team has had x consecutive wins?

a. $f(x) = 300 \cdot 1.5^x$
b. $f(x) = 1.5 \cdot 300^x$
c. $f(x) = x \cdot 300^{1.5}$
d. $f(x) = 300 \cdot x^{1.5}$

3. Monica puts $3,000 in a savings account that earns 1% in interest each year. This scenario is represented by the exponential function $f(x) = 3{,}000 \cdot 1.01^x$. Interpret the meaning of 1.01 in this function.

a. 1.01 is the amount of money earned in interest each year.
b. 1.01 is the number of years the money is saved in the account.
c. 1.01 is the initial value, which is the original amount of money in the savings account.
d. 1.01 is the common ratio for the rate of exponential growth.

4. Melissa started a Spanish club at her high school. In its first year, the club had 15 members. Each year after, the number of club members increased by a factor of 1.1. Which function represents the number of members in the Spanish club x years after it started?

a. $f(x) = 1.1 \cdot x^{15}$
b. $f(x) = 15^x \cdot 0.1$
c. $f(x) = 15 \cdot 1.1^x$
d. $f(x) = 1.1 \cdot 15^x$

5. The population of Plainview can be modeled with the exponential function $f(x) = 30,000 \cdot 1.25^x$. Interpret the meaning of 30,000 in this function.

a. 30,000 is the common ratio for the rate of exponential growth.
b. 30,000 is the final value, which is the ending population of Plainview.
c. 30,000 is the number of years during which the population of Plainview increases.
d. 30,000 is the initial value, which is the original population of Plainview for which the model applies.

6. The table below shows the growth of a culture of bacteria. Which exponential function represents the number of bacteria x days after it started growing?

Bacteria	100	300	900	2,700	8,100
Day	0	1	2	3	4

a. $f(x) = 100 \cdot 3^x$
b. $f(x) = 3 \cdot 100^x$
c. $f(x) = x \cdot 100^3$
d. $f(x) = 100 \cdot x^3$

7. The world population in 2010 was approximately 6.92 billion. The annual rate of increase was about 1.2%. Assuming the rate of increase continues to be 1.2%, the exponential function that models the world population is $f(x) = 6.92 \cdot 1.012^x$. Interpret the meaning of x in this function.

a. x stands for the initial value, which is the world population in 2010.
b. x stands for the number of years the population increases after 2010.
c. x stands for the number of people the population increases by.
d. x stands for the common ratio for the rate of exponential growth.

8. The table of ordered pairs shows an exponential relationship. Which exponential function matches the table?

x	0	1	2	3	4
y	10	20	40	80	160

a. $f(x) = 2 \cdot 10^x$
b. $f(x) = 10 \cdot 10^x$
c. $f(x) = 10 \cdot 2^x$
d. $f(x) = 10 \cdot x^2$

9. A city with a population of 50,000 grows at a rate of 4% each year. Which function represents the population at the end of x years?

a. $f(x) = 0.04 \cdot 50{,}000^x$
b. $f(x) = 50{,}000 \cdot 0.04^x$
c. $f(x) = 50{,}000 \cdot 1.04^x$
d. $f(x) = 1.04 \cdot 50{,}000^x$

10. Margo invested $500 in a company's stock at the end of 2021. The value of the stock is predicted to increase at a rate of 10% each year. Which exponential function represents the growth in the value of the stock for *x* years?

a. $f(x) = 1.10 \cdot 500^x$
b. $f(x) = 10 \cdot 500^x$
c. $f(x) = 500 \cdot 0.10^x$
d. $f(x) = 500 \cdot 1.10^x$

11. John buys a car for $17,500. The value of the car depreciates at a rate of 9% per year. Which exponential function represents the value of the car after *x* years?

a. $f(x) = 17{,}500 \cdot 0.91^x$
b. $f(x) = 17{,}500 \cdot 1.09^x$
c. $f(x) = 0.91 \cdot 17{,}500^x$
d. $f(x) = 17{,}500 \cdot 9^x$

12. A new iPhone sells for $829. With each year that passes, the phone loses 20% of its original value. Which exponential function represents the phone's value *x* years after it is sold?

a. $f(x) = 829 \cdot 0.2^x$
b. $f(x) = 829 \cdot 80^x$
c. $f(x) = 829 \cdot 20^x$
d. $f(x) = 829 \cdot 0.8^x$

13. A painting is worth $500 and is expected to increase in value at a rate of 6% each year. Write an exponential function that represents this scenario. Then, find the expected value of the painting in 7 years. Round your answer to the nearest hundredth.

a. $3,000.00
b. $715.36
c. $3,710.00
d. $751.82

14. A laptop valued at $1,300 depreciates at a rate of 13% per year. Write an exponential function that represents this scenario. Then, find the expected value of the laptop after 2 years. Round your answer to the nearest hundredth.

a. $9,700.00
b. $983.97
c. $1,287.00
d. $659.97

15. An investment valued at $60,000 increases at a rate of 7.8% each year. Write an exponential function that represents this scenario. Then, find the value of the investment after 15 years. Round your answer to the nearest hundredth.

a. $60,117.00
b. $70,200.00
c. $185,111.19
d. $17,746.60

9.3 Graphing Exponential Functions and Determining Key Features

Exponential functions increase or decrease at a compounding rate. Unlike a linear function, which produces a straight line when graphed, exponential functions will have a sweeping arc shape when graphed.

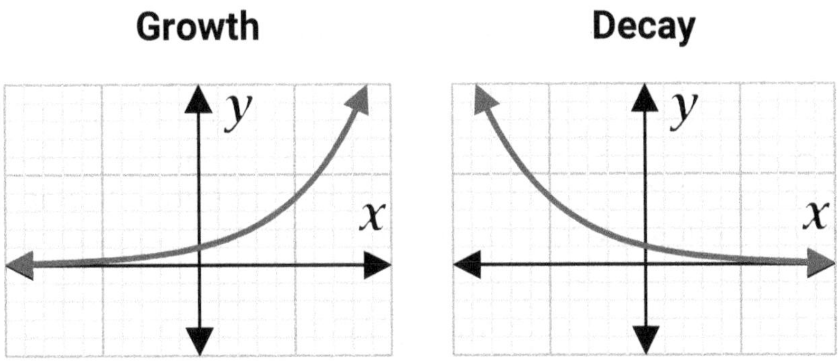

Growth and decay functions have the form $y = ab^x$.

y is the dependent variable.

a represents the original, or initial, value.

b represents the growth factor.

x is the independent variable. This is often a unit of time.

For example, the graph below shows an exponential function. The equation is $y = 2^x$. For any function, the y-intercept can be found by plugging in 0 for x. $y = 2^x$ becomes $y = 2^0$, which simplifies to $y = 1$. The y-intercept of the function is (0,1).

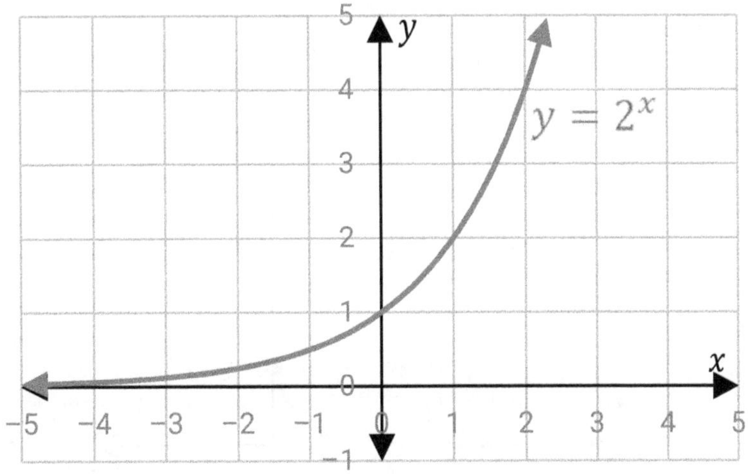

You can tell by the graph that the exponential function is a growth function, not a decay function. Remember, the exponential growth/decay function is $y = ab^x$. Growth functions have b-values greater than 1. Decay functions have b-values less than 1 but greater than 0. Growth functions will grow away from the horizontal asymptote to the right. Decay functions will decay towards the asymptote on the right.

EXAMPLES

EXAMPLE 1:

Graph the exponential function $f(x) = 2(4)^x$. Identify the graph's key features, including the y-intercept, if it is growth or decay, and the horizontal asymptote.

Answer:

The function takes the form $y = a(b)^x + k$, where k is the horizontal asymptote. Here, $k = 0$, so the horizontal asymptote is $y = 0$. To graph an exponential function, start by finding the initial value or y-intercept, which is the value of the function when $x = 0$. Substitute 0 into the equation for x and solve.

$$f(0) = 2(4)^0$$
$$f(0) = 2(1)$$
$$f(0) = 2$$

Since $f(0) = 2$, The y-intercept is (0,2). Graph this point onto the coordinate plane. Next, find the value of the function when $x = 1$. Substitute 1 into the equation and solve. Since $f(1) = 8$, graph the point (1,8) on the coordinate plane.

$$f(1) = 2(4)^1$$
$$f(1) = 2(4)$$
$$f(1) = 8$$

From here, use the points (0,2) and (1,8) to sketch a curve and determine the direction of the slope. Notice that as the value of x increases, the value of $f(x)$ also increases. This tells us that the function shows exponential growth. Finally, analyze the function's end behavior to determine its horizontal asymptote.

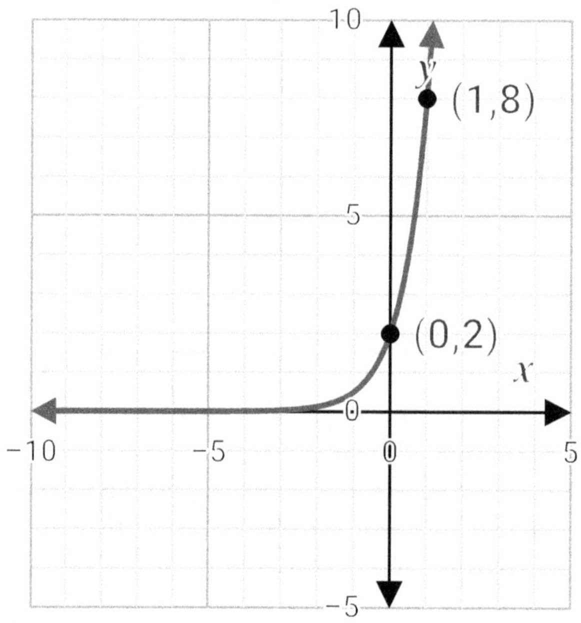

EXAMPLE 2:

Kasia buys a car for \$30,000. The value of the car depreciates at a rate of 15% per year. This exponential relationship is modeled by the function $f(x) = 30{,}000(0.85)^x$, where x stands for time elapsed (in years), and $f(x)$ stands for the value of the car. Graph the exponential function and identify the graph's key features, including the y-intercept, if it is growth or decay, and the horizontal asymptote.

Answer: The function takes the form $y = a(b)^x + k$, where k is the horizontal asymptote. Here, $k = 0$, so the horizontal asymptote is $y = 0$. To graph an exponential function, start by finding the initial value or y-intercept, which is the value of the function when $x = 0$. Substitute 0 into the equation for x and solve. Since $f(0) = 30{,}000$, The y-intercept is (0,30,000). Graph this point onto the coordinate plane.

$$f(0) = 30{,}000(0.85)^0$$
$$f(0) = 30{,}000(1)$$
$$f(0) = 30{,}000$$

Next, find the value of the function when $x = 1$. Substitute 1 into the equation and solve. Since $f(1) = 25{,}500$, graph the point (1,25,500) on the coordinate plane.

$$f(1) = 30{,}000(0.85)^1$$
$$f(1) = 30{,}000(0.85)$$
$$f(1) = 25{,}500$$

From here, use the points (0,30,000) and (1,25,500) to sketch a curve and determine the direction of the slope. Notice that as the value of x increases, the function approaches the horizontal asymptote. This means the function represents exponential decay.

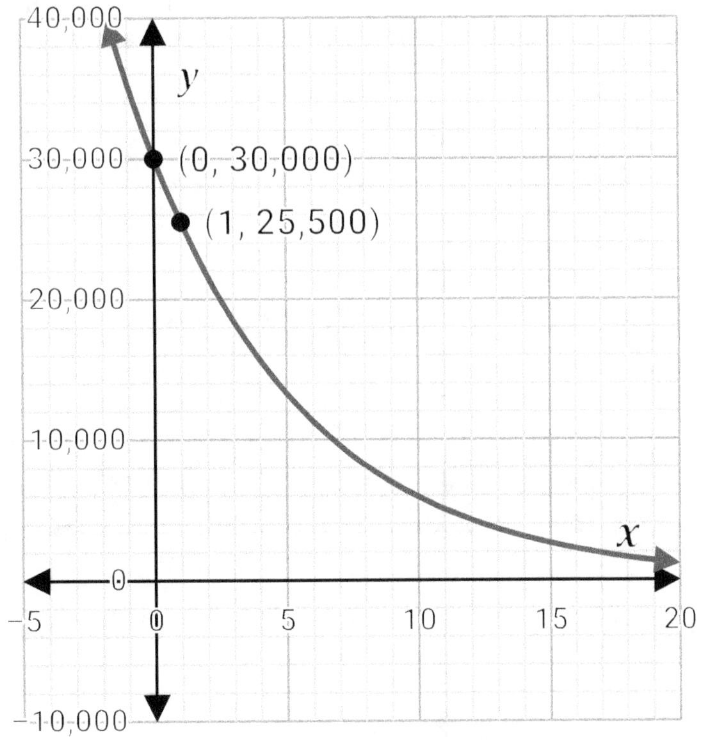

EXAMPLE 3:

The graph of a function $f(x)$ is shown on the coordinate plane below. What is the graph of $f(x) + 4$?

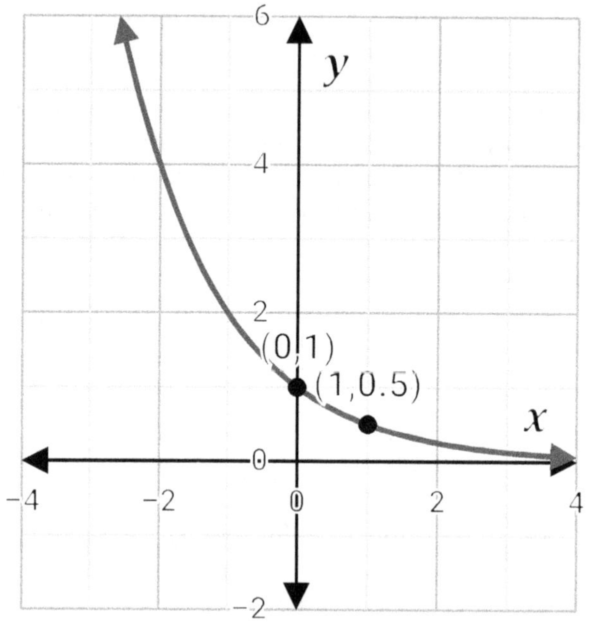

Answer: The horizontal asymptote can be changed by adding a constant term to the function. For the exponential function $f(x) = b^x + k$, the value of y approaches but never reaches a horizontal asymptote of k on one end of the graph. The graph of $f(x)$ has a horizontal asymptote of $y = 0$. The horizontal asymptote of $f(x) + 4$ is $y = 0 + 4$, or $y = 4$. Therefore, the graph can be sketched by shifting the current graph up by 4 units so that the value of y approaches but never reaches 4.

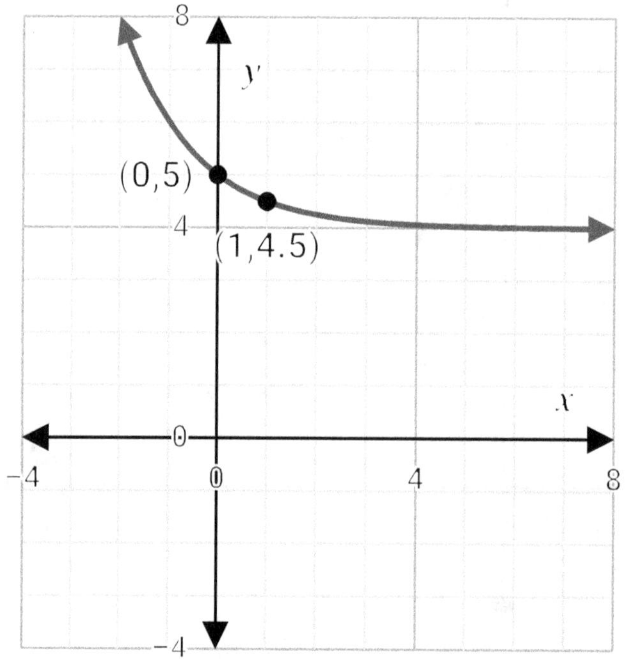

Practice Questions

1. Choose the graph of the exponential function $f(x) = 4 \cdot 2^x$.

a.

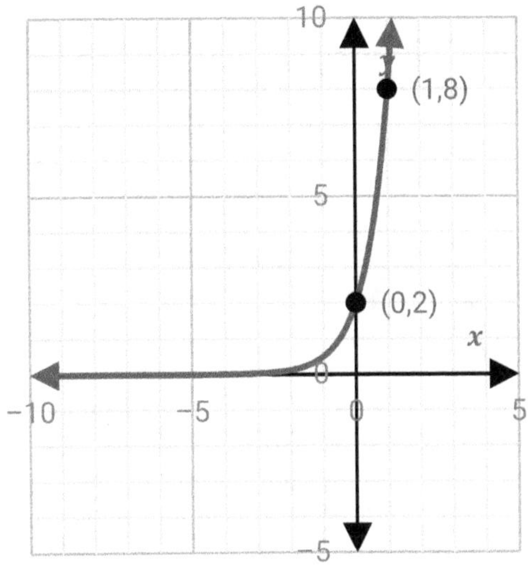

b.

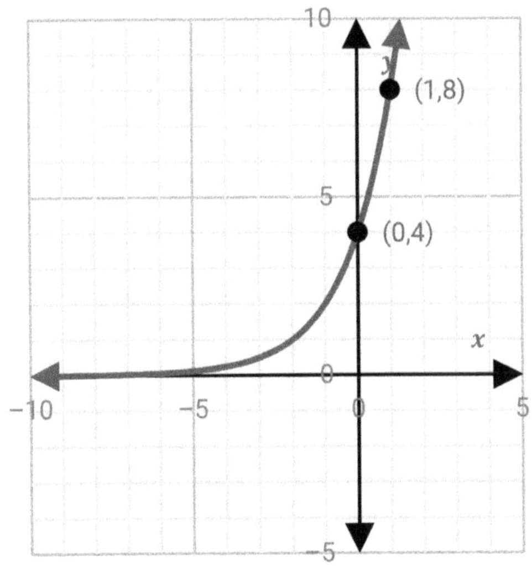

c.

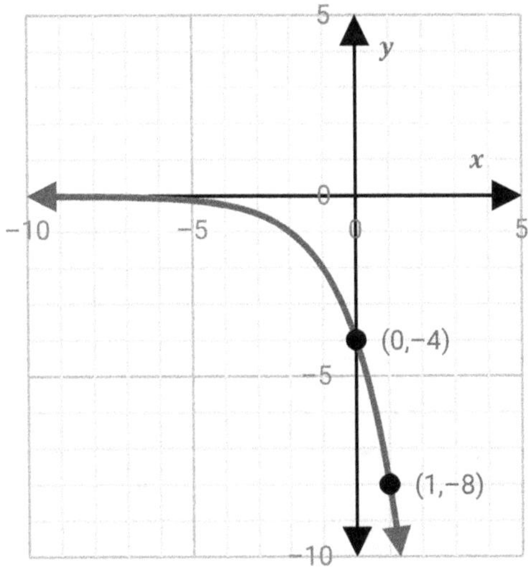

d.

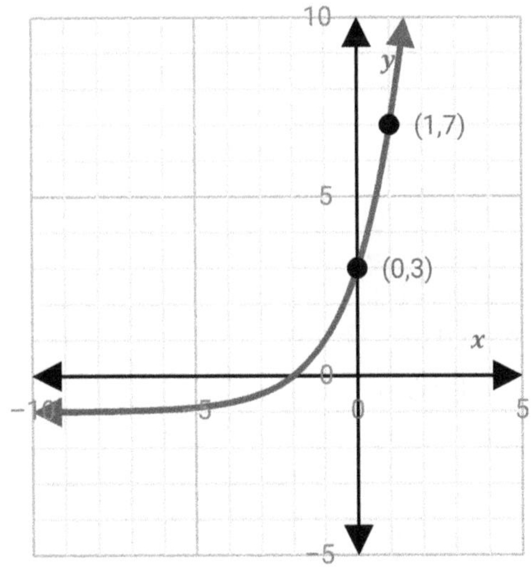

2. What is the y-intercept of the exponential function graphed on the coordinate plane below?

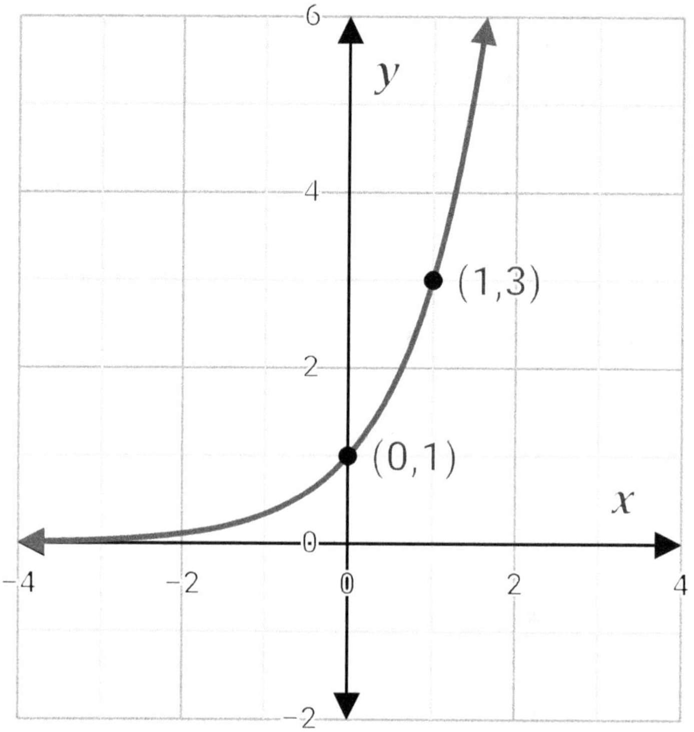

a. (1,3)
b. (0,1)
c. (0,0)
d. (−2,1)

3. Which of the following gives the general shape of the graph of the exponential function $f(x) = -4\left(\frac{2}{5}\right)^x$?

a.

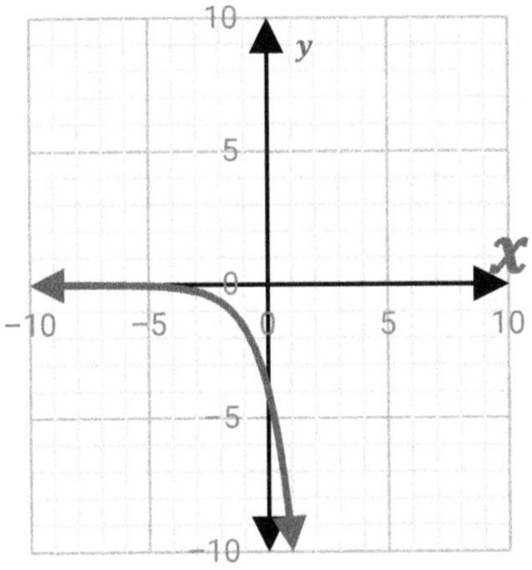

b.

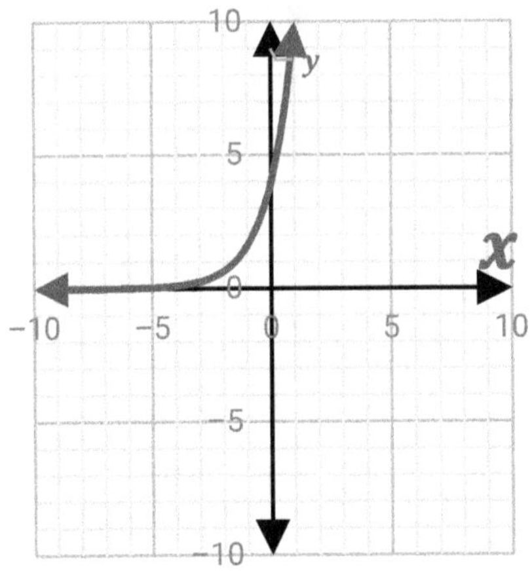

c.

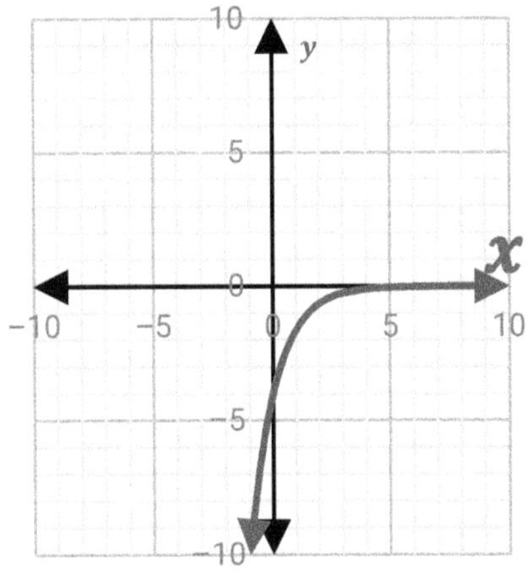

d.

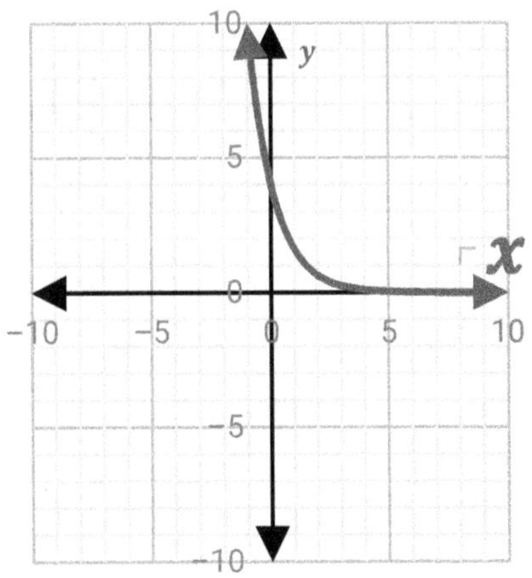

4. An exponential function is graphed on the coordinate plane shown below. What is the horizontal asymptote of the graph?

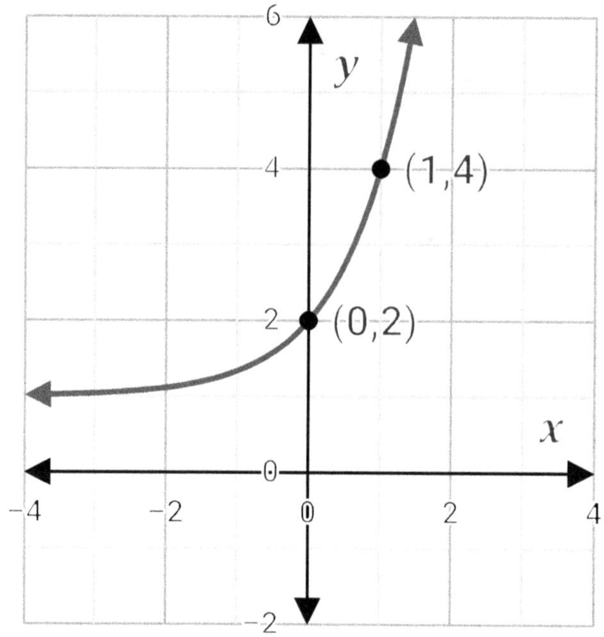

a. $y = 4$
b. $y = 2$
c. $y = 1$
d. $y = 0$

5. Choose the graph of the exponential function $f(x) = 6 \cdot \frac{1^x}{3}$.

a.

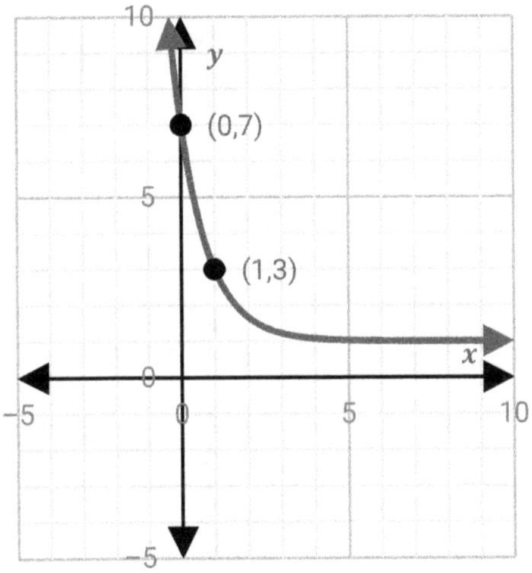

c.

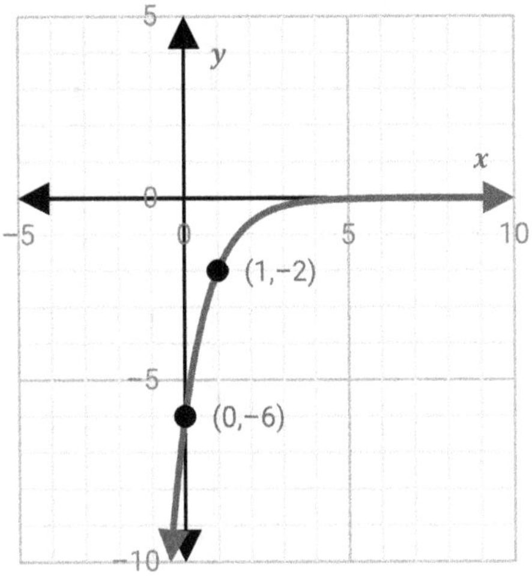

b.

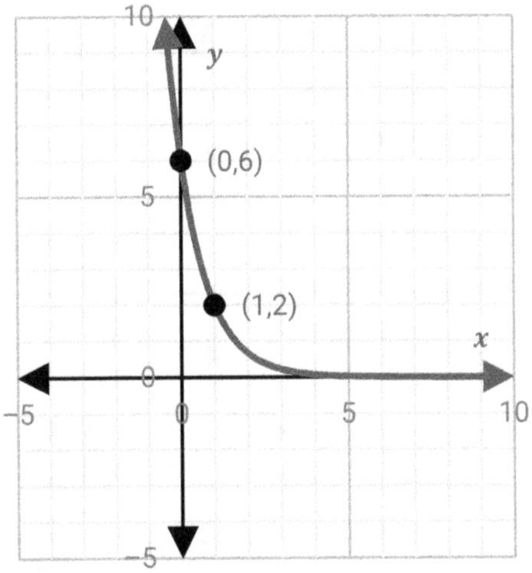

d.

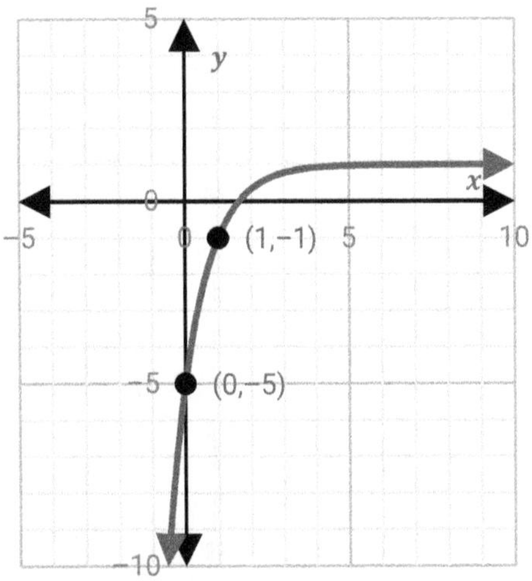

6. Which graph shows an exponential function with an initial value of $\frac{1}{2}$?

a.

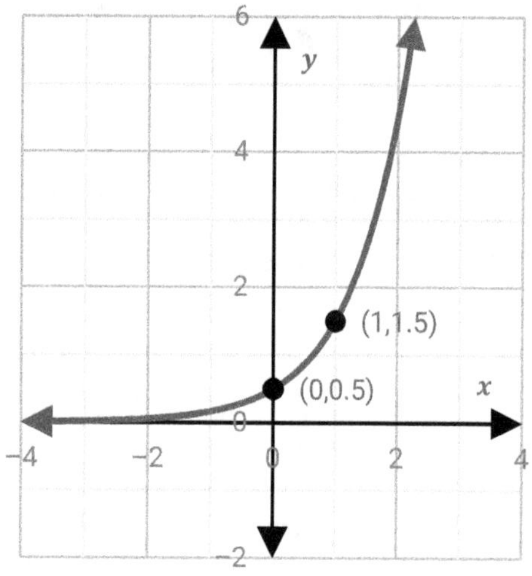

c.

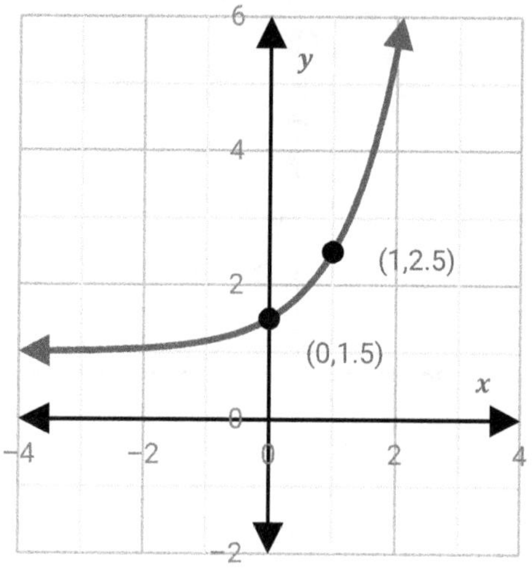

b.

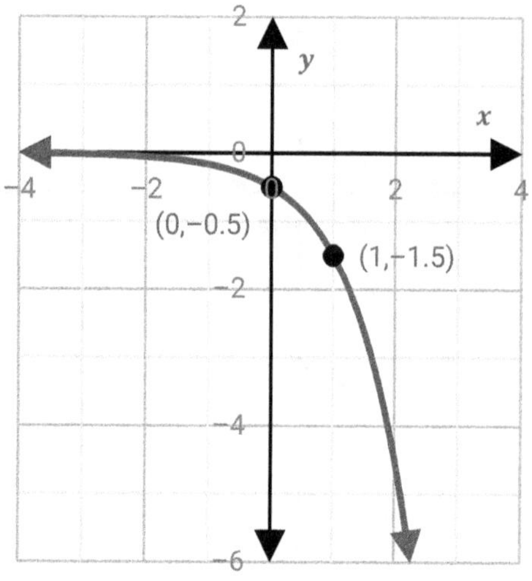

d.

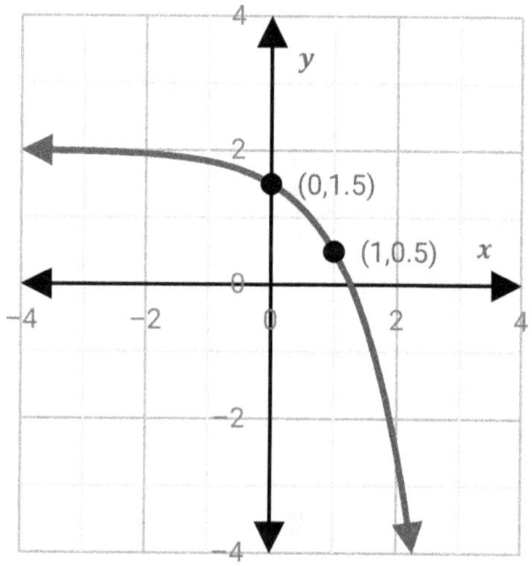

7. Based on the graph of the exponential function below, which statement is true?

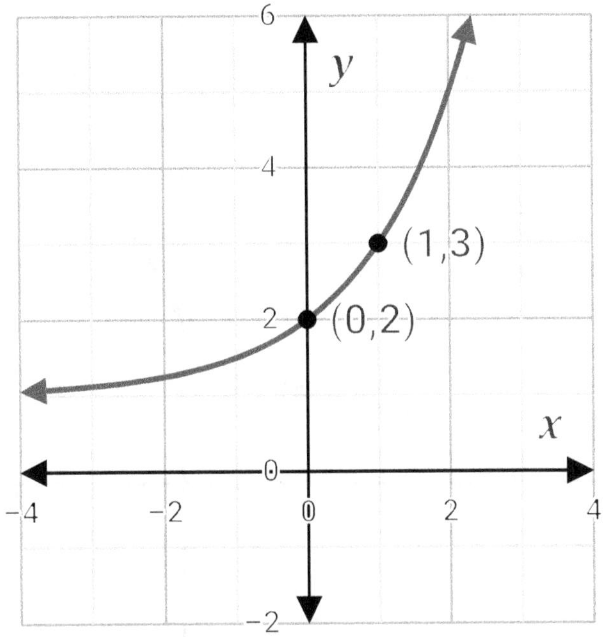

a. The graph is increasing, so the graph shows exponential growth.
b. The graph is increasing, so the graph shows exponential decay.
c. The graph is decreasing, so the graph shows exponential growth.
d. The graph is decreasing, so the graph shows exponential decay.

8. What is the horizontal asymptote of the graph of the exponential function shown?

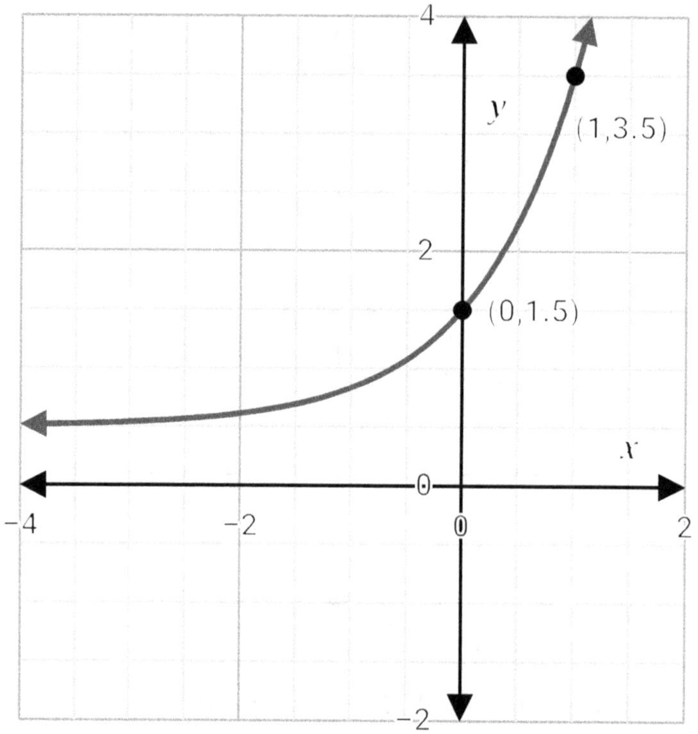

a. $y = 3.5$
b. $y = 1.5$
c. $y = -0.5$
d. $y = 0.5$

9. Choose the graph of the function $f(x) = \frac{2^x}{3}$.

a.

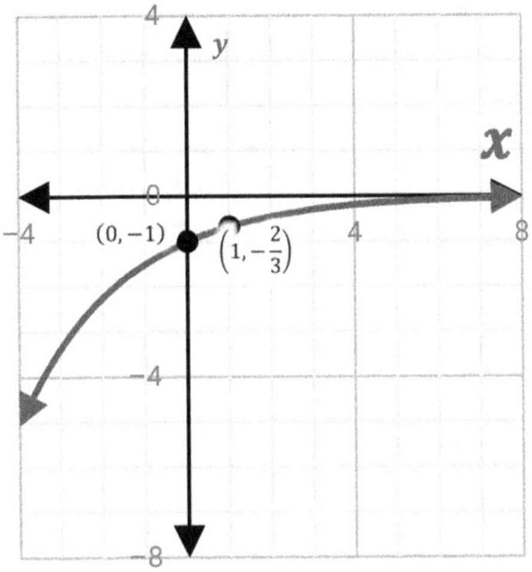

c.

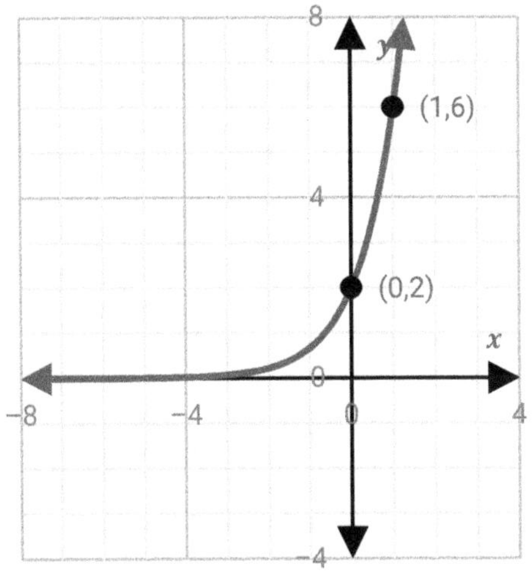

b.

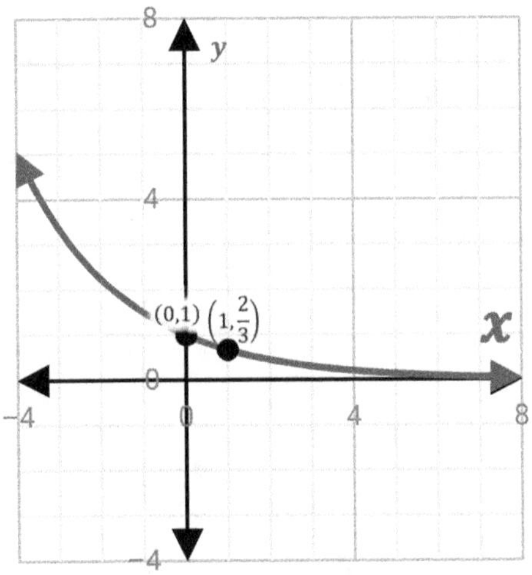

d.

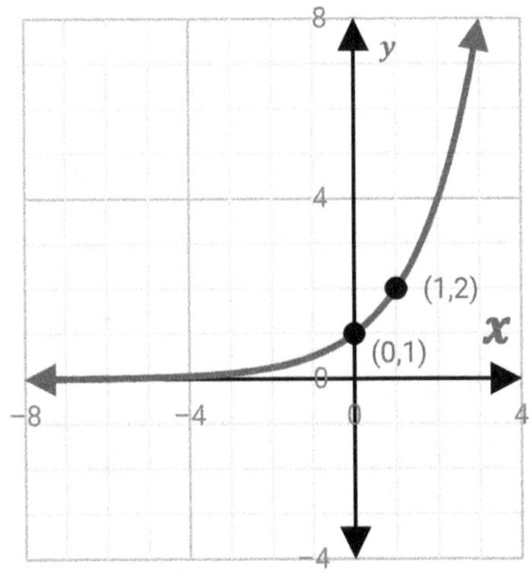

10. What is the y-intercept of the function shown?

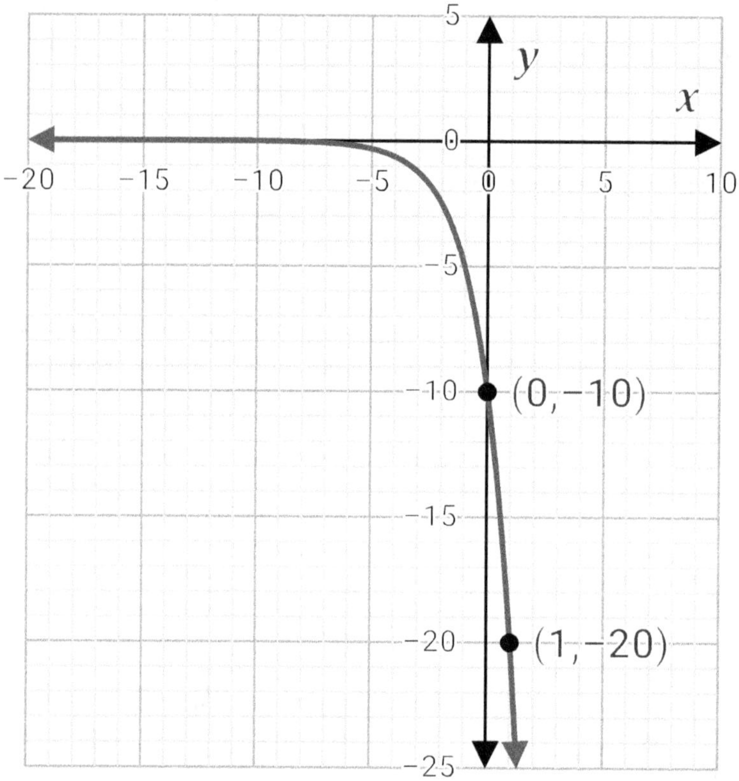

a. (0,0)
b. $(1, -20)$
c. (1,1)
d. $(0, -10)$

11. In the exponential function $f(x) = \frac{1}{4}^x + 2$, the base, b, of the exponent is $\frac{1}{4}$. What does this tell us about the function?

a. $b < 1$, so the function is a growth function.
b. $b < 1$, so the function is a decay function.
c. $0 < b < 1$, so the function is a growth function.
d. $0 < b < 1$, so the function is a decay function.

12. The graph of a function f is shown on the coordinate plane below. Which graph shows $f(x) + 3$?

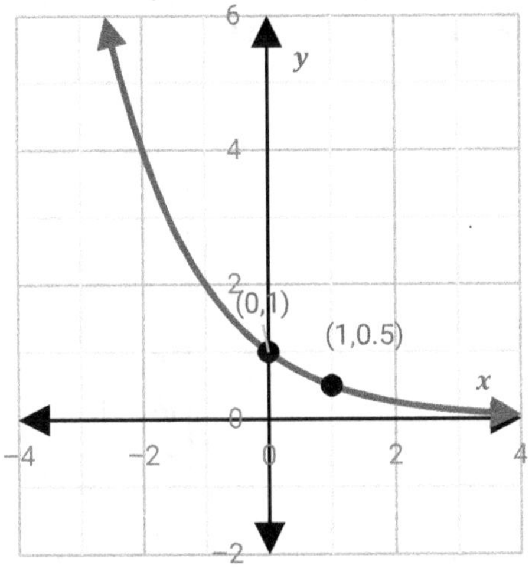

a. 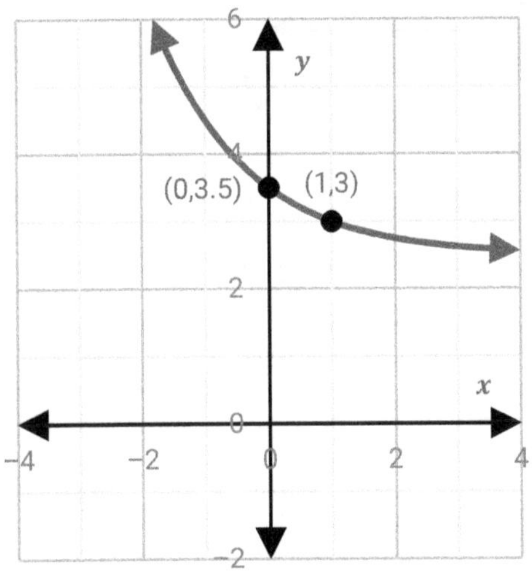

c. 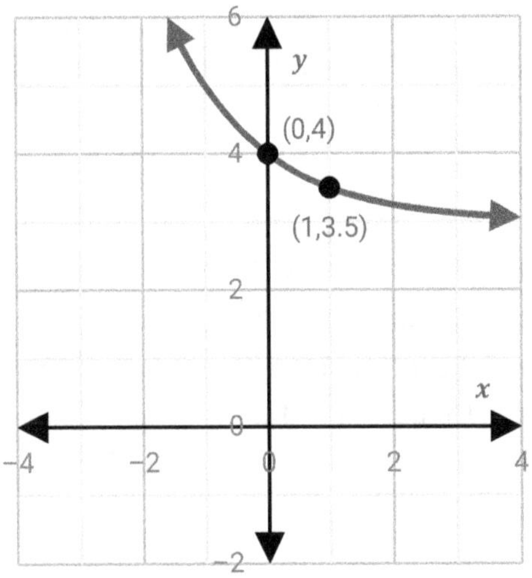

b. 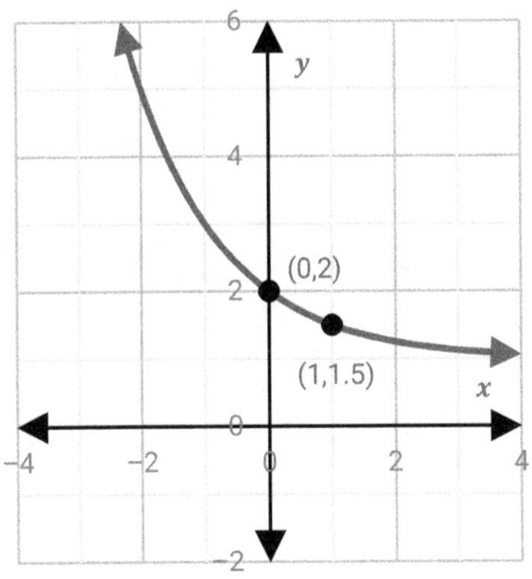

d. 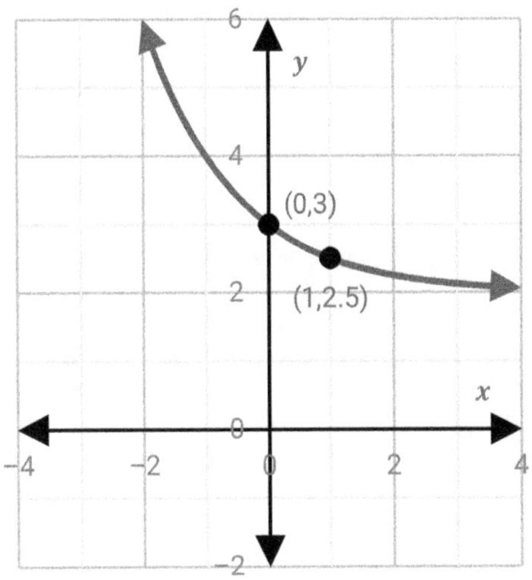

13. The rate at which a culture of bacteria self-replicates is modeled by the exponential function $f(x) = 9 \cdot 2^x$. In this function, x stands for time elapsed in minutes, and $f(x)$ stands for the number of bacteria cells. This function is graphed on the coordinate plane below. Based on this information, how many bacteria cells are present at 2 minutes?

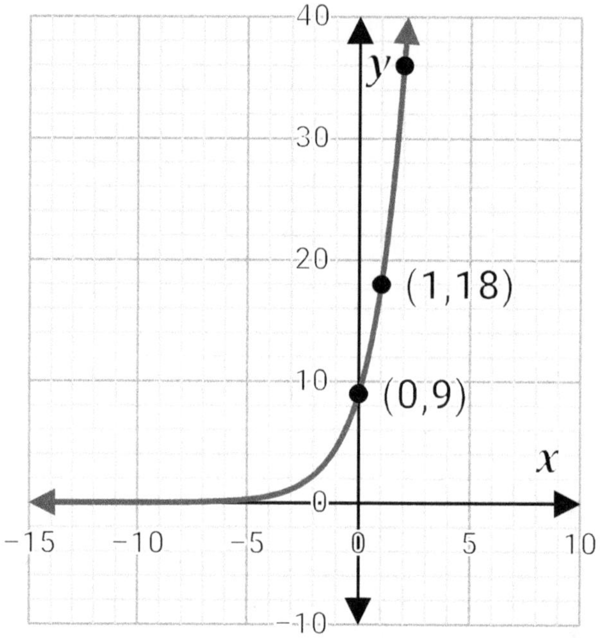

a. 36 bacteria cells
b. 18 bacteria cells
c. 9 bacteria cells
d. 1 bacteria cell

14. A population of 8 beavers tripled every month. This exponential relationship can be modeled by the function $f(x) = 8 \cdot 3^x$, where x stands for time elapsed in months, and $f(x)$ stands for the total number of beavers. Which graph matches this exponential function?

a.

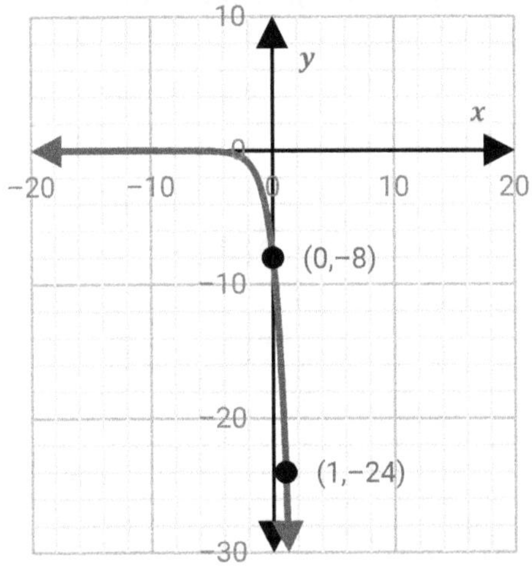

c.

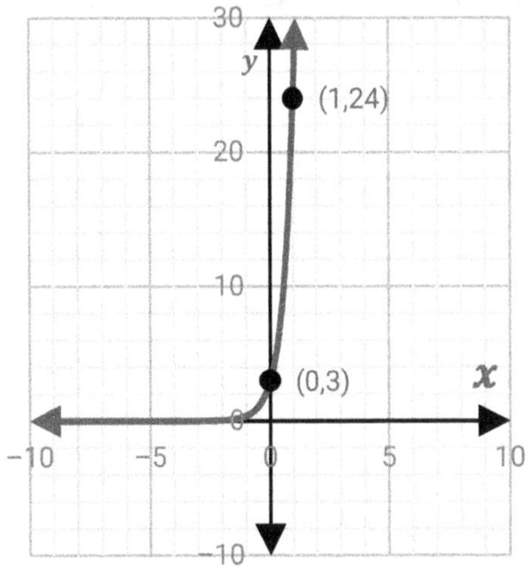

b.

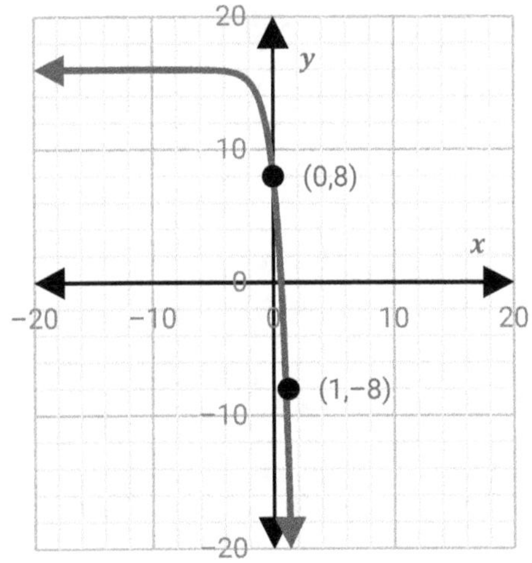

d.

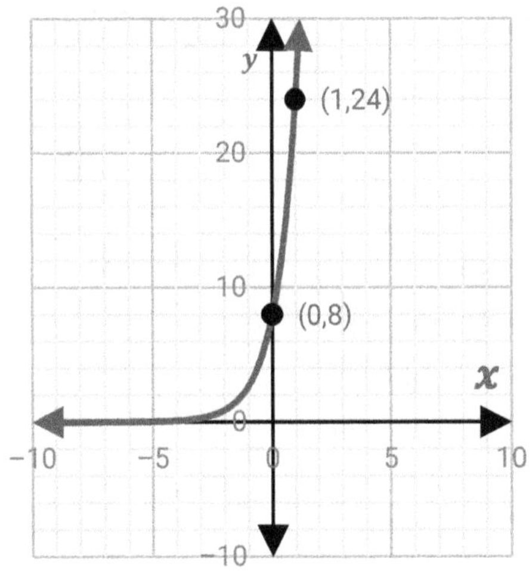

15. On his 18th birthday, Keith opens a bank account with \$2,000. He doubles the amount of money in the account each year. This scenario can be modeled by the exponential function $f(x) = 2,000 \cdot 2^x$, where x stands for time elapsed in years, and $f(x)$ stands for Keith's total amount of money saved. Based on this information, how much money will Keith save after 3 years?

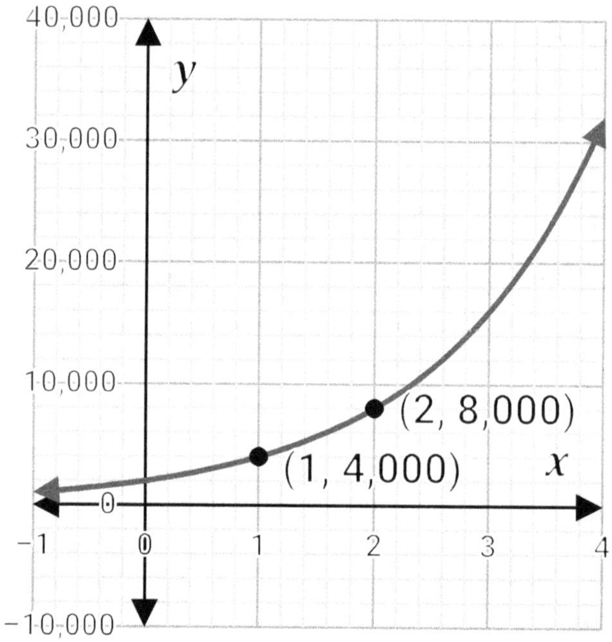

a. \$4,000.00
b. \$8,000.00
c. \$16,000.00
d. \$28,000.00

Chapter 9 Quiz

1. After being heated by a Bunsen burner, a certain type of metal is left to cool. The temperature, in degrees Fahrenheit, of the metal decreases at an exponential rate as it cools. If the temperature, y, of the metal can be modeled by the exponential function $y = 200(0.85)^x + 72$, which of the following is the range of values for the temperature of the metal at time, $x \geq 0$, in minutes?

a. $72 < y \leq 200$
b. $72 < y \leq 272$
c. $y > 72$
d. $y < 200$

2. Find the domain of the exponential function $f(x) = 3 \cdot 2^x$.

a. $x \geq 0$
b. $x \geq 2$
c. $-\infty \leq x \leq \infty$
d. $x \geq 3$

3. A radioactive isotope decays in such a fashion that each week the amount of the isotope is 25% less than the week before. If you start with 8 grams of the isotope, which of the following functions correctly models the mass, $M(t)$, of the isotope after t weeks?

a. $M(t) = 8 - 2t$
b. $M(t) = 8 \cdot 0.75^t$
c. $M(t) = 8 \cdot 1.25^t$
d. $M(t) = 8 \cdot 0.25^t$

4. You begin with 10 grams of a radioactive isotope. It decays in such a fashion that G, the number of grams remaining, is an exponential function of the number of weeks that pass, as shown in the graph. What is the range of the function G?

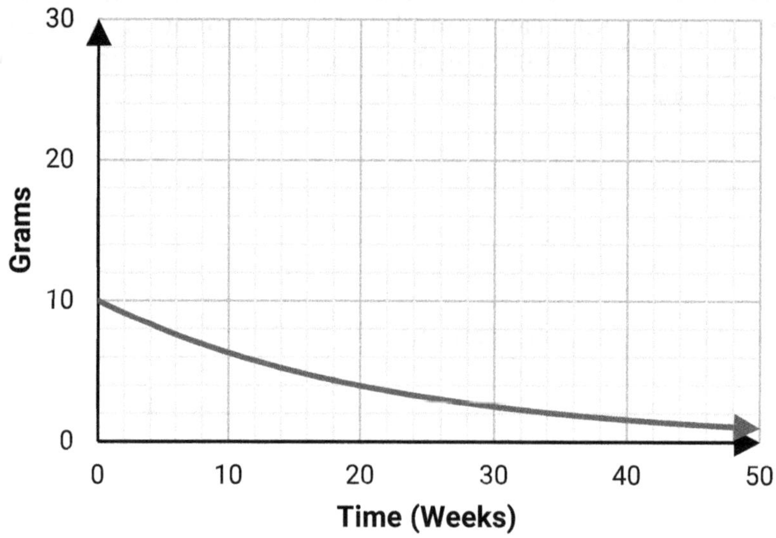

a. $0 < G \leq 10$
b. $G \geq 0$
c. $G \geq 10$
d. $-\infty \leq G \leq \infty$

5. State the range for $y = 5\left(\frac{3}{2}\right)^x - 4$.

a. $y > 5$
b. $y < 4$
c. $y > -4$
d. $y < -4$

6. State the range for $y = \left(\frac{2}{5}\right)^x + 6$.

a. $y > 6$
b. $y < 6$
c. $y > 1$
d. $y < 1$

7. The diagram shows the graph of an exponential function. What is the *y*-intercept of the function?

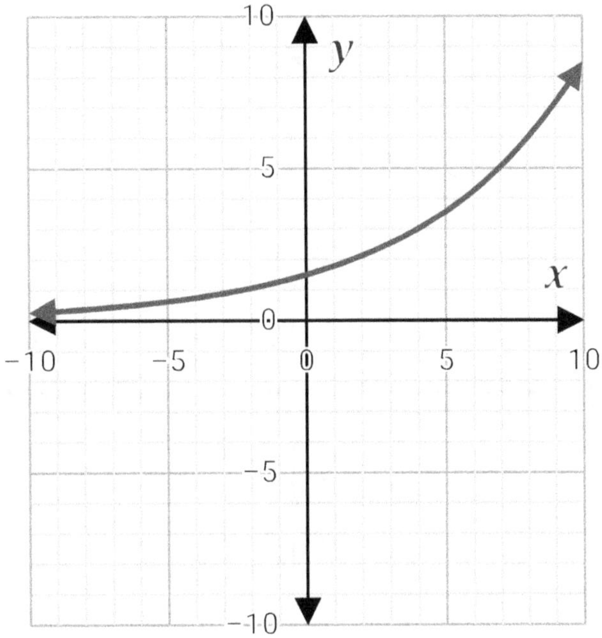

a. (0,1.5)
b. (0,0)
c. (−7,1)
d. (4,1)

8. The diagram shows the graphs of four functions. Which one is the graph of the function $y = 3 \cdot \left(\frac{2}{3}\right)^x$?

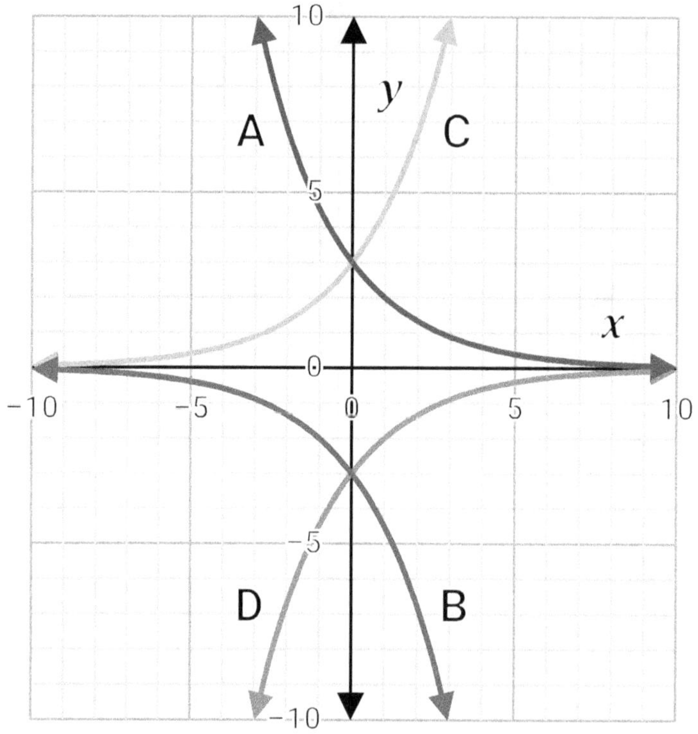

a. Graph A
b. Graph B
c. Graph C
d. Graph D

9. After careful research, you find an investment that promises to increase in value by 4% each year (that is, each year your investment will be worth 4% more than the year before). If you invest \$500 in it, find a function that correctly models the amount, $A(t)$, that your investment is worth after t years.

a. $A(t) = 500 \cdot 0.04^t$
b. $A(t) = 500 \cdot 4^t$
c. $A(t) = 500 + 20t$
d. $A(t) = 500 \cdot 1.04^t$

10. Match the equation to the graph of the exponential function $y = 2^x + 1$.

a. 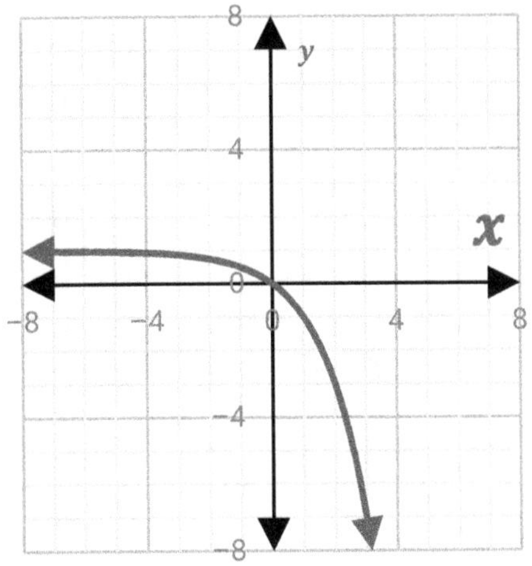

c. 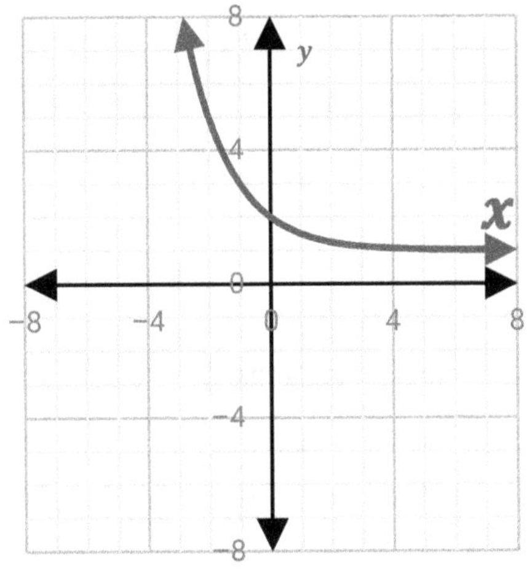

b. 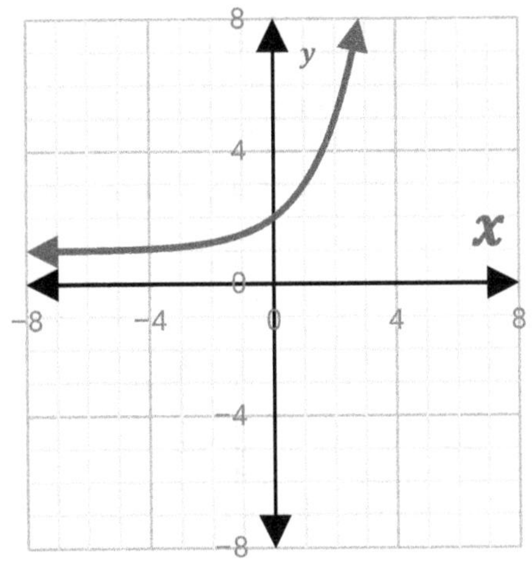

d. 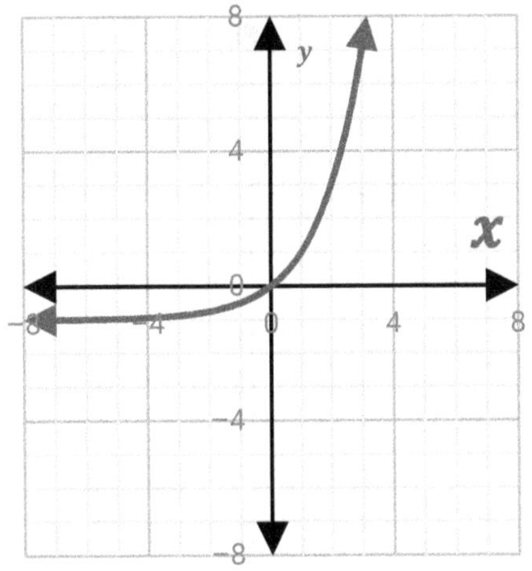

11. Thirteen grams of a salt solution decays in a petri dish at an exponential rate. The amount, A, of salt left in the solution after time, t, in hours can be represented by the exponential function $A = 10\left(\frac{1}{2}\right)^t + 3, t \geq 0$. What is the range of values of salt left in the solution after t hours, in grams?

a. $3 < A \leq 13$
b. $3 < A \leq 10$
c. $A > 13$
d. $A < 13$

12. Match the equation to the graph of the exponential function $y = -4^x + 2$.

a.

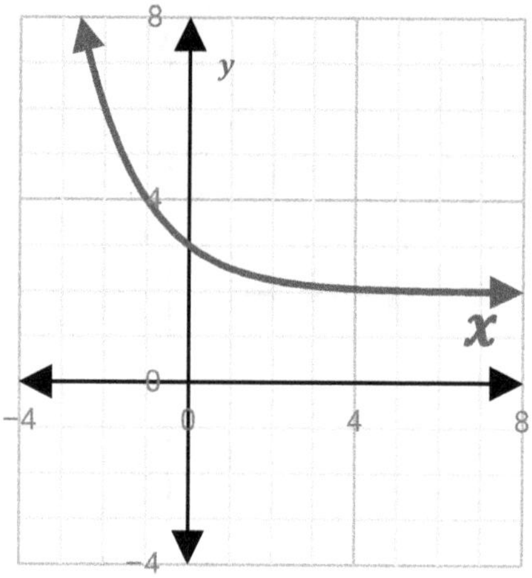

c.

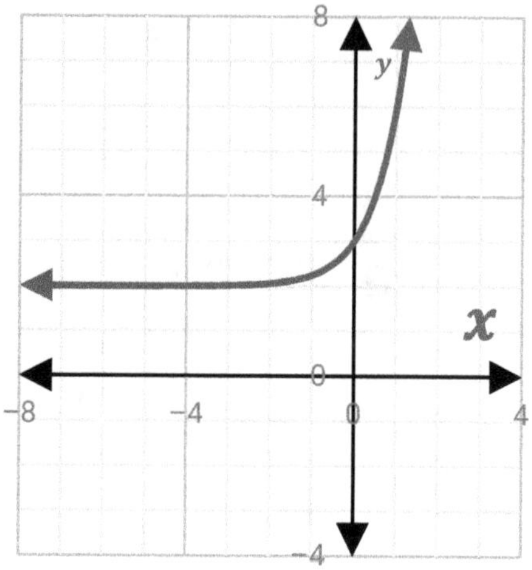

b.

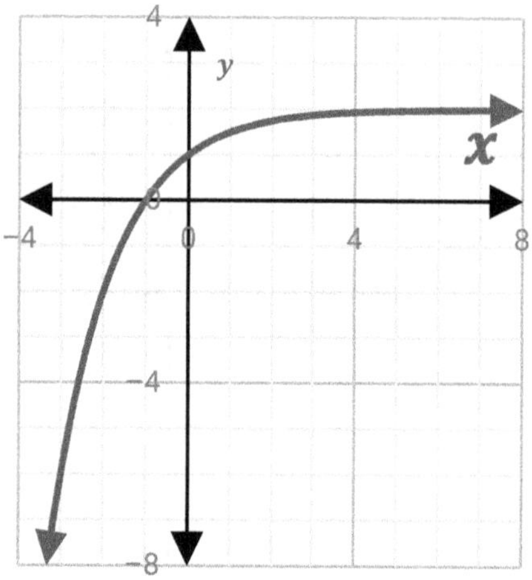

d.

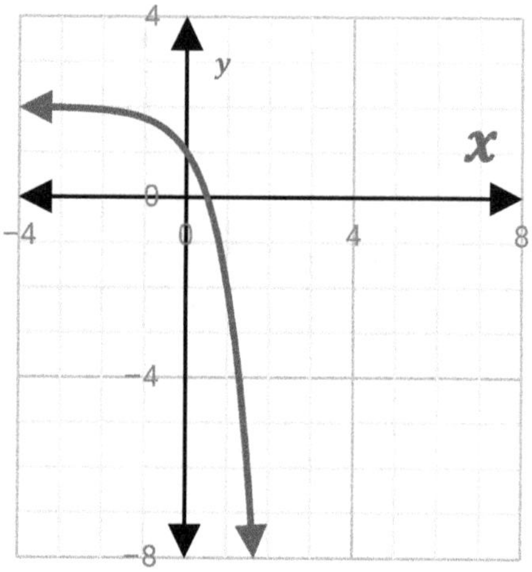

13. Match the equation to the graph of the exponential function $y = 3 \cdot \left(\frac{1}{2}\right)^x$.

a.

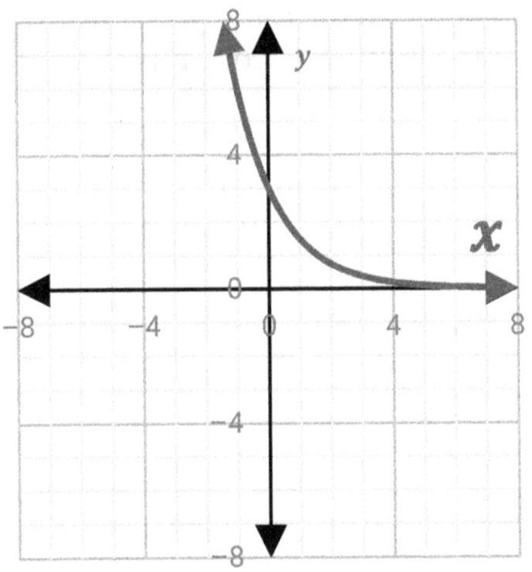

c.

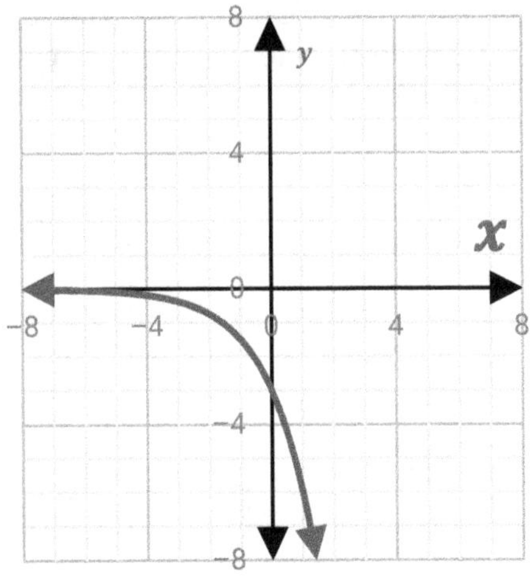

b.

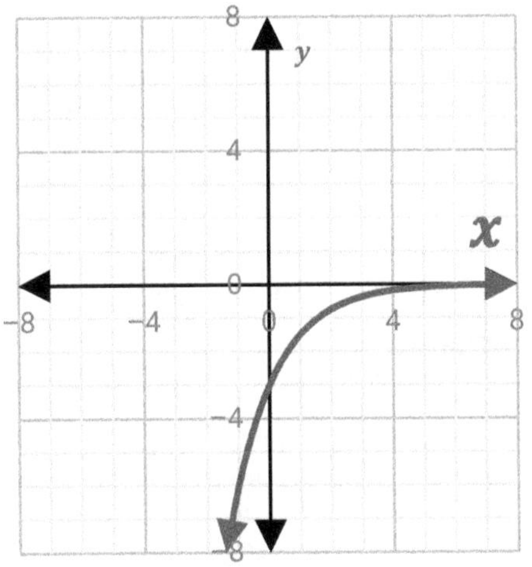

d.

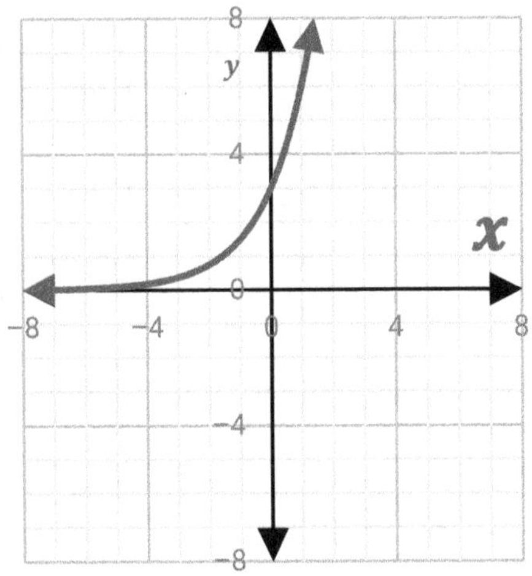

14. State the domain and range for $y = 2^x + 3$.

a. Domain: $-\infty < x < \infty$; Range: $y > 3$

b. Domain: $-\infty < x < \infty$; Range: $y < 3$

c. Domain: $0 < x < \infty$; Range: $y > 5$

d. Domain: $0 < x < \infty$; Range: $y < 5$

15. You invest \$100 in an investment whose value grows by 4% a year. That is, at any moment it is worth 4% more than a year prior. This makes the value, in dollars, an exponential function of time whose graph appears in the diagram. If you leave the money in this investment indefinitely, what is the range of this function, where V is the value of the investment?

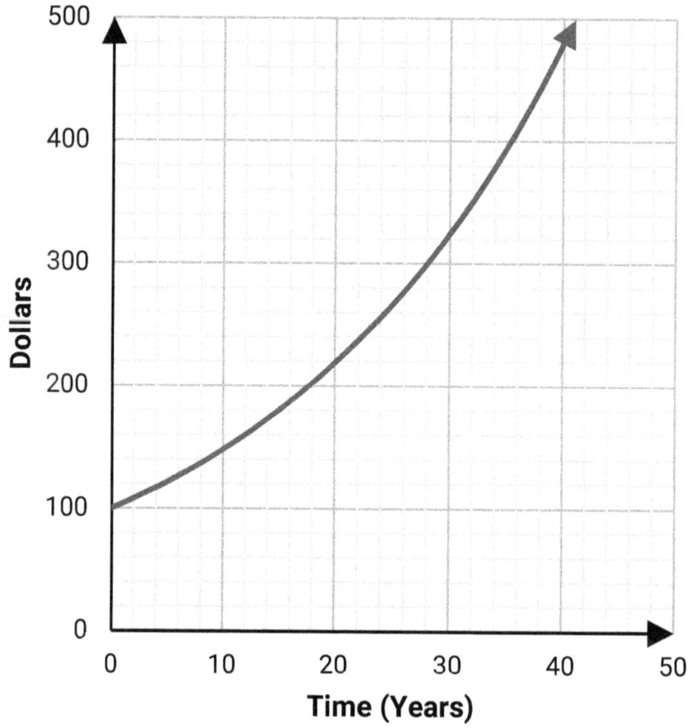

a. $V \geq 4$
b. $-\infty \leq V \leq \infty$
c. $V \geq 0$
d. $V \geq 100$

16. **During your 3-month summer vacation, you work a summer job. You save \$1,500 for the first 2 months. After the third month of work, have earned enough money to invest \$500 dollars into a mutual fund that earns 5% each year. The value, A, of what you have invested plus what you have saved can be modeled by the exponential function $A = 1,500(1.05)^t$, where $t \geq 0$ is time, in years. Which of the following graphs best represents the value of what you have invested plus what you have saved in dollars?**

a.
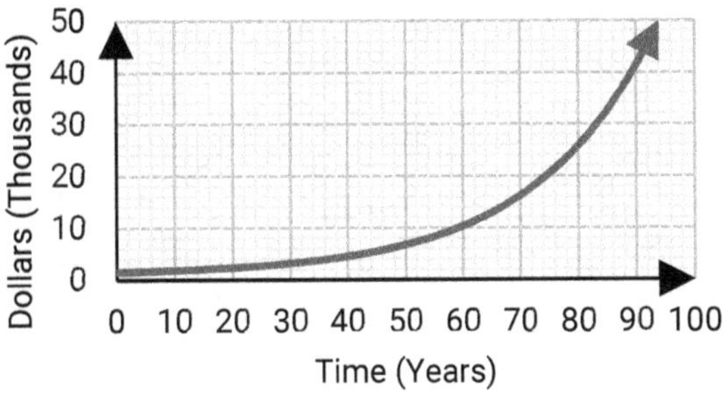

b.
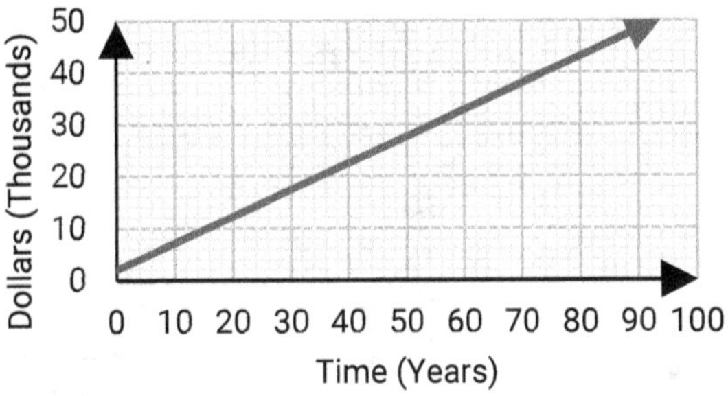

c.
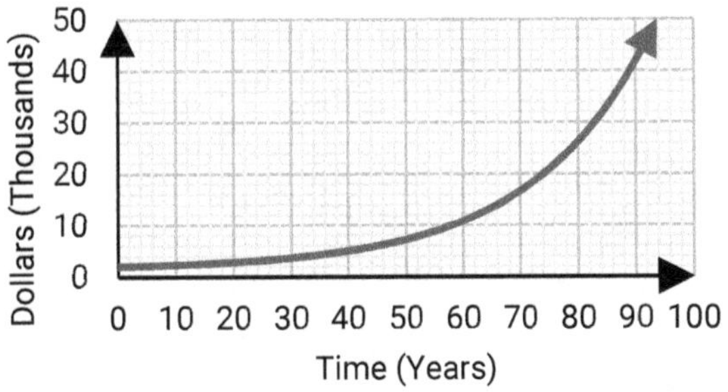

d.
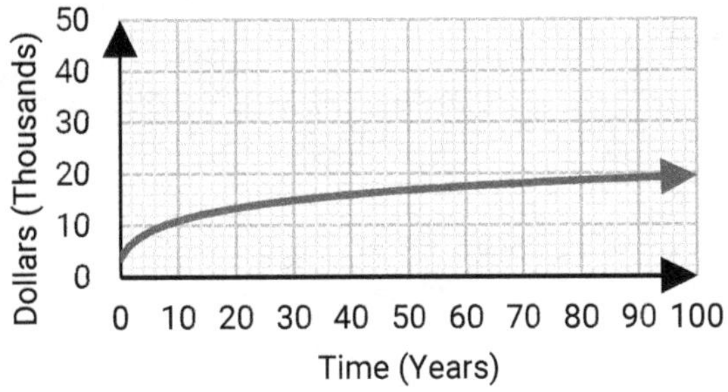

17. You purchase a new car for \$20,000. The car depreciates at an exponential rate of 15% each year. We can write the value of the car, A, as the exponential function $A = 20,000(0.85)^t$, where $t \geq 0$ is time, in years. Which of the following graphs best represents the value of the car in dollars at time t?

a. 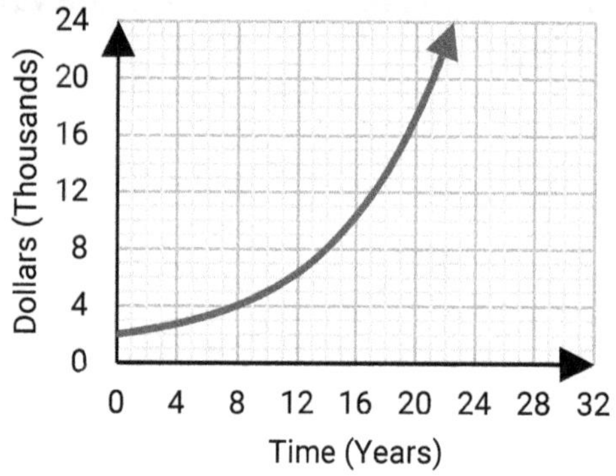

c. 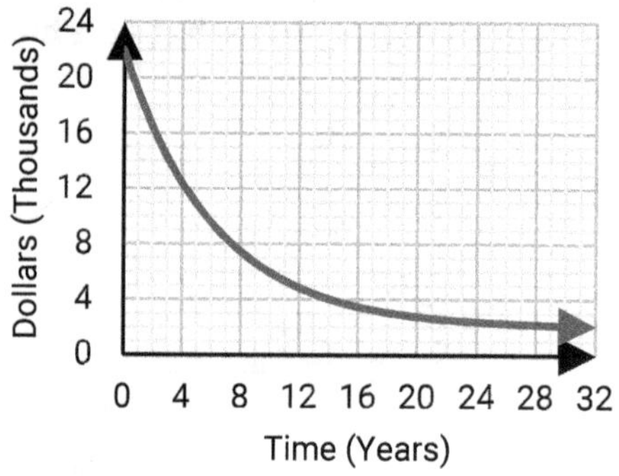

b. 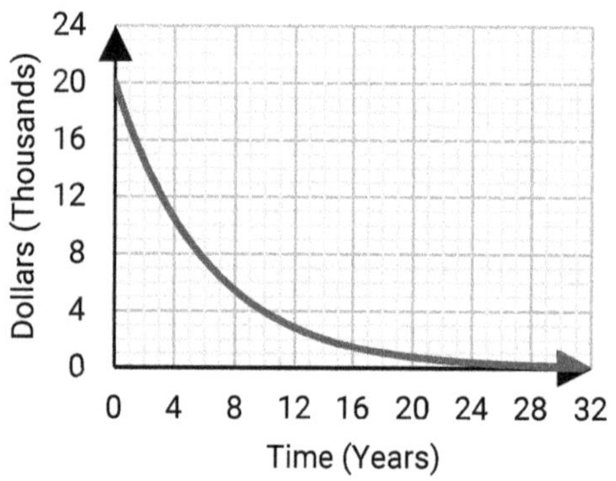

d. 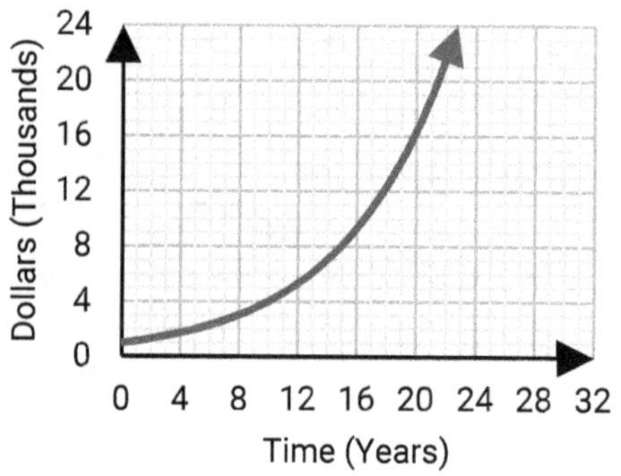

18. State the range for $y = -\left(\frac{4}{5}\right)^x + \frac{2}{3}$.

a. $y > -\frac{4}{5}$
b. $y < -\frac{4}{5}$
c. $y > \frac{2}{3}$
d. $y < \frac{2}{3}$

19. State the domain and range for $y = -4(3)^x$.

a. Domain: $-\infty < x < \infty$; Range: $y > 0$
b. Domain: $-\infty < x < \infty$; Range: $y < 0$
c. Domain: $-4 < x < \infty$; Range: $y > 3$
d. Domain: $-4 < x < \infty$; Range: $y < 3$

20. A researcher studying a colony of 100 bacteria in a nutrient solution under controlled conditions finds that the number of bacteria grows exponentially, doubling every hour. Which of the following functions correctly models the population, $P(t)$, of bacteria after t hours?

a. $P(t) = 100 \cdot 2^t$
b. $P(t) = 100^{2t}$
c. $P(t) = t \cdot 100^2$
d. $P(t) = 100 \cdot t^2$

Chapter 10: Radical Expressions

10.1 Simplifying Radical Expressions

When simplifying radical expressions that are square roots, look for ways to break down numbers into perfect squares. For example, $\sqrt{90}$ contains a perfect square, so it can be simplified. $\sqrt{90}$ is equivalent to $(\sqrt{9})(\sqrt{10})$. $\sqrt{9}$ simplifies to 3, so $(\sqrt{9})(\sqrt{10})$ can be simplified to $3\sqrt{10}$. The same process will apply for expressions with variables.

For example, consider the expression $\sqrt{18x^3}$. Look for ways to take out perfect squares from 18 and x^3. $\sqrt{18}$ is equivalent to $(\sqrt{9})(\sqrt{2})$. 9 is a perfect square. Rewrite $\sqrt{18x^3}$ as $3\sqrt{2x^3}$. Now look at the variable. $\sqrt{x^3}$ is equivalent to $\left(\sqrt{x^2}\right)\left(\sqrt{x}\right)$. $\left(\sqrt{x^2}\right)$ simplifies to x, so $3\sqrt{2x^3}$ becomes $3x\sqrt{2x}$.

Some radical expressions will contain more than one variable. Again, look for perfect squares in each part of the expression. For example, consider the expression $\sqrt{54x^5y}$. Look for ways to break apart the 54, the x^5, and y. If anything is a perfect square, move this out of the radical.

$\sqrt{54x^5y}$ is equivalent to $\sqrt{(9)(6)(x^4)(x)(y)}$. In this expanded form, we can see that the (9) and the (x^4) are perfect squares. Take out the square root of each term. 3 is the square root of 9, so pull out a 3. x^2 is the square root of x^4, so pull out an x^2. $\sqrt{(9)(6)(x^4)(x)(y)}$ becomes $3x^2\sqrt{(6)(x)(y)}$. There are no other perfect squares to remove from under the radical, so the expression is simplified: $3x^2\sqrt{6xy}$.

Some radical expressions will be in the form of a fraction. If a fraction is involved, look for square roots in the numerator and the denominator. For example, consider the expression $\frac{\sqrt{25x^3}}{\sqrt{y^4}}$.

Sometimes it is helpful to write the expression in an expanded form so that the square roots are more visible. $\frac{\sqrt{25x^3}}{\sqrt{y^4}}$ can be rewritten as $\frac{\sqrt{(5^2)(x^2)(x)}}{\sqrt{(y^2)^2}}$, which simplifies to $\frac{5x\sqrt{x}}{y^2}$.

The important thing to remember is to look for square roots within each piece of the expression. Pull out these square roots, and leave everything else under the radical.

EXAMPLES

EXAMPLE 1:

Simplify the radical expression $\sqrt{40}$.

Answer:

$\sqrt{40}$ can be expressed as the product of more than one radical. $\sqrt{40}$ contains a perfect square, $\sqrt{4}$, so use this as one of the factors. $\sqrt{40}$ becomes $\left(\sqrt{4}\right)\left(\sqrt{10}\right)$. The square root of $\sqrt{4}$ is 2. $\left(\sqrt{4}\right)\left(\sqrt{10}\right)$ becomes $2\sqrt{10}$.

Example 2:

Simplify the radical expression $\sqrt{24x^{24}}$.

Answer:

Rewrite $\sqrt{24x^{24}}$ as the product of two radicals. $\sqrt{24x^{24}}$ can be expressed as $\left(\sqrt{4x^{12}x^{12}}\right)\left(\sqrt{6}\right)$, which becomes $2x^{12}\sqrt{6}$.

Practice Questions

1. Simplify the radical expression $\sqrt{98}$.

a. 49

b. $7\sqrt{14}$

c. $7\sqrt{2}$

d. $49\sqrt{2}$

2. Simplify the radical expression $\sqrt{20}$.

a. 10

b. $4\sqrt{5}$

c. $2\sqrt{10}$

d. $2\sqrt{5}$

3. Simplify the radical expression $\sqrt{1,000}$.

a. $2\sqrt{250}$

b. $10\sqrt{10}$

c. $100\sqrt{10}$

d. $5\sqrt{40}$

4. Simplify the radical expression $\sqrt{128}$.

a. $2\sqrt{32}$

b. $4\sqrt{8}$

c. $8\sqrt{2}$

d. $2\sqrt{64}$

5. Simplify the radical expression $\sqrt{180}$.

a. $2\sqrt{45}$

b. $3\sqrt{20}$

c. $6\sqrt{5}$

d. $2\sqrt{90}$

6. Simplify the radical expression $\sqrt{162}$.

a. $3\sqrt{18}$
b. $9\sqrt{2}$
c. $3\sqrt{2}$
d. $81\sqrt{2}$

7. Simplify the radical expression $\sqrt{50x^3}$. Assume x is nonnegative.

a. $25x$
b. $5\sqrt{2x^2}$
c. $x\sqrt{50}$
d. $5x\sqrt{2x}$

8. Simplify the radical expression $\sqrt{64x}$. Assume x is nonnegative.

a. $8\sqrt{x}$
b. $2\sqrt{16x}$
c. $4\sqrt{4x}$
d. $8x$

9. Simplify the radical expression $\sqrt{216x^2}$. Assume x is nonnegative.

a. $3x\sqrt{24}$
b. $6x\sqrt{6}$
c. $6\sqrt{6x^2}$
d. $2x\sqrt{54}$

10. Simplify the radical expression $\sqrt{63x^2y^3}$. Assume x and y are nonnegative.

a. $xy\sqrt{63y}$
b. $3xy\sqrt{7y}$
c. $3x\sqrt{7y^3}$
d. $8xy\sqrt{y}$

11. Simplify the radical expression $\sqrt{30x^5y^7}$. Assume x and y are nonnegative.

a. $x^2y^3\sqrt{30xy}$
b. $2x^2y^3\sqrt{15xy}$
c. $6xy\sqrt{5x^3y^5}$
d. $xy\sqrt{30x^3y^5}$

12. Simplify the radical expression $\sqrt{121x^6y^3}$. Assume x and y are nonnegative.

a. $11x^3y\sqrt{y}$
b. $x^3y\sqrt{121y}$
c. $11x\sqrt{x^3y^3}$
d. $x^3\sqrt{121y^3}$

13. A square has an area of 300 square feet. Find the length of a side of the square. Simplify your answer.

a. $10\sqrt{3}$ feet
b. $50\sqrt{3}$ feet
c. $4\sqrt{75}$ feet
d. 75 feet

14. If a square has sides of length s, then a diagonal of the square has length $\sqrt{2s^2}$. Simplify the expression for the length of the diagonal.

a. $2\sqrt{s}$
b. $s\sqrt{2}$
c. $2\sqrt{s^2}$
d. $2s$

15. When a heavy, dense object is dropped and allowed to fall freely, the time that it takes to fall a distance of d feet is approximately $t = \sqrt{\frac{d}{16}}$ seconds. How long does it take such an object to fall 448 feet? Simplify your answer.

a. $\sqrt{14}$ seconds
b. $2\sqrt{14}$ seconds
c. $8\sqrt{7}$ seconds
d. $2\sqrt{7}$ seconds

10.2 Solving Equations Involving Radicals

A radical equation is one where the unknown variable is under a radical. There can be one or more radicals in the equation. Sometimes the radical can be canceled out right away. For example, if you have an equation like $\sqrt{3x-4}=\sqrt{2x+9}$, you can square both sides so that the radical is eliminated. Squaring a square root eliminates the radical.

$$\sqrt{3x-4}=\sqrt{2x+9}$$
$$3x-4=2x+9$$

At this point, the equation is solved by using inverse operations and combining like terms. Start by adding 4 to both sides of the equation.

$$3x-4+4=2x+9+4$$
$$3x=2x+13$$

Then, subtract $2x$ from both sides.

$$3x-2x=2x-2x+13$$
$$x=13$$

When solving equations with radicals, you always need to check for extraneous solutions. **Extraneous solutions** are ones that result in a false statement when substituted back into the original equation. They are not true solutions.

Substitute 13 back into the original equation.

$$\sqrt{3(13)-4}=\sqrt{2(13)+9}$$
$$\sqrt{39-4}=\sqrt{26+9}$$
$$\sqrt{35}=\sqrt{35}$$

Since this statement is true, $x=13$ is a solution to the equation.

Similarly, some equations will have numbers inside and outside of the radical. For example, consider the equation $2\sqrt{x-1}=\sqrt{3x+4}$. Solving the equation means isolating the variable x. To do so, we need to eliminate the radicals by squaring both sides.

$$\left(2\sqrt{x-1}\right)^2=\left(\sqrt{3x+4}\right)^2$$
$$2^2(x-1)=3x+4$$
$$4(x-1)=3x+4$$

Now, distribute and use inverse operations to solve for x.

$$4x-4=3x+4$$

Add 4 to both sides of the equation.

$$4x=3x+8$$

Then, subtract $3x$ from both sides.

$$x=8$$

Now, check this solution by plugging it back into the original equation.

$$2\sqrt{(8)-1} = \sqrt{3(8)+4}$$
$$2\sqrt{8-1} = \sqrt{24+4}$$
$$2\sqrt{7} = \sqrt{28}$$
$$2\sqrt{7} = \sqrt{4\cdot 7}$$
$$2\sqrt{7} = 2\sqrt{7}$$

Since this statement is true, $x = 8$ is a solution to the equation.

Some equations with radicals need to be rearranged before squaring both sides. For example, consider the equation $2\sqrt{x+5}+8 = 18$. Before squaring the left side of the equation, subtract 8 from both sides.

$$2\sqrt{x+5} = 10$$

Then, divide both sides by 2.

$$\sqrt{x+5} = 5$$

Now, square both sides.

$$\left(\sqrt{x+5}\right)^2 = 5^2$$
$$x+5 = 25$$

Finally, subtract 5 from both sides.

$$x = 20$$

Now, check this solution by plugging it back into the original equation.

$$2\sqrt{(20)+5}+8 = 18$$
$$2\sqrt{25}+8 = 18$$
$$2\cdot 5+8 = 18$$
$$10+8 = 18$$
$$18 = 18$$

Since this statement is true, $x = 20$ is a solution to the equation.

EXAMPLES

EXAMPLE 1:

Solve for x: $\sqrt{2x-3} = 5$.

Answer:

Start by squaring both sides of the equation.

$$\left(\sqrt{2x+3}\right)^2 = 5^2$$

Squaring a square root eliminates the square root symbol.

$$2x - 3 = 25$$

Now it is a two-step equation. Add 3 to both sides.

$$2x = 28$$

Then, divide both sides by 2.

$$x = 14$$

Now, check this solution by plugging it back into the original equation.

$$\begin{aligned} \sqrt{2(14) - 3} &= 5 \\ \sqrt{28 - 3} &= 5 \\ \sqrt{25} &= 5 \\ 5 &= 5 \end{aligned}$$

Since this statement is true, $x = 14$ is a solution to the equation.

EXAMPLE 2:

Solve for x: $2\sqrt{8x + 1} - 8 = 2$

Answer:

Start by adding 8 to both sides.

$$2\sqrt{8x + 1} = 10$$

Now, divide both sides by 2.

$$\sqrt{8x + 1} = 5$$

Eliminate the radical by squaring both sides of the equation.

$$\begin{aligned} \left(\sqrt{8x + 1}\right)^2 &= 5^2 \\ 8x + 1 &= 25 \end{aligned}$$

Now x can be isolated in two steps. Subtract 1 from both sides.

$$8x = 24$$

Then, divide both sides by 8.

$$x = 3$$

Now, check this solution by plugging it back into the original equation.

$$\begin{aligned} 2\sqrt{8(3) + 1} - 8 &= 2 \\ 2\sqrt{24 + 1} - 8 &= 2 \\ 2\sqrt{25} - 8 &= 2 \\ 2 \cdot 5 - 8 &= 2 \end{aligned}$$

$$10 - 8 = 2$$
$$2 = 2$$

Since this statement is true, $x = 3$ is a solution to the equation.

EXAMPLE 3:

Solve for x: $\sqrt{9 - x} = \sqrt{1 - 9x}$

Answer:

Start by squaring both sides of the equation. Squaring a square root eliminates the radical.

$$\left(\sqrt{9 - x}\right)^2 = \left(\sqrt{1 - 9x}\right)^2$$
$$9 - x = 1 - 9x$$

Collect all the variables on one side and the constants on the other side.

$$-x = -8 - 9x$$
$$8x = -8$$

Divide both sides by 8.

$$x = -1$$

Now, check this solution by plugging it back into the original equation.

$$\sqrt{9 - (-1)} = \sqrt{1 - 9(-1)}$$
$$\sqrt{9 + 1} = \sqrt{1 + 9}$$
$$\sqrt{10} = \sqrt{10}$$

Since this statement is true, $x = -1$ is a solution to the equation.

EXAMPLE 4:

Solve for x: $\sqrt{3x + 10} = x + 2$

Answer:

First, undo the radical by squaring both sides of the equation.

$$\left(\sqrt{3x + 10}\right)^2 = (x + 2)^2$$
$$3x + 10 = x^2 + 4x + 4$$

Now, move all terms to the right side of the equation so that there is 0 on the left side.

$$0 = x^2 + x - 6$$

From here, factor the right side of the equation.

$$0 = (x - 2)(x + 3)$$

Now, solve this equation.

$$x = 2, -3$$

Since we are working with a radical equation, substitute each value for x into the original equation and simplify to check for extraneous solutions. Start with $x = 2$.

$$\sqrt{3(2) + 10} = (2) + 2$$
$$\sqrt{6 + 10} = 2 + 2$$
$$\sqrt{16} = 4$$
$$4 = 4$$

Since this statement is true, $x = 2$ is a solution to the equation. Now, check $x = -3$.

$$\sqrt{3(-3) + 10} = (-3) + 2$$
$$\sqrt{-9 + 10} = -3 + 2$$
$$\sqrt{1} = -1$$
$$1 = -1$$

Since this statement is not true, $x = -3$ is an extraneous solution for the equation. Therefore, only $x = 2$ is a true solution.

Practice Questions

1. Solve the equation $\sqrt{3x - 5} = 7$.

a. $x = 4$

b. $x = 16$

c. $x = 18$

d. $x = 21$

2. Solve the equation $\sqrt{2x - 1} = 9$.

a. $x = 5$

b. $x = 18$

c. $x = 24$

d. $x = 41$

3. Solve the equation $4\sqrt{6x - 2} + 5 = 13$.

a. $x = 1$

b. $x = 4$

c. $x = 6$

d. No real solutions

4. Solve the equation $2\sqrt{9x + 4} - 4 = 10$.

a. $x = 49$

b. $x = 45$

c. $x = 7$

d. $x = 5$

5. Solve the equation $3\sqrt{2x+8} - 9 = 27$.

a. $x = 68$

b. $x = 144$

c. $x = 12$

d. $x = 136$

6. Solve the equation $\sqrt{4x+5} = \sqrt{x-10}$.

a. $x = 5$

b. $x = -15$

c. $x = 15$

d. No real solutions

7. Solve the equation $\sqrt{x-9} = 4$.

a. $x = 25$

b. $x = -5$

c. $x = 7$

d. No real solutions

8. Solve the equation $\sqrt{x+17} = 3$.

a. $x = 20$

b. $x = -14$

c. $x = 26$

d. $x = -8$

9. Solve the equation $x - \sqrt{8-2x} = 0$.

a. $x = -4$

b. $x = 2$ or $x = -4$

c. $x = \frac{8}{3}$

d. $x = 2$

10. Solve the equation $\sqrt{2x-6} + 3 = x - 3$.

a. $x = 3$ or $x = 5$

b. $x = 5$

c. $x = -3$ or $x = -5$

d. $x = 3$

11. Solve the equation $\sqrt{4x-13} = \sqrt{3x-1}$.

a. $x = 2$

b. $x = -12$

c. $x = -14$

d. $x = 12$

12. Solve the equation $\sqrt{2x^2 + 5x - 7} - \sqrt{x^2 + 5x - 3} = 0$ for all real solutions.

a. $x = -2$
b. $x = 2$
c. $x = \sqrt{10}$
d. $x = 2$ or $x = -2$

13. Solve the equation: $\sqrt{4x - 4} = 6$.

a. $x = 40$
b. $x = 10$
c. $x = 36$
d. No real solutions

14. Solve the equation: $2\sqrt{3x - 8} + 4 = 12$.

a. $x = 4$
b. $x = 16$
c. $x = 24$
d. $x = 8$

15. You begin with a rectangular piece of paper that has an area of x square centimeters. By cutting off a piece whose area is 11 square centimeters, you produce a square in which each side measures 7 centimeters. Find x, the area of the original rectangle.

a. $x = 60$ square inches
b. $x = 18$ square inches
c. $x = 38$ square inches
d. $x = 2\sqrt{15}$ square inches

10.3 Adding and Subtracting Radical Expressions

The important rule for adding and subtracting radicals is if the radicals are the same, the coefficients can be added or subtracted. Radicals are considered to be the same if they have the same index and the same value under the radical. If the radicals are not the same, they cannot be combined. However, some radicals can be simplified to become like radicals, allowing for adding or subtracting. This is like the rule for adding and subtracting like terms.

For example, $3\sqrt{5} + 2\sqrt{5}$ is an expression with the same radical, $\sqrt{5}$. Since the radicals are the same, we can simply add the coefficients. $3\sqrt{5} + 2\sqrt{5} = 5\sqrt{5}$. Similarly, $7\sqrt{5} - 3\sqrt{5}$ has the same radical, so the coefficients can be subtracted: $7\sqrt{5} - 3\sqrt{5} = 4\sqrt{5}$.

Sometimes the radicals will be different, such as $3\sqrt{8} + 5\sqrt{18}$. In this case, the goal is to simplify the radicals to make them the same.

$3\sqrt{8}$ is equivalent to $3\sqrt{4}\sqrt{2}$, which simplifies to $6\sqrt{2}$.

$5\sqrt{18}$ is equivalent to $5\sqrt{9}\sqrt{2}$, which simplifies to $15\sqrt{2}$.

Now the radicals are the same, so we can add the two terms.

$$6\sqrt{2} + 15\sqrt{2} = 21\sqrt{2}$$

Let's look at an example with larger numbers. For example, $2\sqrt{50} - 6\sqrt{8}$.

Look at the radicals and try to break them down. The goal is to break them down so that the only thing left is a common radical.

$2\sqrt{50}$ is equivalent to $2\sqrt{25}\sqrt{2}$, which simplifies to $10\sqrt{2}$.

$6\sqrt{8}$ is equivalent to $6\sqrt{4}\sqrt{2}$, which simplifies to $12\sqrt{2}$.

Notice how we are left with radicals that are the same, $\sqrt{2}$. That is the goal. Now we can subtract the coefficients.

$$10\sqrt{2} - 12\sqrt{2} = -2\sqrt{2}$$

Review Video: Adding and Subtracting Radical Expressions
Visit mometrix.com/academy and enter code: 752176

Examples

Example 1:

Add the radicals: $2\sqrt{3} + 5\sqrt{3}$.

Answer:

The terms can be combined because the same value is under the radical.

$$2\sqrt{3} + 5\sqrt{3} = 7\sqrt{3}$$

EXAMPLE 2:

Subtract the radicals: $2\sqrt{27} - 5\sqrt{3}$.

Answer:

Simplify the first radical.

$$2\sqrt{9}\sqrt{3} - 5\sqrt{3}$$

The square root of 9 is 3, so pull this out from the radical.

$$2 \cdot 3\sqrt{3} - 5\sqrt{3} = 6\sqrt{3} - 5\sqrt{3}$$

Now the radicals can be combined because the radicals are the same.

$$6\sqrt{3} - 5\sqrt{3} = 1\sqrt{3} = \sqrt{3}$$

EXAMPLE 3:

Subtract the radicals: $6\sqrt{20} - 4\sqrt{18}$.

Answer:

Start by simplifying each radical.

$$6\sqrt{20} = 6\sqrt{4}\sqrt{5} = 6 \cdot 2\sqrt{5} = 12\sqrt{5}$$
$$4\sqrt{18} = 4\sqrt{9}\sqrt{2} = 4 \cdot 3\sqrt{2} = 12\sqrt{2}$$

Now, rewrite the expression.

$$12\sqrt{5} - 12\sqrt{2}$$

Since the radicals are not the same, the expression cannot be simplified any further.

EXAMPLE 4:

Simplify the expression $6x\sqrt{8x} + 2\sqrt{18x^3}$. Assume x is nonnegative.

Answer:

First, simplify each radical.

$$6x\sqrt{4 \cdot 2x} + 2\sqrt{9 \cdot 2 \cdot x^2 \cdot x}$$
$$6x \cdot 2\sqrt{2x} + 2 \cdot 3 \cdot x\sqrt{2x}$$
$$12x\sqrt{2x} + 6x\sqrt{2x}$$

Now, add the coefficients since the radicals are the same.

$$(12x + 6x)\sqrt{2x} = 18x\sqrt{2x}$$

PRACTICE QUESTIONS

1. Simplify the expression $4\sqrt{6} + 8\sqrt{6}$.

a. $12\sqrt{12}$

b. 72

c. $12\sqrt{6}$

d. $24\sqrt{3}$

2. Simplify the expression $20\sqrt{42} + 30\sqrt{42} + 11\sqrt{42}$.

a. $183\sqrt{14}$

b. $103\sqrt{42}$

c. $61\sqrt{126}$

d. $61\sqrt{42}$

3. Simplify the expression $-12\sqrt{17x} + 9\sqrt{17x}$. Assume x is nonnegative.

a. $-51x$

b. $-\sqrt{34x}$

c. $-72\sqrt{17x}$

d. $-3\sqrt{17x}$

4. Simplify the expression $10\sqrt{10} - 5\sqrt{10}$.

a. $5\sqrt{10}$

b. $-50\sqrt{10}$

c. $10\sqrt{5}$

d. 5

5. Simplify the expression $12\sqrt{21} - 16\sqrt{21} + \sqrt{21}$.

a. $-4\sqrt{21}$

b. $-3\sqrt{21}$

c. $-5\sqrt{21}$

d. $-4 + \sqrt{21}$

6. Simplify the expression $(8\sqrt{3} - 10\sqrt{5y}) - (7\sqrt{3} - 12\sqrt{5y})$. Assume y is nonnegative.

a. 3

b. $\sqrt{3} + 2\sqrt{5y}$

c. $\sqrt{6} + 2\sqrt{10y}$

d. $3\sqrt{3 + 5y}$

7. Simplify the expression $14\sqrt{2} + 3\sqrt{50}$.

a. $34\sqrt{13}$
b. $89\sqrt{2}$
c. $29\sqrt{2}$
d. 58

8. Simplify the expression $-15\sqrt{40} + 25\sqrt{10}$.

a. $10\sqrt{30}$
b. $-5\sqrt{10}$
c. $-35\sqrt{10}$
d. $50\sqrt{2}$

9. Simplify the expression $9y\sqrt{3y} + 4\sqrt{27y^3}$. Assume y is nonnegative.

a. $13\sqrt{30(y + y^3)}$
b. $45y\sqrt{3y}$
c. $21y\sqrt{3y}$
d. $117y^2$

10. Simplify the expression $18\sqrt{7} - 8\sqrt{63}$.

a. $-6\sqrt{7}$
b. -6
c. $20\sqrt{14}$
d. $-54\sqrt{7}$

11. Simplify the expression $16\sqrt{132} - 31\sqrt{33}$.

a. $\sqrt{33}$
b. 1
c. $-45\sqrt{11}$
d. $33\sqrt{33}$

12. Simplify the expression $26x^2\sqrt{5x} - 2\sqrt{500x^5}$. Assume x is nonnegative.

a. $6x^2\sqrt{10x}$
b. $6x^2\sqrt{5x}$
c. $6x^2$
d. $-174\sqrt{5x}$

13. A rectangle has length $\sqrt{10}$ cm and width $\sqrt{90}$ cm. Find the perimeter of the rectangle. Simplify your answer.

a. $4\sqrt{10}$ cm

b. $8\sqrt{10}$ cm

c. $20\sqrt{10}$ cm

d. 40 cm

14. You are constructing two small square pens, each with an area of 35 square feet, and one large square pen with an area of 140 square feet. How many feet of fencing will you need altogether? Give an exact answer.

a. $24\sqrt{35}$ ft

b. $4\sqrt{35}$ ft

c. $16\sqrt{35}$ ft

d. $16\sqrt{210}$ ft

15. When a heavy, dense object is dropped and allowed to fall freely, the time that it takes to fall a distance of d feet is approximately $t = \sqrt{\frac{d}{16}}$ seconds. You drop a brick from the top of a tower. How much time passes between the moment when the brick is 32 feet below you and the moment when it is 128 feet below you? Give an exact answer.

a. $\sqrt{2}$ seconds

b. $\sqrt{6}$ seconds

c. 2 seconds

d. $\frac{\sqrt{2}}{4}$ seconds

10.4 Multiplying Radical Expressions

When multiplying radical expressions, there are a few important rules to keep in mind. Remember, the term under the radical symbol is called a **radicand**. When you multiply radicals with the same index, you keep the product under the radical symbol. For example, $\sqrt{7} \times \sqrt{3}$ becomes $\sqrt{21}$. Multiply the radicands and keep the product under the radical symbol.

Some radical expressions will contain a coefficient in front of the radical. When multiplying radicals with coefficients, multiply the coefficient by the coefficient, and the radicand by the radicand. For example, when multiplying $3\sqrt{5} \times 2\sqrt{7}$, multiply 3×2 and then $\sqrt{5} \times \sqrt{7}$.

$$3\sqrt{5} \times 2\sqrt{7} = 6\sqrt{35}$$

Radical expressions can also include binomials. These more advanced radical expressions will require the distributive property, and potentially the FOIL process.

Let's look at a radical expression that requires the distributive property.

$4(3 + \sqrt{7})$ can be simplified by distributing 4 into all terms within the parentheses. This means 4×3 and $4 \times \sqrt{7}$.

$$4 \times 3 = 12$$
$$4 \times \sqrt{7} = 4\sqrt{7}$$

The result is $12 + 4\sqrt{7}$.

A similar process can be used when radical binomials are multiplied. When binomials are multiplied, the FOIL process can be used. Remember, FOIL stands for First, Outer, Inner, Last. Multiply the terms in this order so that the expression becomes fully distributed.

For example, the radical expression $(2 + \sqrt{7})(3 - \sqrt{7})$ can be simplified by FOILing.

F: $2 \times 3 = 6$

O: $2 \times (-\sqrt{7}) = -2\sqrt{7}$

I: $\sqrt{7} \times 3 = 3\sqrt{7}$

L: $\sqrt{7} \times (-\sqrt{7}) = -7$

Remember to incorporate the signs that are in front of each term. This + or − is attached to the term.

Therefore, $(2 + \sqrt{7})(3 - \sqrt{7})$ becomes $6 - 2\sqrt{7} + 3\sqrt{7} - 7$. Now combine like terms, and it becomes $\sqrt{7} - 1$ or $-1 + \sqrt{7}$.

EXAMPLES

EXAMPLE 1:

Multiply. $4\sqrt{3} \times 3\sqrt{2}$

Answer:

Multiply the numbers outside of the radicals and under the radicals.

$$4\sqrt{3} \times 3\sqrt{2} = 12\sqrt{6}$$

EXAMPLE 2:

Multiply. $8\sqrt{3} \times 9\sqrt{27}$

Answer:

Multiply the numbers outside of the radical.

$$8\sqrt{3} \times 9\sqrt{27} = 72\sqrt{3}\sqrt{27}$$

Write the square root of 27 using a perfect square factor.

$$72\sqrt{3}\sqrt{27} = 72\sqrt{3}\sqrt{3}\sqrt{9}$$

The square root of 9 is 3.

$$3 \times 72\sqrt{3}\sqrt{3} = 216\sqrt{3}\sqrt{3}$$

$\sqrt{3}\sqrt{3} = \sqrt{9}$ simplifies to 3, so this expression can be simplified further.

$$216 \times 3 = 648$$

EXAMPLE 3:

Multiply the radical expression: $(3 + \sqrt{2})(4 + \sqrt{2})$.

Answer:

The expression can be simplified by multiplying using the FOIL method.

First: $3 \times 4 = 12$

Outer: $3 \times \sqrt{2} = 3\sqrt{2}$

Inner: $\sqrt{2} \times 4 = 4\sqrt{2}$

Last: $\sqrt{2} \times \sqrt{2} = 2$

$$12 + 3\sqrt{2} + 4\sqrt{2} + 2 = 14 + 7\sqrt{2}$$

PRACTICE QUESTIONS

1. Simplify the product $3\sqrt{11} \cdot 4\sqrt{14}$.

a. $24\sqrt{38}$
b. $12\sqrt{154}$
c. $24\sqrt{39}$
d. 60

2. Simplify the product $5\sqrt{6} \cdot 8\sqrt{14}$.

a. $80\sqrt{21}$
b. $80\sqrt{5}$
c. $160\sqrt{21}$
d. $40\sqrt{84}$

3. Simplify the product $(-4\sqrt{30}) \cdot (2\sqrt{6})$.

a. -48
b. $-48\sqrt{5}$
c. $-288\sqrt{5}$
d. $-16\sqrt{6}$

4. Simplify the product $3\sqrt{2} \cdot \sqrt{6}$.

a. $6\sqrt{3}$
b. $12\sqrt{3}$
c. $8\sqrt{3}$
d. $6\sqrt{2}$

5. Simplify the product $3\sqrt{10} \cdot 2\sqrt{15}$.

a. $150\sqrt{6}$
b. 30
c. $30\sqrt{6}$
d. $25\sqrt{6}$

6. Simplify the product $(5\sqrt{7})(3\sqrt{70})$.

a. $15\sqrt{77}$
b. $515\sqrt{10}$
c. $56\sqrt{10}$
d. $105\sqrt{10}$

7. Simplify the product $(10\sqrt{3}) \cdot (-2\sqrt{21})$.

a. $-180\sqrt{7}$

b. $-60\sqrt{7}$

c. $-40\sqrt{6}$

d. $-60\sqrt{2}$

8. Simplify the product $\sqrt{2}(\sqrt{14} + \sqrt{22})$.

a. $2\sqrt{7} + 2\sqrt{11}$

b. $4 + 2\sqrt{6}$

c. $6\sqrt{2}$

d. $4\sqrt{7} + 4\sqrt{11}$

9. Simplify the product $\sqrt{6}(\sqrt{10} + \sqrt{15})$.

a. $2\sqrt{15} + 3\sqrt{10}$

b. $4\sqrt{15} + 9\sqrt{10}$

c. $\sqrt{60} + \sqrt{90}$

d. $5\sqrt{6}$

10. Simplify the product $(2 + \sqrt{10})(3 + \sqrt{10})$.

a. $6 + 7\sqrt{10}$

b. $16 + 5\sqrt{10}$

c. 16

d. $6 + 15\sqrt{10}$

11. Simplify the product $(6 - \sqrt{15})(5 - \sqrt{15})$.

a. $30 - 9\sqrt{15}$

b. $15 - 11\sqrt{15}$

c. $30 - 13\sqrt{15}$

d. $45 - 11\sqrt{15}$

12. Simplify the product $(7 + \sqrt{2})(7 - \sqrt{2})$.

a. $49 + 14\sqrt{2}$

b. 45

c. 47

d. $49 - 14\sqrt{2}$

13. A rectangular circuit board has a length of $4\sqrt{35}$ mm and a width of $2\sqrt{5}$ mm. Find the exact area of the circuit board. Simplify your answer.

a. $40\sqrt{7}$ mm^2

b. $30\sqrt{7}$ mm^2

c. $200\sqrt{7}$ mm^2

d. $16\sqrt{10}$ mm^2

14. The height of a triangular window from the base to the top of the window is $\sqrt{33}$ feet and the base of the window is $6\sqrt{3}$ feet long. Find the exact area of the window. Simplify your answer.

a. 18 square feet

b. $9\sqrt{11}$ square feet

c. $3\sqrt{99}$ square feet

d. $27\sqrt{11}$ square feet

15. A rectangular sheet of origami paper has a length of $8 + \sqrt{3}$ cm and a width of $6 + \sqrt{3}$ cm. Find the exact area of the sheet. Simplify your answer.

a. $51 + 16\sqrt{3}$ cm^2

b. $51 + 14\sqrt{3}$ cm^2

c. 51 cm^2

d. $14 + 2\sqrt{3}$ cm^2

10.5 Dividing Radical Expressions

When dividing radical expressions, there are a few important rules to consider. When dividing a radical by a radical, both radicals can be combined into one "tall" radical if they have the same index. If the index is not provided, it is assumed that it is 2, or a square root.

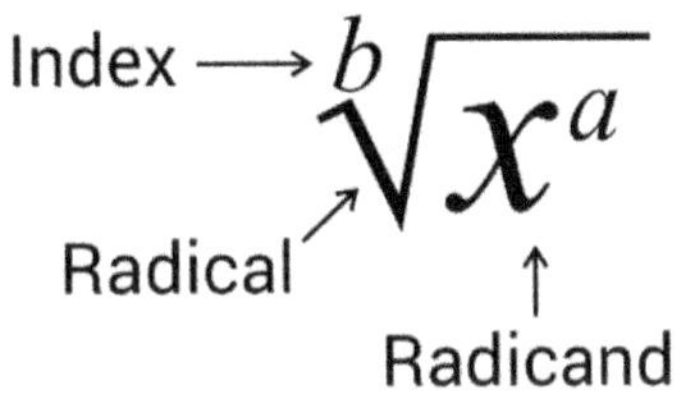

For example, $\frac{\sqrt{6}}{\sqrt{2}}$ has the same index, so it can be combined under one tall radical.

$$\frac{\sqrt{6}}{\sqrt{2}} = \sqrt{\frac{6}{2}}$$

The fraction $\frac{6}{2}$ simplifies to 3.

$$\sqrt{\frac{6}{2}} = \sqrt{3}$$

Let's look at another radical expression with division. Sometimes radicals in the denominator can be eliminated by combining radicals. For example, $\frac{2\sqrt{14}}{\sqrt{7}}$ can be rewritten by combining the radicals.

$$\frac{2\sqrt{14}}{\sqrt{7}} = 2\sqrt{\frac{14}{7}} = 2\sqrt{2}$$

More advanced division expressions will have coefficients and radicals in the numerator and denominator. For example, consider the radical expression $\frac{-\frac{16}{\sqrt{2}}}{\frac{12}{\sqrt{14}}}$. Start by rewriting the division as multiplication by taking the reciprocal of the second fraction.

$$\frac{\left(-\frac{16}{\sqrt{2}}\right)}{\left(\frac{12}{\sqrt{14}}\right)} = \left(-\frac{16}{\sqrt{2}}\right) \div \left(\frac{12}{\sqrt{14}}\right) = \left(-\frac{16}{\sqrt{2}}\right) \times \left(\frac{\sqrt{14}}{12}\right) = -\frac{16\sqrt{14}}{12\sqrt{2}}$$

Now we see that the radicals $\sqrt{2}$ and $\sqrt{14}$ can be combined under one tall radical.

$$-\frac{16\sqrt{14}}{12\sqrt{2}} = -\frac{16}{12}\sqrt{\frac{14}{2}}$$

Reduce each fraction to simplify.

$$-\frac{4}{3}\sqrt{7}$$

Examples

Example 1:

$$\frac{3\sqrt{20}}{2\sqrt{4}}$$

Answer:

Divide the numbers under the radical.

$$\frac{3\sqrt{20}}{2\sqrt{4}} = \frac{3}{2}\sqrt{\frac{20}{4}} = \frac{3}{2}\sqrt{5}$$

Example 2:

$$\frac{\sqrt{15x}}{3\sqrt{10x}}$$

Answer:

Split apart each radical.

$$\frac{\sqrt{3}\sqrt{5x}}{3\sqrt{2}\sqrt{5x}}$$

$\frac{\sqrt{5x}}{\sqrt{5x}}$ divides to 1, or cancels.

$$\frac{\sqrt{3}}{3\sqrt{2}}$$

The proper way to write radicals is by rationalizing the denominator so there is no longer a radical in the denominator. However, this step will be covered in the next section, so for now it is okay to leave the answer unrationalized.

Practice Questions

1. Simplify the quotient $\frac{\sqrt{20}}{10}$.

a. $\frac{2\sqrt{5}}{5}$

b. $\frac{\sqrt{10}}{5}$

c. $\sqrt{2}$

d. $\frac{\sqrt{5}}{5}$

2. Simplify the quotient $\frac{5\sqrt{72}}{9}$.

a. $\frac{10\sqrt{2}}{3}$
b. $20\sqrt{2}$
c. $\frac{5\sqrt{8}}{3}$
d. $10\sqrt{2}$

3. Simplify the quotient $\frac{\sqrt{50x^3}}{15x}$. Assume x is positive.

a. $\frac{x^2\sqrt{10}}{3}$
b. $\frac{x\sqrt{10}}{3}$
c. $\frac{\sqrt{2x}}{3}$
d. $\frac{5\sqrt{2x}}{3}$

4. Simplify the quotient $\frac{14}{\sqrt{28}}$.

a. $\frac{7}{\sqrt{7}}$
b. $\frac{14}{\sqrt{7}}$
c. $\frac{\sqrt{7}}{7}$
d. $7\sqrt{7}$

5. Simplify the quotient $\frac{18}{\sqrt{36}}$. Assume x is positive.

a. $\frac{1}{3}$
b. 18
c. 3
d. $\frac{1}{2}$

6. Simplify the quotient $\frac{3x}{\sqrt{6x^2}}$. Assume x is positive.

a. $\frac{\sqrt{6}}{2}$
b. $\frac{3x}{\sqrt{6}}$
c. $\frac{\sqrt{2}}{2}$
d. $\frac{3}{\sqrt{6}}$

7. Simplify the quotient $\frac{\sqrt{8}+\sqrt{18}}{\sqrt{32}}$.

a. $2\sqrt{2}$

b. $\frac{\sqrt{13}}{4}$

c. $\frac{5}{4}$

d. $\frac{3\sqrt{2}}{2}$

8. Simplify the quotient $\frac{\sqrt{15}}{\sqrt{21}}$.

a. $\frac{5}{\sqrt{7}}$

b. $\frac{\sqrt{5}}{7}$

c. $\frac{\sqrt{315}}{21}$

d. $\frac{\sqrt{5}}{\sqrt{7}}$

9. Simplify the quotient $\frac{2\sqrt{3}}{3\sqrt{6}}$.

a. $\frac{2}{\sqrt{6}}$

b. $\frac{2}{6}$

c. $\frac{2}{3\sqrt{3}}$

d. $\frac{2}{3\sqrt{2}}$

10. Simplify the quotient $\frac{3\sqrt{105}}{8\sqrt{30}}$.

a. $\frac{21}{16}$

b. $\frac{3\sqrt{14}}{8}$

c. $\frac{3\sqrt{7}}{8\sqrt{2}}$

d. $\frac{\sqrt{42}}{16}$

11. Simplify the quotient $\frac{8\sqrt{12}-5\sqrt{27}}{4\sqrt{6}}$.

a. $2\sqrt{2}-15\sqrt{3}$

b. $\frac{1}{4\sqrt{2}}$

c. $\frac{3\sqrt{15}}{4\sqrt{2}}$

d. $-\frac{15\sqrt{2}}{8}$

12. Simplify the quotient $\frac{9\sqrt{12x^3}}{2\sqrt{15x}}$. Assume x is positive.

a. $\frac{9x^2}{\sqrt{5}}$
b. $\frac{18x}{\sqrt{5}}$
c. $\frac{18x^2}{\sqrt{5}}$
d. $\frac{9x}{\sqrt{5}}$

13. The length of one side of the triangle is $\frac{8\sqrt{2}}{\sqrt{16}}$ inches. Simplify this measurement by rationalizing the denominator.

a. $\frac{\sqrt{2}}{2}$ inches
b. $\frac{1}{2}$ inch
c. $2\sqrt{2}$ inches
d. $4\sqrt{2}$ inches

14. Jessica took a math quiz on rationalizing denominators. When she got her graded quiz paper returned, she saw that her teacher wrote her score with a radical in the denominator. Jessica scored $\frac{\sqrt{36}}{\sqrt{49}}$ on her quiz. Simplify this expression by rationalizing the denominator to figure out Jessica's quiz score.

a. Jessica got 6 out of 7 questions correct on her quiz.
b. Jessica got 36 out of 49 questions correct on her quiz.
c. Jessica got 49 out of 36 questions correct on her quiz.
d. Jessica got 7 out of 6 questions correct on her quiz.

15. The two diagonals of a parallelogram have lengths $\sqrt{22}$ cm and $\sqrt{26}$ cm. Find the ratio of the longer diagonal to the shorter diagonal.

a. $\frac{\sqrt{13}}{\sqrt{11}}$
b. $\frac{\sqrt{11}}{\sqrt{13}}$
c. $\sqrt{3}$
d. $\frac{2\sqrt{6}}{11}$

10.6 Rationalizing the Denominator

The phrase **rationalizing the denominator** refers to the process of eliminating any radicals from the denominator of a fraction. When you simplify a radical expression, the radicals should be removed from the denominator. This can be achieved by multiplying the expression by something that will cancel out the radicals. Remember, multiplying a square root by itself eliminates the radical. For example, multiplying $\sqrt{3} \times \sqrt{3}$ gives the result $\sqrt{9}$, which is 3. A square root squared will cancel out the square root.

Consider the radical expression $\frac{3}{\sqrt{5}}$. We have a radical in the denominator, so we need to address this. Think of a value to multiply $\sqrt{5}$ by so that the square root symbol will be eliminated. Multiplying $\sqrt{5}$ by itself will eliminate the radical.

$$\sqrt{5} \cdot \sqrt{5} = 5$$

This means we should multiply the original expression by the fraction $\frac{\sqrt{5}}{\sqrt{5}}$. Remember, to keep the value of our fraction the same, we must multiply by a value of 1, which is why we multiply the numerator and denominator by $\sqrt{5}$.

$$\frac{3 \cdot \sqrt{5}}{\sqrt{5} \cdot \sqrt{5}}$$

The denominator becomes 5, and the numerator becomes $3\sqrt{5}$. The result is $\frac{3\sqrt{5}}{5}$, which cannot be simplified any further.

Let's look at an example with radicals in the numerator and the denominator. For example, $\frac{\sqrt{28}}{\sqrt{3}}$. Notice the radical in the denominator. This needs to be eliminated by rationalizing the denominator. Eliminate the square root multiplying by $\frac{\sqrt{5}}{\sqrt{3}}$.

$$\frac{\sqrt{28} \cdot \sqrt{3}}{\sqrt{3} \cdot \sqrt{3}}$$

The denominator simplifies to 3.

$$\frac{\sqrt{28}\sqrt{3}}{3}$$

The numerator can also be simplified. Factor the perfect square 4 out of 28. The square root of 4 is 2.

$$\frac{\sqrt{4}\sqrt{7}\sqrt{3}}{3} = \frac{2\sqrt{7}\sqrt{3}}{3} = \frac{2\sqrt{21}}{3}$$

There is no longer a radical in the denominator, and the expression is fully simplified.

Examples

Example 1:

$$\frac{4}{\sqrt{5}}$$

Answer:

The radical in the denominator needs to be eliminated. Multiply the expression by $\frac{\sqrt{5}}{\sqrt{5}}$.

$$\frac{4\sqrt{5}}{\sqrt{5}\cdot\sqrt{5}}$$

The square root of 5 times the square root of 5 equals 5.

$$\frac{4\sqrt{5}}{5}$$

Example 2:

$$\frac{7\sqrt{11}}{\sqrt{2}}$$

Answer:

The denominator needs to be rationalized because there is currently a radical in the denominator. Multiply the expression by $\frac{\sqrt{2}}{\sqrt{2}}$.

$$\frac{7\sqrt{11}\cdot\sqrt{2}}{\sqrt{2}\cdot\sqrt{2}}$$

Multiply the numerator and denominator.

$$\frac{7\sqrt{22}}{2}$$

Practice Questions

1. Rationalize the denominator. If possible, simplify the fraction.

$$\frac{7}{\sqrt{2}}$$

a. $\frac{7}{4}$
b. $\frac{\sqrt{2}}{7}$
c. $\frac{7\sqrt{2}}{4}$
d. $\frac{7\sqrt{2}}{2}$

2. Rationalize the denominator. If possible, simplify the fraction.

$$\frac{2}{\sqrt{7}}$$

a. $\frac{2\sqrt{7}}{14}$
b. $\frac{2\sqrt{7}}{7}$
c. $\frac{2\sqrt{7}}{49}$
d. $\frac{14}{49}$

3. Rationalize the denominator. If possible, simplify the fraction.

$$\frac{12}{\sqrt{3}}$$

a. $2\sqrt{3}$
b. $4\sqrt{3}$
c. $\frac{4\sqrt{3}}{3}$
d. 12

4. Rationalize the denominator. If possible, simplify the fraction.

$$\frac{25}{\sqrt{5}}$$

a. $5\sqrt{5}$
b. $\sqrt{5}$
c. $\frac{25\sqrt{5}}{10}$
d. $2.5\sqrt{5}$

5. Rationalize the denominator. If possible, simplify the fraction.

$$\frac{12}{\sqrt{10}}$$

a. $\frac{12\sqrt{5}}{5}$
b. $\frac{12\sqrt{10}}{5}$
c. $\frac{6\sqrt{10}}{5}$
d. $\frac{6\sqrt{5}}{5}$

6. Rationalize the denominator. If possible, simplify the fraction.

$$\frac{1}{3\sqrt{2}}$$

a. $\frac{\sqrt{2}}{6}$
b. $\frac{\sqrt{1}}{3}$
c. $6\sqrt{2}$
d. $2\sqrt{6}$

7. Rationalize the denominator. If possible, simplify the fraction.

$$\frac{9}{\sqrt{15}}$$

a. $\frac{3\sqrt{15}}{5}$
b. $\frac{9\sqrt{15}}{30}$
c. $\frac{9\sqrt{15}}{17}$
d. $9\sqrt{15}$

8. Rationalize the denominator. If possible, simplify the fraction.

$$\frac{5}{2\sqrt{3}}$$

a. $\frac{5\sqrt{3}}{12}$
b. $\frac{5\sqrt{3}}{6}$
c. $\frac{5}{2}$
d. $\frac{5}{4}$

9. Rationalize the denominator. If possible, simplify the fraction.

$$\frac{4}{\sqrt{6}}$$

a. $\frac{\sqrt{6}}{2}$
b. $\frac{\sqrt{6}}{9}$
c. $\frac{\sqrt{6}}{3}$
d. $\frac{2\sqrt{6}}{3}$

10. Rationalize the denominator. If possible, simplify the fraction.

$$\frac{\sqrt{7}}{\sqrt{5}}$$

a. $\frac{\sqrt{35}}{5}$
b. $\frac{\sqrt{7}}{5}$
c. $\frac{\sqrt{7}}{25}$
d. $\frac{\sqrt{35}}{10}$

11. Rationalize the denominator. If possible, simplify the fraction.

$$\frac{5}{\sqrt{10}}$$

a. $\frac{\sqrt{10}}{20}$
b. $\frac{5\sqrt{10}}{12}$
c. $\frac{\sqrt{10}}{2}$
d. $\frac{\sqrt{10}}{4}$

12. Rationalize the denominator. If possible, simplify the fraction.

$$\frac{7}{\sqrt{12}}$$

a. $\frac{7\sqrt{3}}{12}$
b. $\frac{7\sqrt{12}}{24}$
c. $\frac{7\sqrt{3}}{6}$
d. $\frac{7\sqrt{12}}{144}$

13. A rectangular printed circuit board has an area of 100 mm^2 and a width of $\sqrt{70}$ mm. Find the exact length of the circuit board.

a. $10\sqrt{10}$ mm
b. $\frac{10\sqrt{70}}{7}$ mm
c. $100\sqrt{70}$ mm
d. $\frac{10\sqrt{7}}{7}$ mm

14. Christopher's family is eating pizza for dinner. While they're eating, Christopher gives them a math riddle. He tells them that he estimates that they have already eaten $\frac{8}{3\sqrt{10}}$ of the pizza. Simplify this expression by rationalizing the denominator.

a. $\frac{4\sqrt{10}}{30}$

b. $\frac{4\sqrt{10}}{15}$

c. $\frac{8\sqrt{10}}{30}$

d. $\frac{8\sqrt{10}}{15}$

15. A rock dropped from the top of a cliff falls $16t$ feet in $\sqrt{t}$ seconds $(t > 0)$. The average speed of the rock equals the total distance fallen divided by the total time required for the fall. Find the average speed of the rock during this time.

a. $4t$ feet per second

b. $16\sqrt{t}$ feet per second

c. $16t^2$ feet per second

d. $4\sqrt{t}$ feet per second

10.7 Simplifying Radical Expressions with Binomial Denominators

When simplifying radical expressions, the answer cannot contain radicals in the denominator. Some radical expressions will have binomial denominator, such as $\frac{1-\sqrt{3}}{5+\sqrt{3}}$.

Remember, when rationalizing the denominator of simpler expressions, we multiply the expression by a fraction that gets rid of the radicals in the denominator. To do this with a binomial denominator we multiply the expression by its conjugate. The **conjugate** is the result of "flipping the sign." For example, the conjugate of $5+\sqrt{3}$ is $5-\sqrt{3}$. When we multiply an expression by the conjugate, the middle terms cancel out when FOILed and the radicand becomes a perfect square.

For example, notice what happens to the denominator after we FOIL. Let's use the original example, $\frac{1-\sqrt{3}}{5+\sqrt{3}}$.

Multiply the numerator and denominator of the expression by the conjugate of the denominator.

$$\frac{(1-\sqrt{3})(5-\sqrt{3})}{(5+\sqrt{3})(5-\sqrt{3})}$$

When we FOIL the expression, we are left with the following expression.

$$\frac{5-\sqrt{3}-5\sqrt{3}+3}{25+5\sqrt{3}-5\sqrt{3}-3}$$

This seems like a much more complicated expression but notice how the terms will cancel out and simplify.

$$\frac{8-6\sqrt{3}}{25-3}=\frac{8-6\sqrt{3}}{22}$$

Reduce. 8, 6, and 22 can all be divided by 2.

$$\frac{4-3\sqrt{3}}{11}$$

This is now in simplest form with no radicals in the denominator.

EXAMPLES

EXAMPLE 1:

$$\frac{2}{5+\sqrt{2}}$$

Answer:

Recognize that there is a radical in the denominator that needs to be rationalized. Rationalize the denominator by multiplying by the conjugate. The conjugate of the denominator is $(5-\sqrt{2})$.

$$\frac{2(5-\sqrt{2})}{(5+\sqrt{2})(5-\sqrt{2})}$$

Distribute in the numerator and FOIL the denominator.

$$\frac{10 - 2\sqrt{2}}{25 + 5\sqrt{2} - 5\sqrt{2} - 2}$$

Simplify the denominator.

$$\frac{10 - 2\sqrt{2}}{23}$$

EXAMPLE 2:

$$\frac{\sqrt{3}}{\sqrt{3} + 2}$$

Answer:

Multiply the numerator and denominator of the expression by the conjugate of the denominator.

The conjugate of $\sqrt{3} + 2$ is $\sqrt{3} - 2$.

$$\frac{\sqrt{3}(\sqrt{3} - 2)}{(\sqrt{3} + 2)(\sqrt{3} - 2)}$$

Distribute in the numerator and FOIL the denominator.

$$\frac{3 - 2\sqrt{3}}{3 - 2\sqrt{3} + 2\sqrt{3} - 4}$$

Notice how the middle terms in the denominator will cancel out.

$$\frac{3 - 2\sqrt{3}}{-1}$$

Divide by -1 to fully simplify the expression.

$$-3 + 2\sqrt{3} = 2\sqrt{3} - 3$$

PRACTICE QUESTIONS

1. Simplify the expression by rationalizing the denominator.

$$\frac{1}{\sqrt{2} + 1}$$

a. $\sqrt{2} + 1$
b. $\sqrt{2} - 1$
c. $1 - \sqrt{2}$
d. $\sqrt{2}$

2. Simplify the expression by rationalizing the denominator.

$$\frac{4}{3+\sqrt{5}}$$

a. $\sqrt{3}-5$
b. $\sqrt{5}-3$
c. $3-\sqrt{5}$
d. $3+\sqrt{5}$

3. Simplify the expression by rationalizing the denominator.

$$\frac{2}{\sqrt{6}-3}$$

a. $-\frac{2\sqrt{6}+3}{3}$
b. $\frac{2\sqrt{6}+3}{3}$
c. $-\frac{2\sqrt{6}+6}{3}$
d. $\frac{2\sqrt{6}+6}{3}$

4. Simplify the expression by rationalizing the denominator.

$$\frac{4}{\sqrt{2}-2}$$

a. $2\sqrt{2}-4$
b. $-2\sqrt{2}-4$
c. $-2\sqrt{2}+4$
d. $2\sqrt{2}+4$

5. Simplify the expression by rationalizing the denominator.

$$\frac{3}{1+\sqrt{3}}$$

a. $\frac{3\sqrt{3}}{2}$
b. $\frac{3-3\sqrt{3}}{2}$
c. $-\frac{3\sqrt{3}}{2}$
d. $-\frac{3-3\sqrt{3}}{2}$

6. Simplify the expression by rationalizing the denominator.

$$\frac{3}{2-\sqrt{5}}$$

a. $-6-3\sqrt{5}$
b. $6+3\sqrt{5}$
c. $3\sqrt{5}-6$
d. $\frac{6+3\sqrt{5}}{-21}$

7. Simplify the expression by rationalizing the denominator.

$$\frac{5}{\sqrt{2}-\sqrt{3}}$$

a. $-5\sqrt{2}-5\sqrt{3}$
b. $5\sqrt{2}+5\sqrt{3}$
c. $5\sqrt{2}-5\sqrt{3}$
d. $-5\sqrt{2}+5\sqrt{3}$

8. Simplify the expression by rationalizing the denominator.

$$\frac{\sqrt{6}}{\sqrt{2}+3}$$

a. $-\frac{2\sqrt{3}-3\sqrt{6}}{7}$
b. $\frac{2\sqrt{3}-3\sqrt{6}}{7}$
c. $-\frac{3\sqrt{6}-2\sqrt{3}}{5}$
d. $\frac{2\sqrt{3}-3\sqrt{6}}{5}$

9. Simplify the expression by rationalizing the denominator.

$$\frac{2}{\sqrt{5}+2\sqrt{3}}$$

a. $\frac{2\sqrt{5}+4\sqrt{3}}{7}$
b. $-\frac{2\sqrt{5}+4\sqrt{3}}{7}$
c. $\frac{2\sqrt{5}-4\sqrt{3}}{7}$
d. $-\frac{2\sqrt{5}-4\sqrt{3}}{7}$

10. Simplify the expression by rationalizing the denominator.

$$\frac{\sqrt{3}}{5+2\sqrt{3}}$$

a. $\frac{5\sqrt{3}-6}{37}$
b. $\frac{5\sqrt{3}+6}{13}$
c. $\frac{5\sqrt{3}-6}{13}$
d. $\frac{5\sqrt{3}-23}{13}$

11. Simplify the expression by rationalizing the denominator.

$$\frac{-1+\sqrt{3}}{\sqrt{3}-1}$$

a. 2
b. 1
c. $\frac{-\sqrt{3}}{2}$
d. $\frac{\sqrt{3}}{2}$

12. Simplify the expression by rationalizing the denominator.

$$\frac{\sqrt{5}+3}{4-\sqrt{5}}$$

a. $\frac{12\sqrt{5}+17}{11}$
b. $\frac{7\sqrt{5}+17}{11}$
c. $\frac{12\sqrt{5}+60}{11}$
d. $\frac{7\sqrt{5}+60}{11}$

13. The margin of error is a statistic that tells the amount of random sampling error in a survey's results. A poll before a local election predicted that the incumbent candidate would win the election, with a margin of error of $\frac{2-\sqrt{3}}{4+\sqrt{3}}$. Rationalize the binomial denominator.

a. $\frac{11-6\sqrt{3}}{13}$
b. $\frac{11-6\sqrt{3}}{19}$
c. $\frac{11+2\sqrt{3}}{13}$
d. $\frac{11+2\sqrt{3}}{19}$

14. Holly and Theo are working on rationalizing binomial denominators. They each simplify the expression $\frac{\sqrt{3}-2}{\sqrt{3}+1}$, but they have come up with different answers. Holly says that after rationalizing the denominator and simplifying, there are 2 radicals in the numerator. Theo says that his answer only has 1 radical in the numerator. What is the simplified form of this expression, and who is correct?

a. Holly is correct. The simplified form of the expression is $\frac{3-\sqrt{3}-2\sqrt{3}+2}{2}$.

b. Theo is correct. The simplified form of the expression is $\frac{3-\sqrt{3}-2\sqrt{3}+2}{2}$.

c. Holly is correct. The simplified form of the expression is $\frac{5-3\sqrt{3}}{2}$.

d. Theo is correct. The simplified form of the expression is $\frac{5-3\sqrt{3}}{2}$.

15. The length of a rectangle is $\frac{5\sqrt{2}}{2\sqrt{5}-3}$ centimeters. Simplify this measurement by rationalizing the denominator.

a. $\frac{25\sqrt{12}}{36}$ centimeters

b. $\frac{25\sqrt{12}}{11}$ centimeters

c. $\frac{10\sqrt{10}+15\sqrt{2}}{11}$ centimeters

d. $\frac{10\sqrt{10}+15\sqrt{2}}{36}$ centimeters

Chapter 10 Quiz

1. Simplify the expression: $8\sqrt{7} - 6\sqrt{7}$.

a. $\sqrt{7}$
b. $14\sqrt{7}$
c. $-2\sqrt{7}$
d. $2\sqrt{7}$

2. Subtract and simplify: $2\sqrt{50} - 6\sqrt{2}$.

a. $4\sqrt{2}$
b. $2\sqrt{2}$
c. $4\sqrt{50}$
d. $2\sqrt{4}$

3. Simplify the expression: $3\sqrt{18} + 3\sqrt{5} - 2\sqrt{2}$.

a. $3\sqrt{5} + 7\sqrt{2}$
b. $\sqrt{5} + \sqrt{2}$
c. $5\sqrt{5} + 7\sqrt{2}$
d. $3\sqrt{5} + 3\sqrt{2}$

4. Divide the expression: $\frac{4\sqrt{14}}{8\sqrt{2}}$.

a. $\frac{\sqrt{3}}{2}$
b. $\frac{\sqrt{7}}{14}$
c. $\frac{\sqrt{7}}{2}$
d. $\frac{\sqrt{7}}{4}$

5. Simplify the expression: $\frac{\sqrt{24x^2}}{\sqrt{3x}}$.

a. $2\sqrt{2x^2}$
b. $24\sqrt{2x}$
c. $2\sqrt{2x}$
d. $2\sqrt{2}$

6. Which of the following radical expressions is NOT equivalent to $\frac{\sqrt{15}}{5\sqrt{20}}$?

a. $\frac{\sqrt{300}}{(5)(20)}$

b. $\frac{\sqrt{300}}{100}$

c. $\frac{\sqrt{3}}{10}$

d. $\sqrt{10}$

7. Multiply the expression: $\sqrt{3}(\sqrt{3}-5)$.

a. $5\sqrt{3}$

b. $3-5\sqrt{3}$

c. $3\sqrt{3}$

d. $3-\sqrt{3}$

8. Multiply the following expression: $2\sqrt{24}\times 3\sqrt{21}$.

a. $3\sqrt{4}$

b. $6\sqrt{504}$

c. $36\sqrt{4}$

d. $36\sqrt{14}$

9. You are purchasing a small amount of sod for a rectangular dry spot in your backyard. The dimensions of the dry spot are $3\sqrt{10}$ ft by $2\sqrt{6}$ ft. How many square feet of sod do you need to purchase?

a. $12\sqrt{5}\text{ ft}^2$

b. $19\sqrt{5}\text{ ft}^2$

c. $12\sqrt{12}\text{ ft}^2$

d. $12\sqrt{15}\text{ ft}^2$

10. Rationalize the denominator in the fraction $\frac{8}{3\sqrt{11}}$. Simplify your answer.

a. $\frac{8\sqrt{11}}{3}$

b. $\frac{4\sqrt{11}}{33}$

c. $\frac{8\sqrt{11}}{99}$

d. $\frac{8\sqrt{11}}{33}$

11. Rationalize the denominator in the fraction $\frac{10}{\sqrt{5}}$. Simplify your answer.

a. $2\sqrt{5}$
b. $\sqrt{2}$
c. $10\sqrt{5}$
d. $\frac{2\sqrt{5}}{5}$

12. A rectangle has an area of 180 square inches and a length of $8\sqrt{15}$ square inches. Find the exact width of the rectangle. Rationalize the denominator and simplify your answer.

a. $\frac{\sqrt{15}}{10}$ inches
b. $\frac{3\sqrt{15}}{2}$ inches
c. $1{,}440\sqrt{15}$ inches
d. $\frac{3}{2}$ inches

13. Simplify the following expression: $\sqrt{125x}$.

a. $5\sqrt{x}$
b. $\sqrt{5x}$
c. $5\sqrt{5x}$
d. $5\sqrt{12x}$

14. Simplify the radical expression: $\sqrt{50}$.

a. $5\sqrt{2}$
b. $2\sqrt{2}$
c. $5\sqrt{5}$
d. $\sqrt{2}$

15. Two students are measuring the wingspan of butterflies in their science class. Student A says the wingspan of the butterfly is $\sqrt{28}$ cm. Student B says the wingspan is $\sqrt{4}\sqrt{7}$ cm. Their teacher says the wingspan should measure $2\sqrt{7}$ cm. Which student is correct?

a. Student A is correct.
b. Student B is correct.
c. Both students are correct.
d. Neither student is correct.

16. Rationalize the denominator of the following expression and simplify your answer.

$$\frac{7}{4+\sqrt{11}}$$

a. $\frac{28-\sqrt{11}}{5}$
b. $\frac{28+7\sqrt{11}}{27}$
c. $\frac{7\sqrt{11}}{15}$
d. $\frac{28-7\sqrt{11}}{5}$

17. Rationalize the denominator of the following expression and simplify your answer as much as possible:

$$\frac{10}{5-\sqrt{5}}$$

a. $\frac{5-\sqrt{5}}{2}$
b. $\frac{5+\sqrt{5}}{2}$
c. $3\sqrt{5}$
d. $\frac{5+10\sqrt{5}}{2}$

18. A miniature robotic car travels 6 meters in a straight line at a constant speed of $7-\sqrt{47}$ meters per second. How long does it take the car to make this trip? Simplify your answer as much as possible and rationalize the denominator if applicable.

a. $21+3\sqrt{47}$ seconds
b. $\frac{\sqrt{47}}{9}$ seconds
c. $21-3\sqrt{47}$ seconds
d. $\frac{7+\sqrt{47}}{9}$ seconds

19. What value of x makes the following equation true?

$$4\sqrt{2x-6}+7=31$$

a. $x=21$
b. $x=24$
c. $x=36$
d. $x=42$

20. What value of x makes the following equation true?

$$3\sqrt{3x+9} - 11$$

a. $x = 9$
b. $x = 16$
c. $x = 24$
d. $x = 81$

Practice Test #1

1. Solve the inequality and graph your answer: $-2(3x - 5) \geq 58$.

a. $x < -8$

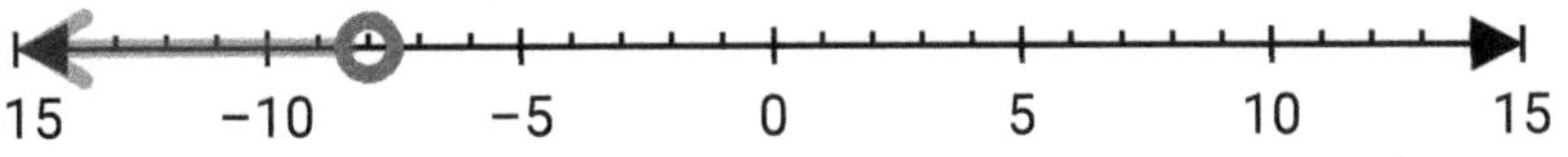

b. $x \leq -8$

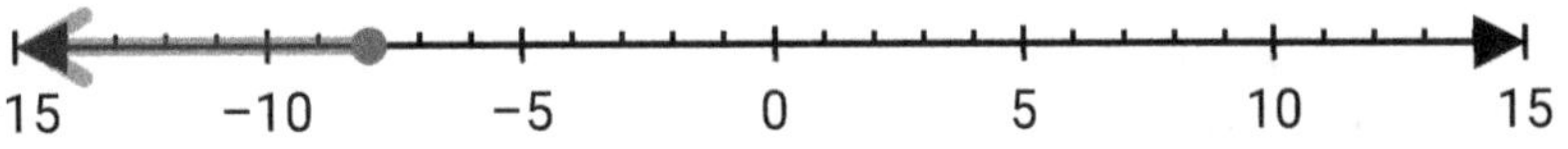

c. $x \geq -8$

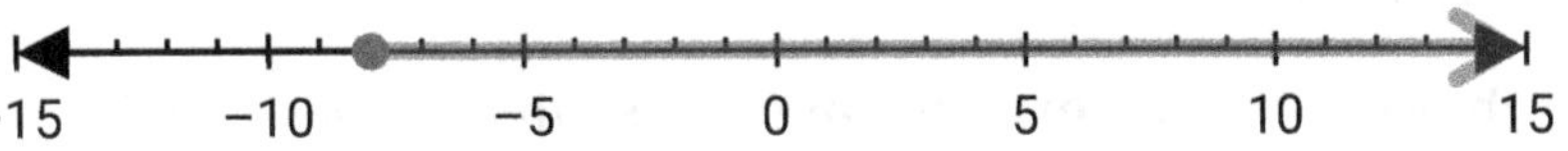

d. $x > -8$

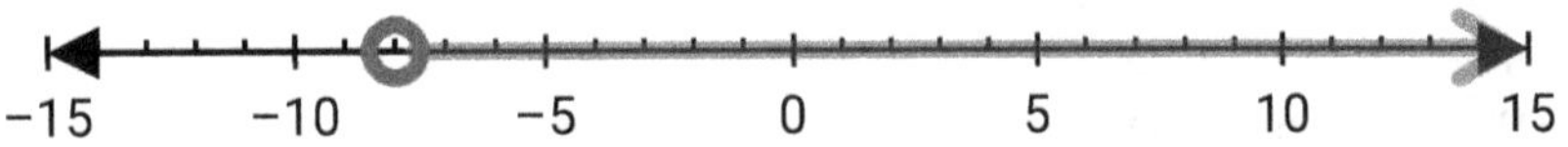

2. What is the value of x for the linear function $y = 3x - 7$ when the value of y is 5?

a. -2
b. 1
c. 4
d. 8

3. Rationalize the denominator in the fraction $\frac{30}{\sqrt{42}}$. Simplify your answer.

a. $\frac{5\sqrt{42}}{7}$
b. $\frac{5\sqrt{21}}{7}$
c. $5\sqrt{6}$
d. $\frac{10\sqrt{21}}{7}$

4. Simplify $\frac{9xy^5}{12x^4y^3}$.

a. $\frac{3y^8}{4x^5}$
b. $\frac{3y^2}{4x^4}$
c. $\frac{3y^2}{4x^3}$
d. $\frac{3x^3}{4y^2}$

5. Which of the following statements describes a variable relationship that shows correlation but not causation?

a. An elementary student's weight and the size of their vocabulary.
b. The speed of a vehicle and the amount of time it takes to drive 60 miles with no traffic.
c. The amount of time spent exercising and the number of calories burned.
d. The dimensions of a rectangular garden and the amount of square footage available for planting.

6. Juan bought a pack of gum at the corner store. He gave 13 pieces of gum to his friend and has 11 pieces left. How many pieces of gum did the pack originally contain?

a. 14 pieces of gum
b. 18 pieces of gum
c. 22 pieces of gum
d. 24 pieces of gum

7. Which choice is NOT an example of the distributive property?

a. $8(x - 5) = 8x - 40$
b. $-2(y - 1) = -2y + 2$
c. $3(4 + x) = 12 + 3x$
d. $6(9 - y) = 54 + 9y$

8. Write a linear equation in standard form that represents the graph shown below.

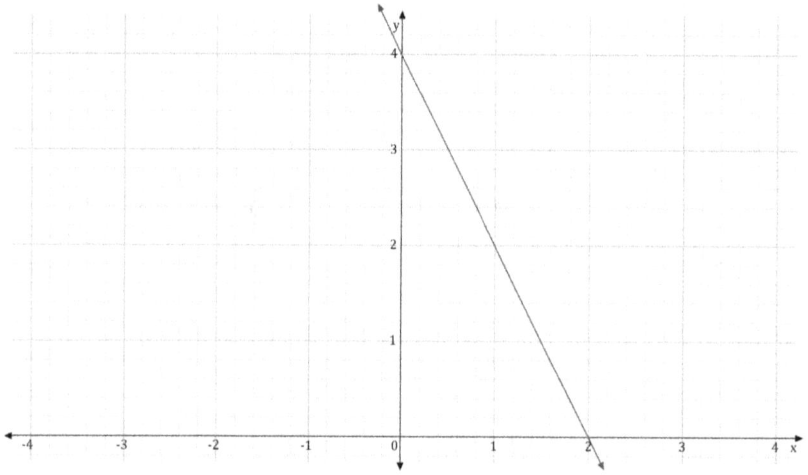

a. $2x + 4 = y$
b. $4x + y = 2$
c. $x + 2y = 4$
d. $2x + y = 4$

9. The square of 7 more than twice some number is 169. Which of the following could be the values of the number?

a. −3 or 10
b. −10 or 3
c. −13 or 13
d. −9 or 9

10. Solve the following quadratic using the complete the square method: $x^2 - 12x + 35 = 0$.

a. $x = -2$ or $x = 6$
b. $x = 2$ or $x = 6$
c. $x = 5$ or $x = 7$
d. $x = -7$ or $x = -5$

11. Poppy's Pizza charges \$10 for each large pizza, plus a delivery fee of \$3. The graph of this linear function passes through point $(3, 33)$ and has a slope of $m = 10$. Which graph shows the total cost for an order at Poppy's Pizza in relation to the number of pizzas ordered?

a.

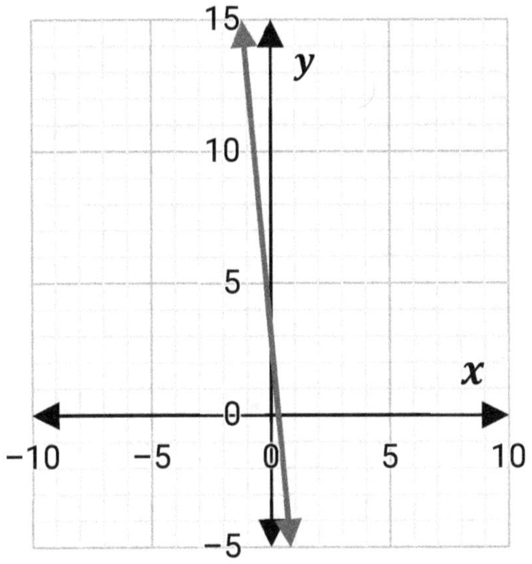

b.

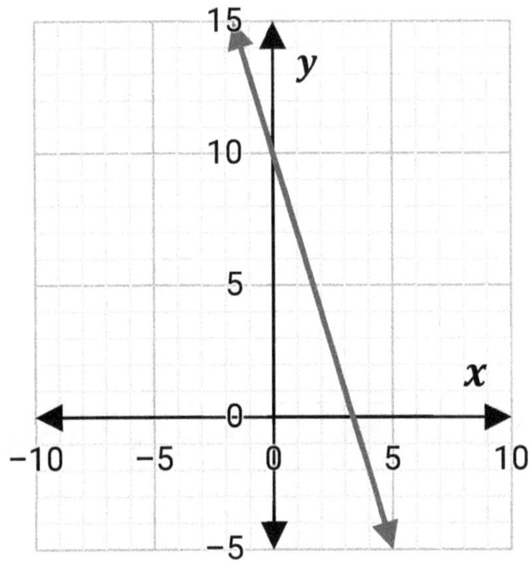

c.

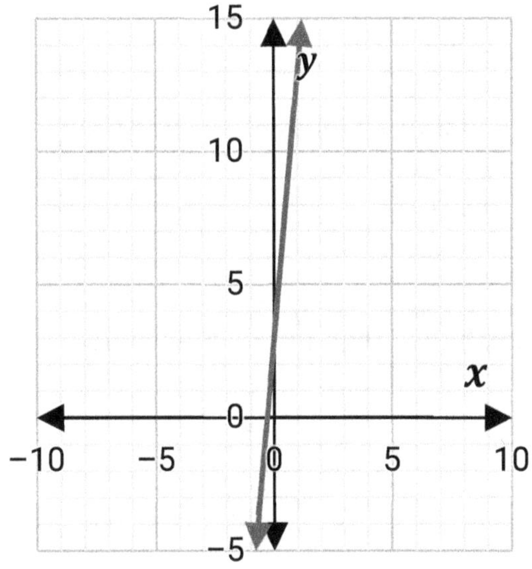

d.

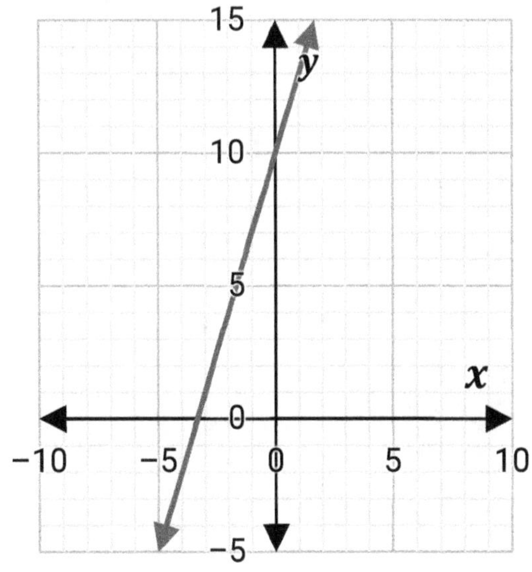

12. A rectangular pen has an area of $20x^2 + 55x$ square feet. The width of the pen is $5x$ feet. Find the length of the pen.

a. $100x^3 + 275x^2$ feet
b. $59x$ feet
c. $4x + 11$ feet
d. $4x^2 + 55$ feet

13. Simplify the expression: $\sqrt{50} + 5\sqrt{18}$.

a. $15\sqrt{2}$
b. $20\sqrt{2}$
c. $5\sqrt{2}$
d. $5\sqrt{68}$

14. Seven times a number x is less than 56. What values for the number make this true?

a. $x < 8$
b. $x \leq 8$
c. $x \geq 8$
d. $x > 8$

15. Graph the inequality: $y > -x + 1$.

a.
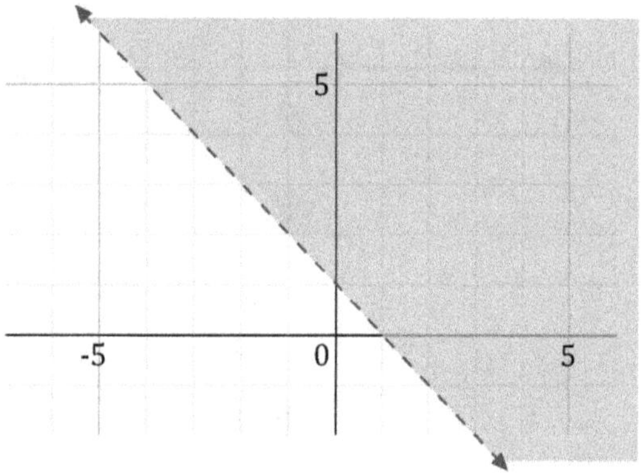

c.
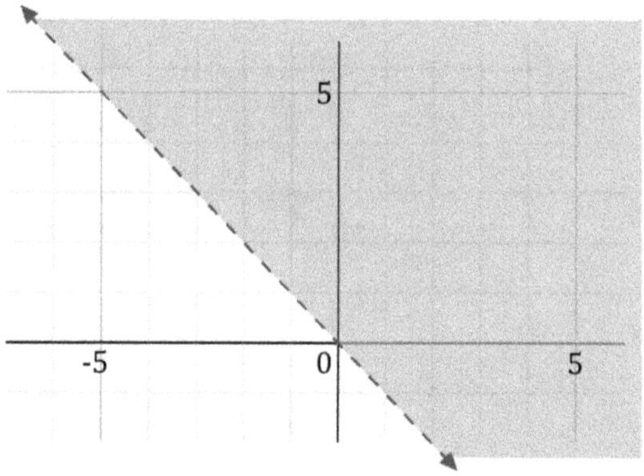

b.
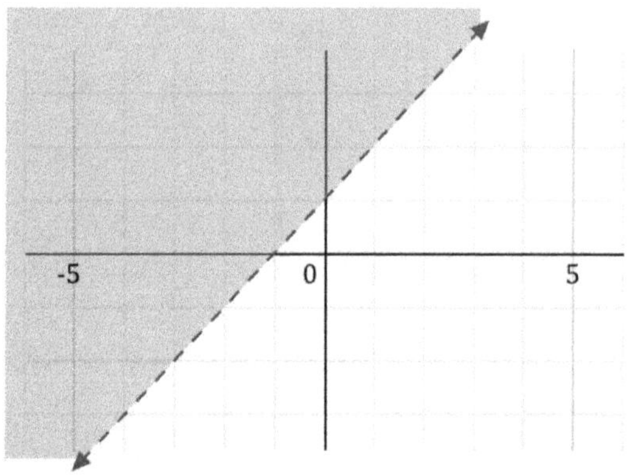

d.
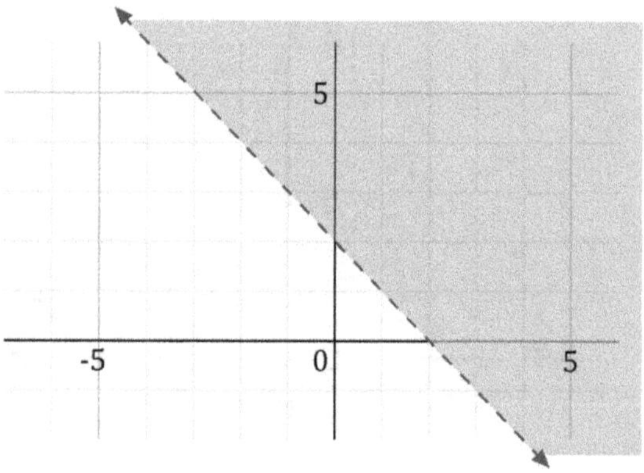

16. You have a square garden plot whose length and width are x meters. If you expand this into a rectangular plot by making it 6 meters longer and 4 meters wider, what will the area of the expanded plot be?

a. $24x^2$ square meters
b. $x^2 + 24x + 10$ square meters
c. $x^2 + 10x + 24$ square meters
d. $x^2 + 24$ square meters

17. Subtract $4xy - 8x^2 + 6$ from $2x^2 - 6xy + 4$.

a. $2x^2 - 2x + 2$
b. $10x^2 - 10xy - 2$
c. $2x^2 - 12xy + 4$
d. $10x^2 - 2 + 4$

18. Solve the following system of equations using the graphing method.

$$\begin{cases} 3x - 2y = -2 \\ x + 2y = 18 \end{cases}$$

a. $(2,8)$
b. $(-2,-2)$
c. $(-4,-5)$
d. $(4,7)$

19. Simplify the equation by combining like terms.

$$9x^2 - 6x - 11x - 16x^2 = -62$$

a. $-7x^2 - 17x = -62$
b. $7x^2 - 17x = 62$
c. $-7x^2 - 10x = -62$
d. $25x^2 + 10x = 62$

20. Combine the following algebraic fractions, and simplify the answer if possible: $\frac{10x+3}{2x+6} - \frac{4x-5}{2x+6}$.

a. $\frac{-x+15}{2x+6}$
b. $\frac{5x-1}{2x+6}$
c. $\frac{3x+4}{x+3}$
d. $\frac{x-1}{x+3}$

21. The function $f(x) = -8x^2 + 32x + 32$ models the height of a ball when thrown from the top of a building that is 32 feet high. What is the function of the height of the ball, $g(x)$, if it is thrown into the air from the ground floor?

a. $g(x) = -8x^2 + 32x$
b. $g(x) = -8x^2 + 32x + 15$
c. $g(x) = -28x^2 + 52x + 15$
d. $g(x) = -8x^2 + 32$

22. Evaluate the product: $\left(-10ab^5\right)\left(-\frac{1}{2}a^8b\right)$.

a. $-5a^8b^5$
b. $5a^8b^5$
c. $-5a^9b^6$
d. $5a^9b^6$

23. To better understand their shoppers and increase sales, Sasha's Boutique surveyed its customer base to find out the most common ages of the people who shop there. Many of their shoppers were one of two ages, which are represented by x in the equation $|x - 23| = 3$. Solve for x to determine the two ages.

a. $x = 20, 23$
b. $x = 20$
c. $x = 20, 26$
d. $x = 26$

24. Which choice shows an incorrect match between the inequality and the written description?

a. $x > 3$: "x is greater than 3"
b. $x < 3$: "x is less than 3"
c. $x \geq 3$: "x is greater than or equal to 3"
d. $x \leq 3$: "x is greater than or equal to 3"

25. Factor completely: $3a^2b^3 + 12ab^5$.

a. $ab^3(3a + 12b^2)$
b. $ab^3(3a + 12ab^2)$
c. $3ab^3(a + 4b^2)$
d. $3b^3(a^2 + 4ab^2)$

26. The height of the cone is represented by the variable h. If $h = \sqrt{12}$, what is the height of the cone in simplest form?

a. $2\sqrt{3}$
b. $3\sqrt{3}$
c. $2\sqrt{5}$
d. $12\sqrt{3}$

27. Simplify $-5a^2(a^3 - 2a^2 + 9a - 3)$.

a. $-4a^6 - 7a^4 + 4a^2 - 8a^2$
b. $-4a^5 - 7a^4 + 4a^3 - 8a^2$
c. $-5a^6 + 10a^4 - 45a^2 + 15a^2$
d. $-5a^5 + 10a^4 - 45a^3 + 15a^2$

28. Simplify the following expression:

$$(6b^2 + 4b + 2d) - (4b^2 + 18b + d)$$

a. $10b^2 + 14b + 3d$
b. $10b^2 + 14b + d$
c. $2b^2 - 14b + d$
d. $2b^2 - 14b + 3d$

29. A teacher gives a 100-point test to a large class. The student with the lowest score gets 30 points and the student with the highest score gets 80 points. The teacher considers two schemes for raising the scores by "curving" them. The first scheme raises the 30 to 40, the 80 to 100, and other scores to fit the line between these two extremes. The second scheme raises the 30 to 60, the 80 to 90, and other scores to fit the line between these extremes. Let x be a student's original score and y be the corresponding curved score. Write a system of equations that models these two schemes.

a. $\begin{cases} 6x - 5y = -20 \\ 3x - 5y = -210 \end{cases}$
b. $\begin{cases} 6x - 5y = -20 \\ 5x - 3y = -30 \end{cases}$
c. $\begin{cases} 5x - 6y = -90 \\ 3x - 5y = -210 \end{cases}$
d. $\begin{cases} 5x - 6y = -90 \\ 5x - 3y = -30 \end{cases}$

30. Solve the quadratic equation by using square roots: $(3x + 6)^2 - 144 = 0$.

a. $x = -12$ or $x = -12$
b. $x = -6$ or $x = 8$
c. $x = -6$ or $x = 2$
d. $x = -2$ or $x = 6$

31. The weight of an object on Earth varies directly as the same object's weight on the Moon. A 100-pound object on Earth weighs 16.5 pounds on the Moon. How much would an object weighing 20 pounds on Earth weigh on the Moon? Round your answer to the nearest tenth of a pound.

a. 13.9 pounds
b. 121.2 pounds
c. 3.3 pounds
d. 0.3 pounds

32. Factor $x^2 - 16x + 64$.

a. $(x - 16)(x + 4)$
b. $(x + 16)(x - 4)$
c. $(x - 8)^2$
d. $(x + 8)^2$

33. Write a function f in standard form, given that f is a quadratic function where $f(-4) = 0, f(2) = 0$, and $f(3) = -7$.

a. $f(x) = -x^2 - 2x + 8$
b. $f(x) = x^2 + 2x - 8$
c. $f(x) = \frac{1}{9}x^2 - 2x + 8$
d. $f(x) = \frac{1}{9}x^2 - \frac{2}{9}x - \frac{8}{9}$

34. Kristen is walking up a hill that is approximately 400 feet tall. She gains a constant elevation of 5 feet per second as she climbs the hill. Use a linear function to calculate the number of seconds it will take Kristen to reach the top of the hill.

a. It will take Kristen 44 seconds to climb the hill.
b. It will take Kristen 56 seconds to climb the hill.
c. It will take Kristen 66 seconds to climb the hill.
d. It will take Kristen 80 seconds to climb the hill.

35. Find the value of d in the equation $\frac{d}{3} = 14$.

a. $d = 14$
b. $d = 17$
c. $d = 24$
d. $d = 42$

36. Which graph correctly shows the solution set for $|x + 3| > 7$?

a.

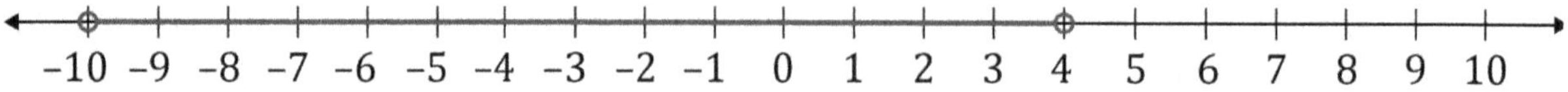

b.

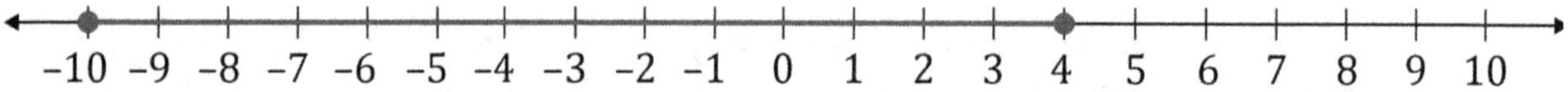

c.

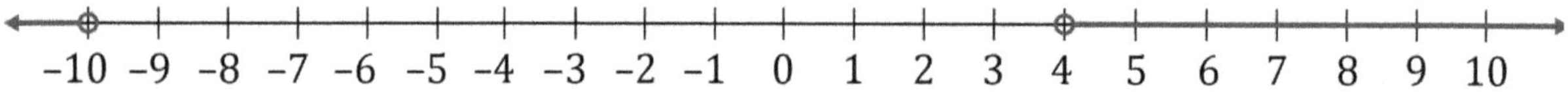

d.

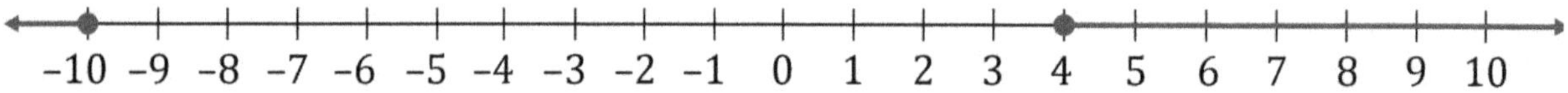

37. Given the data below, which statement best describes the correlation coefficient (r), rounded to the nearest hundredth, of the number of police cars on a turnpike and the average speed of truck drivers?

Number of Patrol Cars	Average Speed (miles per hour)
3	63
4	62
5	59
6	52
7	55
8	51

a. There is a strong negative linear association between the two variables.
b. There is a moderate negative linear association between the two variables.
c. There is a strong positive linear association between the two variables.
d. There is no association between the two variables.

38. Solve the inequality: $-2x + 5 < 1$ or $x - 3 \geq 5$.

a. $x < 2$ or $x \geq 8$
b. $x > 2$
c. $x \geq 8$
d. $2 < x \leq 8$

39. The total revenue for selling a television can be represented by the expression $\frac{3n^2+5n+2}{2n-3}$. If the number of televisions sold can be expressed by $\frac{n^2+2n+1}{8n-12}$, which of the following expressions represents the unit price for a television set in terms of n?

a. $\frac{n+1}{12n+8}$
b. $\frac{12n+4}{n+1}$
c. $\frac{12n+8}{2n+1}$
d. $\frac{12n+8}{n+1}$

40. Solve the equation: $-2 + \sqrt{4x} = 6$.

a. $x = 21$
b. $x = 24$
c. $x = 16$
d. $x = 12$

41. Lamar and Ashley are playing a math game. One player needs to write a linear equation, and the other player needs to write an equation for a line that is perpendicular to the first one. Lamar starts by writing the equation $f(x) = -3x + 4$. If Ashley needs to write an equation for a line that is perpendicular to Lamar's, which equation could she write?

a. $g(x) = -\frac{1}{3}x + 4$
b. $g(x) = -4x + 3$
c. $g(x) = -3x + 8$
d. $g(x) = \frac{1}{3}x + 2$

42. The sum of one-half a number, t, and one-fourth of two more than the number is 20. What is the value of t?

a. $t = 26$
b. $t = 15$
c. $t = \frac{26}{2}$
d. $t = \frac{15}{2}$

43. Find the value of y in the equation $21 - 8y = 7y - 9$.

a. $y = -30$
b. $y = 30$
c. $y = 2$
d. $y = -2$

44. A fence is being built in your neighborhood park. One-third of the fence has been constructed. If the completed fence will be $24\sqrt{12}$ feet, how much of the fence has been built?

a. $9\sqrt{3}$ feet
b. $16\sqrt{3}$ feet
c. $16\sqrt{12}$ feet
d. $8\sqrt{4}$ feet

45. Determine whether the linear equations below are parallel. If so, what is their slope?

$$\mathbf{16x + 8y = 24}$$
$$\mathbf{2x + y = 7}$$

a. These lines are not parallel because their slopes are different.
b. These lines are not parallel because the slope of the line $2x + y = 7$ is undefined.
c. These lines are parallel. The slope of both lines is 2.
d. These lines are parallel. The slope of both lines is –2.

46. Use the distributive property to simplify the expression $3(b + 5) - 11$.

a. $3b - 6$
b. $3b + 4$
c. $15b - 11$
d. $15b + 4$

47. What is the graph of the function $y = -x^2 + 3x + 10$?

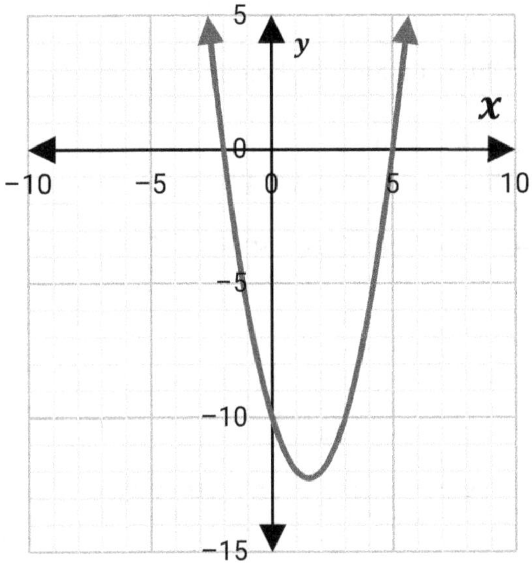

a.

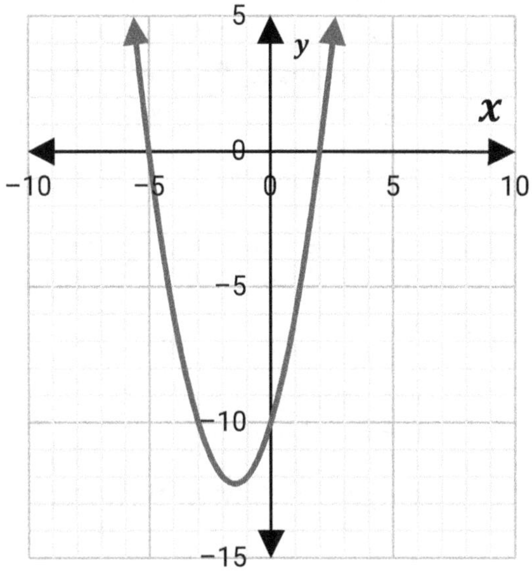

c.

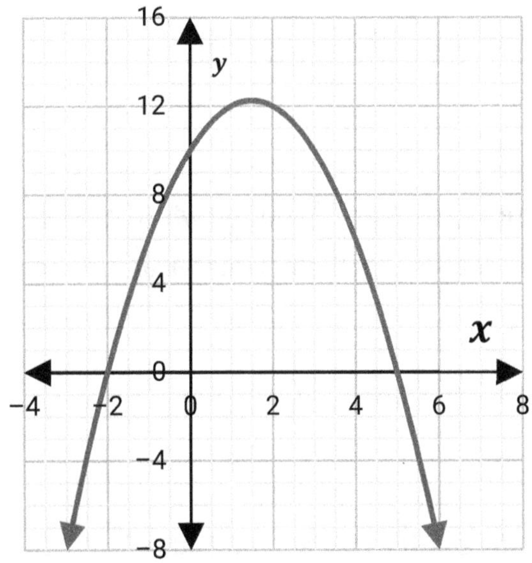

b.

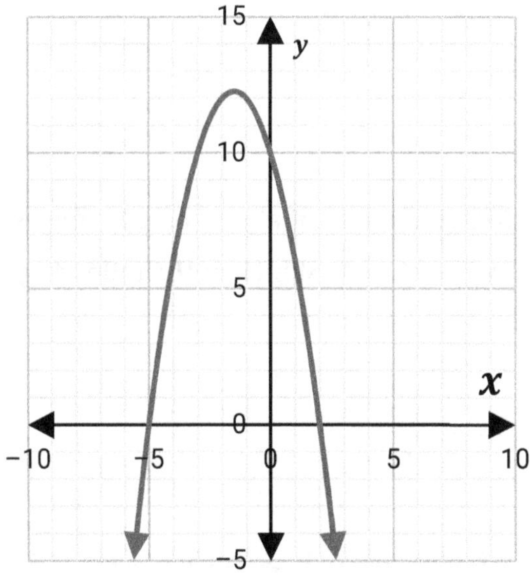

d.

48. A ball is thrown from 10 feet above the ground in the air with an initial velocity of 25 feet per second. The height, h, of the ball above the ground can be modeled by the quadratic equation $h = -16t^2 + 25t + 10$ where t is time, in seconds. To the nearest tenth of a second, how long will it take the ball to hit ground?

a. 0.5 seconds
b. 2.1 seconds
c. 1.9 seconds
d. 0.3 seconds

49. The volume of a rectangular prism with a square base can be expressed by the polynomial $3n^3 + 24n^2 + 48n$. What are the lengths of the sides of the square base and the height of the prism in terms of n?

a. The lengths of the sides of the square base are $3n + 12$, and the height of prism is n.
b. The lengths of the sides of the square base are $n^2 + 12$, and the height of prism is 3.
c. The lengths of the sides of the square base are $n + 4$, and the height of prism is $3n$.
d. The lengths of the sides of the square base are n, and the height of prism is $n + 4$.

50. Which of the following is an equation of a linear function in point-slope form that contains the points $(3, -6)$ and $(4, -3)$?

a. $3x + 4y = 12$
b. $y + 4 = 3(x + 3)$
c. $y = 3x - 9$
d. $y + 3 = 3(x - 4)$

51. Simplify the following algebraic expression. Leave your answer in factored form.

$$\frac{10x^2 - 5x}{15x}$$

a. $\frac{2x^2-1}{3x}$
b. $\frac{2x-1}{3}$
c. $\frac{1}{3}$
d. $\frac{x}{3}$

52. The length of a rectangle is represented by the expression $\frac{10}{n}$ and the width by the expression $\frac{5}{4n}$. Which of the following expressions represents the perimeter of the rectangle?

a. $\frac{15}{4n^3}$
b. $\frac{15}{4n^2}$
c. $\frac{90}{n}$
d. $\frac{45}{2n}$

53. At a baseball game, Jake's concession stand sells hotdogs for \$3 and sodas for \$1.50. At the end of the night, Jake earned 300 dollars in sales. The equation $3x + 1.50y = 300$ can be used to determine the number of hotdogs and sodas sold in one night. In this equation, x represents the number of hotdogs sold, and y represents the number of sodas sold at the baseball game. Convert the equation from standard form to slope-intercept form.

a. $y = -2x + 200$
b. $y = -3x + 300$
c. $y = 2x - 200$
d. $y = 3x - 300$

54. Divide $2a^2 + 3a + 1$ by $a - 2$.

a. $2a + 7 + \frac{15}{a-2}$
b. $2a + 22$
c. $2a + 7 - \frac{13}{a-2}$
d. $2a - 6$

55. Mia opened a bank account in January and deposited \$10. Each month, she plans to double the amount deposited in the previous month. Based on this information, how much money will Mia deposit in August?

a. \$1,380
b. \$1,280
c. \$138
d. \$128

56. What is the equation of a linear function in standard form that contains the points $(-1, 3)$ and $(4, -2)$?

a. $x - y = 2$
b. $x + y = -2$
c. $-x + 2 = y$
d. $x + y = 2$

57. You have just turned onto a highway. The table shows the mile marker M you can reach without speeding after t hours of driving at a constant rate under optimal traffic conditions. Write a linear inequality that represents the possible values of t and M under all traffic conditions.

t	1	2	3	5
M	73	133	193	313

a. $60t + M \leq 133$
b. $60t - M \leq -13$
c. $60t + M \geq 133$
d. $60t - M \geq -13$

58. Use the discriminant to determine how many times the graph of $y = x^2 + 2x - 8$ crosses the x-axis on the coordinate plane.

a. The graph will touch the x-axis twice because the discriminant is positive.
b. The graph will touch the x-axis once because the discriminant is 0.
c. The graph does not touch the x-axis because the discriminant is negative.
d. It cannot be determined how many times the graph crosses the x-axis using the discriminant.

59. State the range for the quadratic function $y = x^2 - 2x - 3$.

a. $-\infty < y < \infty$
b. $y \geq 1$
c. $y \geq -4$
d. $y \leq -4$

60. Factor $16s^2 + 121$.

a. $(4s - 11)(4s + 11)$
b. $(4s + 11)^2$
c. $(4s - 11)^2$
d. Cannot be factored

61. Which equation of a line contains point $(9, -1)$ and is perpendicular to the graph of the line $x + 3y = 6$?

a. $y = 3x - 28$
b. $y = 3x + 26$
c. $y = -\frac{1}{3}x - 28$
d. $y = -\frac{1}{3} + 26$

62. Factor $3x^2 + 11x + 10$.

a. $(x - 1)(3x + 11)$
b. $(x + 1)(3x + 11)$
c. $(x + 5)(3x + 2)$
d. $(x + 2)(3x + 5)$

63. Solve the following inequality: $\frac{x}{2} > 3$.

a. $x < 6$
b. $x > 6$
c. $x \leq 6$
d. $x \geq 6$

64. Which equation of a line is perpendicular to the x-axis, and what is its slope?

a. $y = 8; m = 0$
b. $y = 8; m = \text{undefined}$
c. $x = 8; m = 0$
d. $x = 8; m = \text{undefined}$

65. The length of a rectangle is $x + 4$, and the width is $x + 2$. If the area of the rectangle is 63 square feet, what is the value for x?

a. $x = 4$
b. $x = 7$
c. $x = 5$
d. $x = 2$

66. If $f(x) = x^2 - 3x$, then find the value of $\frac{f(8)}{f(2)}$.

a. 20
b. -20
c. 4
d. -4

67. Find the value of a in the equation $7a - 21 = 28$.

a. $a = 49$
b. $a = 42$
c. $a = 1$
d. $a = 7$

68. Aaron is looking for a new apartment for himself and his cat, Omar. He's narrowed his search down to two options. Apartment A costs $1,200 each month and has a one-time pet deposit of $300. Apartment B costs $1,050 each month and has a one-time pet deposit of $600. How many months will it take for the total costs of both apartments to be the same? Write and solve an equation using the variable m to represent months.

a. 4 months
b. 3 months
c. 2 months
d. 1 month

69. Sue is at the pet store looking for her dog's favorite brand of dog kibble. She has $80 to spend at the store, and she needs to buy 5 bags of food in order to stock up for a few months. Which of the following statements is true about the bags of food Sue can buy?

a. Each bag of food needs to be less than or equal to $3.
b. Each bag of food needs to be less than or equal to $3.20.
c. Each bag of food needs to be less than or equal to $16.
d. Each bag of food needs to be less than or equal to $5.

70. What is the value of y for the linear function $y = 5x + 3$ when the value of x is 8?

a. 11
b. 16
c. 40
d. 43

71. Convert $y = 3(x - 2)^2 + 1$ into standard form.

a. $y = 3(x - 2)^2$
b. $y = 3(x + 1)(x + 13)$
c. $y = 3x^2 - 12x - 11$
d. $y = 3x^2 - 12x + 13$

72. From the ground, a projectile is launched into the air. The projectile reaches its maximum height of 126 feet after 3 seconds. If the projectile follows a parabolic path, which of the following is the equation for the path of the projectile in terms of its height, *h* in feet, from the ground and time, *t* in seconds. State your answer in vertex form.

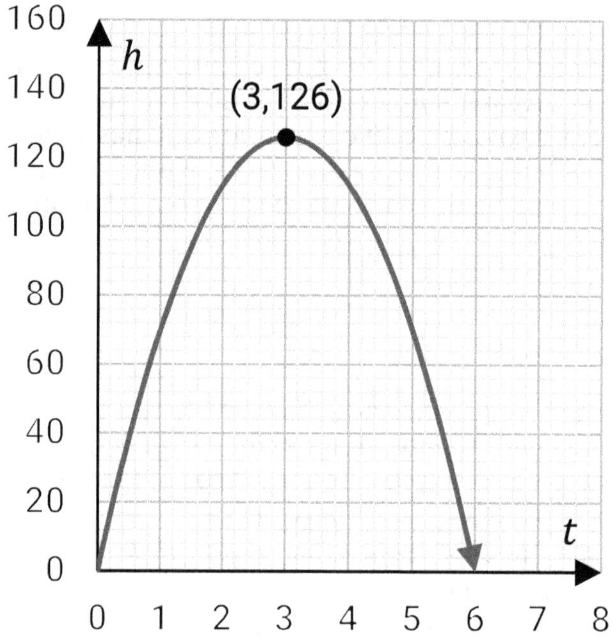

a. $h = -14(t-3)^2 + 126$
b. $h = 14(t-3)^2 + 126$
c. $h = -\frac{3}{126}(t-126)^2 + 3$
d. $h = -(t-126)^2 + 3$

73. Kayla goes for a walk every morning. The distance Kayla walks can be modeled by the equation $d = 3h$, where d is the distance walked in miles and h is the number of hours walked. Which graph best represents the relationship between x hours and y distance?

a.

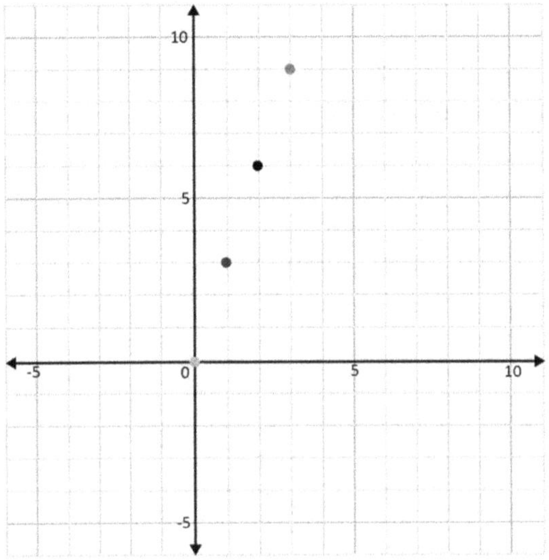

c.

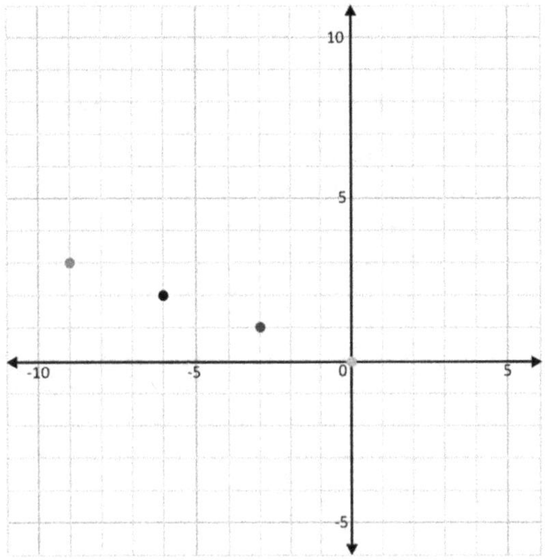

b.

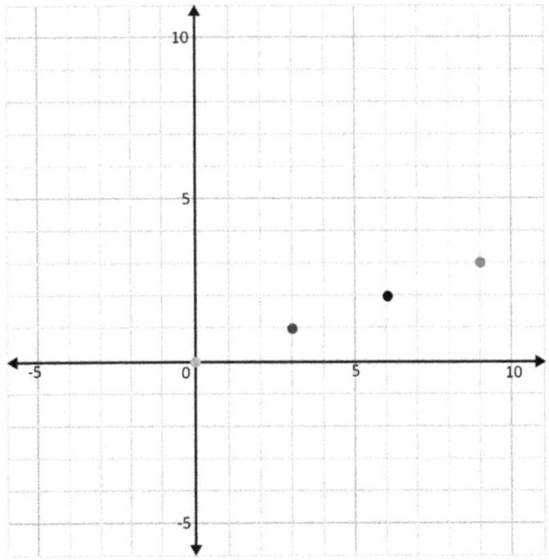

d.

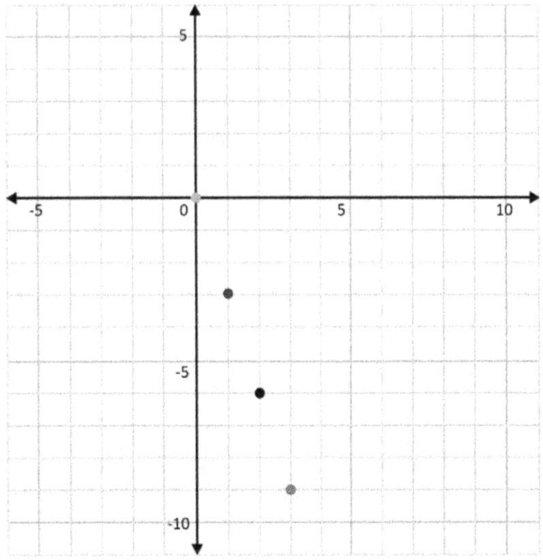

74. Solve the equation $-x + 8y = -14$ for y.

a. $y = \frac{8}{-14-x}$
b. $y = \frac{-14-x}{8}$
c. $y = \frac{-14+x}{8}$
d. $y = \frac{8}{-14+x}$

75. The product of three times the square of a number n and the sum of four times its cube and one. Which of the following represents the product in terms of n?

a. $12n^6 + 43n^2$
b. $7n^6 + 4n^2$
c. $12n^5 + 3n^2$
d. $7n^5 + 4n^3$

Answer Key and Explanations

1. B: To solve the inequality, we must isolate the variable on one side of the inequality. In this problem, we can start by applying the distributive property on the left side of the equation, multiplying the –2 by the $3x$ and then by the –5.

$$-6x + 10 \geq 58$$

The next step is to subtract 10 from both sides.

$$-6x \geq 48$$

Finally, we divide both sides by –6. Because we are dividing by a negative number, we need to "flip" the inequality.

$$x \leq -8$$

To graph $x \leq -8$, we start with a solid dot at –8 because our inequality is less than or equal to. The "or equal to" part tells us we need a solid dot. Then, because x is on the left, we follow the direction of the less than arrow by drawing our ray to the left. The arrow on the end of our ray indicates that all answers to the left on the number line satisfy the inequality.

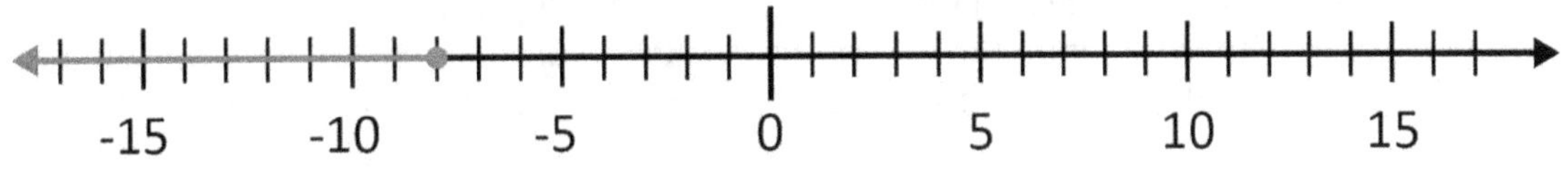

2. C: To find the value of x for the linear function when the value of y is 5, substitute 5 for y and evaluate the equation to find the value of x.

$$5 = 3x - 7$$

To solve for x, start by adding 7 to both sides.

$$12 = 3x$$

Then, divide both sides by 3.

$$x = 4$$

3. A: To rationalize the denominator, multiply by $\frac{\sqrt{42}}{\sqrt{42}}$.

$$\frac{30}{\sqrt{42}} = \frac{30}{\sqrt{42}} \cdot \frac{\sqrt{42}}{\sqrt{42}} = \frac{30\sqrt{42}}{42}$$

Then, simplify by dividing the common factor of 6 in the numerator and denominator

$$\frac{6 \cdot 5\sqrt{42}}{6 \cdot 7} = \frac{5\sqrt{42}}{7}$$

4. C: Use the rules for exponents to simplify.

$\frac{9xy^5}{12x^4y^3} = \frac{9}{12} \cdot x^{1-4} \cdot y^{5-3}$	Rewrite the expression using the rule $\frac{a^m}{a^n} = a^{m-n}$.
$= \frac{3}{4} \cdot x^{-3} \cdot y^2$	Reduce $\frac{9}{12}$ to $\frac{3}{4}$ and subtract the exponents for x and y.
$= \frac{3y^2}{4x^3}$	Rewrite the answer with no negative exponents using the rule $a^{-n} = \frac{1}{a^n}$.

5. A: Correlation refers to the strength of a linear relationship between two variables. Causation refers to instances in which one variable directly causes the outcome of another. In other words, correlation means that there is a relationship or connection between two variables, whereas causation means that one event explicitly causes another event to occur. Correlation and causation can exist at the same time, but correlation does not imply causation. In this question, we need to identify which variable relationship demonstrates correlation but not causation. There is no causation in the relationship between an elementary student's weight and the size of their vocabulary. While the two variables are correlated, one does not cause the other to happen. Therefore, the correct answer is A.

6. D: To start, analyze the problem and write an equation that represents the scenario. Since we don't know how many pieces of gum Juan started with, use a variable to represent this quantity. In this explanation, the variable g is used to represent the amount of gum Juan started with.

$$g$$

Since he gave 13 pieces of gum to his friend, subtract 13 from the original amount, g.

$$g - 13$$

Juan has 11 pieces of gum left, so set the expression equal to 11.

$$g - 13 = 11$$

To find the answer, solve for g in the equation. First, isolate the variable by doing inverse operations. Since the opposite of -13 is $+13$, add 13 to both sides of the equation.

$$g - 13 + 13 = 11 + 13$$

Combine like terms.

$$g = 24$$

To check your work, substitute the solution back into the original equation.

$$(24) - 13 = 11$$
$$11 = 11 \checkmark$$

7. D: When $6(9 - y)$ is multiplied using the distributive property, it becomes $54 - 6y$, not $54 + 9y$.

8. D: First, identify the y–intercept. The line intersects the y–axis at point (0,4), so the y–intercept, b, of this linear equation is 4. From here, we need to identify the slope of the line, m. Use another point from the graph to find the slope. In this explanation, we will use point (1,2) as well as the y-intercept, point (0,4). To find the slope when given the coordinates of two points, find the change in y–values over the change in x–values. To do so, use the slope formula $m = \frac{y_2 - y_1}{x_2 - x_1}$. Substitute the x- and y–values from the two coordinates given.

$$m = \frac{4-2}{0-1} = \frac{2}{-1} = -2$$

The slope of this line is –2.

Since we now know the slope and the y–intercept, we can write an equation in slope–intercept form. The slope–intercept form of the equation is $y = mx + b$, where m is the slope, and b is the y-intercept. The slope–intercept equation of the line is $y = -2x + 4$.

Now that the equation is in slope–intercept form, change it into standard form, which is $Ax + By = C$. Move the x–term from the right side of the equation to the left by adding $2x$ to both sides of the equation. Therefore, the linear equation in two variables is $2x + y = 4$.

9. B: Let x represent the number. Then, the square of 7 more than twice x is represented by the expression $(7 + 2x)^2$. Since this expression equals to 169, we have the following quadratic equation.

$$(7 + 2x)^2 = 169$$

Take the square root of both sides to solve for x.

$$\begin{aligned}
(7 + 2x)^2 &= 169 \\
\sqrt{(7 + 2x)^2} &= \sqrt{169} \\
7 + 2x &= \pm 13 \\
2x &= -7 \pm 13 \\
\frac{2x}{2} &= \frac{-7 \pm 13}{2} \\
x &= \frac{-7-13}{2} \text{ or } x = \frac{-7+13}{2} \\
x &= \frac{-20}{2} \text{ or } x = \frac{6}{2} \\
x &= -10 \text{ or } x = 3
\end{aligned}$$

So, x can be either –10 or 3.

10. C: First, complete the square to put the equation into the form $(ax + b)^2 = c$.

$x^2 - 12x = -35$	Move the constant term to the right-hand side of the equation.
$x^2 - 12x = -35$	Locate the coefficient to the linear term.
$\left(\frac{-12}{2}\right)^2 = (-6)^2 = 36$	Take one-half of this coefficient and square it.

$x^2 - 12x + 36 = -35 + 36$ Add the squared value to both sides of the equation.

$(x - 6)^2 = 1$ Factor the perfect square trinomial $x^2 - 12x + 36$.

Since the converted equation is in the form $(ax + b)^2 = c$, we can take the square root of both sides to solve for x.

$$(x - 6)^2 = 1$$
$$\sqrt{(x - 6)^2} = \sqrt{1}$$
$$x - 6 = \pm 1$$
$$x - 6 + 6 = 6 \pm 1$$
$$x = 6 \pm 1$$
$$x = 6 - 1 \text{ or } x = 6 + 1$$
$$x = 5 \text{ or } x = 7$$

11. C: To graph this linear function given the slope and a point the line passes through, determine the slope-intercept equation of the line. The slope-intercept equation of a line is $y = mx + b$, where m stands for the slope, and b stands for the y-intercept. Since we know the slope, m, is 10, start by replacing m with 10 in the equation.

$$y = 10x + b$$

Next, substitute the x- and y-coordinates into the equation. Since the y-coordinate is 33, replace y with 33 in the equation. Since the x-coordinate is 3, replace x with 3 in the equation.

$$(33) = 10(3) + b$$

From here, solve the equation for b to identify the y-intercept. Simplify the equation by multiplying 10 and 3.

$$33 = 30 + b$$

Next, solve for b by subtracting 30 from both sides.

$$b = 3$$

Finally, substitute 3 into the equation for b. The slope-intercept equation of a line that passes through point (3,33) and has a slope of $m = 10$ is $y = 10x + 3$.

Now that the linear equation is in slope-intercept form, graph the function. Since the y-intercept is 3, the line will intersect the y-axis at 3, or (0,3). Plot a point at this location. From here, use the slope to move to the next point. Since the slope is 10, move 10 units up and 1 unit right to find the next point on the graph. The coordinate pair for this point is (1,13). Repeat this process to continue plotting points on the line of this linear function. When the value of x is 3, the value of y will be 33. The graph in choice C shows all these points on the line.

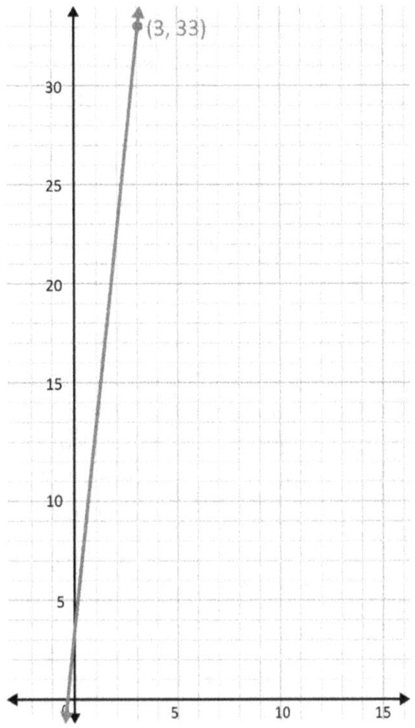

12. C: The area of a rectangle is the product of its length and width, $A = lw$. We can rearrange this to calculate the length as the area divided by the width, $l = \frac{A}{w}$. The area is $20x^2 + 55x$ square feet and the width is $5x$ feet. Substitute the expressions into the formula.

$$l = \frac{A}{w} = \frac{20x^2 + 55x}{5x}$$

Next, we separate the fraction into two fractions, putting each term of the numerator over the denominator. Then we divide coefficients by coefficients and variables by variables, recalling that $\frac{x^m}{x^n} = x^{m-n}$ and that a fraction with the same numerator and denominator equals 1.

$$\frac{20x^2 + 55x}{5x} = \frac{20x^2}{5x} + \frac{55x}{5x} = \frac{20}{5} \cdot \frac{x^2}{x} + \frac{55}{5} \cdot \frac{x}{x} = 4x^{2-1} + 11 \cdot 1 = 4x + 11$$

Therefore, the length of the pen is $4x + 11$ feet.

13. B: Simplify the expression so that the radicals are the same.

$\sqrt{50}$ can be simplified to $\sqrt{25}\sqrt{2}$, or $5\sqrt{2}$.

$5\sqrt{18}$ can be simplified to $5\sqrt{9}\sqrt{2}$, or $15\sqrt{2}$.

Rewrite $\sqrt{50} + 5\sqrt{18}$ as $5\sqrt{2} + 15\sqrt{2}$. Now that the radicals are the same, the coefficients can be added.

$$5\sqrt{2} + 15\sqrt{2} = 20\sqrt{2}$$

14. A: To solve this problem, we must first translate the written problem into a math inequality. "Seven times a number x" translates to $7x$. "Is less than 56" translates to the less than symbol ($<$) and the constant 56.

$$7x < 56$$

To solve this, divide both sides of the inequality by 7.

$$x < 8$$

15. A: The inequality $y > -x + 1$ is in slope-intercept form. To graph the inequality, treat it as though it is an equation and identify the y-intercept and the slope. If we treat $y > -x + 1$ as though it is $y = -x + 1$, we know that the y-intercept is 1, and the slope is –1. Now the line can be graphed. The line will be dashed because the inequality does not include "equal to." The shaded region will be above the line because the inequality reads, "y is greater than," so shade the y-values greater than the line.

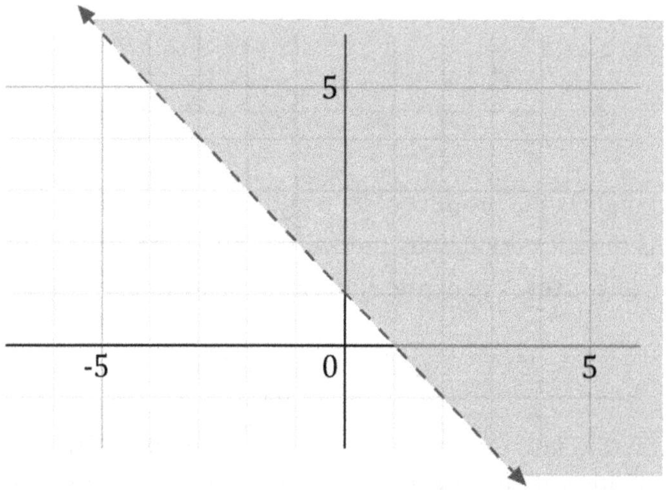

The shaded region above the dashed line means that any value in the red area is considered a solution for the inequality. The dashed line means that values on the line are not solutions.

16. C: Increasing the original length of x by 6 produces a new length of $(x + 6)$ meters. Similarly, the new width will be $(x + 4)$ meters. Since the area of a rectangle is the product of its length and width, The area of the new rectangle will be $(x + 6)(x + 4)$ square meters. Calculate this product using the FOIL method and combine like terms.

$$\begin{aligned}(x+6)(x+4) &= \overbrace{x \cdot x}^{F} + \overbrace{x \cdot 4}^{O} + \overbrace{6 \cdot x}^{I} + \overbrace{6 \cdot 4}^{L} \\ &= x^2 + 4x + 6x + 24 \\ &= x^2 + 10x + 24\end{aligned}$$

The area of the expanded plot will be $x^2 + 10x + 24$ square meters.

17. B: Polynomials can be subtracted vertically. Line up the like terms in the first polynomial under the second polynomial.

$$\begin{array}{rrrr} & 2x^2 & -6xy & +4 \\ - & \underline{(-8x^2} & \underline{+4xy} & \underline{+6)} \end{array}$$

Take the additive inverse of the bottom polynomial by distributing the negative sign to the bottom polynomial. Then remove the parentheses from both polynomials before combining them.

$$\begin{array}{rrrr} & 2x^2 & -6xy & +4 \\ + & 8x^2 & -4xy & -6 \\ \hline & 10x^2 & -10xy & -2 \end{array}$$

18. D: To solve this system by the graphing method, we graph both lines by any convenient method. For instance, we can solve the first equation for y, putting it in slope-intercept form, $y = mx + b$.

$$\begin{aligned} 3x - 2y &= -2 \\ -2y &= -3x - 2 \\ y &= \frac{3}{2}x + 1 \end{aligned}$$

The line has a y-intercept of $b = 1$ and therefore passes through the point (0,1). The slope $m = \frac{3}{2}$ indicates that moving along the line, we rise 3 units for every 2 units we move to the right. If we do this starting at the point (0,1), we end up at the point (2,4). To graph the line, we plot these two points and draw the line through them.

Similarly, we put the second equation in slope-intercept form.

$$\begin{aligned} x + 2y &= 18 \\ 2y &= -x + 18 \\ y &= -\frac{1}{2}x + 9 \end{aligned}$$

The line has a y-intercept of $b = 9$ and therefore passes through the point (0,9). The slope $m = -\frac{1}{2}$ indicates that moving along the line, we rise –1 unit (that is, we drop 1 unit) for every 2 units we move to the right. If we do this starting at the point (0,9), we end up at the point (2,8). Again, we plot the points and draw the line through them. Here is the graph of both lines.

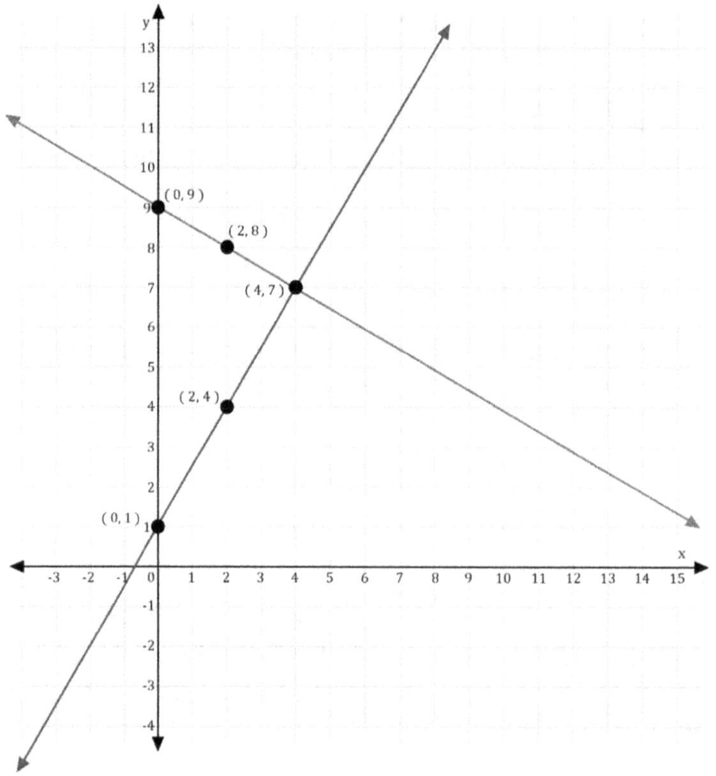

Since each line represents all the solutions to one equation, their intersection point—the point they have in common—solves both equations. That is, it solves the system. These lines intersect at the point (4,7).

19. A: To combine like terms, look for terms that have the same variable part. Once identified, combine them into a single term by adding or subtracting their coefficients.

$$9x^2 - 6x - 11x - 16x^2 = -62$$

Start by combining the terms $9x^2$ and $-16x^2$ on the left side of the equation.

$$-7x^2 - 6x - 11x = -62$$

Next, combine the terms $-6x$ and $-11x$ on the left side of the equation.

$$-7x^2 - 17x = -62$$

20. C: Since the denominators are the same, combine the numerators to write the expression as one fraction.

$$\frac{10x + 3}{2x + 6} - \frac{4x - 5}{2x + 6} = \frac{10x + 3 - (4x - 5)}{2x + 6} = \frac{6x + 8}{2x + 6}$$

Simplifying the answer by factoring out a 2 in both the numerator and denominator.

$$\frac{6x+8}{2x+6} = \frac{2(3x+4)}{2(x+3)} = \frac{3x+4}{x+3}$$

21. A: Since the ball will now be tossed from 32 feet below the original throw, the entire graph will move down by 32 units, which is the same as subtracting 32 from the original function.

$$g(x) = -8x^2 + 32x + 32 - 32$$
$$g(x) = -8x^2 + 32x$$

22. D: Using commutativity and associativity of multiplication and the law of exponents that states $x^m \cdot x^n = x^{m+n}$, we get $(-10ab^5)\left(-\frac{1}{2}a^8b\right) = (-10)\left(-\frac{1}{2}\right) \cdot a \cdot a^8 \cdot b^5 \cdot b = 5a^{1+8}b^{5+1} = 5a^9b^6$.

23. C: For this problem, we are given the equation $|x - 23| = 3$. Since the expression in absolute value bars containing x is already by itself on the left side, the first step is dropping the absolute value bars by writing $\pm$ in front of the 3.

$$x - 23 = \pm 3$$

Now, add 23 to both sides to get x by itself.

$$x = \pm 3 + 23$$

If 3 is positive, then $x = 3 + 23 = 26$, and if 3 is negative, then $x = -3 + 23 = 20$. Most of the shoppers at Sasha's Boutique are either 20 or 26 years old.

24. D: Choice D gives the inequality, $x \leq 3$. By definition, this inequality is read as, "x is less than or equal to 3." This contradicts the statement in the answer choice.

25. C: First, determine the greatest common factor (GCF) for the terms by decomposing each of them into their prime factors.

$$3a^2b^3 = 3 \cdot a \cdot a \cdot b \cdot b \cdot b \text{ and } 12ab^5 = 2 \cdot 2 \cdot 3 \cdot a \cdot b \cdot b \cdot b \cdot b \cdot b$$

To find the GCF, compare common prime factors for each term (bolded below).

$$\mathbf{3} \cdot \boldsymbol{a} \cdot a \cdot \boldsymbol{b} \cdot \boldsymbol{b} \cdot \boldsymbol{b} \text{ and } 2 \cdot 2 \cdot \mathbf{3} \cdot \boldsymbol{a} \cdot \boldsymbol{b} \cdot \boldsymbol{b} \cdot \boldsymbol{b} \cdot b \cdot b$$

The GCF is the product of the common prime factors.

$$3 \cdot a \cdot a \cdot b \cdot b \cdot b = 3ab^3$$

Factoring the GCF from each term we have:

$$3a^2b^3 + 12ab^5 = 3ab^3(a + 4b^2)$$

We can always check our answer by taking the product of the GCF and the expression inside the parentheses by the distribution property of multiplication.

$$3ab^3(a + 4b^2) = 3ab^3 \cdot a + 3ab^3 \cdot 4b^2 = 3a^2b^3 + 12ab^5$$

26. A: $\sqrt{12}$ can be expressed as the product of two radicals: $\sqrt{4}\sqrt{3}$. The square root of 4 is 2, so $\sqrt{4}\sqrt{3}$ becomes $2\sqrt{3}$. The height of the cone in simplest form is $2\sqrt{3}$.

27. D: Use the distributive property for multiplication to simplify.

$$\begin{aligned} -5a^2(a^3 - 2a^2 + 9a - 3) &= (-5a^2) \cdot a^3 - (-5a^2) \cdot 2a^2 + (-5a^2) \cdot 9a - (-5a^2) \cdot 3 \\ &= -5a^5 + 10a^4 - 45a^3 + 15a^2 \end{aligned}$$

28. C: Start by removing the parentheses from the first portion of the expression.

$$6b^2 + 4b + 2d - (4b^2 + 18b + d)$$

Next, use the distributive property to distribute the negative sign to each term in the second set of parentheses. To do so, multiply each term in parentheses by –1.

$$6b^2 + 4b + 2d - 4b^2 - 18b - d$$

From here, combine like terms. To do so, look for terms that have the same variable part. Once identified, combine them into a single term by adding or subtracting their coefficients. Start with $6b^2$ and $-4b^2$.

$$2b^2 + 4b + 2d - 18b - d$$

Next, combine the terms $4b$ and $-18b$.

$$2b^2 - 14b + 2d - d$$

Finally, combine the terms $2d$ and $-d$.

$$2b^2 - 14b + d$$

29. A: Under the first scheme, a score of 30 becomes 40 and a score of 80 becomes 100, so the points (30,40) and (80,100) define the line for the first scheme. Substituting these two points into the slope formula, we find the slope of this line:

$$m = \frac{y_2 - y_1}{x_2 - x_1} = \frac{100 - 40}{80 - 30} = \frac{60}{50} = \frac{6}{5}$$

Now we substitute this slope and one of the points, say (30,40), into the point-slope form of a linear equation, $y - y_1 = m(x - x_1)$, and simplify, getting an equation for the first scheme.

$$\begin{aligned} y - 40 &= \frac{6}{5}(x - 30) \\ y - 40 &= \frac{6}{5}x - 36 \\ -\frac{6}{5}x + y &= 4 \\ 6x - 5y &= -20 \end{aligned}$$

Next, we repeat these steps for the second scheme. The points (30,60) and (80,90) define the line for this scheme. Substituting these two points into the slope formula, we find the slope of this line is:

$$m = \frac{90 - 60}{80 - 30} = \frac{30}{50} = \frac{3}{5}$$

Again, we substitute this slope and one of the points, say (30,60), into the point-slope form of a linear equation and simplify, getting an equation for the second scheme.

$$\begin{aligned} y - 60 &= \frac{3}{5}(x - 30) \\ y - 60 &= \frac{3}{5}x - 18 \\ -\frac{3}{5}x + y &= 42 \\ 3x - 5y &= -210 \end{aligned}$$

The equations for the first and second schemes form the desired system of equations.

30. C: First, put the quadratic equation in the form $(ax + b)^2 = c$ by adding 144 to both sides.

$$\begin{aligned} (3x + 6)^2 - 144 &= 0 \\ (3x + 6)^2 &= 144 \end{aligned}$$

Now, take the square root of both sides to solve for x.

$$\begin{aligned} 3x + 6 &= \pm 12 \\ 3x &= -6 \pm 12 \\ x &= \frac{-6 - 12}{3} \text{ or } x = \frac{-6 + 12}{3} \\ x &= \frac{-18}{3} \text{ or } x = \frac{6}{3} \\ x &= -6 \text{ or } x = 2 \end{aligned}$$

31. C: Since the weight varies directly, set up the following proportion and solve for the weight of the object on the Moon.

$$\begin{aligned} \frac{100}{16.5} &= \frac{20}{M} \\ 330 &= 100M \\ 3.3 &= M \end{aligned}$$

Additionally, we can use the equation $y = kx$ to solve direct variation problems. This problem refers to the relationship between the weight of an object on Earth and the Moon. Instead of using the variables x and y, we'll use the variable E to represent the weight of the object on Earth, and the variable M to represent the weight of the object on the Moon.

$$y = kx \rightarrow E = kM$$

First, rewrite the equation by substituting 100 for E and 16.5 for M. Then, solve for k to find the constant variation.

$$100 = k(16.5)$$

$$\frac{100}{16.5} = \frac{16.5k}{16.5}$$

$$6.06 \approx k$$

Next, substitute 6.06 in the equation $E = kM$ for k.

$$E = 6.06\ M$$

To find the value of M when $E = 20$, substitute 20 into the equation for E. Then, solve for M.

$$20 = 6.06\ M$$
$$\frac{20}{6.06} = \frac{6.06M}{6.06}$$
$$M \approx 3.3$$

Therefore, an object weighing 20 pounds on Earth would weigh about 3.3 pounds on the moon.

32. C: A factorable quadratic trinomial in the form $a^2 - 2ab + b^2$ is called a perfect square trinomial, where a^2 and b^2 are perfect squares, and factor to $(a - b)(a - b) = (a - b)^2$.

For our trinomial, x^2 and 64 are perfect squares so, $a = x$ and $b = 8$. Since $2ab = 2x \cdot 8 = 16x$, we have the following factorization.

$$x^2 - 16x + 64 = (x - 8)^2$$

33. A: In factored form, a quadratic function having two real zeros can be written as $f(x) = a(x - m)(x - n)$, where m and n are the zeros of the function, and a is a real number that does not equal 0.

Since $f(-4) = 0$ and $f(2) = 0$, the values –4 and 2 are the zeros for the function. Substitute the zeros into the factored form equation.

$$f(x) = a\big(x - (-4)\big)(x - 2)$$
$$f(x) = a(x + 4)(x - 2)$$

Then, substitute $f(3) = -7$ and solve for a.

$$-7 = a(3 + 4)(3 - 2)$$
$$-7 = a(7)(1)$$
$$-7 = 7a$$
$$a = -1$$

Finally, substitute the value for a and expand the factored form equation to write the quadratic function in standard form.

$$\begin{aligned} f(x) &= -1(x+4)(x-2) \\ &= -1(x^2 - 2x + 4x - 8) \\ &= -1(x^2 + 2x - 8) \\ &= -x^2 - 2x + 8 \end{aligned}$$

So, the quadratic function is $f(x) = -x^2 - 2x + 8$.

34. D: A linear function can be written to describe Kristen's walk up the hill. Her elevation depends on how many seconds she has been walking, so the variable, x, will represent time in seconds, and the variable, y, will represent the elevation in feet, expressed as $y = 5x$. The hill is 400 feet, so $400 = 5x$ becomes $x = 80$. It will take Kristen 80 seconds to climb the hill.

35. D: Solve for d in the equation.

$$\frac{d}{3} = 14$$

Isolate the variable by doing inverse operations. Multiply both sides of the equation by 3.

$$d = 42$$

36. C: To solve, isolate the variable x in the inequality. Since this inequality involves absolute value, rewrite this inequality as two separate inequalities without absolute value bars. $|x + 3| > 7$ becomes $x + 3 > 7$ and $x + 3 < -7$. Notice that the sign is flipped in $x + 3 < -7$. From here, isolate x in the first inequality by doing inverse operations. Subtract 3 from both sides of the inequality.

$$\begin{aligned} x + 3 - 3 &> 7 - 3 \\ x &> 4 \end{aligned}$$

Next, isolate x in the second inequality by doing inverse operations. Subtract 3 from both sides of the inequality.

$$\begin{aligned} x + 3 - 3 &< -7 - 3 \\ x &< -10 \end{aligned}$$

From here, graph $x > 4$ or $x < -10$ on the same number line. The first inequality states that x is greater than 4, so the number line must show an open circle at 4 with a line extending to the right. The second inequality states that x is less than -10, so the number line must also show an open circle at -10 with a line extending to the left.

37. A: The correlation coefficient, represented by the variable r, is a measurement of the strength of the linear relationship between two quantitative variables or sets of data. It is a number ranging from -1.0 to 1.0, with 1.0 indicating a perfect positive linear association, 0 indicating no association, and -1.0 indicating a perfect negative linear association. In this scenario, the correlation coefficient represents the relationship between the number of patrol cars on the turnpike and the average speed of truck drivers. The correlation coefficient (r) is found by dividing the covariance by the product of the standard deviations of each variable. The correlation coefficient of this set of data can be calculated using a calculator and is approximately -0.92. Since -0.92 is a negative number, it indicates a negative association. Since -0.92 is very close to -1.0, a correlation coefficient of -0.92

indicates a strong negative linear association. When this data is plotted on the coordinate plane, it shows a nearly perfect straight line, pointing in a negative direction.

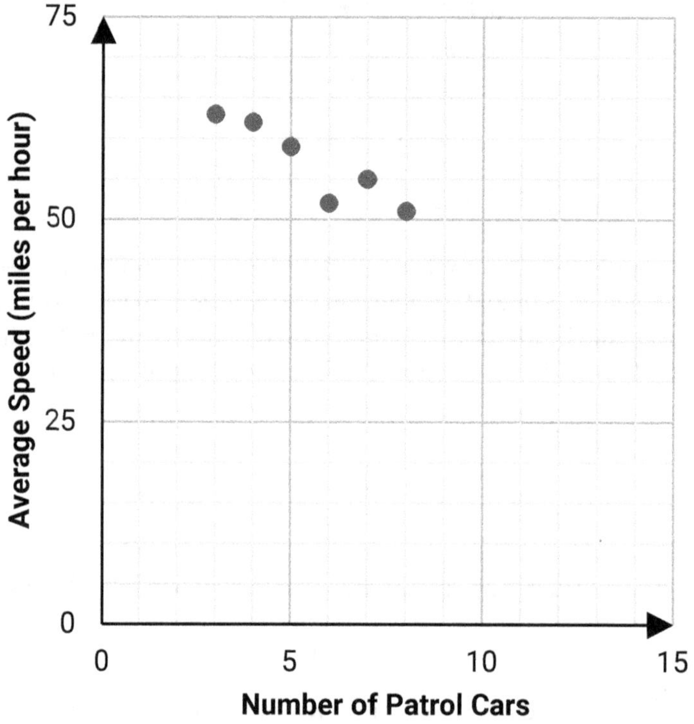

38. B: To solve an "or" inequality, we must solve each part of the inequality separately. We can solve the first part by subtracting 5 from both sides and then dividing both sides by –2.

$$-2x + 5 < 1$$
$$-2x < -4$$
$$x > 2$$

We can solve the second part by adding 3 to both sides.

$$x - 3 \geq 5$$
$$x \geq 8$$

Now we check to see if the two parts overlap. The answer can be expressed as $x > 2$, since any value that satisfies at least one of the two parts counts as a solution. We can see this illustrated on the number line below.

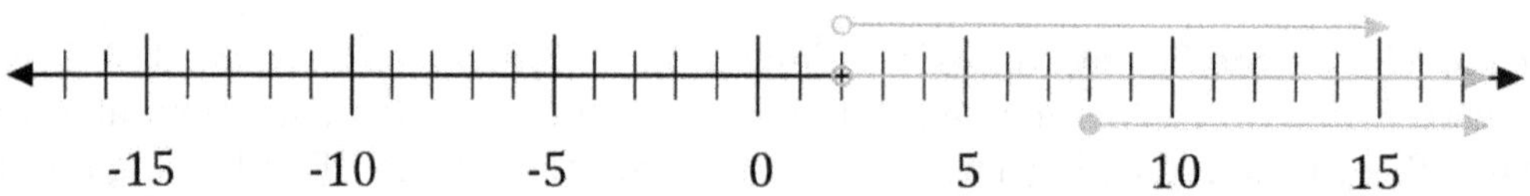

39. D: The unit price for the televisions can be found using the equation unit price = total revenue ÷ number of units sold. Substituting the given values, we have the following equation.

$$\text{unit price} = \frac{3n^2 + 5n + 2}{2n - 3} \div \frac{n^2 + 2n + 1}{8n - 12}$$

To divide, multiply by the reciprocal of the second fraction.

$$\text{unit price} = \frac{3n^2 + 5n + 2}{2n - 3} \cdot \frac{8n - 12}{n^2 + 2n + 1}$$

To simplify, first completely factor the numerator and denominator of each fraction.

$$\frac{(3n + 2)(n + 1)}{2n - 3} \cdot \frac{4(2n - 3)}{(n + 1)(n + 1)}$$

Next, cancel out common factors.

$$\frac{(3n + 2)\cancel{(n + 1)}}{\cancel{2n - 3}} \cdot \frac{4\cancel{(2n - 3)}}{\cancel{(n + 1)}(n + 1)}$$

Finally, multiply the remaining factors.

$$\frac{3n + 2}{1} \cdot \frac{4}{n + 1} = \frac{12n + 8}{n + 1}$$

In terms of n, the unit price of a television is $\frac{12n+8}{n+1}$.

40. C: Isolate the radical by adding 2 to each side.

$$-2 + 2 + \sqrt{4x} = 6 + 2$$
$$\sqrt{4x} = 8$$

Get rid of the radical by squaring both sides of the equation.

$$\left(\sqrt{4x}\right)^2 = (8)^2$$
$$4x = 64$$

Divide both sides by 4.

$$\frac{4x}{4} = \frac{64}{4}$$
$$x = 16$$

Since both sides of the equation were squared when solving, we need to check for extraneous solutions. However, since no solution is not an answer choice, you can assume $x = 16$ is not an extraneous solution.

41. D: The given function and answer choices are in the form $f(x) = mx + b$, where m is the slope and b is the y-intercept. Perpendicular lines have slopes that are the opposite reciprocal of each other. Start by identifying the slope, m, in the original equation, $f(x) = -3x + 4$. From looking at the equation, we see that the slope is –3. Next, identify the opposite reciprocal of –3. The opposite reciprocal of -3 is $\frac{1}{3}$. An equation for a line that is perpendicular to Lamar's line must have a slope of $\frac{1}{3}$. The equation $g(x) = \frac{1}{3}x + 2$ is the only answer option given with a slope of $\frac{1}{3}$.

42. A: One-half of t can be represented by the expression $\frac{t}{2}$, and one-fourth of two more than the number can be represented by $\frac{t+2}{4}$. Since the sum of the two expressions is 20, we have the following equation.

$$\frac{t}{2} + \frac{t+2}{4} = 20$$

To solve a rational equation, multiply both sides of the equation by the lowest common denominator (LCD), to clear out any denominators. The LCD for the denominators 2 and 4 is 4, so multiply both sides of the equation by 4.

$$4\left(\frac{t}{2} + \frac{t+2}{4}\right) = 4 \cdot 20$$
$$2t + t + 2 = 80$$

Now that the denominators are cleared out, solve the equation to find the value for t.

$$2t + t + 2 = 80$$
$$3t + 2 = 80$$
$$3t = 78$$
$$t = 26$$

43. C: Solve for y in the equation.

$$21 - 8y = 7y - 9$$

Since there are variables on both sides of the equal sign, move all variables to one side and combine like terms. Add $8y$ to both sides of the equation.

$$21 - 8y + 8y = 7y + 8y - 9$$
$$21 = 15y - 9$$

Next, add 9 to both sides of the equation.

$$21 + 9 = 15y - 9 + 9$$
$$30 = 15y$$

Finally, divide both sides of the equation by 15.

$$\frac{30}{15} = \frac{15y}{15}$$
$$2 = y$$

44. B: The total length of the fence will be $24\sqrt{12}$ feet. One-third of the fence has been completed, so divide $24\sqrt{12}$ by 3. $\frac{24\sqrt{12}}{3}$ can be simplified to $8\sqrt{12}$. The square root of 12 can be split into the factors 4 and 3, giving the expression $8\sqrt{4}\sqrt{3}$, and then $16\sqrt{3}$. Thus, $16\sqrt{3}$ feet of fence has been built.

45. D: Two lines are parallel if they have the same slope and different y–intercepts. To determine whether the linear equations are parallel, start by identifying the slope, m, of each line. To find the slope, write each equation in slope–intercept form, $y = mx + b$.

$$16x + 8y = 24$$

First, move the x–term to the right side of the equation by subtracting $16x$ from both sides of the equation.

$$\begin{aligned} 16x - 16x + 8y &= 24 - 16x \\ 8y &= 24 - 16x \end{aligned}$$

Next, isolate the variable y by dividing both sides of the equation by 8.

$$\begin{aligned} \frac{8y}{8} &= \frac{24 - 16x}{8} \\ y &= 3 - 2x \\ y &= -2x + 3 \end{aligned}$$

Since the value of m in the equation is -2, the slope of this line is -2.

Now, look at the second equation. Write this equation in slope–intercept form, $y = mx + b$.

$$2x + y = 7$$

First, move the x–term to the right side of the equation by subtracting $2x$ from both sides of the equation.

$$\begin{aligned} 2x - 2x + y &= 7 - 2x \\ y &= 7 - 2x \\ y &= -2x + 7 \end{aligned}$$

Since the value of m in the equation is -2, the slope of the line is -2.

Written in slope–intercept form, both equations have the same slope, –2, but different y–intercepts. Therefore, these two linear equations are parallel.

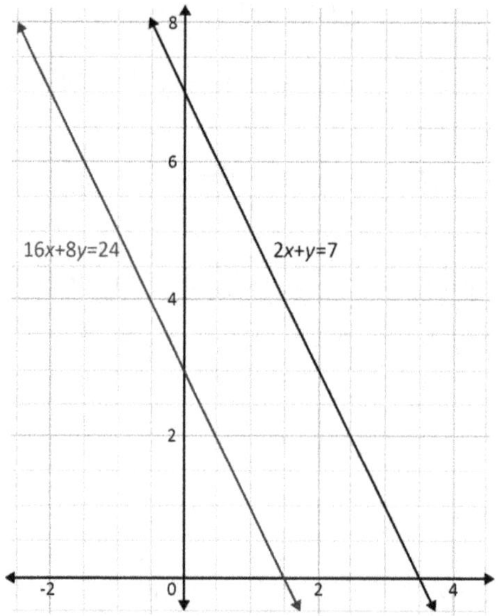

46. B: Apply the distributive property to the expression by multiplying each term inside of the parentheses by 3.

$$3(b + 5) - 11$$

$$3b + 15 - 11$$

Combine like terms to simplify the expression.

$$3b + 4$$

47. B: A quadratic function in standard form, $y = ax^2 + bx + c$, where a, b, and c are real numbers and $a \neq 0$, opens up when $a > 0$ and opens down when $a < 0$. Since $a = -1$ for our quadratic function, its graph opens down.

The x-coordinate for the vertex for a quadratic function can be found using the formula $x = -\frac{b}{2a}$. The y-coordinate for the vertex can be found by substituting $x = -\frac{b}{2a}$ into the function.

$$x = -\frac{3}{2(-1)} = -\frac{3}{-2} = \frac{3}{2} = 1.5$$

Therefore, the x-coordinate of the vertex is 1.5. Next, substitute $x = \frac{3}{2}$ into the function.

$$y = -\left(\frac{3}{2}\right)^2 + 3\left(\frac{3}{2}\right) + 10$$
$$y = -\frac{9}{4} + \frac{9}{2} + 10$$
$$y = \frac{49}{4}$$
$$y = 12.25$$

So, the vertex is $\left(\frac{3}{2}, \frac{49}{4}\right)$, or (1.5,12.25).

Then, locate additional points on the graph by identifying the x-intercepts of the graph. Substitute $y = 0$ into the equation to find the x-intercepts.

$$0 = -x^2 + 3x + 10$$
$$-1 \cdot 0 = -1(-x^2 + 3x + 10)$$
$$0 = x^2 - 3x - 10$$
$$0 = (x + 2)(x - 5)$$

$$x + 2 = 0 \qquad\qquad x - 5 = 0$$

$$x = -2 \qquad\qquad x = 5$$

So, the x-intercepts are (−2,0) and (5,0). Use all these points to find the graph of the function in choice B.

48. C: A quadratic equation can be solved for x when it is written in standard form by using the quadratic formula. The standard form equation of a quadratic is $ax^2 + bx + c = 0$, where a, b, and c are constants and $a \neq 0$.

The quadratic formula is $x = \frac{-b \pm \sqrt{b^2 - 4ac}}{2a}$.

To find the time when the ball hits the ground, set $h = 0$.

$$0 = -16t^2 + 25t + 10$$

Now the quadratic equation is in standard form where $a = -16$, $b = 25$, and $c = 10$. Substituting our values into the quadratic formula and solving for t, we get the following equation.

$$t = \frac{-(25) \pm \sqrt{(25)^2 - 4(-16)(10)}}{2(-16)}$$
$$= \frac{-25 \pm \sqrt{625 + 640}}{-32}$$
$$= \frac{-25 \pm \sqrt{1,265}}{-32}$$
$$= \frac{25 \pm \sqrt{1,265}}{32}$$

We can separate our two solutions as follows.

$$t = \frac{25\sqrt{1,265}}{32} = -0.3 \text{ or } t = \frac{25 + \sqrt{1,265}}{32} = 1.9$$

While both solutions satisfy the quadratic equation, time cannot have a negative value so we can disregard our answer of –0.3. Thus, to the nearest tenth of a second, it takes 1.9 seconds for the ball to hit the ground.

49. C: The volume, V, of a rectangular prism with a square base is $V = s^2h$, where s is the length of a side of the square and h is the height of the prism.

Completely factor the polynomial.

$$3n^3 + 24n^2 + 48n = 3n(n^2 + 8n + 16)$$

Factor out the greatest common factor (GCF) of $3n$ from each term in the polynomial.

$$3n(n^2 + 8n + 16) = 3n(n^2 + 2 \cdot n \cdot 4 + 4^2)$$
$$= 3n(n + 4)^2$$

The quadratic trinomial is a perfect square trinomial that can be factored as $a^2 + 2ab + b^2 = (a + b)^2$.

So, the factored form for the volume of the prism can be modeled by $V = 3n(n + 4)^2$. Thus, $h = 3n$, and $s = n + 4$.

50. D: The point-slope equation of a line is $y - y_1 = m(x - x_1)$, where x_1 and y_1 are the coordinates of a point on the line (x_1, y_1), and m is the slope. Start by identifying the slope using the slope formula $m = \frac{y_2 - y_1}{x_2 - x_1}$. Substitute the x- and y-values from the given coordinates.

$$m = \frac{y_2 - y_1}{x_2 - x_1}$$
$$m = \frac{-3 - (-6)}{4 - 3}$$
$$m = \frac{3}{1} = 3$$

Since we know the slope is 3, replace m with 3 in the equation.

$$y - y_1 = m(x - x_1)$$
$$y - y_1 = 3(x - x_1)$$

Next, substitute the x- and y-coordinates from one of the points into the equation. In this explanation, we will use point $(4, -3)$.

$$y - (-3) = 3(x - 4)$$

51. B: Factor the numerator and denominator completely.

$$\frac{10x^2 - 5x}{15x} = \frac{5x(2x - 1)}{5(3x)}$$

Divide common factors from the numerator and denominator.

$$\frac{\cancel{5x}(2x-1)}{\cancel{5}(3\cancel{x})} = \frac{2x-1}{3}$$

52. D: The perimeter of a rectangle can be found by taking twice the sum of the length and width.

$$P = 2(l + w)$$
$$P = 2\left(\frac{10}{n} + \frac{5}{4n}\right)$$

To combine two rational expressions with unlike denominators, we must first find the least common denominator (LCD). The LCD is the lowest common multiple between the two denominators, which in this case is $4n$. Multiply the first expression in the parentheses by some form of 1 that will give it the LCD.

$$2\left(\frac{10}{n}\cdot\frac{4}{4} + \frac{5}{4n}\right) = 2\left(\frac{40}{4n} + \frac{5}{4n}\right)$$

Combine the numerators for each expression.

$$2\left(\frac{40}{4n} + \frac{5}{4n}\right) = 2\left(\frac{40+5}{4n}\right) = 2\left(\frac{45}{4n}\right) = \frac{90}{4n}$$

Simplify the fraction.

$$\frac{90}{4n} = \frac{45}{2n}$$

The perimeter of the rectangle is $\frac{45}{2n}$.

53. A: The equation $3x + 1.50y = 300$ is written in standard form, we need to convert it into slope-intercept form, $y = mx + b$. First, move the x-term to the right side of the equation by subtracting $3x$ from both sides of the equation.

$$3x + 1.50y = 300$$
$$3x - 3x + 1.50y = 300 - 3x$$

$$1.50y = 300 - 3x$$

Next, isolate the variable y in the term $1.50y$ by dividing both sides of the equation by 1.50.

$$\frac{1.50y}{1.50} = \frac{300-3x}{1.50}$$

$$y = 200 - 2x$$

Finally, switch the order of the two terms on the right side of the equation. The slope-intercept form of the linear equation given is $y = -2x + 200$.

54. A: We will use long division to divide.

$$a-2\overline{)\,2a^2+3a+1}$$

Set up as a long division problem.

$$\begin{array}{r}2a\\ a-2\overline{)\,2a^2+3a+1}\end{array}$$

Divide $2a^2$ by a.

$$\begin{array}{r}2a\\ a-2\overline{)\,2a^2+3a+1}\\ \underline{-(2a^2-4a)}\end{array}$$

Multiply $a-2$ by $2a$ and write beneath $2a^2+3a$.

$$\begin{array}{r}2a\\ a-2\overline{)\,2a^2+3a+1}\\ \underline{-(2a^2-4a)}\\ 7a\end{array}$$

Subtract the result from above to get a remainder of $7a$.

$$\begin{array}{r}2a+7\\ a-2\overline{)\,2a^2+3a+1}\\ \underline{-(2a^2-4a)}\\ 7a+1\end{array}$$

Bring down the 1 and divide a into $7a$.

$$\begin{array}{r}2a+7\\ a-2\overline{)\,2a^2+3a+1}\\ \underline{-(2a^2-4a)}\\ 7a+1\\ \underline{-\ (7a-14)}\end{array}$$

Multiply $a-2$ by 7 and write underneath $7a+1$.

$$\begin{array}{r}2a+7\\ a-2\overline{)\,2a^2+3a+1}\\ \underline{-(2a^2-4a)}\\ 7a+1\\ \underline{-\ (7a-14)}\\ 15\end{array}$$

Subtract the result from above to get a remainder of 15.

Dividing $2a^2-3a+1$ by $a-2$ yields a quotient of $2a+7$ and a remainder of 15 which can be written as $2a+7+\frac{15}{a-2}$.

55. B: To figure out how much money Mia plans to deposit, use the formula $a_n = a_1r^{(n-1)}$. First, identify the values of the variables used in the equation. August will be Mia's eighth deposit, so we are finding the eighth term in the geometric sequence. Since n is the number in the sequence that needs to be identified, replace n with 8 in the formula. The variable a_1 is the first term in the sequence. Mia's first deposit was in January when she opened her bank account with $10. Therefore, the first term in the sequence is 10. Each month, Mia plans to double the amount of money deposited. Each term gets doubled, or multiplied by 2, resulting in the subsequent term in the sequence. Based on this information, the common ratio r in this sequence is 2. Therefore, replace r with 2 in the formula.

$$a_n = a_1r^{(n-1)}$$
$$a_8 = (10)(2)^{(8-1)}$$

Simplify using the order of operations.

$$a_8 = (10)(2)^{(7)}$$
$$a_8 = (10)(128)$$
$$a_8 = 1{,}280$$

Mia plans to deposit \$1,280 into her bank account in August.

56. D: Start by finding the slope, m. To do so, use the slope formula $m = \frac{y_2 - y_1}{x_2 - x_1}$. Substitute the x- and y-values from the two coordinates given.

$$m = \frac{3 - (-2)}{-1 - 4} = \frac{5}{-5} = -1$$

Now that we know the slope is –1, we can write a point-slope equation using the slope and one of the points given. The point-slope equation of a line is $y - y_1 = m(x - x_1)$, where x_1 and y_1 are the coordinates of one point on the line, and m is the slope. Since we know the slope is –1, replace m with –1 in the equation. Next, substitute the x- and y-coordinates from one of the points into the equation. In this explanation, we will use point $(-1,3)$. The point-slope form of the equation is $y - 3 = -1(x - (-1))$.

Rearrange the equation to write it in standard form.

$$y - 3 = -1(x + 1)$$
$$y - 3 + 3 = -1x - 1 + 3$$
$$y = -1x + 2$$
$$x + y = 2$$

57. D: The table shows a linear relationship between t and M. To find an equation of this line, we choose two convenient points from the table, say $(1,73)$ and $(2,133)$, and use them to find the slope of the line: $m = \frac{133-73}{2-1} = \frac{60}{1} = 60$. Then we substitute this slope and either point into point-slope form, which looks like $M - M_1 = m(t - t_1)$ using these variables, and solve for M.

$$M - 73 = 60(t - 1)$$
$$M - 73 = 60t - 60$$
$$M = 60t + 13$$

Since each M-value on this line is the largest that makes the inequality true for the corresponding t-value, the rest are smaller. So, the inequality we seek is $M \leq 60t + 13$. In other words, if we must slow down due to traffic, the number of miles traveled will be less than the miles traveled if there was no traffic. The answer choices are in standard form, so rearrange this inequality to get $60t - M \geq -13$.

58. A: To find the x-intercepts for the graph of a quadratic function, we set $y = 0$ and solve for x. Setting $y = 0$ for our function, we get the quadratic equation, $0 = x^2 + 2x - 8$. The roots for the quadratic equation are also the x-intercepts of the corresponding graph.

While the discriminant for the quadratic formula will not tell you what the roots are, it can be used to determine how many roots exist. The discriminant for a quadratic equation in the standard form

$0 = ax^2 + bx + c$ is $b^2 - 4ac$, where a, b, and c are real numbers, and $a \neq 0$. Three conditions exist to determine how many roots a quadratic equation has using its discriminant.

$b^2 - 4ac > 0$	There are two distinct real roots.
$b^2 - 4ac = 0$	There is one real root (often referred to as a double root).
$b^2 - 4ac < 0$	No real roots exist.

For our quadratic equation $0 = x^2 + 2x - 8$, $a = 1$, $b = 2$, and $c = -8$. Find the discriminant.

$$b^2 - 4ac = (2)^2 - 4(1)(-8) = 4 + 32 = 36$$

Since the discriminant is positive in value, the quadratic equation has two distinct real roots, so the graph of the quadratic function $y = x^2 + 2x - 8$ crosses the x-axis two times.

59. C: The y-coordinate of the vertex for the graph of a quadratic function is either the minimum value or maximum value for the function. To find the vertex of the quadratic function, use the formula $x = -\frac{b}{2a}$ to find the x-coordinate, then plug this value into the original equation to find the y-coordinate.

$$x = \frac{-(-2)}{2(1)} = \frac{2}{2} = 1$$

Now, plug this in for x in the original equation and solve for y.

$$y = (1)^2 - 2(1) - 3 = 1 - 2 - 3 = -4$$

Therefore, the vertex of the parabola is at $(1, -4)$. Since the value of a is positive, the parabola opens up. This means the range of the function is $y \geq -4$.

60. D: A quadratic binomial in the form $a^2 - b^2$ is called a difference of two squares, where a^2 and b^2 are perfect squares, and factors to $(a - b)(a + b)$.

The two terms in our binomial are perfect squares and can be written as $(4s)^2 + (11)^2$, but it is not a difference of two perfect squares. It is a sum that cannot be factored.

61. A: Two lines are perpendicular if their slopes are opposite reciprocals of one another. To identify an equation of a line perpendicular to $x + 3y = 6$, start by identifying the slope, m, of the given line. To do so, write the equation in slope–intercept form, $y = mx + b$.

First, move the x–term to the right side of the equation by subtracting x from both sides of the equation.

$$3y = 6 - x$$

Next, divide both sides of the equation by 3.

$$\frac{3y}{3} = \frac{6 - x}{3}$$
$$y = 2 - \frac{1}{3}x$$

Switch the order of the two terms on the right side of the equation. The slope-intercept form of the linear equation given is $y = -\frac{1}{3}x + 2$.

Now that the equation is written in slope-intercept form, identify the slope. Since the value of m in this equation is $-\frac{1}{3}$, the slope of a perpendicular line must be the opposite reciprocal of $-\frac{1}{3}$, which is 3.

Write a new equation of a line in slope-intercept form, $y = mx + b$, and replace m with 3 in the equation.

$$y = 3x + b$$

Next, substitute the point $(9, -1)$ into the equation $y = 3x + b$.

$$(-1) = 3(9) + b$$

Simplify the equation by multiplying 3 and 9.

$$-1 = 27 + b$$

From here, solve for b. Subtract 27 from both sides of the equation.

$$-28 = b$$

Now that we know the slope is 3 and the y-intercept is -28, write the equation for this line in slope-intercept form, $y = mx + b$.

$$y = 3x - 28$$

The equation of a line that is perpendicular to the graph of $x + 3y = 6$ and contains point $(9, -1)$ is $y = 3x - 28$.

62. D: Decompose $11x$ to the sum of two terms so that one has a common factor with $3x^2$ and the other has a common factor with 10.

$$3x^2 + 6x + 5x + 10$$

Now, factor the greatest common factor (GCF) from the first two terms and the second two terms.

$$3x(x + 2) + 5(x + 2)$$

Finally, factor the GCF of $(x + 2)$ from the two groups.

$$(x + 2)(3x + 5)$$

63. B: $\frac{x}{2} > 3$ is an inequality that can be solved in one step. Treat the inequality as an equation, and isolate the variable using inverse operations. Multiply both sides by 2.

$$2\left(\frac{x}{2}\right) > 3(2)$$

$$x > 6$$

64. D: The equation of a line perpendicular to the x-axis runs vertically, passing through the x-axis. The x-coordinate of each point on this line is constant, represented by the variable k. Therefore, an equation of a line perpendicular to the x-axis is $x = k$. Based on the answer choices, k is equal to 8. Therefore, $x = 8$.

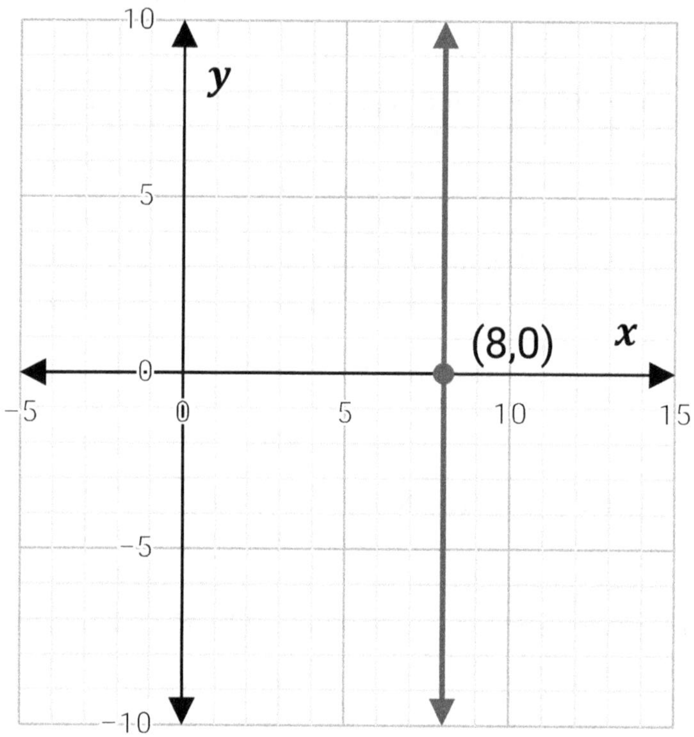

A vertical line always has an undefined slope. Recall that slope is rise over run, which is the change in y over the change in x. A vertical line will always have a change in x equal to 0. Since we can't divide by 0, its slope is undefined. Therefore, the slope of the line $x = 8$ is undefined.

65. C: The area, A, of a rectangle is $A = l \times w$, where l is the length of the rectangle, and w is the width of the rectangle.

Substituting the given values for the length, width, and area, we have the following equation.

$$63 = (x + 4)(x + 2)$$
$$63 = x^2 + 2x + 4x + 8$$
$$63 = x^2 + 6x + 8$$

This equation can be solved by factoring. First set the equation equal to 0 by subtracting 63 from both sides.

$$0 = x^2 + 6x - 55$$

Factor the quadratic on the right-hand side, then solve for x.

$$0 = x^2 + 6x - 55$$
$$0 = (x + 11)(x - 5)$$
$$x + 11 = 0 \text{ and } x - 5 = 0$$
$$x = -11 \text{ and } x = 5$$

While both answers satisfy the quadratic equation, only 5 satisfies the x-value for the possible dimensions of the rectangle since $x = 5$ yields positive values for the length and width of the rectangle, but $x = -11$ yields a length of –7 feet and a width of –9 feet, which is not possible.

66. B: To solve, start by finding the value of $f(8)$. Substitute 8 into the function $f(x) = x^2 - 3x$ by replacing the variable x with 8 and simplifying the expression.

$$f(x) = x^2 - 3x$$
$$f(8) = (8)^2 - 3(8)$$
$$f(8) = 64 - 3(8)$$
$$f(8) = 64 - 24$$
$$f(8) = 40$$

Now, find the value of $f(2)$. Substitute 2 into the function $f(x) = x^2 - 3x$ by replacing the variable x with 2, then simplify the expression.

$$f(x) = x^2 - 3x$$
$$f(2) = (2)^2 - 3(2)$$
$$f(2) = 4 - 3(2)$$
$$f(2) = 4 - 6$$
$$f(2) = -2$$

Now that we know the values of $f(8)$ and $f(2)$ in relation to the function $f(x) = x^2 - 3x$, we can find the value of $\frac{f(8)}{f(2)}$.

$$\frac{f(8)}{f(2)} = \frac{40}{-2} = -20$$

67. D: To find the answer, solve for a in the equation by isolating it on one side of the equation.

$$7a - 21 = 28$$

Start by adding 21 to both sides of the equation.

$$7a - 21 + 21 = 28 + 21$$
$$7a = 49$$

Next, divide both sides of the equation by 7.

$$\frac{7a}{7} = \frac{49}{7}$$
$$a = 7$$

68. C: The equation that represents the cost of Apartment A is $1{,}200m + 300$, and the equation that represents Apartment B is $1{,}050m + 600$. To find how many months before the cost is the same, set the equations equal to one another and solve for m.

$$1{,}200m + 300 = 1{,}050m + 600$$

First, subtract $1{,}050m$ from both sides of the equation.

$$1{,}200m - 1{,}050m + 300 = 1{,}050m - 1{,}050m + 600$$
$$150m + 300 = 600$$

Then, subtract 300 from both sides of the equation.

$$150m + 300 - 300 = 600 - 300$$
$$150m = 300$$

Finally, divide both sides of the equation by 150.

$$\frac{150m}{150} = \frac{300}{150}$$
$$m = 2$$

It will take 2 months for the costs of both apartments to be the same.

69. C: Let the variable b represent the price of each bag of food. Sue only has $80 to spend, so the total cost of her dog food purchase needs to be less than or equal to 80. The inequality $5b \leq 80$ can be used to determine the maximum price per bag. To solve for the price per bag, divide both sides by 5.

$$b \leq 16$$

This means that Sue can afford 5 bags of dog food as long as each bag is less than or equal to $16.

70. D: To find the value of y for the linear function when the value of x is 8, substitute 8 in for x and evaluate to find the value of y.

$$y = 5(8) + 3$$
$$y = 40 + 3$$
$$y = 43$$

71. D: A quadratic function in standard form is $y = ax^2 + bx + c$, where a, b, and c are real numbers and $a \neq 0$.

$y = 3(x - 2)(x - 2) + 1$	Square the binomial.
$y = 3(x^2 - 4x + 4) + 1$	Distribute using the FOIL method.
$y = 3x^2 - 12x + 12 + 1$	Distribute the coefficient of 3.
$y = 3x^2 - 12x + 13$	Combine like terms.

72. A: The equation of a parabola in vertex form is $y = a(x - h)^2 + k$, where (h, k) is the vertex of the parabola, and a is a real number that cannot be 0.

Since the projectile follows a parabolic path, its maximum height represents the y-coordinate of the vertex of the parabola, and the time at which this occurs represents the x-coordinate of the vertex. Substituting the vertex (3,126) into the vertex form, we get the following equation.

$$h = a(t - 3)^2 + 126$$

Also, since the projectile is launched from the ground, its height is 0 at time $t = 0$, so the point (0,0) is another point on the graph of the parabola. Substitute (0,0) in for x and y respectively to find the value for a.

$$\begin{aligned} 0 &= a(0-3)^2 + 126 \\ 0 &= a(-3)^2 + 126 \\ 0 &= 9a + 126 \\ -126 &= 9a \\ \frac{-126}{9} &= a \\ -14 &= a \end{aligned}$$

In vertex form, the equation for the path of the projectile is $h = -14(t - 3)^2 + 126$. Note the graph of the parabola opens downward, verifying the value for a is less than 0.

73. A: Start by creating a table that shows the relationship between hours walked (h) and distance (d).

Hours walked (h)	Distance in miles (d)
0	0
1	3
2	6
3	9

As you can see, the table represents the equation $d = 3h$. The independent variable is hours and the dependent variable is the distance walked in miles. From here, we can use the data in the table to graph points on the coordinate plane. The x-axis represents the independent variable, which is hours walked, and the y-axis represents the dependent variable, which is the distance walked in miles.

Rewrite the data from the table as points on the coordinate plane, (x y): (0,0), (1,3), (2,6), (3,9). The only graph with all four points is choice A.

74. C: Solve the equation for y by isolating the variable y on one side of the equation.

$$-x + 8y = -14$$

First, move x from the left side of the equation to the right by adding x to both sides of the equation.

$$\begin{aligned} -x + x + 8y &= -14 + x \\ 8y &= -14 + x \end{aligned}$$

Next, divide both sides of the equation by 8.

$$\frac{8y}{8} = \frac{-14 + x}{8}$$
$$y = \frac{-14 + x}{8}$$

75. C: First write the statement as an expression.

$$3n^2(4n^3 + 1)$$

Now, multiply the left side of the equation using the distributive property of multiplication.

$$3n^2(4n^3 + 1)$$

$$3n^2 \cdot 4n^3 + 3n^2 \cdot 1$$

$$12n^{2+3} + 3n^2$$

$$12n^5 + 3n^2$$

So, the product in terms of n is $12n^5 + 3n^2$.

Practice Tests #2 and #3

To take these additional practice tests, visit the bonus page for this product:
mometrix.com/bonus948/algebra1tb

Lesson and Chapter Quiz Answer Key

For worked out answer explanations, please visit the bonus page for this product at **mometrix.com/bonus948/algebra1tb** or scan the QR code.

Lesson Quizzes

Chapter 1: Solving Equations

	1.1	1.2	1.3	1.4	1.5	1.6	1.7	1.8	1.9	1.10
1	C	C	D	A	C	B	D	A	B	C
2	A	D	B	A	B	B	B	B	D	C
3	A	A	C	B	A	D	C	C	D	C
4	B	A	D	A	A	A	D	A	B	D
5	D	B	A	C	B	C	C	C	C	B
6	A	A	A	B	A	D	A	D	D	A
7	D	D	B	A	B	D	A	B	B	C
8	B	C	B	B	D	A	C	C	A	D
9	B	D	C	B	B	C	D	C	D	A
10	A	D	A	B	D	D	A	B	A	C
11	D	B	B	A	A	A	C	C	B	D
12	B	A	A	A	D	A	B	B	D	B
13	A	B	D	B	C	C	B	B	C	A
14	B	C	D	B	B	B	A	C	D	B
15	D	A	D	C	C	A	D	B	A	A

CHAPTER 2: FUNCTIONS AND SEQUENCES

	2.1	2.2	2.3	2.4	2.5	2.6
1	A	C	A	B	C	D
2	D	B	B	A	B	A
3	B	A	B	C	B	A
4	A	D	A	C	C	D
5	A	A	C	B	A	A
6	B	B	C	B	C	A
7	B	A	B	C	A	B
8	C	C	A	B	C	B
9	A	B	B	D	D	A
10	D	D	D	C	B	D
11	D	B	C	A	B	C
12	B	B	C	A	C	A
13	C	A	A	B	C	C
14	B	C	D	C	D	A
15	C	C	D	A	D	B

CHAPTER 3: LINEAR EQUATIONS

	3.1	3.2	3.3	3.4	3.5	3.6	3.7	3.8	3.9	3.10	3.11	3.12
1	D	C	A	D	D	B	A	A	B	C	B	D
2	A	B	A	D	B	B	A	C	B	C	C	A
3	D	B	B	A	D	B	D	D	C	B	D	D
4	A	A	A	C	A	C	C	C	D	C	A	A
5	C	D	D	A	C	A	B	D	C	D	B	B
6	C	A	C	B	C	B	C	A	A	A	B	B
7	B	C	C	D	B	B	C	D	A	D	C	C
8	A	D	D	D	C	C	A	A	D	B	A	A
9	C	C	B	D	D	C	D	B	C	B	D	B
10	B	B	A	C	A	D	B	B	D	D	A	D
11	D	A	C	A	D	A	B	C	B	C	A	B
12	D	A	D	D	C	A	C	B	B	C	C	C
13	B	D	B	B	C	D	A	A	A	A	B	C
14	C	C	B	D	B	B	D	D	C	B	D	A
15	A	B	A	D	D	C	C	C	C	B	D	D

Chapter 4: Inequalities

	4.1	4.2	4.3	4.4	4.5	4.6	4.7
1	C	A	A	C	A	A	A
2	D	A	B	A	B	A	C
3	C	D	A	D	D	D	D
4	B	D	D	A	B	A	A
5	B	B	C	B	A	C	B
6	A	C	C	D	C	B	C
7	C	B	D	C	C	B	A
8	D	D	A	B	D	A	C
9	B	C	B	B	B	D	D
10	D	C	B	C	D	C	A
11	A	B	D	A	B	B	C
12	B	D	B	D	B	D	B
13	B	B	D	B	C	C	D
14	C	A	A	C	D	A	C
15	C	B	C	A	D	C	B

Chapter 5: Systems of Equations

	5.1	5.2	5.3	5.4	5.5
1	A	D	A	A	B
2	C	B	C	C	B
3	C	C	B	D	D
4	B	D	D	B	C
5	C	C	C	D	B
6	B	A	A	A	A
7	D	B	D	A	D
8	A	C	D	C	B
9	B	C	B	B	C
10	C	B	A	D	A
11	A	D	C	C	D
12	D	D	C	D	A
13	B	A	B	B	C
14	C	B	A	C	B
15	B	C	B	A	A

CHAPTER 6: POLYNOMIALS

	6.1	6.2	6.3	6.4	6.5	6.6	6.7	6.8	6.9	6.10	6.11	6.12
1	C	B	C	D	B	B	C	B	C	C	D	B
2	A	C	D	B	C	D	C	A	A	B	A	B
3	D	A	C	B	D	A	B	B	D	D	D	A
4	B	C	D	A	D	B	A	A	B	B	D	D
5	D	D	C	B	B	C	C	A	A	C	A	C
6	B	B	B	D	C	A	D	A	D	D	C	C
7	B	A	B	D	B	D	A	D	D	C	A	B
8	B	D	D	D	C	A	D	C	C	A	B	C
9	B	C	B	A	D	C	B	B	B	C	D	B
10	C	A	C	B	D	B	C	A	A	B	C	A
11	D	B	D	C	B	C	C	B	B	B	B	B
12	A	D	B	A	A	D	C	D	D	D	A	C
13	B	C	B	C	C	A	D	B	D	C	C	C
14	A	B	D	D	C	C	D	A	C	D	B	D
15	D	A	B	B	D	D	B	B	C	C	C	A

CHAPTER 7: ALGEBRAIC FRACTIONS

	7.1	7.2	7.3	7.4	7.5	7.6	7.7
1	D	C	A	D	B	B	D
2	A	B	C	A	B	D	B
3	B	D	D	A	C	B	A
4	C	A	C	C	A	A	C
5	D	B	C	A	D	C	D
6	D	A	B	B	C	C	C
7	A	B	A	C	C	D	C
8	D	D	A	C	A	B	A
9	C	A	A	D	D	A	B
10	D	D	B	B	D	A	D
11	A	B	A	C	B	A	B
12	B	D	C	D	C	C	B
13	A	C	A	A	D	B	C
14	B	C	C	A	A	D	D
15	C	B	C	C	D	D	C

CHAPTER 8: QUADRATIC FUNCTIONS

	8.1	8.2	8.3	8.4	8.5	8.6	8.7	8.8	8.9	8.10	8.11
1	A	B	B	D	A	B	D	B	C	B	B
2	A	D	C	A	B	A	B	C	B	C	D
3	C	A	B	D	C	D	B	C	D	D	B
4	C	C	D	A	C	D	A	D	B	C	A
5	C	B	B	C	D	B	C	A	D	A	A
6	B	B	A	B	C	B	A	D	A	D	C
7	B	D	A	B	B	C	B	D	A	B	B
8	D	C	C	A	D	A	D	A	A	D	A
9	A	D	A	D	A	C	A	B	C	D	B
10	C	A	D	C	A	D	A	D	B	C	D
11	D	A	D	C	C	B	D	C	C	C	C
12	D	B	B	C	D	D	C	C	C	A	A
13	B	B	C	B	C	A	C	B	A	B	D
14	A	D	C	A	B	A	A	B	C	C	D
15	C	A	D	D	B	C	B	D	C	D	A

CHAPTER 9: EXPONENTIAL FUNCTIONS

	9.1	9.2	9.3
1	C	B	B
2	A	A	B
3	D	D	C
4	D	C	C
5	B	D	B
6	C	A	A
7	D	B	A
8	B	C	D
9	B	C	B
10	D	D	D
11	D	A	D
12	C	D	C
13	C	D	A
14	A	B	D
15	A	C	C

CHAPTER 10: RADICAL EXPRESSIONS

	10.1	10.2	10.3	10.4	10.5	10.6	10.7
1	C	C	C	B	D	D	B
2	D	D	D	A	A	B	C
3	B	A	D	B	C	B	C
4	C	D	A	A	A	A	B
5	C	A	B	C	C	C	D
6	B	D	B	D	D	A	A
7	D	A	C	B	C	A	A
8	A	D	B	A	D	B	A
9	B	D	C	A	D	D	D
10	B	A	A	B	C	A	C
11	A	D	A	D	B	C	B
12	A	B	B	C	D	C	B
13	A	B	B	A	C	B	A
14	B	D	C	B	A	B	D
15	D	A	A	B	A	B	C

Chapter Quizzes

	Ch1	Ch2	Ch3	Ch4	Ch5	Ch6	Ch7	Ch8	Ch9	Ch10
1	A	B	C	B	B	A	D	C	B	D
2	B	A	D	B	B	A	D	A	C	A
3	B	A	C	D	C	B	A	D	B	A
4	D	C	B	D	A	B	C	C	A	C
5	B	D	D	C	D	D	D	B	C	C
6	C	B	D	D	D	A	A	A	A	D
7	C	A	A	A	A	B	D	B	A	B
8	D	D	C	A	D	C	B	A	A	D
9	D	C	A	B	D	D	B	B	D	D
10	B	C	C	A	C	A	B	B	B	D
11	A	D	C	D	C	D	C	B	A	A
12	D	C	A	B	C	C	B	B	D	B
13	A	C	C	D	D	D	C	C	A	C
14	D	B	B	C	D	D	C	B	A	A
15	A	B	D	A	C	B	A	A	D	C
16	D	D	C	D	A	A	C	B	A	D
17	D	D	B	C	A	C	C	A	B	B
18	D	B	B	A	C	B	B	A	D	A
19	B	A	B	A	C	C	B	B	B	A
20	D	C	C	D	C	C	A	A	A	C

For worked out answer explanations, please visit the bonus page for this product at **mometrix.com/bonus948/algebra1tb** or scan the QR code.

How to Overcome Test Anxiety

Just the thought of taking a test is enough to make most people a little nervous. A test is an important event that can have a long-term impact on your future, so it's important to take it seriously and it's natural to feel anxious about performing well. But just because anxiety is normal, that doesn't mean that it's helpful in test taking, or that you should simply accept it as part of your life. Anxiety can have a variety of effects. These effects can be mild, like making you feel slightly nervous, or severe, like blocking your ability to focus or remember even a simple detail.

If you experience test anxiety—whether severe or mild—it's important to know how to beat it. To discover this, first you need to understand what causes test anxiety.

Causes of Test Anxiety

While we often think of anxiety as an uncontrollable emotional state, it can actually be caused by simple, practical things. One of the most common causes of test anxiety is that a person does not feel adequately prepared for their test. This feeling can be the result of many different issues such as poor study habits or lack of organization, but the most common culprit is time management. Starting to study too late, failing to organize your study time to cover all of the material, or being distracted while you study will mean that you're not well prepared for the test. This may lead to cramming the night before, which will cause you to be physically and mentally exhausted for the test. Poor time management also contributes to feelings of stress, fear, and hopelessness as you realize you are not well prepared but don't know what to do about it.

Other times, test anxiety is not related to your preparation for the test but comes from unresolved fear. This may be a past failure on a test, or poor performance on tests in general. It may come from comparing yourself to others who seem to be performing better or from the stress of living up to expectations. Anxiety may be driven by fears of the future—how failure on this test would affect your educational and career goals. These fears are often completely irrational, but they can still negatively impact your test performance.

Elements of Test Anxiety

As mentioned earlier, test anxiety is considered to be an emotional state, but it has physical and mental components as well. Sometimes you may not even realize that you are suffering from test anxiety until you notice the physical symptoms. These can include trembling hands, rapid heartbeat, sweating, nausea, and tense muscles. Extreme anxiety may lead to fainting or vomiting. Obviously, any of these symptoms can have a negative impact on testing. It is important to recognize them as soon as they begin to occur so that you can address the problem before it damages your performance.

The mental components of test anxiety include trouble focusing and inability to remember learned information. During a test, your mind is on high alert, which can help you recall information and stay focused for an extended period of time. However, anxiety interferes with your mind's natural processes, causing you to blank out, even on the questions you know well. The strain of testing during anxiety makes it difficult to stay focused, especially on a test that may take several hours. Extreme anxiety can take a huge mental toll, making it difficult not only to recall test information but even to understand the test questions or pull your thoughts together.

Effects of Test Anxiety

Test anxiety is like a disease—if left untreated, it will get progressively worse. Anxiety leads to poor performance, and this reinforces the feelings of fear and failure, which in turn lead to poor performances on subsequent tests. It can grow from a mild nervousness to a crippling condition. If allowed to progress, test anxiety can have a big impact on your schooling, and consequently on your future.

Test anxiety can spread to other parts of your life. Anxiety on tests can become anxiety in any stressful situation, and blanking on a test can turn into panicking in a job situation. But fortunately, you don't have to let anxiety rule your testing and determine your grades. There are a number of relatively simple steps you can take to move past anxiety and function normally on a test and in the rest of life.

Physical Steps for Beating Test Anxiety

While test anxiety is a serious problem, the good news is that it can be overcome. It doesn't have to control your ability to think and remember information. While it may take time, you can begin taking steps today to beat anxiety.

Just as your first hint that you may be struggling with anxiety comes from the physical symptoms, the first step to treating it is also physical. Rest is crucial for having a clear, strong mind. If you are tired, it is much easier to give in to anxiety. But if you establish good sleep habits, your body and mind will be ready to perform optimally, without the strain of exhaustion. Additionally, sleeping well helps you to retain information better, so you're more likely to recall the answers when you see the test questions.

Getting good sleep means more than going to bed on time. It's important to allow your brain time to relax. Take study breaks from time to time so it doesn't get overworked, and don't study right before bed. Take time to rest your mind before trying to rest your body, or you may find it difficult to fall asleep.

Along with sleep, other aspects of physical health are important in preparing for a test. Good nutrition is vital for good brain function. Sugary foods and drinks may give a burst of energy but this burst is followed by a crash, both physically and emotionally. Instead, fuel your body with protein and vitamin-rich foods.

Also, drink plenty of water. Dehydration can lead to headaches and exhaustion, especially if your brain is already under stress from the rigors of the test. Particularly if your test is a long one, drink water during the breaks. And if possible, take an energy-boosting snack to eat between sections.

Along with sleep and diet, a third important part of physical health is exercise. Maintaining a steady workout schedule is helpful, but even taking 5-minute study breaks to walk can help get your blood pumping faster and clear your head. Exercise also releases endorphins, which contribute to a positive feeling and can help combat test anxiety.

When you nurture your physical health, you are also contributing to your mental health. If your body is healthy, your mind is much more likely to be healthy as well. So take time to rest, nourish your body with healthy food and water, and get moving as much as possible. Taking these physical steps will make you stronger and more able to take the mental steps necessary to overcome test anxiety.

Mental Steps for Beating Test Anxiety

Working on the mental side of test anxiety can be more challenging, but as with the physical side, there are clear steps you can take to overcome it. As mentioned earlier, test anxiety often stems from lack of preparation, so the obvious solution is to prepare for the test. Effective studying may be the most important weapon you have for beating test anxiety, but you can and should employ several other mental tools to combat fear.

First, boost your confidence by reminding yourself of past success—tests or projects that you aced. If you're putting as much effort into preparing for this test as you did for those, there's no reason you should expect to fail here. Work hard to prepare; then trust your preparation.

Second, surround yourself with encouraging people. It can be helpful to find a study group, but be sure that the people you're around will encourage a positive attitude. If you spend time with others who are anxious or cynical, this will only contribute to your own anxiety. Look for others who are motivated to study hard from a desire to succeed, not from a fear of failure.

Third, reward yourself. A test is physically and mentally tiring, even without anxiety, and it can be helpful to have something to look forward to. Plan an activity following the test, regardless of the outcome, such as going to a movie or getting ice cream.

When you are taking the test, if you find yourself beginning to feel anxious, remind yourself that you know the material. Visualize successfully completing the test. Then take a few deep, relaxing breaths and return to it. Work through the questions carefully but with confidence, knowing that you are capable of succeeding.

Developing a healthy mental approach to test taking will also aid in other areas of life. Test anxiety affects more than just the actual test—it can be damaging to your mental health and even contribute to depression. It's important to beat test anxiety before it becomes a problem for more than testing.

Study Strategy

Being prepared for the test is necessary to combat anxiety, but what does being prepared look like? You may study for hours on end and still not feel prepared. What you need is a strategy for test prep. The next few pages outline our recommended steps to help you plan out and conquer the challenge of preparation.

Step 1: Scope Out the Test

Learn everything you can about the format (multiple choice, essay, etc.) and what will be on the test. Gather any study materials, course outlines, or sample exams that may be available. Not only will this help you to prepare, but knowing what to expect can help to alleviate test anxiety.

Step 2: Map Out the Material

Look through the textbook or study guide and make note of how many chapters or sections it has. Then divide these over the time you have. For example, if a book has 15 chapters and you have five days to study, you need to cover three chapters each day. Even better, if you have the time, leave an extra day at the end for overall review after you have gone through the material in depth.

If time is limited, you may need to prioritize the material. Look through it and make note of which sections you think you already have a good grasp on, and which need review. While you are studying, skim quickly through the familiar sections and take more time on the challenging parts.

Write out your plan so you don't get lost as you go. Having a written plan also helps you feel more in control of the study, so anxiety is less likely to arise from feeling overwhelmed at the amount to cover.

Step 3: Gather Your Tools

Decide what study method works best for you. Do you prefer to highlight in the book as you study and then go back over the highlighted portions? Or do you type out notes of the important information? Or is it helpful to make flashcards that you can carry with you? Assemble the pens, index cards, highlighters, post-it notes, and any other materials you may need so you won't be distracted by getting up to find things while you study.

If you're having a hard time retaining the information or organizing your notes, experiment with different methods. For example, try color-coding by subject with colored pens, highlighters, or post-it notes. If you learn better by hearing, try recording yourself reading your notes so you can listen while in the car, working out, or simply sitting at your desk. Ask a friend to quiz you from your flashcards, or try teaching someone the material to solidify it in your mind.

Step 4: Create Your Environment

It's important to avoid distractions while you study. This includes both the obvious distractions like visitors and the subtle distractions like an uncomfortable chair (or a too-comfortable couch that makes you want to fall asleep). Set up the best study environment possible: good lighting and a comfortable work area. If background music helps you focus, you may want to turn it on, but otherwise keep the room quiet. If you are using a computer to take notes, be sure you don't have any other windows open, especially applications like social media, games, or anything else that could distract you. Silence your phone and turn off notifications. Be sure to keep water close by so you stay hydrated while you study (but avoid unhealthy drinks and snacks).

Also, take into account the best time of day to study. Are you freshest first thing in the morning? Try to set aside some time then to work through the material. Is your mind clearer in the afternoon or evening? Schedule your study session then. Another method is to study at the same time of day that you will take the test, so that your brain gets used to working on the material at that time and will be ready to focus at test time.

Step 5: Study!

Once you have done all the study preparation, it's time to settle into the actual studying. Sit down, take a few moments to settle your mind so you can focus, and begin to follow your study plan. Don't give in to distractions or let yourself procrastinate. This is your time to prepare so you'll be ready to fearlessly approach the test. Make the most of the time and stay focused.

Of course, you don't want to burn out. If you study too long you may find that you're not retaining the information very well. Take regular study breaks. For example, taking five minutes out of every hour to walk briskly, breathing deeply and swinging your arms, can help your mind stay fresh.

As you get to the end of each chapter or section, it's a good idea to do a quick review. Remind yourself of what you learned and work on any difficult parts. When you feel that you've mastered the material, move on to the next part. At the end of your study session, briefly skim through your notes again.

But while review is helpful, cramming last minute is NOT. If at all possible, work ahead so that you won't need to fit all your study into the last day. Cramming overloads your brain with more information than it can process and retain, and your tired mind may struggle to recall even previously learned information when it is overwhelmed with last-minute study. Also, the urgent nature of cramming and the stress placed on your brain contribute to anxiety. You'll be more likely to go to the test feeling unprepared and having trouble thinking clearly.

So don't cram, and don't stay up late before the test, even just to review your notes at a leisurely pace. Your brain needs rest more than it needs to go over the information again. In fact, plan to finish your studies by noon or early afternoon the day before the test. Give your brain the rest of the day to relax or focus on other things, and get a good night's sleep. Then you will be fresh for the test and better able to recall what you've studied.

Step 6: Take a Practice Test

Many courses offer sample tests, either online or in the study materials. This is an excellent resource to check whether you have mastered the material, as well as to prepare for the test format and environment.

Check the test format ahead of time: the number of questions, the type (multiple choice, free response, etc.), and the time limit. Then create a plan for working through them. For example, if you have 30 minutes to take a 60-question test, your limit is 30 seconds per question. Spend less time on the questions you know well so that you can take more time on the difficult ones.

If you have time to take several practice tests, take the first one open book, with no time limit. Work through the questions at your own pace and make sure you fully understand them. Gradually work up to taking a test under test conditions: sit at a desk with all study materials put away and set a timer. Pace yourself to make sure you finish the test with time to spare and go back to check your answers if you have time.

After each test, check your answers. On the questions you missed, be sure you understand why you missed them. Did you misread the question (tests can use tricky wording)? Did you forget the information? Or was it something you hadn't learned? Go back and study any shaky areas that the practice tests reveal.

Taking these tests not only helps with your grade, but also aids in combating test anxiety. If you're already used to the test conditions, you're less likely to worry about it, and working through tests until you're scoring well gives you a confidence boost. Go through the practice tests until you feel comfortable, and then you can go into the test knowing that you're ready for it.

Test Tips

On test day, you should be confident, knowing that you've prepared well and are ready to answer the questions. But aside from preparation, there are several test day strategies you can employ to maximize your performance.

First, as stated before, get a good night's sleep the night before the test (and for several nights before that, if possible). Go into the test with a fresh, alert mind rather than staying up late to study.

Try not to change too much about your normal routine on the day of the test. It's important to eat a nutritious breakfast, but if you normally don't eat breakfast at all, consider eating just a protein bar. If you're a coffee drinker, go ahead and have your normal coffee. Just make sure you time it so that the caffeine doesn't wear off right in the middle of your test. Avoid sugary beverages, and drink enough water to stay hydrated but not so much that you need a restroom break 10 minutes into the test. If your test isn't first thing in the morning, consider going for a walk or doing a light workout before the test to get your blood flowing.

Allow yourself enough time to get ready, and leave for the test with plenty of time to spare so you won't have the anxiety of scrambling to arrive in time. Another reason to be early is to select a good seat. It's helpful to sit away from doors and windows, which can be distracting. Find a good seat, get out your supplies, and settle your mind before the test begins.

When the test begins, start by going over the instructions carefully, even if you already know what to expect. Make sure you avoid any careless mistakes by following the directions.

Then begin working through the questions, pacing yourself as you've practiced. If you're not sure on an answer, don't spend too much time on it, and don't let it shake your confidence. Either skip it and come back later, or eliminate as many wrong answers as possible and guess among the remaining ones. Don't dwell on these questions as you continue—put them out of your mind and focus on what lies ahead.

Be sure to read all of the answer choices, even if you're sure the first one is the right answer. Sometimes you'll find a better one if you keep reading. But don't second-guess yourself if you do immediately know the answer. Your gut instinct is usually right. Don't let test anxiety rob you of the information you know.

If you have time at the end of the test (and if the test format allows), go back and review your answers. Be cautious about changing any, since your first instinct tends to be correct, but make sure you didn't misread any of the questions or accidentally mark the wrong answer choice. Look over any you skipped and make an educated guess.

At the end, leave the test feeling confident. You've done your best, so don't waste time worrying about your performance or wishing you could change anything. Instead, celebrate the successful completion of this test. And finally, use this test to learn how to deal with anxiety even better next time.

Review Video: Test Anxiety
Visit mometrix.com/academy and enter code: 100340

Important Qualification

Not all anxiety is created equal. If your test anxiety is causing major issues in your life beyond the classroom or testing center, or if you are experiencing troubling physical symptoms related to your anxiety, it may be a sign of a serious physiological or psychological condition. If this sounds like your situation, we strongly encourage you to seek professional help.

Additional Bonus Material

Due to our efforts to try to keep this book to a manageable length, we've created a link that will give you access to all of your additional bonus material:

mometrix.com/bonus948/algebra1tb

www.ingramcontent.com/pod-product-compliance
Lightning Source LLC
LaVergne TN
LVHW062133200926
848484LV00007B/1805